List of the Elements with Their Atomic Symbols and Atomic Weights

Name	Symbol	Atomic Number	Atomic Weight	Name	Symbol	Atomic Number	Atomic Weight
Actinium	Ac	89	227.028	Meitnerium	Mt	109	(268)
Aluminum	Al	13	26.9815	Mendelevium	Md	101	(258)
Americium	Am	95	(243)	Mercury	Hg	80	200.59
Antimony	Sb	51	121.76	Molybdenum	Mo	42	95.94
Argon	Ar	18	39.948	Neodymium	Nd	60	144.24
Arsenic	As	33	74.9216	Neon	Ne	10	20.1797
Astatine	At	85	(210)	Neptunium	Np	93	237.048
Barium	Ba	56	137.327	Nickel	Ni	28	58.693
Berkelium	Bk	97	(247)	Niobium	Nb	41	92.9064
Beryllium	Be	4	9.01218	Nitrogen	N	7	14.0067
Bismuth	Bi	83	208.980	Nobelium	No	102	(259)
Bohrium	Bh	107	(264)	Osmium	Os	76	190.23
Boron	B	5	10.811	Oxygen	O	8	15.9994
Bromine	Br	35	79.904	Palladium	Pd	46	106.42
Cadmium	Cd	48	112.411	Phosphorus	P	15	30.9738
Calcium	Ca	20	40.078	Platinum	Pt	78	195.08
Californium	Cf	98	(251)	Plutonium	Pu	94	(244)
Carbon	C	6	12.011	Polonium	Po	84	(209)
Cerium	Ce	58	140.115	Potassium	K	19	39.0983
Cesium	Cs	55	132.905	Praseodymium	Pr	59	140.908
Chlorine	Cl	17	35.4527	Promethium	Pm	61	(145)
Chromium	Cr	24	51.9961	Protactinium	Pa	91	231.036
Cobalt	Co	27	58.9332	Radium	Ra	88	226.025
Copernicium	Cn	112	(285)	Radon	Rn	86	(222)
Copper	Cu	29	63.546	Rhenium	Re	75	186.207
Curium	Cm	96	(247)	Rhodium	Rh	45	102.906
Darmstadtium	Ds	110	(271)	Roentgenium	Rg	111	(272)
Dubnium	Db	105	(262)	Rubidium	Rb	37	85.4678
Dysprosium	Dy	66	162.50	Ruthenium	Ru	44	101.07
Einsteinium	Es	99	(252)	Rutherfordium	Rf	104	(261)
Erbium	Er	68	167.26	Samarium	Sm	62	150.36
Europium	Eu	63	151.965	Scandium	Sc	21	44.9559
Fermium	Fm	100	(257)	Seaborgium	Sg	106	(266)
Fluorine	F	9	18.9984	Selenium	Se	34	78.96
Francium	Fr	87	(223)	Silicon	Si	14	28.0855
Gadolinium	Gd	64	157.25	Silver	Ag	47	107.868
Gallium	Ga	31	69.723	Sodium	Na	11	22.9898
Germanium	Ge	32	72.61	Strontium	Sr	38	87.62
Gold	Au	79	196.967	Sulfur	S	16	32.066
Hafnium	Hf	72	178.49	Tantalum	Ta	73	180.948
Hassium	Hs	108	(269)	Technetium	Tc	43	(98)
Helium	He	2	4.00260	Tellurium	Te	52	127.60
Holmium	Ho	67	164.930	Terbium	Tb	65	158.925
Hydrogen	H	1	1.00794	Thallium	Tl	81	204.383
Indium	In	49	114.818	Thorium	Th	90	232.038
Iodine	I	53	126.904	Thulium	Tm	69	168.934
Iridium	Ir	77	192.22	Tin	Sn	50	118.710
Iron	Fe	26	55.847	Titanium	Ti	22	47.88
Krypton	Kr	36	83.80	Tungsten	W	74	183.84
Lanthanum	La	57	138.906	Uranium	U	92	238.029
Lawrencium	Lr	103	(260)	Vanadium	V	23	50.9415
Lead	Pb	82	207.2	Xenon	Xe	54	131.29
Lithium	Li	3	6.941	Ytterbium	Yb	70	173.04
Lutetium	Lu	71	174.967	Yttrium	Y	39	88.9059
Magnesium	Mg	12	24.3050	Zinc	Zn	30	65.39
Manganese	Mn	25	54.9381	Zirconium	Zr	40	91.224

PEARSON   ALWAYS LEARNING

John McMurry • David S. Ballantine • Carl A. Hoeger
Virginia E. Peterson

Fundamentals of General Chemistry
Volume I

Fourth Custom Edition for the Community Colleges of Spokane

Taken from:
Fundamentals of General, Organic, and Biological Chemistry,
Seventh Edition
by John McMurry, David S. Ballantine, Carl A. Hoeger,
and Virginia E. Peterson

Cover Art: Courtesy of Photodisc/Getty Images.

Taken from:

Fundamentals of General, Organic, and Biological Chemistry, Seventh Edition
by John McMurry, David S. Ballantine, Carl A. Hoeger, and Virginia E. Peterson
Copyright © 2013, 2010, 20087, 2003, 1999, 1996, 1992 by Pearson Education, Inc.
Published by Pearson
Glenview, IL 60025

All rights reserved. No part of this book may be reproduced, in any form or by any means, without permission in writing from the publisher.

This special edition published in cooperation with Pearson Learning Solutions.

All trademarks, service marks, registered trademarks, and registered service marks are the property of their respective owners and are used herein for identification purposes only.

Pearson Learning Solutions, 501 Boylston Street, Suite 900, Boston, MA 02116
A Pearson Education Company
www.pearsoned.com

Printed in the United States of America

1 2 3 4 5 6 7 8 9 10 V092 16 15 14 13 12

0002000010271674130

CW

ISBN 10: 1-256-78508-3
ISBN 13: 978-1-256-78508-8

About the Authors

John McMurry, educated at Harvard and Columbia, has taught approximately 17,000 students in general and organic chemistry over a 30-year period. A professor of chemistry at Cornell University since 1980, Dr. McMurry previously spent 13 years on the faculty at the University of California at Santa Cruz. He has received numerous awards, including the Alfred P. Sloan Fellowship (1969–71), the National Institute of Health Career Development Award (1975–80), the Alexander von Humboldt Senior Scientist Award (1986–87), and the Max Planck Research Award (1991).

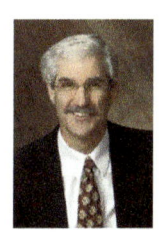

David S. Ballantine received his B.S. in Chemistry in 1977 from the College of William and Mary in Williamsburg, VA, and his Ph.D. in Chemistry in 1983 from the University of Maryland at College Park. After several years as a researcher at the Naval Research Labs in Washington, DC, he joined the faculty in the Department of Chemistry and Biochemistry of Northern Illinois University, where he has been a professor since 1989. He was awarded the Excellence in Undergraduate Teaching Award in 1998 and has been departmental Director of Undergraduate Studies since 2008. In addition, he is the coordinator for the Introductory and General Chemistry programs and is responsible for supervision of the laboratory teaching assistants.

Carl A. Hoeger received his B.S. in Chemistry from San Diego State University and his Ph.D. in Organic Chemistry from the University of Wisconsin, Madison in 1983. After a postdoctoral stint at the University of California, Riverside, he joined the Peptide Biology Laboratory at the Salk Institute in 1985, where he ran the NIH Peptide Facility while doing basic research in the development of peptide agonists and antagonists. During this time he also taught general, organic, and biochemistry at San Diego City College, Palomar College, and Miramar College. He joined the teaching faculty at University of California, San Diego, in 1998. Dr. Hoeger has been teaching chemistry to undergraduates for over 20 years, where he continues to explore the use of technology in the classroom; his current project involves the use of videopodcasts as adjuncts to live lectures. In 2004, he won the Paul and Barbara Saltman Distinguished Teaching Award from UCSD. He is deeply involved with the General Chemistry program at UCSD and also shares partial responsibility for the training and guidance of teaching assistants in the Chemistry and Biochemistry departments.

Virginia E. Peterson received her B.S. in Chemistry in 1967 from the University of Washington in Seattle and her Ph.D. in Biochemistry in 1980 from the University of Maryland at College Park. Between her undergraduate and graduate years she worked in lipid, diabetes, and heart disease research at Stanford University. Following her Ph.D. she took a position in the Biochemistry Department at the University of Missouri in Columbia and is now Professor Emerita. When she retired in 2011 she had been the Director of Undergraduate Advising for the department for 8 years and had taught both senior capstone classes and biochemistry classes for nonscience majors. Although retired, Dr. Peterson continues to advise undergraduates and teach classes. Awards include both the college-level and the university-wide Excellence in Teaching Award and, in 2006, the University's Outstanding Advisor Award and the State of Missouri Outstanding University Advisor Award. Dr. Peterson believes in public service and in 2003 received the Silver Beaver Award for service from the Boy Scouts of America.

Brief Contents

Features ix

Preface x

1 Matter and Measurements 2

2 Atoms and the Periodic Table 44

3 Ionic Compounds 72

4 Molecular Compounds 98

5 Classification and Balancing of Chemical Reactions 132

6 Chemical Reactions: Mole and Mass Relationships 158

7 Chemical Reactions: Energy, Rates, and Equilibrium 178

8 Gases, Liquids, and Solids 212

9 Solutions 252

10 Acids and Bases 290

11 Nuclear Chemistry 328

12 Introduction to Organic Chemistry: Alkanes 356

Appendices A-1

Glossary A-6

Answers to Selected Problems A-13

Photo Credits C-1

Index I-1

Contents

Features ix
Preface x

1 Matter and Measurements 2

1.1 Chemistry: The Central Science 3
1.2 States of Matter 5
1.3 Classification of Matter 6
   CHEMISTRY IN ACTION: *Aspirin—A Case Study* 8
1.4 Chemical Elements and Symbols 9
1.5 Elements and the Periodic Table 11
1.6 Chemical Reactions: An Example of Chemical Change 14
1.7 Physical Quantities 14
   CHEMISTRY IN ACTION: *Mercury and Mercury Poisoning* 15
1.8 Measuring Mass, Length, and Volume 17
1.9 Measurement and Significant Figures 19
1.10 Scientific Notation 21
1.11 Rounding Off Numbers 23
1.12 Problem Solving: Unit Conversions and Estimating Answers 25
1.13 Temperature, Heat, and Energy 29
   CHEMISTRY IN ACTION: *Temperature–Sensitive Materials* 31
1.14 Density and Specific Gravity 33
   CHEMISTRY IN ACTION: *A Measurement Example: Obesity and Body Fat* 35

2 Atoms and the Periodic Table 44

2.1 Atomic Theory 45
   CHEMISTRY IN ACTION: *Are Atoms Real?* 48
2.2 Elements and Atomic Number 48
2.3 Isotopes and Atomic Weight 50
2.4 The Periodic Table 52
2.5 Some Characteristics of Different Groups 54
   CHEMISTRY IN ACTION: *The Origin of Chemical Elements* 56
2.6 Electronic Structure of Atoms 56
2.7 Electron Configurations 59
2.8 Electron Configurations and the Periodic Table 62
2.9 Electron-Dot Symbols 65
   CHEMISTRY IN ACTION: *Atoms and Light* 66

3 Ionic Compounds 72

3.1 Ions 73
3.2 Periodic Properties and Ion Formation 75
3.3 Ionic Bonds 77
3.4 Some Properties of Ionic Compounds 77
   CHEMISTRY IN ACTION: *Ionic Liquids* 78
3.5 Ions and the Octet Rule 79
3.6 Ions of Some Common Elements 80
3.7 Naming Ions 82
   CHEMISTRY IN ACTION: *Salt* 83
3.8 Polyatomic Ions 85
   CHEMISTRY IN ACTION: *Biologically Important Ions* 86
3.9 Formulas of Ionic Compounds 86
3.10 Naming Ionic Compounds 89
3.11 H^+ and OH^- Ions: An Introduction to Acids and Bases 91
   CHEMISTRY IN ACTION: *Osteoporosis* 93

4 Molecular Compounds 98

4.1 Covalent Bonds 99
4.2 Covalent Bonds and the Periodic Table 101
4.3 Multiple Covalent Bonds 104
4.4 Coordinate Covalent Bonds 106
4.5 Characteristics of Molecular Compounds 107
4.6 Molecular Formulas and Lewis Structures 108
4.7 Drawing Lewis Structures 108
   CHEMISTRY IN ACTION: *CO and NO: Pollutants or Miracle Molecules?* 113
4.8 The Shapes of Molecules 114
   CHEMISTRY IN ACTION: *VERY Big Molecules* 118
4.9 Polar Covalent Bonds and Electronegativity 119
4.10 Polar Molecules 121
4.11 Naming Binary Molecular Compounds 123
   CHEMISTRY IN ACTION: *Damascenone by Any Other Name Would Smell as Sweet* 125

v

5 Classification and Balancing of Chemical Reactions 132

- 5.1 Chemical Equations 133
- 5.2 Balancing Chemical Equations 135
- 5.3 Classes of Chemical Reactions 138
- 5.4 Precipitation Reactions and Solubility Guidelines 139
  - **CHEMISTRY IN ACTION:** *Gout and Kidney Stones: Problems in Solubility* 140
- 5.5 Acids, Bases, and Neutralization Reactions 141
- 5.6 Redox Reactions 142
  - **CHEMISTRY IN ACTION:** *Batteries* 147
- 5.7 Recognizing Redox Reactions 148
- 5.8 Net Ionic Equations 150

6 Chemical Reactions: Mole and Mass Relationships 158

- 6.1 The Mole and Avogadro's Number 159
- 6.2 Gram–Mole Conversions 163
  - **CHEMISTRY IN ACTION:** *Did Ben Franklin Have Avogadro's Number? A Ballpark Calculation* 164
- 6.3 Mole Relationships and Chemical Equations 165
- 6.4 Mass Relationships and Chemical Equations 167
- 6.5 Limiting Reagent and Percent Yield 169
  - **CHEMISTRY IN ACTION:** *Anemia—A Limiting Reagent Problem?* 172

7 Chemical Reactions: Energy, Rates, and Equilibrium 178

- 7.1 Energy and Chemical Bonds 179
- 7.2 Heat Changes during Chemical Reactions 180
- 7.3 Exothermic and Endothermic Reactions 181
  - **CHEMISTRY IN ACTION:** *Energy from Food* 185
- 7.4 Why Do Chemical Reactions Occur? Free Energy 186
- 7.5 How Do Chemical Reactions Occur? Reaction Rates 190
- 7.6 Effects of Temperature, Concentration, and Catalysts on Reaction Rates 192
  - **CHEMISTRY IN ACTION:** *Regulation of Body Temperature* 195
- 7.7 Reversible Reactions and Chemical Equilibrium 195
- 7.8 Equilibrium Equations and Equilibrium Constants 196
- 7.9 Le Châtelier's Principle: The Effect of Changing Conditions on Equilibria 200
  - **CHEMISTRY IN ACTION:** *Coupled Reactions* 204

8 Gases, Liquids, and Solids 212

- 8.1 States of Matter and Their Changes 213
- 8.2 Intermolecular Forces 216
- 8.3 Gases and the Kinetic–Molecular Theory 220
- 8.4 Pressure 221
  - **CHEMISTRY IN ACTION:** *Greenhouse Gases and Global Warming* 224
- 8.5 Boyle's Law: The Relation between Volume and Pressure 225
  - **CHEMISTRY IN ACTION:** *Blood Pressure* 228
- 8.6 Charles's Law: The Relation between Volume and Temperature 228
- 8.7 Gay-Lussac's Law: The Relation between Pressure and Temperature 230
- 8.8 The Combined Gas Law 231
- 8.9 Avogadro's Law: The Relation between Volume and Molar Amount 232
- 8.10 The Ideal Gas Law 233
- 8.11 Partial Pressure and Dalton's Law 236
- 8.12 Liquids 237
- 8.13 Water: A Unique Liquid 239
- 8.14 Solids 240
- 8.15 Changes of State 242
  - **CHEMISTRY IN ACTION:** CO_2 *as an Environmentally Friendly Solvent* 245

9 Solutions 252

- 9.1 Mixtures and Solutions 253
- 9.2 The Solution Process 255
- 9.3 Solid Hydrates 257
- 9.4 Solubility 258
- 9.5 The Effect of Temperature on Solubility 258
- 9.6 The Effect of Pressure on Solubility: Henry's Law 260
- 9.7 Units of Concentration 262
  - **CHEMISTRY IN ACTION:** *Breathing and Oxygen Transport* 263
- 9.8 Dilution 270
- 9.9 Ions in Solution: Electrolytes 272
- 9.10 Electrolytes in Body Fluids: Equivalents and Milliequivalents 273

CONTENTS vii

11 Nuclear Chemistry 328
- 11.1 Nuclear Reactions 329
- 11.2 The Discovery and Nature of Radioactivity 330
- 11.3 Stable and Unstable Isotopes 331
- 11.4 Nuclear Decay 332
- 11.5 Radioactive Half-Life 337
  - CHEMISTRY IN ACTION: *Medical Uses of Radioactivity* 338
- 11.6 Radioactive Decay Series 340
- 11.7 Ionizing Radiation 341
- 11.8 Detecting Radiation 343
- 11.9 Measuring Radiation 344
  - CHEMISTRY IN ACTION: *Irradiated Food* 345
- 11.10 Artificial Transmutation 347
  - CHEMISTRY IN ACTION: *Body Imaging* 348
- 11.11 Nuclear Fission and Nuclear Fusion 349

- 9.11 Properties of Solutions 275
  - CHEMISTRY IN ACTION: *Electrolytes, Fluid Replacement, and Sports Drinks* 276
- 9.12 Osmosis and Osmotic Pressure 279
- 9.13 Dialysis 283
  - CHEMISTRY IN ACTION: *Timed-Release Medications* 284

10 Acids and Bases 290
- 10.1 Acids and Bases in Aqueous Solution 291
- 10.2 Some Common Acids and Bases 292
- 10.3 The Brønsted–Lowry Definition of Acids and Bases 293
- 10.4 Acid and Base Strength 296
  - CHEMISTRY IN ACTION: *GERD—Too Much Acid or Not Enough?* 299
- 10.5 Acid Dissociation Constants 301
- 10.6 Water as Both an Acid and a Base 302
- 10.7 Measuring Acidity in Aqueous Solution: pH 303
- 10.8 Working with pH 306
- 10.9 Laboratory Determination of Acidity 308
- 10.10 Buffer Solutions 308
  - CHEMISTRY IN ACTION: *Buffers in the Body: Acidosis and Alkalosis* 312
- 10.11 Acid and Base Equivalents 313
- 10.12 Some Common Acid–Base Reactions 316
- 10.13 Titration 317
  - CHEMISTRY IN ACTION: *Acid Rain* 320
- 10.14 Acidity and Basicity of Salt Solutions 321

12 Introduction to Organic Chemistry: Alkanes 356
- 12.1 The Nature of Organic Molecules 357
- 12.2 Families of Organic Molecules: Functional Groups 359
- 12.3 The Structure of Organic Molecules: Alkanes and Their Isomers 364
- 12.4 Drawing Organic Structures 367
- 12.5 The Shapes of Organic Molecules 372
- 12.6 Naming Alkanes 374
- 12.7 Properties of Alkanes 380
- 12.8 Reactions of Alkanes 381
  - MASTERING REACTIONS: *Organic Chemistry and the Curved Arrow Formalism* 382
- 12.9 Cycloalkanes 383
  - CHEMISTRY IN ACTION: *Surprising Uses of Petroleum* 385
- 12.10 Drawing and Naming Cycloalkanes 386

Appendices A-1
Glossary A-6
Answers to Selected Problems A-13
Photo Credits C-1
Index I-1

Features

CHEMISTRY IN ACTION

Aspirin—A Case Study 8
Mercury and Mercury Poisoning 15
Temperature-Sensitive Materials 31
A Measurement Example: Obesity and Body Fat 35
Are Atoms Real? 48
The Origin of Chemical Elements 56
Atoms and Light 66
Ionic Liquids 78
Salt 83
Biologically Important Ions 86
Osteoporosis 93
CO and NO: Pollutants or Miracle Molecules? 113
VERY Big Molecules 118
Damascenone by Any Other Name Would Smell as Sweet 125
Gout and Kidney Stones: Problems in Solubility 140
Batteries 147
Did Ben Franklin Have Avogadro's Number? A Ballpark Calculation 164
Anemia—A Limiting Reagent Problem? 172
Energy from Food 185
Regulation of Body Temperature 195
Coupled Reactions 204
Greenhouse Gases and Global Warming 224
Blood Pressure 228
CO_2 as an Environmentally Friendly Solvent 245
Breathing and Oxygen Transport 263
Electrolytes, Fluid Replacement, and Sports Drinks 276
Timed-Release Medications 284
GERD—Too Much Acid or Not Enough? 299
Buffers in the Body: Acidosis and Alkalosis 312
Acid Rain 320
Medical Uses of Radioactivity 338
Irradiated Food 345
Body Imaging 348
Surprising Uses of Petroleum 385
The Chemistry of Vision and Color 406
Polycyclic Aromatic Hydrocarbons and Cancer 420
Ethyl Alcohol as a Drug and a Poison 446
Phenols as Antioxidants 449
Inhaled Anesthetics 453
Knowing What You Work With: Material Safety Data Sheets 465
Organic Compounds in Body Fluids and the "Solubility Switch" 473
Toxicology 478
Chemical Warfare among the Insects 489
How Toxic Is Toxic? 499
Acids for the Skin 528
Kevlar: A Life-Saving Polymer 538
Proteins in the Diet 564
Protein Analysis by Electrophoresis 568

Collagen—A Tale of Two Diseases 576
Prions: Proteins That Cause Disease 579
Extremozymes—Enzymes from the Edge 595
Enzymes in Medical Diagnosis 601
Enzyme Inhibitors as Drugs 607
Vitamins, Minerals, and Food Labels 615
Life without Sunlight 627
Basal Metabolism 636
Plants and Photosynthesis 649
Chirality and Drugs 663
Cell-Surface Carbohydrates and Blood Type 672
Carbohydrates and Fiber in the Diet 679
Cell Walls: Rigid Defense Systems 685
Tooth Decay 701
Microbial Fermentations: Ancient and Modern 703
Diagnosis and Monitoring of Diabetes 709
The Biochemistry of Running 712
Lipids in the Diet 728
Detergents 731
Butter and Its Substitutes 740
Lipids and Atherosclerosis 757
Fat Storage: A Good Thing or Not? 760
The Liver, Clearinghouse for Metabolism 767
It's a Ribozyme! 790
Viruses and AIDS 794
Influenza—Variations on a Theme 799
One Genome To Represent Us All? 808
Serendipity and the Polymerase Chain Reaction 815
DNA Fingerprinting 817
Gout: When Biochemistry Goes Awry 833
The Importance of Essential Amino Acids and Effects of Deficiencies 836
Homeostasis 845
Plant Hormones 855
The Blood–Brain Barrier 878
Automated Clinical Laboratory Analysis 889

MASTERING REACTIONS

Organic Chemistry and the Curved Arrow Formalism 382
How Addition Reactions Occur 414
How Eliminations Occur 441
Carbonyl Additions 506

Preface

This textbook and its related digital resources provide students in the allied health sciences with a needed background in chemistry and biochemistry while offering a general context for chemical concepts to ensure that students in other disciplines gain an appreciation of the importance of chemistry in everyday life.

To teach chemistry all the way from "What is an atom?" to "How do we get energy from glucose?" is a challenge. Throughout our general chemistry and organic chemistry coverage, the focus is on concepts fundamental to the chemistry of living things and everyday life. In our biochemistry coverage we strive to meet the further challenge of providing a context for the application of those concepts in biological systems. Our goal is to provide enough detail for thorough understanding while avoiding so much detail that students are overwhelmed. Many practical and relevant examples are included to illustrate the concepts and enhance student learning.

The material covered is ample for a two-term introduction to general, organic, and biological chemistry. While the general and early organic chapters contain concepts that are fundamental to understanding the material in biochemistry, the later chapters can be covered individually and in an order that can be adjusted to meet the needs of the students and the duration of the course.

The writing style is clear and concise and punctuated with practical and familiar examples from students' personal experience. Art work, diagrams, and molecular models are used extensively to provide graphical illustration of concepts to enhance student understanding. Since the true test of knowledge is the ability to apply that knowledge appropriately, we include numerous worked examples that incorporate consistent problem-solving strategies.

Regardless of their career paths, all students will be citizens in an increasingly technological society. When they recognize the principles of chemistry at work not just in their careers but in their daily lives, they are prepared to make informed decisions on scientific issues based on a firm understanding of the underlying concepts.

New to This Edition

The major theme of this revision is *making connections*, which is accomplished in a variety of ways:

- **NEW and updated *Chemistry in Action* boxes** highlight and strengthen the connections between general, organic, and biological chemistry.
- **NEW *Mastering Reactions* boxes** discuss, in some depth, the "how" behind a number of organic reactions.
- **NEW in-chapter questions specifically related to *Chemistry in Action* applications and *Mastering Reactions*** reinforce the connection between the chapter content and practical applications.
- **NEW Concept Maps** added to certain chapters, draw connections between general, organic, and biological chemistry—in particular those chapters dealing with intermolecular forces, chemical reactions and energy, acid–base chemistry, and relationships between functional groups, proteins, and their properties.
- **NEW and updated Concept Links offer** visual reminders for students that indicate when new material builds on concepts from previous chapters. **Updated questions in the End of Chapter section build on Concept Links** and require students to recall information learned in previous chapters.
- **NEW and updated end-of-chapter (EOC) problems:** approximately 20–25% of the end-of-chapter problems have been revised to enhance clarity.
- **All Chapter Goals tied to EOC problem sets:** chapter summaries include a list of EOC problems that correspond to the chapter goals for a greater connection between problems and concepts.

- **Chapters 1 and 2** have been restructured to place a greater emphasis on building math skills.
- **Chapter 6 (Chemical Reactions)** has been reorganized into two chapters: Chapter 5 (Classification and Balancing of Chemical Reactions) and Chapter 6 (Chemical Reactions: Mole and Mass Relationships) to allow student to narrow their focus; Chapter 5 focuses on the qualitative aspect of reactions, while Chapter 6 focuses on calculations.

Organization

General Chemistry: Chapters 1–11 The introduction to elements, atoms, the periodic table, and the quantitative nature of chemistry (Chapters 1 and 2) is followed by chapters that individually highlight the nature of ionic and molecular compounds (Chapters 3 and 4. The next three chapters discuss chemical reactions and their stoichiometry, energies, rates, and equilibria (Chapters 5, 6, and 7). Topics relevant to the chemistry of life follow: Gases, Liquids, and Solids (Chapter 8); Solutions (Chapter 9); and Acids and Bases (Chapter 10). Nuclear Chemistry (Chapter 11) closes the general chemistry sequence.

Organic Chemistry: Chapters 12–17 These chapters concisely focus on what students must know in order to understand biochemistry. The introduction to hydrocarbons (Chapters 12 and 13) includes the basics of nomenclature, which is thereafter kept to a minimum. Discussion of functional groups with single bonds to oxygen, sulfur, or a halogen (Chapter 14) is followed by a short chapter on amines, which are so important to the chemistry of living things and drugs (Chapter 15). After introducing aldehydes and ketones (Chapter 16), the chemistry of carboxylic acids and their derivatives (including amides) is covered (Chapter 17), with a focus on similarities among the derivatives. More attention to the mechanisms by which organic reactions occur and the vernacular used to describe them has been incorporated into this edition.

Biological Chemistry: Chapters 18–29 Rather than proceed through the complexities of protein, carbohydrate, lipid, and nucleic acid structure before getting to the roles of these compounds in the body, structure and function are integrated in this text. Protein structure (Chapter 18) is followed by enzyme and coenzyme chemistry (Chapter 19). With enzymes introduced, the central pathways and themes of biochemical energy production can be described (Chapter 20). If the time you have available to cover biochemistry is limited, stop with Chapter 20 and your students will have an excellent preparation in the essentials of metabolism. The following chapters cover carbohydrate chemistry (Chapters 21 and 22), then lipid chemistry (Chapters 23 and 24). Next we discuss nucleic acids and protein synthesis (Chapter 25) and genomics (Chapter 26). The last three chapters cover protein and amino acid metabolism (Chapter 27), the function of hormones and neurotransmitters, and the action of drugs (Chapter 28), and provide an overview of the chemistry of body fluids (Chapter 29).

Chapter by Chapter Changes

COVERAGE OF GENERAL CHEMISTRY

The major revisions in this section involve reorganization or revision of content to strengthen the connections between concepts and to provide a more focused coverage of specific concepts. In order to reinforce the relationship between topics, Concept Maps have been included in several chapters to illustrate the connections between concepts.

Specific changes to chapters are provided below:

Chapter 1

- Chapters 1 and 2 from the sixth edition have been combined; a greater emphasis is placed on math skills. Goals were revised and updated to reflect the combined chapter.

- The concept of homogeneous and heterogeneous mixtures is introduced (previously in Chapter 9).
- There are several new references to the Application boxes (now titled *Chemistry in Action*), both in the text and in the problems. Four Application boxes were updated to provide more current connections to everyday life and the health fields.

Chapter 2

- Chapter 3 from the sixth edition has become Chapter 2 in the seventh edition: Atoms and the Periodic Table.
- Information on the periodic table has been updated (the 117th element has been discovered, no longer considered a metalloid; 112th element has been named).
- Application boxes (*Chemistry in Action*) have been modified to enhance clarity, relevance to the student, and connection to the text.

Chapter 3

- Chapter 3 in this edition was Chapter 4 in the sixth edition: Ionic Compounds.
- There is a new Application (*Chemistry in Action*) box titled "Ionic Liquids."
- Changes have been made to the boxes to enhance clarity, relevance to the student, and connection to the text.

Chapter 4

- Chapter 4 in this edition was Chapter 5 in the sixth edition: Molecular Compounds.
- Section 11 (Characteristics of Molecular Compounds) has been moved; it is now Section 5.

Chapter 5

- Chapter 5 in this edition, Classification and Balancing of Chemical Reactions, is a portion of Chapter 6 from the sixth edition (6e Sections 6.1–6.2 and 6.8–6.13).
- There are several new references to the Application (*Chemistry in Action*) boxes, both in the text and in the problems.

Chapter 6

- Chapter 6 in this edition, Chemical Reactions: Mole and Mass Relationships, is a portion of Chapter 6 from the sixth edition (6e Sections 6.3 – 6.7).
- There are several new references to the Applications boxes, both in the text and in the problems.
- A new concept map has been added, relating topics in Chapters 3 and 4 to topics in Chapters 5 and 6 and to topics in Chapters 7 and 10.

Chapter 7

- An explanation of bond energies has been added to show how the energy of chemical reactions is related to the covalent bonds in reactants and products.
- Bond and reaction energies in units of both kcal and kJ have been consistently included.
- A new concept map has been added at the end of chapter that shows how energy, rates, and equilibrium are related.
- There is a new *Chemistry in Action* application box titled "Coupled Reactions."

Chapter 8

- Section 8.11 (Intermolecular Forces) has been moved to Section 8.2 to help students make the connection between these forces and the physical states and properties of matter that are discussed in the subsequent sections.
- Chemistry in Action application boxes have been revised to strengthen the connection with chapter content.
- There is a new Concept Map relating molecular shape and polarity (Chapter 4) and the energy of chemical and physical changes (Chapter 7) to intermolecular forces and the physical states of matter.

Chapter 9

- Section 9.7 (Units of Concentration) has been reorganized to add mass/mass units and improve connections between units.
- A new Concept Map has been added to show the relationship between intermolecular forces (Chapter 8) and the formation of solutions and between concentration units of molarity and mole/mass relationships of reactions in solution.

Chapter 10

- Section 10.4 (Water as Both Acid and Base) and Section 10.6 (Dissociation of Water) have been combined to strengthen the connection between these concepts.
- Section 10.11 (Buffer Solutions) and Section 10.12 (Buffers in the Body) have been combined to strengthen the connection between these concepts and reduce redundancy of content in later chapters.
- Content in the *Chemistry in Action* application boxes has been combined and revised to strengthen connections between concepts and practical applications.
- New Concept Map has been added to show the relationships between strong/weak electrolytes (Chapter 9) and the extent of formation of H^+ and OH^- ions in acid/base solutions, and between equilibrium (Chapter 7) and strong/weak acids.

Chapter 11

- One *Chemistry in Action* application box was eliminated and others were revised to strengthen the connections between chapter content and practical applications.

COVERAGE OF ORGANIC CHEMISTRY

A major emphasis in this edition was placed on making the fundamental reactions that organic molecules undergo much clearer to the reader, with particular attention on those reactions encountered again in biochemical transformations. Also new to this edition is the expanded use and evaluation of line-angle structure for organic molecules, which are so important when discussing biomolecules. Most of the Application boxes (*Chemistry in Action*) have been updated to reflect current understanding and research. A number of instructors have asked for an increased discussion of the mechanisms of organic reactions; however, since many that teach this class did not want it to be integrated directly into the text we developed a completely new feature titled *Mastering Reactions*. This boxed feature discusses in relative depth the "how" behind a number of organic reactions. We have designed *Mastering Reactions* so that they may be integrated into an instructor's lecture or simply left out with no detriment to the material in the text itself.

Other specific changes to chapters are provided below:

Chapter 12

- There is a new feature box called *Mastering Reactions* that explains curved-arrow formalism used in organic mechanisms.
- There is a functional group scheme map that will aid in classifying functional groups.
- Table 1 has been substantially reworked to include line structures and sulfur compounds.

Chapter 13

- Sixth edition section 13.7 has been converted into a *Mastering Reactions* box (How Addition Reactions Occur). The content of *Mastering Reactions* box includes expanded discussion of Markovnikov's Rule.
- Chapter 13 now includes in-text references to *Chemistry in Action* boxes, including in-text problems related to them. There are also several cross-references to the *Mastering Reactions* boxes.

Chapter 14

- The language used to describe the classification of alcohols has been adjusted to make it clearer for the reader.
- A *Mastering Reactions* box (How Eliminations Occur) has been added. Discussion of Zaitsev's Rule and its mechanistic explanation are included.

Chapter 15

- A new *Chemistry in Action* box (Knowing What You Work With: Material Safety Data Sheets) has been added.

Chapter 16

- A *Mastering Reactions* box (Carbonyl Additions) has been added, with an emphasis on hemiacetal and acetal formation.
- The discussion of formation of cyclic hemiacetals and acetals has been adjusted to make it more clear to the reader.

Chapter 17

- The colors used in many of the illustrations were corrected and/or modified to allow students to easily follow which atoms come from which starting materials in the formation and degradation of the various carboxylic acid derivatives.

Chapter 18

- There are new references to the Chemistry in Action boxes, both in the text and in the problems.
- There is an expanded discussion of isoelectric points.
- There is a new Concept Map illustrating the organizing principles of protein structure, types of proteins, and amino acids.

Chapter 19

- There is an expanded discussion of minerals, including a new table.
- A clarification of the definition of uncompetitive inhibition (previously noncompetitive inhibition) has been added.

Chapter 20

- A new Concept Map relating biochemical energy to chemical energy concepts discussed in earlier chapters has been added.
- Energy calculations are in both kcalories and kjoules.
- The discussion of "uncouplers" has been integrated into the text.

Chapter 21

- A new *Chemistry in Action* box was added, combining and updating concepts from earlier applications discussing aspects of dietary carbohydrates.
- Many ribbon molecules were made clearer by floating the model on white rather than black backgrounds.
- A new worked example was added to clarify how to analyze a complex molecule for its component structures.

Chapter 22

- The text discussion was made more readable by reducing the jargon present in this chapter.
- The discussion of glucose metabolism in diabetes and metabolic syndrome was freshened.

Chapter 23

- The discussion of cholesterol and bile acids was moved from Chapter 28 to this chapter.
- Dietary and obesity statistics were updated.
- Text information about medical uses of liposomes was added.

Chapter 24

- Jargon was removed and concepts were clarified by a more thorough explanation of reactions.
- A clearer explanation of how triacylglycerides are digested, absorbed, and moved through the body to destination cells was added.
- The discussion of energy yields from fat metabolism was extended for clarity.

Chapter 25

- The retrovirus information has been updated to focus on retroviruses in general.
- The influenza information focuses on the nature of the common influenza viruses and new research directions.

Chapter 26

- This chapter, Genomics, was Chapter 27 in the sixth edition. It has been updated to reflect the current state of genome mapping.
- The *Chemistry In Action* box, DNA Fingerprinting, has been updated to include PCR fingerprinting.

Chapter 27

- This chapter, Protein and Amino Acid Metabolism, was Chapter 28 in the sixth edition.
- Changes have been made to enhance clarity, relevance to the student, and connection to the text.

Chapter 28

- The chapter is now focused only on the messenger aspect of these peptides, amino acid derivatives, and steroids.
- Discussions were made clearer by spelling-out terms instead of defining abbreviations.
- The steroid-abuse section was revamped to increase relevance and enhance clarity for the student.

Chapter 29

- Changes were made to enhance clarity, relevance to the student, and connection to the text.

KEY FEATURES

Focus on Learning

Worked Examples Most Worked Examples include an **Analysis** section that precedes the **Solution**. The Analysis lays out the approach to solving a problem of the given type. When appropriate, a **Ballpark Estimate** gives students an overview of the relationships needed to solve the problem and provides an intuitive approach to arrive at a rough estimate of the answer. The Solution presents the worked-out example using the strategy laid out in the Analysis and, in many cases, includes expanded discussion to enhance student understanding. When applicable, following the Solution there is a Ballpark Check that compares the calculated answer to the Ballpark Estimate and verifies that the answer makes chemical and physical sense.

Worked Example 1.11 Factor Labels: Unit Conversions

A child is 21.5 inches long at birth. How long is this in centimeters?

ANALYSIS This problem calls for converting from inches to centimeters, so we will need to know how many centimeters are in an inch and how to use this information as a conversion factor.

BALLPARK ESTIMATE It takes about 2.5 cm to make 1 in., and so it should take two and a half times as many centimeters to make a distance equal to approximately 20 in., or about 20 in. × 2.5 = 50 cm.

SOLUTION

STEP 1: Identify given information. Length = 21.5 in.

STEP 2: Identify answer and units. Length = ?? cm

STEP 3: Identify conversion factor. 1 in. = 2.54 cm → $\dfrac{2.54 \text{ cm}}{1 \text{ in.}}$

STEP 4: Solve. Multiply the known length (in inches) by the conversion factor so that units cancel, providing the answer (in centimeters).

21.5 in. × $\dfrac{2.54 \text{ cm}}{1 \text{ in.}}$ = 54.6 cm (Rounded off from 54.61)

BALLPARK CHECK How does this value compare with the ballpark estimate we made at the beginning? Are the final units correct? 54.6 cm is close to our original estimate of 50 cm.

Key Concept Problems are integrated throughout the chapters to focus attention on the use of essential concepts, as do the **Understanding Key Concepts problems** at the end of each chapter. Understanding Key Concepts problems are designed to test students' mastery of the core principles developed in the chapter. Students thus have an opportunity to ask "Did I get it?" before they proceed. Most of these Key Concept Problems use graphics or molecular-level art to illustrate the core principles and will be particularly useful to visual learners.

KEY CONCEPT PROBLEM 6.4

What is the molecular weight of cytosine, a component of DNA (deoxyribonucleic acid)? (black = C, blue = N, red = O, white = H.)

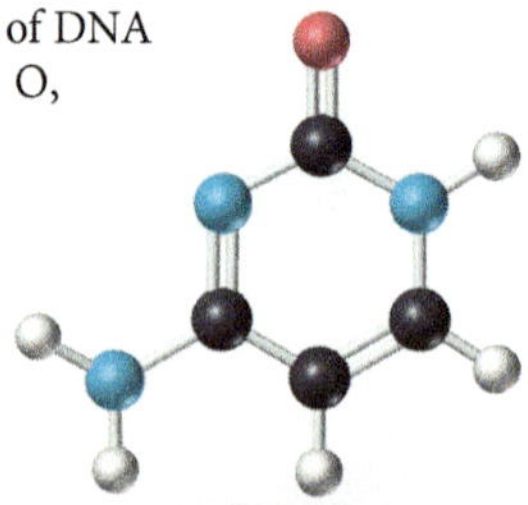

Cytosine

Problems The problems within the chapters, for which brief answers are given in an appendix, cover every skill and topic to be understood. One or more problems follow each Worked Example and others stand alone at the ends of sections.

PROBLEM 1.18
Write appropriate conversion factors and carry out the following conversions:
(a) 16.0 oz = ? g (b) 2500 mL = ? L (c) 99.0 L = ? qt

PROBLEM 1.19
Convert 0.840 qt to milliliters in a single calculation using more than one conversion factor.

More Color-Keyed, Labeled Equations It is entirely too easy to skip looking at a chemical equation while reading the text. We have used color extensively to call attention to the aspects of chemical equations and structures under discussion, a continuing feature of this book that has been judged to be very helpful.

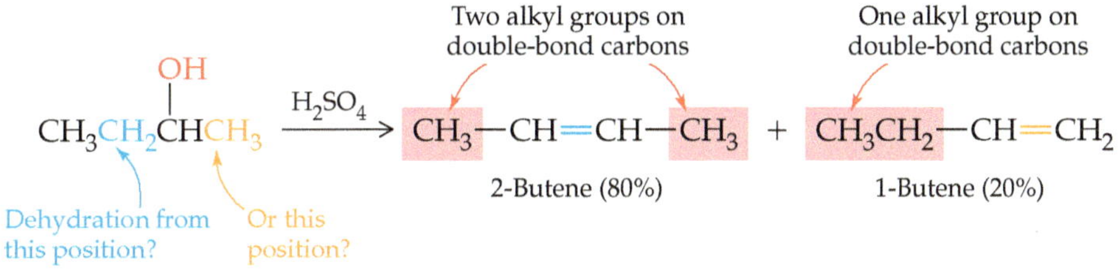

$$CH_3CH_2CHCH_3 \xrightarrow{H_2SO_4} CH_3-CH=CH-CH_3 + CH_3CH_2-CH=CH_2$$

(with OH on the second carbon of the starting material)

Two alkyl groups on double-bond carbons — 2-Butene (80%)
One alkyl group on double-bond carbons — 1-Butene (20%)

Dehydration from this position? Or this position?

Key Words Every key term is boldfaced on its first use, fully defined in the margin adjacent to that use, and listed at the end of the chapter. These are the terms students must understand to continue with the subject at hand. Definitions of all Key Words are collected in the Glossary.

Focus on Relevancy

Chemistry is often considered to be a difficult and tedious subject. But when students make a connection between a concept in class and an application in their daily lives, the chemistry comes alive, and they get excited about the subject. The applications in this book strive to capture student interest and emphasize the relevance of the scientific concepts. The use of relevant applications makes the concepts more accessible and increases understanding.

Applications—now titled *Chemistry in Action*—are both integrated into the discussions in the text and set off from the text. Each boxed application provides sufficient information for reasonable understanding and, in many cases, extends the concepts discussed in the text in new ways. The boxes end with a cross-reference to end-of-chapter problems that can be assigned by the instructor.

CHEMISTRY IN ACTION

Anemia – A Limiting Reagent Problem?

Anemia is the most commonly diagnosed blood disorder, with symptoms typically including lethargy, fatigue, poor concentration, and sensitivity to cold. Although anemia has many causes, including genetic factors, the most common cause is insufficient dietary intake or absorption of iron.

Hemoglobin (abbreviated Hb), the iron-containing protein found in red blood cells, is responsible for oxygen transport throughout the body. Low iron levels in the body result in decreased production and incorporation of Hb into red blood cells. In addition, blood loss due to injury or to menstruation in women increases the body's demand for iron in order to replace lost Hb. In the United States, nearly 20% of women of child-bearing age suffer from iron-deficiency anemia compared to only 2% of adult men.

The recommended minimum daily iron intake is 8 mg for adult men and 18 mg for premenopausal women. One way to ensure sufficient iron intake is a well-balanced diet that includes iron-fortified grains and cereals, red meat, egg yolks, leafy green vegetables, tomatoes, and raisins. Vegetarians should pay extra attention to their diet, because the iron in fruits and vegetables is not as readily absorbed by the body as the iron

▲ Can cooking in cast iron pots decrease anemia?

in meat, poultry, and fish. Vitamin supplements containing folic acid and either ferrous sulfate or ferrous gluconate can decrease iron deficiencies, and vitamin C increases the absorption of iron by the body.

However, the simplest way to increase dietary iron may be to use cast iron cookware. Studies have demonstrated that the iron content of many foods increases when cooked in an iron pot. Other studies involving Ethiopian children showed that those who ate food cooked in iron cookware were less likely to suffer from iron-deficiency anemia than their playmates who ate similar foods prepared in aluminum cookware.

See Chemistry in Action Problems 6.59 and 6.60 at the end of the chapter.

NEW Feature box in this edition—*Mastering Reactions* include How Addition Reactions Occur, How Elimination Reactions Occur, and Carbonyl Additions and discuss how these important organic transformations are believed to occur. This new feature allows instructors to easily introduce discussions of mechanism into their coverage of organic chemistry.

MASTERING REACTIONS

Organic Chemistry and the Curved Arrow Formalism

Starting with this chapter and continuing on through the remainder of this text, you will be exploring the world of organic chemistry and its close relative, biochemistry. Both of these areas of chemistry are much more "visual" than those you have been studying; organic chemists, for example, look at how and why reactions occur by examining the flow of electrons. For example, consider the following reaction of 2-iodopropane with sodium cyanide:

This seemingly simple process (known as a *substitution reaction*, discussed in Chapter 13) is not adequately described by the equation. To help to understand what may really be going on, organic chemists use what is loosely described as "electron pushing" and have adopted what is known as *curved arrow formalism* to represent it. The movement of electrons is depicted using curved arrows, where the number of electrons corresponds to the head of the arrow. Single-headed arrows represent movement of one electron, while a double-headed arrow indicates

The convention is to show the movement *from* an area of high electron density (the start of the arrow) *to* one of lower electron density (the head of the arrow). Using curved arrow formalism, we can examine the reaction of 2-iodopropane with sodium cyanide in more detail. There are two distinct paths by which this reaction can occur:

Path 1

Path 2

Notice that while both pathways lead ultimately to the same product, the curved arrow formalism shows us that they have significantly different ways of occurring. Although it is not important right now to understand which of the two paths

Focus on Making Connections

This can be a difficult course to teach. Much of what students are interested in lies in the last part of the course, but the material they need to understand the biochemistry is found in the first two-thirds. It is easy to lose sight of the connections among general, organic, and biological chemistry, so we use a feature—**Concepts to Review**—to call attention to these connections. From Chapter 4 on, the Concepts to Review section at the beginning of the chapter lists topics covered in earlier chapters that form the basis for what is discussed in the current chapter.

We have also retained the successful Concept Link icons and Looking Ahead notes.

Concept Link icons ▶▶ are used extensively to indicate places where previously covered material is relevant to the discussion at hand. These links provide cross-references and also serve to highlight important chemical themes as they are revisited.

LOOKING AHEAD ▶▶ notes call attention to connections between just-covered material and discussions in forthcoming chapters. These notes are designed to illustrate to the students why what they are learning will be useful in what lies ahead.

NEW Concept Maps are used to illustrate and reinforce the connections between concepts discussed in each chapter and concepts in previous or later chapters.

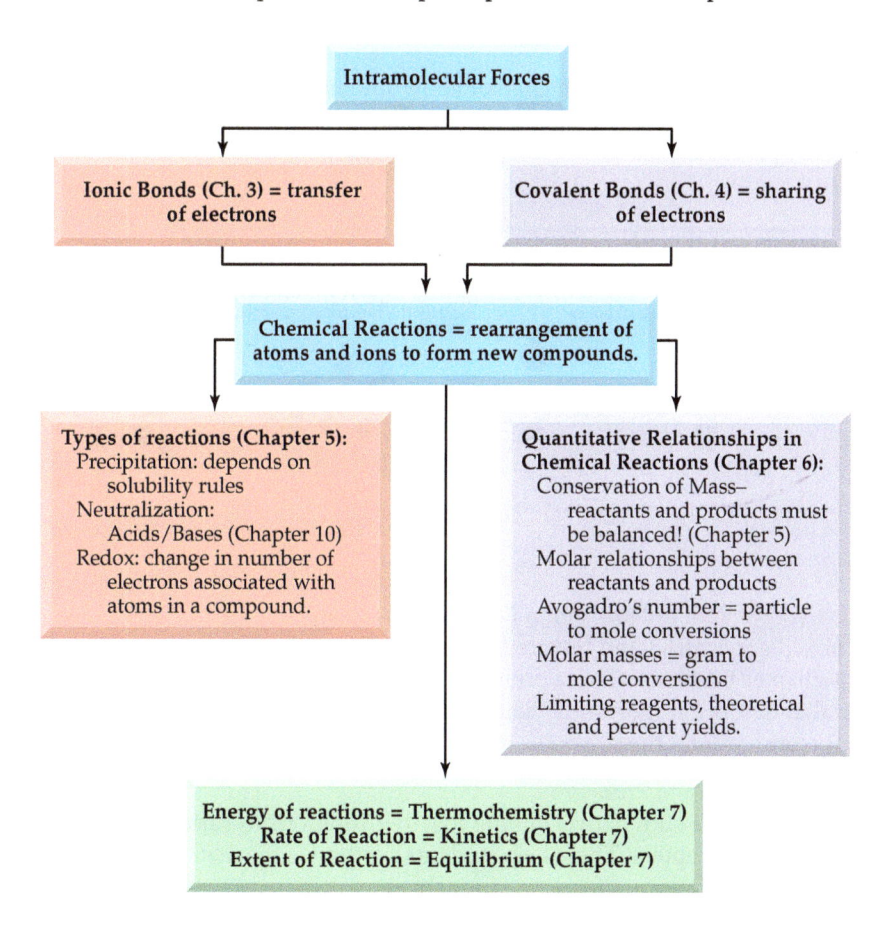

```
                        Intramolecular Forces
                         /                \
          Ionic Bonds (Ch. 3) = transfer    Covalent Bonds (Ch. 4) = sharing
                  of electrons                     of electrons
                         \                /
              Chemical Reactions = rearrangement of
              atoms and ions to form new compounds.
                         /                \
    Types of reactions (Chapter 5):        Quantitative Relationships in
      Precipitation: depends on              Chemical Reactions (Chapter 6):
        solubility rules                      Conservation of Mass–
      Neutralization:                           reactants and products must
        Acids/Bases (Chapter 10)                be balanced! (Chapter 5)
      Redox: change in number of              Molar relationships between
        electrons associated with               reactants and products
        atoms in a compound.                  Avogadro's number = particle
                                                to mole conversions
                                              Molar masses = gram to
                                                mole conversions
                                              Limiting reagents, theoretical
                                                and percent yields.

              Energy of reactions = Thermochemistry (Chapter 7)
              Rate of Reaction = Kinetics (Chapter 7)
              Extent of Reaction = Equilibrium (Chapter 7)
```

Focus on Studying

End of Chapter Section

Summary: Revisiting the Chapter Goals
The Chapter Summary revisits the Chapter Goals that open the chapter. Each of the questions posed at the start of the chapter is answered by a summary of the essential information needed to attain the corresponding goal.

SUMMARY: REVISITING THE CHAPTER GOALS

1. What are the basic properties of organic compounds? Compounds made up primarily of carbon atoms are classified as organic. Many organic compounds contain carbon atoms that are joined in long chains by a combination of single (C—C), double (C=C), or triple (C≡C) bonds. In this chapter, we focused primarily on *alkanes*, hydrocarbon compounds that contain only single bonds between all C atoms (see *Problems 29, 31, 32*).

is represented by lines and the locations of C and H atoms are understood (see *Problems 22–24, 44, 45, 48, 49–51*).

5. What are alkanes and cycloalkanes, and how are they named? Compounds that contain only carbon and hydrogen are called *hydrocarbons*, and hydrocarbons that have only single bonds are called *alkanes*. A *straight-chain alkane* has all its carbons connected in a row, a *branched-chain alkane* has a

---

Key Words
All of the chapter's boldface terms are listed in alphabetical order and are cross-referenced to the page where it appears in the text.

Understanding Key Concepts
The problems at the end of each chapter allow students to test their mastery of the core principles developed in the chapter. Students have an opportunity to ask "Did I get it?" before they proceed.

UNDERSTANDING KEY CONCEPTS

12.22 How many hydrogen atoms are needed to complete the hydrocarbon formulas for the following carbon backbones?

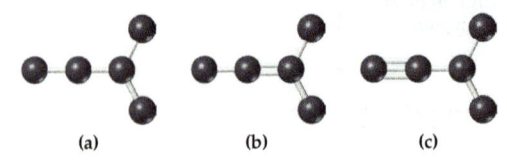

(a)   (b)   (c)

12.23 Convert the following models into condensed structures (black = C; white = H; red = O):

12.25 Convert the following models into line drawings and identify the functional groups in each:

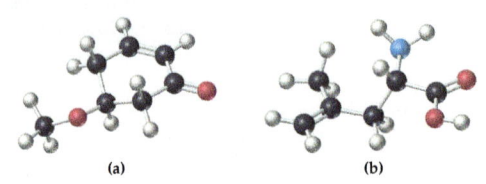

(a)   (b)

12.26 Give systematic names for the following alkanes:

Chemistry in Action and Mastering Reactions Problems
Each boxed application and feature throughout the text ends with a cross-reference to end-of-chapter problems. These problems help students test their understanding of the material and, more importantly, help students see the connection between chemistry and the world around them.

General Questions and Problems
These problems are cumulative, pulling together topics from various parts of the chapter and previous chapters. These help students synthesize the material just learned while helping them review topics from previous chapters.

Acknowledgments

Although this text is now in its seventh edition, each revision has aspired to improve the quality and accuracy of the content and emphasize its relevance to the student users. Achieving this goal requires the coordinated efforts of a dedicated team of editors and media experts. Without them, this textbook would not be possible.

On behalf of all my coauthors, I would like to thank Adam Jaworski (Editor in Chief) and Jeanne Zalesky (Executive Editor) for building an excellent team for this project. Thanks also to Jared Sterzer (Production Manager), Wendy Perez (Project Manager), Eric Schrader (Photo Researcher), Lisa Tarabokjia (Editorial Assistant), and Connie Long (Art Specialist) for their attention to detail as we moved forward. Erica Frost, our developmental editor, deserves special recognition for providing invaluable feedback—her painstaking perusal of each chapter and her eye for details have contributed greatly to the accessibility and relevance of the text. Very special thanks also to Lisa Pierce, Assistant Editor, who patiently guided the process and worked closely with us—thank you for your flexibility and dedication to the success of this project.

The value of this text has also been enhanced by the many individuals who have worked to improve the ancillary materials. Particular thanks to Susan McMurry for her efforts to ensure the accuracy of the answers to problems provided in the text and her revisions of the solutions manuals. Thanks to Ashley Eklund, Miriam Adrianowicz, and Lauren Layn for their work on the media supplements. Thanks also to Margaret Trombley, Kristin Mayo, and Damon Botsakos for their efforts to expand and improve Mastering Chemistry.

Finally, thank you to the many instructors and students who have used the sixth edition and have provided valuable insights and feedback to improve the accuracy of the current edition. We gratefully acknowledge the following reviewers for their contributions to the seventh edition.

Accuracy Reviewers of the Seventh Edition

Sheikh Ahmed, *West Virginia University*
Danae R. Quirk Dorr, *Minnesota State University, Mankato*
Karen Ericson, *Indiana University-Purdue University, Fort Wayne*
Barbara Mowery, *York College of Pennsylvania*
Susan Thomas, *University of Texas, San Antonio*
Richard Triplett, *Des Moines Area Community College*

Reviewers of the Seventh Edition

Francis Burns, *Ferris State University*
Lisa L. Crozier, *Northeast Wisconsin Technical Center*
Robert P. Dixon, *Southern Illinois University, Edwardsville*
Luther Giddings, *Salt Lake Community College*
Arlene Haffa, *University of Wisconsin, Oshkosh*
L. Jaye Hopkins, *Spokane Community College*
Mohammad Mahroof, *Saint Cloud State University*
Gregory Marks, *Carroll University*
Van Quach, *Florida State University*
Douglas Raynie, *South Dakota State University*

Reviewers of the Previous Editions

Sheikh Ahmed, *West Virginia University*
Stanley Bajue, *CUNY-Medgar Evers College*
Daniel Bender, *Sacramento City College*
Dianne A. Bennett, *Sacramento City College*
Alfredo Castro, *Felician College*
Gezahegn Chaka, *Louisiana State University, Alexandria*
Michael Columbia, *Indiana University-Purdue University, Fort Wayne*
Rajeev B. Dabke, *Columbus State University*
Danae R. Quirk Dorr, *Minnesota State University, Mankato*

Pamela S. Doyle, *Essex County College*
Marie E. Dunstan, *York College of Pennsylvania*
Karen L. Ericson, *Indiana University-Purdue University, Fort Wayne*
Charles P. Gibson, *University of Wisconsin, Oshkosh*
Clifford Gottlieb, *Shasta College*
Mildred V. Hall, *Clark State Community College*
Meg Hausman, *University of Southern Maine*
Ronald Hirko, *South Dakota State University*
L. Jaye Hopkins, *Spokane Community College*
Margaret Isbell, *Sacramento City College*
James T. Johnson, *Sinclair Community College*
Margaret G. Kimble, *Indiana University-Purdue University Fort Wayne*
Grace Lasker, *Lake Washington Technical College*
Ashley Mahoney, *Bethel University*
Matthew G. Marmorino, *Indiana University, South Bend*
Diann Marten, *South Central College, Mankato*
Barbara D. Mowery, *York College of Pennsylvania*
Tracey Arnold Murray, *Capital University*
Andrew M. Napper, *Shawnee State University*
Lisa Nichols, *Butte Community College*
Glenn S. Nomura, *Georgia Perimeter College*
Douglas E. Raynie, *South Dakota State University*
Paul D. Root, *Henry Ford Community College*
Victor V. Ryzhov, *Northern Illinois University*
Karen Sanchez, *Florida Community College, Jacksonville-South*
Mir Shamsuddin, *Loyola University, Chicago*
Jeanne A. Stuckey, *University of Michigan*
John Sullivan, *Highland Community College*
Deborah E. Swain, *North Carolina Central University*
Susan T. Thomas, *University of Texas, San Antonio*
Yakov Woldman, *Valdosta State University*

The authors are committed to maintaining the highest quality and accuracy and look forward to comments from students and instructors regarding any aspect of this text and supporting materials. Questions or comments should be directed to the lead co-author.

David S. Ballantine
dballant@niu.edu

Resources in Print and Online

Name of Supplement	Available in Print	Available Online	Instructor or Student Supplement	Description
MasteringChemistry® (www.masteringchemistry.com)		✓	Supplement for Instructors and Students	MasteringChemistry from Pearson has been designed and refined with a single purpose in mind: to help educators create those moments of understanding with their students. The Mastering platform delivers engaging, dynamic learning opportunities—focused on your course objectives and responsive to each student's progress—that are proven to help students absorb course material and understand difficult concepts. By complementing your teaching with our engaging technology and content, you can be confident your students will arrive at those moments—moments of true understanding. The seventh edition will feature 20 new general, organic, and biological (GOB) specific tutorials, totaling over 100 GOB tutorials.
Instructor Resource Manual (isbn: 0321765427)	✓	✓	Supplement for Instructors	The manual features lecture outlines with presentation suggestions, teaching tips, suggested in-class demonstrations, and topics for classroom discussion.
Test Item File (isbn: 0321765435)	✓	✓	Supplement for Instructors	This has been updated to reflect the revisions in this text and contains questions in a bank of more than 2,000 multiple-choice questions.
Instructor Resource Center on DVD (isbn: 0321776119)		✓	Supplement for Instructors	This DVD provides an integrated collection of resources designed to help you make efficient and effective use of your time. The DVD features art from the text, including figures and tables in PDF format for high-resolution printing, as well as pre-built PowerPoint™ presentations. The first presentation contains the images, figures, and tables embedded within the PowerPoint slides, while the second includes a complete, modifiable, lecture outline. The final two presentations contain worked in-chapter sample exercises and questions to be used with Classroom Response Systems. This DVD also contains animations, as well as the TestGen version of the Test Item File, which allows you to create and tailor exams to your needs.
Study Guide and Full Solutions Manual (isbn: 032177616X) Study Guide and Selected Solutions Manual (isbn: 0321776100)	✓		Supplement for Students	**Study Guide and Full Solutions Manual** and **Study Guide and Selected Solutions Manual**, both by Susan McMurry. The selected version provides solutions only to those problems that have a short answer in the text's Selected Answer Appendix. Both versions explain in detail how the answers to the in-text and end-of-chapter problems are obtained. They also contain chapter summaries, study hints, and self-tests for each chapter.
Chemistry and Life in the Laboratory: Experiments, 6e (isbn: 0321751604)	✓		Supplement for Laboratory	**Chemistry and Life in the Laboratory, sixth edition**, by Victor L. Heasley, Val J. Christensen, Gene E. Heasley. Written specifically to accompany any fundamentals of general, organic and biological chemistry text, this manual contains 34 comprehensive and accessible experiments specifically for GOB students.
Catalyst: The Pearson Custom Laboratory Program for Chemistry		✓	Supplement for Laboratory	This program allows you to custom-build a chemistry lab manual that matches your content needs and course organization. You can either write your own labs using the Lab Authoring Kit tool or you can select from the hundreds of labs available at http://www.pearsonlearningsolutions.com/custom-library/catalyst. This program also allows you to add your own course notes, syllabi, or other materials.

Personalized Coaching and Feedback At Your Fingertips

MasteringChemistry®

MasteringChemistry™ has been designed and refined with a single purpose in mind: to help educators create that moment of understanding with their students. The Mastering platform delivers engaging, dynamic learning opportunities—focused on your course objectives and responsive to each student's progress—that are proven to help students absorb course material and understand difficult concepts.

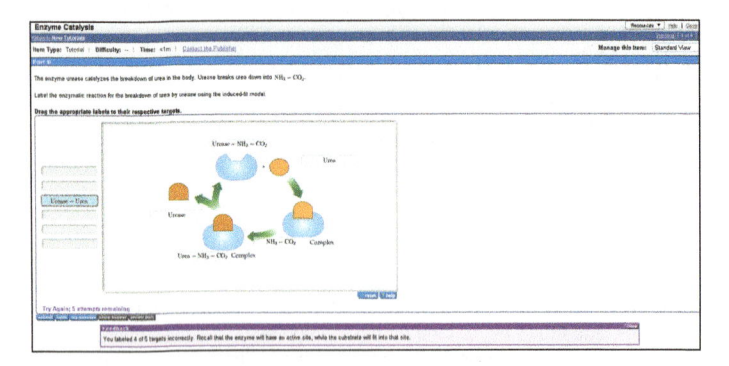

NEW! Chemistry Tutorials

MasteringChemistry® self-paced tutorials are designed to coach students with hints and feedback specific to their individual misconceptions. For the Seventh Edition, new tutorials have been created to guide students through the most challenging General, Organic, and Biological Chemistry topics and help them make connections between different concepts.

Unmatched Gradebook Capability

MasteringChemistry is the only system to capture the step-by-step work of each student in your class, including wrong answers submitted, hints requested, and time taken on every step. This data powers an unprecedented gradebook.

Gradebook Diagnostics

Instructors can identify at a glance students who are having difficulty with the color-coded gradebook. With a single click, charts summarize the most difficult problems in each assignment, vulnerable students, grade distribution, and even score improvement over the course.

Extend Learning Beyond The Classroom

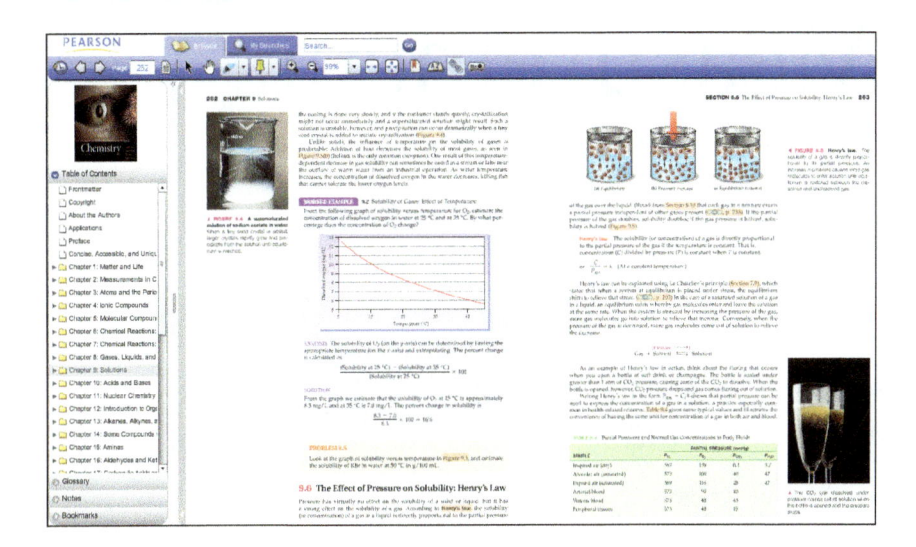

Pearson eText

Pearson eText gives students access to the text whenever and wherever they can access the Internet. The eText pages look exactly like the printed text and include powerful interactive and customization functions.

- Students can create notes, highlight text in different colors, create bookmarks, zoom, click hyperlinked words and phrases to view definitions, and view in single-page or two-page view.
- Students can link directly to associated media files, enabling them to view an animation as they read the text.
- It is possible to perform a full-text search and have the ability to save and export notes.

Instructors can share their notes and highlights with students and can also hide chapters that they do not want their students to read.

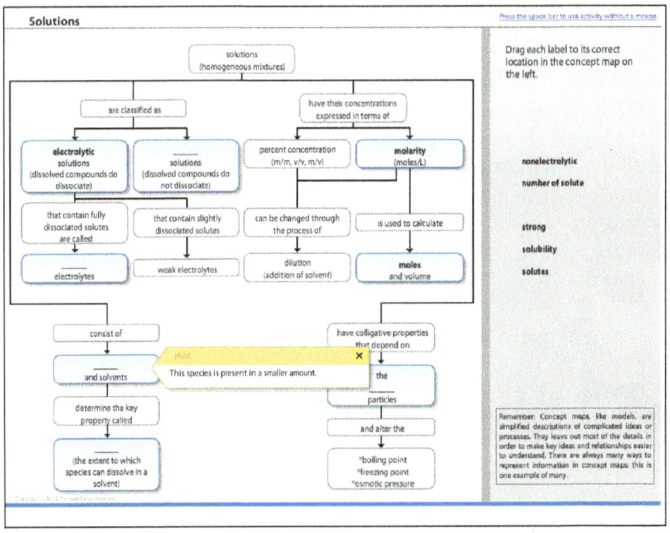

NEW! Concept Map problems

These interactive maps help students synthesize material they learned in previous chapters and demonstrate their understanding of interrelatedness of concepts in general, organic, and biological chemistry.

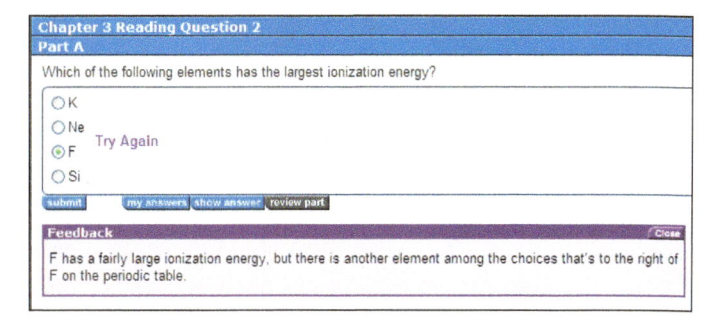

Reading Quizzes

Chapter-specific quizzes and activities focus on important, hard-to-grasp chemistry concepts.

xxv

CHAPTER 1

Matter and Measurements

CONTENTS

1.1 Chemistry: The Central Science
1.2 States of Matter
1.3 Classification of Matter
1.4 Chemical Elements and Symbols
1.5 Elements and the Periodic Table
1.6 Chemical Reaction: An Example of a Chemical Change
1.7 Physical Quantities
1.8 Measuring Mass, Length, and Volume
1.9 Measurement and Significant Figures
1.10 Scientific Notation
1.11 Rounding Off Numbers
1.12 Problem Solving: Unit Conversions and Estimating Answers
1.13 Temperature, Heat, and Energy
1.14 Density and Specific Gravity

◀ Increasing our knowledge of the chemical and physical properties of matter depends on our ability to make measurements that are precise and accurate.

CHAPTER GOALS

1. **What is matter and how is it classified?**
   THE GOAL: Be able to discuss the properties of matter, describe the three states of matter, distinguish between mixtures and pure substances, and distinguish between elements and compounds.

2. **How are chemical elements represented?**
   THE GOAL: Be able to name and give the symbols of common elements.

3. **What kinds of properties does matter have?**
   THE GOAL: Be able to distinguish between chemical and physical properties.

4. **What units are used to measure properties, and how can a quantity be converted from one unit to another?**
   THE GOAL: Be able to name and use the metric and SI units of measurement for mass, length, volume, and temperature and be able to convert quantities from one unit to another using conversion factors.

5. **How good are the reported measurements?**
   THE GOAL: Be able to interpret the number of significant figures in a measurement and round off numbers in calculations involving measurements.

6. **How are large and small numbers best represented?**
   THE GOAL: Be able to interpret prefixes for units of measurement and express numbers in scientific notation.

7. **What techniques are used to solve problems?**
   THE GOAL: Be able to analyze a problem, use the factor-label method to solve the problem, and check the result to ensure that it makes sense chemically and physically.

8. **What are temperature, specific heat, density, and specific gravity?**
   THE GOAL: Be able to define these quantities and use them in calculations.

Earth, air, fire, water—the ancient philosophers believed that all matter was composed of these four fundamental substances. We now know that matter is much more complex, made up of nearly 100 naturally occurring fundamental substances, or elements, in millions of unique combinations. Everything you see, touch, taste, and smell is made of chemicals formed from these elements. Many chemicals occur naturally, but others are synthetic, including the plastics, fibers, and medicines that are so critical to modern life. Just as everything you see is made of chemicals, many of the natural changes you see taking place around you are the result of *chemical reactions*—the change of one chemical into another. The crackling fire of a log burning in the fireplace, the color change of a leaf in the fall, and the changes that a human body undergoes as it grows and ages are all results of chemical reactions. To understand these and other natural processes, you must have a basic understanding of chemistry.

As you might expect, the chemistry of living organisms is complex, and it is not possible to understand all concepts without a proper foundation. Thus, the general plan of this book is to gradually increase in complexity, beginning in the first 11 chapters with a grounding in the scientific fundamentals that govern all of chemistry. In the following six chapters, we look at the nature of the carbon-containing substances, or *organic chemicals*, that compose all living things. In the final 12 chapters, we apply what we have learned in the first part of the book to the study of biological chemistry.

We begin in Chapter 1 with an examination of the states and properties of matter and an introduction to the systems of measurement that are essential to our understanding of matter and its behavior.

1.1 Chemistry: The Central Science

Chemistry is often referred to as "the central science" because it is crucial to nearly all other sciences. In fact, as more and more is learned, the historical dividing lines between chemistry, biology, and physics are fading, and current research is more interdisciplinary. Figure 1.1 diagrams the relationship of chemistry and biological chemistry to other fields of scientific study. Whatever the discipline in which you are most interested, the study of chemistry builds the necessary foundation.

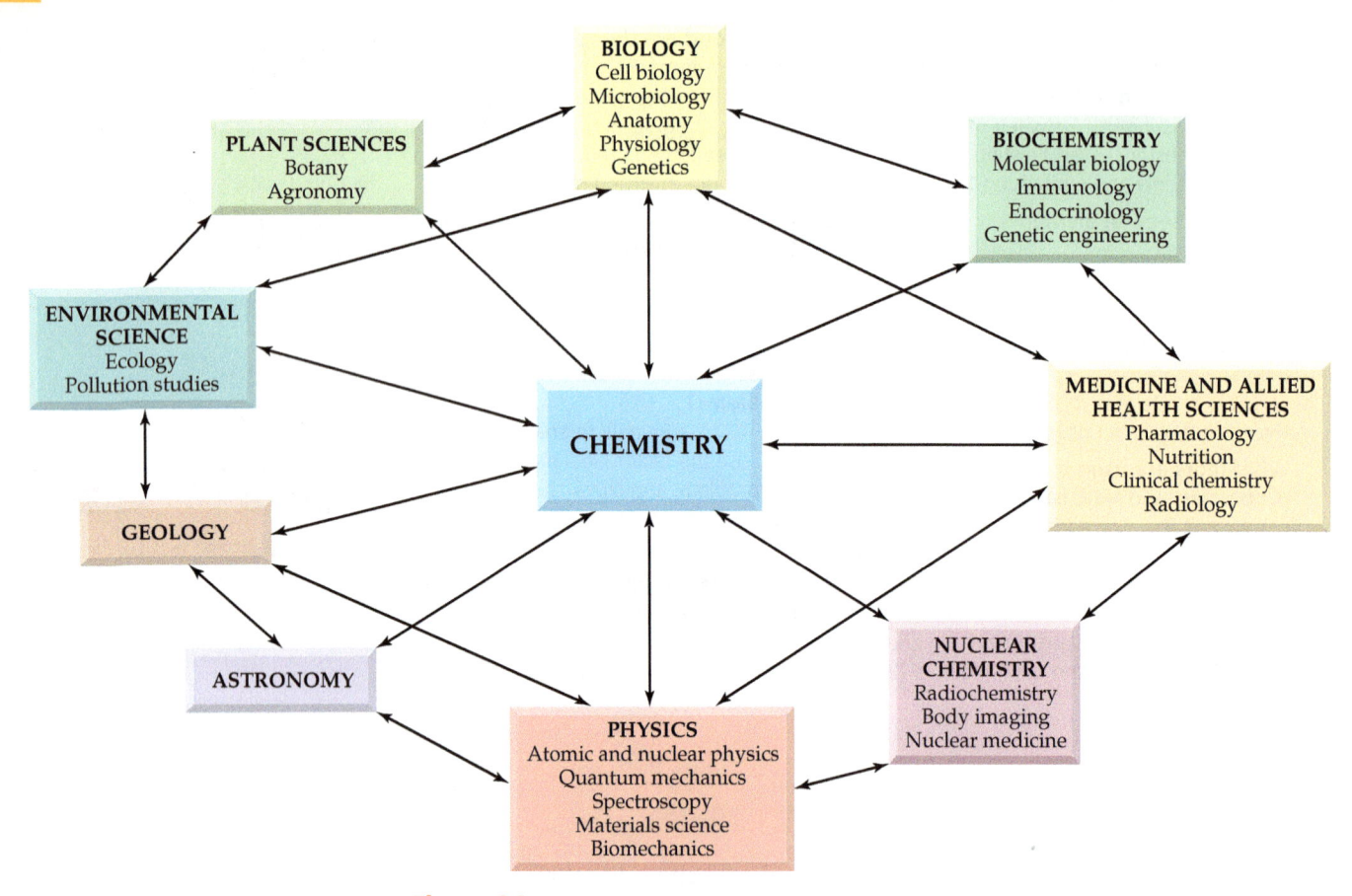

CHAPTER 1 Matter and Measurements

Figure 1.1 diagram:

- **BIOLOGY**: Cell biology, Microbiology, Anatomy, Physiology, Genetics
- **PLANT SCIENCES**: Botany, Agronomy
- **BIOCHEMISTRY**: Molecular biology, Immunology, Endocrinology, Genetic engineering
- **ENVIRONMENTAL SCIENCE**: Ecology, Pollution studies
- **CHEMISTRY**
- **MEDICINE AND ALLIED HEALTH SCIENCES**: Pharmacology, Nutrition, Clinical chemistry, Radiology
- **GEOLOGY**
- **ASTRONOMY**
- **NUCLEAR CHEMISTRY**: Radiochemistry, Body imaging, Nuclear medicine
- **PHYSICS**: Atomic and nuclear physics, Quantum mechanics, Spectroscopy, Materials science, Biomechanics

▲ Figure 1.1
Some relationships between chemistry—the central science—and other scientific and health-related disciplines.

Chemistry The study of the nature, properties, and transformations of matter.

Matter The physical material that makes up the universe; anything that has mass and occupies space.

Scientific Method The systematic process of observation, hypothesis, and experimentation used to expand and refine a body of knowledge.

Property A characteristic useful for identifying a substance or object.

Physical change A change that does not affect the chemical makeup of a substance or object.

Chemical change A change in the chemical makeup of a substance.

Chemistry is the study of matter—its nature, properties, and transformations. **Matter**, in turn, is a catchall word used to describe anything physically real—anything you can see, touch, taste, or smell. In more scientific terms, matter is anything that has mass and volume. As with our knowledge of all the other sciences, our knowledge of chemistry has developed by application of a process called the **scientific method** (see Chemistry in Action on p. 8). Starting with observations and measurements of the physical world, we form hypotheses to explain what we have observed. These hypotheses can then be tested by more observations and measurements, or experiments, to improve our understanding.

How might we describe different kinds of matter more specifically? Any characteristic that can be used to describe or identify something is called a **property**; size, color, and temperature are all familiar examples. Less familiar properties include *chemical composition*, which describes what matter is made of, and *chemical reactivity*, which describes how matter behaves. Rather than focus on the properties themselves, however, it is often more useful to think about *changes* in properties. Changes are of two types: *physical* and *chemical*. A **physical change** is one that does not alter the chemical makeup of a substance, whereas a **chemical change** is one that *does* alter a substance's chemical makeup. The melting of solid ice to give liquid water, for instance, is a physical change because the water changes only in form but not in chemical makeup. The rusting of an iron bicycle left in the rain, however, is a chemical change because iron combines with oxygen and moisture from the air to give a new substance, rust.

Table 1.1 lists some chemical and physical properties of several familiar substances—water, table sugar (sucrose), and baking soda (sodium bicarbonate). Note in Table 1.1 that the changes occurring when sugar and baking soda are heated are chemical changes, because new substances are produced.

SECTION 1.2 States of Matter 5

TABLE 1.1 Some Properties of Water, Sugar, and Baking Soda

Water	Sugar (Sucrose)	Baking Soda (Sodium Bicarbonate)
Physical properties		
Colorless liquid	White crystals	White powder
Odorless	Odorless	Odorless
Melting point: 0 °C	Begins to decompose at 160 °C, turning black and giving off water.	Decomposes at 270 °C, giving off water and carbon dioxide.
Boiling point: 100 °C	—	—
Chemical properties		
Composition:*	Composition:*	Composition:*
11.2% hydrogen	6.4% hydrogen	27.4% sodium
88.8% oxygen	42.1% carbon	1.2% hydrogen
	51.5% oxygen	14.3% carbon
		57.1% oxygen
Does not burn.	Burns in air.	Does not burn.

*Compositions are given by mass percent.

▲ Burning of potassium in water is an example of a chemical change.

PROBLEM 1.1
Identify each of the following as a physical change or a chemical change:
- **(a)** Grinding of a metal
- **(b)** Fruit ripening
- **(c)** Wood burning
- **(d)** A rain puddle evaporating

1.2 States of Matter

Matter exists in three forms: solid, liquid, and gas. A **solid** has a definite volume and a definite shape that does not change regardless of the container in which it is placed; for example, a wooden block, marbles, or a cube of ice. A **liquid**, by contrast, has a definite volume but an indefinite shape. The volume of a liquid, such as water, does not change when it is poured into a different container, but its shape does. A **gas** is different still, having neither a definite volume nor a definite shape. A gas expands to fill the volume and take the shape of any container it is placed in, such as the helium in a balloon or steam formed by boiling water (Figure 1.2).

Solid A substance that has a definite shape and volume.

Liquid A substance that has a definite volume but assumes the shape of its container.

Gas A substance that has neither a definite volume nor a definite shape.

◀ Figure 1.2
The three states of matter—solid, liquid, and gas.

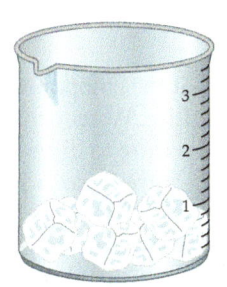

(a) Ice: A solid has a definite volume and a definite shape independent of its container.

(b) Water: A liquid has a definite volume but a variable shape that depends on its container.

(c) Steam: A gas has both variable volume and shape that depend on its container.

6 CHAPTER 1 Matter and Measurements

State of matter The physical state of a substance as a solid, liquid, or gas.

Change of state The conversion of a substance from one state to another—for example, from liquid to gas.

Many substances, such as water, can exist in all three phases, or **states of matter**—the solid state, the liquid state, and the gaseous state—depending on the temperature. The conversion of a substance from one state to another is known as a **change of state**. The melting of a solid, the freezing or boiling of a liquid, and the condensing of a gas to a liquid are familiar to everyone.

Worked Example 1.1 Identifying States of Matter

▶▶▶ The symbol °C means degrees Celsius and will be discussed in Section 1.13.

Formaldehyde is a disinfectant, a preservative, and a raw material for the manufacturing of plastics. Its melting point is −92 °C and its boiling point is −19.5 °C. Is formaldehyde a gas, a liquid, or a solid at room temperature (25 °C)?

ANALYSIS The state of matter of any substance depends on its temperature. How do the melting point and boiling point of formaldehyde compare with room temperature?

SOLUTION
Room temperature (25 °C) is above the boiling point of formaldehyde (−19.5 °C), and so the formaldehyde is a gas.

PROBLEM 1.2
Acetic acid, which gives the sour taste to vinegar, has a melting point of 16.7 °C and a boiling point of 118 °C. Predict the physical state of acetic acid when the ambient temperature is 10 °C.

1.3 Classification of Matter

The first question a chemist asks about an unknown substance is whether it is a pure substance or a mixture. Every sample of matter is one or the other. Water and sugar alone are pure substances, but stirring some sugar into a glass of water creates a *mixture*.

Pure substance A substance that has a uniform chemical composition throughout.

Mixture A blend of two or more substances, each of which retains its chemical identity.

Homogeneous mixture A uniform mixture that has the same composition throughout.

Heterogeneous mixture A non-uniform mixture that has regions of different composition.

▶▶▶ We'll revisit the properties of mixtures in Section 9.1 when we discuss solutions.

What is the difference between a pure substance and a mixture? One difference is that a **pure substance** is uniform in its chemical composition and its properties all the way down to the microscopic level. Every sample of water, sugar, or baking soda, regardless of source, has the composition and properties listed in Table 1.1. A **mixture**, however, can vary in both composition and properties, depending on how it is made. A **homogeneous mixture** is a blend of two or more pure substances having a uniform composition at the microscopic level. Sugar dissolved in water is one example. You cannot always distinguish between a pure substance and a homogeneous mixture just by looking. The sugar–water mixture *looks* just like pure water but differs on a molecular level. The amount of sugar dissolved in a glass of water will determine the sweetness, boiling point, and other properties of the mixture. A **heterogeneous mixture**, by contrast, is a blend of two or more pure substances having non-uniform composition, such as a vegetable stew in which each spoonful is different. It is relatively easy to distinguish heterogeneous mixtures from pure substances.

Another difference between a pure substance and a mixture is that the components of a mixture can be separated without changing their chemical identities. Water can be separated from a sugar–water mixture, for example, by boiling the mixture to drive off the steam and then condensing the steam to recover the pure water. Pure sugar is left behind in the container.

Element A fundamental substance that cannot be broken down chemically into any simpler substance.

▶▶▶ Elements are explored in the next section of this chapter (Section 1.4).

Pure substances are themselves classified into two groups: those that can undergo a chemical breakdown to yield simpler substances and those that cannot. A pure substance that cannot be broken down chemically into simpler substances is called an **element**. Examples include hydrogen, oxygen, aluminum, gold, and sulfur. At the time this book was printed, 118 elements had been identified, although only 91 of these occur naturally. All the millions of other substances in the universe are derived from them.

SECTION 1.3 Classification of Matter   7

Any pure material that *can* be broken down into simpler substances by a chemical change is called a **chemical compound**. The term *compound* implies "more than one" (think "compound fracture"). A chemical compound, therefore, is formed by combining two or more elements to make a new substance. Water, for example, can be chemically changed by passing an electric current through it to produce hydrogen and oxygen. In writing this chemical change, the initial substance, or **reactant** (water), is written on the left; the new substances, or **products** (hydrogen and oxygen), are written on the right; and an arrow connects the two parts to indicate a chemical change, or **chemical reaction**. The conditions necessary to bring about the reaction are written above and below the arrow.

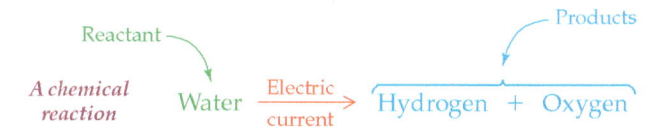

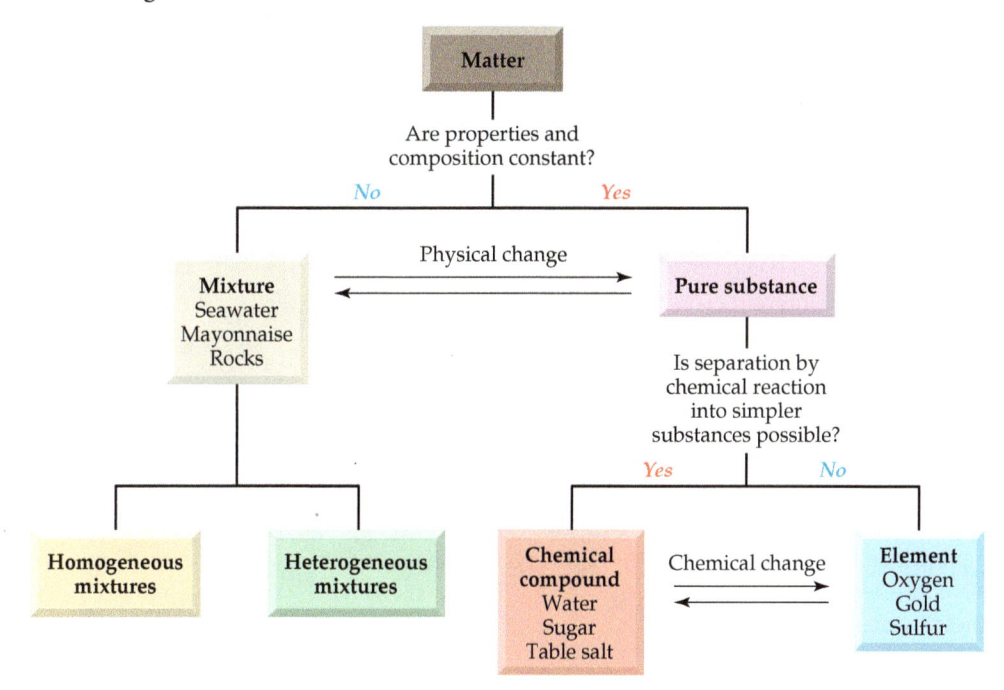

Chemical compound A pure substance that can be broken down into simpler substances by chemical reactions.

Reactant A starting substance that undergoes change during a chemical reaction.

Product A substance formed as the result of a chemical reaction.

Chemical reaction A process in which the identity and composition of one or more substances are changed.

▶▶ We will discuss how chemical reactions are represented in more detail in Section 1.6, and how reactions are classified in Chapter 5.

A chemical reaction   Water $\xrightarrow{\text{Electric current}}$ Hydrogen + Oxygen

(Reactant; Products)

The classification of matter into mixtures, pure compounds, and elements is summarized in Figure 1.3.

Figure 1.3
◀ A scheme for the classification of matter.

Matter
— Are properties and composition constant?
  - No → Mixture (Seawater, Mayonnaise, Rocks)
    - Homogeneous mixtures
    - Heterogeneous mixtures
  - Yes → Pure substance
    - Is separation by chemical reaction into simpler substances possible?
      - Yes → Chemical compound (Water, Sugar, Table salt)
      - No → Element (Oxygen, Gold, Sulfur)

Physical change connects Mixture ↔ Pure substance. Chemical change connects Chemical compound ↔ Element.

Worked Example 1.2 Classifying Matter

Classify each of the following as a mixture or a pure substance. If a mixture, classify it as heterogeneous or homogeneous. If a pure substance, identify it as an element or a compound.

(a) Vanilla ice cream   (b) Sugar

ANALYSIS Refer to the definitions of pure substances and mixtures. Is the substance composed of more than one kind of matter? Is the composition uniform?

SOLUTION

(a) Vanilla ice cream is composed of more than one substance—cream, sugar, and vanilla flavoring. The composition appears to be uniform throughout, so this is a homogeneous mixture.

(b) Sugar is composed of only one kind of matter—pure sugar. This is a pure substance. It can be converted to some other substance by a chemical change (see Table 1.1), so it is not an element. It must be a compound.

PROBLEM 1.3
Classify each of the following as a mixture or a pure substance. If a mixture, classify it as heterogeneous or homogeneous. If a pure substance, identify it as an element or a compound.
(a) Concrete (b) The helium in a balloon (c) A lead weight (d) Wood

PROBLEM 1.4
Classify each of the following as a physical change or a chemical change:
(a) Dissolving sugar in water
(b) Producing carbon dioxide gas and solid lime by heating limestone
(c) Frying an egg
(d) The conversion of salicylic acid to acetylsalicylic acid (see the following Chemistry in Action)

▶▶ Prostaglandins are discussed in Section 24.9.

CHEMISTRY IN ACTION

Aspirin—A Case Study

Acetylsalicylic acid, more commonly known as aspirin, is perhaps the first true wonder drug. It is used as an analgesic to reduce fevers and to relieve headaches and body pains. It possesses anticoagulant properties, which in low doses can help prevent heart attacks and minimize the damage caused by strokes. But how was it discovered, and how does it work? The "discovery" of aspirin is a combination of serendipity and a process known as the scientific method: observation, evaluation of data, formation of a hypothesis, and the design of experiments to test the hypothesis and further our understanding.

The origins of aspirin can be traced back to the ancient Greek physician Hippocrates in 400 B.C., who prescribed the bark and leaves of the willow tree to relieve pain and fever. His knowledge of the therapeutic properties of these substances was the result of systematic observations and the evaluation of folklore—knowledge of the common people obtained through trial and error. The development of aspirin took another step forward in 1828 when scientists isolated a bitter-tasting yellow extract, called salicin, from willow bark. Experimental evidence identified salicin as the active ingredient responsible for the observed medical effects. Salicin could be easily converted by chemical reaction to salicylic acid (SA), which by the late 1800s was being mass-produced and marketed. SA had an unpleasant taste, however, and often caused stomach irritation and indigestion.

Further experiments were performed to convert salicylic acid to a substance that retained the therapeutic activity of SA, but without the unpleasant side effects. The discovery of acetylsalicylic acid (ASA), a derivative of SA, has often been attributed to Felix Hoffman, a chemist working for the Bayer pharmaceutical labs, but the first synthesis of ASA was actually reported by a French chemist, Charles Gerhardt, in 1853. Nevertheless, Hoffman obtained a patent for ASA in 1900, and Bayer marketed the new drug, now called aspirin, in water-soluble tablets.

▲ *Hippocrates.* The ancient Greek physician prescribed a precursor of aspirin found in willow bark to relieve pain.

But, how does aspirin work? Once again, experimental data provided insights into the therapeutic activity of aspirin. In 1971, the British pharmacologist John Vane discovered that aspirin suppresses the body's production of prostaglandins, which are responsible for the pain and swelling that accompany inflammation. The discovery of this mechanism led to the development of new analgesic drugs.

Research continues to explore aspirin's potential for preventing colon cancer, cancer of the esophagus, and other diseases.

See Chemistry in Action Problem 1.96 at the end of the chapter.

> **KEY CONCEPT PROBLEM 1.5**
>
> In the image below, red spheres represent element A and blue spheres represent element B. Identify the process illustrated in the image as a chemical change or a physical change. Explain your answer.

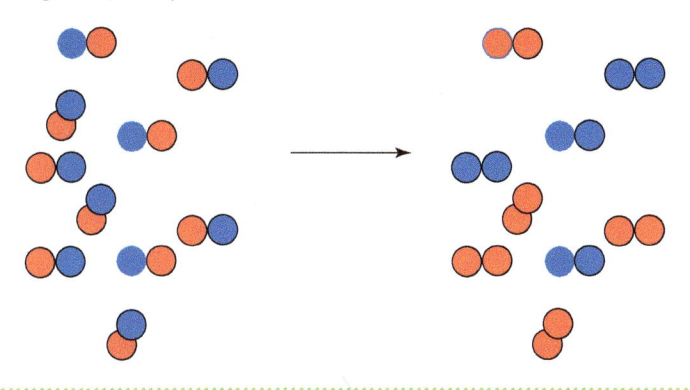

1.4 Chemical Elements and Symbols

As of the date this book was printed, 118 chemical elements have been identified. Some are certainly familiar to you—oxygen, helium, iron, aluminum, copper, and gold, for example—but many others are probably unfamiliar—rhenium, niobium, thulium, and promethium. Rather than write out the full names of elements, chemists use a shorthand notation in which elements are referred to by one- or two-letter symbols. The names and symbols of some common elements are listed in Table 1.2, and a complete alphabetical list is given inside the front cover of this book.

Note that all two-letter symbols have only their first letter capitalized, whereas the second letter is always lowercase. The symbols of most common elements are the first one or two letters of the elements' commonly used names, such as H (hydrogen) and Al (aluminum). Pay special attention, however, to the elements grouped in the last column to the right in Table 1.2. The symbols for these elements are derived from their original Latin names, such as Na for sodium, once known as *natrium*. The only way to learn these symbols is to memorize them; fortunately, they are few in number.

Only 91 of the elements occur naturally; the remaining elements have been produced artificially by chemists and physicists. Each element has its own distinctive properties, and just about all of the first 95 elements have been put to use in some way that takes advantage

▶▶ We will discuss the creation of new elements by nuclear bombardment in Chapter 11.

TABLE 1.2 Names and Symbols for Some Common Elements

Elements with Symbols Based on Modern Names							Elements with Symbols Based on Latin Names	
Al	Aluminum	Co	Cobalt	N	Nitrogen	Cu	Copper (*cuprum*)
Ar	Argon	F	Fluorine	O	Oxygen	Au	Gold (*aurum*)
Ba	Barium	He	Helium	P	Phosphorus	Fe	Iron (*ferrum*)
Bi	Bismuth	H	Hydrogen	Pt	Platinum	Pb	Lead (*plumbum*)
B	Boron	I	Iodine	Rn	Radon	Hg	Mercury (*hydrargyrum*)
Br	Bromine	Li	Lithium	Si	Silicon	K	Potassium (*kalium*)
Ca	Calcium	Mg	Magnesium	S	Sulfur	Ag	Silver (*argentum*)
C	Carbon	Mn	Manganese	Ti	Titanium	Na	Sodium (*natrium*)
Cl	Chlorine	Ni	Nickel	Zn	Zinc	Sn	Tin (*stannum*)

TABLE 1.3 Elemental Composition of the Earth's Crust and the Human Body*

Earth's Crust		Human Body	
Oxygen	46.1%	Oxygen	61%
Silicon	28.2%	Carbon	23%
Aluminum	8.2%	Hydrogen	10%
Iron	5.6%	Nitrogen	2.6%
Calcium	4.1%	Calcium	1.4%
Sodium	2.4%	Phosphorus	1.1%
Magnesium	2.3%	Sulfur	0.20%
Potassium	2.1%	Potassium	0.20%
Titanium	0.57%	Sodium	0.14%
Hydrogen	0.14%	Chlorine	0.12%

*Mass percent values are given.

Chemical formula A notation for a chemical compound using element symbols and subscripts to show how many atoms of each element are present.

▶▶ We'll learn more about the structure of atoms and how they form compounds in Chapter 2.

of those properties. As indicated in Table 1.3, which shows the approximate elemental composition of the earth's crust and the human body, the naturally occurring elements are not equally abundant. Oxygen and silicon together account for nearly 75% of the mass in the earth's crust; oxygen, carbon, and hydrogen account for nearly all the mass of a human body.

Just as elements combine to form chemical compounds, symbols are combined to produce **chemical formulas**, which show by subscripts how many *atoms* (the smallest fundamental units) of each element are in a given chemical compound. For example, the formula H_2O represents water, which contains 2 hydrogen atoms combined with 1 oxygen atom. Similarly, the formula CH_4 represents methane (natural gas), and the formula $C_{12}H_{22}O_{11}$ represents table sugar (sucrose). When no subscript is given for an element, as for carbon in the formula CH_4, a subscript of "1" is understood.

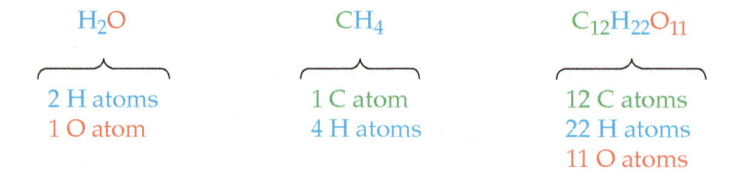

H_2O — 2 H atoms, 1 O atom

CH_4 — 1 C atom, 4 H atoms

$C_{12}H_{22}O_{11}$ — 12 C atoms, 22 H atoms, 11 O atoms

PROBLEM 1.6
Match the names of the elements described below (a–f) with their elemental symbols (1–6).
(a) Sodium, a major component in table salt
(b) Tungsten, a metal used in light bulb filaments
(c) Strontium, used to produce brilliant red colors in fireworks
(d) Titanium, used in artificial hips and knee-replacement joints
(e) Fluorine, added to municipal water supplies to strengthen tooth enamel
(f) Tin, a metal used in solder

(1) W　　(2) Na　　(3) Sn　　(4) F　　(5) Ti　　(6) Sr

PROBLEM 1.7
Identify the elements represented in each of the following chemical formulas, and tell the number of atoms of each element:
(a) NH_3 (ammonia)　　(b) $NaHCO_3$ (sodium bicarbonate)
(c) C_8H_{18} (octane, a component of gasoline)　　(d) $C_6H_8O_6$ (vitamin C)

1.5 Elements and the Periodic Table

The symbols of the known elements are normally presented in a tabular format called the **periodic table**, as shown in Figure 1.4 and the inside front cover of this book. We will have much more to say about the periodic table and how it is numbered later, but will note for now that it is the most important organizing principle in chemistry. An enormous amount of information is embedded in the periodic table, information that gives chemists the ability to explain known chemical behavior of elements and to predict new behavior. The elements can be roughly divided into three groups: *metals*, *nonmetals*, and *metalloids* (sometimes called *semimetals*).

Periodic table A tabular format listing all known elements.

▶▶ The organization of the periodic table will be discussed in Chapter 2.

1A 1	2A 2	3B 3	4B 4	5B 5	6B 6	7B 7	8B 8	8B 9	8B 10	1B 11	2B 12	3A 13	4A 14	5A 15	6A 16	7A 17	8A 18
1 H 1.00794																	2 He 4.00260
3 Li 6.941	4 Be 9.01218											5 B 10.81	6 C 12.011	7 N 14.0067	8 O 15.9994	9 F 18.9984	10 Ne 20.1797
11 Na 22.98977	12 Mg 24.305											13 Al 26.98154	14 Si 28.0855	15 P 30.9738	16 S 32.066	17 Cl 35.4527	18 Ar 39.948
19 K 39.0983	20 Ca 40.078	21 Sc 44.9559	22 Ti 47.88	23 V 50.9415	24 Cr 51.996	25 Mn 54.9380	26 Fe 55.847	27 Co 58.9332	28 Ni 58.69	29 Cu 63.546	30 Zn 65.39	31 Ga 69.72	32 Ge 72.61	33 As 74.9216	34 Se 78.96	35 Br 79.904	36 Kr 83.80
37 Rb 85.4678	38 Sr 87.62	39 Y 88.9059	40 Zr 91.224	41 Nb 92.9064	42 Mo 95.94	43 Tc (98)	44 Ru 101.07	45 Rh 102.9055	46 Pd 106.42	47 Ag 107.8682	48 Cd 112.41	49 In 114.82	50 Sn 118.710	51 Sb 121.757	52 Te 127.60	53 I 126.9045	54 Xe 131.29
55 Cs 132.9054	56 Ba 137.33	57 *La 138.9055	72 Hf 178.49	73 Ta 180.9479	74 W 183.85	75 Re 186.207	76 Os 190.2	77 Ir 192.22	78 Pt 195.08	79 Au 196.9665	80 Hg 200.59	81 Tl 204.383	82 Pb 207.2	83 Bi 208.9804	84 Po (209)	85 At (210)	86 Rn (222)
87 Fr (223)	88 Ra 226.0254	89 †Ac 227.0278	104 Rf (261)	105 Db (262)	106 Sg (266)	107 Bh (264)	108 Hs (269)	109 Mt (268)	110 Ds (271)	111 Rg (272)	112 Cn (285)	113 (284)	114 (289)	115 (288)	116 (292)	117 (293)	118 (294)

58 Ce 140.12	59 Pr 140.9077	60 Nd 144.24	61 Pm (145)	62 Sm 150.36	63 Eu 151.965	64 Gd 157.25	65 Tb 158.9254	66 Dy 162.50	67 Ho 164.9304	68 Er 167.26	69 Tm 168.9342	70 Yb 173.04	71 Lu 174.967
90 Th 232.0381	91 Pa 231.0399	92 U 238.0289	93 Np 237.048	94 Pu (244)	95 Am (243)	96 Cm (247)	97 Bk (247)	98 Cf (251)	99 Es (252)	100 Fm (257)	101 Md (258)	102 No (259)	103 Lr (262)

Metals Metalloids Nonmetals

▲ **Figure 1.4**
The periodic table of the elements.
Metals appear on the left, nonmetals on the right, and metalloids in a zigzag band between metals and nonmetals. The numbering system is explained in Section 2.4.

Ninety-four of the currently known elements are metals—aluminum, gold, copper, and zinc, for example. **Metals** are solid at room temperature (except for mercury), usually have a lustrous appearance when freshly cut, are good conductors of heat and electricity, and are malleable rather than brittle. That is, metals can be pounded into different shapes rather than shattering when struck. Note that metals occur on the left side of the periodic table.

Metal A malleable element, with a lustrous appearance, that is a good conductor of heat and electricity.

Eighteen elements are **nonmetals**. All are poor conductors of heat and electricity. Eleven are gases at room temperature, six are brittle solids, and one is a liquid. Oxygen and nitrogen, for example, are gases present in air; sulfur is a solid found in large underground deposits. Bromine is the only liquid nonmetal. Note that nonmetals occur on the right side of the periodic table.

Nonmetal An element that is a poor conductor of heat and electricity.

Only six elements are **metalloids**, so named because their properties are intermediate between those of metals and nonmetals. Boron, silicon, and arsenic are examples. Pure silicon has a lustrous or shiny surface, like a metal, but it is brittle, like a nonmetal, and its electrical conductivity lies between that of metals and nonmetals. Note that metalloids occur in a zigzag band between metals on the left and nonmetals on the right side of the periodic table.

Metalloid An element whose properties are intermediate between those of a metal and a nonmetal.

CHAPTER 1 Matter and Measurements

(a) (b) (c)

▲ **Metals: Gold, zinc, and copper.**
(a) Known for its beauty, gold is very unreactive and is used primarily in jewelry and in electronic components. (b) Zinc, an essential trace element in our diets, has industrial uses ranging from the manufacture of brass, to roofing materials, to batteries. (c) Copper is widely used in electrical wiring, in water pipes, and in coins.

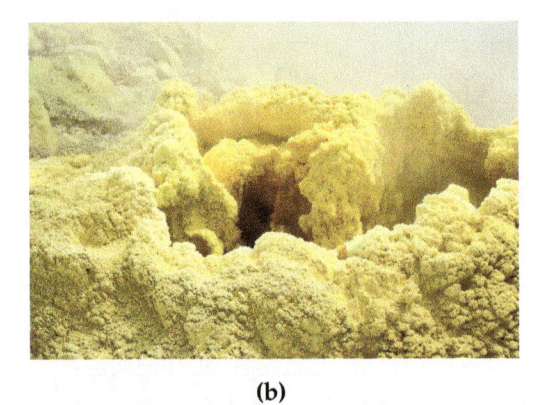

(a) (b) (c)

▲ **Nonmetals: Nitrogen, sulfur, and iodine.**
(a) Nitrogen, (b) sulfur, and (c) iodine are essential to all living things. Pure nitrogen, which constitutes almost 80% of air, is a gas at room temperature and does not condense to a liquid until it is cooled to −328 °C. Sulfur, a yellow solid, is found in large underground deposits in Texas and Louisiana. Iodine is a dark violet crystalline solid that was first isolated from seaweed.

(a) (b)

▲ **Metalloids: Boron and silicon.**
(a) Boron is a strong, hard metalloid used in making the composite materials found in military aircraft. (b) Silicon is well known for its use in making computer chips.

Those elements essential for human life are listed in Table 1.4. In addition to the well-known elements carbon, hydrogen, oxygen, and nitrogen, less familiar elements such as molybdenum and selenium are also important.

TABLE 1.4 Elements Essential for Human Life*

Element	Symbol	Function
Carbon	C	These four elements are present in all living organisms
Hydrogen	H	
Oxygen	O	
Nitrogen	N	
Arsenic	As	May affect cell growth and heart function
Boron	B	Aids in the use of Ca, P, and Mg
Calcium*	Ca	Necessary for growth of teeth and bones
Chlorine*	Cl	Necessary for maintaining salt balance in body fluids
Chromium	Cr	Aids in carbohydrate metabolism
Cobalt	Co	Component of vitamin B_{12}
Copper	Cu	Necessary to maintain blood chemistry
Fluorine	F	Aids in the development of teeth and bones
Iodine	I	Necessary for thyroid function
Iron	Fe	Necessary for oxygen-carrying ability of blood
Magnesium*	Mg	Necessary for bones, teeth, and muscle and nerve action
Manganese	Mn	Necessary for carbohydrate metabolism and bone formation
Molybdenum	Mo	Component of enzymes necessary for metabolism
Nickel	Ni	Aids in the use of Fe and Cu
Phosphorus*	P	Necessary for growth of bones and teeth; present in DNA/RNA
Potassium*	K	Component of body fluids; necessary for nerve action
Selenium	Se	Aids vitamin E action and fat metabolism
Silicon	Si	Helps form connective tissue and bone
Sodium*	Na	Component of body fluids; necessary for nerve and muscle action
Sulfur*	S	Component of proteins; necessary for blood clotting
Zinc	Zn	Necessary for growth, healing, and overall health

*C, H, O, and N are present in most foods. Other elements listed vary in their distribution in different foods. Those marked with an asterisk are *macronutrients*, essential in the diet at more than 100 mg/day; the rest, other than C, H, O, and N, are *micronutrients*, essential at 15 mg or less per day.

LOOKING AHEAD ▶▶ The elements listed in Table 1.4 are not present in our bodies in their free forms. Instead, they are combined into many thousands of different chemical compounds. We will talk about some compounds formed by metals in Chapter 3 and compounds formed by nonmetals in Chapter 4.

PROBLEM 1.8
The six metalloids are boron (B), silicon (Si), germanium (Ge), arsenic (As), antimony (Sb), and tellurium (Te). Locate them in the periodic table, and tell where they appear with respect to metals and nonmetals.

PROBLEM 1.9
Locate the element Hg (discussed in the Chemisty in Action on p. 15) in the periodic table. Is it a metal, nonmetal, or metalloid? What physical and chemical properties contribute to the toxicity of mercury and compounds containing mercury?

1.6 Chemical Reactions: An Example of Chemical Change

If we take a quick look at an example of a chemical reaction, we can reinforce some of the ideas discussed in the previous sections. The element *nickel* is a hard, shiny metal, and the compound *hydrogen chloride* is a colorless gas that dissolves in water to give a solution called *hydrochloric acid*. When pieces of nickel are added to hydrochloric acid in a test tube, the nickel is slowly eaten away, the colorless solution turns green, and a gas bubbles out of the test tube. The change in color, the dissolving of the nickel, and the appearance of gas bubbles are indications that a chemical reaction is taking place, as shown in Figure 1.5.

Overall, the reaction of nickel with hydrochloric acid can be either written in words or represented in a shorthand notation using symbols to represent the elements or compounds involved as reactants and products, as shown below.

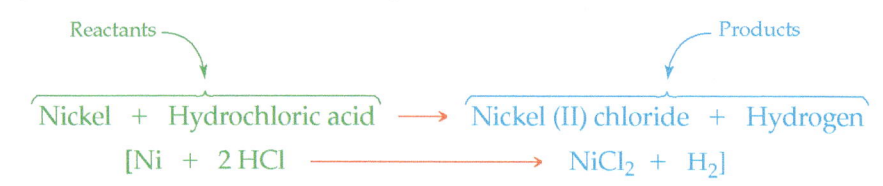

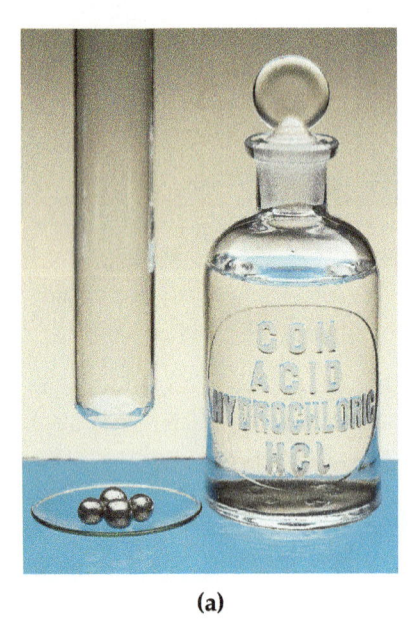

$$\underbrace{\text{Nickel} + \text{Hydrochloric acid}}_{\text{Reactants}} \longrightarrow \underbrace{\text{Nickel (II) chloride} + \text{Hydrogen}}_{\text{Products}}$$

$$[\text{Ni} + 2\,\text{HCl} \longrightarrow \text{NiCl}_2 + \text{H}_2]$$

(a)   (b)   (c)

▲ **Figure 1.5**
Reactants and products of a chemical reaction.
(a) The reactants: The flat dish contains pieces of nickel, an element that is a typical lustrous metal. The bottle contains hydrochloric acid, a solution of the chemical compound hydrogen chloride in water. These reactants are about to be combined in the test tube. (b) The reaction: As the chemical reaction occurs, the colorless solution turns green when water-insoluble nickel metal slowly changes into the water-soluble chemical compound nickel (II) chloride. Gas bubbles of the element hydrogen are produced and rise slowly through the green solution. (c) The product: Hydrogen gas can be collected as it bubbles from the solution. Removal of water from the solution leaves behind the other product, a solid green chemical compound known as nickel (II) chloride.

1.7 Physical Quantities

Our understanding of matter depends on our ability to measure the changes in physical properties associated with physical and chemical change. Mass, volume, temperature, density, and other physical properties that can be measured are called **physical quantities** and are described by both a number and a **unit** of defined size:

Physical quantity A physical property that can be measured.

Unit A defined quantity used as a standard of measurement.

$$\underset{\text{Number}}{61.2} \; \underset{\text{Unit}}{\text{kilograms}}$$

CHEMISTRY IN ACTION

Mercury and Mercury Poisoning

Mercury, the only metallic element that is liquid at room temperature, has fascinated people for millennia. Egyptian kings were buried in their pyramids along with containers of mercury, alchemists during the Middle Ages used mercury to dissolve gold, and Spanish galleons carried loads of mercury to the New World in the 1600s for use in gold and silver mining. Even its symbol, Hg, from the Latin *hydrargyrum,* meaning "liquid silver," hints at mercury's uniqueness.

Much of the recent interest in mercury has concerned its toxicity, but there are some surprises. For example, the mercury compound Hg_2Cl_2 (called *calomel*) is nontoxic and has a long history of medical use as a laxative, yet it is also used as a fungicide and rat poison. Dental amalgam, a solid alloy of approximately 50% elemental mercury, 35% silver, 13% tin, 1% copper, and trace amounts of zinc, has been used by dentists for many years to fill tooth cavities, with little or no adverse effects except in individuals with a hypersensitivity to mercury. Yet exposure to elemental mercury *vapor* for long periods leads to mood swings, headaches, tremors, and loss of hair and teeth. The widespread use of mercuric nitrate, a mercury compound employed to make the felt used in hats, exposed many hatters of the eighteenth and nineteenth centuries to toxic levels of mercury. The eccentric behavior displayed by hatters suffering from mercury poisoning led to the phrase "mad as a hatter."

Why is mercury toxic in some forms but not in others? It turns out that the toxicity of mercury and its compounds is related to solubility. Only soluble mercury compounds are toxic, because they can be transported through the bloodstream to all parts of the body, where they react with different enzymes and interfere with various biological processes. Elemental mercury and insoluble mercury compounds become toxic only when converted into soluble compounds, reactions that are extremely slow in the body. Calomel, for example, is an insoluble mercury compound that passes through the body long before it is converted into any soluble compounds. Mercury alloys were considered safe for dental use because mercury does not evaporate readily from the alloys and it neither reacts with nor dissolves in saliva. Mercury vapor, however, remains in the lungs when breathed, until it is slowly converted into soluble compounds. Soluble organic forms of mercury can be particularly toxic. Trace amounts are found in nearly all seafood, but some larger species such as king mackerel and swordfish contain higher levels of mercury. Because mercury can affect the developing brain and nervous system of a fetus, pregnant women are often advised to avoid consuming them.

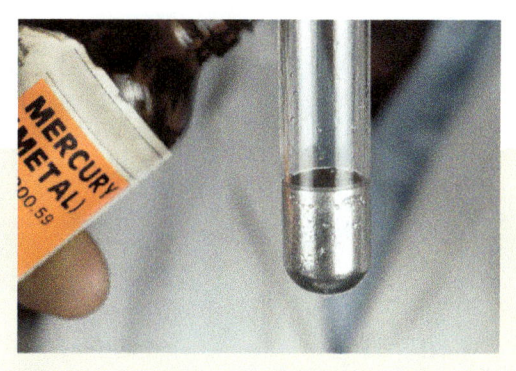

▲ Elemental Mercury, a liquid at room temperature, forms many toxic compounds.

Recent events have raised new concerns regarding the safe use of mercury in some other applications. Perhaps the most controversial example is the use of thimerosal, an organic mercury compound, as a preservative in flu vaccines. While there is anecdotal evidence suggesting a link between thimerosal and autism in children, most scientific data seem to refute this claim. In response to these concerns, preservative-free versions of the influenza vaccine are available for use in infants, children, and pregnant women.

See Chemistry in Action Problem 1.97 at the end of the chapter.

The number alone is not much good without a unit. If you asked how much blood an accident victim had lost, the answer "three" would not tell you much. Three drops? Three milliliters? Three pints? Three liters? (By the way, an adult human has only 5–6 liters of blood.)

Any physical quantity can be measured in many different units. For example, a person's height might be measured in inches, feet, yards, centimeters, or many other units. To avoid confusion, scientists from around the world have agreed on a system of standard units, called by the French name *Système International d'Unites* (International System of Units), abbreviated *SI*. **SI units** for some common physical quantities are given in Table 1.5. Mass is measured in *kilograms* (kg), length is measured in *meters* (m), volume is measured in *cubic meters* (m^3), temperature is measured in *kelvins* (K), and time is measured in *seconds* (s, not sec).

SI units are closely related to the more familiar *metric units* used in all industrialized nations of the world except the United States. If you compare the SI and metric units shown in Table 1.5, you will find that the basic metric unit of mass is the *gram* (g) rather than the kilogram (1 g = 1/1000 kg), the metric unit of volume is the *liter* (L) rather than the cubic meter (1 L = 1/1000 m^3), and the metric unit of temperature

SI units Units of measurement defined by the International System of Units.

TABLE 1.5 Some SI and Metric Units and Their Equivalents

Quantity	SI Unit (Symbol)	Metric Unit (Symbol)	Equivalents
Mass	Kilogram (kg)	Gram (g)	1 kg = 1000 g = 2.205 lb
Length	Meter (m)	Meter (m)	1 m = 3.280 ft
Volume	Cubic meter (m^3)	Liter (L)	1 m^3 = 1000 L = 264.2 gal
Temperature	Kelvin (K)	Celsius degree (°C)	See Section 1.13
Time	Second (s)	Second (s)	—

is the *Celsius degree* (°C) rather than the kelvin. The meter is the unit of length, and the second is the unit of time in both systems. Although SI units are now preferred in scientific research, metric units are still used in some fields. You will probably find yourself working with both.

In addition to the units listed in Table 1.5, many other widely used units are derived from them. For instance, units of *meters per second* (m/s) are often used for *speed*—the distance covered in a given time. Similarly, units of *grams per cubic centimeter* (g/cm^3) are often used for *density*—the mass of substance in a given volume. We will see other such derived units in future chapters.

One problem with any system of measurement is that the sizes of the units often turn out to be inconveniently large or small for the problem at hand. A biologist describing the diameter of a red blood cell (0.000 006 m) would find the meter to be an inconveniently large unit, but an astronomer measuring the average distance from the earth to the sun (150,000,000,000 m) would find the meter to be inconveniently small. For this reason, metric and SI units can be modified by prefixes to refer to either smaller or larger quantities. For instance, the SI unit for mass—the kilogram—differs by the prefix *kilo-* from the metric unit gram. *Kilo-* indicates that a kilogram is 1000 times as large as a gram:

$$1 \text{ kg} = (1000)(1 \text{ g}) = 1000 \text{ g}$$

Small quantities of active ingredients in medications are often reported in *milligrams* (mg). The prefix *milli-* shows that the unit gram has been divided by 1000, which is the same as multiplying by 0.001:

$$1 \text{ mg} = \left(\frac{1}{1000}\right)(1 \text{ g}) = (0.001)(1 \text{ g}) = 0.001 \text{ g}$$

▶▶▶ The use of exponents is reviewed in Section 1.10.

A list of prefixes is given in Table 1.6, with the most common ones displayed in color. Note that the exponents are multiples of 3 for *mega-* (10^6), *kilo-* (10^3), *milli-* (10^{-3}), *micro-* (10^{-6}), *nano-* (10^{-9}), and *pico-* (10^{-12}). The prefixes *centi-*, meaning 1/100, and *deci-*, meaning 1/10, indicate exponents that are not multiples of 3. *Centi-* is seen most often in the length unit *centimeter* (1 cm = 0.01 m), and *deci-* is used most often in clinical chemistry, where the concentrations of blood components are given in milligrams per deciliter (1 dL = 0.1 L). These prefixes allow us to compare the magnitudes of different numbers by noting how the prefixes modify a common unit.

For example,

$$1 \text{ meter} = 10 \text{ dm} = 100 \text{ cm} = 1000 \text{ mm} = 1{,}000{,}000 \text{ } \mu\text{m}$$

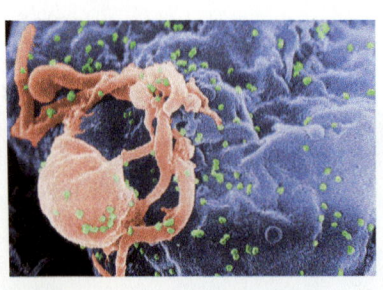

▲ The HIV-1 virus particles (in green) budding from the surface of a lymphocyte have an approximate diameter of 0.000 000 120 m.

Such comparisons will be useful when we start performing calculations involving units in Section 1.12. Note also in Table 1.6 that numbers having five or more digits to the right of the decimal point are shown with thin spaces every three digits for convenience—0.000 001, for example. This manner of writing numbers is becoming more common and will be used throughout this book.

SECTION 1.8 Measuring Mass, Length, and Volume 17

TABLE 1.6 Some Prefixes for Multiples of Metric and SI Units

Prefix	Symbol	Base Unit Multiplied By*	Example
mega	M	$1{,}000{,}000 = 10^6$	1 megameter (Mm) = 10^6 m
kilo	k	$1000 = 10^3$	1 kilogram (kg) = 10^3 g
hecto	h	$100 = 10^2$	1 hectogram (hg) = 100 g
deka	da	$10 = 10^1$	1 dekaliter (daL) = 10 L
deci	d	$0.1 = 10^{-1}$	1 deciliter (dL) = 0.1 L
centi	c	$0.01 = 10^{-2}$	1 centimeter (cm) = 0.01 m
milli	m	$0.001 = 10^{-3}$	1 milligram (mg) = 0.001 g
micro	μ	$0.000\,001 = 10^{-6}$	1 micrometer (μm) = 10^{-6} m
nano	n	$0.000\,000\,001 = 10^{-9}$	1 nanogram (ng) = 10^{-9} g
pico	p	$0.000\,000\,000\,001 = 10^{-12}$	1 picogram (pg) = 10^{-12} g
femto	f	$0.000\,000\,000\,000\,001 = 10^{-15}$	1 femtogram (fg) = 10^{-15} g

*The scientific notation method of writing large and small numbers (for example, 10^6 for 1,000,000) is explained in Section 1.10.

PROBLEM 1.10
Give the full name of the following units and express the quantities in terms of the basic unit (for example, 1 mL = 1 milliliter = 0.001 L):
(a) 1 cm   (b) 1 dg   (c) 1 km   (d) 1 μs   (e) 1 ng

1.8 Measuring Mass, Length, and Volume

The terms *mass* and *weight*, though often used interchangeably, really have quite different meanings. **Mass** is a measure of the amount of matter in an object, whereas **weight** is a measure of the gravitational pull that the earth, moon, or other large body exerts on an object. Clearly, the amount of matter in an object does not depend on location. Whether you are standing on the earth or standing on the moon, the mass of your body is the same. On the other hand, the weight of an object *does* depend on location. Your weight on earth might be 140 lb, but it would only be 23 lb on the moon because the pull of gravity there is only about one-sixth as great.

At the same location, two objects with identical masses have identical weights; that is, gravity pulls equally on both. Thus, the *mass* of an object can be determined by comparing the *weight* of the object to the weight of a known reference standard. Much of the confusion between mass and weight is simply due to a language problem: We speak of "weighing" when we really mean that we are measuring mass by comparing two weights. Figure 1.6 shows a two-pan balance in which the mass of objects are measured by comparison with the known masses of standard materials, such as brass weights.

Mass A measure of the amount of matter in an object.

Weight A measure of the gravitational force that the earth or other large body exerts on an object.

◀ **Figure 1.6**
The two-pan balance is used to measure the mass of objects, such as the pennies on the left pan, by comparing them with the mass of standard objects, such as the brass weights on the right pan.

One kilogram, the SI unit for mass, is equal to 2.205 lb—too large a quantity for many purposes in chemistry and medicine. Thus, smaller units of mass such as the gram, milligram (mg), and microgram (μg), are more commonly used. Table 1.7 shows the relationships between metric and common units for mass.

The meter is the standard measure of length, or distance, in both the SI and metric systems. One meter is 39.37 inches (about 10% longer than a yard), a length that is much too large for most measurements in chemistry and medicine. Other, more commonly used measures of length are the *centimeter* (cm; 1/100 m) and the *millimeter* (mm; 1/1000 m). One centimeter is a bit less than half an inch—0.3937 inch to be exact. A millimeter, in turn, is 0.03937 inch, or about the thickness of a dime. Table 1.8 lists the relationships of these units.

Volume is the amount of space occupied by an object. The SI unit for volume—the cubic meter, m^3—is so large that the liter (1 L = 0.001 m^3 = 1 dm^3) is much more commonly used in chemistry and medicine. One liter has the volume of a cube 10 cm (1 dm) on edge and is a bit larger than one U.S. quart. Each liter is further divided into

TABLE 1.7 Units of Mass

Unit	Equivalent	Unit	Equivalent
1 kilogram (kg)	= 1000 grams = 2.205 pounds	1 ton	= 2000 pounds = 907.03 kilograms
1 gram (g)	= 0.001 kilogram = 1000 milligrams = 0.035 27 ounce	1 pound (lb)	= 16 ounces = 0.454 kilogram = 454 grams
1 milligram (mg)	= 0.001 gram = 1000 micrograms	1 ounce (oz)	= 0.028 35 kilogram = 28.35 grams
1 microgram (μg)	= 0.000 001 gram = 0.001 milligram		= 28,350 milligrams

TABLE 1.8 Units of Length

Unit	Equivalent
1 kilometer (km)	= 1000 meters = 0.6214 mile
1 meter (m)	= 100 centimeters = 1000 millimeters = 1.0936 yards = 39.37 inches
1 centimeter (cm)	= 0.01 meter = 10 millimeters = 0.3937 inch
1 millimeter (mm)	= 0.001 meter = 0.1 centimeter
1 mile (mi)	= 1.609 kilometers = 1609 meters
1 yard (yd)	= 0.9144 meter = 91.44 centimeters
1 foot (ft)	= 0.3048 meter = 30.48 centimeters
1 inch (in)	= 2.54 centimeters = 25.4 millimeters

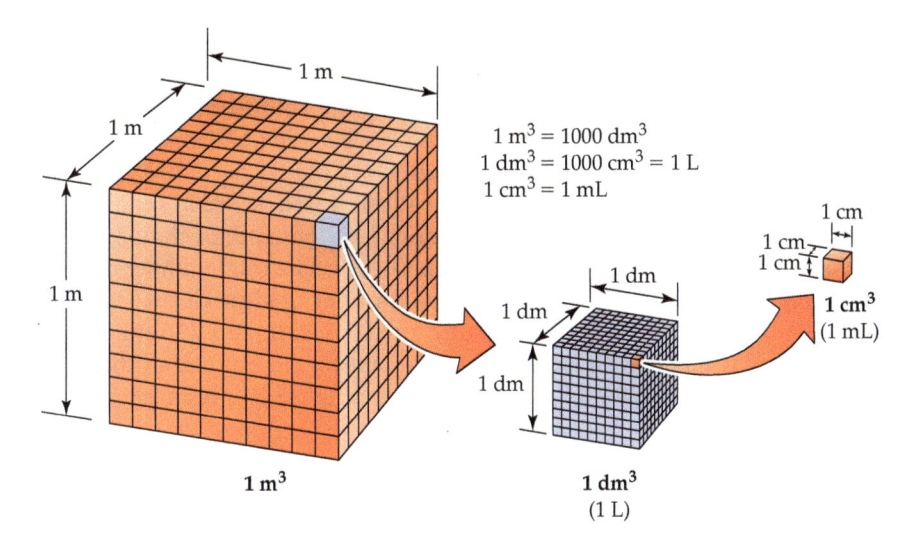

◀ Figure 1.7
A cubic meter is the volume of a cube 1 m on edge. Each cubic meter contains 1000 cubic decimeters (liters), and each cubic decimeter contains 1000 cubic centimeters (milliliters). Thus, there are 1000 mL in a liter and 1000 L in a cubic meter.

$1 m^3 = 1000 dm^3$
$1 dm^3 = 1000 cm^3 = 1 L$
$1 cm^3 = 1 mL$

TABLE 1.9 Units of Volume

Unit	Equivalent
1 cubic meter (m³)	= 1000 liters
	= 264.2 gallons
1 liter (L)	= 0.001 cubic meter
	= 1000 milliliters
	= 1.057 quarts
1 deciliter (dL)	= 0.1 liter
	= 100 milliliters
1 milliliter (mL)	= 0.001 liter
	= 1000 microliters
1 microliter (μL)	= 0.001 milliliter
1 gallon (gal)	= 3.7854 liters
1 quart (qt)	= 0.9464 liter
	= 946.4 milliliters
1 fluid ounce (fl oz)	= 29.57 milliliters

1000 *milliliters* (mL), with 1 mL being the size of a cube 1 cm on edge, or 1 cm³. In fact, the milliliter is often called a *cubic centimeter* (cm³ or cc) in medical work. Figure 1.7 shows the divisions of a cubic meter, and Table 1.9 shows the relationships among units of volume.

1.9 Measurement and Significant Figures

How much does a tennis ball weigh? If you put a tennis ball on an ordinary bathroom scale, the scale would probably register 0 lb (or 0 kg if you have a metric scale). If you placed the same tennis ball on a common laboratory balance, however, you might get a reading of 54.07 g. Trying again by placing the ball on an expensive analytical balance like those found in clinical and research laboratories, you might find a mass of 54.071 38 g. Clearly, the precision of your answer depends on the equipment used for the measurement.

Every experimental measurement, no matter how precise, has a degree of uncertainty to it because there is always a limit to the number of digits that can be determined. An analytical balance, for example, might reach its limit in measuring mass to the fifth decimal place, and weighing the tennis ball several times might produce

▲ The tennis ball weighs 54.07 g on this common laboratory balance, which is capable of determining mass to about 0.01 g.

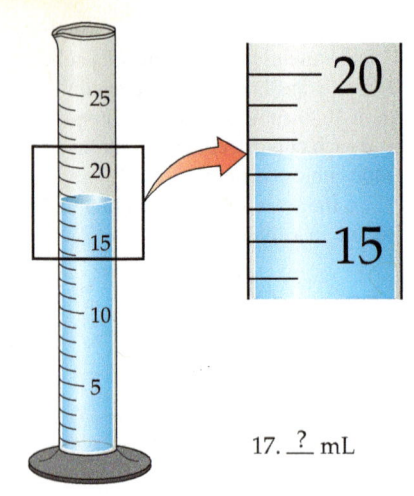

Figure 1.8
What is the volume of liquid in this graduated cylinder?

Significant figures The number of meaningful digits used to express a value.

slightly different readings, such as 54.071 39 g, 54.071 38 g, and 54.071 37 g. Also, different people making the same measurement might come up with slightly different answers. How, for instance, would you record the volume of the liquid shown in Figure 1.8? It is clear that the volume of liquid lies between 17.0 and 18.0 mL, but the exact value of the last digit must be estimated.

To indicate the precision of a measurement, the value recorded should use all the digits known with certainty, plus one additional estimated digit that is usually considered uncertain by plus or minus 1 (written as ±1). The total number of digits used to express such a measurement is called the number of **significant figures**. Thus, the quantity 54.07 g has four significant figures (5, 4, 0, and 7), and the quantity 54.071 38 g has seven significant figures. *Remember:* All but one of the significant figures are known with certainty; the last significant figure is only an estimate accurate to ±1.

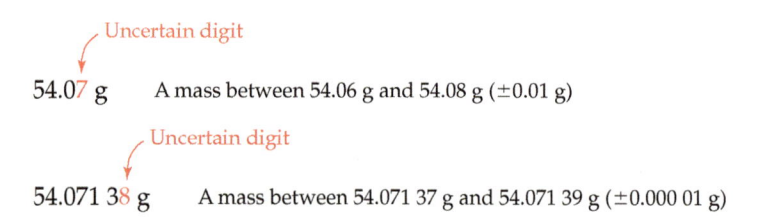

Uncertain digit
54.07 g   A mass between 54.06 g and 54.08 g (±0.01 g)

Uncertain digit
54.071 38 g   A mass between 54.071 37 g and 54.071 39 g (±0.000 01 g)

Deciding the number of significant figures in a given measurement is usually simple, but it can be troublesome when zeros are involved. Depending on the circumstances, a zero might be significant or might be just a space-filler to locate the decimal point. For example, how many significant figures does each of the following measurements have?

94.072 g	Five significant figures (9, 4, 0, 7, 2)
0.0834 cm	Three significant figures (8, 3, 4)
0.029 07 mL	Four significant figures (2, 9, 0, 7)
138.200 m	Six significant figures (1, 3, 8, 2, 0, 0)
23,000 kg	*Anywhere* from two (2, 3) to five (2, 3, 0, 0, 0) significant figures

The following rules are helpful for determining the number of significant figures when zeros are present:

RULE 1: Zeros in the middle of a number are like any other digit; they are always significant. Thus, 94.072 g has five significant figures.

RULE 2: Zeros at the beginning of a number are not significant; they act only to locate the decimal point. Thus, 0.0834 cm has three significant figures, and 0.029 07 mL has four.

RULE 3: Zeros at the end of a number and *after* the decimal point are significant. It is assumed that these zeros would not be shown unless they were significant. Thus, 138.200 m has six significant figures. If the value were known to only four significant figures, we would write 138.2 m.

RULE 4: Zeros at the end of a number and *before* an implied decimal point may or may not be significant. We cannot tell whether they are part of the measurement or whether they act only to locate the unwritten but implied decimal point. Thus, 23,000 kg may have two, three, four, or five significant figures. Adding a decimal point at the end would indicate that all five numbers are significant.

Often, however, a little common sense is useful. A temperature reading of 20 °C probably has two significant figures rather than one, because one significant figure would imply a temperature anywhere from 10 °C to 30 °C and would be of little use. Similarly, a volume given as 300 mL probably has three significant figures. On the other hand, a figure of 150,000,000 km for the distance between the earth and the sun has only two or three significant figures because the distance is variable. We will see a better way to deal with this problem in the next section.

One final point about significant figures: some numbers, such as those obtained when counting objects and those that are part of a definition, are *exact* and effectively have an unlimited number of significant figures. Thus, a class might have *exactly* 32 students (not 31.9, 32.0, or 32.1), and 1 foot is defined to have *exactly* 12 inches.

▲ The number of seats in this auditorium is an exact number with an unlimited number of significant figures.

Worked Example 1.3 Significant Figures of Measurements

How many significant figures do the following measurements have?
(a) 2730.78 m   (b) 0.0076 mL   (c) 3400 kg   (d) 3400.0 m^2

ANALYSIS All nonzero numbers are significant; the number of significant figures will then depend on the status of the zeros in each case. (Hint: which rule applies in each case?)

SOLUTION
(a) Six (rule 1)   (b) Two (rule 2)
(c) Two, three, or four (rule 4)   (d) Five (rule 3)

PROBLEM 1.11
How many significant figures do the following measurements have?
(a) 3.45 m   (b) 0.1400 kg
(c) 10.003 L   (d) 35 cents

KEY CONCEPT PROBLEM 1.12
How would you record the temperature reading on the following Celsius thermometer? How many significant figures do you have in your answer?

1.10 Scientific Notation

Rather than write very large or very small numbers in their entirety, it is more convenient to express them using *scientific notation*. A number is written in **scientific notation** as the product of a number between 1 and 10, times the number 10 raised to a power. Thus, 215 is written in scientific notation as 2.15×10^2:

$$215 = 2.15 \times 100 = 2.15(10 \times 10) = 2.15 \times 10^2$$

Notice that in this case, where the number is *larger* than 1, the decimal point has been moved *to the left* until it follows the first digit. The exponent on the 10 tells how many places we had to move the decimal point to position it just after the first digit:

$$215. = 2.15 \times 10^2$$

Decimal point is moved two places to the left, so exponent is 2.

Scientific notation A number expressed as the product of a number between 1 and 10, times the number 10 raised to a power.

To express a number *smaller* than 1 in scientific notation, we have to move the decimal point *to the right* until it follows the first digit. The number of places moved is the negative exponent of 10. For example, the number 0.002 15 can be rewritten as 2.15×10^{-3}:

$$0.002\ 15 = 2.15 \times \frac{1}{1000} = 2.15 \times \frac{1}{10 \times 10 \times 10} = 2.15 \times \frac{1}{10^3} = 2.15 \times 10^{-3}$$

$$0.002\ 15 = 2.15 \times 10^{-3}$$

Decimal point is moved three places to the right, so exponent is −3.

To convert a number written in scientific notation to standard notation, the process is reversed. For a number with a *positive* exponent, the decimal point is moved to the *right* a number of places equal to the exponent:

$$3.7962 \times 10^4 = 37,962$$

Positive exponent of 4, so decimal point is moved to the right four places.

For a number with a *negative* exponent, the decimal point is moved to the *left* a number of places equal to the exponent:

$$1.56 \times 10^{-8} = 0.000\ 000\ 015\ 6$$

Negative exponent of −8, so decimal point is moved to the left eight places.

Scientific notation is particularly helpful for indicating how many significant figures are present in a number that has zeros at the end but to the left of a decimal point. If we read, for instance, that the distance from the earth to the sun is 150,000,000 km, we do not really know how many significant figures are indicated. Some of the zeros might be significant, or they might merely act to locate the decimal point. Using scientific notation, however, we can indicate how many of the zeros are significant. Rewriting 150,000,000 as 1.5×10^8 indicates two significant figures, whereas writing it as 1.500×10^8 indicates four significant figures. Scientific notation is not ordinarily used for numbers that are easily written, such as 10 or 175, although it is sometimes helpful in doing arithmetic.

▶▶ Rules for doing arithmetic with numbers written in scientific notation are reviewed in Appendix A.

▲ How many molecules are in this 1 g pile of table sugar?

Worked Example 1.4 Significant Figures and Scientific Notation

There are 1,760,000,000,000,000,000,000 molecules of sucrose (table sugar) in 1 g. Use scientific notation to express this number with four significant figures.

ANALYSIS Because the number is larger than 1, the exponent will be positive. You will have to move the decimal point 21 places to the left.

SOLUTION
The first four digits—1, 7, 6, and 0—are significant, meaning that only the first of the 19 zeros is significant. Because we have to move the decimal point 21 places to the left to put it after the first significant digit, the answer is 1.760×10^{21}.

Worked Example 1.5 Scientific Notation

The rhinovirus responsible for the common cold has a diameter of 20 nm, or 0.000 000 020 m. Express this number in scientific notation.

ANALYSIS The number is smaller than 1, and so the exponent will be negative. You will have to move the decimal point eight places to the right.

SOLUTION
There are only two significant figures, because zeros at the beginning of a number are not significant. We have to move the decimal point 8 places to the right to place it after the first digit, so the answer is 2.0×10^{-8} m.

Worked Example 1.6 Scientific Notation and Unit Conversions

A clinical laboratory found that a blood sample contained 0.0026 g of phosphorus and 0.000 101 g of iron.

(a) Give these quantities in scientific notation.
(b) Give these quantities in the units normally used to report them—milligrams for phosphorus and micrograms for iron.

ANALYSIS Is the number larger or smaller than 1? How many places do you have to move the decimal point?

SOLUTION

(a) 0.0026 g phosphorus = 2.6×10^{-3} g phosphorus

0.000 101 g iron = 1.01×10^{-4} g iron

(b) We know from Table 1.6 that 1 mg = 1×10^{-3} g, where the exponent is -3. Expressing the amount of phosphorus in milligrams is straightforward because the amount in grams (2.6×10^{-3} g) already has an exponent of -3. Thus, 2.6×10^{-3} g = 2.6 mg of phosphorus.

$$(2.6 \times 10^{-3} \text{ g}) \left(\frac{1 \text{ mg}}{1 \times 10^{-3} \text{ g}} \right) = 2.6 \text{ mg}$$

We know from Table 1.6 that 1 μg = 1×10^{-6} g where the exponent is -6. Expressing the amount of iron in micrograms thus requires that we restate the amount in grams so that the exponent is -6. We can do this by moving the decimal point six places to the right:

$$0.000\ 101 \text{ g iron} = 101 \times 10^{-6} \text{ g iron} = 101 \ \mu\text{g iron}$$

PROBLEM 1.13
Convert the following values to scientific notation:
(a) 0.058 g   (b) 46,792 m   (c) 0.006 072 cm   (d) 345.3 kg

PROBLEM 1.14
Convert the following values from scientific notation to standard notation:
(a) 4.885×10^4 mg   (b) 8.3×10^{-6} m   (c) 4.00×10^{-2} m

PROBLEM 1.15
Rewrite the following numbers in scientific notation as indicated:
(a) 630,000 with five significant figures
(b) 1300 with three significant figures
(c) 794,200,000,000 with four significant figures

1.11 Rounding Off Numbers

It often happens, particularly when doing arithmetic on a pocket calculator, that a quantity appears to have more significant figures than are really justified. For example, you might calculate the gas mileage of your car by finding that it takes 11.70 gallons of gasoline to drive 278 miles:

$$\text{Mileage} = \frac{\text{Miles}}{\text{Gallons}} = \frac{278 \text{ mi}}{11.70 \text{ gal}} = 23.760\ 684 \text{ mi/gal (mpg)}$$

24 CHAPTER 1 Matter and Measurements

Rounding off A procedure used for deleting nonsignificant figures.

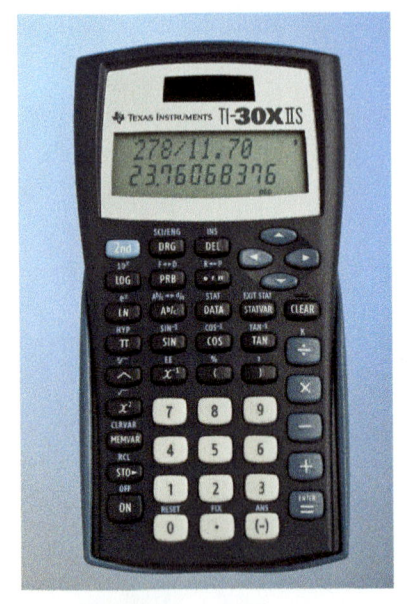

Although the answer on a calculator has eight digits, your calculated result is really not as precise as it appears. In fact, as we will see below, your answer is good to only three significant figures and should be **rounded off** to 23.8 mi/gal.

How do you decide how many digits to keep? The full answer to this question is a bit complex and involves a mathematical treatment called *error analysis*, but for many purposes, a simplified procedure using just two rules is sufficient:

RULE 1: In carrying out a multiplication or division, the answer cannot have more significant figures than either of the original numbers. This is just a common-sense rule if you think about it. After all, if you do not know the number of miles you drove to better than three significant figures (278 could mean 277, 278, or 279), you certainly cannot calculate your mileage to more than the same number of significant figures.

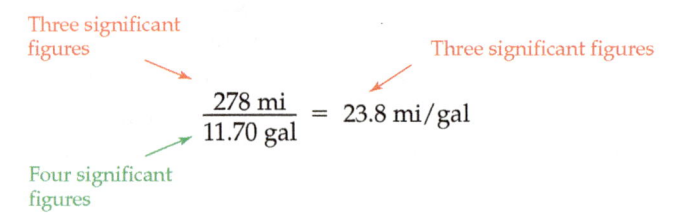

Three significant figures　　　　　　　　Three significant figures

$$\frac{278 \text{ mi}}{11.70 \text{ gal}} = 23.8 \text{ mi/gal}$$

Four significant figures

RULE 2: In carrying out an addition or subtraction, the answer cannot have more digits after the decimal point than either of the original numbers. For example, if you have 3.18 L of water and you add 0.013 15 L more, you now have 3.19 L. Again, this rule is just common sense. If you do not know the volume you started with past the second decimal place (it could be 3.17, 3.18, or 3.19), you cannot know the total of the combined volumes past the same decimal place.

Volume of water at start → 3.18? ?? L ← Two digits after decimal point
Volume of water added → + 0.013 15 L ← Five digits after decimal point
Total volume of water → 3.19? ?? L ← Two digits after decimal point

▲ Calculators often display more digits than are justified by the precision of the data.

If a calculation has several steps, it is generally best to round off at the end after all the steps have been carried out, keeping the number of significant figures determined by the least precise number in your calculations. Once you decide how many digits to retain for your answer, the rules for rounding off numbers are straightforward:

RULE 1: If the first digit you remove is 4 or less, drop it and all following digits. Thus, 2.4271 becomes 2.4 when rounded off to two significant figures because the first of the dropped digits (a 2) is 4 or less.

RULE 2: If the first digit you remove is 5 or greater, round the number up by adding a 1 to the digit to the left of the one you drop. Thus, 4.5832 becomes 4.6 when rounded off to two significant figures because the first of the dropped digits (an 8) is 5 or greater.

Worked Example 1.7 Significant Figures and Calculations: Addition/Subtraction

Suppose that you weigh 124 lb before dinner. How much will you weigh after dinner if you eat 1.884 lb of food?

ANALYSIS When performing addition or subtraction, the number of significant figures you report in the final answer is determined by the number of digits in the least precise number in the calculation.

SECTION 1.12 Problem Solving: Unit Conversions and Estimating Answers

SOLUTION
Your after-dinner weight is found by adding your original weight to the weight of the food consumed:

$$\begin{array}{r} 124 \text{ lb} \\ \underline{1.884 \text{ lb}} \\ 125.884 \text{ lb (Unrounded)} \end{array}$$

Because the value of your original weight has no significant figures after the decimal point, your after-dinner weight also must have no significant figures after the decimal point. Thus, 125.884 lb must be rounded off to 126 lb.

Worked Example 1.8 Significant Figures and Calculations: Multiplication/Division

To make currant jelly, 13.75 cups of sugar was added to 18 cups of currant juice. How much sugar was added per cup of juice?

ANALYSIS For calculations involving multiplication or division, the final answer cannot have more significant figures than either of the original numbers.

SOLUTION
The quantity of sugar must be divided by the quantity of juice:

$$\frac{13.75 \text{ cups sugar}}{18 \text{ cups juice}} = 0.763\,888\,89 \; \frac{\text{cup sugar}}{\text{cup juice}} \; (\text{Unrounded})$$

The number of significant figures in the answer is limited to two by the quantity 18 cups in the calculation and must be rounded to 0.76 cup of sugar per cup of juice.

PROBLEM 1.16
Round off the following quantities to the indicated number of significant figures:
(a) 2.304 g (three significant figures)
(b) 188.3784 mL (five significant figures)
(c) 0.008 87 L (one significant figure)
(d) 1.000 39 kg (four significant figures)

PROBLEM 1.17
Carry out the following calculations, rounding each result to the correct number of significant figures:
(a) 4.87 mL + 46.0 mL
(b) 3.4 × 0.023 g
(c) 19.333 m − 7.4 m
(d) 55 mg − 4.671 mg + 0.894 mg
(e) 62,911 ÷ 611

1.12 Problem Solving: Unit Conversions and Estimating Answers

Many activities in the laboratory and in medicine—measuring, weighing, preparing solutions, and so forth—require converting a quantity from one unit to another. For example: "These pills contain 1.3 grains of aspirin, but I need 200 mg. Is one pill enough?" Converting between units is not mysterious; we all do it every day. If you run 9 laps around a 400-meter track, for instance, you have to convert between the distance unit "lap" and the distance unit "meter" to find that you have run 3600 m (9 laps times

▲ Currency exchange between the US$ and Euros is another activity that requires a unit conversion.

26 CHAPTER 1 Matter and Measurements

400 m/lap). If you want to find how many miles that is, you have to convert again to find that 3600 m = 2.237 mi.

The simplest way to carry out calculations involving different units is to use the **factor-label method**. In this method, a quantity in one unit is converted into an equivalent quantity in a different unit by using a **conversion factor** that expresses the relationship between units:

> **Factor-label method** A problem-solving procedure in which equations are set up so that unwanted units cancel and only the desired units remain.
>
> **Conversion factor** An expression of the numerical relationship between two units.

$$\text{Starting quantity} \times \text{Conversion factor} = \text{Equivalent quantity}$$

As an example, we learned from Table 1.8 that 1 km = 0.6214 mi. Writing this relationship as a fraction restates it in the form of a conversion factor, either kilometers per mile or miles per kilometer.

Since 1 km = 0.6214 mi, then:

Conversion factors between kilometers and miles

$$\frac{1 \text{ km}}{0.6214 \text{ mi}} = 1 \quad \text{or} \quad \frac{0.6214 \text{ mi}}{1 \text{ km}} = 1$$

Note that this and all other conversion factors are numerically equal to 1 because the value of the quantity above the division line (the numerator) is equal in value to the quantity below the division line (the denominator). Thus, multiplying by a conversion factor is equivalent to multiplying by 1 and so does not change the value of the quantity being multiplied:

These two quantities are the same.
$$\frac{1 \text{ km}}{0.6214 \text{ mi}} \quad \text{or} \quad \frac{0.6214 \text{ mi}}{1 \text{ km}}$$
These two quantities are the same.

The key to the factor-label method of problem solving is that units are treated like numbers and can thus be multiplied and divided (though not added or subtracted) just as numbers can. When solving a problem, the idea is to set up an equation so that all unwanted units cancel, leaving only the desired units. Usually, it is best to start by writing what you know and then manipulating that known quantity. For example, if you know there are 26.22 mi in a marathon and want to find how many kilometers that is, you could write the distance in miles and multiply by the conversion factor in kilometers per mile. The unit "mi" cancels because it appears both above and below the division line, leaving "km" as the only remaining unit.

$$26.22 \text{ mi} \times \frac{1 \text{ km}}{0.6214 \text{ mi}} = 42.20 \text{ km}$$

Starting quantity — Conversion factor — Equivalent quantity

The factor-label method gives the right answer only if the equation is set up so that the unwanted unit (or units) cancel. If the equation is set up in any other way, the units will not cancel and you will not get the right answer. Thus, if you selected the incorrect conversion factor (miles per kilometer) for the above problem, you would end up with an incorrect answer expressed in meaningless units:

$$\text{Incorrect} \quad 26.22 \text{ mi} \times \frac{0.6214 \text{ mi}}{1 \text{ km}} = 16.29 \frac{\text{mi}^2}{\text{km}} \quad \text{Incorrect}$$

The main drawback to using the factor-label method is that it is possible to get an answer without really understanding what you are doing. It is therefore best when solving a problem to first think through a rough estimate, or *ballpark estimate*, as a check on your work. If your ballpark estimate is not close to the final calculated solution, there is a misunderstanding somewhere and you should think the problem through again. If, for example, you came up with the answer 5.3 cm^3 when calculating the volume of a human cell, you should realize that such an answer could not possibly be right. Cells are too tiny to be distinguished with the naked eye, but a volume of 5.3 cm^3 is about the size

SECTION 1.12 Problem Solving: Unit Conversions and Estimating Answers

of a walnut. The Worked Examples 1.11, 1.12, and 1.13 at the end of this section show how to estimate the answers to simple unit-conversion problems.

The factor-label method and the use of ballpark estimates are techniques that will help you solve problems of many kinds, not just unit conversions. Problems sometimes seem complicated, but you can usually sort out the complications by analyzing the problem properly:

STEP 1: Identify the information given, including units.

STEP 2: Identify the information needed in the answer, including units.

STEP 3: Find the relationship(s) between the known information and unknown answer, and plan a series of steps, including conversion factors, for getting from one to the other.

STEP 4: Solve the problem.

BALLPARK CHECK Make a ballpark estimate at the beginning and check it against your final answer to be sure the value and the units of your calculated answer are reasonable.

Worked Example 1.9 Factor Labels: Unit Conversions

Write conversion factors for the following pairs of units (use Tables 1.7–1.9):
(a) Deciliters and milliliters
(b) Pounds and grams

ANALYSIS Start with the appropriate equivalency relationship and rearrange to form conversion factors.

SOLUTION

(a) Since 1 dL = 0.1 L and 1 mL = 0.001 L, then 1 dL = $(0.1 \text{ L})\left(\dfrac{1 \text{ mL}}{0.001 \text{ L}}\right)$ = 100 mL. The conversion factors are

$$\dfrac{1 \text{ dL}}{100 \text{ mL}} \quad \text{and} \quad \dfrac{100 \text{ mL}}{1 \text{ dL}}$$

(b) $\dfrac{1 \text{ lb}}{454 \text{ g}} \quad \text{and} \quad \dfrac{454 \text{ g}}{1 \text{ lb}}$

Worked Example 1.10 Factor Labels: Unit Conversions

(a) Convert 0.75 lb to grams.
(b) Convert 0.50 qt to deciliters.

ANALYSIS Start with conversion factors and set up equations so that units cancel appropriately.

SOLUTION

(a) Select the conversion factor from Worked Example 1.9(b) so that the "lb" units cancel and "g" remains:

$$0.75 \text{ lb} \times \dfrac{454 \text{ g}}{1 \text{ lb}} = 340 \text{ g}$$

(b) In this, as in many problems, it is convenient to use more than one conversion factor. As long as the unwanted units cancel correctly, two or more conversion factors can be strung together in the same calculation. In this case, we can convert first between quarts and milliliters, and then between milliliters and deciliters:

$$0.50 \text{ qt} \times \dfrac{946.4 \text{ mL}}{1 \text{ qt}} \times \dfrac{1 \text{ dL}}{100 \text{ mL}} = 4.7 \text{ dL}$$

CHAPTER 1 Matter and Measurements

Worked Example 1.11 Factor Labels: Unit Conversions

A child is 21.5 inches long at birth. How long is this in centimeters?

ANALYSIS This problem calls for converting from inches to centimeters, so we will need to know how many centimeters are in an inch and how to use this information as a conversion factor.

BALLPARK ESTIMATE It takes about 2.5 cm to make 1 in., and so it should take two and a half times as many centimeters to make a distance equal to approximately 20 in., or about 20 in. × 2.5 = 50 cm.

SOLUTION

STEP 1: **Identify given information.**   Length = 21.5 in.

STEP 2: **Identify answer and units.**   Length = ?? cm

STEP 3: **Identify conversion factor.**   1 in. = 2.54 cm → $\frac{2.54 \text{ cm}}{1 \text{ in.}}$

STEP 4: **Solve.** Multiply the known length (in inches) by the conversion factor so that units cancel, providing the answer (in centimeters).

$$21.5 \text{ in.} \times \frac{2.54 \text{ cm}}{1 \text{ in.}} = 54.6 \text{ cm (Rounded off from 54.61)}$$

BALLPARK CHECK How does this value compare with the ballpark estimate we made at the beginning? Are the final units correct? 54.6 cm is close to our original estimate of 50 cm.

Worked Example 1.12 Factor Labels: Concentration to Mass

A patient requires an injection of 0.012 g of a pain killer available as a 15 mg/mL solution. How many milliliters of solution should be administered?

ANALYSIS Knowing the amount of pain killer in 1 mL allows us to use the concentration as a conversion factor to determine the volume of solution that would contain the desired amount.

BALLPARK ESTIMATE One milliliter contains 15 mg of the pain killer, or 0.015 g. Since only 0.012 g is needed, a little less than 1.0 mL should be administered.

▲ How many milliliters should be injected?

SOLUTION

STEP 1: **Identify known information.**
Dosage = 0.012 g
Concentration = 15 mg/mL

STEP 2: **Identify answer and units.**
Volume to administer = ?? mL

STEP 3: **Identify conversion factors.** Two conversion factors are needed. First, g must be converted to mg. Once we have the mass in mg, we can calculate mL using the conversion factor of mL/mg.

$$1 \text{ mg} = .001 \text{ g} \Rightarrow \frac{1 \text{ mg}}{0.001 \text{ g}}$$

$$15 \text{ mg/mL} \Rightarrow \frac{1 \text{ mL}}{15 \text{ mg}}$$

STEP 4: **Solve.** Starting from the desired dosage, we use the conversion factors to cancel units, obtaining the final answer in mL.

$$(0.012 \text{ g})\left(\frac{1 \text{ mg}}{0.001 \text{ g}}\right)\left(\frac{1 \text{ mL}}{15 \text{ mg}}\right) = 0.80 \text{ mL}$$

BALLPARK CHECK Consistent with our initial estimate of a little less than 1 mL.

SECTION 1.13 Temperature, Heat, and Energy

Worked Example 1.13 Factor Labels: Multiple Conversion Calculations

Administration of digitalis to control atrial fibrillation in heart patients must be carefully regulated because even a modest overdose can be fatal. To take differences between patients into account, dosages are sometimes prescribed in micrograms per kilogram of body weight ($\mu g/kg$). Thus, two people may differ greatly in weight, but both will receive the proper dosage. At a dosage of 20 $\mu g/kg$ body weight, how many milligrams of digitalis should a 160 lb patient receive?

ANALYSIS Knowing the patient's body weight (in kg) and the recommended dosage (in $\mu g/kg$), we can calculate the appropriate amount of digitalis.

BALLPARK ESTIMATE Since a kilogram is roughly equal to 2 lb, a 160 lb patient has a mass of about 80 kg. At a dosage of 20 $\mu g/kg$, an 80 kg patient should receive 80 × 20 μg, or about 1600 μg of digitalis, or 1.6 mg.

SOLUTION

STEP 1: Identify known information.

Patient weight = 160 lb
Prescribed dosage = 20 μg digitalis/kg body weight
Delivered dosage = ?? mg digitalis

STEP 2: Identify answer and units.

STEP 3: Identify conversion factors. Two conversions are needed. First, convert the patient's weight in pounds to weight in kg. The correct dose can then be determined based on μg digitalis/kg of body weight. Finally, the dosage in μg is converted to mg.

$$1 \text{ kg} = 2.205 \text{ lb} \rightarrow \frac{1 \text{ kg}}{2.205 \text{ lb}}$$

$$1 \text{ mg} = (0.001 \text{ g})\left(\frac{1 \mu g}{10^{-6} \text{ g}}\right) = 1000 \mu g$$

STEP 4: Solve. Use the known information and the conversion factors so that units cancel, obtaining the answer in mg.

$$160 \text{ lb} \times \frac{1 \text{ kg}}{2.205 \text{ lb}} \times \frac{20 \mu g \text{ digitalis}}{1 \text{ kg}} \times \frac{1 \text{ mg}}{1000 \mu g}$$

$$= 1.5 \text{ mg digitalis (Rounded off)}$$

BALLPARK CHECK Close to our estimate of 1.6 mg.

PROBLEM 1.18
Write appropriate conversion factors and carry out the following conversions:
(a) 16.0 oz = ? g   (b) 2500 mL = ? L   (c) 99.0 L = ? qt

PROBLEM 1.19
Convert 0.840 qt to milliliters in a single calculation using more than one conversion factor.

PROBLEM 1.20
One international nautical mile is defined as exactly 6076.1155 ft, and a speed of 1 knot is defined as one international nautical mile per hour. What is the speed in meters per second of a boat traveling at a speed of 14.3 knots? (Hint: what conversion factor is needed to convert from feet to meters? From hours to seconds?)

PROBLEM 1.21
Calculate the dosage in milligrams per kilogram body weight for a 135 lb adult who takes two aspirin tablets containing 0.324 g of aspirin each. Calculate the dosage for a 40 lb child who also takes two aspirin tablets.

1.13 Temperature, Heat, and Energy

All chemical reactions are accompanied by a change in **energy**, which is defined in scientific terms as *the capacity to do work or supply heat* (Figure 1.9). Detailed discussion of the various kinds of energy will be included in Chapter 7, but for now we will look at the various units used to describe energy and heat, and how heat energy can be gained or lost by matter.

Temperature, the measure of the amount of heat energy in an object, is commonly reported either in Fahrenheit (°F) or Celsius (°C) units. The SI unit for reporting temperature, however, is the *kelvin* (K). (Note that we say only "kelvin," not "degrees kelvin".)

The kelvin and the celsius degree are the same size—both are 1/100 of the interval between the freezing point of water and the boiling point of water at atmospheric pressure.

Energy The capacity to do work or supply heat.

Temperature The measure of the amount of heat energy in an object.

Thus, a change in temperature of 1 °C is equal to a change of 1 K. The only difference between the Kelvin and Celsius temperature scales is that they have different zero points. The Celsius scale assigns a value of 0 °C to the freezing point of water, but the Kelvin scale assigns a value of 0 K to the coldest possible temperature, sometimes called *absolute zero*, which is equal to −273.15 °C. Thus, 0 K = −273.15 °C, and +273.15 K = 0 °C. For example, a warm spring day with a temperature of 25 °C has a Kelvin temperature of 298 K (for most purposes, rounding off to 273 is sufficient):

$$\text{Temperature in K} = \text{Temperature in °C} + 273.15$$
$$\text{Temperature in °C} = \text{Temperature in K} - 273.15$$

For practical applications in medicine and clinical chemistry, the Fahrenheit and Celsius scales are used almost exclusively. The Fahrenheit scale defines the freezing point of water as 32 °F and the boiling point of water as 212 °F, whereas 0 °C and 100 °C are the freezing and boiling points of water on the Celsius scale. Thus, it takes 180 Fahrenheit degrees to cover the same range encompassed by only 100 celsius degrees, and a Celsius degree is therefore exactly 180/100 = 9/5 = 1.8 times as large as a Fahrenheit degree. In other words, a change in temperature of 1.0 °C is equal to a change of 1.8 °F. Figure 1.10 gives a comparison of all three scales.

Converting between the Fahrenheit and Celsius scales is similar to converting between different units of length or volume, but is a bit more complex because two corrections need to be made—one to adjust for the difference in degree size and one to adjust for the different zero points. The degree-size correction is made by using the relationship 1 °C = (9/5) °F and 1 °F = (5/9) °C. The zero-point correction is made by remembering that the freezing point is higher by 32 on the Fahrenheit scale than on the Celsius scale. These corrections are incorporated into the following formulas, which show the conversion methods:

Celsius to Fahrenheit: $\quad °F = \left(\dfrac{9\,°F}{5\,°C} \times °C \right) + 32\,°F$

Fahrenheit to Celsius: $\quad °C = \dfrac{5\,°C}{9\,°F} \times (°F - 32\,°F)$

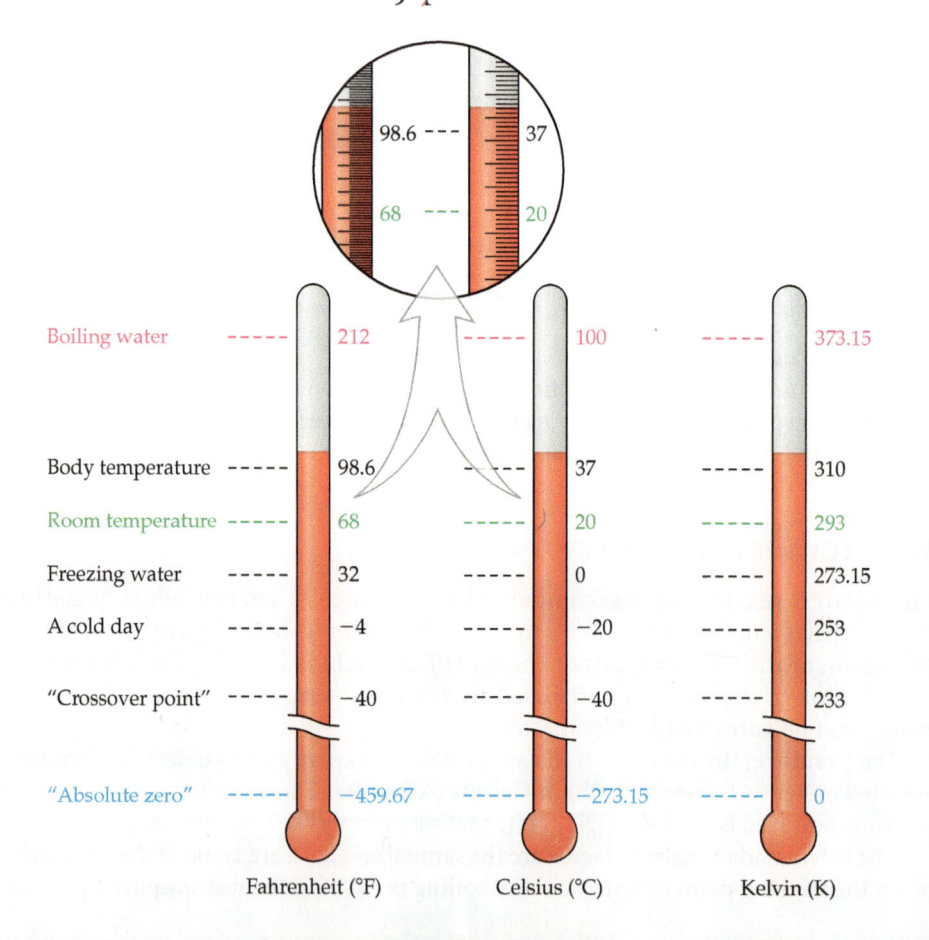

▲ **Figure 1.9**
The reaction of aluminum with bromine releases energy in the form of heat.
When the reaction is complete, the products undergo no further change.

▶ **Figure 1.10**
A comparison of the Fahrenheit, Celsius, and Kelvin temperature scales.
One Fahrenheit degree is 5/9 the size of a kelvin or a celsius degree.

	Fahrenheit (°F)	Celsius (°C)	Kelvin (K)
Boiling water	212	100	373.15
Body temperature	98.6	37	310
Room temperature	68	20	293
Freezing water	32	0	273.15
A cold day	−4	−20	253
"Crossover point"	−40	−40	233
"Absolute zero"	−459.67	−273.15	0

CHEMISTRY IN ACTION

Temperature–Sensitive Materials

Wouldn't it be nice to be able to tell if the baby's formula bottle is too hot without touching it? Or to easily determine if the package of chicken you are buying for dinner has been stored appropriately? Temperature-sensitive materials are already being used in these and other applications. Although these materials have been used previously in many popular "fads," like mood rings or clothes that changed color at different temperatures, more practical applications are emerging.

Most current applications use substances known as thermochromic materials that change color as their temperature increases, and they change from the liquid phase to a semi-crystalline ordered state. These "liquid crystals" can be incorporated into plastics or paints and can be used to monitor the temperature of the products or packages in which they are incorporated. For example, some meat packaging now includes a temperature strip that darkens when the meat is stored above a certain temperature, which makes the meat unsafe to eat. Some beverage containers turn color to indicate when the beverage has reached its optimal temperature for consumption. Hospitals and other medical facilities now routinely use strips that, when placed under the tongue or applied to the forehead, change color to indicate the patient's body temperature. In the future, we may even see road signs that change color to warn us of dangerous icy road conditions.

See Chemistry in Action Problems 1.98 and 1.99 at the end of the chapter.

Energy is represented in SI units by the unit *joule* (J; pronounced "jool"), but the metric unit *calorie* (cal) is still widely used in medicine. In most of this text we will present energy values in both units of calories and joules. One calorie is the amount of heat necessary to raise the temperature of 1 g of water by 1 °C. A *kilocalorie* (kcal), often called a *large calorie (Cal)* or *food calorie* by nutritionists, equals 1000 cal:

$$1000 \text{ cal} = 1 \text{ kcal} \qquad 1000 \text{ J} = 1 \text{ kJ}$$
$$1 \text{ cal} = 4.184 \text{ J} \qquad 1 \text{ kcal} = 4.184 \text{ kJ}$$

Not all substances have their temperatures raised to the same extent when equal amounts of heat energy are added. One calorie raises the temperature of 1 g of water by 1 °C but raises the temperature of 1 g of iron by 10 °C. The amount of heat needed to raise the temperature of 1 g of a substance by 1 °C is called the **specific heat** of the substance. It is measured in units of cal/(g · °C).

$$\text{Specific heat} = \frac{\text{calories}}{\text{grams} \times °C}$$

Specific heats vary greatly from one substance to another, as shown in Table 1.10. The specific heat of water, 1.00 cal/(g · °C) (or 4.184 J/g °C) is higher than that of most other substances, which means that a large transfer of heat is required to change the temperature of a given amount of water by a given number of degrees. One consequence is that the human body, which is about 60% water, is able to withstand changing outside conditions.

Knowing the mass and specific heat of a substance makes it possible to calculate how much heat must be added or removed to accomplish a given temperature change, as shown in Worked Example 1.15.

$$\text{Heat (cal)} = \text{Mass (g)} \times \text{Temperature change (°C)} \times \text{Specific heat}\left(\frac{\text{cal}}{\text{g} \cdot °C}\right)$$

Specific heat The amount of heat that will raise the temperature of 1 g of a substance by 1 °C.

TABLE 1.10 Specific Heats of Some Common Substances

Substance	Specific Heat [cal/g °C]; [J/g °C]
Ethanol	0.59; 2.5
Gold	0.031; 0.13
Iron	0.106; 0.444
Mercury	0.033; 0.14
Sodium	0.293; 1.23
Water	1.00; 4.18

CHAPTER 1 Matter and Measurements

Worked Example 1.14 Temperature Conversions: Fahrenheit to Celsius

A body temperature above 107 °F can be fatal. What does 107 °F correspond to on the Celsius scale?

ANALYSIS Using the temperature (in °F) and the appropriate temperature conversion equation we can convert from the Fahrenheit scale to the Celsius scale.

BALLPARK ESTIMATE Note in Figure 1.10 that normal body temperature is 98.6 °F, or 37 °C. A temperature of 107 °F is approximately 8 °F above normal; since 1 °C is nearly 2 °F, then 8 °F is about 4 °C. Thus, the 107 °F body temperature is 41 °C.

SOLUTION

STEP 1: Identify known information. Temperature = 107 °F

STEP 2: Identify answer and units. Temperature = ?? °C

STEP 3: Identify conversion factors. We can convert from °F to °C using this equation.

$$°C = \frac{5\ °C}{9\ °F} \times (°F - 32\ °F)$$

STEP 4: Solve. Substitute the known temperature (in °F) into the equation.

$$°C = \frac{5\ °C}{9\ °F} \times (107\ °F - 32\ °F) = 42\ °C^*$$

(Rounded off from 41.666 667 °C)

BALLPARK CHECK Close to our estimate of 41 °C.

*It is worth noting that the 5/9 conversion factor in the equation is an exact conversion, and so does not impact the number of significant figures in the final answer.

Worked Example 1.15 Specific Heat: Mass, Temperature, and Energy

Taking a bath might use about 95 kg of water. How much energy (in calories and Joules) is needed to heat the water from a cold 15 °C to a warm 40 °C?

ANALYSIS From the amount of water being heated (95 kg) and the amount of the temperature change (40 °C − 15 °C = 25 °C), the total amount of energy needed can be calculated by using specific heat [1.00 cal/(g·°C)] as a conversion factor.

BALLPARK ESTIMATE The water is being heated 25 °C (from 15 °C to 40 °C), and it therefore takes 25 cal to heat each gram. The tub contains nearly 100,000 g (95 kg is 95,000 g), and so it takes about 25 × 100,000 cal, or 2,500,000 cal, to heat all the water in the tub.

SOLUTION

STEP 1: Identify known information.

Mass of water = 95 kg
Temperature change = 40 °C − 15 °C = 25 °C

STEP 2: Identify answer and units. Heat = ?? cal

STEP 3: Identify conversion factors. The amount of energy (in cal) can be calculated using the specific heat of water (cal/g · °C), and will depend on both the mass of water (in g) to be heated and the total temperature change (in °C). In order for the units in specific heat to cancel correctly, the mass of water must first be converted from kg to g.

$$\text{Specific heat} = \frac{1.0\ \text{cal}}{\text{g} \cdot °C}$$

$$1\ \text{kg} = 1000\ \text{g} \rightarrow \frac{1000\ \text{g}}{1\ \text{kg}}$$

STEP 4: Solve. Starting with the known information, use the conversion factors to cancel unwanted units.

$$95\ \text{kg} \times \frac{1000\ \text{g}}{\text{kg}} \times \frac{1.00\ \text{cal}}{\text{g} \cdot °C} \times 25\ °C = 2,400,000\ \text{cal}$$

$$= 2.4 \times 10^6\ \text{cal (or } 1.0 \times 10^7\ \text{J)}$$

BALLPARK CHECK Close to our estimate of 2.5×10^6 cal.

PROBLEM 1.22
The highest land temperature ever recorded was 136 °F in Al Aziziyah, Libya, on September 13, 1922. What is this temperature on the kelvin scale?

PROBLEM 1.23
The patient in the photo in the Chemistry in Action on page 31 has a temperature of 39 °C. What is the body temperature of the patient in °F?

PROBLEM 1.24
Assuming that Coca-Cola has the same specific heat as water, how much energy in calories is removed when 350 g of Coca-Cola (about the contents of one 12 oz can) is cooled from room temperature (25 °C) to refrigerator temperature (3 °C)?

PROBLEM 1.25
What is the specific heat of aluminum if it takes 161 cal (674 J) to raise the temperature of a 75 g aluminum bar by 10.0 °C?

1.14 Density and Specific Gravity

One further physical quantity that we will take up in this chapter is **density**, which relates the mass of an object to its volume. Density is usually expressed in units of grams per cubic centimeter (g/cm^3) for solids and grams per milliliter (g/mL) for liquids. Thus, if we know the density of a substance, we know both the mass of a given volume and the volume of a given mass. The densities of some common materials are listed in Table 1.11.

$$\text{Density} = \frac{\text{Mass (g)}}{\text{Volume (mL or cm}^3\text{)}}$$

Although most substances contract when cooled and expand when heated, water behaves differently. Water contracts when cooled from 100 °C to 3.98 °C, but below this temperature it begins to *expand* again. The density of liquid water is at its maximum of 1.0000 g/mL at 3.98 °C but decreases to 0.999 87 g/mL at 0 °C. When freezing occurs, the density drops still further to a value of 0.917 g/cm^3 for ice at 0 °C. Since a less dense substance will float on top of a more dense fluid, ice and any other substance with a density less than that of water will float in water. Conversely, any substance with a density greater than that of water will sink in water.

Knowing the density of a liquid is useful because it is often easier to measure a liquid's volume rather than its mass. Suppose, for example, that you need 1.50 g of

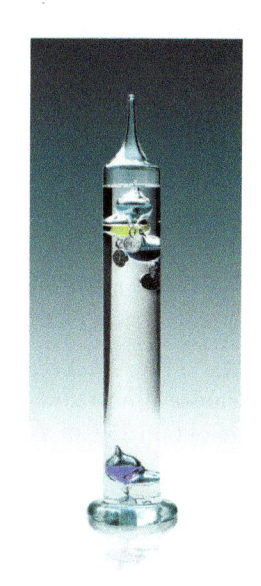

Density The physical property that relates the mass of an object to its volume; mass per unit volume.

▲ The Galileo thermometer contains several weighted bulbs which rise or fall as the density of the liquid changes with temperature.

TABLE 1.11 Densities of Some Common Materials at 25 °C

Substance	Density*	Substance	Density*
Gases		Solids	
Helium	0.000 194	Ice (0 °C)	0.917
Air	0.001 185	Gold	19.3
Liquids		Human fat	0.94
Water (3.98 °C)	1.0000	Cork	0.22–0.26
Urine	1.003–1.030	Table sugar	1.59
Blood plasma	1.027	Balsa wood	0.12
		Earth	5.54

*Densities are in g/cm^3 for solids and g/mL for liquids and gases.

Specific gravity The density of a substance divided by the density of water at the same temperature.

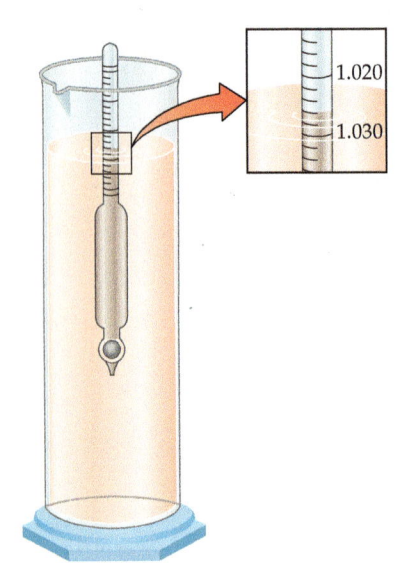

▲ **Figure 1.11**
A hydrometer for measuring specific gravity.
The instrument has a weighted bulb at the end of a calibrated glass tube. The depth to which the hydrometer sinks in a liquid indicates the liquid's specific gravity.

ethanol. Rather than use a dropper to weigh out exactly the right amount, it would be much easier to look up the density of ethanol (0.7893 g/mL at 20 °C) and measure the correct volume (1.90 mL) with a syringe or graduated cylinder. Thus, density acts as a conversion factor between mass (g) and volume (mL).

$$1.50 \text{ g ethanol} \times \frac{1 \text{ mL ethanol}}{0.7893 \text{ g ethanol}} = 1.90 \text{ mL ethanol}$$

For many purposes, ranging from winemaking to medicine, it is more convenient to use *specific gravity* than density. The **specific gravity** (sp gr) of a substance (usually a liquid) is simply the density of the substance divided by the density of water at the same temperature. Because all units cancel, specific gravity is unitless:

$$\text{Specific gravity} = \frac{\text{Density of substance (g/mL)}}{\text{Density of water at the same temperature (g/mL)}}$$

At typical temperatures, the density of water is very close to 1 g/mL. Thus, the specific gravity of a substance is numerically equal to its density and is used in the same way.

The specific gravity of a liquid can be measured using an instrument called a *hydrometer*, which consists of a weighted bulb on the end of a calibrated glass tube, as shown in Figure 1.11. The depth to which the hydrometer sinks when placed in a fluid indicates the fluid's specific gravity: the lower the bulb sinks, the lower the specific gravity of the fluid.

In medicine, a hydrometer called a *urinometer* is used to indicate the amount of solids dissolved in urine. Although the specific gravity of normal urine is about 1.003–1.030, conditions such as diabetes mellitus or a high fever cause an abnormally high urine specific gravity, indicating either excessive elimination of solids or decreased elimination of water. Abnormally low specific gravity is found in individuals using diuretics—drugs that increase water elimination.

Worked Example 1.16 Density: Mass-to-Volume Conversion

What volume of isopropyl alcohol (rubbing alcohol) would you use if you needed 25.0 g? The density of isopropyl alcohol is 0.7855 g/mL at 20 °C.

ANALYSIS The known information is the mass of isopropyl alcohol needed (25.0 g). The density (0.7855 g/mL) acts as a conversion factor between mass and the unknown volume of isopropyl alcohol.

BALLPARK ESTIMATE Because 1 mL of isopropyl alcohol contains only 0.7885 g of the alcohol, obtaining 1 g of alcohol would require almost 20% more than 1 mL, or about 1.2 mL. Therefore, a volume of about 25 × 1.2 mL = 30 mL is needed to obtain 25 g of alcohol.

SOLUTION

STEP 1: Identify known information.

Mass of rubbing alcohol = 25.0 g
Density of rubbing alcohol = 0.7855 g/mL

STEP 2: Identify answer and units.

Volume of rubbing alcohol = ?? mL

STEP 3: Identify conversion factors. Starting with the mass of isopropyl alcohol (in g), the corresponding volume (in mL) can be calculated using density (g/mL) as the conversion factor.

Density = g/mL → 1/density = mL/g

STEP 4: Solve. Starting with the known information, set up the equation with conversion factors so that unwanted units cancel.

$$25.0 \text{ g alcohol} \times \frac{1 \text{ mL alcohol}}{0.7855 \text{ g alcohol}} = 31.8 \text{ mL alcohol}$$

BALLPARK CHECK Our estimate was 30 mL.

Chemistry in Action

A Measurement Example: Obesity and Body Fat

According to the U.S. Centers for Disease Control and Prevention, the U.S. population is suffering from a fat epidemic. Over the last 25 years, the percentage of adults 20 years or older identified as obese increased from 15% in the late 1970s to nearly 33% in 2008. Even children and adolescents are gaining too much weight: The number of overweight children in all age groups increased by nearly a factor of 3, with the biggest increase seen among teenagers (from 5% to 18.1%). Of particular concern is the fact that 80% of children who were overweight as teenagers were identified as obese at age 25. Obesity increases the risk for many adverse health conditions, including type 2 diabetes and heart disease.

How do we define obesity, however, and how is it measured? Obesity is defined by reference to *body mass index* (BMI), which is equal to a person's mass in kilograms divided by the square of his or her height in meters. BMI can also be calculated by dividing a person's weight in pounds by the square of her or his height in inches multiplied by 703. For instance, someone 5 ft 7 in. (67 inches; 1.70 m) tall weighing 147 lb (66.7 kg) has a BMI of 23:

$$\text{BMI} = \frac{\text{weight (kg)}}{[\text{height (m)}]^2}, \quad \text{or} \quad \frac{\text{weight (lb)}}{[\text{height (in.)}]^2} \times 703$$

A BMI of 25 or above is considered overweight, and a BMI of 30 or above is obese. By these standards, approximately 61% of the U.S. population is overweight. Health professionals are concerned by the rapid rise in obesity in the United States because of the link between BMI and health problems. Many reports have documented the correlation between health and BMI, including a recent study on more than 1 million adults. The lowest death risk from any cause, including cancer and heart disease, is associated with a BMI between 22 and 24. Risk increases steadily as BMI increases, more than doubling for a BMI above 29.

An individual's percentage of body fat is most easily measured by the skinfold-thickness method. The skin at several locations on the arm, shoulder, and waist is pinched, and the thickness of the fat layer beneath the skin is measured with calipers. Comparing the measured results to those in a standard table gives an estimation of percentage body fat. As an alternative to skinfold measurement, a more accurate assessment of body fat can be made by underwater immersion. The person's underwater body weight is less than her or his weight on land because water gives the body buoyancy. The higher the percentage of body fat, the more buoyant the person and the greater the difference between land weight and underwater body weight. Checking the observed buoyancy on a standard table then gives an estimation of body fat.

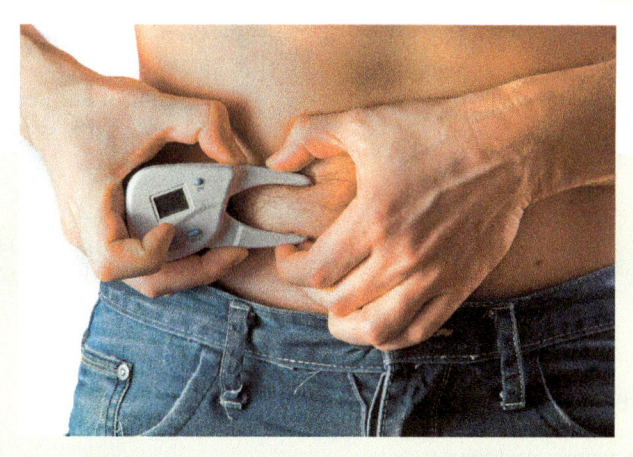

▲ A person's percentage body fat can be estimated by measuring the thickness of the fat layer under the skin.

See Chemistry in Action Problems 1.100 and 1.101 at the end of the chapter.

Weight (lb)

Height	110	115	120	125	130	135	140	145	150	155	160	165	170	175	180	185	190	195	200
5'0"	21	22	23	24	25	26	27	28	29	30	31	32	33	34	35	36	37	38	39
5'2"	20	21	22	23	24	25	26	27	27	28	29	30	31	32	33	34	35	36	37
5'4"	19	20	21	21	22	23	24	25	26	27	27	28	29	30	31	32	33	33	34
5'6"	18	19	19	20	21	22	23	23	24	25	26	27	27	28	29	30	31	31	32
5'8"	17	17	18	19	20	21	21	22	23	24	24	25	26	27	27	28	29	30	30
5'10"	16	17	17	18	19	19	20	21	22	22	23	24	24	25	26	27	27	28	29
6'0"	15	16	16	17	18	18	19	20	20	21	22	22	23	24	24	25	26	26	27
6'2"	14	15	15	16	17	17	18	19	19	20	21	21	22	22	23	24	24	25	26
6'4"	13	14	15	15	16	16	17	18	18	19	19	20	21	21	22	23	23	24	24

Body Mass Index (numbers in boxes)

36 CHAPTER 1 Matter and Measurements

The specific gravity of urine, measured by a urinometer, is used to diagnose conditions such as diabetes.

PROBLEM 1.26
A sample of pumice, a porous volcanic rock, weighs 17.4 grams and has a volume of 27.3 cm^3. If this sample is placed in a container of water, will it sink or will it float? Explain.

PROBLEM 1.27
Chloroform, once used as an anesthetic agent, has a density of 1.474 g/mL. What volume would you use if you needed 12.37 g?

PROBLEM 1.28
The sulfuric acid solution in an automobile battery typically has a specific gravity of about 1.27. Is battery acid more dense or less dense than pure water?

SUMMARY: REVISITING THE CHAPTER GOALS

1. What is matter and how is it classified? *Matter* is anything that has mass and occupies volume—that is, anything physically real. Matter can be classified by its physical state as *solid*, *liquid*, or *gas*. A solid has a definite volume and shape, a liquid has a definite volume but indefinite shape, and a gas has neither a definite volume nor a definite shape. Matter can also be classified by composition as being either *pure* or a *mixture*. Every pure substance is either an *element* or a *chemical compound*. Elements are fundamental substances that cannot be chemically changed into anything simpler. A chemical compound, by contrast, can be broken down by chemical change into simpler substances. Mixtures are composed of two or more pure substances and can be separated into component parts by physical means (*see Problems 40–45, 96, 103*).

2. How are chemical elements represented? Elements are represented by one- or two-letter symbols, such as H for hydrogen, Ca for calcium, Al for aluminum, and so on. Most symbols are the first one or two letters of the element name, but some symbols are derived from Latin names—Na (sodium), for example. All the known elements are commonly organized into a form called the *periodic table*. Most elements are *metals*, 18 are *nonmetals*, and 6 are *metalloids* (*see Problems 29–31, 48–57, 96, 102, 103*).

3. What kinds of properties does matter have? A *property* is any characteristic that can be used to describe or identify something: *physical* properties can be seen or measured without changing the chemical identity of the substance (that is, color, melting point), while *chemical* properties can only be seen or measured when the substance undergoes a *chemical change*, such as a chemical reaction (*see Problems 37–39, 42–44, 47, 97, 102, 103*).

4. What units are used to measure properties, and how can a quantity be converted from one unit to another? A property that can be measured is called a *physical quantity* and is described by both a number and a label, or *unit*. The preferred units are either those of the International System of Units (*SI units*) or the *metric system*. Mass, the amount of matter an object contains, is measured in *kilograms* (kg) or *grams* (g). Length is measured in *meters* (m). Volume is measured in *cubic meters* (m^3) in the SI system and in *liters* (L) or *milliliters* (mL) in the metric system. Temperature is measured in *kelvins* (K) in the SI system and in *degrees celsius* (°C) in the metric system. A measurement in one unit can be converted to another unit by multiplying by a *conversion factor* that expresses the exact relationship between the units (*see Problems 58–63, 72–82, 100, 101, 104, 105, 107–109, 121*).

5. How good are the reported measurements? When measuring physical quantities or using them in calculations, it is important to indicate the exactness of the measurement by *rounding off* the final answer using the correct number of *significant figures*. All but one of the significant figures in a number is known with certainty; the final digit is estimated to ±1 (*see Problems 32–35, 64–71, 104, 112*).

6. How are large and small numbers best represented? Measurements of small and large quantities are usually written in *scientific notation* as the product of a number between 1 and 10, times a power of 10. Numbers greater than 10 have a positive exponent, and numbers less than 1 have a negative exponent. For example, $3562 = 3.562 \times 10^3$, and $0.00391 = 3.91 \times 10^{-3}$ (*see Problems 64–71, 75, 82, 108*).

7. What techniques are used to solve problems? Problems are best solved by applying the *factor-label method*, in which units can be multiplied and divided just as numbers can. The idea is to set up an equation so that all unwanted units cancel, leaving only the desired units. Usually it is best to start by identifying the known and needed information, then decide how to convert the known information to the answer, and finally check to make sure the answer is reasonable both chemically and physically (*see Problems 76–82, 101, 106, 107, 109, 110–112, 114, 115, 118–123*).

8. What are temperature, specific heat, density, and specific gravity? *Temperature* is a measure of how hot or cold an object is. The *specific heat* of a substance is the amount of heat necessary to raise the temperature of 1 g of the substance by 1 °C (1 cal/g °C or 4.184 J/g °C). Water has an unusually high specific heat, which helps our bodies to maintain an even temperature. *Density*, the physical property that relates mass to volume, is expressed in units of grams per milliliter (g/mL) for a liquid or grams per cubic centimeter (g/cm^3) for a solid. The *specific gravity* of a liquid is the density of the liquid divided by the density of water at the same temperature. Because the density of water is approximately 1 g/mL, specific gravity and density have the same numerical value (*see Problems 32, 36, 42, 43, 83–89, 90–95, 98, 99, 106, 109, 113, 118–120, 122, 123*).

Understanding Key Concepts

KEY WORDS

Change of state, *p. 6*
Chemical change, *p. 4*
Chemical compound, *p. 7*
Chemical formula, *p. 10*
Chemical reaction, *p. 7*
Chemistry, *p. 4*
Conversion factor, *p. 26*
Density, *p. 33*
Element, *p. 6*
Energy, *p. 29*
Factor-label method, *p. 26*
Gas, *p. 5*
Heterogeneous mixture, *p. 6*
Homogeneous mixture, *p. 6*
Liquid, *p. 5*
Mass, *p. 17*
Matter, *p. 4*
Metal, *p. 11*
Metalloid, *p. 11*
Mixture, *p. 6*

Nonmetal, *p. 11*
Periodic table, *p. 11*
Physical change, *p. 4*
Physical quantity, *p. 14*
Product, *p. 7*
Property, *p. 4*
Pure substance, *p. 6*
Reactant, *p. 7*
Rounding off, *p. 24*
Scientific Method, *p. 4*
Scientific notation, *p. 21*
SI units, *p. 15*
Significant figures, *p. 20*
Solid, *p. 5*
Specific gravity, *p. 34*
Specific heat, *p. 31*
State of matter, *p. 6*
Temperature, *p. 29*
Unit, *p. 14*
Weight, *p. 17*

UNDERSTANDING KEY CONCEPTS

The problems in this section are intended as a bridge between the Chapter Summary and the Additional Problems that follow. Primarily visual in nature, they are designed to help you test your grasp of the chapter's most important principles before attempting to solve quantitative problems. Answers to all Key Concept problems are at the end of the book following the appendixes.

1.29 The six elements in blue at the far right of the periodic table are gases at room temperature. The red elements in the middle of the table are the so-called coinage metals. Identify each of these elements using the periodic table inside the front cover of this book.

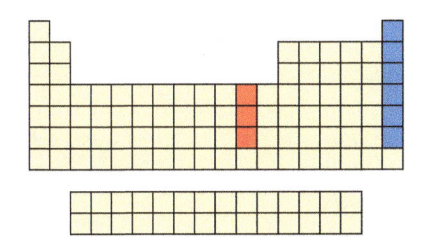

1.30 Identify the three elements indicated on the following periodic table and tell which is a metal, which is a nonmetal, and which is a metalloid.

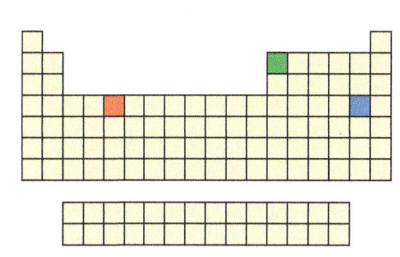

1.31 The radioactive element indicated on the following periodic table is used in smoke detectors. Identify it, and tell whether it is a metal, a nonmetal, or a metalloid.

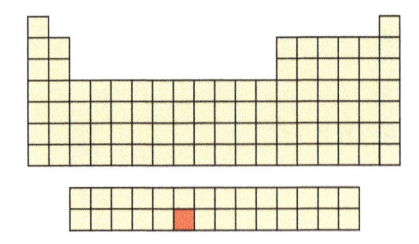

1.32 (a) What is the specific gravity of the following solution?
(b) How many significant figures does your answer have?
(c) Is the solution more dense or less dense than water?

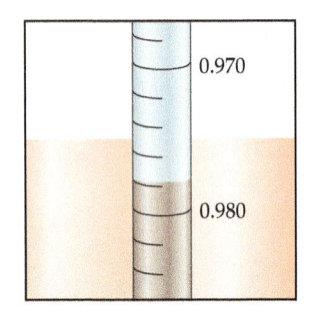

1.33 Assume that you have two graduated cylinders, one with a capacity of 5 mL (a) and the other with a capacity of 50 mL (b). Draw a line in each showing how much liquid you would add if you needed to measure 2.64 mL of water. Which cylinder do you think is more precise? Explain.

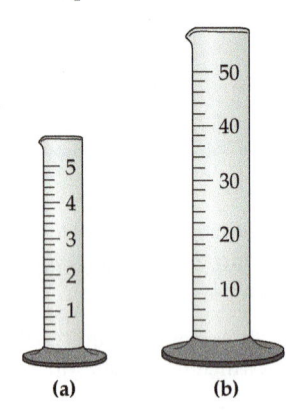

1.34 State the length of the pencil depicted in the accompanying figure in both inches and centimeters using appropriate numbers of significant figures.

1.35 Assume that you are delivering a solution sample from a pipette. Figures (a) and (b) show the volume level before and after dispensing the sample, respectively. State the liquid level (in mL) before and after dispensing the sample, and calculate the volume of the sample.

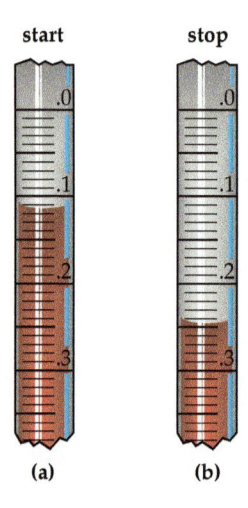

1.36 Assume that identical hydrometers are placed in ethanol (sp gr 0.7893) and in chloroform (sp gr 1.4832). In which liquid will the hydrometer float higher? Explain.

ADDITIONAL PROBLEMS

These exercises are divided into sections by topic. Each section begins with review and conceptual questions, followed by numerical problems of varying levels of difficulty. Many of the problems dealing with more difficult concepts or skills are presented in pairs, with each even-numbered problem followed by an odd-numbered one requiring similar skills. The final section consists of unpaired General Questions and Problems that draw on various parts of the chapter and, in future chapters, may even require the use of concepts from previous chapters. Answers to all even-numbered problems are given at the end of the book following the appendixes.

CHEMISTRY AND THE PROPERTIES OF MATTER

1.37 What is the difference between a physical change and a chemical change?

1.38 Which of the following is a physical change and which is a chemical change?
(a) Boiling water
(b) Decomposing water by passing an electric current through it

(c) Exploding of potassium metal when placed in water

(d) Breaking of glass

1.39 Which of the following is a physical change and which is a chemical change?

(a) Making lemonade (lemons + water + sugar)

(b) Frying eggs

(c) Burning a candle

(d) Whipping cream

(e) Leaves changing color

STATES AND CLASSIFICATION OF MATTER

1.40 Name and describe the three states of matter.

1.41 Name two changes of state, and describe what causes each to occur.

1.42 Sulfur dioxide is a compound produced when sulfur burns in air. It has a melting point of $-72.7\,°C$ and a boiling point of $-10\,°C$. In what state does it exist at room temperature (298 K)? (refer to Figure 1.10).

1.43 Butane (C_4H_8) is an easily compressible gas used in cigarette lighters. It has a melting point of $-138.4\,°C$ and a boiling point of $-0.5\,°C$. Would you expect a butane lighter to work in winter when the temperature outdoors is 25 °F? Why or why not? (refer to Figure 1.10).

1.44 Classify each of the following as a mixture or a pure substance:

(a) Pea soup (b) Seawater

(c) The contents of a propane tank

(d) Urine (e) Lead

(f) A multivitamin tablet

1.45 Which of these terms, (i) mixture, (ii) solid, (iii) liquid, (iv) gas, (v) chemical element, (vi) chemical compound, applies to the following substances at room temperature?

(a) Gasoline (b) Iodine

(c) Water (d) Air

(e) Blood (f) Sodium bicarbonate

(g) Gaseous ammonia (h) Silicon

1.46 Hydrogen peroxide, often used in solutions to cleanse cuts and scrapes, breaks down to yield water and oxygen:

Hydrogen peroxide ⟶ Water + Oxygen

(a) Identify the reactants and products.

(b) Which of the substances are chemical compounds, and which are elements?

1.47 When sodium metal is placed in water, the following change occurs:

Sodium + Water ⟶ Hydrogen + Sodium hydroxide

(a) Identify the reactants and products.

(b) Which of the substances are elements, and which are chemical compounds?

ELEMENTS AND THEIR SYMBOLS

1.48 Describe the general properties of metals, nonmetals, and metalloids.

1.49 What is the most abundant element in the earth's crust? In the human body? List the name and symbol for each.

1.50 What are the symbols for the following elements?

(a) Gadolinium (used in color TV screens)

(b) Germanium (used in semiconductors)

(c) Technetium (used in biomedical imaging)

(d) Arsenic (used in pesticides)

(e) Cadmium (used in rechargeable batteries)

1.51 Supply the missing names or symbols for the elements in the spaces provided:

(a) N _____ (b) K _____

(c) Cl _____ (d) _____ Calcium

(e) _____ Phosphorus (f) _____ Manganese

1.52 Correct the following statements.

(a) The symbol for bromine is BR.

(b) The symbol for manganese is Mg.

(c) The symbol for carbon is Ca.

(d) The symbol for potassium is Po.

1.53 Correct the following statements.

(a) Carbon dioxide has the formula CO2.

(b) Carbon dioxide has the formula Co_2.

(c) Table salt, NaCl, is composed of nitrogen and chlorine.

1.54 The amino acid glycine has the formula $C_2H_5NO_2$. Which elements are present in glycine? What is the total number of atoms represented by the formula?

1.55 Glucose, a form of sugar, has the formula $C_6H_{12}O_6$. Which elements are included in this compound, and how many atoms of each are present?

1.56 Write the formula for ibuprofen: 13 carbons, 18 hydrogens, and 2 oxygens.

1.57 Given the physical properties of the following elements classify each one as a metal, nonmetal, or metalloid:

(a) a hard, shiny, very dense solid that conducts electricity

(b) a brittle, gray solid that conducts electricity poorly

(c) a brown, crystalline solid that does not conduct electricity

(d) a colorless, odorless gas

PHYSICAL QUANTITIES: DEFINITIONS AND UNITS

1.58 What is the difference between a physical quantity and a number?

40 CHAPTER 1 Matter and Measurements

1.59 What are the units used in the SI system to measure mass, volume, length, and temperature? In the metric system?

1.60 Give the full name of the following units:

(a) cc (b) dm (c) mm
(d) nL (e) mg (f) m³

1.61 Write the symbol for the following units:

(a) nanogram (b) centimeter
(c) microliter (d) micrometer
(e) milligram

1.62 How many picograms are in 1 mg? In 35 ng?

1.63 How many microliters are in 1 L? In 20 mL?

SCIENTIFIC NOTATION AND SIGNIFICANT FIGURES

1.64 Express the following numbers in scientific notation with the correct number of significant figures:

(a) 9457 (b) 0.000 07
(c) 20,000,000,000 (four significant figures)
(d) 0.012 345 (e) 652.38

1.65 Convert the following numbers from scientific notation to standard notation:

(a) 5.28×10^3 (b) 8.205×10^{-2}
(c) 1.84×10^{-5} (d) 6.37×10^4

1.66 How many significant figures does each of the following numbers have?

(a) 237,401 (b) 0.300 (c) 3.01
(d) 244.4 (e) 50,000 (f) 660

1.67 How many significant figures are there in each of the following quantities?

(a) Distance from New York City to Wellington, New Zealand, 14,397 km
(b) Average body temperature of a crocodile, 25.6 °C
(c) Melting point of gold, 1064 °C
(d) Diameter of an influenza virus, 0.000 01 mm
(e) Radius of a phosphorus atom, 0.110 nm

1.68 The diameter of the earth at the equator is 7926.381 mi.

(a) Round off the earth's diameter to four significant figures, to two significant figures, and to six significant figures.
(b) Express the earth's diameter in scientific notation.

1.69 Round off each of the numbers in Problem 1.67 to two significant figures, and express them in scientific notation.

1.70 Carry out the following calculations, express each answer to the correct number of significant figures, and include units in the answers.

(a) 9.02 g + 3.1 g (b) 88.80 cm + 7.391 cm

(c) 362 mL − 99.5 mL
(d) 12.4 mg + 6.378 mg + 2.089 mg

1.71 Carry out the following calculations, express the answers to the correct numbers of significant figures, and include units in the answers.

(a) $5280 \dfrac{\text{ft}}{\text{mi}} \times 6.2 \text{ mi}$

(b) 4.5 m × 3.25 m

(c) $2.50 \text{ g} \div 8.3 \dfrac{\text{g}}{\text{cm}^3}$

(d) 4.70 cm × 6.8 cm × 2.54 cm

UNIT CONVERSIONS AND PROBLEM SOLVING

1.72 Carry out the following conversions:

(a) 3.614 mg to centigrams
(b) 12.0 kL to megaliters
(c) 14.4 μm to millimeters
(d) 6.03×10^{-6} cg to nanograms
(e) 174.5 mL to deciliters
(f) 1.5×10^{-2} km to centimeters

1.73 Carry out the following conversions. Consult Tables 1.7–1.9 as needed.

(a) 56.4 mi to kilometers and to megameters
(b) 2.0 L to quarts and to fluid ounces
(c) 7 ft 2.0 in. to centimeters and to meters
(d) 1.35 lb to kilograms and to decigrams

1.74 Express the following quantities in more convenient units by using SI unit prefixes:

(a) 9.78×10^4 g (b) 1.33×10^{-4} L
(c) 0.000 000 000 46 g (d) 2.99×10^8 cm

1.75 Fill in the blanks to complete the equivalencies either with appropriate units prefixes or with the appropriate scientific notation. The first blank is filled in as an example.

(a) 125 km = 1.25×10^5 m
(b) 6.285×10^3 mg = _____? _____ kg
(c) 47.35 dL = 4.735 × _____? _____ mL
(d) 67.4 cm = 6.7×10^{-4} _____? _____

1.76 The speed limit in Canada is 100 km/h.

(a) How many miles per hour is this?
(b) How many feet per second?

1.77 The muzzle velocity of a projectile fired from a 9 mm handgun is 1200 ft/s.

(a) How many miles per hour is this?
(b) How many meters per second?

1.78 The diameter of a red blood cell is 6×10^{-6} m.

(a) How many centimeters is this?

(b) How many red blood cells are needed to make a line 1 cm long? 1 in. long?

1.79 The Willis Tower in Chicago has an approximate floor area of 418,000 m². How many square feet of floor space is this?

1.80 A normal value for blood cholesterol is 200 mg/dL of blood. If a normal adult has a total blood volume of 5 L, how much total cholesterol is present?

1.81 The recommended daily dose of calcium for an 18-year-old male is 1200 mg. If 1.0 cup of whole milk contains 290 mg of calcium and milk is his only calcium source, how much milk should an 18-year-old male drink each day?

1.82 The white blood cell concentration in normal blood is approximately 12,000 cells/mm³ of blood. How many white blood cells does a normal adult with 5 L of blood have? Express the answer in scientific notation.

ENERGY, HEAT, AND TEMPERATURE

1.83 The boiling point of liquid nitrogen, used in the removal of warts and in other surgical applications, is −195.8 °C. What is this temperature in kelvins and in degrees Fahrenheit? (3.74 J/g °C)

1.84 Diethyl ether, a substance once used as a general anesthetic, has a specific heat of 0.895 cal/(g °C). How many calories and how many kilocalories of heat are needed to raise the temperature of 30.0 g of diethyl ether from 10.0 °C to 30.0 °C? How many Joules and kiloJoules?

1.85 Aluminum has a specific heat of 0.215 cal/(g °C). When 25.7 cal (108.5 J) of heat is added to 18.4 g of aluminum at 20.0°, what is the final temperature of the aluminum?

1.86 Calculate the specific heat of copper if it takes 23 cal (96 J) to heat a 5.0 g sample from 25 °C to 75 °C.

1.87 The specific heat of fat is 0.45 cal/(g·°C) (1.9 J/g °C) and the density of fat is 0.94 g/cm³. How much energy (in calories and joules) is needed to heat 10 cm³ of fat from room temperature (25 °C) to its melting point (35 °C)?

1.88 A 150 g sample of mercury and a 150 g sample of iron are at an initial temperature of 25.0 °C. If 250 cal (1050 J) of heat is applied to each sample, what is the final temperature of each? (See Table 1.10.)

1.89 When 100 cal (418 J) of heat is applied to a 125 g sample, the temperature increases by 28 °C. Calculate the specific heat of the sample and compare your answer to the values in Table 1.10. What is the identity of the sample?

DENSITY AND SPECIFIC GRAVITY

1.90 Aspirin has a density of 1.40 g/cm³. What is the volume in cubic centimeters of a tablet weighing 250 mg?

1.91 Gaseous hydrogen has a density of 0.0899 g/L at 0 °C. How many liters would you need if you wanted 1.0078 g of hydrogen?

1.92 What is the density of lead (in g/cm³) if a rectangular bar measuring 0.500 cm in height, 1.55 cm in width, and 25.00 cm in length has a mass of 220.9 g?

1.93 What is the density of lithium metal (in g/cm³) if a cube measuring 0.82 cm $\times$ 1.45 cm $\times$ 1.25 cm has a mass of 0.794 g?

1.94 Ethanol produced by fermentation has a specific gravity of 0.787 at 25 °C. What is the volume of 125 g of ethanol at this temperature? (The density of water at 25 °C is 0.997 g/mL.)

1.95 Ethylene glycol, commonly used as automobile antifreeze, has a specific gravity of 1.1088 at room temperature (25 °C). What is the mass of 1.00 L of ethylene glycol at this temperature?

CHEMISTRY IN ACTION

1.96 The active ingredient in aspirin, acetylsalicylic acid (ASA), has the formula $C_9H_8O_4$ and melts at 140 °C. Identify the elements and how many atoms of each are present in ASA. Is it a solid or a liquid at room temperature? [*Aspirin—A Case Study, p. 8*]

1.97 Calomel (Hg_2Cl_2) is not toxic but methyl mercury chloride (CH_3HgCl) is highly toxic. What physical property explains this difference in toxicity? [*Mercury and Mercury Poisoning, p. 15*]

1.98 A thermochromic plastic chip included in a shipping container for beef undergoes an irreversible color change if the storage temperature exceeds 28 °F. What is this temperature on the Celsius and Kelvin scales? [*Temperature-Sensitive Materials, p. 31*]

1.99 A temperature-sensitive bath toy undergoes several color changes in the temperature range from 37 °C to 47 °C. What is the corresponding temperature range on the Fahrenheit scale? [*Temperature-Sensitive Materials, p. 31*]

1.100 Calculate the BMI for an individual who is

(a) 5 ft 1 in. tall and weighs 155 lb

(b) 5 ft 11 in. tall and weighs 170 lb

(c) 6 ft 3 in. tall and weighs 195 lb

Which of these individuals is likely to have increased health risks? [*A Measurement Example: Obesity and Body Fat, p. 35*]

1.101 Liposuction is a technique for removing fat deposits from various areas of the body. How many liters of fat would have to be removed to result in a 5.0 lb weight loss? The density of human fat is 0.94 g/mL. [*A Measurement Example: Obesity and Body Fat, p. 35*]

GENERAL QUESTIONS AND PROBLEMS

1.102 The most recently discovered element is number 117, Unuseptium. Based on its location in the periodic table, classify it as a metal, nonmetal, or metalloid and discuss the physical properties (physical state, conductivity, etc.) you would expect it to exhibit.

1.103 A white solid with a melting point of 730 °C is melted. When electricity is passed through the resultant liquid, a brown gas and a molten metal are produced. Neither the metal nor the gas can be broken down into anything simpler by chemical means. Classify each—the white solid, the molten metal, and the brown gas—as a mixture, a compound, or an element.

1.104 Refer to the pencil in Problem 1.34. Using the equivalent values in Table 1.8 as conversion factors, convert the length measured in inches to centimeters. Compare the calculated length in centimeters to the length in centimeters measured using the metric ruler. How do the two values compare? Explain any differences.

1.105 Gemstones are weighed in carats, where 1 carat = 200 mg exactly. What is the mass in grams of the Hope diamond, the world's largest blue diamond, at 44.4 carats?

1.106 The relationship between the nutritional unit for energy and the metric unit is 1 Calorie = 1 kcal.

(a) One donut contains 350 Calories. Convert this to calories and joules.

(b) If the energy in one donut was used to heat 35.5 kg of water, calculate the increase in temperature of the water (in °C).

1.107 Drug dosages are typically prescribed in units of milligrams per kilogram of body weight. A new drug has a recommended dosage of 9 mg/kg.

(a) How many mgs would a 130 lb woman have to take to obtain this dosage?

(b) How many 125 mg tablets should a 40 lb child take to receive the recommended dosage?

1.108 A clinical report gave the following data from a blood analysis: iron, 39 mg/dL; calcium, 8.3 mg/dL; cholesterol, 224 mg/dL. Express each of these quantities in grams per deciliter, writing the answers in scientific notation.

1.109 The Spirit of America Goodyear blimp has a volume of 2.027×10^5 ft^3.

(a) Convert this volume to L.

(b) When in operation it is filled with helium gas. If the density of helium at room temperature is 0.179 g/L, calculate the mass of helium in the blimp.

(c) What is the mass of air occupying the same volume? The density of air at room temperature is 1.20 g/L.

1.110 Approximately 75 mL of blood is pumped by a normal human heart at each beat. Assuming an average pulse of 72 beats per minute, how many milliliters of blood are pumped in one day?

1.111 A doctor has ordered that a patient be given 15 g of glucose, which is available in a concentration of 50.00 g glucose/1000.0 mL of solution. What volume of solution should be given to the patient?

1.112 Reconsider the volume of the sample dispensed by pipette in Problem 1.35. Assuming that the solution in the pipette has a density of 0.963 g/mL, calculate the mass of solution dispensed in the problem to the correct number of significant figures.

1.113 Today, thermometers containing mercury are used less frequently than in the past because of concerns regarding the toxicity of mercury and because of its relatively high melting point (−39 °C). This means that mercury thermometers cannot be used in very cold environments because the mercury is a solid under such conditions. Alcohol thermometers, however, can be used over a temperature range from −115 °C (the melting point of alcohol) to 78.5 °C (the boiling point of alcohol).

(a) What is the effective temperature range of the alcohol thermometer in °F?

(b) The densities of alcohol and mercury are 0.79 g/mL and 13.6 g/mL, respectively. If the volume of liquid in a typical laboratory thermometer is 1.0 mL, what mass of alcohol is contained in the thermometer? What mass of mercury?

1.114 In a typical person, the level of blood glucose (also known as blood sugar) is about 85 mg/100 mL of blood. If an average body contains about 11 pints of blood, how many grams and how many pounds of glucose are present in the blood?

1.115 A patient is receiving 3000 mL/day of a solution that contains 5 g of dextrose (glucose) per 100 mL of solution. If glucose provides 4 kcal/g of energy, how many kilocalories per day is the patient receiving from the glucose?

1.116 A rough guide to fluid requirements based on body weight is 100 mL/kg for the first 10 kg of body weight, 50 mL/kg for the next 10 kg, and 20 mL/kg for weight over 20 kg. What volume of fluid per day is needed by a 55 kg woman? Give the answer with two significant figures.

1.117 Chloral hydrate, a sedative and sleep-inducing drug, is available as a solution labeled 10.0 gr/fluidram. What volume in milliliters should be administered to a patient who is meant to receive 7.5 gr per dose? (1 gr = 64.8 mg ; 1 fluidram = 3.72 mL)

1.118 When 1.0 tablespoon of butter is burned or used by our body, it releases 100 kcal (100 food Calories or 418. 4 kJ) of energy. If we could use all the energy provided, how many tablespoons of butter would have to be burned to raise the temperature of 3.00 L of water from 18.0 °C to 90.0 °C?

1.119 An archeologist finds a 1.62 kg goblet that she believes to be made of pure gold. When 1350 cal (5650 J) of heat is added to the goblet, its temperature increases by 7.8 °C. Calculate the specific heat of the goblet. Is it made of gold? Explain.

1.120 In another test, the archeologist in Problem 1.119 determines that the volume of the goblet is 205 mL. Calculate the density of the goblet and compare it with the density of gold (19.3 g/mL), lead (11.4 g/mL), and iron (7.86 g/mL). What is the goblet probably made of?

1.121 Sulfuric acid (H_2SO_4, density = 1.83 g/mL) is produced in larger amounts than any other chemical: 2.01×10^{11} lb worldwide in 2004. What is the volume of this amount in liters?

1.122 Imagine that you place a piece of cork measuring 1.30 cm × 5.50 cm × 3.00 cm in a pan of water and that on top of the cork you place a small cube of lead measuring 1.15 cm on each edge. The density of cork is 0.235 g/cm^3 and the density of lead is 11.35 g/cm^3. Will the combination of cork plus lead float or sink?

1.123 At a certain point, the Celsius and Fahrenheit scales "cross," and at this point the numerical value of the Celsius temperature is the same as the numerical value of the Fahrenheit temperature. At what temperature does this crossover occur?

CHAPTER 2

Atoms and the Periodic Table

CONTENTS

2.1 Atomic Theory
2.2 Elements and Atomic Number
2.3 Isotopes and Atomic Weight
2.4 The Periodic Table
2.5 Some Characteristics of Different Groups
2.6 Electronic Structure of Atoms
2.7 Electron Configurations
2.8 Electron Configurations and the Periodic Table
2.9 Electron-Dot Symbols

◀ These basaltic columns at the Devil's Post-pile National Monument in northern California are one example of repeating patterns that can be found in nature.

CHAPTER GOALS

1. **What is the modern theory of atomic structure?**
   THE GOAL: Be able to explain the major assumptions of atomic theory.

2. **How do atoms of different elements differ?**
   THE GOAL: Be able to explain the composition of different atoms according to the number of protons, neutrons, and electrons they contain.

3. **What are isotopes, and what is atomic weight?**
   THE GOAL: Be able to explain what isotopes are and how they affect an element's atomic weight.

4. **How is the periodic table arranged?**
   THE GOAL: Be able to describe how elements are arranged in the periodic table, name the subdivisions of the periodic table, and relate the position of an element in the periodic table to its electronic structure.

5. **How are electrons arranged in atoms?**
   THE GOAL: Be able to explain how electrons are distributed in shells and subshells around the nucleus of an atom, how valence electrons can be represented as electron-dot symbols, and how the electron configurations can help explain the chemical properties of the elements.

Chemistry must be studied on two levels. In the previous chapter we dealt with chemistry on the large-scale, or *macroscopic*, level, looking at the properties and transformations of matter that we can see and measure. Now we are ready to look at the sub-microscopic, or atomic level, studying the behavior and properties of individual atoms. Although scientists have long been convinced of their existence, only within the past 20 years have powerful new instruments made it possible to see individual atoms. In this chapter, we will look at modern atomic theory and how the structure of atoms influences macroscopic properties.

2.1 Atomic Theory

Take a piece of aluminum foil, and cut it in two. Then, take one of the pieces and cut *it* in two, and so on. Assuming that you have extremely small scissors and extraordinary dexterity, how long can you keep dividing the foil? Is there a limit, or is matter infinitely divisible into ever smaller and smaller pieces? Historically, this argument can be traced as far back as the ancient Greek philosophers. Aristotle believed that matter could be divided infinitely, while Democritus argued (correctly) that there is a limit. The smallest and simplest bit that aluminum (or any other element) can be divided and still be identifiable as aluminum is called an **atom**, a word derived from the Greek *atomos*, meaning "indivisible."

Chemistry is founded on four fundamental assumptions about atoms and matter, which together make up modern **atomic theory**:

- All matter is composed of atoms.
- The atoms of a given element differ from the atoms of all other elements.
- Chemical compounds consist of atoms combined in specific ratios. That is, only whole atoms can combine—one A atom with one B atom, or one A atom with two B atoms, and so on. The enormous diversity in the substances we see around us is based on the vast number of ways that atoms can combine with one another.
- Chemical reactions change only the way that atoms are combined in compounds. The atoms themselves are unchanged.

Atoms are extremely small, ranging from about 7.4×10^{-11} m in diameter for a hydrogen atom to 5.24×10^{-10} m for a cesium atom. In mass, atoms vary from 1.67×10^{-24} g for hydrogen to 3.95×10^{-22} g for uranium, one of the heaviest naturally occurring atoms. It is difficult to appreciate just how small atoms are, although it might help if you realize that a fine pencil line is about 3 million atoms across and that even the smallest speck of dust contains about 10^{16} atoms. Our current understanding

Atom The smallest and simplest particle of an element.

Atomic theory A set of assumptions proposed by the English scientist John Dalton to explain the chemical behavior of matter.

▶▶ We will further explore the topics of chemical compounds in Chapters 3 and 4, and chemical reactions in Chapters 5 and 6.

45

46 CHAPTER 2 Atoms and the Periodic Table

TABLE 2.1 A Comparison of Subatomic Particles

Name	Symbol	Mass (Grams)	Mass (amu)	Charge (Charge Units)
Proton	p	$1.672\,622 \times 10^{-24}$	$1.007\,276$	$+1$
Neutron	n	$1.674\,927 \times 10^{-24}$	$1.008\,665$	0
Electron	e^-	$9.109\,328 \times 10^{-28}$	$5.485\,799 \times 10^{-4}$	-1

Subatomic particles Three kinds of fundamental particles from which atoms are made: protons, neutrons, and electrons.

Proton A positively charged subatomic particle.

Neutron An electrically neutral subatomic particle.

Electron A negatively charged subatomic particle.

Atomic mass unit (amu) A convenient unit for describing the mass of an atom; 1 amu = $\frac{1}{12}$ the mass of a carbon-12 atom.

▲ The relative size of a nucleus in an atom is the same as that of a pea in the middle of this stadium.

Nucleus The dense, central core of an atom that contains protons and neutrons.

▶ Figure 2.1
The structure of an atom.
Protons and neutrons are packed together in the nucleus, whereas electrons move about in the large surrounding volume. Virtually all the mass of an atom is concentrated in the nucleus.

of atomic structure is the result of many experiments performed in the late 1800s and early 1900s (see Chemistry in Action on p. 48).

Atoms are composed of tiny **subatomic particles** called *protons, neutrons,* and *electrons.* A **proton** has a mass of $1.672\,622 \times 10^{-24}$ g and carries a positive $(+)$ electrical charge; a **neutron** has a mass similar to that of a proton $(1.674\,927 \times 10^{-24}$ g$)$ but is electrically neutral; and an **electron** has a mass that is only $1/1836$ that of a proton $(9.109\,328 \times 10^{-28}$ g$)$ and carries a negative $(-)$ electrical charge. In fact, electrons are so much lighter than protons and neutrons that their mass is usually ignored. Table 2.1 compares the properties of the three fundamental subatomic particles.

The masses of atoms and their constituent subatomic particles are so small when measured in grams that it is more convenient to express them on a *relative* mass scale. That is, one atom is assigned a mass, and all others are measured relative to it. The process is like deciding that a golf ball (46.0 g) will be assigned a mass of 1. A baseball (149 g), which is $149/46.0 = 3.24$ times heavier than a golf ball, would then have a mass of about 3.24; a volleyball (270 g) would have a mass of $270/46.0 = 5.87$; and so on.

The basis for the relative atomic mass scale is an atom of carbon that contains 6 protons and 6 neutrons. Such an atom is assigned a mass of exactly 12 **atomic mass units** (**amu**; also called a *dalton* in honor of the English scientist John Dalton, who proposed most of atomic theory as we know it), where 1 amu = $1.660\,539 \times 10^{-24}$ g. Thus, for all practical purposes, both a proton and a neutron have a mass of 1 amu (Table 2.1). Hydrogen atoms are only about $\frac{1}{12}$th as heavy as carbon atoms and have a mass close to 1 amu, magnesium atoms are about twice as heavy as carbon atoms and have a mass close to 24 amu, and so forth.

Subatomic particles are not distributed at random throughout an atom. Rather, the protons and neutrons are packed closely together in a dense core called the **nucleus**. Surrounding the nucleus, the electrons move about rapidly through a large, mostly empty volume of space (Figure 2.1). Measurements show that the diameter of a nucleus is only about 10^{-15} m, whereas that of the atom itself is about 10^{-10} m. For comparison, if an atom were the size of a large domed stadium, the nucleus would be approximately the size of a small pea in the center of the playing field.

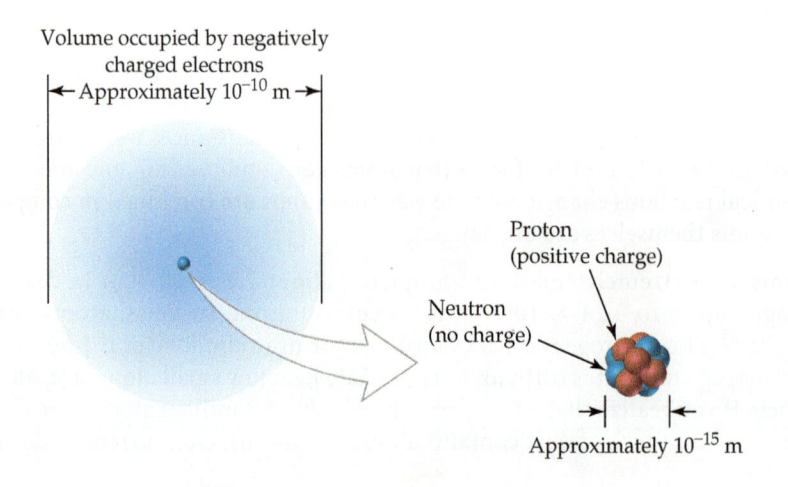

Volume occupied by negatively charged electrons
← Approximately 10^{-10} m →

Proton (positive charge)

Neutron (no charge)

Approximately 10^{-15} m

The structure of the atom is determined by an interplay of different attractive and repulsive forces. Because unlike charges attract one another, the negatively charged electrons are held near the positively charged nucleus. But because like charges repel one another, the electrons also try to get as far away from one another as possible, accounting for the relatively large volume they occupy. The positively charged protons in the nucleus also repel one another, but are nevertheless held together by a unique attraction called the *nuclear strong force*, which we will discuss further in Chapter 11.

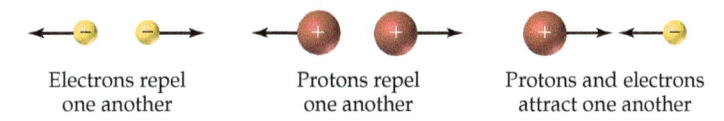

Electrons repel one another

Protons repel one another

Protons and electrons attract one another

Worked Example 2.1 Atomic Mass Units: Gram-to-Atom Conversions

How many atoms are in a small piece of aluminum foil with a mass of 0.100 g? The mass of an atom of aluminum is 27.0 amu.

ANALYSIS We know the sample mass in grams and the mass of one atom in atomic mass units. To find the number of atoms in the sample, two conversions are needed, the first between grams and atomic mass units and the second between atomic mass units and the number of atoms. The conversion factor between atomic mass units and grams is 1 amu = $1.660\,539 \times 10^{-24}$ g.

BALLPARK ESTIMATE An atom of aluminum has a mass of 27.0 amu; since 1 amu ~ 10^{-24} g, the mass of a single aluminum atom is very small ($\approx 10^{-23}$ g). A very *large* number of atoms, therefore, (10^{22} ?) is needed to obtain a mass of 0.100 g.

SOLUTION

STEP 1: Identify known information.

Mass of aluminum foil = 0.100 g
1 Al atom = 27.0 amu

STEP 2: Identify unknown answer and units.

Number of Al atoms = ?

STEP 3: Identify needed conversion factors. Knowing the mass of foil (in g) and the mass of individual atoms (in amu) we need to convert from atoms/amu to atoms/g.

1 amu = $1.660\,539 \times 10^{-24}$ g

$$\rightarrow \frac{1 \text{ amu}}{1.660\,539 \times 10^{-24} \text{ g}}$$

STEP 4: Solve. Set up an equation using known information and conversion factors so that unwanted units cancel.

$$(0.100 \text{ g})\left(\frac{1 \text{ amu}}{1.660\,539 \times 10^{-24} \text{ g}}\right)\left(\frac{1 \text{ Al atom}}{27.0 \text{ amu}}\right)$$
$$= 2.23 \times 10^{21} \text{ Al atoms}$$

BALLPARK CHECK Our estimate was 10^{22}, which is within a factor of 10.

PROBLEM 2.1
What is the mass in grams of 150×10^{12} iron atoms, each having a mass of 56 amu?

PROBLEM 2.2
How many atoms are in each of the following?
(a) 1.0 g of hydrogen atoms, each of mass 1.0 amu
(b) 12.0 g of carbon atoms, each of mass 12.0 amu
(c) 23.0 g of sodium atoms, each of mass 23.0 amu

PROBLEM 2.3
What pattern do you see in your answers to Problem 2.2? (We will return to this very important pattern in Chapter 6.)

PROBLEM 2.4
The atoms in the gold foil used in Rutherford's experiments have an estimated radius of 1.44×10^{-10} m (see Chemistry in Action on p. 48). If we assume that the radius of the nucleus of a gold atom is 1.5×10^{-15} m, what fraction of the volume of the atom is occupied by the nucleus? (Volume = $4/3\,\pi r^3$)

48 CHAPTER 2 Atoms and the Periodic Table

CHEMISTRY IN ACTION

Are Atoms Real?

Chemistry rests on the premise that matter is composed of the tiny particles we call atoms. Every chemical reaction and every physical law that governs the behavior of matter is explained by chemists in terms of atomic theory. But how do we know that atoms are real and not just an imaginary concept? And how do we know the structure of the atom? The development of our understanding of atomic structure is another example of the scientific method at work.

Dalton's atomic theory was originally published in 1808, but many prominent scientists dismissed it. Over the next century, however, several unrelated experiments provided insight into the nature of matter and the structure of the atom. Nineteenth-century investigations into electricity, for example, demonstrated that matter was composed of charged particles—rubbing a glass rod with a silk cloth would generate "static electricity," the same phenomenon that shocks you when you walk across a carpet and then touch a metal surface. It was also known that passing electricity through certain substances, such as water, decomposed the compounds into their constituent elements (hydrogen and oxygen, in the case of water). Several hypotheses were proposed to explain the nature and origin of these charged particles, but our current understanding of atomic structure developed incrementally from several key experiments.

Experiments performed in 1897 by J. J. Thomson demonstrated that matter contained negatively charged particles that were 1000 times lighter than H^+, the lightest positively charged particles found in aqueous solution, and that the mass-to-charge ratio of these particles was the same regardless of the material used to produce the particles (Section 6.10 and Chapter 10). This result implied that atoms were not the smallest particles of matter but that they could be divided into even smaller particles. In 1909, Robert Millikan determined that the charge associated with the "electron," as these particles were now called, was 1.6×10^{-19} coulombs.

But where did the electron fit in the overall structure of matter? The pieces to this puzzle fell into place as a result of experiments performed in 1910 by Ernest Rutherford. He bombarded a gold foil with positively charged "alpha" particles emitted from radium during radioactive decay. The majority of these particles passed straight through the foil, but a small fraction of them were deflected, and a few even bounced back. From these results, Rutherford deduced that an atom consists mostly of empty space (occupied by the negatively charged electrons) and that most of the mass and all of the positive charges are contained in a relatively small, dense region that he called the "nucleus."

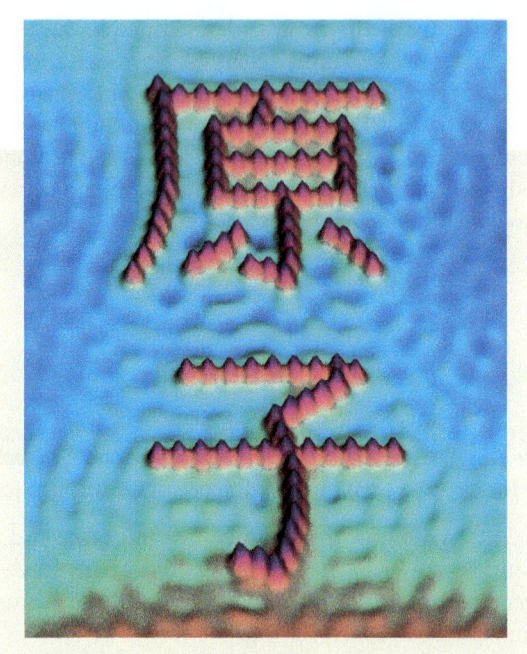

▲ STM image of the Kanji characters for "atom" formed by iron atoms (radius = 126 pm) deposited on a copper metal surface.

We can now actually "see" and manipulate individual atoms through the use of a device called a *scanning tunneling microscope*, or STM. With the STM, invented in 1981 by a research team at the IBM Corporation, magnifications of up to 10 million have been achieved, allowing chemists to look directly at atoms. The accompanying photograph shows a computer-enhanced representation of iron atoms that have been deposited on a copper surface.

Most present uses of the STM involve studies of surface chemistry, such as the events accompanying the corrosion of metals and the ordering of large molecules in polymers. Work is also underway using the STM to determine the structures of complex biological molecules, such as immunoglobulin G and streptavidin.

See Chemistry in Action Problems 2.84 and 2.85 at the end of the chapter.

2.2 Elements and Atomic Number

Atomic number (Z) The number of protons in atoms of a given element; the number of electrons in atoms of a given element.

Atoms of different elements differ from one another according to how many protons they contain, a value called the element's **atomic number (Z)**. Thus, if we know the number of protons in an atom, we can identify the element. Any atom with 6 protons, for example, is a carbon atom because the atomic number for carbon is 6 ($Z = 6$).

Atoms are neutral overall and have no net charge because the number of positively charged protons in an atom is the same as the number of negatively charged electrons. Thus, the atomic number also equals the number of electrons in every atom of a given element. Hydrogen, $Z = 1$, has only 1 proton and 1 electron; carbon, $Z = 6$, has 6 protons and 6 electrons; sodium, $Z = 11$, has 11 protons and 11 electrons; and so on, up to the element with the largest known atomic number ($Z = 118$). In a periodic table, elements are listed in order of increasing atomic number, beginning at the upper left and ending at the lower right.

The sum of the protons and neutrons in an atom is called the atom's **mass number (A)**. Hydrogen atoms with 1 proton and no neutrons have mass number 1, carbon atoms with 6 protons and 6 neutrons have mass number 12, sodium atoms with 11 protons and 12 neutrons have mass number 23, and so on. Except for hydrogen, atoms generally contain at least as many neutrons as protons and frequently contain more. There is no simple way to predict how many neutrons a given atom will have.

Mass number (A) The total number of protons and neutrons in an atom.

Worked Example 2.2 Atomic Structure: Protons, Neutrons, and Electrons

Phosphorus has the atomic number $Z = 15$. How many protons, electrons, and neutrons are there in phosphorus atoms, which have mass number $A = 31$?

ANALYSIS The atomic number gives the number of protons, which is the same as the number of electrons, and the mass number gives the total number of protons plus neutrons.

SOLUTION
Phosphorus atoms, with $Z = 15$, have 15 protons and 15 electrons. To find the number of neutrons, subtract the atomic number from the mass number:

Mass number (sum of protons and neutrons) − Atomic number (number of protons)

$31 - 15 = 16$ neutrons

Worked Example 2.3 Atomic Structure: Atomic Number and Atomic Mass

An atom contains 28 protons and has $A = 60$. Give the number of electrons and neutrons in the atom, and identify the element.

ANALYSIS The number of protons and the number of electrons are the same and are equal to the atomic number Z, 28 in this case. Subtracting the number of protons (28) from the total number of protons plus neutrons (60) gives the number of neutrons.

SOLUTION
The atom has 28 electrons and $60 - 28 = 32$ neutrons. The list of elements inside the front cover shows that the element with atomic number 28 is nickel (Ni).

PROBLEM 2.5
Use the list inside the front cover to identify the following elements:
(a) $A = 186$, with 111 neutrons
(b) $A = 59$, with 21 neutrons
(c) $A = 127$, with 75 neutrons

PROBLEM 2.6
The cobalt used in cancer treatments has $Z = 27$ and $A = 60$. How many protons, neutrons, and electrons are in these cobalt atoms?

2.3 Isotopes and Atomic Weight

All atoms of a given element have the same number of protons, equal to the atomic number (Z) characteristic of that element. But, different atoms of an element can have different numbers of neutrons and therefore different mass numbers. Atoms with identical atomic numbers but different mass numbers are called **isotopes**. Hydrogen, for example, has three isotopes. The most abundant hydrogen isotope, called *protium*, has no neutrons and thus has a mass number of 1. A second hydrogen isotope, called *deuterium*, has one neutron and a mass number of 2; and a third isotope, called *tritium*, has two neutrons and a mass number of 3. Tritium is unstable and does not occur naturally in significant amounts, although it can be made in nuclear reactors.

Isotopes Atoms with identical atomic numbers but different mass numbers.

▶▶ We will see that isotopes of the same element have the same *chemical* behavior (Chapter 5), but very different *nuclear* behavior (Chapter 11).

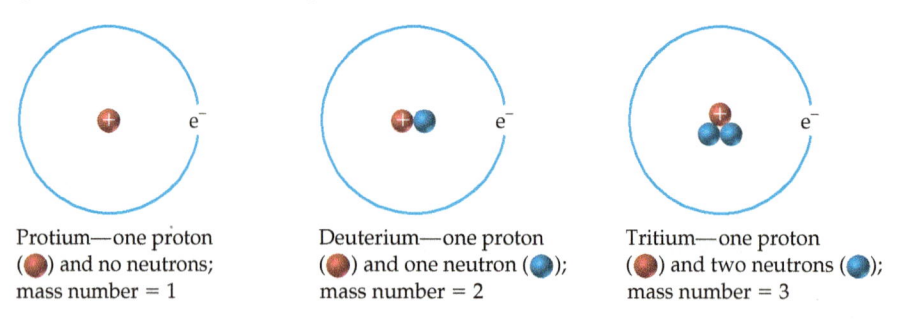

Protium—one proton (●) and no neutrons; mass number = 1

Deuterium—one proton (●) and one neutron (●); mass number = 2

Tritium—one proton (●) and two neutrons (●); mass number = 3

A specific isotope is represented by showing its mass number (A) as a superscript and its atomic number (Z) as a subscript in front of the atomic symbol, for example, $^A_Z X$, where X represents the symbol for the element. Thus, protium is $^1_1 H$, deuterium is $^2_1 H$, and tritium is $^3_1 H$.

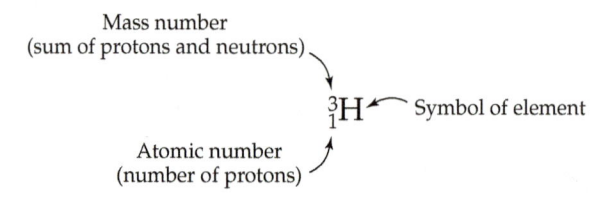

$$^3_1 H$$

Mass number (sum of protons and neutrons) — Symbol of element
Atomic number (number of protons)

Unlike the three isotopes of hydrogen, the isotopes of most elements do not have distinctive names. Instead, the mass number of the isotope is given after the name of the element. The $^{235}_{92}U$ isotope used in nuclear reactors, for example, is usually referred to as uranium-235, or U-235.

▶▶ We will discuss nuclear reactors in Section 11.11.

Most naturally occurring elements are mixtures of isotopes. In a large sample of naturally occurring hydrogen atoms, for example, 99.985% have mass number $A = 1$ (protium) and 0.015% have mass number $A = 2$ (deuterium). Therefore, it is useful to know the *average* mass of the atoms in a large sample, a value called the element's **atomic weight**. For hydrogen, the atomic weight is 1.008 amu. Atomic weights for all elements are given on the inside of the front cover of this book.

Atomic weight The weighted average mass of an element's atoms.

To calculate the atomic weight of an element, the individual masses of the naturally occurring isotopes and the percentage of each must be known. The atomic weight can then be calculated as the sum of the masses of the individual isotopes for that element, or

$$\text{Atomic weight} = \Sigma \left[(\text{isotopic abundance}) \times (\text{isotopic mass}) \right]$$

where the Greek symbol Σ indicates the mathematical summing of terms.

Chlorine, for example, occurs on earth as a mixture of 75.77% Cl-35 atoms (mass = 34.97 amu) and 24.23% Cl-37 atoms (mass = 36.97 amu). The atomic weight is found by calculating the percentage of the mass contributed by each isotope. For chlorine, the calculation is done in the following way (to four significant figures), giving an atomic weight of 35.45 amu:

$$\text{Contribution from } ^{35}Cl: (0.7577)(34.97 \text{ amu}) = 26.4968 \text{ amu}$$
$$\text{Contribution from } ^{37}Cl: (0.2423)(36.97 \text{ amu}) = \underline{8.9578 \text{ amu}}$$
$$\text{Atomic weight} = 35.4546 = 35.45 \text{ amu}$$
(rounded to four significant figures)

The final number of significant figures in this case (four) was determined by the atomic masses. Note that the final rounding to four significant figures was not done until *after* the final answer was obtained.

Worked Example 2.4 Average Atomic Mass: Weighted-Average Calculation

Gallium is a metal with a very low melting point—it will melt in the palm of your hand. It has two naturally occurring isotopes: 60.4% is Ga-69 (mass = 68.9257 amu), and 39.6% is Ga-71 (mass = 70.9248 amu). Calculate the atomic weight for gallium.

ANALYSIS We can calculate the average atomic mass for the element by summing up the contributions from each of the naturally occurring isotopes.

BALLPARK ESTIMATE The masses of the two naturally occurring isotopes of gallium differ by 2 amu (68.9 and 70.9 amu). Since slightly more than half of the Ga atoms are the lighter isotope (Ga-69), the average mass will be slightly less than halfway between the two isotopic masses; estimate = 69.8 amu.

SOLUTION

STEP 1: **Identify known information.**

Ga-69 (60.4% at 68.9257 amu)
Ga-71 (39.6% at 70.9248 amu)

STEP 2: **Identify the unknown answer and units.**

Atomic weight for Ga (in amu) = ?

STEP 3: **Identify conversion factors or equations.** This equation calculates the average atomic weight as a weighted average of all naturally occurring isotopes.

Atomic weight = $\Sigma[$ (isotopic abundance) $\times$ (isotopic mass) $]$

STEP 4: **Solve.** Substitute known information and solve.

Atomic weight = $(0.604) \times (68.9257 \text{ amu}) = 41.6311$ amu
$+ (0.396) \times (70.9248 \text{ amu}) = 28.0862$ amu

Atomic weight = 69.7 amu (3 significant figures)

BALLPARK CHECK Our estimate (69.8 amu) is close!

Worked Example 2.5 Identifying Isotopes from Atomic Mass and Atomic Number

Identify element X in the symbol $^{194}_{78}X$, and give its atomic number, mass number, number of protons, number of electrons, and number of neutrons.

ANALYSIS The identity of the atom corresponds to the atomic number—78.

SOLUTION
Element X has Z = 78, which shows that it is platinum. (Look inside the front cover for the list of elements.) The isotope $^{194}_{78}Pt$ has a mass number of 194, and we can subtract the atomic number from the mass number to get the number of neutrons. This platinum isotope therefore has 78 protons, 78 electrons, and 194 − 78 = 116 neutrons.

PROBLEM 2.7
Potassium (K) has two naturally occurring isotopes: K-39 (93.12%; mass = 38.9637 amu) and K-41 (6.88%; 40.9618 amu). Calculate the atomic weight for potassium. How does your answer compare with the atomic weight given in the list inside the front cover of this book?

PROBLEM 2.8
Bromine, an element present in compounds used as sanitizers and fumigants (for example, ethylene bromide), has two naturally occurring isotopes, with mass numbers 79 and 81. Write the symbols for both, including their atomic numbers and mass numbers.

PROBLEM 2.9
An element used to sanitize water supplies has two naturally occurring isotopes with mass numbers of 35 and 37, and 17 electrons. Write the symbols for both isotopes, including their atomic numbers and mass numbers.

2.4 The Periodic Table

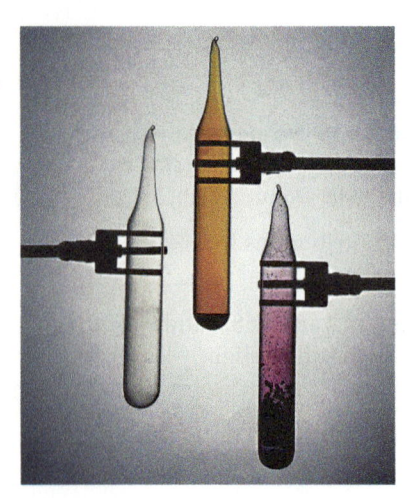

Ten elements have been known since the beginning of recorded history: antimony (Sb), carbon (C), copper (Cu), gold (Au), iron (Fe), lead (Pb), mercury (Hg), silver (Ag), sulfur (S), and tin (Sn). It is worth noting that the symbols for many of these elements are derived from their Latin names, a reminder that they have been known since the time when Latin was the language used for all scholarly work. The first "new" element to be found in several thousand years was arsenic (As), discovered in about 1250. In fact, only 24 elements were known up to the time of the American Revolution in 1776.

As the pace of discovery quickened in the late 1700s and early 1800s, chemists began to look for similarities among elements that might make it possible to draw general conclusions. Particularly important was Johann Döbereiner's observation in 1829 that there were several *triads*, or groups of three elements, that appeared to have similar chemical and physical properties. For example, lithium, sodium, and potassium were all known to be silvery metals that react violently with water; chlorine, bromine, and iodine were all known to be colored nonmetals with pungent odors.

▲ Samples of chlorine, bromine, and iodine, one of Döbereiner's triads of elements with similar chemical properties.

Numerous attempts were made in the mid-1800s to account for the similarities among groups of elements, but the great breakthrough came in 1869 when the Russian chemist Dmitri Mendeleev organized the elements in order of increasing mass and then grouped elements into columns based on similarities in chemical behavior. His table is a forerunner of the modern periodic table, introduced previously in Section 1.5 and shown again in Figure 2.2. The table has boxes for each element that give the symbol, atomic number, and atomic mass of the element:

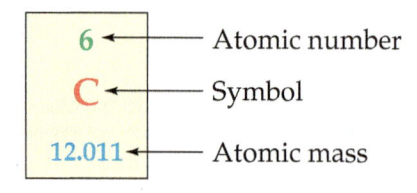

```
6        ← Atomic number
C        ← Symbol
12.011   ← Atomic mass
```

Period One of the 7 horizontal rows of elements in the periodic table.

Group One of the 18 vertical columns of elements in the periodic table.

Beginning at the upper left corner of the periodic table, elements are arranged by increasing atomic number into seven horizontal rows, called **periods**, and 18 vertical columns, called **groups**. When organized in this way, *the elements in a given group have similar chemical properties.* Lithium, sodium, potassium, and the other elements in group 1A behave similarly. Chlorine, bromine, iodine, and the other elements in group 7A behave similarly, and so on throughout the table.

Note that different periods (rows) contain different numbers of elements. The first period contains only 2 elements, hydrogen and helium; the second and third periods each contain 8 elements; the fourth and fifth periods each contain 18; the sixth and seventh periods contain 32. Note also that the 14 elements following lanthanum (the *lanthanides*) and the 14 following actinium (the *actinides*) are pulled out and shown below the others.

Main group element An element in one of the 2 groups on the left or the 6 groups on the right of the periodic table.

Transition metal element An element in one of the 10 smaller groups near the middle of the periodic table.

Inner transition metal element An element in one of the 14 groups shown separately at the bottom of the periodic table.

Groups are numbered in two ways, both shown in Figure 2.2. The 2 large groups on the far left and the 6 on the far right are called the **main group elements** and are numbered 1A through 8A. The 10 smaller groups in the middle of the table are called the **transition metal elements** and are numbered 1B through 8B. Alternatively, all 18 groups are numbered sequentially from 1 to 18. The 14 groups shown separately at the bottom of the table are called the **inner transition metal elements** and are not numbered.

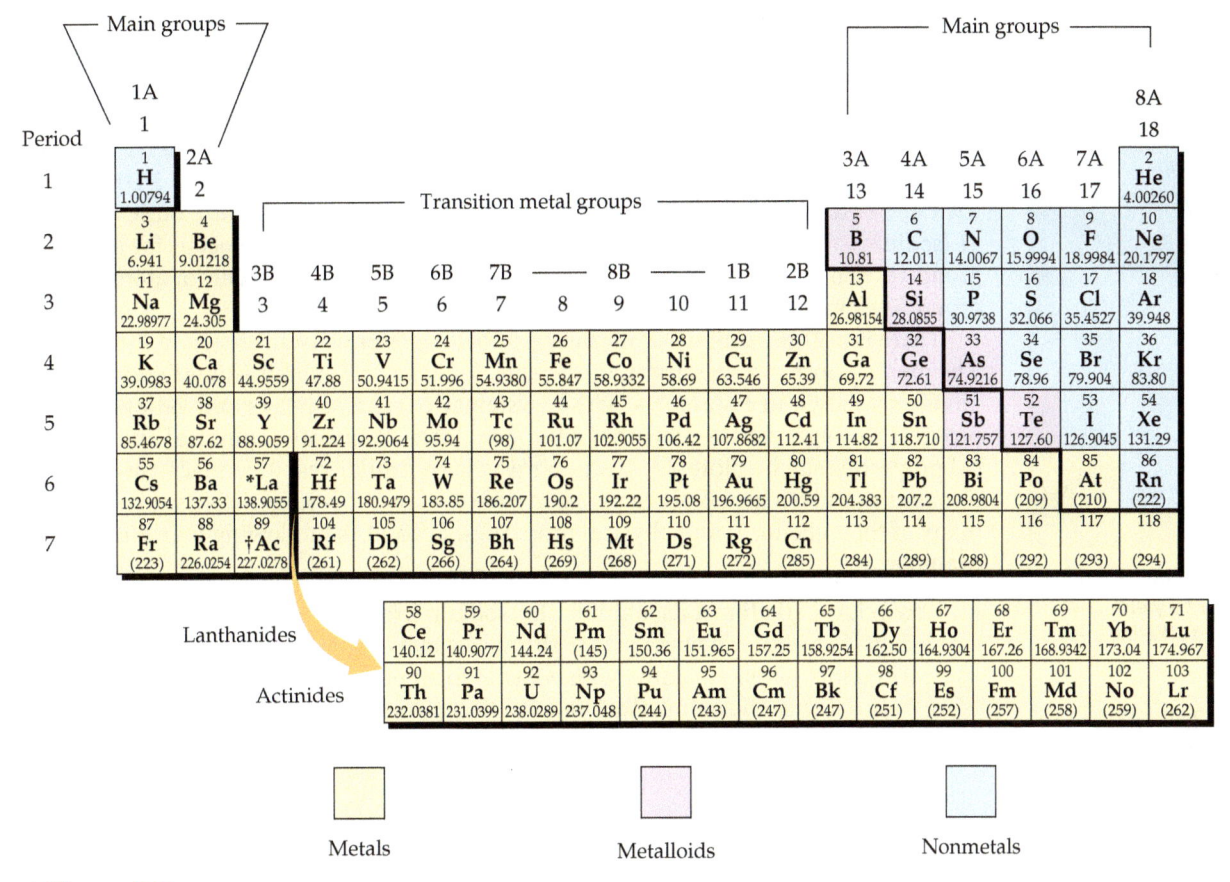

SECTION 2.4 The Periodic Table 53

▲ Figure 2.2
The periodic table of the elements.
Each element is identified by a one- or two-letter symbol and is characterized by an *atomic number*. The table begins with hydrogen (H, atomic number 1) in the upper left-hand corner and continues to the yet unnamed element with atomic number 118. The 14 elements following lanthanum (La, atomic number 57) and the 14 elements following actinium (Ac, atomic number 89) are pulled out and shown below the others.

Elements are organized into 18 vertical columns, or *groups*, and 7 horizontal rows, or *periods*. The 2 groups on the left and the 6 on the right are the *main groups*; the 10 in the middle are the *transition metal groups*. The 14 elements following lanthanum are the *lanthanides*, and the 14 elements following actinium are the *actinides*; together these are known as the *inner transition metals*. Two systems for numbering the groups are explained in the text.

Those elements (except hydrogen) on the left-hand side of the black zigzag line running from boron (B) to tellurium (Te) are *metals* (yellow), those elements to the right of the line are *nonmetals* (blue), and most elements abutting the line are *metalloids* (purple).

PROBLEM 2.10
Locate aluminum in the periodic table, and give its group number and period number.

PROBLEM 2.11
Identify the group 1B element in period 5 and the group 2A element in period 4.

PROBLEM 2.12
There are five elements in group 5A of the periodic table. Identify them, and give the period of each.

54 CHAPTER 2 Atoms and the Periodic Table

2.5 Some Characteristics of Different Groups

To see why the periodic table has the name it does, look at the graph of atomic radius versus atomic number in Figure 2.3. The graph shows an obvious *periodicity*—a repeating rise-and-fall pattern. Beginning on the left with atomic number 1 (hydrogen), the sizes of the atoms increase to a maximum at atomic number 3 (lithium), then decrease to a minimum, then increase again to a maximum at atomic number 11 (sodium), then decrease, and so on. It turns out that the maxima occur for atoms of group 1A elements—Li, Na, K, Rb, Cs, and Fr—and the minima occur for atoms of the group 7A elements.

There is nothing unique about the periodicity of atomic radii shown in Figure 2.3. The melting points of the first 100 elements, for example, exhibit similar periodic behavior, as shown in Figure 2.4, with a systematic trend of peaks and valleys as you progress through the elements in the periodic table. Many other physical and chemical properties can be plotted in a similar way with similar results. In fact, the various elements in a given group of the periodic table usually show remarkable similarities in many of their chemical and physical properties. Look at the following four groups, for example:

- **Group 1A—Alkali metals:** Lithium (Li), sodium (Na), potassium (K), rubidium (Rb), cesium (Cs), and francium (Fr) are shiny, soft metals with low melting points. All react rapidly (often violently) with water to form products that are highly alkaline, or basic—hence the name **alkali metals**. Because of their high reactivity, the alkali metals are never found in nature in the pure state but only in combination with other elements.

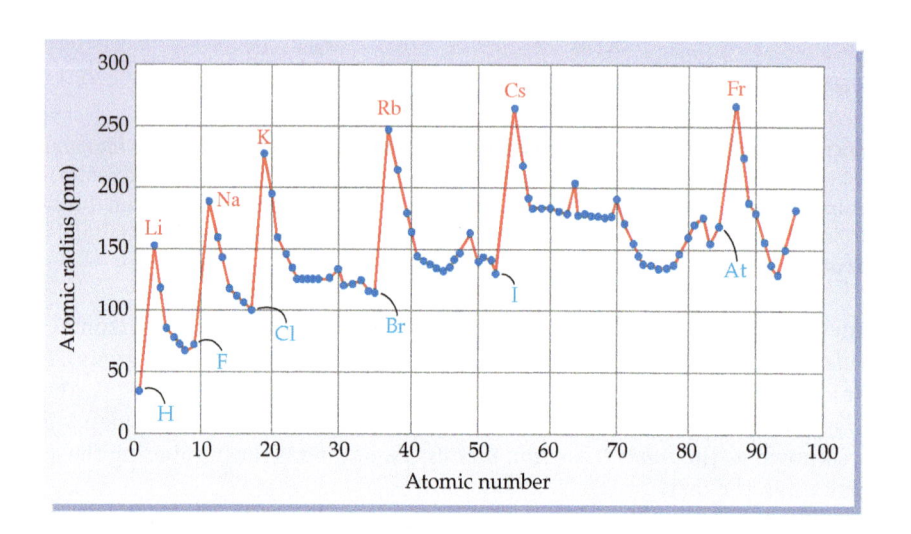

▲ Sodium, an alkali metal, reacts violently with water to yield hydrogen gas and an alkaline (basic) solution.

Alkali metal An element in group 1A of the periodic table.

▶ **Figure 2.3**
A graph of atomic radius in picometers (pm) versus atomic number shows a periodic rise-and-fall pattern.
The maxima occur for atoms of the group 1A elements (Li, Na, K, Rb, Cs, Fr, in red); the minima occur for atoms of the group 7A elements (blue). Accurate data are not available for the group 8A elements.

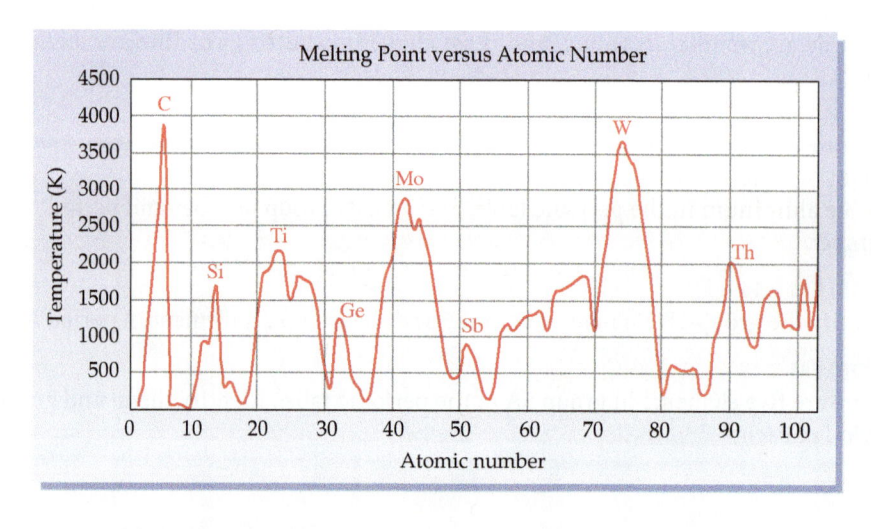

▶ **Figure 2.4**
A graph of melting point versus atomic number shows periodic properties similar to the trend in Figure 2.3.
While the maxima and minima are not as sharp as in Figure 2.3, the change in melting points of the elements still show a similar periodic trend.

- **Group 2A—Alkaline earth metals:** Beryllium (Be), magnesium (Mg), calcium (Ca), strontium (Sr), barium (Ba), and radium (Ra) are also lustrous, silvery metals, but are less reactive than their neighbors in group 1A. Like the alkali metals, the alkaline earths are never found in nature in the pure state.
- **Group 7A—Halogens:** Fluorine (F), chlorine (Cl), bromine (Br), iodine (I), and astatine (At) are colorful and corrosive nonmetals. All are found in nature only in combination with other elements, such as with sodium in table salt (sodium chloride, NaCl). In fact, the group name **halogen** is taken from the Greek word *hals*, meaning salt.
- **Group 8A—Noble gases:** Helium (He), neon (Ne), argon (Ar), krypton (Kr), xenon (Xe), and radon (Rn) are colorless gases. The elements in this group were labeled the "noble" gases because of their lack of chemical reactivity—helium, neon, and argon don't combine with any other elements, whereas krypton and xenon combine with a very few.

Alkaline earth metal An element in group 2A of the periodic table.

Halogen An element in group 7A of the periodic table.

Noble gas An element in group 8A of the periodic table.

▶▶ The reason for the similarity in chemical properties of elements within each group will be explained in Section 2.8.

Although the resemblances are not as pronounced as they are within a single group, *neighboring* elements often behave similarly as well. Thus, as noted in Section 1.5 and indicated in Figure 2.2, the periodic table can be divided into three major classes of elements—*metals*, *nonmetals*, and *metalloids* (metal-like). Metals, the largest category of elements, are found on the left side of the periodic table, bounded on the right by a zigzag line running from boron (B) at the top to astatine (At) at the bottom. Nonmetals are found on the right side of the periodic table, and six of the elements adjacent to the zigzag boundary between metals and nonmetals are metalloids.

LOOKING AHEAD ▶▶ Carbon, the element on which life is based, is a group 4A nonmetal near the top right of the periodic table. Clustered near carbon are other elements often found in living organisms, including oxygen, nitrogen, phosphorus, and sulfur. We will look at the subject of *organic chemistry*—the chemistry of carbon compounds—in Chapters 12–17, and move on to *biochemistry*—the chemistry of living things—in Chapters 18–29.

PROBLEM 2.13
Identify the following elements as metals, nonmetals, or metalloids:
(a) Ti   (b) Te
(c) Se   (d) Sc
(e) At   (f) Ar

PROBLEM 2.14
Locate (a) krypton, (b) strontium, (c) nitrogen, and (d) cobalt in the periodic table. Indicate which categories apply to each: (i) metal, (ii) nonmetal, (iii) transition element, (iv) main group element, (v) noble gas.

PROBLEM 2.15
Heavier elements were formed in stars by the fusion of hydrogen and helium nuclei (see Chemistry in Action on p. 56). How many He-4 nuclei would be needed to form a Fe-56 nucleus? What additional particles would be needed?

KEY CONCEPT PROBLEM 2.16
Identify the elements whose nuclei are shown below. For each, tell its group number, its period number, and whether it is a metal, nonmetal, or metalloid.

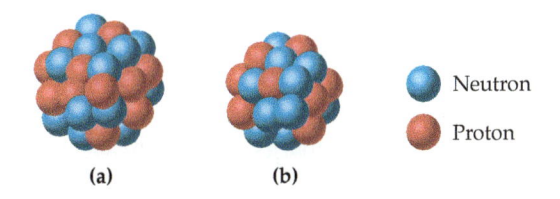

CHEMISTRY IN ACTION

The Origin of Chemical Elements

Astronomers believe that the universe began some 15 billion years ago in an extraordinary moment they call the "big bang." Initially, the temperature must have been inconceivably high, but after 1 second, it had dropped to about 10^{10} K and subatomic particles began to form: protons, neutrons, and electrons. After 3 minutes, the temperature had dropped to 10^9 K, and protons began fusing with neutrons to form helium nuclei, $^4_2$He.

Matter remained in this form for many millions of years, until the expanding universe had cooled to about 10,000 K and electrons were then able to bind to protons and to helium nuclei, forming stable hydrogen and helium atoms.

The attractive force of gravity acting on regions of higher-than-average density of hydrogen and helium atoms slowly produced massive local concentrations of matter and ultimately formed billions of galaxies, each with many billions of stars. As the gas clouds of hydrogen and helium condensed under gravitational attraction and stars formed, their temperatures reached 10^7 K, and their densities reached 100 g/cm³. Protons and neutrons again fused to yield helium nuclei, generating vast amounts of heat and light.

Most of these early stars probably burned out after a few billion years, but a few were so massive that, as their nuclear fuel diminished, gravitational attraction caused a rapid contraction leading to still higher core temperatures and higher densities—up to 5×10^8 K and 5×10^5 g/cm³. Under such extreme conditions, larger nuclei were formed, including carbon, oxygen, silicon, magnesium, and iron. Ultimately, the stars underwent a gravitational collapse resulting in the synthesis of still heavier elements and an explosion visible throughout the universe as a *supernova*.

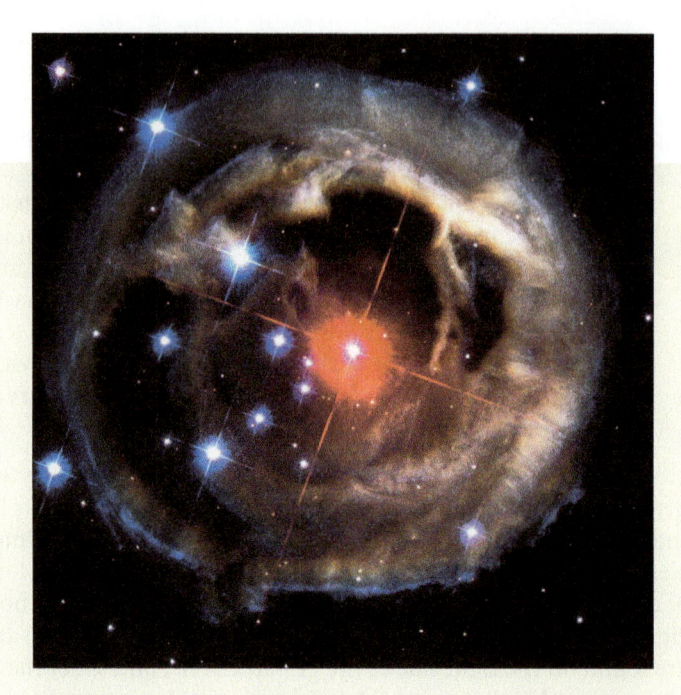

▲ "Light echoes" illuminate dust around the supergiant star V838 monocerotis, as seen from the Hubble telescope.

Matter from exploding supernovas was blown throughout the galaxy, forming a new generation of stars and planets. Our own sun and solar system formed about 4.5 billion years ago from matter released by former supernovas. Except for hydrogen and helium, all the atoms in our bodies and our entire solar system were created more than 5 billion years ago in exploding stars. We and our world are made from the ashes of dying stars.

See Chemistry in Action Problems 2.86 and 2.87 at the end of this chapter.

2.6 Electronic Structure of Atoms

Why does the periodic table have the shape it does, with periods of different length? Why are periodic variations observed in atomic radii and in so many other characteristics of the elements? And why do elements in a given group of the periodic table show similar chemical behavior? These questions occupied the thoughts of chemists for more than 50 years after Mendeleev, and it was not until well into the 1920s that the answers were established. Today, we know that *the properties of the elements are determined by the arrangement of electrons in their atoms.*

Our current understanding of the electronic structure of atoms is based on the now accepted *quantum mechanical model*, developed by Austrian physicist Erwin Schrödinger in 1926. One of the fundamental assumptions of the model is that electrons have both particle-like and wave-like properties, and that the behavior of electrons can be described using a mathematical equation called a wave function. One consequence of this assumption is that electrons are not perfectly free to move about in an atom. Instead, each electron is restricted to a certain region of space within the atom, depending on the energy level of the electron. Different electrons have different amounts of energy and thus occupy different regions within the atom.

Furthermore, the energies of electrons are *quantized*, or restricted to having only certain values.

To understand the idea of quantization, think about the difference between stairs and a ramp. A ramp is *not* quantized because it changes height continuously. Stairs, by contrast, *are* quantized because they change height only by a fixed amount. You can climb one stair or two stairs, but you cannot climb 1.5 stairs. In the same way, the energy values available to electrons in an atom change only in steps rather than continuously.

The wave functions derived from the quantum mechanical model also provide important information about the location of electrons in an atom. Just as a person can be found by giving his or her address within a state, an electron can be found by giving its "address" within an atom. Furthermore, just as a person's address is composed of several successively narrower categories—city, street, and house number—an electron's address is also composed of successively narrower categories—*shell, subshell,* and *orbital,* which are defined by the quantum mechanical model.

The electrons in an atom are grouped around the nucleus into **shells**, roughly like the layers in an onion, according to the energy of the electrons. The farther a shell is from the nucleus, the larger it is, the more electrons it can hold, and the higher the energies of those electrons. The first shell (the one nearest the nucleus) can hold only 2 electrons, the second shell can hold 8, the third shell can hold 18, and the fourth shell can hold 32 electrons.

▲ Stairs are *quantized* because they change height in discrete amounts. A ramp, by contrast, is not quantized because it changes height continuously.

Shell (electron) A grouping of electrons in an atom according to energy.

Shell number:	1	2	3	4
Electron capacity:	2	8	18	32

Within shells, electrons are further grouped into **subshells** of four different types, identified in order of increasing energy by the letters *s, p, d,* and *f.* The first shell has only one subshell, of the *s* type. The second shell has two subshells: an *s* subshell and a *p* subshell. The third shell has an *s,* a *p,* and a *d* subshell. The fourth shell has an *s,* a *p,* a *d,* and an *f* subshell. Of the four types, we will be concerned mainly with *s* and *p* subshells because most of the elements found in living organisms use only these. A specific subshell is symbolized by writing the number of the shell, followed by the letter for the subshell. For example, the designation 3*p* refers to the *p* subshell in the third shell. Note that the number of subshells in a given shell is equal to the shell number. For example, shell number 3 has 3 subshells.

Subshell (electron) A grouping of electrons in a shell according to the shape of the region of space they occupy.

Finally, within each subshell, electrons are grouped into **orbitals**, regions of space within an atom where the specific electrons are most likely to be found. There are different numbers of orbitals within the different kinds of subshells. A given *s* subshell has only 1 orbital, a *p* subshell has 3 orbitals, a *d* subshell has 5 orbitals, and an *f* subshell has 7 orbitals. Each orbital can hold only two electrons, which differ in a property known as *spin*. If one electron in an orbital has a clockwise spin, the other electron in the same orbital must have a counterclockwise spin. The configuration of shells, subshells, and orbitals is summarized in the figure below.

Orbital A region of space within an atom where an electron in a given subshell can be found.

Shell number:	1	2	3	4
Subshell designation:	s	s , p	s , p , d	s , p , d , f
Number of orbitals:	1	1 , 3	1 , 3 , 5	1 , 3 , 5 , 7

Different orbitals have different shapes and orientations, which are described by the quantum mechanical model. Orbitals in *s* subshells are spherical regions centered about the nucleus, whereas orbitals in *p* subshells are roughly dumbbell-shaped regions (Figure 2.5). As shown in Figure 2.5(b), the three *p* orbitals in a given subshell are oriented at right angles to one another.

The overall electron distribution within an atom is summarized in Table 2.2 and in the following list:

- The first shell holds only 2 electrons. The 2 electrons have different spins and are in a single 1*s* orbital.

▶ **Figure 2.5**
The shapes of *s* and *p* orbitals.
(a) The *s* orbitals and **(b)** the *p* orbitals. The three *p* orbitals in a given subshell are oriented at right angles to one another. Each orbital can hold only two electrons.

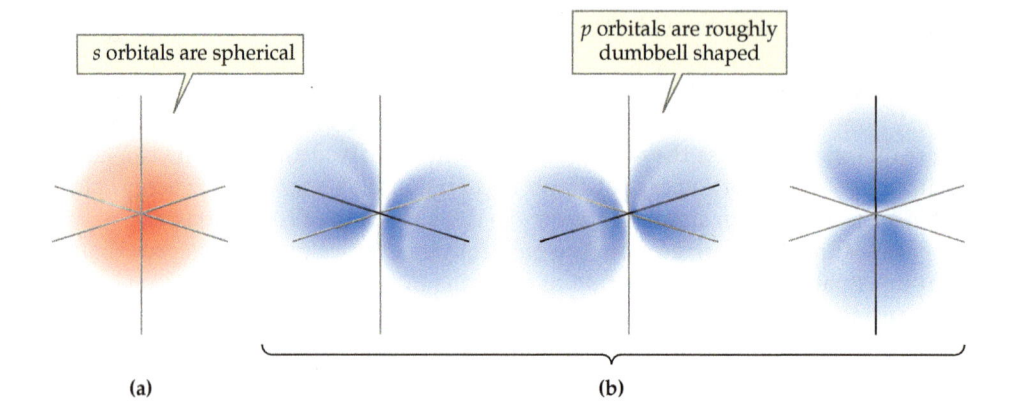

s orbitals are spherical

p orbitals are roughly dumbbell shaped

(a)   (b)

- The second shell holds 8 electrons. Two are in a 2*s* orbital, and 6 are in the three different 2*p* orbitals (two per 2*p* orbital).
- The third shell holds 18 electrons. Two are in a 3*s* orbital, 6 are in three 3*p* orbitals, and 10 are in five 3*d* orbitals.
- The fourth shell holds 32 electrons. Two are in a 4*s* orbital, 6 are in three 4*p* orbitals, 10 are in five 4*d* orbitals, and 14 are in seven 4*f* orbitals.

Worked Example 2.6 Atomic Structure: Electron Shells

How many electrons are present in an atom that has its first and second shells filled and has 4 electrons in its third shell? Name the element.

ANALYSIS The number of electrons in the atom is calculated by adding the total electrons in each shell. We can identify the element from the number of protons in the nucleus, which is equal to the number of electrons in the atom.

SOLUTION
The first shell of an atom holds 2 electrons in its 1*s* orbital, and the second shell holds 8 electrons (2 in a 2*s* orbital and 6 in three 2*p* orbitals). Thus, the atom has a total of 2 + 8 + 4 = 14 electrons and must be silicon (Si).

PROBLEM 2.17
How many electrons are present in an atom in which the first and second shells and the 3*s* subshell are filled? Name the element.

PROBLEM 2.18
An element has completely filled $n = 1$ and $n = 2$ shells and has 6 electrons in the $n = 3$ shell. Identify the element and its major group (i.e., main group, transition, etc.). Is it a metal or a nonmetal? Identify the orbital in which the last electron is found.

TABLE 2.2 Electron Distribution in Atoms

SHELL NUMBER:	1	2	3	4
Subshell designation:	s	s , p	s , p , d	s , p , d , f
Number of orbitals:	1	1 , 3	1 , 3 , 5	1 , 3 , 5 , 7
Number of electrons:	2	2 , 6	2 , 6 , 10	2 , 6 , 10 , 14
Total electron capacity:	2	8	18	32

2.7 Electron Configurations

The exact arrangement of electrons in an atom's shells and subshells is called the atom's **electron configuration** and can be predicted by applying three rules:

RULE 1: Electrons occupy the lowest-energy orbitals available, beginning with 1s and continuing in the order shown in Figure 2.6a. Within each shell, the orbital energies increase in the order *s, p, d, f*. The overall ordering is complicated, however, by the fact that some "crossover" of energies occurs between orbitals in different shells above the 3*p* level. For example, the 4*s* orbital is lower in energy than the 3*d* orbitals, and is therefore filled first. The energy level diagram can be used to predict the order in which orbitals are filled, but it may be hard to remember. The schematic in Figure 2.6b may also be used and is easier to remember.

RULE 2: Each orbital can hold only two electrons, which must be of opposite spin.

RULE 3: Two or more orbitals with the same energy—the three *p* orbitals or the five *d* orbitals in a given shell, for example—are each half-filled by one electron before any one orbital is completely filled by addition of the second electron.

Electron configurations of the first 20 elements are shown in Table 2.3. Notice that the number of electrons in each subshell is indicated by a superscript. For example, the notation $1s^2\ 2s^2\ 2p^6\ 3s^2$ for magnesium means that magnesium atoms have 2 electrons in the first shell, 8 electrons in the second shell, and 2 electrons in the third shell.

Mg (atomic number 12): $1s^2\ 2s^2\ 2p^6\ 3s^2$
- 2 electrons in first shell
- 8 electrons in second shell
- 2 electrons in third shell

Electron configuration The specific arrangement of electrons in an atom's shells and subshells.

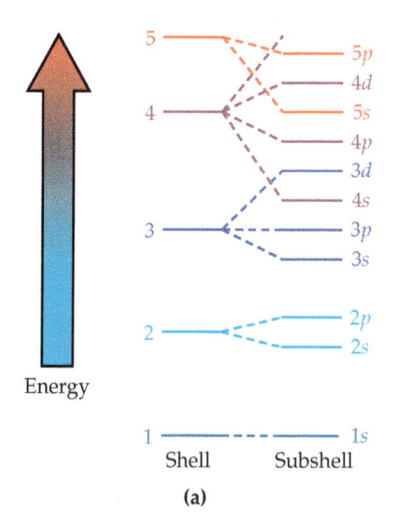

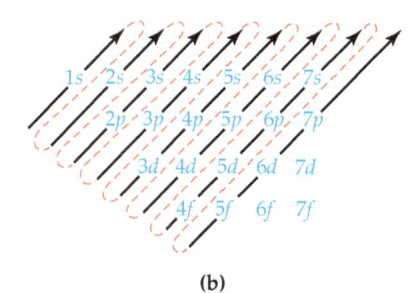

▲ **Figure 2.6**
Order of orbital energy levels.
(a) An energy-level diagram shows the order in which orbitals will be filled within each shell. Above the 3*p* level, there is some crossover of energies among orbitals in different shells.
(b) A simple scheme to remember the order in which the orbitals are filled.

TABLE 2.3 Electron Configurations of the First 20 Elements

	Element	Atomic Number	Electron Configuration
H	Hydrogen	1	$1s^1$
He	Helium	2	$1s^2$
Li	Lithium	3	$1s^2\ 2s^1$
Be	Beryllium	4	$1s^2\ 2s^2$
B	Boron	5	$1s^2\ 2s^2\ 2p^1$
C	Carbon	6	$1s^2\ 2s^2\ 2p^2$
N	Nitrogen	7	$1s^2\ 2s^2\ 2p^3$
O	Oxygen	8	$1s^2\ 2s^2\ 2p^4$
F	Fluorine	9	$1s^2\ 2s^2\ 2p^5$
Ne	Neon	10	$1s^2\ 2s^2\ 2p^6$
Na	Sodium	11	$1s^2\ 2s^2\ 2p^6\ 3s^1$
Mg	Magnesium	12	$1s^2\ 2s^2\ 2p^6\ 3s^2$
Al	Aluminum	13	$1s^2\ 2s^2\ 2p^6\ 3s^2\ 3p^1$
Si	Silicon	14	$1s^2\ 2s^2\ 2p^6\ 3s^2\ 3p^2$
P	Phosphorus	15	$1s^2\ 2s^2\ 2p^6\ 3s^2\ 3p^3$
S	Sulfur	16	$1s^2\ 2s^2\ 2p^6\ 3s^2\ 3p^4$
Cl	Chlorine	17	$1s^2\ 2s^2\ 2p^6\ 3s^2\ 3p^5$
Ar	Argon	18	$1s^2\ 2s^2\ 2p^6\ 3s^2\ 3p^6$
K	Potassium	19	$1s^2\ 2s^2\ 2p^6\ 3s^2\ 3p^6\ 4s^1$
Ca	Calcium	20	$1s^2\ 2s^2\ 2p^6\ 3s^2\ 3p^6\ 4s^2$

As you read through the following electron configurations, check the atomic number and the location of each element in the periodic table (Figure 2.2). See if you can detect the relationship between electron configuration and position in the table.

- **Hydrogen ($Z = 1$):** The single electron in a hydrogen atom is in the lowest-energy, 1s, level. The configuration can be represented in either of two ways:

$$H \quad 1s^1 \quad \text{or} \quad \frac{\uparrow}{1s^1}$$

In the written representation, the superscript in the notation $1s^1$ means that the 1s orbital is occupied by one electron. In the graphic representation, the 1s orbital is indicated by a line and the single electron in this orbital is shown by an up arrow ($\uparrow$). A single electron in an orbital is often referred to as being *unpaired*.

- **Helium ($Z = 2$):** The two electrons in helium are both in the lowest-energy, 1s, orbital, and their spins are *paired*, as represented by up and down arrows ($\uparrow\downarrow$):

$$He \quad 1s^2 \quad \text{or} \quad \frac{\uparrow\downarrow}{1s^2}$$

- **Lithium ($Z = 3$):** With the first shell full, the second shell begins to fill. The third electron goes into the 2s orbital:

$$Li \quad 1s^2\,2s^1 \quad \text{or} \quad \frac{\uparrow\downarrow}{1s^2} \; \frac{\uparrow}{2s^1}$$

Because [He] has the configuration of a filled $1s^2$ orbital, it is sometimes substituted for the $1s^2$ orbital in depictions of electron pairing. Using this alternative shorthand notation, the electron configuration for Li is written [He] $2s^1$.

- **Beryllium ($Z = 4$):** An electron next pairs up to fill the 2s orbital:

$$Be \quad 1s^2\,2s^2 \quad \text{or} \quad \frac{\uparrow\downarrow}{1s^2} \; \frac{\uparrow\downarrow}{2s^2} \quad \text{or} \quad [He]\,2s^2$$

- **Boron ($Z = 5$), Carbon ($Z = 6$), Nitrogen ($Z = 7$):** The next three electrons enter the three 2p orbitals, one at a time. Note that representing the configurations with lines and arrows gives more information than the alternative written notations because the filling and pairing of electrons in individual orbitals within the p subshell is shown.

$$B \quad 1s^2\,2s^2\,2p^1 \quad \text{or} \quad \frac{\uparrow\downarrow}{1s^2} \; \frac{\uparrow\downarrow}{2s^2} \; \underbrace{\frac{\uparrow}{}\,\frac{}{}\,\frac{}{}}_{2p^1} \quad \text{or} \quad [He]\,2s^2\,2p^1$$

$$C \quad 1s^2\,2s^2\,2p^2 \quad \text{or} \quad \frac{\uparrow\downarrow}{1s^2} \; \frac{\uparrow\downarrow}{2s^2} \; \underbrace{\frac{\uparrow}{}\,\frac{\uparrow}{}\,\frac{}{}}_{2p^2} \quad \text{or} \quad [He]\,2s^2\,2p^2$$

$$N \quad 1s^2\,2s^2\,2p^3 \quad \text{or} \quad \frac{\uparrow\downarrow}{1s^2} \; \frac{\uparrow\downarrow}{2s^2} \; \underbrace{\frac{\uparrow}{}\,\frac{\uparrow}{}\,\frac{\uparrow}{}}_{2p^3} \quad \text{or} \quad [He]\,2s^2\,2p^3$$

- **Oxygen ($Z = 8$), Fluorine ($Z = 9$), Neon ($Z = 10$):** Electrons now pair up one by one to fill the three 2p orbitals and fully occupy the second shell:

$$O \quad 1s^2\,2s^2\,2p^4 \quad \text{or} \quad \frac{\uparrow\downarrow}{1s^2} \; \frac{\uparrow\downarrow}{2s^2} \; \underbrace{\frac{\uparrow\downarrow}{}\,\frac{\uparrow}{}\,\frac{\uparrow}{}}_{2p^4} \quad \text{or} \quad [He]\,2s^2\,2p^4$$

$$F \quad 1s^2\,2s^2\,2p^5 \quad \text{or} \quad \frac{\uparrow\downarrow}{1s^2} \; \frac{\uparrow\downarrow}{2s^2} \; \underbrace{\frac{\uparrow\downarrow}{}\,\frac{\uparrow\downarrow}{}\,\frac{\uparrow}{}}_{2p^5} \quad \text{or} \quad [He]\,2s^2\,2p^5$$

$$Ne \quad 1s^2\,2s^2\,2p^6 \quad \text{or} \quad \frac{\uparrow\downarrow}{1s^2} \; \frac{\uparrow\downarrow}{2s^2} \; \underbrace{\frac{\uparrow\downarrow}{}\,\frac{\uparrow\downarrow}{}\,\frac{\uparrow\downarrow}{}}_{2p^6}$$

At this point, we may use the shorthand notation [Ne] to represent the electron configuration for a completely filled set of orbitals in the second shell.

- **Sodium to Calcium (Z = 11 − 20):** The pattern seen for lithium through neon is seen again for sodium ($Z = 11$) through argon ($Z = 18$) as the 3s and 3p subshells fill up. For elements having a third filled shell, we may use [Ar] to represent a completely filled third shell. After argon, however, the first crossover in subshell energies occurs. As indicated in Figure 2.6, the 4s subshell is lower in energy than the 3d subshell and is filled first. Potassium ($Z = 19$) and calcium ($Z = 20$) therefore have the following electron configurations:

  K $1s^2\ 2s^2\ 2p^6\ 3s^2\ 3p^6\ 4s^1$ or [Ar]$4s^1$ **Ca** $1s^2\ 2s^2\ 2p^6\ 3s^2\ 3p^6\ 4s^2$ or [Ar]$4s^2$

Worked Example 2.7 Atomic Structure: Electron Configurations

Show how the electron configuration of magnesium can be assigned.

ANALYSIS Magnesium, $Z = 12$, has 12 electrons to be placed in specific orbitals. Assignments are made by putting 2 electrons in each orbital, according to the order shown in Figure 2.6.

- The first 2 electrons are placed in the 1s orbital ($1s^2$).
- The next 2 electrons are placed in the 2s orbital ($2s^2$).
- The next 6 electrons are placed in the three available 2p orbitals ($2p^6$).
- The remaining 2 electrons are both put in the 3s orbital ($3s^2$).

SOLUTION
Magnesium has the configuration $1s^2\ 2s^2\ 2p^6\ 3s^2$ or [Ne]$3s^2$.

Worked Example 2.8 Electron Configurations: Orbital-Filling Diagrams

Write the electron configuration of phosphorus, $Z = 15$, using up and down arrows to show how the electrons in each orbital are paired.

ANALYSIS Phosphorus has 15 electrons, which occupy orbitals according to the order shown in Figure 2.6.

- The first 2 are paired and fill the first shell ($1s^2$).
- The next 8 fill the second shell ($2s^2\ 2p^6$). All electrons are paired.
- The remaining 5 electrons enter the third shell, where 2 fill the 3s orbital ($3s^2$) and 3 occupy the 3p subshell, one in each of the three p orbitals.

SOLUTION

P ↑↓ ↑↓ ↑↓ ↑↓ ↑↓ ↑↓ ↑ ↑ ↑
   $1s^2$ $2s^2$ $2p^6$ $3s^2$ $3p^3$

PROBLEM 2.19
Write electron configurations for the following elements. (You can check your answers in Table 2.3.)
(a) C   (b) P   (c) Cl   (d) K

PROBLEM 2.20
For an atom containing 33 electrons, identify the incompletely filled subshell, and show the paired and/or unpaired electrons in this subshell using up and down arrows.

KEY CONCEPT PROBLEM 2.21
Identify the atom with the following orbital-filling diagram.

$1s^2\ 2s^2\ 2p^6\ 3s^2\ 3p^6$ ↑↓   ↑↓ ↑↓ ↑↓ ↑↓ ↑↓   ↑ __ __
                                 4s   3d   4p

2.8 Electron Configurations and the Periodic Table

How is an atom's electron configuration related to its chemical behavior, and why do elements with similar behavior occur in the same group of the periodic table? As shown in Figure 2.7, the periodic table can be divided into four regions, or *blocks*, of elements according to the electron shells and subshells occupied by *the subshell filled last*.

- The main group 1A and 2A elements on the left side of the table (plus He) are called the **s-block elements** because an *s* subshell is filled last in these elements.
- The main group 3A–8A elements on the right side of the table (except He) are the **p-block elements** because a *p* subshell is filled last in these elements.
- The transition metals in the middle of the table are the **d-block elements** because a *d* subshell is filled last in these elements.
- The inner transition metals detached at the bottom of the table are the **f-block elements** because an *f* subshell is filled last in these elements.

Thinking of the periodic table as outlined in Figure 2.7 provides a simple way to remember the order of orbital filling shown previously in Figure 2.6. Beginning at the top left corner of the periodic table, the first row contains only two elements (H and He) because only two electrons are required to fill the *s* orbital in the first shell, $1s^2$. The second row begins with two *s*-block elements (Li and Be) and continues with six *p*-block elements (B through Ne), so electrons fill the next available *s* orbital (2*s*) and then the first available *p* orbitals (2*p*). The third row is similar to the second row, so the 3*s* and 3*p* orbitals are filled next. The fourth row again starts with 2 *s*-block elements (K and Ca) but is then followed by 10 *d*-block elements (Sc through Zn) and 6 *p*-block elements (Ga through Kr). Thus, the order of orbital filling is 4*s* followed by the first available *d* orbitals (3*d*) followed by 4*p*. Continuing through successive rows of the periodic table gives the entire filling order, identical to that shown in Figure 2.6.

$$1s \rightarrow 2s \rightarrow 2p \rightarrow 3s \rightarrow 3p \rightarrow 4s \rightarrow 3d \rightarrow 4p \rightarrow 5s \rightarrow$$
$$4d \rightarrow 5p \rightarrow 6s \rightarrow 4f \rightarrow 5d \rightarrow 6p \rightarrow 7s \rightarrow 5f \rightarrow 6d \rightarrow 7p$$

But why do the elements in a given group of the periodic table have similar properties? The answer emerges when you look at Table 2.4, which gives electron configurations for elements in the main groups 1A, 2A, 7A, and 8A. Focusing only on the

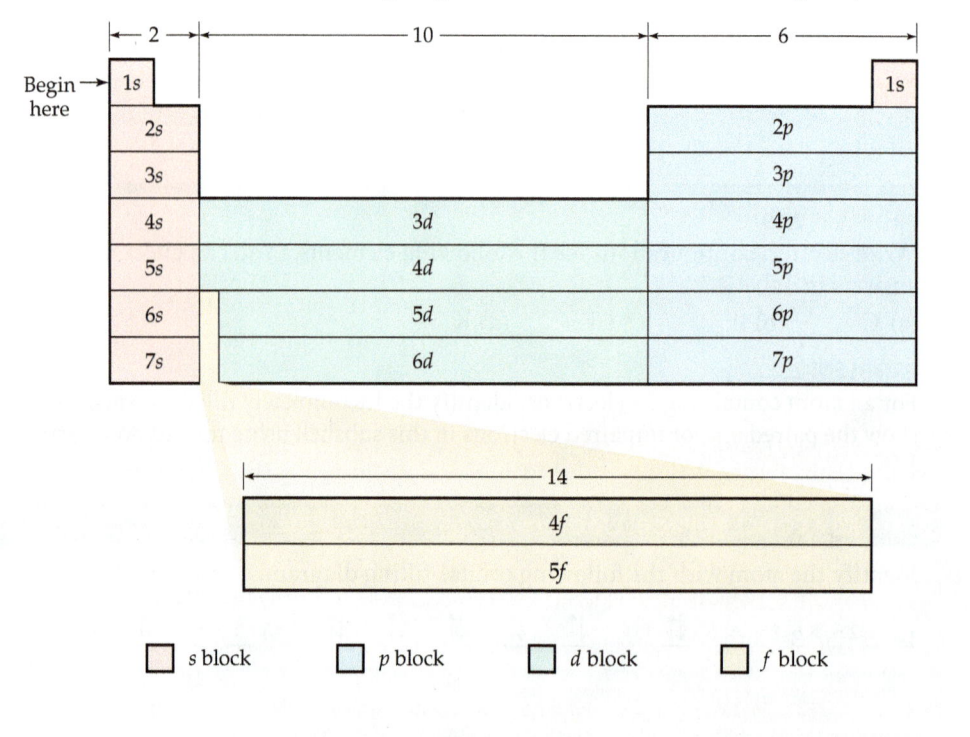

---

s-Block element A main group element that results from the filling of an *s* orbital.

p-Block element A main group element that results from the filling of *p* orbitals.

d-Block element A transition metal element that results from the filling of *d* orbitals.

f-Block element An inner transition metal element that results from the filling of *f* orbitals.

▶ **Figure 2.7**
The blocks of elements in the periodic table correspond to filling the different types of subshells.
Beginning at the top left and going across successive rows of the periodic table provides a method for remembering the order of orbital filling:
$1s \rightarrow 2s \rightarrow 2p \rightarrow 3s \rightarrow 3p \rightarrow 4s \rightarrow 3d \rightarrow 4p$, and so on.

Table 2.4 Valence-Shell Electron Configurations for Group 1A, 2A, 7A, and 8A Elements

Group	Element	Atomic Number	Valence-Shell Electron Configuration
1A	Li (lithium)	3	$2s^1$
	Na (sodium)	11	$3s^1$
	K (potassium)	19	$4s^1$
	Rb (rubidium)	37	$5s^1$
	Cs (cesium)	55	$6s^1$
2A	Be (beryllium)	4	$2s^2$
	Mg (magnesium)	12	$3s^2$
	Ca (calcium)	20	$4s^2$
	Sr (strontium)	38	$5s^2$
	Ba (barium)	56	$6s^2$
7A	F (fluorine)	9	$2s^2\,2p^5$
	Cl (chlorine)	17	$3s^2\,3p^5$
	Br (bromine)	35	$4s^2\,4p^5$
	I (iodine)	53	$5s^2\,5p^5$
8A	He (helium)	2	$1s^2$
	Ne (neon)	10	$2s^2\,2p^6$
	Ar (argon)	18	$3s^2\,3p^6$
	Kr (krypton)	36	$4s^2\,4p^6$
	Xe (xenon)	54	$5s^2\,5p^6$

electrons in the outermost shell, or **valence shell**, *elements in the same group of the periodic table have similar electron configurations in their valence shells.* The group 1A elements, for example, all have one **valence electron**, ns^1 (where n represents the number of the valence shell: $n = 2$ for Li; $n = 3$ for Na; $n = 4$ for K; and so on). The group 2A elements have two valence electrons (ns^2); the group 7A elements have seven valence electrons ($ns^2\,np^5$); and the group 8A elements (except He) have eight valence electrons ($ns^2\,np^6$). You might also notice that the group numbers from 1A through 8A give the numbers of valence electrons for the elements in each main group. It is worth noting that the valence electrons are those in the outermost shell—not necessarily in the orbitals that were filled last!

What is true for the main group elements is also true for the other groups in the periodic table: atoms within a given group have the same number of valence electrons and have similar electron configurations. *Because the valence electrons are the most loosely held, they are the most important in determining an element's properties.* Similar electron configurations thus explain why the elements in a given group of the periodic table have similar chemical behavior.

Valence shell The outermost electron shell of an atom.

Valence electron An electron in the valence shell of an atom.

LOOKING AHEAD ▶▶ We have seen that elements in a given group have similar chemical behavior because they have similar valence electron configurations, and that many chemical properties exhibit periodic trends across the periodic table. The *chemical* behavior of nearly all the elements can be predicted based on their position in the periodic table, and this will be examined in more detail in Chapters 3 and 4. Similarly, the *nuclear* behavior of the different isotopes of a given element is related to the configuration of the nucleus (that is, the number of neutrons and protons) and will be examined in Chapter 11.

Worked Example 2.9 Electron Configurations: Valence Electrons

Write the electron configuration for the following elements, using both the complete and the shorthand notations. Indicate which electrons are the valence electrons.

(a) Na   (b) Cl   (c) Zr

ANALYSIS Locate the row and the block in which each of the elements is found in Figure 2.7. The location can be used to determine the complete electron configuration and to identify the valence electrons.

SOLUTION

(a) Na (sodium) is located in the third row and in the first column of the *s*-block. Therefore, all orbitals up to the 3*s* are completely filled, and there is one electron in the 3*s* orbital.

$$\text{Na: } 1s^2\, 2s^2\, 2p^6\, \underline{3s^1} \quad \text{or} \quad [\text{Ne}]\, \underline{3s^1} \quad \text{(valence electrons are underlined)}$$

(b) Cl (chlorine) is located in the third row and in the fifth column of the *p*-block.

$$\text{Cl: } 1s^2\, 2s^2\, 2p^6\, \underline{3s^2\, 3p^5} \quad \text{or} \quad [\text{Ne}]\, \underline{3s^2\, 3p^5}$$

(c) Zr (zirconium) is located in the fifth row and in the second column of the *d*-block. All orbitals up to the 4*d* are completely filled, and there are 2 electrons in the 4*d* orbitals. Note that the 4*d* orbitals are filled after the 5*s* orbitals in both Figures 2.6 and 2.7.

$$\text{Zr: } 1s^2\, 2s^2\, 2p^6\, 3s^1\, 3p^6\, 4s^2\, 3d^{10}\, 4p^6\, \underline{5s^2\, 4d^2} \quad \text{or} \quad [\text{Kr}]\, \underline{5s^2\, 4d^2}$$

Worked Example 2.10 Electron Configurations: Valence-Shell Configurations

Using *n* to represent the number of the valence shell, write a general valence-shell configuration for the elements in group 6A.

ANALYSIS The elements in group 6A have 6 valence electrons. In each element, the first two of these electrons are in the valence *s* subshell, giving ns^2, and the next four electrons are in the valence *p* subshell, giving np^4.

SOLUTION
For group 6A, the general valence-shell configuration is $ns^2\, np^4$.

Worked Example 2.11 Electron Configurations: Inner Shells versus Valence Shell

How many electrons are in a tin atom? Give the number of electrons in each shell. How many valence electrons are there in a tin atom? Write the valence-shell configuration for tin.

ANALYSIS The total number of electrons will be the same as the atomic number for tin $(Z = 50)$. The number of valence electrons will equal the number of electrons in the valence shell.

SOLUTION
Checking the periodic table shows that tin has atomic number 50 and is in group 4A. The number of electrons in each shell is

Shell number: 1   2   3   4   5
Number of electrons: 2   8   18   18   4

As expected from the group number, tin has 4 valence electrons. They are in the 5*s* and 5*p* subshells and have the configuration $5s^2\, 5p^2$.

PROBLEM 2.22
Write the electron configuration for the following elements, using both the complete and the shorthand notations. Indicate which electrons are the valence electrons.
(a) F   (b) Al   (c) As

PROBLEM 2.23
Identify the group in which all the elements have the valence-shell configuration ns^2.

PROBLEM 2.24
For chlorine, identify the group number, give the number of electrons in each occupied shell, and write its valence-shell configuration.

KEY CONCEPT PROBLEM 2.25
Identify the group number, and write the general valence-shell configuration (for example, ns^1 for group 1A elements) for the elements indicated in red in the following periodic table.

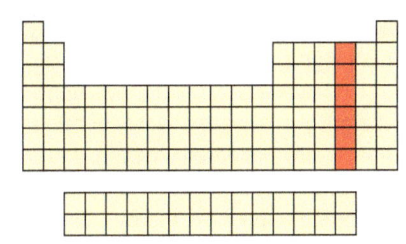

2.9 Electron-Dot Symbols

Valence electrons play such an important role in the behavior of atoms that it is useful to have a method for including them with atomic symbols. In an **electron-dot symbol**, dots are placed around the atomic symbol to indicate the number of valence electrons present. A group 1A atom, such as sodium, has a single dot; a group 2A atom, such as magnesium, has two dots; a group 3A atom, such as boron, has three dots; and so on.

Table 2.5 gives electron-dot symbols for atoms of the first few elements in each main group. As shown, the dots are distributed around the four sides of the element symbol, singly at first until each of the four sides has one dot. As more electron dots are added they will form pairs, with no more than two dots on a side. Note that helium differs from other noble gases in having only two valence electrons rather than eight. Nevertheless, helium is considered a member of group 8A because its properties resemble those of the other noble gases and because its highest occupied subshell is filled ($1s^2$).

Electron-dot symbol An atomic symbol with dots placed around it to indicate the number of valence electrons.

TABLE 2.5 Electron-Dot Symbols for Some Main Group Elements

1A	2A	3A	4A	5A	6A	7A	NOBLE GASES
H·							He:
Li·	·Be·	·B·	·C·	·N:	·O:	·F:	:Ne:
Na·	·Mg·	·Al·	·Si·	·P:	·S:	·Cl:	:Ar:
K·	·Ca·	·Ga·	·Ge·	·As:	·Se:	·Br:	:Kr:

CHEMISTRY IN ACTION

Atoms and Light

What we see as *light* is really a wave of energy moving through space. The shorter the length of the wave (the *wavelength*), the higher the energy; the longer the wavelength, the lower the energy.

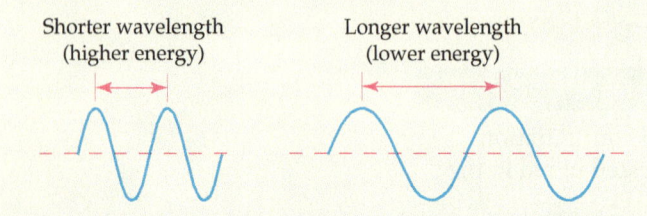

Shorter wavelength (higher energy)

Longer wavelength (lower energy)

Visible light has wavelengths in the range 400–800 nm, but that is just one small part of the overall *electromagnetic spectrum*, shown in the accompanying figure. Although we cannot see the other wavelengths of electromagnetic energy, we use them for many purposes and their names may be familiar to you: gamma rays, X rays, ultraviolet (UV) rays, infrared (IR) rays, microwaves, and radio waves.

What happens when a beam of electromagnetic energy collides with an atom? Remember that electrons are located in orbitals based on their energy levels. An atom with its electrons in their usual, lowest-energy locations is said to be in its *ground state*. If the amount of electromagnetic energy is just right, an electron can be kicked up from its usual energy level to a higher one. Energy from an electrical discharge or in the form of heat can also boost electrons to higher energy levels. With one of its electrons promoted to a higher energy, an atom is said to be *excited*. The excited state does not last long, though, because the electron quickly drops back to its more stable, ground-state energy level, releasing its extra energy in the process. If the released energy falls in the range of visible light, we can see the result. Many practical applications, from neon lights to fireworks, are the result of this phenomenon.

▲ The brilliant colors of fireworks are due to the release of the energy from excited atoms as electrons fall from higher to lower energy levels.

In "neon" lights, noble gas atoms are excited by an electric discharge, giving rise to a variety of colors that depend on the gas—red from neon, white from krypton, and blue from argon—as electrons release energy and return to their ground states. Similarly, mercury or sodium atoms excited by electrical energy are responsible for the intense bluish or yellowish light, respectively, provided by some street lamps. In the same manner, metal atoms excited by heat are responsible for the spectacular colors of fireworks—red from strontium, green from barium, and blue from copper, for example.

The concentration of certain biologically important metals in body fluids, such as blood or urine, is measured by sensitive instruments relying on the same principle of electron excitation that we see in fireworks. These instruments measure the intensity of color produced in a flame by lithium (red), sodium (yellow), and potassium (violet), to determine the concentrations of these metals, which are included in most clinical lab reports.

See Chemistry in Action Problems 2.88 and 2.89 at the end of the chapter.

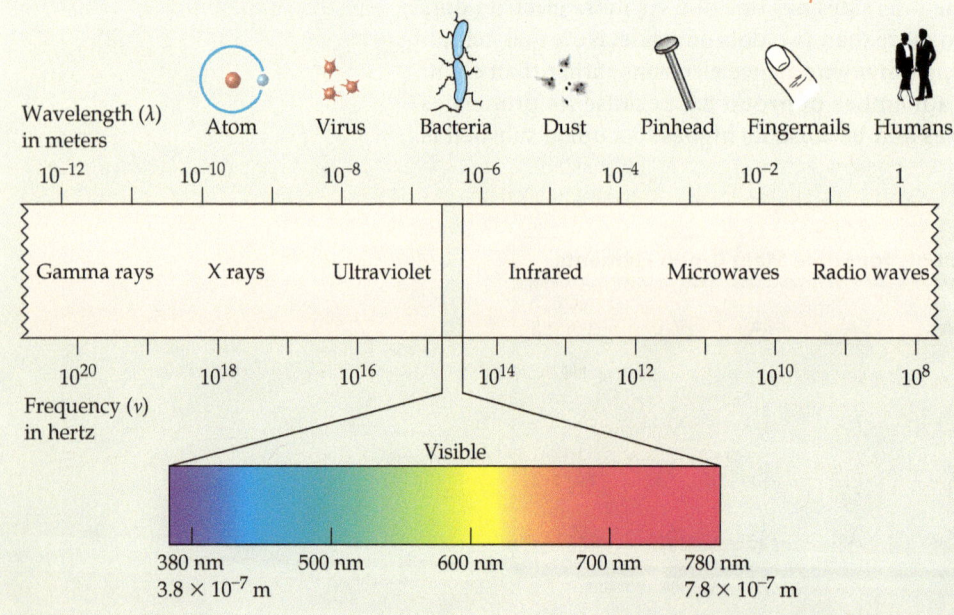

Wavelength (λ) in meters							
	Atom	Virus	Bacteria	Dust	Pinhead	Fingernails	Humans
10^{-12}	10^{-10}	10^{-8}	10^{-6}	10^{-4}	10^{-2}		1

Gamma rays — X rays — Ultraviolet — Infrared — Microwaves — Radio waves

| 10^{20} | 10^{18} | 10^{16} | 10^{14} | 10^{12} | 10^{10} | 10^{8} |

Frequency (ν) in hertz

Visible

380 nm — 500 nm — 600 nm — 700 nm — 780 nm
3.8×10^{-7} m 7.8×10^{-7} m

◀ The electromagnetic spectrum consists of a continuous range of wavelengths, with the familiar visible region accounting for only a small portion near the middle of the range.

Worked Example 2.12 Electron Configurations: Electron-Dot Symbols

Write the electron-dot symbol for any element X in group 5A.

ANALYSIS The group number, 5A, indicates 5 valence electrons. The first four are distributed singly around the four sides of the element symbol, and any additional are placed to form electron pairs.

SOLUTION

·Ẍ: (5 electrons)

PROBLEM 2.26
Write the electron-dot symbol for any element X in group 3A.

PROBLEM 2.27
Write electron-dot symbols for radon, lead, xenon, and radium.

PROBLEM 2.28
When an electron in a strontium atom drops from the excited state to the ground state, it emits red light, as explained in the Chemistry in Action on p. 66. When an electron in a copper atom drops from the excited state to the ground state, it emits blue light. What are the approximate wavelengths of the red light and the blue light? Which color is associated with higher energy?

SUMMARY: REVISITING THE CHAPTER GOALS

1. What is the modern theory of atomic structure? All matter is composed of *atoms*. An atom is the smallest and simplest unit into which a sample of an element can be divided while maintaining the properties of the element. Atoms are made up of subatomic particles called *protons, neutrons,* and *electrons*. Protons have a positive electrical charge, neutrons are electrically neutral, and electrons have a negative electrical charge. The protons and neutrons in an atom are present in a dense, positively charged central region called the *nucleus*. Electrons are situated a relatively large distance away from the nucleus, leaving most of the atom as empty space (*see Problems 34, 42, 43*).

2. How do atoms of different elements differ? Elements differ according to the number of protons their atoms contain, a value called the element's *atomic number* (Z). All atoms of a given element have the same number of protons and an equal number of electrons. The number of neutrons in an atom is not predictable but is generally as great or greater than the number of protons. The total number of protons plus neutrons in an atom is called the atom's *mass number* (A) (*see Problems 35, 44, 46, 86, 87, 92*).

3. What are isotopes, and what is atomic weight? Atoms with identical numbers of protons and electrons but different numbers of neutrons are called *isotopes*. The atomic weight of an element is the weighted average mass of atoms of the element's naturally occurring isotopes measured in *atomic mass units* (amu) (*see Problems 36–41, 45–53, 92, 96, 97*).

4. How is the periodic table arranged? Elements are organized into the *periodic table*, consisting of 7 rows, or *periods*, and 18 columns, or *groups*. The 2 groups on the left side of the table and the 6 groups on the right are called the *main group elements*. The 10 groups in the middle are the *transition metal groups*, and the 14 groups pulled out and displayed below the main part of the table are called the *inner transition metal groups*. Within a given group in the table, elements have the same number of valence electrons in their valence shell and similar electron configurations (*see Problems 29, 30, 54–65, 90, 91, 93*).

5. How are electrons arranged in atoms? The electrons surrounding an atom are grouped into layers, or *shells*. Within each shell, electrons are grouped into *subshells*, and within each subshell into *orbitals*—regions of space in which electrons are most likely to be found. The *s* orbitals are spherical, and the *p* orbitals are dumbbell-shaped.

Each shell can hold a specific number of electrons. The first shell can hold 2 electrons in an *s* orbital ($1s^2$); the second shell can hold 8 electrons in one *s* and three *p* orbitals ($2s^2\ 2p^6$); the third shell can hold 18 electrons in one *s*, three *p*, and five *d* orbitals ($3s^2\ 3p^6\ 3d^{10}$); and so on. The *electron configuration* of an element is predicted by assigning the element's electrons into orbitals, beginning with the lowest-energy orbital. The electrons in the outermost shell, or *valence shell*, can be represented using electron-dot symbols (*see Problems 31–33, 54, 55, 57, 66–83, 91, 94, 95, 98–100, 102–106*).

KEY WORDS

Alkali metal, *p. 54*
Alkaline earth metal, *p. 55*
Atom, *p. 45*
Atomic mass unit (amu), *p. 46*
Atomic number (Z), *p. 48*
Atomic theory, *p. 45*
Atomic weight, *p. 50*
d-Block element, *p. 62*
Electron, *p. 46*
Electron configuration, *p. 59*
Electron-dot symbol, *p. 65*
f-Block element, *p. 62*
Group, *p. 52*
Halogen, *p. 55*
Inner transition metal element, *p. 52*
Isotopes, *p. 50*

Main group element, *p. 52*
Mass number (A), *p. 49*
Neutron, *p. 46*
Noble gas, *p. 55*
Nucleus, *p. 46*
Orbital, *p. 57*
p-Block element, *p. 62*
Period, *p. 52*
Proton, *p. 46*
s-Block element, *p. 62*
Shell (electron), *p. 57*
Subatomic particles, *p. 46*
Subshell (electron), *p. 57*
Transition metal element, *p. 52*
Valence electron, *p. 63*
Valence shell, *p. 63*

UNDERSTANDING KEY CONCEPTS

2.29 Where on the following outline of a periodic table do the indicated elements or groups of elements appear?

(a) Alkali metals
(b) Halogens
(c) Alkaline earth metals
(d) Transition metals
(e) Hydrogen
(f) Helium
(g) Metalloids

2.30 Is the element marked in red on the following periodic table likely to be a gas, a liquid, or a solid? What is the atomic number of the element in blue? Name at least one other element that is likely to be similar to the element in green.

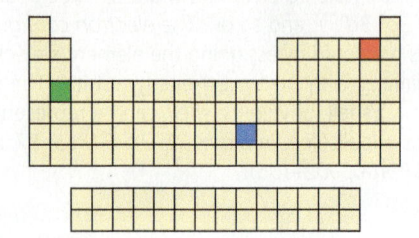

2.31 Use the blank periodic table below to show where the elements matching the following descriptions appear.

(a) Elements with the valence-shell electron configuration $ns^2 np^5$
(b) An element whose third shell contains two *p* electrons
(c) Elements with a completely filled valence shell

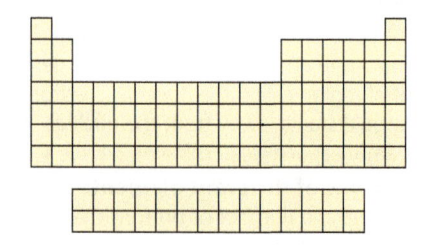

2.32 What atom has the following orbital-filling diagram?

$1s^2\ 2s^2\ 2p^6\ 3s^2\ 3p^6$ ⇅   ⇅ ⇅ ⇅ ⇅ ⇅   ⇅ ↓ ↓
                                    4s   3d   4p

2.33 Use the orbital-filling diagram below to show the electron configuration for As:

$1s^2\ 2s^2\ 2p^6\ 3s^2\ 3p^6$ __   __ __ __ __ __   __ __ __
                                    4s   3d   4p

ADDITIONAL PROBLEMS

ATOMIC THEORY AND THE COMPOSITION OF ATOMS

2.34 What four fundamental assumptions about atoms and matter make up modern atomic theory?

2.35 How do atoms of different elements differ?

2.36 Find the mass in grams of one atom of the following elements:

(a) Bi, atomic weight 208.9804 amu

(b) Xe, atomic weight 131.29 amu

(c) He, atomic weight 4.0026 amu

2.37 Find the mass in atomic mass units of the following:

(a) 1 O atom, with a mass of 2.66×10^{-23} g

(b) 1 Br atom, with a mass of 1.31×10^{-22} g

2.38 What is the mass in grams of 6.022×10^{23} N atoms of mass 14.01 amu?

2.39 What is the mass in grams of 6.022×10^{23} O atoms of mass 16.00 amu?

2.40 How many O atoms of mass 15.99 amu are in 15.99 g of oxygen?

2.41 How many C atoms of mass 12.00 amu are in 12.00 g of carbon?

2.42 What are the names of the three subatomic particles? What are their approximate masses in atomic mass units, and what electrical charge does each have?

2.43 Where within an atom are the three types of subatomic particles located?

2.44 Give the number of neutrons in each naturally occurring isotope of argon: argon-36, argon-38, argon-40.

2.45 Give the number of protons, neutrons, and electrons in the following isotopes:

(a) Al-27   (b) $^{28}_{14}Si$

(c) B-11   (d) $^{115}_{47}Ag$

2.46 Which of the following symbols represent isotopes of the same element?

(a) $^{19}_{9}X$ (b) $^{19}_{10}X$

(c) $^{21}_{9}X$ (d) $^{21}_{12}X$

2.47 Give the name and the number of neutrons in each isotope listed in Problem 2.46.

2.48 Write the symbols for the following isotopes:

(a) Its atoms contain 6 protons and 8 neutrons.

(b) Its atoms have mass number 39 and contain 19 protons.

(c) Its atoms have mass number 20 and contain 10 electrons.

2.49 Write the symbols for the following isotopes:

(a) Its atoms contain 50 electrons and 70 neutrons.

(b) Its atoms have $A = 56$ and $Z = 26$.

(c) Its atoms have $A = 226$ and contain 88 electrons.

2.50 There are three naturally occurring isotopes of carbon, with mass numbers of 12, 13, and 14. How many neutrons does each have? Write the symbol for each isotope, indicating its atomic number and mass number.

2.51 One of the most widely used isotopes in medical diagnostics is technicium-99*m* (the *m* indicates that it is a *metastable* isotope). Write the symbol for this isotope, indicating both mass number and atomic number.

2.52 Naturally occurring copper is a mixture of 69.17% Cu-63 with a mass of 62.93 amu and 30.83% Cu-65 with a mass of 64.93 amu. What is the atomic weight of copper?

2.53 Naturally occurring lithium is a mixture of 92.58% Li-7 with a mass of 7.016 amu and 7.42% Li-6 with a mass of 6.015 amu. What is the atomic weight of lithium?

THE PERIODIC TABLE

2.54 Why does the third period in the periodic table contain eight elements?

2.55 Why does the fourth period in the periodic table contain 18 elements?

2.56 Americium, atomic number 95, is used in household smoke detectors. What is the symbol for americium? Is americium a metal, a nonmetal, or a metalloid?

2.57 What subshell is being filled for the metalloid elements?

2.58 Answer the following questions for the elements from scandium through zinc:

(a) Are they metals or nonmetals?

(b) To what general class of elements do they belong?

(c) What subshell is being filled by electrons in these elements?

2.59 Answer the following questions for the elements from cerium through lutetium:

(a) Are they metals or nonmetals?

(b) To what general class of elements do they belong?

(c) What subshell is being filled by electrons in these elements?

2.60 For (a) rubidium (b) tungsten, (c) germanium, and (d) krypton, which of the following terms apply? (i) metal, (ii) nonmetal (iii) metalloid (iv) transition element, (v) main group element, (vi) noble gas, (vii) alkali metal, (viii) alkaline earth metal.

2.61 For (a) calcium, (b) palladium, (c) carbon, and (d) radon, which of the following terms apply? (i) metal, (ii) nonmetal, (iii) metalloid (iv) transition element, (v) main group element, (vi) noble gas, (vii) alkali metal, (viii) alkaline earth metal.

2.62 Name an element in the periodic table that you would expect to be chemically similar to sulfur.

2.63 Name an element in the periodic table that you would expect to be chemically similar to potassium.

2.64 What elements in addition to lithium make up the alkali metal family?

2.65 What elements in addition to fluorine make up the halogen family?

ELECTRON CONFIGURATIONS

2.66 What is the maximum number of electrons that can go into an orbital?

2.67 What are the shapes and locations within an atom of *s* and *p* orbitals?

2.68 What is the maximum number of electrons that can go into the first shell? The second shell? The third shell?

2.69 What is the total number of orbitals in the third shell? The fourth shell?

2.70 How many subshells are there in the third shell? The fourth shell? The fifth shell?

2.71 How many orbitals would you expect to find in the last subshell of the fifth shell? How many electrons would you need to fill this subshell?

2.72 How many electrons are present in an atom with its 1*s*, 2*s*, and 2*p* subshells filled? What is this element?

2.73 How many electrons are present in an atom with its 1*s*, 2*s*, 2*p*, 3*s*, 3*p*, and 4*s* subshells filled and with two electrons in the 3*d* subshell? What is this element?

2.74 Use arrows to show electron pairing in the valence *p* subshell of

(a) Sulfur
(b) Bromine
(c) Silicon

2.75 Use arrows to show electron pairing in the 5*s* and 4*d* orbitals of

(a) Rubidum
(b) Niobium
(c) Rhodium

2.76 Determine the number of unpaired electrons for each of the atoms in Problems 2.74 and 2.75.

2.77 Without looking back in the text, write the electron configurations for the following:

(a) Titanium $Z = 22$
(b) Phosphorus, $Z = 15$
(c) Argon, $Z = 18$
(d) Lanthanum, $Z = 57$

2.78 How many electrons does the element with $Z = 12$ have in its valence shell? Write the electron-dot symbol for this element.

2.79 How many valence electrons do group 4A elements have? Explain. Write a generic electron-dot symbol for elements in this group.

2.80 Identify the valence subshell occupied by electrons in beryllium and arsenic atoms.

2.81 What group in the periodic table has the valence-shell configuration $ns^2 \, np^3$?

2.82 Give the number of valence electrons and draw electron-dot symbols for atoms of the following elements:

(a) Kr
(b) C
(c) Ca
(d) K
(e) B
(f) Cl

2.83 Using *n* for the number of the valence shell, write a general valence-shell configuration for the elements in group 6A and in group 2A.

CHEMISTRY IN ACTION

2.84 What is the advantage of using a scanning tunneling microscope rather than a normal light microscope? [*Are Atoms Real?* p. 48]

2.85 For the Kanji character in the lower portion of the figure on p. 48: (a) How wide is the character in terms of iron atoms? (b) Given the radius of an iron atom is 126 pm, calculate the width of this character in centimeters. [*Are Atoms Real?* p. 48]

2.86 What are the first two elements that are made in stars? [*The Origin of Chemical Elements*, p. 56]

2.87 How are elements heavier than iron made? [*The Origin of Chemical Elements*, p. 56]

2.88 Which type of electromagnetic energy in the following pairs is of higher energy? [*Atoms and Light*, p. 66]

(a) Infrared, ultraviolet
(b) Gamma waves, microwaves
(c) Visible light, X rays

2.89 Why do you suppose ultraviolet rays from the sun are more damaging to the skin than visible light? [*Atoms and Light*, p. 66]

GENERAL QUESTIONS AND PROBLEMS

2.90 What elements in addition to helium make up the noble gas family?

2.91 Hydrogen is placed in group 1A on many periodic charts, even though it is not an alkali metal. On other periodic charts, however, hydrogen is included with group 7A even though it is not a halogen. Explain. (Hint: draw electron-dot symbols for H and for the 1A and 7A elements.)

2.92 Tellurium ($Z = 52$) has a *lower* atomic number than iodine ($Z = 53$), yet it has a *higher* atomic weight (127.60 amu for Te versus 126.90 amu for I). How is this possible?

2.93 What is the atomic number of the yet-undiscovered element directly below francium (Fr) in the periodic table?

2.94 Give the number of electrons in each shell for lead.

2.95 Identify the highest-energy occupied subshell in atoms of the following elements:

(a) Iodine
(b) Scandium
(c) Arsenic
(d) Aluminum

2.96 What is the atomic weight of naturally occurring bromine, which contains 50.69% Br-79 of mass 78.92 amu and 49.31% Br-81 of mass 80.91 amu?

2.97 (a) What is the mass (in amu and in grams) of a single atom of Carbon-12?
(b) What is the mass (in grams) of 6.02×10^{23} atoms of Carbon-12?
(c) Based on your answer to part (b), what would be the mass of 6.02×10^{23} atoms of Sodium-23?

2.98 An unidentified element is found to have an electron configuration by shell of 2 8 18 8 2. To what group and period does this element belong? Is the element a metal or a nonmetal? How many protons does an atom of the element have? What is the name of the element? Write its electron-dot symbol.

2.99 Germanium, atomic number 32, is used in building semiconductors for microelectronic devices, and has an electron configuration by shell of 2 8 18 4.

(a) Write the electronic configuration for germanium.
(b) In what shell and orbitals are the valence electrons?

2.100 Tin, atomic number 50, is directly beneath germanium (Problem 2.99) in the periodic table. What electron configuration by shell would you expect tin to have? Is tin a metal or a nonmetal?

2.101 A blood sample is found to contain 8.6 mg/dL of Ca. How many atoms of Ca are present in 8.6 mg? The atomic weight of Ca is 40.08 amu.

2.102 What is wrong with the following electron configurations?
(a) Ni $1s^2 \, 2s^2 \, 2p^6 \, 3s^2 \, 3p^6 \, 3d^{10}$
(b) N $1s^2 \, 2p^5$
(c) Si $1s^2 \, 2s^2 \, 2p$ ⇅ __ __
(d) Mg $1s^2 \, 2s^2 \, 2p^6 \, 3s$ ↑↑

2.103 Not all elements follow exactly the electron-filling order described in Figure 2.7. Atoms of which elements are represented by the following electron configurations?

(a) $1s^2 \, 2s^2 \, 2p^6 \, 3s^2 \, 3p^6 \, 3d^5 \, 4s^1$
(b) $1s^2 \, 2s^2 \, 2p^6 \, 3s^2 \, 3p^6 \, 3d^{10} \, 4s^1$
(c) $1s^2 \, 2s^2 \, 2p^6 \, 3s^2 \, 3p^6 \, 3d^{10} \, 4s^2 \, 4p^6 \, 4d^5 \, 5s^1$
(d) $1s^2 \, 2s^2 \, 2p^6 \, 3s^2 \, 3p^6 \, 3d^{10} \, 4s^2 \, 4p^6 \, 4d^{10} \, 5s^1$

2.104 What similarities do you see in the electron configurations for the atoms in Problem 2.103? How might these similarities explain their anomalous electron configurations?

2.105 Based on the identity of the elements whose electron configurations are given in Problem 2.103, write the electron configurations for the element with atomic number $Z = 79$.

2.106 What orbital is filled last in the most recently discovered element 117?

CHAPTER 3

Ionic Compounds

CONTENTS

- **3.1** Ions
- **3.2** Periodic Properties and Ion Formation
- **3.3** Ionic Bonds
- **3.4** Some Properties of Ionic Compounds
- **3.5** Ions and the Octet Rule
- **3.6** Ions of Some Common Elements
- **3.7** Naming Ions
- **3.8** Polyatomic Ions
- **3.9** Formulas of Ionic Compounds
- **3.10** Naming Ionic Compounds
- **3.11** H^+ and OH^- Ions: An Introduction to Acids and Bases

◀ Stalagmites and stalactites, such as these in a cave in the Nangu Stone Forest in China, are composed of the ionic compounds calcium carbonate, $CaCO_3$, and magnesium carbonate, $MgCO_3$.

CHAPTER GOALS

1. **What is an ion, what is an ionic bond, and what are the general characteristics of ionic compounds?**
   THE GOAL: Be able to describe ions and ionic bonds, and give the general properties of compounds that contain ionic bonds.

2. **What is the octet rule, and how does it apply to ions?**
   THE GOAL: Be able to state the octet rule, and use it to predict the electron configurations of ions of main group elements. (◀◀ B.)

3. **What is the relationship between an element's position in the periodic table and the formation of its ion?**
   THE GOAL: Be able to predict what ions are likely to be formed by atoms of a given element. (◀◀ A, B.)

4. **What determines the chemical formula of an ionic compound?**
   THE GOAL: Be able to write formulas for ionic compounds, given the identities of the ions.

5. **How are ionic compounds named?**
   THE GOAL: Be able to name an ionic compound from its formula or give the formula of a compound from its name.

6. **What are acids and bases?**
   THE GOAL: Be able to recognize common acids and bases.

CONCEPTS TO REVIEW
◀◀

A. The Periodic Table
(Sections 2.4 and 2.5)

B. Electron Configurations
(Sections 2.7 and 2.8)

There are more than 19 million known chemical compounds, ranging in size from small *diatomic* (two-atom) substances like carbon monoxide, CO, to deoxyribonucleic acid (DNA), which can contain several *billion* atoms linked together in a precise way. Clearly, there must be some force that holds atoms together in compounds; otherwise, the atoms would simply drift apart and no compounds could exist. The forces that hold atoms together in compounds are called *chemical bonds* and are of two major types: *ionic bonds* and *covalent bonds*. In this chapter, we look at ionic bonds and at the substances formed by them. In the next chapter, we will look at covalent bonds.

All chemical bonds result from the electrical attraction between opposite charges—between positively charged nuclei and negatively charged electrons. As a result, the way that different elements form bonds is related to their different electron configurations and the changes that take place as each atom tries to achieve a more stable electron configuration.

3.1 Ions

A general rule noted by early chemists is that metals, on the left side of the periodic table, tend to form compounds with nonmetals, on the right side of the table. The alkali metals of group 1A, for instance, react with the halogens of group 7A to form a variety of compounds. Sodium chloride (table salt), formed by the reaction of sodium with chlorine, is a familiar example. The names and chemical formulas of some other compounds containing elements from groups 1A and 7A include:

Potassium iodide, KI	Added to table salt to provide the iodide ion that is needed by the thyroid gland
Sodium fluoride, NaF	Added to many municipal water supplies to provide fluoride ion for the prevention of tooth decay
Sodium iodide, NaI	Used in laboratory scintillation counters to detect radiation (See Section 11.8)

The compositions and the properties of these alkali metal–halogen compounds are similar. For instance, the two elements always combine in a 1:1 ratio: one alkali metal atom for every halogen atom. Each compound has a high melting point (all are over 500 °C); each is a stable, white, crystalline solid; and each is soluble in water.

74 CHAPTER 3 Ionic Compounds

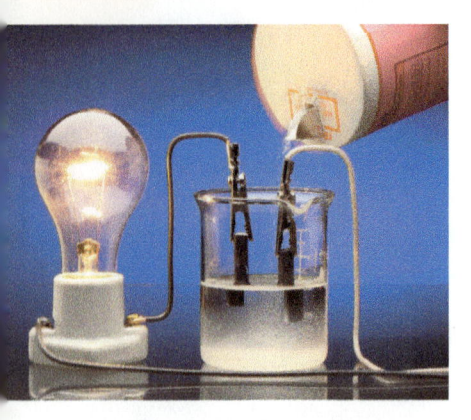

A solution of sodium chloride in water conducts electricity, allowing the bulb to light.

Ion An electrically charged atom or group of atoms.

Cation A positively charged ion.

Anion A negatively charged ion.

Furthermore, a water solution containing each compound conducts electricity, a property that gives a clue to the kind of chemical bond holding the atoms together.

Electricity can only flow through a medium containing charged particles that are free to move. The electrical conductivity of metals, for example, results from the movement of negatively charged electrons through the metal. But what charged particles might be present in the water solutions of alkali metal–halogen compounds? To answer this question, think about the composition of atoms. Atoms are electrically neutral because they contain equal numbers of protons and electrons. By gaining or losing one or more electrons, however, an atom can be converted into a charged particle called an **ion**.

The *loss* of one or more electrons from a neutral atom gives a *positively* charged ion called a **cation** (cat-ion). As we saw in Section 2.8, sodium and other alkali metal atoms have a single electron in their valence shell and an electron configuration symbolized as ns^1, where n represents the shell number. By losing this electron, an alkali metal is converted to a positively charged cation.

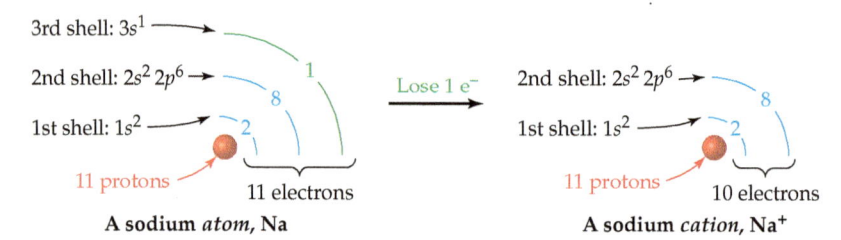

3rd shell: $3s^1$
2nd shell: $2s^2\,2p^6$
1st shell: $1s^2$
11 protons — 11 electrons
A sodium *atom*, Na

Lose 1 e⁻ →

2nd shell: $2s^2\,2p^6$
1st shell: $1s^2$
11 protons — 10 electrons
A sodium *cation*, Na⁺

Conversely, the *gain* of one or more electrons by a neutral atom gives a *negatively* charged ion called an **anion** (an-ion). Chlorine and other halogen atoms have ns^2np^5 valence electrons and can easily gain an additional electron to fill out their valence subshell, thereby forming negatively charged anions.

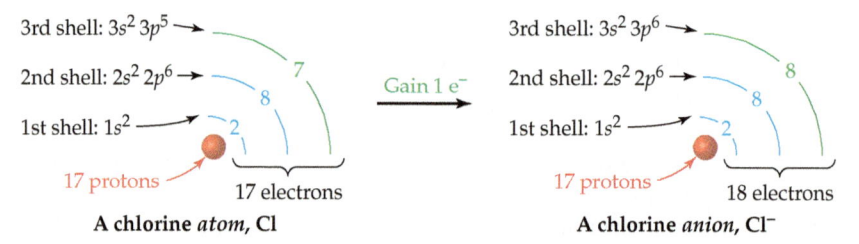

3rd shell: $3s^2\,3p^5$
2nd shell: $2s^2\,2p^6$
1st shell: $1s^2$
17 protons — 17 electrons
A chlorine *atom*, Cl

Gain 1 e⁻ →

3rd shell: $3s^2\,3p^6$
2nd shell: $2s^2\,2p^6$
1st shell: $1s^2$
17 protons — 18 electrons
A chlorine *anion*, Cl⁻

The symbol for a cation is written by adding the positive charge as a superscript to the symbol for the element; an anion symbol is written by adding the negative charge as a superscript. If one electron is lost or gained, the charge is +1 or −1 but the number 1 is omitted in the notation, as in Na⁺ and Cl⁻. If two or more electrons are lost or gained, however, the charge is ±2 or greater and the number *is* used, as in Ca^{2+} and N^{3-}.

PROBLEM 3.1
Magnesium atoms lose two electrons when they react. Write the symbol of the ion that is formed. Is it a cation or an anion?

PROBLEM 3.2
Sulfur atoms gain two electrons when they react. Write the symbol of the ion that is formed. Is it a cation or an anion?

KEY CONCEPT PROBLEM 3.3
Write the symbol for the ion depicted here. Is it a cation or an anion?

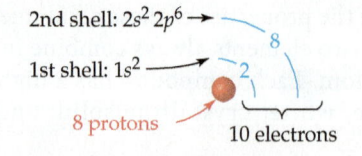

2nd shell: $2s^2\,2p^6$
1st shell: $1s^2$
8 protons — 10 electrons

3.2 Periodic Properties and Ion Formation

The ease with which an atom loses an electron to form a positively charged cation is measured by a property called the atom's **ionization energy**, defined as the energy required to remove one electron from a single atom in the gaseous state. Conversely, the ease with which an atom *gains* an electron to form a negatively charged anion is measured by a property called **electron affinity**, defined as the energy released on adding an electron to a single atom in the gaseous state.

> **Ionization energy** The energy required to remove one electron from a single atom in the gaseous state.
>
> **Electron affinity** The energy released on adding an electron to a single atom in the gaseous state.

Ionization energy (energy is added)   Atom + Energy $\xrightarrow{\text{Gain } e^-}$ Cation + Electron

Electron affinity (energy is relased)   Atom + Electron $\xrightarrow{\text{Lose } e^-}$ Anion + Energy

The relative magnitudes of ionization energies and electron affinities for elements in the first four rows of the periodic table are shown in Figure 3.1. Because ionization energy measures the amount of energy that must be *added* to pull an electron away from a neutral atom, the small values shown in Figure 3.1 for alkali metals (Li, Na, K) and other elements on the left side of the periodic table mean that these elements lose an electron easily. Conversely, the large values shown for halogens (F, Cl, Br) and noble gases (He, Ne, Ar, Kr) on the right side of the periodic table mean that these elements do not lose an electron easily. Electron affinities, however, measure the amount of energy *released* when an atom gains an electron. Although electron affinities are small compared to ionization energies, the halogens nevertheless have the largest values and therefore gain an electron most easily, whereas metals have the smallest values and do not gain an electron easily:

Alkali metal
- Small ionization energy—electron easily lost
- Small electron affinity—electron not easily gained
- *Net result*: Cation formation is favored

Halogen
- Large ionization energy—electron not easily lost
- Large electron affinity—electron easily gained
- *Net result*: Anion formation is favored

You might also note in Figure 3.1 that main group elements near the *middle* of the periodic table—boron ($Z = 5$, group 3A) carbon ($Z = 6$, group 4A), and nitrogen ($Z = 7$, group 5A)—neither lose nor gain electrons easily and thus do not form ions easily. In the next chapter, we will see that these elements tend not to form ionic bonds but form covalent bonds instead.

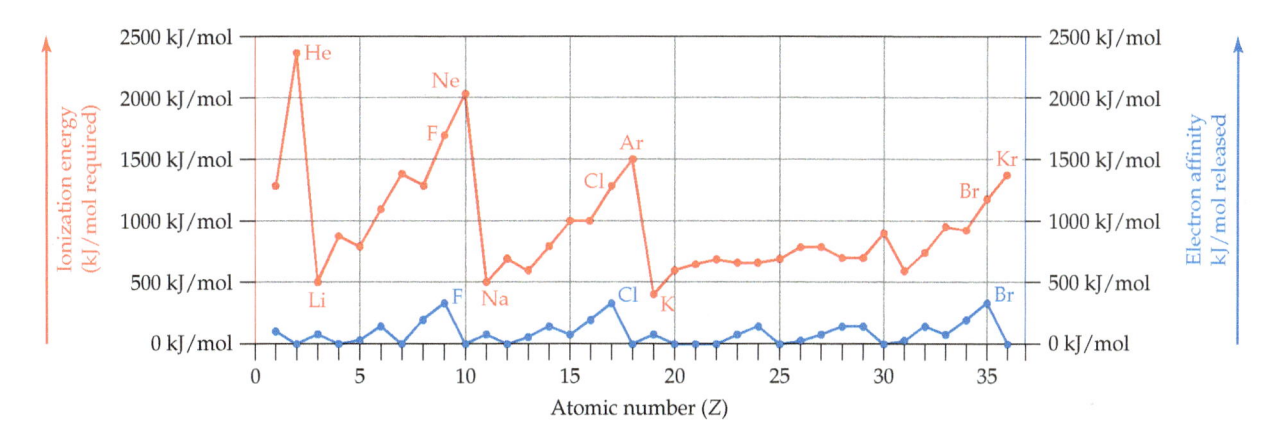

▲ **Figure 3.1**
Relative ionization energies (red) and electron affinities (blue) for elements in the first four rows of the periodic table.
Those elements having a value of zero for electron affinity do not accept an electron. Note that the alkali metals (Li, Na, K) have the lowest ionization energies and lose an electron most easily, whereas the halogens (F, Cl, Br) have the highest electron affinities and gain an electron most easily. The noble gases (He, Ne, Ar, Kr) neither gain nor lose an electron easily.

76 CHAPTER 3 Ionic Compounds

Because alkali metals such as sodium tend to lose an electron, and halogens such as chlorine tend to gain an electron, these two elements (sodium and chlorine) will react with each other by transfer of an electron from the metal to the halogen (Figure 3.2). The product that results—sodium chloride (NaCl)—is electrically neutral because the positive charge of each Na$^+$ ion is balanced by the negative charge of each Cl$^-$ ion.

▶ **Figure 3.2**
(a) Chlorine is a toxic green gas, sodium is a reactive metal, and sodium chloride is a harmless white solid. (b) Sodium metal burns with an intense yellow flame when immersed in chlorine gas, yielding white sodium chloride "smoke."

(a)   (b)

Worked Example 3.1 Periodic Trends: Ionization Energy

Look at the periodic trends in Figure 3.1, and predict where the ionization energy of rubidium is likely to fall on the chart.

ANALYSIS Identify the group number of rubidium (group 1A), and find where other members of the group appear in Figure 3.1.

SOLUTION
Rubidium (Rb) is the alkali metal below potassium (K) in the periodic table. Since the alkali metals Li, Na, and K all have ionization energies near the bottom of the chart, the ionization energy of rubidium is probably similar.

Worked Example 3.2 Periodic Trends: Formation of Anions and Cations

Which element is likely to lose an electron more easily, Mg or S?

ANALYSIS Identify the group numbers of the elements, and find where members of those groups appear in Figure 3.1.

SOLUTION
Magnesium, a group 2A element on the left side of the periodic table, has a relatively low ionization energy, and loses an electron easily. Sulfur, a group 6A element on the right side of the table, has a higher ionization energy, and loses an electron less easily.

PROBLEM 3.4
Look at the periodic trends in Figure 3.1, and predict approximately where the ionization energy of xenon is likely to fall.

PROBLEM 3.5
Which element in the following pairs is likely to lose an electron more easily?
(a) Be or B   **(b)** Ca or Co   **(c)** Sc or Se

PROBLEM 3.6
Which element in the following pairs is likely to gain an electron more easily?
(a) H or He   **(b)** S or Si   **(c)** Cr or Mn

3.3 Ionic Bonds

When sodium reacts with chlorine, the product is sodium chloride, a compound completely unlike either of the elements from which it is formed. Sodium is a soft, silvery metal that reacts violently with water, and chlorine is a corrosive, poisonous, green gas (Figure 3.2a). When chemically combined, however, they produce our familiar table salt containing Na$^+$ ions and Cl$^-$ ions. Because opposite electrical charges attract each other, the positive Na$^+$ ion and negative Cl$^-$ ion are said to be held together by an **ionic bond**.

When a vast number of sodium atoms transfer electrons to an equally vast number of chlorine atoms, a visible crystal of sodium chloride results. In this crystal, equal numbers of Na$^+$ and Cl$^-$ ions are packed together in a regular arrangement. Each positively charged Na$^+$ ion is surrounded by six negatively charged Cl$^-$ ions, and each Cl$^-$ ion is surrounded by six Na$^+$ ions (Figure 3.3). This packing arrangement allows each ion to be stabilized by the attraction of unlike charges on its six nearest-neighbor ions, while being as far as possible from ions of like charge.

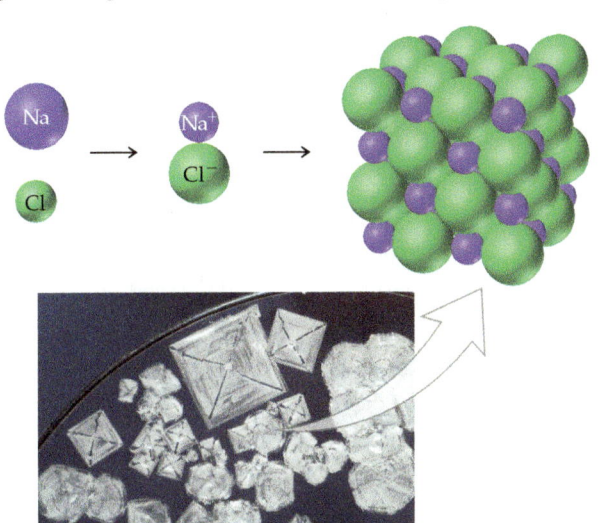

Ionic bond The electrical attractions between ions of opposite charge in a crystal.

◄ **Figure 3.3**
The arrangement of Na$^+$ and Cl$^-$ ions in a sodium chloride crystal. Each positively charged Na$^+$ ion is surrounded by six negatively charged Cl$^-$ ions, and each Cl$^-$ ion is surrounded by six Na$^+$ ions. The crystal is held together by ionic bonds—the attraction between oppositely charged ions.

Because of the three-dimensional arrangement of ions in a sodium chloride crystal, we cannot speak of specific ionic bonds between specific pairs of ions. Rather, there are many ions attracted by ionic bonds to their nearest neighbors. We therefore speak of the whole NaCl crystal as being an **ionic solid** and of such compounds as being **ionic compounds**. The same is true of all compounds composed of ions.

Ionic solid A crystalline solid held together by ionic bonds.

Ionic compound A compound that contains ionic bonds.

3.4 Some Properties of Ionic Compounds

Like sodium chloride, ionic compounds are usually crystalline solids. Different ions vary in size and charge, therefore, they are packed together in crystals in different ways. The ions in each compound settle into a pattern that efficiently fills space and maximizes ionic bonding.

Because the ions in an ionic solid are held rigidly in place by attraction to their neighbors, they cannot move about. Once an ionic solid is dissolved in water, however, the ions can move freely, thereby accounting for the electrical conductivity of these compounds in solution.

The high melting points and boiling points observed for ionic compounds are also accounted for by ionic bonding. The attractive force between oppositely charged particles is extremely strong, and the ions need to gain a large amount of energy by being heated to high temperatures for them to loosen their grip on one another. Sodium chloride, for example, melts at 801 °C and boils at 1413 °C; potassium iodide melts at 681 °C and boils at 1330 °C.

▲ The melting point of sodium chloride is 801 °C.

Despite the strength of ionic bonds, ionic solids shatter if struck sharply. A blow disrupts the orderly arrangement of cations and anions, forcing particles of like electrical charge closer together. The close proximity of like charges creates repulsive energies that split the crystal apart.

Ionic compounds dissolve in water if the attraction between water and the ions overcomes the attraction of the ions for one another. Compounds like sodium chloride are very soluble in water and can be dissolved to make solutions of high concentration. Do not be misled, however, by the ease with which sodium chloride and other familiar ionic compounds dissolve in water. Many other ionic compounds, such as magnesium hydroxide or barium sulfate, are not water-soluble, because the attractive forces between these ions and water is not sufficient to overcome the ionic attractions in the crystals.

PROBLEM 3.7
Consider the ionic liquids described in the Chemistry in Action below. How are the properties of these ionic liquids different from other common ionic substances?

CHEMISTRY IN ACTION

Ionic Liquids

Imagine a substance that could help solve the problems of nuclear waste, make solar energy more efficient, revolutionize the development of biomass-based renewable energies, serve as a solvent for enzyme-based biochemical transformations, and act as a major component in a spinning-liquid mirror telescope stationed on the moon. Ionic liquids can do all that—and more! When discussing ionic substances, most of us think of hard, crystalline materials like common table salt (see Chemistry in Action, p. 83), with high melting points. But ionic liquids have very different properties, including low melting points, high viscosity, low-to-moderate electrical conductivity, and low volatility, which make them suitable for the widely varied uses described previously.

Although the details of the discovery of ionic liquids are in dispute, one of the first *room temperature ionic liquids* (or RTILs), ethylammonium nitrate, was synthesized in 1914 by Paul Walden. Most RTILs developed since then consist of a bulky, asymmetric organic cation (see Organic Chemistry in Chapters 12–19), combined with a variety of anions. The bulky cations cannot pack together in an ordered fashion, and so these substances do not condense into a solid at ambient temperatures. Rather, they tend to form highly viscous liquids that exhibit low volatility, ideal properties for a large-diameter spinning liquid mirror in a low-pressure environment like the moon. The viscous liquid can be covered with a thin metallic film that will form a parabolic reflective surface to collect long-wavelength infrared light. And the cost of the spinning-liquid mirror is about 1% of a conventional lens, which must be ground and polished.

The bulky cations also provide unique solvent properties, enabling them to dissolve substances that are not very soluble in more conventional solvents. Their low volatility also makes them attractive as "green," or environmentally friendly, solvents. Consider the practice of using biomass as a fuel source. One common approach is to convert sugar or starch (from corn, beets, or cane sugar) into ethanol by the process of fermentation. But the major component of these and most other plants is cellulose. Cellulose is a polymer (see Chemistry in Action on pp. 118 and 538) composed of many sugars joined together in a long chain. Cellulose is chemically similar to starch but is neither highly soluble in most solvents nor subject to fermentation. RTILs, however, such as that illustrated in the figure below, can be used to dissolve cellulose at moderate temperatures and facilitate its breakdown into simple fermentable sugars. At a volume of nearly 700 billion tons of the earth's biomass, cellulose represents an important renewable energy source. The ability to convert cellulose into fuel will certainly help meet our expanding energy needs.

See Chemistry in Action Problems 3.80 and 3.81 at the end of the chapter.

Benzylmethylimidazolium chloride

◀ Pine wood fibers dissolving in an ionic liquid solvent consisting of benzyl methyl imidazolium chloride, whose structural formula is shown.

3.5 Ions and the Octet Rule

We have seen that alkali metal atoms have a single valence-shell electron, ns^1. The electron-dot symbol X· is consistent with this valence electron configuration. Halogens, having seven valence electrons, ns^2np^5, can be represented using :Ẍ· as the electron-dot symbol. Noble gases can be represented as :Ẍ:, since they have eight valence electrons, ns^2np^6. Both the alkali metals and the halogens are extremely reactive, undergoing many chemical reactions and forming many compounds. The noble gases, however, are quite different. They are the least reactive of all elements.

Now look at sodium chloride and similar ionic compounds. When sodium or any other alkali metal reacts with chlorine or any other halogen, the metal transfers an electron from its valence shell to the valence shell of the halogen. Sodium thereby changes its valence-shell electron configuration from $2s^22p^63s^1$ in the atom to $2s^22p^6(3s^0)$ in the Na$^+$ ion, and chlorine changes from $3s^23p^5$ in the atom to $3s^23p^6$ in the Cl$^-$ ion. *As a result, both sodium and chlorine gain noble gas electron configurations, with 8 valence electrons.* The Na$^+$ ion has 8 electrons in the $n = 2$ shell, matching the electron configuration of neon. The Cl$^-$ ion has 8 electrons in the $n = 3$ shell, matching the electron configuration of argon.

$$\text{Na} + \text{Cl} \longrightarrow \text{Na}^+ + \text{Cl}^-$$
$$1s^2\,2s^2\,2p^6\,3s^1 \quad 1s^2\,2s^2\,2p^6\,3s^2\,3p^5 \quad \underbrace{1s^2\,2s^2\,2p^6 3s^0}_{\text{Neon configuration}} \quad \underbrace{1s^2\,2s^2\,2p^6\,3s^2\,3p^6}_{\text{Argon configuration}}$$

$$\text{Na·} + \text{·C̈l:} \longrightarrow \text{Na}^+ + \text{:C̈l:}^-$$

Evidently there is something special about having 8 valence electrons (filled s and p subshells) that leads to stability and lack of chemical reactivity. In fact, observations of many chemical compounds have shown that main group elements frequently combine in such a way that each winds up with 8 valence electrons, a so-called *electron octet*. This conclusion is summarized in a statement called the **octet rule**:

Octet rule Main group elements tend to undergo reactions that leave them with 8 valence electrons.

Put another way, main group *metals* tend to lose electrons when they react so that they attain an electron configuration like that of the noble gas just *before* them in the periodic table, and reactive main group *nonmetals* tend to gain electrons when they react so that they attain an electron configuration like that of the noble gas just *after* them in the periodic table. In both cases, the product ions have filled s and p subshells in their valence electron shell.

Worked Example 3.3 Electron Configurations: Octet Rule for Cations

Write the electron configuration of magnesium ($Z = 12$). Show how many electrons a magnesium atom must lose to form an ion with a filled shell (8 electrons), and write the configuration of the ion. Explain the reason for the ion's charge, and write the ion's symbol.

ANALYSIS Write the electron configuration of magnesium as described in Section 2.7 and count the number of electrons in the valence shell.

SOLUTION
Magnesium has the electron configuration $1s^22s^22p^63s^2$. Since the second shell contains an octet of electrons ($2s^22p^6$) and the third shell is only partially filled ($3s^2$), magnesium can achieve a valence-shell octet by losing the 2 electrons in the

3s subshell. The result is formation of a doubly charged cation, Mg^{2+}, with the neon configuration:

$$Mg^{2+} \quad 1s^2 2s^2 2p^6 \text{ (Neon configuration, or [Ne])}$$

A neutral magnesium atom has 12 protons and 12 electrons. With the loss of 2 electrons, there is an excess of 2 protons, accounting for the +2 charge of the ion, Mg^{2+}.

Worked Example 3.4 Electron Configurations: Octet Rule for Anions

How many electrons must a nitrogen atom, $Z = 7$, gain to attain a noble gas configuration? Write the electron-dot and ion symbols for the ion formed.

ANALYSIS Write the electron configuration of nitrogen, and identify how many more electrons are needed to reach a noble gas configuration.

SOLUTION
Nitrogen, a group 5A element, has the electron configuration $1s^2 2s^2 2p^3$. The second shell contains 5 electrons ($2s^2 2p^3$) and needs 3 more to reach an octet. The result is formation of a triply charged anion, N^{3-}, with 8 valence electrons, matching the neon configuration:

$$N^{3-} \quad 1s^2 2s^2 2p^6 \quad \text{(Neon configuration)} \quad :\ddot{\underset{..}{N}}:^{3-}$$

PROBLEM 3.8
Write the electron configuration of potassium, $Z = 19$, and show how a potassium atom can attain a noble gas configuration.

PROBLEM 3.9
How many electrons must an aluminum atom, $Z = 13$, lose to attain a noble gas configuration? Write the symbol for the ion formed.

KEY CONCEPT PROBLEM 3.10

Which atom in the reaction depicted here gains electrons, and which loses electrons? Draw the electron-dot symbols for the resulting ions.

$$X: \;+\; \cdot \ddot{Y} \cdot \;\longrightarrow\; ?$$

3.6 Ions of Some Common Elements

The periodic table is the key to understanding and remembering which elements form ions and which do not. As shown in Figure 3.4, atoms of elements in the same group tend to form ions of the same charge. The metals of groups 1A and 2A, for example, form only +1 and +2 ions, respectively. The ions of these elements

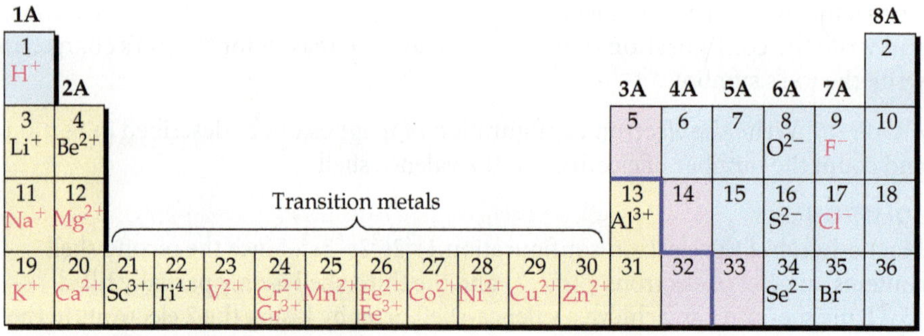

▶ **Figure 3.4**
Common ions formed by elements in the first four periods.
Ions important in biological chemistry are shown in red.

all have noble gas configurations as a result of electron loss from their valence *s* subshells. (Note in the following equations that the electrons being lost are shown as products.)

Group 1A: M· → M⁺ + e⁻
(M = Li, Na, K, Rb, or Cs)

Group 2A: M: → M²⁺ + 2e⁻
(M = Be, Mg, Ca, Sr, Ba, or Ra)

Four of these ions, Na^+, K^+, Mg^{2+}, and Ca^{2+}, are present in body fluids, where they play extremely important roles in biochemical processes.

The only group 3A element commonly encountered in ionic compounds is aluminum, which forms Al^{3+} by loss of three electrons from its valence *s* and *p* subshells. Aluminum is not thought to be an essential element in the human diet, although it is known to be present in some organisms.

The first three elements in groups 4A (C, Si, Ge) and 5A (N, P, As) do not ordinarily form cations or anions, because either too much energy is required to remove an electron or not enough energy is released by adding an electron to make the process energetically favorable. The bonding of these elements is largely covalent and will be described in the next chapter. Carbon, in particular, is the key element on which life is based. Together with hydrogen, nitrogen, phosphorus, and oxygen, carbon is present in all the essential biological compounds that we will be describing throughout the latter half of this book.

The group 6A elements, oxygen and sulfur, form large numbers of compounds, some of which are ionic and some of which are covalent. Their ions have noble gas configurations, achieved by gaining two electrons:

Group 6A: ·Ö· + 2 e⁻ ⟶ :Ö:²⁻

·S̈· + 2 e⁻ ⟶ :S̈:²⁻

The halogens are present in many compounds as ions formed by gaining one electron:

Group 7A: ·Ẍ: + e⁻ ⟶ :Ẍ:⁻
(X = F, Cl, Br, I)

Transition metals lose electrons to form cations, some of which are present in the human body. The charges of transition metal cations are not as predictable as those of main group elements, however, because many transition metal atoms can lose one or more *d* electrons in addition to losing valence *s* electrons. For example, iron (... $3s^2 3p^6 3d^6 4s^2$) forms Fe^{2+} by losing two electrons from the 4*s* subshell and also forms Fe^{3+} by losing an additional electron from the 3*d* subshell. Looking at the electron configuration for iron shows why the octet rule is limited to main group elements: transition metal cations generally do not have noble gas configurations because they would have to lose *all* their *d* electrons.

Important Points about Ion Formation and the Periodic Table:

- **Metals form cations by losing one or more electrons.**
  - Group 1A and 2A metals form +1 and +2 ions, respectively (for example, Li^+ and Mg^{2+}) to achieve a noble gas configuration.
  - Transition metals can form cations of more than one charge (for example, Fe^{2+} and Fe^{3+}) by losing a combination of valence-shell *s* electrons and inner-shell *d* electrons.
- **Reactive nonmetals form anions by gaining one or more electrons to achieve a noble gas configuration.**
  - Group 6A nonmetals oxygen and sulfur form the anions O^{2-} and S^{2-}.
  - Group 7A elements (the halogens) form −1 ions; for example, F^- and Cl^-.

- Group 8A elements (the noble gases) are unreactive.
- Ionic charges of main group elements can be predicted using the group number and the octet rule.
  - For 1A and 2A metals: cation charge = group number
  - For nonmetals in groups 5A, 6A, and 7A: anion charge = 8 − (group number)

Worked Example 3.5 Formation of Ions: Gain/Loss of Valence Electrons

Which of the following ions is likely to form?

(a) S^{3-} (b) Si^{2+} (c) Sr^{2+}

ANALYSIS Count the number of valence electrons in each ion. For main group elements, only ions with a valence octet of electrons are likely to form.

SOLUTION

(a) Sulfur is in group 6A, has 6 valence electrons, and needs only 2 more to reach an octet. Gaining 2 electrons gives an S^{2-} ion with a noble gas configuration, but gaining 3 electrons does not. The S^{3-} ion is, therefore, unlikely to form.

(b) Silicon is a nonmetal in group 4A. Like carbon, it does not form ions because it would have to gain or lose too many electrons (4) to reach a noble gas electron configuration. The Si^{2+} ion does not have an octet and will not form.

(c) Strontium, a metal in group 2A, has only 2 outer-shell electrons and can lose both to reach a noble gas configuration. The Sr^{2+} ion has an octet and, therefore, forms easily.

PROBLEM 3.11

Is molybdenum more likely to form a cation or an anion? Why?

PROBLEM 3.12

Write symbols, both with and without electron dots, for the ions formed by the following processes:

(a) Gain of 2 electrons by selenium   (b) Loss of 2 electrons by barium
(c) Gain of 1 electron by bromine

PROBLEM 3.13

By mass, seawater contains 3.5% NaCl, or table salt (see Chemistry in Action, p. 83). If one liter of seawater contains 35 g of NaCl, how many gallons of water must be evaporated to produce one pound of NaCl?

3.7 Naming Ions

Main group metal cations in groups 1A, 2A, and 3A are named by identifying the metal, followed by the word "ion," as in the following examples:

K^+ Mg^{2+} Al^{3+}
Potassium ion   Magnesium ion   Aluminum ion

It is sometimes a little confusing to use the same name for both a metal and its ion, and you may occasionally have to stop and think about what is meant. For example, it is common practice in nutrition and health-related fields to talk about sodium or potassium in the bloodstream. Because both sodium and potassium *metals* react violently with water, however, they cannot possibly be present in blood. The references are to dissolved sodium and potassium *ions*.

Transition metals, such as iron or chromium, and many metals found in the *p*-block, such as tin and lead, can form more than one type of cation. To avoid confusion, a method is needed to differentiate between ions of these metals. Two systems are used. The first is an old system that gives the ion with the smaller charge the word ending *-ous* and the ion with the larger charge the ending *-ic*.

CHEMISTRY IN ACTION

Salt

If you are like most people, you feel a little guilty about reaching for the salt shaker at mealtime. The notion that high salt intake and high blood pressure go hand in hand is surely among the most highly publicized pieces of nutritional lore ever to appear.

Salt has not always been held in such disrepute. Historically, salt has been prized since the earliest recorded times as a seasoning and a food preservative. Words and phrases in many languages reflect the importance of salt as a life-giving and life-sustaining substance. We refer to a kind and generous person as "the salt of the earth," for instance, and we speak of being "worth one's salt." In Roman times, soldiers were paid in salt; the English word "salary" is derived from the Latin word for paying salt wages (*salarium*).

Salt is perhaps the easiest of all minerals to obtain and purify. The simplest method, used for thousands of years throughout the world in coastal climates where sunshine is abundant and rainfall is scarce, is to evaporate seawater. Though the exact amount varies depending on the source, seawater contains an average of about 3.5% by mass of dissolved substances, most of which is sodium chloride. It has been estimated that evaporation of all the world's oceans would yield approximately *4.5 million cubic miles* of NaCl.

Only about 10% of current world salt production comes from evaporation of seawater. Most salt is obtained by mining the vast deposits of *halite*, or *rock salt*, formed by evaporation of ancient inland seas. These salt beds vary in thickness up to hundreds of meters and vary in depth from a few meters to thousands of meters below the earth's surface. Salt mining has gone on for at least 3400 years, and the Wieliczka mine in Galicia, Poland, has been worked continuously from A.D. 1000 to the present.

What about the link between dietary salt intake and high blood pressure? Although sodium is a macronutrient that we need—it plays a critical role in charge balance and ion transport in cell membranes—too much sodium has been linked to both hypertension and kidney ailments. The recommended daily intake (RDI) for sodium is 2300 mg, which translates to roughly 4 g of salt. However, the average adult in most industrialized countries consumes over twice this amount, with most of it coming from processed foods.

What should an individual do? The best answer, as in so many things, is to use moderation and common sense. People with hypertension should make a strong effort to lower their

▲ In many areas of the world, salt is still harvested by evaporation of ocean or tidal waters.

- 5% added while cooking
- 6% added while eating
- 12% from natural sources
- 77% from processed and prepared foods

sodium intake; others might be well advised to choose unsalted snacks, monitor their consumption of processed food, and read nutrition labels for sodium content.

See Chemistry in Action Problem 3.82 at the end of the chapter.

The second is a newer system in which the charge on the ion is given as a Roman numeral in parentheses right after the metal name. For example:

	Cr^{2+}	Cr^{3+}
Old name:	Chrom*ous* ion	Chrom*ic* ion
New name:	Chromium(II) ion	Chromium(III) ion

We will generally emphasize the new system in this book, but it is important to understand both systems because the old system is often found on labels of commercially supplied chemicals. The small differences between the names in either system illustrate the importance of reading a name very carefully before using a chemical. There are significant differences between compounds consisting of the same two elements but having different charges on the cation. In treating iron-deficiency anemia, for example, iron(II) compounds are preferable because the body absorbs them considerably better than iron(III) compounds.

The names of some common transition metal cations are listed in Table 3.1. Notice that the old names of the copper, iron, and tin ions are derived from their Latin names (*cuprum, ferrum,* and *stannum*).

TABLE 3.1 Names of Some Transition Metal Cations

Element	Symbol	Old Name	New Name
Chromium	Cr^{2+}	Chromous	Chromium(II)
	Cr^{3+}	Chromic	Chromium(III)
Copper	Cu^{+}	Cuprous	Copper(I)
	Cu^{2+}	Cupric	Copper(II)
Iron	Fe^{2+}	Ferrous	Iron(II)
	Fe^{3+}	Ferric	Iron(III)
Mercury	*Hg_2^{2+}	Mercurous	Mercury(I)
	Hg^{2+}	Mercuric	Mercury(II)
Tin	Sn^{2+}	Stannous	Tin(II)
	Sn^{4+}	Stannic	Tin(IV)

*This cation is composed of two mercury atoms, each of which has an average charge of +1.

Anions are named by replacing the ending of the element name with *-ide*, followed by the word "ion" (Table 3.2). For example, the anion formed by fluor*ine* is the fluor*ide* ion, and the anion formed by sul*fur* is the sul*fide* ion.

TABLE 3.2 Names of Some Common Anions

Element	Symbol	Name
Bromine	Br^{-}	Bromide ion
Chlorine	Cl^{-}	Chloride ion
Fluorine	F^{-}	Fluoride ion
Iodine	I^{-}	Iodide ion
Oxygen	O^{2-}	Oxide ion
Sulfur	S^{2-}	Sulfide ion

PROBLEM 3.14
Name the following ions:
(a) Cu^{2+} (b) F^{-} (c) Mg^{2+} (d) S^{2-}

PROBLEM 3.15
Write the symbols for the following ions:
(a) Silver(I) ion (b) Iron(II) ion (c) Cuprous ion (d) Telluride ion

PROBLEM 3.16
Ringer's solution, which is used intravenously to adjust ion concentrations in body fluids, contains the ions of sodium, potassium, calcium, and chlorine. Give the names and symbols of these ions.

3.8 Polyatomic Ions

Ions that are composed of more than one atom are called **polyatomic ions**. Most polyatomic ions contain oxygen and another element, and their chemical formulas include subscripts to show how many of each type of atom are present. Sulfate ion, for example, is composed of 1 sulfur atom and 4 oxygen atoms and has a -2 charge: SO_4^{2-}. The atoms in a polyatomic ion are held together by covalent bonds of the sort discussed in the next chapter, and the entire group of atoms acts as a single unit. A polyatomic ion is charged because it contains a total number of electrons different from the total number of protons in the combined atoms.

The most common polyatomic ions are listed in Table 3.3. Note that the ammonium ion, NH_4^+, and the hydronium ion, H_3O^+, are the only cations; all the others are anions. These ions are encountered so frequently in chemistry, biology, and medicine that there is no alternative but to memorize their names and formulas. Fortunately, there are only a few of them.

Polyatomic ion An ion that is composed of more than one atom.

TABLE 3.3 Some Common Polyatomic Ions

Name	Formula	Name	Formula
Hydronium ion	H_3O^+	Nitrate ion	NO_3^-
Ammonium ion	NH_4^+	Nitrite ion	NO_2^-
Acetate ion	$CH_3CO_2^-$	Oxalate ion	$C_2O_4^{2-}$
Carbonate ion	CO_3^{2-}	Permanganate ion	MnO_4^-
Hydrogen carbonate ion (bicarbonate ion)	HCO_3^-	Phosphate ion	PO_4^{3-}
Chromate ion	CrO_4^{2-}	Hydrogen phosphate ion (biphosphate ion)	HPO_4^{2-}
Dichromate ion	$Cr_2O_7^{2-}$	Dihydrogen phosphate ion	$H_2PO_4^-$
Cyanide ion	CN^-	Sulfate ion	SO_4^{2-}
Hydroxide ion	OH^-	Hydrogen sulfate ion (bisulfate ion)	HSO_4^-
Hypochlorite ion	OCl^-	Sulfite ion	SO_3^{2-}

Note in Table 3.3 that several pairs of ions—CO_3^{2-} and HCO_3^-, for example—are related by the presence or absence of a hydrogen ion, H^+. In such instances, the ion with the hydrogen is sometimes named using the prefix *bi-*. Thus, CO_3^{2-} is the carbonate ion, and HCO_3^- is the bicarbonate ion; similarly, SO_4^{2-} is the sulfate ion, and HSO_4^- is the bisulfate ion.

PROBLEM 3.17
Name the following ions:
(a) NO_3^-
(b) CN^-
(c) OH^-
(d) HPO_4^{2-}

PROBLEM 3.18
Which of the biologically important ions (see Chemistry in Action, p. 86) belong to Group 1A? To Group 2A? To the transition metals? To the halogens?

CHAPTER 3 Ionic Compounds

CHEMISTRY IN ACTION

Biologically Important Ions

The human body requires many different ions for proper functioning. Several of these ions, such as Ca^{2+}, Mg^{2+}, and HPO_4^{2-}, are used as structural materials in bones and teeth in addition to having other essential functions. Although 99% of Ca^{2+} is contained in bones and teeth, small amounts in body fluids play a vital role in transmission of nerve impulses. Other ions, including essential transition metal ions such as Fe^{2+}, are required for specific chemical reactions in the body. And still others, such as K^+, Na^+, and Cl^-, are present in fluids throughout the body.

In order to maintain charge neutrality in solution, the total negative charge (from anions) must balance the total positive charge (from cations). Several monatomic anions, and several polyatomic anions, especially HCO_3^- and HPO_4^{2-}, are present in body fluids where they help balance the cation charges. Some of the most important ions and their functions are shown in the accompanying table.

See Chemistry in Action Problems 3.83, 3.84, and 3.85 at the end of the chapter.

Some Biologically Important Ions

Ion	Location	Function	Dietary source
Ca^{2+}	Outside cell; 99% of Ca^{2+} is in bones and teeth as $Ca_3(PO_4)_2$ and $CaCO_3$	Bone and tooth structure; necessary for blood clotting, muscle contraction, and transmission of nerve impulses	Milk, whole grains, leafy vegetables
Fe^{2+}	Blood hemoglobin	Transports oxygen from lungs to cells	Liver, red meat, leafy green vegetables
K^+	Fluids inside cells	Maintain ion concentrations in cells; regulate insulin release and heartbeat	Milk, oranges, bananas, meat
Na^+	Fluids outside cells	Protect against fluid loss; necessary for muscle contraction and transmission of nerve impulses	Table salt, seafood
Mg^{2+}	Fluids inside cells; bone	Present in many enzymes; needed for energy generation and muscle contraction	Leafy green plants, seafood, nuts
Cl^-	Fluids outside cells; gastric juice	Maintain fluid balance in cells; help transfer CO_2 from blood to lungs	Table salt, seafood
HCO_3^-	Fluids outside cells	Control acid–base balance in blood	By-product of food metabolism
HPO_4^{2-}	Fluids inside cells; bones and teeth	Control acid–base balance in cells	Fish, poultry, milk

3.9 Formulas of Ionic Compounds

Since all chemical compounds are neutral, it is relatively easy to figure out the formulas of ionic compounds. Once the ions are identified, all we need to do is decide how many ions of each type give a total charge of zero. Thus, the chemical formula of an ionic compound tells the ratio of anions and cations.

If the ions have the same charge, only one of each ion is needed:

$$K^+ \text{ and } F^- \text{ form } KF$$
$$Ca^{2+} \text{ and } O^{2-} \text{ form } CaO$$

This makes sense when we look at how many electrons must be gained or lost by each atom in order to satisfy the octet rule:

$$K\cdot + \cdot\ddot{\underset{\cdot\cdot}{F}}\colon \longrightarrow K^+ + \colon\ddot{\underset{\cdot\cdot}{F}}\colon^-$$
$$\cdot Ca\cdot + \cdot\ddot{O}\cdot \longrightarrow Ca^{2+} + \colon\ddot{\underset{\cdot\cdot}{O}}\colon^{2-}$$

If the ions have different charges, however, unequal numbers of anions and cations must combine in order to have a net charge of zero. When potassium and oxygen combine, for example, it takes two K^+ ions to balance the -2 charge of the O^{2-} ion. Put another way, it takes two K atoms to provide the two electrons needed in order to complete the octet for the O atom:

$$2 \text{K}\cdot + \cdot\ddot{\text{O}}\cdot \longrightarrow 2 \text{K}^+ + :\ddot{\text{O}}:^{2-}$$

$$2 \text{K}^+ \text{ and } \text{O}^{2-} \text{ form } \text{K}_2\text{O}$$

The situation is reversed when a Ca^{2+} ion reacts with a Cl^- ion. One Ca atom can provide two electrons; each Cl atom requires only one electron to achieve a complete octet. Thus, there is one Ca^{2+} cation for every two Cl^- anions:

$$\cdot\text{Ca}\cdot + 2\cdot\ddot{\text{Cl}}: \longrightarrow \text{Ca}^{2+} + 2:\ddot{\text{Cl}}:^-$$

$$\text{Ca}^{2+} \text{ and } 2\text{Cl}^- \text{ form } \text{CaCl}_2$$

It sometimes helps when writing the formulas for an ionic compound to remember that, when the two ions have different charges, the number of one ion is equal to the charge on the other ion. In magnesium phosphate, for example, the charge on the magnesium ion is +2 and the charge on the polyatomic phosphate ion is −3. Thus, there must be 3 magnesium ions with a total charge of $3 \times (+2) = +6$, and 2 phosphate ions with a total charge of $2 \times (-3) = -6$ for overall neutrality:

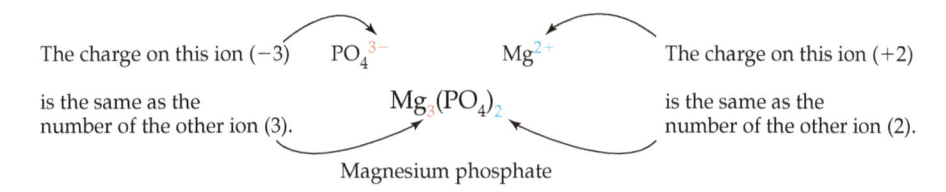

The charge on this ion (−3) PO_4^{3-} Mg^{2+} The charge on this ion (+2)
is the same as the $\text{Mg}_3(\text{PO}_4)_2$ is the same as the
number of the other ion (3). number of the other ion (2).

Magnesium phosphate

The formula of an ionic compound shows the lowest possible ratio of atoms in the compound and is thus known as a *simplest formula*. Because there is no such thing as a single neutral *particle* of an ionic compound, however, we use the term **formula unit** to identify the smallest possible neutral *unit* (Figure 3.5). For NaCl, the formula unit is 1 Na^+ ion and 1 Cl^- ion; for K_2SO_4, the formula unit is 2 K^+ ions and 1 SO_4^{2-} ion; for CaF_2, the formula unit is 1 Ca^{2+} ion and 2 F^- ions; and so on.

Formula unit The formula that identifies the smallest neutral unit of an ionic compound.

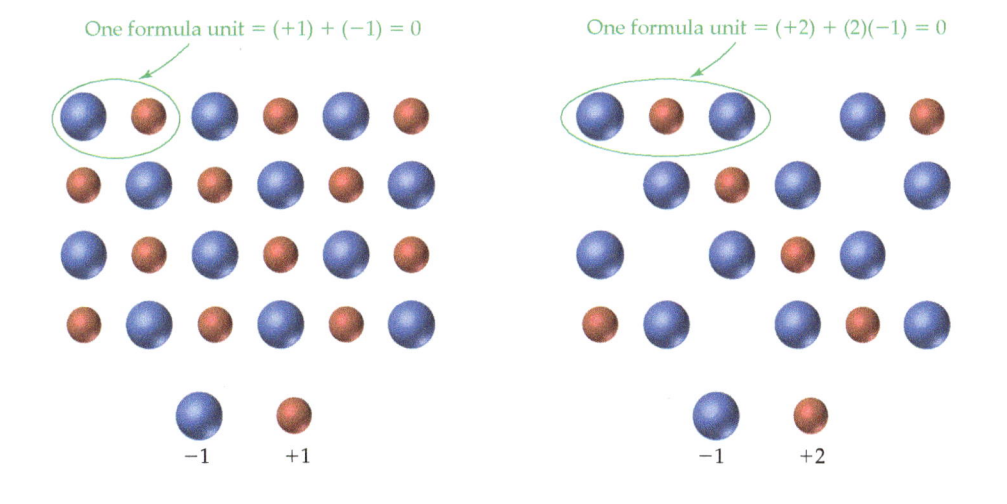

One formula unit = (+1) + (−1) = 0   One formula unit = (+2) + (2)(−1) = 0

−1   +1   −1   +2

Figure 3.5
Formula units of ionic compounds.
The sum of charges on the ions in a formula unit equals zero.

Once the numbers and kinds of ions in a compound are known, the formula is written using the following rules:

- List the cation first and the anion second; for example, NaCl rather than ClNa.
- Do not write the charges of the ions; for example, KF rather than K^+F^-.
- Use parentheses around a polyatomic ion formula if it has a subscript; for example, $\text{Al}_2(\text{SO}_4)_3$ rather than $\text{Al}_2\text{SO}_{43}$.

Worked Example 3.6 Ionic Compounds: Writing Formulas

Write the formula for the compound formed by calcium ions and nitrate ions.

ANALYSIS Knowing the formula and charges on the cation and anion (Figure 3.4), we determine how many of each are needed to yield a neutral formula for the ionic compound.

SOLUTION
The two ions are Ca^{2+} and NO_3^-. Two nitrate ions, each with a -1 charge, will balance the $+2$ charge of the calcium ion.

$$Ca^{2+} \quad \text{Charge} = 1 \times (+2) = +2$$
$$2NO_3^- \quad \text{Charge} = 2 \times (-1) = -2$$

Since there are 2 ions, the nitrate formula must be enclosed in parentheses:

$$Ca(NO_3)_2 \quad \text{Calcium nitrate}$$

PROBLEM 3.19
Write the formulas for the ionic compounds that silver(I) forms with each of the following:
(a) Iodide ion   (b) Oxide ion   (c) Phosphate ion

PROBLEM 3.20
Write the formulas for the ionic compounds that sulfate ion forms with the following:
(a) Sodium ion   (b) Iron(II) ion   (c) Chromium(III) ion

PROBLEM 3.21
The ionic compound containing ammonium ion and carbonate ion gives off the odor of ammonia, a property put to use in smelling salts for reviving someone who has fainted. Write the formula for this compound.

PROBLEM 3.22
An *astringent* is a compound that causes proteins in blood, sweat, and other body fluids to coagulate, a property put to use in antiperspirants. Two safe and effective astringents are the ionic compounds of aluminum with sulfate ion and with acetate ion. Write the formulas of both.

KEY CONCEPT PROBLEM 3.23

Three ionic compounds are represented on this periodic table—red cation with red anion, blue cation with blue anion, and green cation with green anion. Give a likely formula for each compound.

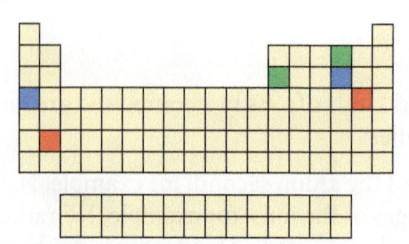

KEY CONCEPT PROBLEM 3.24

The ionic compound calcium nitride is represented here. What is the formula for calcium nitride, and what are the charges on the calcium and nitride ions?

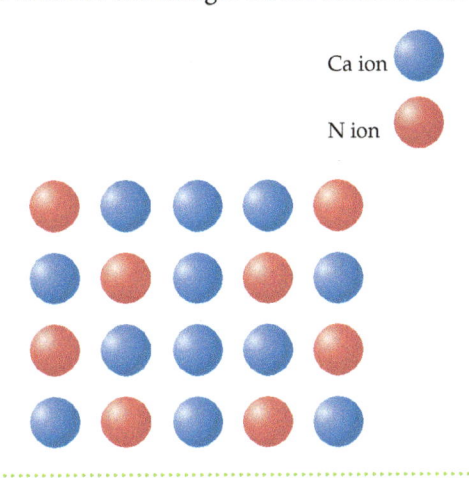

Ca ion ●
N ion ●

3.10 Naming Ionic Compounds

Just as in writing formulas for ionic compounds, these compounds are named by citing first the cation and then the anion, with a space between words. There are two kinds of ionic compounds, and the rules for naming them are slightly different.

Type I: Ionic compounds containing cations of main group elements (1A, 2A, aluminum). Since the charges on these cations do not vary, we do not need to specify the charge on the cation as discussed in Section 3.7. For example, NaCl is sodium chloride and $MgCO_3$ is magnesium carbonate.

Type II: Ionic compounds containing metals that can exhibit more than one charge. Since some metals, including the transition metals, often form more than one ion, we need to specify the charge on the cation in these compounds. Either the old (-ous, -ic) or the new (Roman numerals) system described in Section 3.7 can be used. Thus, $FeCl_2$ is called iron(II) chloride (or ferrous chloride), and $FeCl_3$ is called iron(III) chloride (or ferric chloride). Note that we do *not* name these compounds iron *di*chloride or iron *tri*chloride—once the charge on the metal is known, the number of anions needed to yield a neutral compound is also known and does not need to be included as part of the compound name. Table 3.4 lists some common ionic compounds and their uses.

TABLE 3.4 Some Common Ionic Compounds and Their Applications

Chemical Name (Common Name)	Formula	Applications
Ammonium carbonate	$(NH_4)_2CO_3$	Smelling salts
Calcium hydroxide (hydrated lime)	$Ca(OH)_2$	Mortar, plaster, whitewash
Calcium oxide (lime)	CaO	Lawn treatment, industrial chemical
Lithium carbonate ("lithium")	Li_2CO_3	Treatment of bipolar disorder
Magnesium hydroxide (milk of magnesia)	$Mg(OH)_2$	Antacid
Magnesium sulfate (Epsom salts)	$MgSO_4$	Laxative, anticonvulsant
Potassium permanganate	$KMnO_4$	Antiseptic, disinfectant*
Potassium nitrate (saltpeter)	KNO_3	Fireworks, matches, and desensitizer for teeth
Silver nitrate	$AgNO_3$	Antiseptic, germicide
Sodium bicarbonate (baking soda)	$NaHCO_3$	Baking powder, antacid, mouthwash, deodorizer
Sodium hypochlorite	NaOCl	Disinfectant; active ingredient in household bleach
Zinc oxide	ZnO	Skin protection, in calamine lotion

*Antiseptics and disinfectants can also be harmful/toxic to non-harmful microorganisms, but are used specifically to prevent infection from harmful microorganisms.

CHAPTER 3 Ionic Compounds

LOOKING AHEAD ▶▶▶ Because the formula unit for an ionic compound must be neutral, we can unambiguously write the formula from the name of the compound, and vice versa. As we shall see in Chapter 4, covalent bonding between atoms can produce a much greater variety of compounds. The rules for naming covalent compounds must be able to accommodate multiple combinations of elements (for example, CO and CO_2).

Worked Example 3.7 Ionic Compounds: Formulas Involving Polyatomic Ions

Magnesium carbonate is used as an ingredient in Bufferin (buffered aspirin) tablets. Write its formula.

ANALYSIS Since magnesium is a main group metal, we can determine its ionic compound formula by identifying the charges and formulas for the anion and the cation, remembering that the overall formula must be neutral.

SOLUTION
Look at the cation and the anion parts of the name separately. Magnesium, a group 2A element, forms the doubly positive Mg^{2+} cation; carbonate anion is doubly negative, CO_3^{2-} Because the charges on the anion and cation are equal, a formula of $MgCO_3$ will be neutral.

Worked Example 3.8 Ionic Compounds: Formulas and Ionic Charges

Sodium and calcium both form a wide variety of ionic compounds. Write formulas for the following compounds:

(a) Sodium bromide and calcium bromide
(b) Sodium sulfide and calcium sulfide
(c) Sodium phosphate and calcium phosphate

ANALYSIS Using the formulas and charges for the cations and the anions (from Tables 3.2 and 3.3), we determine how many of each cation and anion are needed to yield a formula that is neutral.

SOLUTION
(a) Cations = Na^+ and Ca^{2+}; anion = Br^-: NaBr and $CaBr_2$
(b) Cations = Na^+ and Ca^{2+}; anion = S^{2-}: Na_2S and CaS
(c) Cations = Na^+ and Ca^{2+}; anion = PO_4^{3-}: Na_3PO_4 and $Ca_3(PO_4)_2$

Worked Example 3.9 Naming Ionic Compounds

Name the following compounds, using Roman numerals to indicate the charges on the cations where necessary:

(a) KF   (b) $MgCl_2$ (c) $AuCl_3$ (d) Fe_2O_3

ANALYSIS For main group metals, the charge is determined from the group number, and no Roman numerals are necessary. For transition metals, the charge on the metal can be determined from the total charge(s) on the anion(s).

SOLUTION
(a) Potassium fluoride. No Roman numeral is necessary because a group 1A metal forms only one cation.
(b) Magnesium chloride. No Roman numeral is necessary because magnesium (group 2A) forms only Mg^{2+}.

SECTION 3.11 H⁺ and OH⁻ Ions: An Introduction to Acids and Bases

(c) Gold(III) chloride. The 3 Cl⁻ ions require a +3 charge on the gold for a neutral formula. Since gold is a transition metal that can form other ions, the Roman numeral is necessary to specify the +3 charge.

(d) Iron(III) oxide. Because the 3 oxide anions (O^{2-}) have a total negative charge of -6, the 2 iron cations must have a total charge of +6. Thus, each is Fe^{3+}, and the charge on each is indicated by the Roman numeral (III).

PROBLEM 3.25
The compound Ag_2S is responsible for much of the tarnish found on silverware. Name this compound, and give the charge on the silver ion.

PROBLEM 3.26
Name the following compounds:
(a) SnO_2 (b) $Ca(CN)_2$ (c) Na_2CO_3
(d) Cu_2SO_4 (e) $Ba(OH)_2$ (f) $Fe(NO_3)_2$

PROBLEM 3.27
Write formulas for the following compounds:
(a) Lithium phosphate   (b) Copper(II) carbonate
(c) Aluminum sulfite   (d) Cuprous fluoride
(e) Ferric sulfate   (f) Ammonium chloride

KEY CONCEPT PROBLEM 3.28
The ionic compound, formed between chromium and oxygen is shown here. Name the compound, and write its formula.

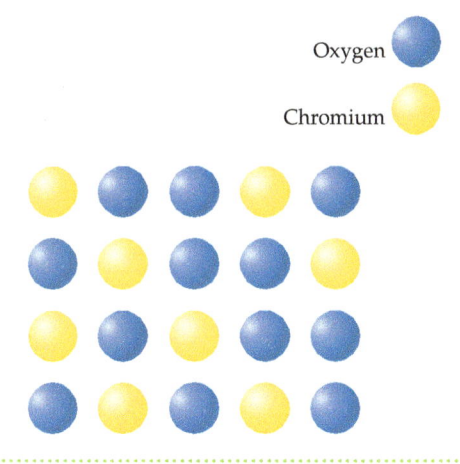

3.11 H⁺ and OH⁻ Ions: An Introduction to Acids and Bases

Two of the most important ions we will be discussing in the remainder of this book are the hydrogen cation (H^+) and the hydroxide anion (OH^-). Since a hydrogen *atom* contains one proton and one electron, a hydrogen *cation* is simply a proton. When an acid dissolves in water, the proton typically attaches to a molecule of water to form the hydronium ion (H_3O^+), but chemists routinely use the H^+ and H_3O^+ ions interchangeably. A hydroxide anion, by contrast, is a polyatomic ion in which an oxygen atom is covalently bonded to a hydrogen atom. Although much of Chapter 10 is devoted to the chemistry of H^+ and OH^- ions, it is worth taking a preliminary look now.

◀◀ In Chapter 10 we will look at the chemical behavior of acids and bases and their importance in many areas of chemistry.

Acid A substance that provides H⁺ ions in water;

Base A substance that provides OH⁻ ions in water

The importance of the H⁺ cation and the OH⁻ anion is that they are fundamental to the concepts of *acids* and *bases*. In fact, one definition of an **acid** is a substance that provides H⁺ ions when dissolved in water; for example, HCl, HNO₃, H₂SO₄, H₃PO₄. One definition of a **base** is a substance that provides OH⁻ ions when dissolved in water; for example, NaOH, KOH, Ba(OH)₂.

Hydrochloric acid (HCl), nitric acid (HNO₃) sulfuric acid (H₂SO₄), and phosphoric acid (H₃PO₄) are among the most common acids. When any of these substances is dissolved in water, H⁺ ions are formed along with the corresponding anion (Table 3.5).

TABLE 3.5 Some Common Acids and the Anions Derived from Them

Acids		Anions	
Acetic acid	CH₃COOH	Acetate ion	*CH₃COO⁻
Carbonic acid	H₂CO₃	Hydrogen carbonate ion (bicarbonate ion) Carbonate ion	CO₃²⁻
Hydrochloric acid	HCl	Chloride ion	Cl⁻
Nitric acid	HNO₃	Nitrate ion	NO₃⁻
Nitrous acid	HNO₂	Nitrite ion	NO₂⁻
Phosphoric acid	H₃PO₄	Dihydrogen phosphate ion Hydrogen phosphate ion Phosphate ion	H₂PO₄⁻ HPO₄²⁻ PO₄³⁻
Sulfuric acid	H₂SO₄	Hydrogen sulfate ion Sulfate ion	HSO₄⁻ SO₄²⁻

*Sometimes written C₂H₃O₂⁻ or as CH₃CO₂⁻.

Different acids can provide different numbers of H⁺ ions per acid molecule. Hydrochloric acid, for instance, provides one H⁺ ion per acid molecule; sulfuric acid can provide two H⁺ ions per acid molecule; and phosphoric acid can provide three H⁺ ions per acid molecule.

▶▶ The behavior of polyprotic acids, or acids that provide more than one H⁺ ion per acid molecule, will be discussed in more detail in Chapter 10.

Sodium hydroxide (NaOH; also known as *lye* or *caustic soda*), potassium hydroxide (KOH; also known as *caustic potash*), and barium hydroxide [Ba(OH)₂] are examples of bases. When any of these compounds dissolves in water, OH⁻ anions go into solution along with the corresponding metal cation. Sodium hydroxide and potassium hydroxide provide one OH⁻ ion per formula unit; barium hydroxide provides two OH⁻ ions per formula unit, as indicated by its formula, Ba(OH)₂.

PROBLEM 3.29
Which of the following compounds are acids, and which are bases? Explain.
(a) HF   (b) Ca(OH)₂   (c) LiOH   (d) HCN

KEY CONCEPT PROBLEM 3.30
One of these pictures represents a solution of HCl, and one represents a solution of H₂SO₄. Which is which?

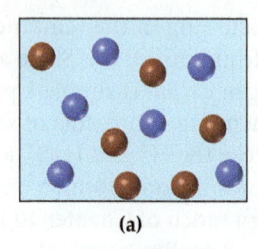

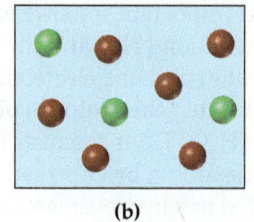

CHEMISTRY IN ACTION

Osteoporosis

Bone consists primarily of two components, one mineral and one organic. About 70% of bone is the ionic compound *hydroxyapatite,* $Ca_{10}(PO_4)_6(OH)_2$, called the *trabecular,* or spongy, bone. This mineral component is intermingled in a complex matrix with about 30% by mass of fibers of the protein *collagen,* called the *cortical,* or compact, bone. Hydroxyapatite gives bone its hardness and strength, whereas collagen fibers add flexibility and resistance to breaking.

Total bone mass in the body increases from birth until reaching a maximum in the mid-30s. By the early 40s, however, an age-related decline in bone mass begins to occur in both sexes. Bone density decreases, and the microarchitecture of bones is disrupted, resulting in weakening of bone structure, particularly in the wrists, hips, and spine. Should this thinning of bones become too great and the bones become too porous and brittle, a clinical condition called *osteoporosis* can result. Osteoporosis is, in fact, the most common of all bone diseases, affecting approximately 25 million people in the United States. Approximately 1.5 million bone fractures each year are caused by osteoporosis, at an estimated healthcare cost of $14 billion.

Although both sexes are affected by osteoporosis, the condition is particularly common in postmenopausal women, who undergo bone loss at a rate of 2–3% per year over and above that of the normal age-related loss. The cumulative lifetime bone loss, in fact, may approach 40–50% in women versus 20–30% in men. It has been estimated that half of all women over age 50 will have an osteoporosis-related bone fracture at some point in their life. Other risk factors, in addition to sex,

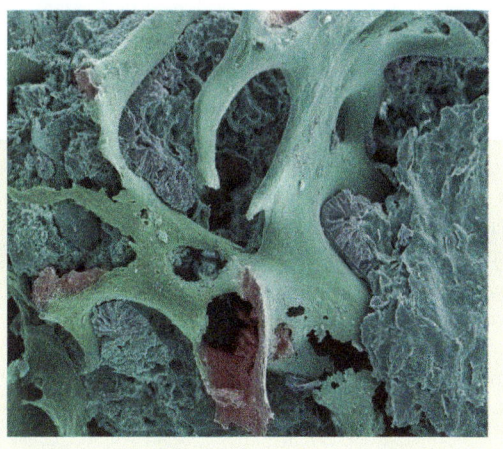

▲ Normal bone is strong and dense; a bone affected by osteoporosis, shown here, is weak and spongy in appearance.

include being thin, being sedentary, having a family history of osteoporosis, smoking, and having a diet low in calcium.

No cure exists for osteoporosis, but treatment for its prevention and management includes estrogen-replacement therapy for postmenopausal women as well as several approved medications called *bisphosphonates* that bind to the calcium in bone, slowing down bone loss by inhibiting the action of *osteoclasts,* or cells that break down bone tissue. Calcium supplements are also recommended, as is appropriate weight-bearing exercise. In addition, treatment with sodium fluoride is under active investigation and shows considerable promise. Fluoride ion reacts with hydroxyapatite to give *fluorapatite,* in which OH^- ions are replaced by F^-, increasing both bone strength and density.

$$Ca_{10}(PO_4)_6(OH)_2 + 2\,F^- \longrightarrow Ca_{10}(PO_4)_6F_2$$
Hydroxyapatite → Fluorapatite

See Chemistry in Action Problems 3.86 and 3.87 at the end of the chapter.

SUMMARY: REVISITING THE CHAPTER GOALS

1. What is an ion, what is an ionic bond, and what are the general characteristics of ionic compounds? Atoms are converted into *cations* by the loss of one or more electrons and into *anions* by the gain of one or more electrons. Ionic compounds are composed of cations and anions held together by *ionic bonds,* which result from the attraction between opposite electrical charges. Ionic compounds conduct electricity when dissolved in water, and they are generally crystalline solids with high melting points and high boiling points (see Problems 33, 35, 38–41, 80, 81, 95–97).

2. What is the octet rule, and how does it apply to ions? A valence-shell electron configuration of 8 electrons in filled *s* and *p* subshells leads to stability and lack of reactivity, as typified by the noble gases in group 8A. According to the *octet rule,* atoms of main group elements tend to form ions in which they have gained or lost the appropriate number of electrons to reach a noble gas configuration (see Problems 42–49, 88, 89).

3. What is the relationship between an element's position in the periodic table and the formation of its ion? Periodic variations in *ionization energy,* the amount of energy that must be supplied to remove an electron from an atom, show that metals lose electrons more easily than nonmetals. As a result, metals usually form cations. Similar periodic variations in *electron affinity,* the amount of energy released on adding an electron to an atom, show that reactive nonmetals gain electrons more easily than metals. As a result, reactive nonmetals usually form anions. The ionic charge can be predicted from the group number and the octet rule. For main group metals, the charge on the cation

94 CHAPTER 3 Ionic Compounds

is equal to the group number. For nonmetals, the charge on the anion is equal to 8 − (group number) (see *Problems 31, 32, 36, 40, 41, 50–57, 88, 89, 96*).

4. What determines the chemical formula of an ionic compound? Ionic compounds contain appropriate numbers of anions and cations to maintain overall neutrality, thereby providing a means of determining their chemical formulas (see *Problems 36, 37, 64, 65, 68, 69, 86, 87, 90, 92, 94*).

5. How are ionic compounds named? Cations have the same name as the metal from which they are derived. Monatomic anions have the name ending *-ide*. For metals that form more than one ion, a Roman numeral equal to the charge on the ion is added to the name of the cation. Alternatively, the ending *-ous* is added to the name of the cation with the lesser charge and the ending *-ic* is added to the name of the cation with the greater charge. To name an ionic compound, the cation name is given first, with the charge of the metal ion indicated if necessary, and the anion name is given second (see *Problems 36, 58-63, 66, 68, 70–75, 93, 94*).

6. What are acids and bases? The hydrogen ion (H^+) and the hydroxide ion (OH^-) are among the most important ions in chemistry because they are fundamental to the idea of acids and bases. According to one common definition, an *acid* is a substance that yields H^+ ions when dissolved in water, and a *base* is a substance that yields OH^- ions when dissolved in water (see *Problems 76–79, 91*).

KEY WORDS

Acid, *p. 92*
Anion, *p. 74*
Base, *p. 92*
Cation, *p. 74*
Electron affinity, *p. 75*

Formula unit, *p. 87*
Ion, *p. 74*
Ionic bond, *p. 77*
Ionic compound, *p. 77*
Ionic solid, *p. 77*

Ionization energy, *p. 75*
Octet rule, *p. 79*
Polyatomic ion, *p. 85*

UNDERSTANDING KEY CONCEPTS

3.31 Where on the blank outline of the periodic table are the following elements found?

(a) Elements that commonly form only one type of cation
(b) Elements that commonly form anions
(c) Elements that can form more than one type of cation
(d) Elements that do not readily form either anions or cations

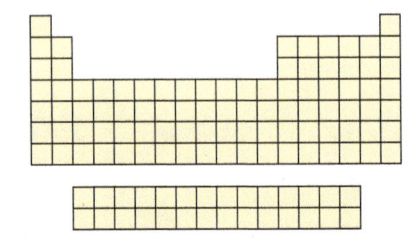

3.32 Where on the blank outline of the periodic table are the following elements found?

(a) Elements that commonly form +2 ions
(b) Elements that commonly form −2 ions
(c) An element that forms a +3 ion

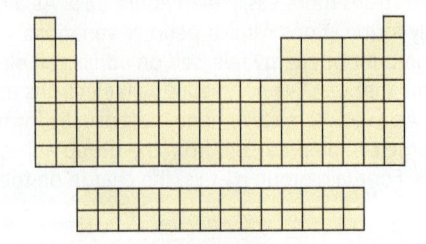

3.33 Write the symbols for the ions represented in the following drawings.

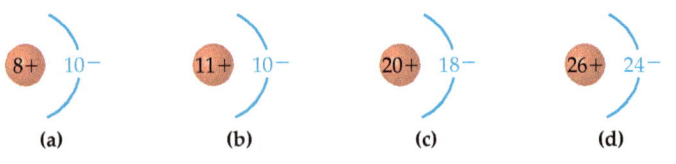

8+ 10−   11+ 10−   20+ 18−   26+ 24−
 (a)   (b)   (c)   (d)

3.34 One of these drawings represents an Na atom, and one represents an Na^+ ion. Tell which is which, and explain why there is a difference in size.

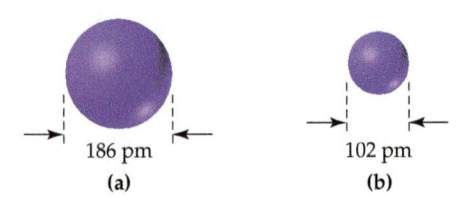

186 pm   102 pm
  (a)   (b)

3.35 One of these drawings represents a Cl atom, and one represents a Cl^- ion. Tell which is which, and explain why there is a difference in size.

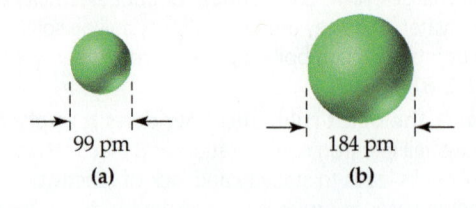

99 pm   184 pm
 (a)   (b)

3.36 The elements in red in the periodic table can form cations having more than one charge. Write the formulas and names of the compounds that are formed between the red cations and the blue anions depicted in the periodic table.

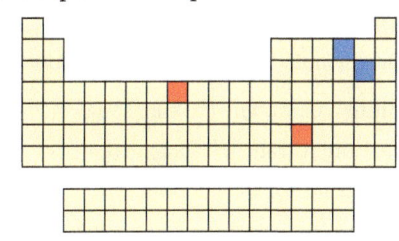

3.37 Each of these drawings (a)–(d) represents one of the following ionic compounds: PbBr$_2$, ZnS, CrF$_3$, Al$_2$O$_3$. Which is which?

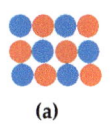

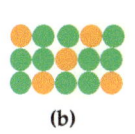

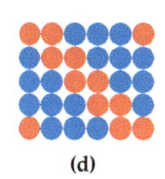

(a) (b) (c)

(d)

ADDITIONAL PROBLEMS

IONS AND IONIC BONDING

3.38 Write equations for loss or gain of electrons by atoms that result in formation of the following ions:

(a) Ca^{2+} (b) Au$^+$
(c) F$^-$ (d) Cr^{3+}

3.39 Write electronic configurations and symbols for the ions formed by the following:

(a) Gain of 3 electrons by phosphorus
(b) Loss of 1 electron by lithium
(c) Loss of 2 electrons by cobalt
(d) Loss of 3 electrons by thallium

3.40 Tell whether each statement about ions is true or false. If a statement is false, explain why.

(a) A cation is formed by addition of one or more electrons to an atom.
(b) Group 4A elements tend to lose 4 electrons to yield ions with a +4 charge.
(c) Group 4A elements tend to gain 4 electrons to yield ions with a −4 charge.
(d) The individual atoms in a polyatomic ion are held together by covalent bonds.

3.41 Tell whether each statement about ionic solids is true or false. If a statement is false, explain why.

(a) Ions are randomly arranged in ionic solids.
(b) All ions are the same size in ionic solids.
(c) Ionic solids can often be shattered by a sharp blow.
(d) Ionic solids have low boiling points.

IONS AND THE OCTET RULE

3.42 What is the *octet rule*?

3.43 Why do H and He not obey the octet rule?

3.44 Write the symbol for an ion that contains 34 protons and 36 electrons.

3.45 What is the charge of an ion that contains 21 protons and 19 electrons?

3.46 Identify the element X in the following ions, and tell which noble gas has the same electron configuration.

(a) X^{2+}, a cation with 36 electrons
(b) X$^-$, an anion with 36 electrons

3.47 Element Z forms an ion Z^{3+}, which contains 31 protons. What is the identity of Z, and how many electrons does Z^{3+} have?

3.48 Write the electron configuration for the following ions:

(a) Rb$^+$ (b) Br$^-$
(c) S^{2-} (d) Ba^{2+}
(e) Al^{3+}

3.49 Based on the following atomic numbers and electronic configurations, write the symbols for the following ions:

(a) Z = 20; 1s^2 2s^2 2p^6 3s^2 3p^6
(b) Z = 8; 1s^2 2s^2 2p^6
(c) Z = 22; 1s^2 2s^2 2p^6 3s^2 3p^6 3d^2
(d) Z = 19; 1s^2 2s^2 2p^6 3s^2 3p^6
(e) Z = 13; 1s^2 2s^2 2p^6

PERIODIC PROPERTIES AND ION FORMATION

3.50 Looking only at the periodic table, tell which member of each pair of atoms has the larger ionization energy and thus loses an electron less easily:

(a) Li and O (b) Li and Cs
(c) K and Zn (d) Mg and N

3.51 Looking only at the periodic table, tell which member of each pair of atoms has the larger electron affinity and thus gains an electron more easily:

(a) Li and S
(b) Ba and I
(c) Ca and Br

CHAPTER 3 Ionic Compounds

3.52 Which of the following ions are likely to form? Explain.
(a) Li^{2+} (b) K^-
(c) Mn^{3+} (d) Zn^{4+}
(e) Ne^+

3.53 What is the charge on the cation formed from the following elements? For those elements that form more than one cation, indicate the ionic charges most commonly observed.
(a) Magnesium (b) Tin
(c) Mercury (d) Aluminum

3.54 Write the electron configurations of Cr^{2+} and Cr^{3+}.

3.55 Write the electron configurations of Co, Co^{2+} and Co^{3+}.

3.56 Would you expect the ionization energy of Li^+ to be less than, greater than, or the same as the ionization energy of Li? Explain.

3.57 (a) Write equations for the loss of an electron by a K atom and the gain of an electron by a K^+ ion.
(b) What is the relationship between the equations?
(c) What is the relationship between the ionization energy of a K atom and the electron affinity of a K^+ ion?

SYMBOLS, FORMULAS, AND NAMES FOR IONS

3.58 Name the following ions:
(a) S^{2-} (b) Sn^{2+} (c) Sr^{2+}
(d) Mg^{2+} (e) Au^+

3.59 Name the following ions in both the old and the new systems:
(a) Cr^{2+} (b) Fe^{3+} (c) Hg^{2+}

3.60 Write symbols for the following ions:
(a) Selenide ion (b) Oxide ion
(c) Silver(I) ion

3.61 Write symbols for the following ions:
(a) Ferrous ion (b) Tin(IV) ion
(c) Lead(II) ion (d) Chromic ion

3.62 Write formulas for the following ions:
(a) Hydroxide ion (b) Bisulfate ion
(c) Acetate ion (d) Permanganate ion
(e) Hypochlorite ion (f) Nitrate ion
(g) Carbonate ion (h) Dichromate ion

3.63 Name the following ions:
(a) NO_2^- (b) CrO_4^{2-} (c) NH_4^+ (d) HPO_4^{2-}

NAMES AND FORMULAS FOR IONIC COMPOUNDS

3.64 Write formulas for the compounds formed by the sulfate ion with the following cations:
(a) Aluminum (b) Silver(I)
(c) Zinc (d) Barium

3.65 Write formulas for the compounds formed by the carbonate ion with the following cations:
(a) Strontium (b) Fe(III)
(c) Ammonium (d) Sn(IV)

3.66 Write the formula for the following substances:
(a) Sodium bicarbonate (baking soda)
(b) Potassium nitrate (a backache remedy)
(c) Calcium carbonate (an antacid)
(d) Ammonium nitrate (first aid cold packs)

3.67 Write the formula for the following compounds:
(a) Calcium hypochlorite, used as a swimming pool disinfectant
(b) Copper(II) sulfate, used to kill algae in swimming pools
(c) Sodium phosphate, used in detergents to enhance cleaning action

3.68 Complete the table by writing in the formula of the compound formed by each pair of ions:

	S^{2-}	Cl^-	PO_4^{3-}	CO_3^{2-}
Copper(II)	CuS			
Ca^{2+}				
NH_4^+				
Ferric ion				

3.69 Complete the table by writing in the formula of the compound formed by each pair of ions:

	O^{2-}	HSO_4^-	HPO_4^{2-}	$C_2O_4^{2-}$
K^+	K_2O			
Ni^{2+}				
NH_4^+				
Chromous				

3.70 Write the name of each compound in the table for Problem 3.68.

3.71 Write the name of each compound in the table for Problem 3.69.

3.72 Name the following substances:
(a) $MgCO_3$ (b) $Ca(CH_3CO_2)_2$
(c) AgCN (d) $Na_2Cr_2O_7$

3.73 Name the following substances:
(a) $Fe(OH)_2$ (b) $KMnO_4$
(c) Na_2CrO_4 (d) $Ba_3(PO_4)_2$

3.74 Which of the following formulas is most likely to be correct for calcium phosphate?
(a) Ca_2PO_4 (b) $CaPO_4$
(c) $Ca_2(PO_4)_3$ (d) $Ca_2(PO_4)_2$

3.75 Fill in the missing information to give the correct formula for each compound:

(a) Al$_?$(SO$_4$)$_?$
(b) (NH$_4$)$_?$(PO$_4$)$_?$
(c) Rb$_?$(SO$_4$)$_?$

ACIDS AND BASES

3.76 What is the difference between an acid and a base?

3.77 Identify the following substances as either an acid or a base:

(a) H$_2$CO$_3$
(b) HCN
(c) Mg(OH)$_2$
(d) KOH

3.78 Write equations to show how the substances listed in Problem 3.77 give ions when dissolved in water.

3.79 Name the anions that result when the acids in Problem 3.77 are dissolved in water.

CHEMISTRY IN ACTION

3.80 Most ionic substances are solids at room temperature. Explain why the RTILs discussed in this application are liquids rather than solids. [*Ionic Liquids, p. 78*]

3.81 Ionic liquids are being evaluated for use in a moon-based spinning-liquid telescope. Which properties of ionic liquids make them particularly well-suited for this application? [*Ionic Liquids, p. 78*]

3.82 What is the RDI for sodium for adults, and what amount of table salt (in grams) contains this quantity of sodium? [*Salt, p. 83*]

3.83 Where are most of the calcium ions found in the body? [*Biologically Important Ions, p. 86*]

3.84 Excess sodium ion is considered hazardous, but a certain amount is necessary for normal body functions. What is the purpose of sodium in the body? [*Biologically Important Ions, p. 86*]

3.85 Before a person is allowed to donate blood, a drop of the blood is tested to be sure that it contains a sufficient amount of iron (men, 41 µg/dL; women, 38 µg/dL). What is the biological role of iron, and which ion of iron is involved? [*Biologically Important Ions, p. 86*]

3.86 Name each ion in hydroxyapatite, Ca$_{10}$(PO$_4$)$_6$(OH)$_2$; give its charge; and show that the formula represents a neutral compound. [*Osteoporosis, p. 93*]

3.87 Sodium fluoride reacts with hydroxyapatite to give fluorapatite. What is the formula of fluorapatite? [*Osteoporosis, p. 93*]

GENERAL QUESTIONS AND PROBLEMS

3.88 Explain why the hydride ion, H$^-$, has a noble gas configuration.

3.89 The H$^-$ ion (Problem 3.88) is stable, but the Li$^-$ ion is not. Explain.

3.90 Many compounds containing a metal and a nonmetal are not ionic, yet they are named using the Roman numeral system for ionic compounds described in Section 3.7. Write the chemical formulas for the following such compounds.

(a) Chromium(VI) oxide
(b) Vanadium(V) chloride
(c) Manganese(IV) oxide
(d) Molybdenum(IV) sulfide

3.91 The arsenate ion has the formula AsO$_4^{3-}$. Write the formula of the corresponding acid that contains this anion.

3.92 One commercially available calcium supplement contains calcium gluconate, a compound that is also used as an anticaking agent in instant coffee.

(a) If this compound contains 1 calcium ion for every 2 gluconate ions, what is the charge on a gluconate ion?

(b) What is the ratio of iron ions to gluconate ions in iron(III) gluconate, a commercial iron supplement?

3.93 The names given for the following compounds are incorrect. Write the correct name for each compound.

(a) Cu$_3$PO$_4$, copper(III) phosphate
(b) Na$_2$SO$_4$, sodium sulfide
(c) MnO$_2$, manganese(II) oxide
(d) AuCl$_3$, gold chloride
(e) Pb(CO$_3$)$_2$, lead(II) acetate
(f) Ni$_2$S$_3$, nickel(II) sulfide

3.94 The formulas given for the following compounds are incorrect. Write the correct formula for each compound.

(a) Cobalt(II) cyanide, CoCN$_2$
(b) Uranium(VI) oxide, UO$_6$
(c) Tin(II) sulfate, Ti(SO$_4$)$_2$
(d) Manganese(IV) oxide; MnO$_4$
(e) Potassium phosphate, K$_2$PO$_4$
(f) Calcium phosphide, CaP
(g) Lithium bisulfate, Li(SO$_4$)$_2$
(h) Aluminum hydroxide; Al$_2$(OH)$_3$

3.95 How many protons, electrons, and neutrons are in each of these ions?

(a) ^{16}O^{2-}
(b) ^{89}Y^{3+}
(c) ^{133}Cs$^+$
(d) ^{81}Br$^-$

3.96 Element X reacts with element Y to give a product containing X^{3+} ions and Y^{2-} ions.

(a) Is element X likely to be a metal or a nonmetal?
(b) Is element Y likely to be a metal or a nonmetal?
(c) What is the formula of the product?
(d) What groups of the periodic table are elements X and Y likely to be in?

3.97 Identify each of the ions having the following charges and electron configurations:

(a) X^{4+}; [Ar] $4s^0 3d^3$
(b) X$^+$; [Ar] $4s^0 3d^{10}$
(c) X^{4+}; [Ar] $4s^0 3d^0$

CHAPTER 4

Molecular Compounds

CONTENTS

- 4.1 Covalent Bonds
- 4.2 Covalent Bonds and the Periodic Table
- 4.3 Multiple Covalent Bonds
- 4.4 Coordinate Covalent Bonds
- 4.5 Characteristics of Molecular Compounds
- 4.6 Molecular Formulas and Lewis Structures
- 4.7 Drawing Lewis Structures
- 4.8 The Shapes of Molecules
- 4.9 Polar Covalent Bonds and Electronegativity
- 4.10 Polar Molecules
- 4.11 Naming Binary Molecular Compounds

◀ The Atomium monument in Brussels, Belgium, provides an artistic image of the binding forces between atoms.

CHAPTER GOALS

1. **What is a covalent bond?**
   THE GOAL: Be able to describe the nature of covalent bonds and how they are formed. (◀◀◀ A, B, C.)

2. **How does the octet rule apply to covalent bond formation?**
   THE GOAL: Be able to use the octet rule to predict the numbers of covalent bonds formed by common main group elements. (◀◀◀ B, C.)

3. **What are the major differences between ionic and molecular compounds?**
   THE GOAL: Be able to compare the structures, compositions, and properties of ionic and molecular compounds.

4. **How are molecular compounds represented?**
   THE GOAL: Be able to interpret molecular formulas and draw Lewis structures for molecules. (◀◀◀ D.)

5. **What is the influence of valence-shell electrons on molecular shape?**
   THE GOAL: Be able to use Lewis structures to predict molecular geometry. (◀◀◀ D.)

6. **When are bonds and molecules polar?**
   THE GOAL: Be able to use electronegativity and molecular geometry to predict bond and molecular polarity. (◀◀◀ A, D.)

CONCEPTS TO REVIEW

A. The Periodic Table (Sections 2.4 and 2.5)

B. Electron Configurations (Sections 2.7 and 2.8)

C. The Octet Rule (Section 3.5)

D. Electron-Dot Symbols (Section 2.9)

We saw in the preceding chapter that ionic compounds are crystalline solids composed of positively and negatively charged ions. Not all substances, however, are ionic. In fact, with the exception of table salt (NaCl), baking soda (NaHCO$_3$), lime for the garden (CaO), and a few others, most of the compounds around us are *not* crystalline, brittle, high-melting ionic solids. We are much more likely to encounter gases (like those in air), liquids (such as water), low-melting solids (such as butter), and flexible solids (like plastics). All these materials are composed of *molecules* rather than ions, all contain *covalent* bonds rather than ionic bonds, and all consist primarily of nonmetal atoms rather than metals.

4.1 Covalent Bonds

How do we describe the bonding in carbon dioxide, water, polyethylene, and the many millions of nonionic compounds that make up our bodies and much of the world around us? Simply put, the bonds in such compounds are formed by the *sharing* of electrons between atoms (in contrast to ionic bonds, which involve the complete transfer of electrons from one atom to another). The bond formed when atoms share electrons is called a **covalent bond**, and the group of atoms held together by covalent bonds is called a **molecule**. A single molecule of water, for example, contains 2 hydrogen atoms and 1 oxygen atom covalently bonded to one another. We might visualize a water molecule using a space-filling model as shown here:

2 hydrogen atoms + 1 oxygen atom Combine to give 1 water molecule (H$_2$O)

Covalent bond A bond formed by sharing electrons between atoms.

Molecule A group of atoms held together by covalent bonds.

Recall that according to the *octet rule* (Section 3.5), main group elements tend to undergo reactions that leave them with completed outer subshells with 8 valence electrons (or 2 for hydrogen), so that they have a noble gas electron configuration. Although metals and reactive nonmetals can achieve an electron octet by gaining or losing an appropriate number of electrons to form ions, the nonmetals can also achieve an electron octet by *sharing* an appropriate number of electrons in covalent bonds.

As an example of how covalent bond formation occurs, let us look first at the bond between 2 hydrogen atoms in a hydrogen molecule, H$_2$. Recall that a hydrogen *atom* consists of a positively charged nucleus and a single, negatively charged 1s valence electron, which we represent as H· using the electron-dot symbol. When 2 hydrogen atoms come together, electrical interactions occur. Some of these interactions are repulsive—the 2 positively charged nuclei repel each other, and the 2 negatively charged electrons repel each other. Other interactions, however, are attractive—each nucleus attracts both electrons, and each electron attracts both nuclei (Figure 4.1). Because the attractive forces are stronger than the repulsive forces, a covalent bond is formed, and the hydrogen atoms stay together.

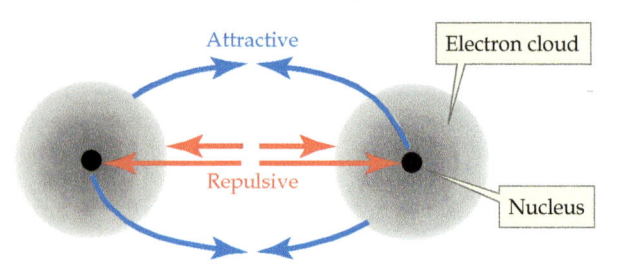

▲ **Figure 4.1**
A covalent H—H bond is the net result of attractive and repulsive forces.
The nucleus–electron attractions (blue arrows) are greater than the nucleus–nucleus and electron–electron repulsions (red arrows), resulting in a net attractive force that holds the atoms together to form an H_2 molecule.

In essence, the electrons act as a kind of "glue" to bind the 2 nuclei together into an H_2 molecule. Both nuclei are simultaneously attracted to the same electrons and are held together, much as two tug-of-war teams pulling on the same rope are held together.

Covalent bond formation in the H—H molecule can be visualized by imagining that the spherical 1s orbitals from the two individual atoms blend together and *overlap* to give an egg-shaped region in the H_2 molecule. The 2 electrons in the H—H covalent bond occupy the central region between the nuclei, giving both atoms a share in 2 valence electrons and the $1s^2$ electron configuration of the noble gas helium. For simplicity, the shared pair of electrons in a covalent bond is often represented as a line between atoms. Thus, the symbols H—H, H:H, and H_2 all represent a hydrogen molecule.

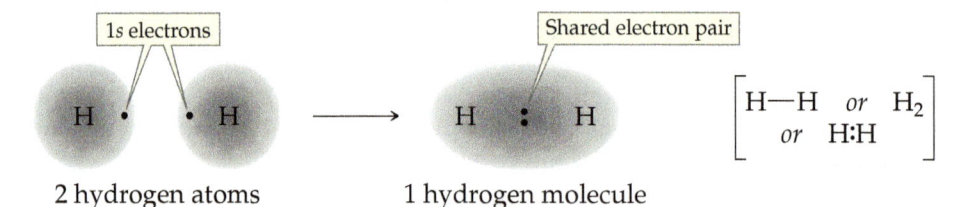

▲ The two teams are joined together because both are holding onto the same rope. In a similar way, two atoms are bonded together when both hold onto the same electrons.

Bond length The optimum distance between nuclei in a covalent bond.

As you might imagine, the magnitudes of the various attractive and repulsive forces between nuclei and electrons in a covalent bond depend on how close the atoms are to each other. If the atoms are too far apart, the attractive forces are small and no bond exists. If the atoms are too close, the repulsive interaction between nuclei is so strong that it pushes the atoms apart. Thus, there is an optimum point where net attractive forces are maximized and where the H_2 molecule is most stable. This optimum distance between nuclei is called the **bond length** and is 74 pm (7.4×10^{-11} m) in the H_2 molecule.

As another example of covalent bond formation, look at the chlorine molecule, Cl_2. An individual chlorine atom has 7 valence electrons and the valence-shell electron configuration $3s^2 3p^5$. Using the electron-dot symbols for the valence electrons, each Cl atom can be represented as :C̈l·. The 3s orbital and 2 of the three 3p orbitals are filled by 2 electrons each, but the third 3p orbital holds only 1 electron. When 2 chlorine atoms approach each other, the unpaired 3p electrons are shared by both atoms in a covalent bond. Each chlorine atom in the resultant Cl_2 molecule now "owns" 6 outer-shell electrons and "shares" 2 more, giving each a valence-shell octet like that of the noble gas argon. We can represent the formation of a covalent bond between chlorine atoms as

$$:\ddot{\text{Cl}}\cdot + \cdot\ddot{\text{Cl}}: \longrightarrow :\ddot{\text{Cl}}:\ddot{\text{Cl}}:$$

Such bond formation can also be pictured as the overlap of the 3p orbitals containing the single electrons, with resultant formation of a region of high electron density between the nuclei:

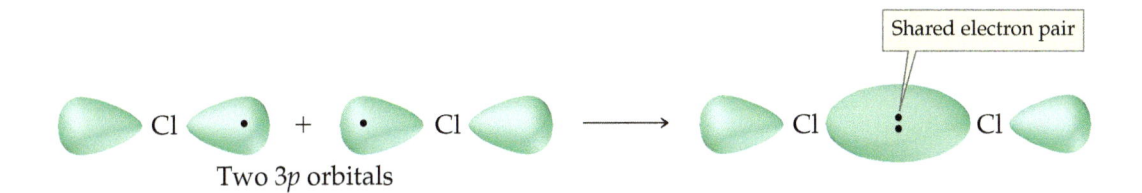

In addition to H_2 and Cl_2, five other elements always exist as *diatomic* (2-atom) molecules (Figure 4.2): nitrogen (N_2) and oxygen (O_2) are colorless, odorless, nontoxic gases present in air; fluorine (F_2) is a pale yellow, highly reactive gas; bromine (Br_2) is a dark red, toxic liquid; and iodine (I_2) is a violet crystalline solid.

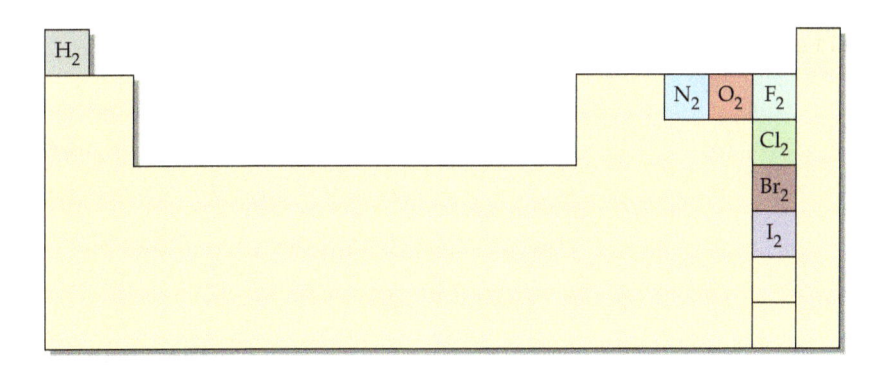

◂ **Figure 4.2**
Diatomic elements in the periodic table.

PROBLEM 4.1
Draw the iodine molecule using electron-dot symbols, and indicate the shared electron pair. What noble gas configuration do the iodine atoms have in an (I_2) molecule?

4.2 Covalent Bonds and the Periodic Table

Covalent bonds can form between unlike atoms as well as between like atoms, making possible a vast number of **molecular compounds**. Water molecules, for example, consist of 2 hydrogen atoms joined by covalent bonds to an oxygen atom, H_2O; ammonia molecules consist of 3 hydrogen atoms covalently bonded to a nitrogen atom, NH_3; and methane molecules consist of 4 hydrogen atoms covalently bonded to a carbon atom, CH_4.

Molecular compound A compound that consists of molecules rather than ions.

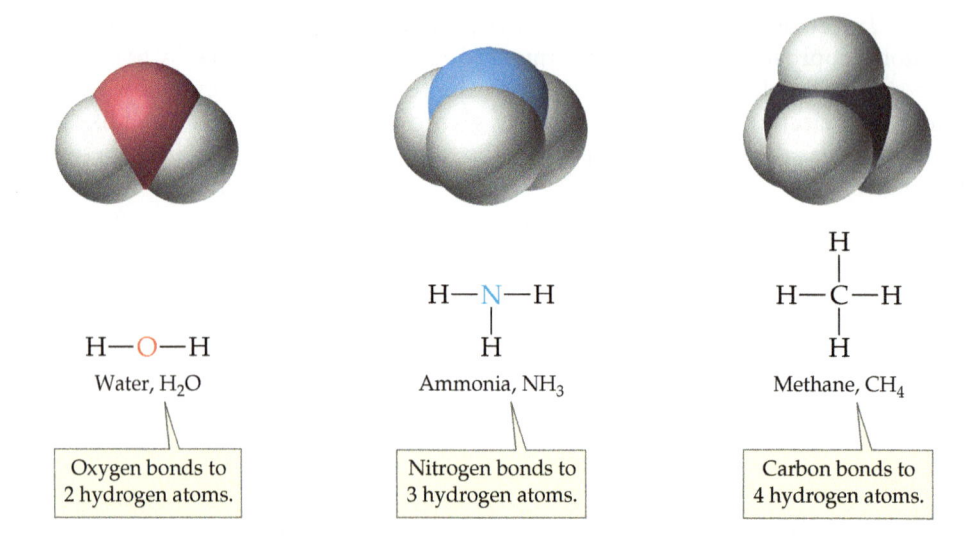

102 CHAPTER 4 Molecular Compounds

H—O—H
Water, H₂O

Oxygen bonds to 2 hydrogen atoms.

H—N—H
   |
   H
Ammonia, NH₃

Nitrogen bonds to 3 hydrogen atoms.

    H
    |
H—C—H
    |
    H
Methane, CH₄

Carbon bonds to 4 hydrogen atoms.

Note that in all these examples, each atom shares enough electrons to achieve a noble gas configuration: 2 electrons for hydrogen and octets for oxygen, nitrogen, and carbon. Hydrogen, with 1 valence electron (H·), needs one more electron to achieve a noble gas configuration (that of helium, $1s^2$) and thus forms 1 covalent bond. Oxygen, with 6 valence electrons (·Ö·), needs two more electrons to have an octet; this happens when oxygen forms 2 covalent bonds. Nitrogen, with 5 valence electrons (·N̈·), needs three more electrons to achieve an octet and thus forms 3 covalent bonds. Carbon, with 4 valence electrons (·C̈·), needs four more electrons and thus forms 4 covalent bonds. Figure 4.3 summarizes the number of covalent bonds typically formed by common main group elements.

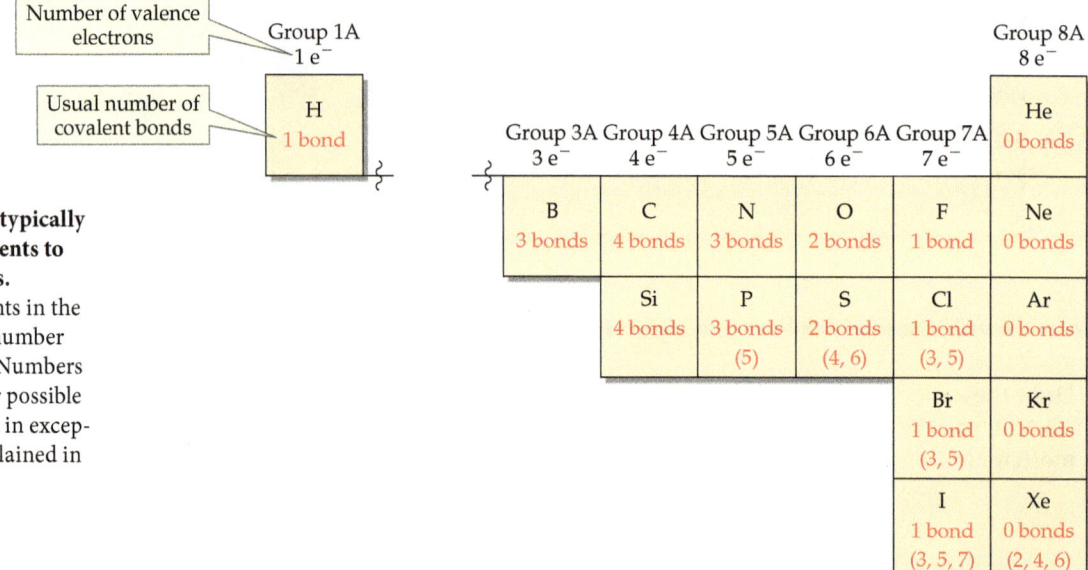

▶ **Figure 4.3**
Numbers of covalent bonds typically formed by main group elements to achieve octet configurations. For P, S, Cl, and other elements in the third period and below, the number of covalent bonds may vary. Numbers in parentheses indicate other possible numbers of bonds that result in exceptions to the octet rule, as explained in the text.

Number of valence electrons

Group 1A
1 e⁻

Usual number of covalent bonds

H
1 bond

	Group 3A 3 e⁻	Group 4A 4 e⁻	Group 5A 5 e⁻	Group 6A 6 e⁻	Group 7A 7 e⁻	Group 8A 8 e⁻
						He 0 bonds
	B 3 bonds	C 4 bonds	N 3 bonds	O 2 bonds	F 1 bond	Ne 0 bonds
		Si 4 bonds	P 3 bonds (5)	S 2 bonds (4, 6)	Cl 1 bond (3, 5)	Ar 0 bonds
					Br 1 bond (3, 5)	Kr 0 bonds
					I 1 bond (3, 5, 7)	Xe 0 bonds (2, 4, 6)

The octet rule is a useful guideline, but it has numerous exceptions. Boron, for example, has only 3 valence electrons it can share (·Ḃ·) and thus forms compounds in which it has only 3 covalent bonds and 6 electrons, such as BF₃. Exceptions to the octet rule are also seen with elements in the third row of the periodic table and below because these elements have vacant *d* orbitals that can be used for bonding. Phosphorus sometimes forms 5 covalent bonds (using 10 bonding electrons); sulfur sometimes forms 4 or 6 covalent bonds (using 8 and 12 bonding electrons, respectively); and chlorine, bromine, and iodine sometimes form 3, 5, or 7 covalent bonds. Phosphorus and sulfur, for example, form molecules such as PCl$_5$, SF$_4$, and SF$_6$.

BF$_3$
Boron trifluoride
(6 valence electrons on B)

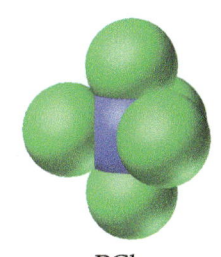

PCl$_5$
Phosphorus pentachloride
(10 valence electrons on P)

SF$_6$
Sulfur hexafluoride
(12 valence electrons on S)

Worked Example 4.1 Molecular Compounds: Octet Rule and Covalent Bonds

Look at Figure 4.3 and tell whether the following molecules are likely to exist.

(a) Br—C—Br with Br above and below
   CBr$_3$

(b) I—Cl
   ICl

(c) H—F—H with H above and below
   FH$_4$

(d) H—S—H
   H$_2$S

ANALYSIS Count the number of covalent bonds formed by each element and see if the numbers correspond to those shown in Figure 4.3.

SOLUTION
(a) No. Carbon needs 4 covalent bonds but has only 3 in CBr$_3$.
(b) Yes. Both iodine and chlorine have 1 covalent bond in ICl.
(c) No. Fluorine only needs 1 covalent bond to achieve an octet. It cannot form more than 1 covalent bond because it is in the second period and does not have valence d orbitals to use for bonding.
(d) Yes. Sulfur, which is in group 6A like oxygen, often forms 2 covalent bonds.

Worked Example 4.2 Molecular Compounds: Electron-Dot Symbols

Using electron-dot symbols, show the reaction between a hydrogen atom and a fluorine atom.

ANALYSIS The electron-dot symbols show the valence electrons for the hydrogen and fluorine atoms. A covalent bond is formed by the sharing of unpaired valence electrons between the 2 atoms.

SOLUTION
Draw the electron-dot symbols for the H and F atoms, showing the covalent bond as a shared electron pair.

H· + ·F̈: ⟶ H:F̈:

Worked Example 4.3 Molecular Compounds: Predicting Number of Bonds

What are likely formulas for the following molecules?
(a) SiH₂Cl?   (b) HBr?   (c) PBr?

ANALYSIS The numbers of covalent bonds formed by each element should be those shown in Figure 4.3.

SOLUTION
(a) Silicon typically forms 4 bonds: SiH₂Cl₂
(b) Hydrogen forms only 1 bond: HBr
(c) Phosphorus typically forms 3 bonds: PBr₃

PROBLEM 4.2
How many covalent bonds are formed by each atom in the following molecules? Draw molecules using the electron-dot symbols and lines to show the covalent bonds.
(a) PH₃   (b) H₂Se   (c) HCl   (d) SiF₄

PROBLEM 4.3
Lead forms both ionic and molecular compounds. Using Figure 4.3, the periodic table, and electronic configurations, predict which of the following lead compounds is more likely to be ionic and which is more likely to be molecular: PbCl₂, PbCl₄.

PROBLEM 4.4
What are likely formulas for the following molecules?
(a) CH₂Cl?   (b) BH?   (c) NI?   (d) SiCl?

4.3 Multiple Covalent Bonds

The bonding in some molecules cannot be explained by the sharing of only 2 electrons between atoms. For example, the carbon and oxygen atoms in carbon dioxide (CO₂) and the nitrogen atoms in the N₂ molecule cannot have electron octets if only 2 electrons are shared:

·Ö· + ·C· + ·Ö·   ·N· + ·N·
   ↓   ↓
  ·Ö:C:Ö·   ·N:N·

UNSTABLE—Carbon has only 6 electrons; each oxygen has only 7.

UNSTABLE—Each nitrogen has only 6 electrons.

The only way the atoms in CO₂ and N₂ can have outer-shell electron octets is by sharing *more* than 2 electrons, resulting in the formation of *multiple* covalent bonds. Only if the carbon atom shares 4 electrons with each oxygen atom do all atoms in CO₂ have electron octets, and only if the 2 nitrogen atoms share 6 electrons do both have electron octets. A bond formed by sharing 2 electrons (one pair) is a **single bond**, a bond formed by sharing 4 electrons (two pairs) is a **double bond**, and a bond formed by sharing 6 electrons (three pairs) is a **triple bond**. Just as a single bond is represented by a single line between atoms, a double bond is represented by two lines between atoms and a triple bond by three lines:

Single bond A covalent bond formed by sharing one electron pair.
Double bond A covalent bond formed by sharing two electron pairs.
Triple bond A covalent bond formed by sharing three electron pairs.

Double bonds
:Ö::C::Ö: or   :Ö=C=Ö:

A triple bond
:N:::N: or   :N≡N:

The carbon atom in CO_2 has 2 double bonds (4e⁻ each) for a total of 8 electrons. Each oxygen atom also has a complete octet: a double bond (4e⁻) plus two sets of lone pairs. Similarly, formation of a triple bond in N_2 allows each nitrogen to obtain a complete octet: 6 electrons from the triple bond plus a lone pair.

Carbon, nitrogen, and oxygen are the elements most often present in multiple bonds. Carbon and nitrogen form both double and triple bonds; oxygen forms double bonds. Multiple covalent bonding is particularly common in *organic* molecules, which consist predominantly of the element carbon. For example, ethylene, a simple compound used commercially to induce ripening in fruit, has the formula C_2H_4. The only way for the 2 carbon atoms to have octets is for them to share 4 electrons in a carbon-carbon double bond:

Ethylene—the carbon atoms share 4 electrons in a double bond.

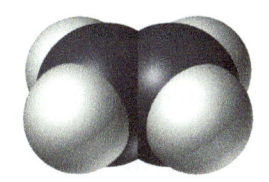

As another example, acetylene, the gas used in welding, has the formula C_2H_2. To achieve octets, the 2 acetylene carbons share 6 electrons in a carbon-carbon triple bond:

Acetylene—the carbon atoms share 6 electrons in a triple bond.

Note that in compounds with multiple bonds like ethylene and acetylene, each carbon atom still forms a total of 4 covalent bonds.

Worked Example 4.4 Molecular Compounds: Multiple Bonds

The compound 1-butene contains a multiple bond. In the following representation, however, only the connections between atoms are shown; the multiple bond is not specifically indicated. Identify the position of the multiple bond.

1-Butene

ANALYSIS Look for 2 adjacent atoms that appear to have fewer than the typical number of covalent bonds, and connect those atoms by a double or triple bond. Refer to Figure 4.3 to see how many bonds will typically be formed by hydrogen and carbon atoms.

SOLUTION

Worked Example 4.5 Multiple Bonds: Electron-Dot and Line Structures

Draw the oxygen molecule (a) using the electron-dot symbols, and (b) using lines rather than dots to indicate covalent bonds.

ANALYSIS Each oxygen atom has 6 valence electrons and will tend to form 2 covalent bonds to reach an octet. Thus, each oxygen will need to share 4 electrons to form a double bond.

SOLUTION

$$:\ddot{O}::\ddot{O}: \quad \text{or} \quad :\ddot{O}=\ddot{O}:$$

PROBLEM 4.5

Acetic acid, the organic constituent of vinegar, can be drawn using electron-dot symbols as shown below. How many outer-shell electrons are associated with each atom? Draw the structure using lines rather than dots to indicate covalent bonds.

PROBLEM 4.6

Identify the positions of all double bonds in caffeine, a stimulant found in coffee and many soft drinks and as an additive in several over-the-counter drugs, such as aspirin.

4.4 Coordinate Covalent Bonds

In the covalent bonds we have seen thus far, the shared electrons have come from different atoms. That is, the bonds result from the overlap of 2 singly occupied valence orbitals, 1 from each atom. Sometimes, though, a bond is formed by the overlap of a filled orbital on 1 atom with a vacant orbital on another atom so that both electrons come from the *same* atom. The bond that results in this case is called a **coordinate covalent bond**.

Coordinate covalent bond The covalent bond that forms when both electrons are donated by the same atom.

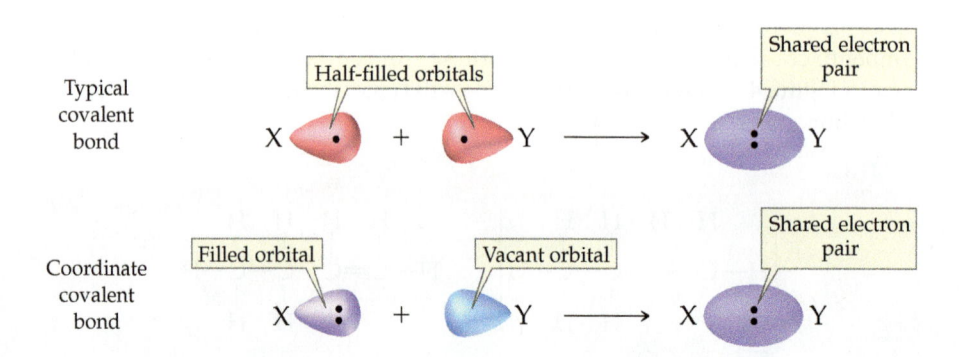

The ammonium ion, NH_4^+, is an example of a species with a coordinate covalent bond. When ammonia reacts in water solution with a hydrogen ion, H^+, the nitrogen atom donates 2 electrons from a filled valence orbital to form a coordinate covalent bond to the hydrogen ion, which has a vacant 1s orbital.

$$H^+ + H-\underset{H}{\overset{H}{\underset{..}{N}}}-H \longrightarrow \left[H-\underset{H}{\overset{H}{\underset{H}{N}}}-H \right]^+$$

Once formed, a coordinate covalent bond contains two shared electrons and is no different from any other covalent bond. All four covalent bonds in NH_4^+ are identical, for example. Note, however, that formation of a coordinate covalent bond often results in unusual bonding patterns, such as an N atom with four covalent bonds rather than the usual three, or an oxygen atom with three bonds rather than the usual two (H_3O^+). An entire class of substances is based on the ability of transition metals to form coordinate covalent bonds with nonmetals. Called *coordination compounds*, many of these substances have important roles in living organisms. For example, toxic metals can be removed from the bloodstream by the formation of water-soluble coordination compounds.

▶▶ As another example, we will see in Chapter 19 that essential metal ions are held in enzyme molecules by coordinate covalent bonds.

4.5 Characteristics of Molecular Compounds

We saw in Section 3.4 that ionic compounds have high melting and boiling points because the attractive forces between oppositely charged ions are so strong that the ions are held tightly together. But *molecules* are neutral, so there is no strong electrostatic attraction between molecules. There are, however, several weaker forces between molecules, called *intermolecular forces*, which we will look at in more detail in Chapter 8.

When intermolecular forces are very weak, molecules of a substance are so weakly attracted to one another that the substance is a gas at ordinary temperatures. If the forces are somewhat stronger, the molecules are pulled together into a liquid; and if the forces are still stronger, the substance becomes a molecular solid. Even so, the melting points and boiling points of molecular solids are usually lower than those of ionic solids.

In addition to having lower melting points and boiling points, molecular compounds differ from ionic compounds in other ways. Most molecular compounds are insoluble in water, for instance, because they have little attraction to the strongly polar water molecules. In addition, they do not conduct electricity when melted because they have no charged particles. Table 4.1 provides a comparison of the properties of ionic and molecular compounds.

TABLE 4.1 A Comparison of Ionic and Molecular Compounds

Ionic Compounds	Molecular Compounds
Smallest components are ions (e.g., Na^+, Cl^-)	Smallest components are molecules (e.g., CO_2, H_2O)
Usually composed of metals combined with nonmetals	Usually composed of nonmetals with nonmetals
Crystalline solids	Gases, liquids, or low-melting-point solids
High melting points (e.g., NaCl = 801 °C)	Low melting points (H_2O = 0.0 °C)
High boiling points (above 700 °C) (e.g., NaCl = 1413 °C)	Low boiling points (e.g., H_2O = 100 °C; CH_3CH_2OH = 76 °C)
Conduct electricity when molten or dissolved in water	Do not conduct electricity
Many are water-soluble	Relatively few are water-soluble
Not soluble in organic liquids	Many are soluble in organic liquids

108 CHAPTER 4 Molecular Compounds

> **PROBLEM 4.7**
> Aluminum chloride ($AlCl_3$) has a melting point of 190 °C, whereas aluminum oxide (Al_2O_3) has a melting point of 2070 °C. Explain why the melting points of the two compounds are so different.

4.6 Molecular Formulas and Lewis Structures

Formulas such as H_2O, NH_3, and CH_4, which show the numbers and kinds of atoms in one molecule of a compound, are called **molecular formulas**. Though important, molecular formulas are limited in their use because they do not provide information about how the atoms in a given molecule are connected.

Much more useful are **structural formulas**, which use lines to show how atoms are connected, and **Lewis structures**, which show both the connections among atoms and the placement of unshared valence electrons. In a water molecule, for instance, the oxygen atom shares two electron pairs in covalent bonds with 2 hydrogen atoms and has two other pairs of valence electrons that are not shared in bonds. Such unshared pairs of valence electrons are called **lone pairs**. In an ammonia molecule, three electron pairs are used in bonding, and there is one lone pair. In methane, all four electron pairs are bonding.

Molecular formula A formula that shows the numbers and kinds of atoms in 1 molecule of a compound.

Structural formula A molecular representation that shows the connections among atoms by using lines to represent covalent bonds.

Lewis structure A molecular representation that shows both the connections among atoms and the locations of lone-pair valence electrons.

Lone pair A pair of electrons that is not used for bonding.

Lewis structures:

H—Ö—H (Water)

H—N̈—H with H below (Ammonia)

H—C—H with H above and below (Methane)

Electron lone pairs are indicated on O and N.

Note how a molecular formula differs from an ionic formula, described previously in Section 3.9. A *molecular* formula gives the number of atoms that are combined in one molecule of a compound, whereas an *ionic* formula gives only a ratio of ions (Figure 4.4). The formula C_2H_4 for ethylene, for example, says that every ethylene molecule consists of 2 carbon atoms and 4 hydrogen atoms. The formula NaCl for sodium chloride, however, says only that there are equal numbers of Na^+ and Cl^- ions in the crystal; the formula says nothing about how the ions interact with one another.

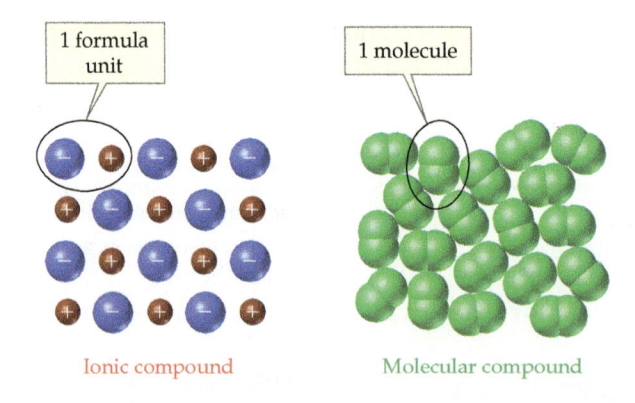

▶ **Figure 4.4**
The distinction between ionic and molecular compounds.
In ionic compounds, the smallest particle is an ion. In molecular compounds, the smallest particle is a molecule.

1 formula unit — Ionic compound
1 molecule — Molecular compound

4.7 Drawing Lewis Structures

To draw a Lewis structure, you first need to know the connections among atoms. Sometimes the connections are obvious. Water, for example, can only be H—O—H because only oxygen can be in the middle and form 2 covalent bonds. Other times, you will have to be told how the atoms are connected.

Two approaches are used for drawing Lewis structures once the connections are known. The first is particularly useful for organic molecules like those found in living

Lewis Structures for Molecules Containing C, N, O, X (Halogen), and H

As summarized in Figure 4.3, carbon, nitrogen, oxygen, halogen, and hydrogen atoms usually maintain consistent bonding patterns:

- C forms 4 covalent bonds and often bonds to other carbon atoms.
- N forms 3 covalent bonds and has one lone pair of electrons.
- O forms 2 covalent bonds and has two lone pairs of electrons.
- Halogens (X = F, Cl, Br, I) form 1 covalent bond and have three lone pairs of electrons.
- H forms 1 covalent bond.

Carbon	Nitrogen	Oxygen	Halogen	Hydrogen
4 bonds	3 bonds	2 bonds	1 bond	1 bond

Relying on these common bonding patterns simplifies the writing of Lewis structures. In ethane (C_2H_6), a constituent of natural gas, for example, 3 of the 4 covalent bonds of each carbon atom are used in bonds to hydrogen, and the fourth is a carbon–carbon bond. There is no other arrangement in which all 8 atoms can have their usual bonding patterns. In acetaldehyde (C_2H_4O), a substance used in manufacturing perfumes, dyes, and plastics, 1 carbon has 3 bonds to hydrogen, while the other has 1 bond to hydrogen and a double bond to oxygen.

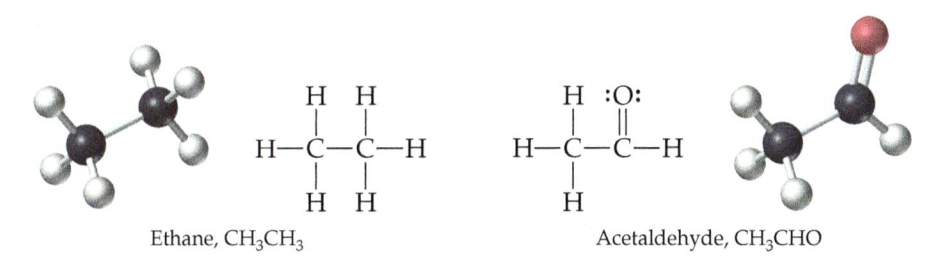

Ethane, CH₃CH₃

Acetaldehyde, CH₃CHO

Because Lewis structures are awkward for larger organic molecules, ethane is more frequently written as a **condensed structure** in which the bonds are not specifically shown. In its condensed form, ethane is CH₃CH₃, meaning that each carbon atom has 3 hydrogen atoms bonded to it (CH₃) and the 2 (CH₃) units are bonded to each other. In the same way, acetaldehyde can be written as CH₃CHO. Note that neither the lone-pair electrons nor the C=O double bond in acetaldehyde is shown explicitly. You will get a lot more practice with such condensed structures in later chapters.

Many of the computer-generated pictures we will be using from now on will be *ball-and-stick models* rather than the space-filling models used previously. Space-filling models are more realistic, but ball-and-stick models do a better job of showing connections and molecular geometry. All models, regardless of type, use a consistent color code in which C is dark gray or black, H is white or ivory, O is red, N is blue, S is yellow, P is dark blue, F is light green, Cl is green, Br is brownish red, and I is purple.

Space-filling

Ball-and-stick

Condensed structure A molecular representation in which bonds are not specifically shown but rather are understood by the order in which atoms are written.

▶▶ Condensed structures are used extensively to represent molecular structures in organic chemistry (Chapters 12–17).

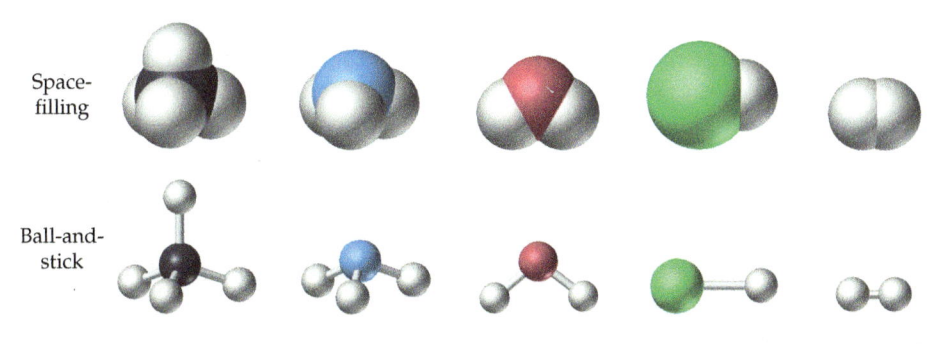

A General Method for Drawing Lewis Structures

A Lewis structure can be drawn for any molecule or polyatomic ion by following a five-step procedure. Take PCl₃, for example, a substance in which 3 chlorine atoms surround the central phosphorus atom.

STEP 1: Find the total number of valence electrons of all atoms in the molecule or ion. In PCl₃, for example, phosphorus (group 5A) has 5 valence electrons and chlorine (group 7A) has 7 valence electrons, giving a total of 26:

$$P + (3 \times Cl) = PCl_3$$
$$5e^- + (3 \times 7e^-) = 26e^-$$

For a polyatomic ion, add 1 electron for each negative charge or subtract 1 for each positive charge. In OH⁻, the total is 8 electrons (6 from oxygen, 1 from hydrogen, plus 1 for the negative charge). In NH₄⁺, the total is 8 (5 from nitrogen, 1 from each of 4 hydrogens, minus 1 for the positive charge).

STEP 2: Draw a line between each pair of connected atoms to represent the two electrons in a covalent bond. Remember that elements in the second row of the periodic table form the number of bonds discussed earlier in this section, whereas elements in the third row and beyond can use more than 8 electrons and form more than the "usual" number of bonds (Figure 4.3). A particularly common pattern is that an atom in the third row (or beyond) occurs as the central atom in a cluster. In PCl₃, for example, the phosphorus atom is in the center with the 3 chlorine atoms bonded to it:

$$\begin{array}{c} Cl \\ | \\ Cl-P-Cl \end{array}$$

STEP 3: Using the remaining electrons, add lone pairs so that each atom connected to the central atom (except H) gets an octet. In PCl₃, 6 of the 26 valence electrons were used to make the covalent bonds. From the remaining 20 electrons, each Cl atom needs three lone pairs to complete the octet:

$$\begin{array}{c} :\ddot{\text{C}}\text{l}: \\ | \\ :\ddot{\text{C}}\text{l}-\text{P}-\ddot{\text{C}}\text{l}: \end{array}$$

STEP 4: Place any remaining electrons in lone pairs on the central atom. In PCl₃, we have used 24 of the 26 available electrons—6 in three single bonds and 18 in the three lone pairs on each chlorine atom. This leaves 2 electrons for one lone pair on phosphorus:

$$\begin{array}{c} :\ddot{\text{C}}\text{l}: \\ | \\ :\ddot{\text{C}}\text{l}-\ddot{\text{P}}-\ddot{\text{C}}\text{l}: \end{array}$$

STEP 5: If the central atom does not yet have an octet after all electrons have been assigned, take a lone pair from a neighboring atom, and form a multiple bond to the central atom. In PCl₃, each atom has an octet, all 26 available electrons have been used, and the Lewis structure is finished.

Worked Examples 4.6–4.8 shows how to deal with cases where this fifth step is needed.

Worked Example 4.6 Multiple Bonds: Electron Dots and Valence Electrons

Draw a Lewis structure for the toxic gas hydrogen cyanide, HCN. The atoms are connected in the order shown in the preceding sentence.

ANALYSIS Follow the procedure outlined in the text.

SOLUTION

STEP 1: Find the total number of valence electrons:

H = 1, C = 4, N = 5   Total number of valence electrons = 10

STEP 2: Draw a line between each pair of connected atoms to represent bonding electron pairs:

$$H-C-N \quad 2 \text{ bonds} = 4 \text{ electrons}; 6 \text{ electrons remaining}$$

STEP 3: Add lone pairs so that each atom (except H) has a complete octet:

$$H-C-\ddot{N}\!:$$

STEP 4: All valence electrons have been used, and so step 4 is not needed. H and N have filled valence shells, but C does not.

STEP 5: If the central atom (C in this case) does not yet have an octet, use lone pairs from a neighboring atom (N) to form multiple bonds. This results in a triple bond between the C and N atoms, as shown in the electron dot and ball-and-stick representations below:

$$H-C \equiv N\!:$$

We can check the structure by noting that all 10 valence electrons have been used (in 4 covalent bonds and one lone pair) and that each atom has the expected number of bonds (1 bond for H, 3 for N, and 4 for C).

Worked Example 4.7 Lewis Structures: Location of Multiple Bonds

Draw a Lewis structure for vinyl chloride, C_2H_3Cl, a substance used in making polyvinyl chloride, or PVC, plastic.

ANALYSIS Since H and Cl form only 1 bond each, the carbon atoms must be bonded to each other, with the remaining atoms bonded to the carbons. With only 4 atoms available to bond with them, the carbon atoms cannot have 4 covalent bonds each unless they are joined by a double bond.

SOLUTION

STEP 1: The total number of valence electrons is 18; 4 from each of the 2 C atoms, 1 from each of the 3 H atoms, and 7 from the Cl atom.

STEP 2: Place the 2 C atoms in the center, and divide the 4 other atoms between them:

$$\begin{array}{cc} H & Cl \\ \backslash & / \\ C-C \\ / & \backslash \\ H & H \end{array}$$

The 5 bonds account for 10 valence electrons, with 8 remaining.

STEP 3: Place 6 of the remaining valence electrons around the Cl atom so that it has a complete octet, and place the remaining 2 valence electrons on one of the C atoms (either C, it does not matter):

$$\begin{array}{cc} H & :\ddot{C}l\!: \\ \backslash & / \\ C-\ddot{C}\!: \\ / & \backslash \\ H & H \end{array}$$

When all the valence electrons are distributed, the C atoms still do not have a complete octet; they each need 4 bonds but have only 3.

STEP 5: The lone pair of electrons on the C atom can be used to form a double bond between the C atoms, giving each a total of 4 bonds (8 electrons). Placement of the double bond yields the Lewis structure and ball-and-stick model for vinyl chloride shown below:

$$\begin{array}{c} H \\ \backslash \\ C=C \\ / \backslash \\ H H \end{array} \begin{array}{c} :\ddot{Cl}: \\ \end{array}$$

All 18 valence electrons are accounted for in 6 covalent bonds and three lone pairs, and each atom has the expected number of bonds.

Worked Example 4.8 Lewis Structures: Octet Rule and Multiple Bonds

Draw a Lewis structure for sulfur dioxide, SO_2. The connections are O—S—O.

ANALYSIS Follow the procedure outlined in the text.

SOLUTION

STEP 1: The total number of valence electrons is 18, 6 from each atom:

$$S + (2 \times O) = SO_2$$
$$6e^- + (2 + 6e^-) = 18e^-$$

STEP 2: O—S—O Two covalent bonds use 4 valence electrons.

STEP 3: :Ö—S—Ö: Adding three lone pairs to each oxygen to give each an octet uses 12 additional valence electrons.

STEP 4: :Ö—S̈—Ö: The remaining 2 valence electrons are placed on sulfur, but sulfur still does not have an octet.

STEP 5: Moving one lone pair from a neighboring oxygen to form a double bond with the central sulfur gives sulfur an octet (it does not matter on which side the S=O bond is written):

$$:\ddot{O}-\ddot{S}=\ddot{O}:$$

NOTE: The Lewis structure for SO_2 includes a single bond to one O and a double bond to the other O. It doesn't matter which O has the double bond—both structures are equally acceptable. In reality, however, the S—O bonds in this molecule are actually closer to 1.5, an average between the two possible structures we could draw. This is an example of *resonance structures*, or different Lewis structures that could be used to represent the same molecule.

▶▶ Aromatic compounds, a class of organic compounds discussed in Section 13.9, are an important example of resonance structures.

PROBLEM 4.8
Methylamine, CH_5N, is responsible for the characteristic odor of decaying fish. Draw a Lewis structure of methylamine.

PROBLEM 4.9
Add lone pairs where appropriate to the following structures:

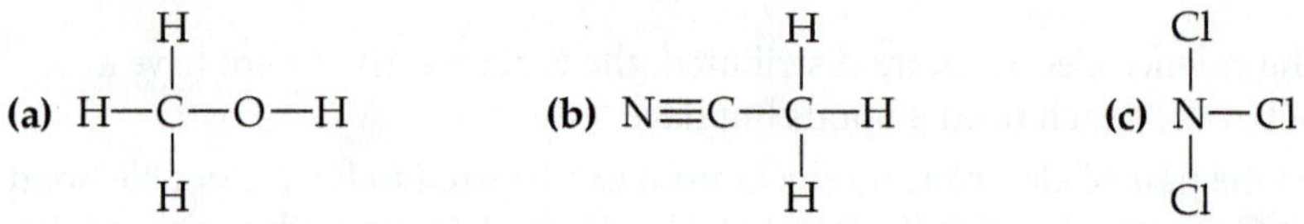

(a)
$$\begin{array}{c} H \\ | \\ H-C-O-H \\ | \\ H \end{array}$$

(b)
$$\begin{array}{c} H \\ | \\ N\equiv C-C-H \\ | \\ H \end{array}$$

(c)
$$\begin{array}{c} Cl \\ | \\ N-Cl \\ | \\ Cl \end{array}$$

PROBLEM 4.10

Draw Lewis structures for the following:

(a) Phosgene, $COCl_2$, a poisonous gas
(b) Hypochlorite ion, OCl^-, present in many swimming pool chemicals
(c) Hydrogen peroxide, H_2O_2
(d) Sulfur dichloride, SCl_2

PROBLEM 4.11

Draw a Lewis structure for nitric acid, HNO_3. The nitrogen atom is in the center, and the hydrogen atom is bonded to an oxygen atom.

CHEMISTRY IN ACTION

CO and NO: Pollutants or Miracle Molecules?

Carbon monoxide (CO) is a killer; everyone knows that. It is to blame for an estimated 3500 accidental deaths and suicides each year in the United States and is the number one cause of all deaths by poisoning. Nitric oxide (NO) is formed in combustion engines and reacts with oxygen to form nitrogen dioxide (NO_2), the reddish-brown gas associated with urban smog. What most people do not know, however, is that our bodies cannot function without these molecules. A startling discovery made in 1992 showed that CO and NO are key chemical messengers in the body, used by cells to regulate critical metabolic processes.

The toxicity of CO in moderate concentration is due to its ability to bind to hemoglobin molecules in the blood, thereby preventing the hemoglobin from carrying oxygen to tissues. The high reactivity of NO leads to the formation of compounds that are toxic irritants. However, low concentrations of CO and NO are produced in cells throughout the body. Both CO and NO are highly soluble in water and can diffuse from one cell to another, where they stimulate production of a substance called *guanylyl cyclase*. Guanylyl cyclase, in turn, controls the production of another substance called *cyclic GMP*, which regulates many cellular functions.

Levels of CO production are particularly high in certain regions of the brain, including those associated with long-term memory. Evidence from experiments with rat brains suggests that a special kind of cell in the brain's hippocampus is signaled by transfer of a molecular messenger from a neighboring cell. The receiving cell responds back to the signaling cell by releasing CO, which causes still more messenger molecules to be sent. After several rounds of this back-and-forth communication, the receiving cell undergoes some sort of change that becomes a memory. When CO production is blocked, possibly in response to a medical condition or exposure to certain toxic metals, long-term memories are no longer stored, and those memories that previously existed are erased. When CO production is stimulated, however, memories are again laid down.

▲ Los Angeles at sunset. Carbon monoxide is a major component of photochemical smog, but it also functions as an essential chemical messenger in our bodies.

NO controls a seemingly limitless range of functions in the body. The immune system uses NO to fight infections and tumors. It is also used to transmit messages between nerve cells and is associated with the processes involved in learning and memory, sleeping, and depression. Its most advertised role, however, is as a *vasodilator*, a substance that allows blood vessels to relax and dilate. This discovery led to the development of a new class of drugs that stimulate production of enzymes called nitric oxide synthases (NOS). These drugs can be used to treat conditions from erectile dysfunction (Viagra) to hypertension. Given the importance of NO in the fields of neuroscience, physiology, and immunology, it is not surprising that it was named "Molecule of the Year" in 1992.

See Chemistry in Action Problems 4.89 and 4.90 at the end of the chapter.

> **KEY CONCEPT PROBLEM 4.12**
>
> The molecular model shown here is a representation of methyl methacrylate, a starting material used to prepare Lucite plastic. Only the connections between atoms are shown; multiple bonds are not indicated.
>
> **(a)** What is the molecular formula of methyl methacrylate?
> **(b)** Indicate the positions of the multiple bonds and lone pairs in methyl methacrylate.

> **PROBLEM 4.13**
> Draw the Lewis dot structures for the molecules CO and NO discussed in the Chemistry in Action box on page 113. How do the Lewis structures provide insight into the reactivity of these molecules?

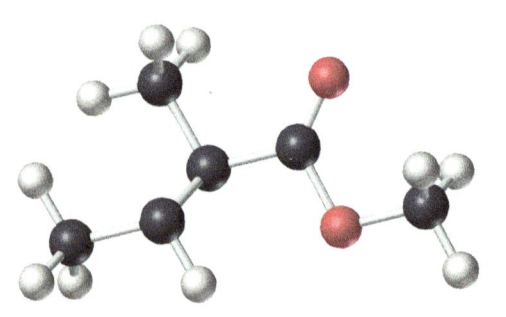

4.8 The Shapes of Molecules

Look back at the computer-generated drawings of molecules in the preceding section, and you will find that the molecules are shown with specific shapes. Acetylene is *linear*, water is *bent*, ammonia is *pyramid-shaped*, methane is *tetrahedral*, and ethylene is flat, or *planar*. What determines such shapes? Why, for example, are the 3 atoms in water connected at an angle of (104.5°) rather than in a straight line? Like so many properties, molecular shapes are related to the numbers and locations of the valence electrons around atoms.

Molecular shapes can be predicted by noting how many bonds and electron pairs surround individual atoms and applying what is called the **valence-shell electron-pair repulsion (VSEPR) model**. The basic idea of the VSEPR model is that the constantly moving valence electrons in bonds and lone pairs make up negatively charged clouds of electrons, which electrically repel one another. The clouds therefore tend to keep as far apart as possible, causing molecules to assume specific shapes. There are three steps to applying the VSEPR model:

STEP 1: Draw a Lewis structure of the molecule, and identify the atom whose geometry is of interest. In a simple molecule like PCl_3 or CO_2, this is usually the central atom.

STEP 2: Count the number of electron charge clouds surrounding the atom of interest. The number of charge clouds is simply the total number of lone pairs plus connections to other atoms. It does not matter whether a connection is a single bond or a multiple bond because we are interested only in the *number* of charge clouds, not in how many electrons each cloud contains. The carbon atom in carbon dioxide, for instance, has 2 double bonds to oxygen (O=C=O), and thus has two charge clouds.

STEP 3: Predict molecular shape by assuming that the charge clouds orient in space so that they are as far away from one another as possible. How they achieve this favorable orientation depends on their number, as summarized in Table 4.2.

If there are only two charge clouds, as occurs on the central atom of CO_2 (2 double bonds) and HCN (1 single bond and 1 triple bond), the clouds are farthest apart when

Valence-shell electron-pair repulsion (VSEPR) model A method for predicting molecular shape by noting how many electron charge clouds surround atoms and assuming that the clouds orient as far away from one another as possible.

SECTION 4.8 The Shapes of Molecules 115

TABLE 4.2 Molecular Geometry Around Atoms with 2, 3, and 4 Charge Clouds

NUMBER OF BONDS	NUMBER OF LONE PAIRS	TOTAL NUMBER OF CHARGE CLOUDS	MOLECULAR GEOMETRY	EXAMPLE
2	0	2	Linear	O=C=O
3	0	3	Trigonal planar	H₂C=O
2	1	3	Bent	O₂S
4	0	4	Tetrahedral	CH₄
3	1	4	Pyramidal	NH₃
2	2	4	Bent	H₂O

they point in opposite directions. Thus, both HCN and CO_2 are linear molecules, with **bond angles** of 180°:

H—C≡N: 180°

$\ddot{O}$=C=$\ddot{O}$ 180°

These molecules are linear, with bond angles of 180°.

Bond angle The angle formed by 3 adjacent atoms in a molecule.

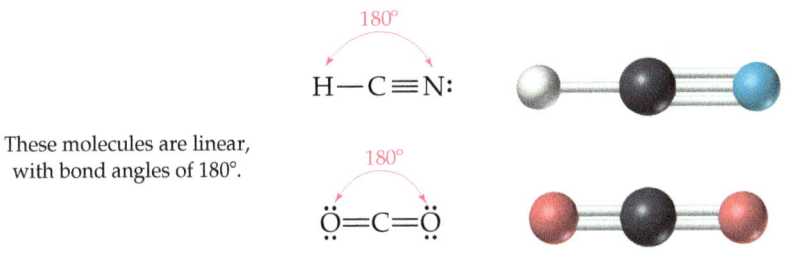

When there are three charge clouds, as occurs on the central atom in formaldehyde (1 single bond and 1 double bond) and SO_2 (1 single bond, 1 double bond, and one lone pair), the clouds will be farthest apart if they lie in a plane and point to the corners of an equilateral triangle. Thus, a formaldehyde molecule is trigonal planar, with all bond angles near 120°. In the same way, an SO_2 molecule has a trigonal planar arrangement of its three electron clouds, but one point of the triangle is occupied by a lone pair. The connection between the 3 atoms is therefore bent rather than linear, with an O—S—O bond angle of approximately 120°:

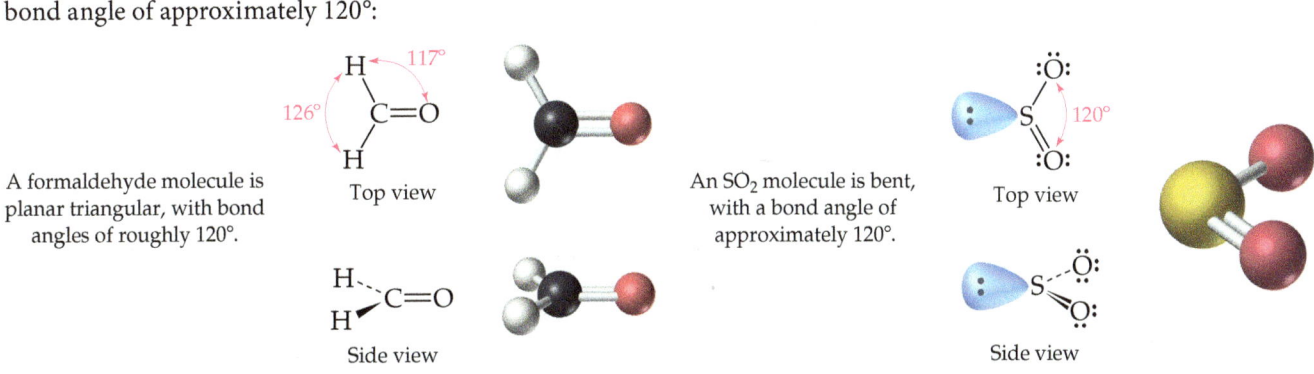

A formaldehyde molecule is planar triangular, with bond angles of roughly 120°.

Top view

Side view

An SO_2 molecule is bent, with a bond angle of approximately 120°.

Top view

Side view

Note how the three-dimensional shapes of molecules like formaldehyde and SO₂ are shown. Solid lines are assumed to be in the plane of the paper; a dashed line recedes behind the plane of the paper away from the viewer; and a dark wedged line protrudes out of the paper toward the viewer. This standard method for showing three-dimensionality will be used throughout the rest of the book.

When there are four charge clouds, as occurs on the central atom in CH₄ (4 single bonds), NH₃ (3 single bonds and one lone pair), and H₂O (2 single bonds and two lone pairs), the clouds can be farthest apart when they extend to the corners of a *regular tetrahedron*. As illustrated in Figure 4.5, a **regular tetrahedron** is a geometric solid whose four identical faces are equilateral triangles. The central atom is at the center of the tetrahedron, the charge clouds point to the corners, and the angle between lines drawn from the center to any two corners is 109.5°.

Regular tetrahedron A geometric figure with four identical triangular faces.

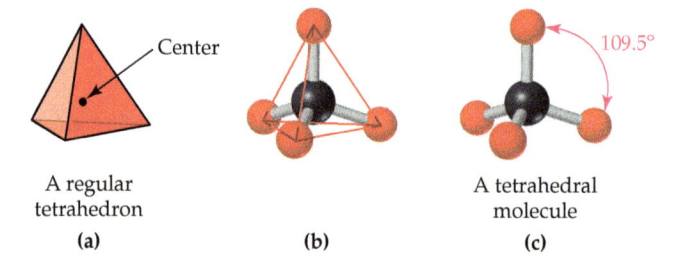

▶ **Figure 4.5**
The tetrahedral geometry of an atom surrounded by four charge clouds. The atom is located at the center of the regular tetrahedron, and the four charge clouds point toward the corners. The bond angle between the center and any two corners is 109.5°.

A regular tetrahedron
(a)

(b)

A tetrahedral molecule
(c)

Because valence-shell electron octets are so common, a great many molecules have geometries based on the tetrahedron. In methane (CH₄), for example, the carbon atom has tetrahedral geometry with H—C—H bond angles of exactly 109.5°. In ammonia (NH₃), the nitrogen atom has a tetrahedral arrangement of its four charge clouds, but one corner of the tetrahedron is occupied by a lone pair, resulting in an overall pyramidal shape for the molecule. Similarly, water, which has two corners of the tetrahedron occupied by lone pairs, has an overall bent shape.

A methane molecule is tetrahedral, with bond angles of 109.5°.

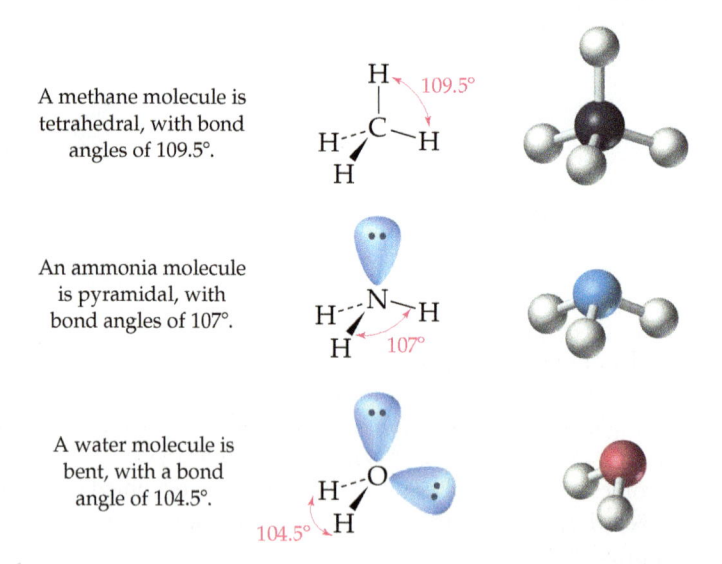

An ammonia molecule is pyramidal, with bond angles of 107°.

A water molecule is bent, with a bond angle of 104.5°.

Note that the H—N—H bond angle in ammonia (107°) and the H—O—H bond angle in water (104.5°) are close to, but not exactly equal to, the ideal 109.5° tetrahedral value. The angles are diminished somewhat from their ideal value because the lone-pair charge clouds repel other electron clouds strongly and compress the rest of the molecule.

The geometry around atoms in larger molecules also derives from the shapes shown in Table 4.2. For example, each of the 2 carbon atoms in ethylene (H₂C=CH₂) has three charge clouds, giving rise to trigonal planar geometry. It turns out that the molecule as a whole is also planar, with H—C—C and H—C—H bond angles of approximately 120°:

The ethylene molecule is planar, with bond angles of 120°.

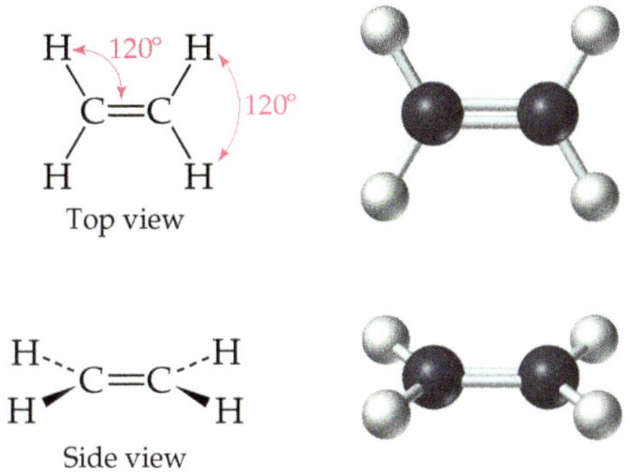

Top view

Side view

Carbon atoms bonded to 4 other atoms are each at the center of a tetrahedron, as shown here for ethane, H₃C—CH₃:

The ethane molecule has tetrahedral carbon atoms, with bond angles of 109.5°.

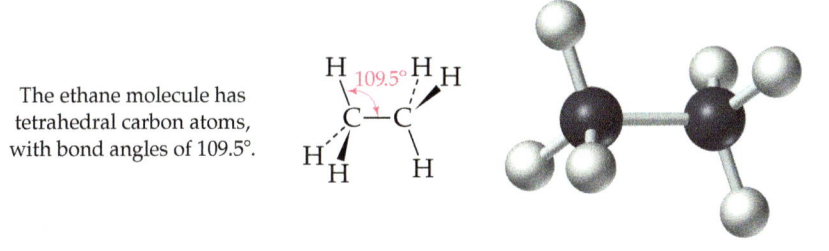

Worked Example 4.9 Lewis Structures: Molecular Shape

What shape would you expect for the hydronium ion, H_3O^+?

ANALYSIS Draw the Lewis structure for the molecular ion, and count the number of charge clouds around the central oxygen atom; imagine the clouds orienting as far away from one another as possible.

SOLUTION
The Lewis structure for the hydronium ion shows that the oxygen atom has four charge clouds (3 single bonds and one lone pair). The hydronium ion is therefore pyramidal with bond angles of approximately 109.5°.

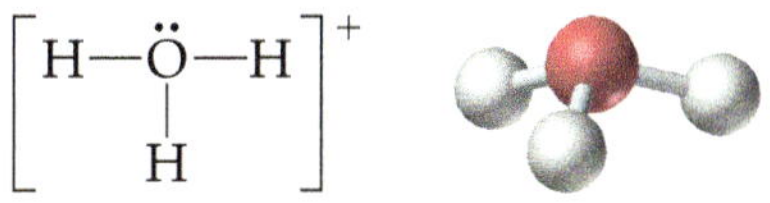

Worked Example 4.10 Lewis Structures: Charge Cloud Geometry

Predict the geometry around each of the carbon atoms in an acetaldehyde molecule, CH_3CHO.

ANALYSIS Draw the Lewis structure and identify the number of charge clouds around each of the central carbon atoms.

SOLUTION
The Lewis structure of acetaldehyde shows that the CH_3 carbon has four charge clouds (4 single bonds) and the CHO carbon atom has three charge clouds (2 single bonds, 1 double bond). Table 4.2 indicates that the CH_3 carbon is tetrahedral, but the CHO carbon is trigonal planar.

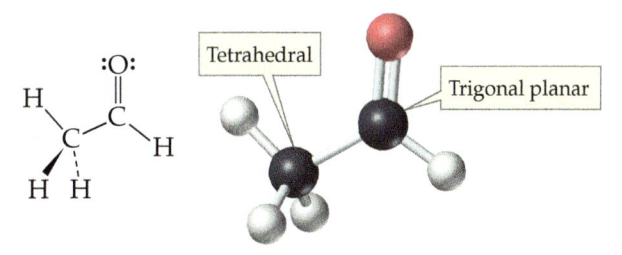

CHEMISTRY IN ACTION

VERY Big Molecules

How big can a molecule be? The answer is very, *very* big. The really big molecules in our bodies and in many items we buy are all *polymers*. Like a string of beads, a polymer is formed of many repeating units connected in a long chain. Each "bead" in the chain comes from a simple molecule that has formed chemical bonds at both ends, linking it to other molecules. The repeating units can be the same:

–a–a–a–a–a–a–a–a–a–a–a–a–

or they can be different. If different, they can be connected in an ordered pattern:

–a–b–a–b–a–b–a–b–a–b–a–b–

or in a random pattern:

–a–b–b–a–b–a–a–a–b–a–b–b–

Furthermore, the polymer chains can have branches, and the branches can have either the same repeating unit as the main chain or a different one:

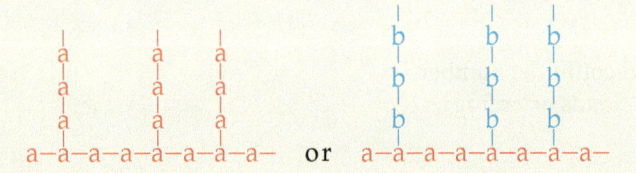

Still other possible variations include complex, three-dimensional networks of "cross-linked" chains. The rubber used in tires, for example, contains polymer chains connected by cross-linking atoms of sulfur to impart greater rigidity.

We all use synthetic polymers every day—we usually call them "plastics." Common synthetic polymers are made by connecting up to several hundred thousand smaller molecules together, producing giant polymer molecules with masses up to several million atomic mass units. Polyethylene, for example, is made by

▲ The protective gear worn by motorcyclists (shown above), firefighters, and security forces are composed of advanced composite materials based on polymers.

combining as many as 50,000 ethylene molecules ($H_2C=CH_2$) to give a polymer with repeating $—CH_2CH—$ units:

Many $H_2C=CH_2$ ⟶ $—CH_2CH_2CH_2CH_2CH_2CH_2—$
Ethylene   Polyethlene

The product is used in such items as chairs, toys, drain pipes, milk bottles, and packaging films. Other examples of polymers include the nylon used in clothing and pantyhose, molded hardware (nuts and bolts), and the Kevlar used in bulletproof vests (see Chemistry in Action on p. 538).

Nature began to exploit the extraordinary variety of polymer properties long before humans did. In fact, despite great progress in recent years, there is still much to be learned about the polymers in living things. Carbohydrates and proteins are polymers, as are the giant molecules of deoxyribonucleic acid (DNA) that govern many cellular processes, including reproduction, in all organisms. Nature's polymer molecules, though, are more complex than any that chemists have yet created.

▶▶ Carbohydrates are polymers composed of sugar molecules linked together in long chains (Chapter 21), while proteins are polymers of smaller molecules called amino acids (Chapter 18). DNA, a polymer of repeating nucleotide subunits, is discussed in Chapter 25.

See Chemistry in Action Problems 4.91 and 4.92 at the end of the chapter.

PROBLEM 4.14
Boron typically only forms 3 covalent bonds because it only has 3 valence electrons, but can form coordinate covalent bonds. Draw the Lewis structure for BF_4^- and predict the molecular shape of the ion.

PROBLEM 4.15
Predict shapes for the organic molecules chloroform, $CHCl_3$, and 1,1-dichloroethylene, $Cl_2C=CH_2$.

PROBLEM 4.16
Polycarbonate, also known as plexiglass, has the basic repeating unit shown below. What is the geometry of the electron clouds for the carbon atoms labeled "a" and "b" in this structure?

PROBLEM 4.17
Hydrogen selenide (H_2Se) resembles hydrogen sulfide (H_2S) in that both compounds have terrible odors and are poisonous. What are their shapes?

> **KEY CONCEPT PROBLEM 4.18**
>
> Draw a structure corresponding to the molecular model of the amino acid methionine shown here, and describe the geometry around the indicated atoms. (Remember the color key discussed in Section 4.7: black = carbon; white = hydrogen; red = oxygen; blue = nitrogen; yellow = sulfur.)
>
> Methionine

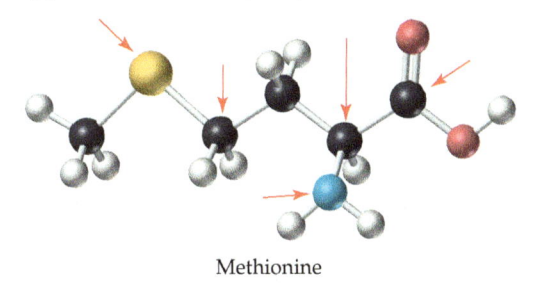

4.9 Polar Covalent Bonds and Electronegativity

Electrons in a covalent bond occupy the region between the bonded atoms. If the atoms are identical, as in H_2 and Cl_2, the electrons are attracted equally to both atoms and are shared equally. If the atoms are *not* identical, however, as in HCl, the bonding electrons may be attracted more strongly by one atom than by the other and may be shared unequally. Such bonds are said to be **polar covalent bonds**. In hydrogen chloride, for example, electrons spend more time near the chlorine atom than near the hydrogen atom. Although the molecule as a whole is neutral, the chlorine is more negative than the hydrogen, resulting in *partial* charges on the atoms. These partial charges are represented by placing a δ− (Greek lowercase *delta*) on the more negative atom and a δ+ on the more positive atom.

A particularly helpful way of visualizing this unequal distribution of bonding electrons is to look at what is called an *electrostatic potential map*, which uses color to portray the calculated electron distribution in a molecule. In HCl, for example, the electron-poor hydrogen is blue, and the electron-rich chlorine is reddish-yellow:

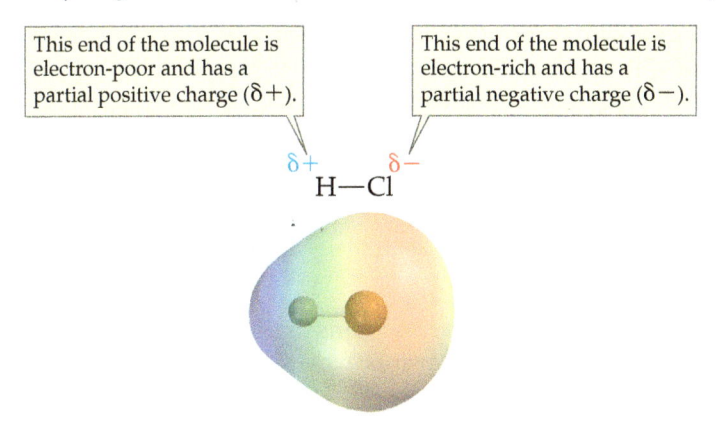

Polar covalent bond A bond in which the electrons are attracted more strongly by one atom than by the other.

This end of the molecule is electron-poor and has a partial positive charge (δ+).

This end of the molecule is electron-rich and has a partial negative charge (δ−).

δ+ δ−
H—Cl

CHAPTER 4   Molecular Compounds

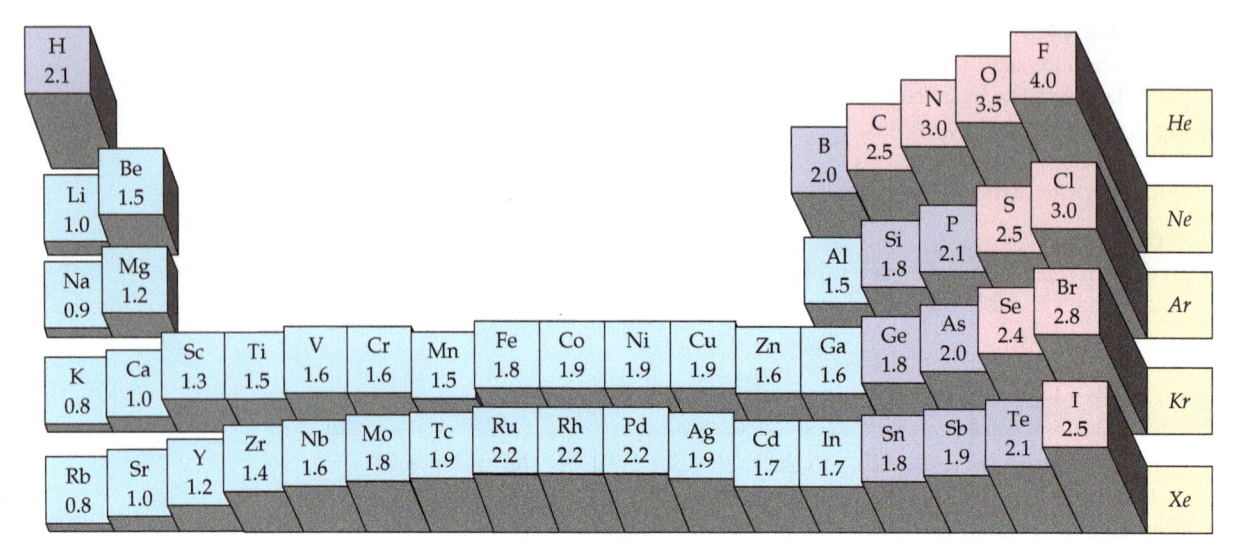

▲ **Figure 4.6**
Electronegativities of several main group and transition metal elements.
Reactive nonmetals at the top right of the periodic table are the most electronegative, and metals at the lower left are the least electronegative. The noble gases are not assigned values.

Electronegativity The ability of an atom to attract electrons in a covalent bond.

The ability of an atom to attract electrons in a covalent bond is called the atom's **electronegativity**. Fluorine, the most electronegative element, is assigned a value of 4, and less electronegative atoms are assigned lower values, as shown in Figure 4.6. Metallic elements on the left side of the periodic table attract electrons only weakly and have lower electronegativities, whereas the halogens and other reactive nonmetal elements on the upper right side of the table attract electrons strongly and have higher electronegativities. Note in Figure 4.6 that electronegativity generally decreases going down the periodic table within a group.

Comparing the electronegativities of bonded atoms makes it possible to compare the polarities of bonds and to predict the occurrence of ionic bonding. Both oxygen (electronegativity 3.5) and nitrogen (3.0), for instance, are more electronegative than carbon (2.5). As a result, both C—O and C—N bonds are polar, with carbon at the positive end. The larger difference in electronegativity values shows that the C—O bond is the more polar of the two:

Less polar
$$^{\delta+}C—N^{\delta-}$$
Electronegativity difference: $3.0 - 2.5 = 0.5$

More polar
$$^{\delta+}C—O^{\delta-}$$
Electronegativity difference: $3.5 - 2.5 = 1.0$

As a rule of thumb, electronegativity differences of less than 0.5 result in nonpolar covalent bonds, differences up to 1.9 indicate increasingly polar covalent bonds, and differences of 2 or more indicate ionic bonds. The electronegativity differences show, for example, that the bond between carbon and fluorine is highly polar covalent, the bond between sodium and chlorine is largely ionic, and the bond between rubidium and fluorine is almost completely ionic:

$$^{\delta+}C—F^{\delta-} \quad Na^+Cl^- \quad Rb^+F^-$$
Electronegativity difference: 1.5   2.1   3.2

E.N difference	Type of bond
0 — 0.4	~ Covalent
0.5 — 1.9	~ Polar covalent
2.0 and above	~ Ionic

Note, though, that there is no sharp dividing line between covalent and ionic bonds; most bonds fall somewhere between two extremes.

LOOKING AHEAD ▶▶▶ The values given in Figure 4.6 indicate that carbon and hydrogen have similar electronegativities. As a result, C—H bonds are nonpolar. We will see in Chapters 12–25 how this fact helps explain the properties of organic and biological compounds, all of which have carbon and hydrogen as their principal constituents.

Worked Example 4.11 Electronegativity: Ionic, Nonpolar, and Polar Covalent Bonds

Predict whether each of the bonds between the following atoms would be ionic, polar covalent, or nonpolar covalent. If polar covalent, which atom would carry the partial positive and negative charges?

(a) C and Br (b) Li and Cl
(c) N and H   (d) Si and I

ANALYSIS Compare the electronegativity values for the atoms and classify the nature of the bonding based on the electronegativity difference.

SOLUTION
(a) The electronegativity for C is 2.5, and for Br is 2.8; the difference is 0.3, indicating nonpolar covalent bonding would occur between these atoms.
(b) The electronegativity for Li is 1.0, and for Cl is 3.0; the difference is 2.0, indicating that ionic bonding would occur between these atoms.
(c) The electronegativity for N is 3.0, and for H is 2.5; the difference is 0.5. Bonding would be polar covalent, with N = $\delta-$ and H = $\delta+$.
(d) The electronegativity for Si is 1.8, and for I is 2.5; the difference is 0.7. Bonding would be polar covalent, with I = $\delta-$, and Si = $\delta+$.

PROBLEM 4.19
The elements H, N, O, P, and S are commonly bonded to carbon in organic compounds. Arrange these elements in order of increasing electronegativity.

PROBLEM 4.20
Use electronegativity differences to classify bonds between the following pairs of atoms as ionic, nonpolar covalent, or polar covalent. For those that are polar, use the symbols $\delta+$ and $\delta-$ to identify the location of the partial charges on the polar covalent bond.

(a) I and Cl   (b) Li and O
(c) Br and Br (d) P and Br

4.10 Polar Molecules

Just as individual bonds can be polar, entire *molecules* can be polar if electrons are attracted more strongly to one part of the molecule than to another. Molecular polarity is due to the sum of all individual bond polarities and lone-pair contributions in the molecule and is often represented by an arrow pointing in the direction that electrons are displaced. The arrow is pointed at the negative end and is crossed at the positive end to resemble a plus sign, $(\delta+) \longmapsto (\delta-)$.

Molecular polarity depends on the shape of the molecule as well as the presence of polar covalent bonds and lone pairs. In water, for example, electrons are displaced away from the less electronegative hydrogen atoms toward the more electronegative oxygen atom so that the net polarity points between the two O—H bonds. In chloromethane, CH_3Cl, electrons are attracted from the carbon/hydrogen part of the molecule toward

122 CHAPTER 4 Molecular Compounds

the electronegative chlorine atom so that the net polarity points along the C—Cl bond. Electrostatic potential maps show these polarities clearly, with electron-poor regions in blue and electron-rich regions in red.

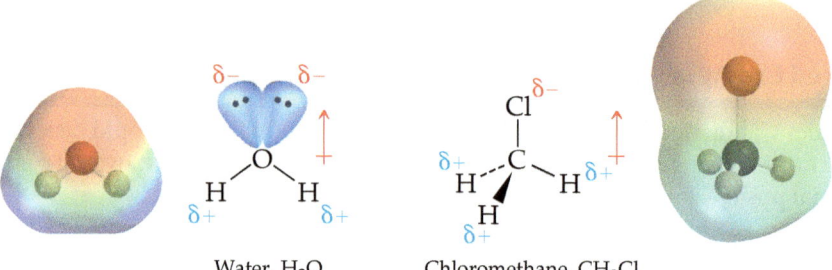

Water, H$_2$O   Chloromethane, CH$_3$Cl

Furthermore, just because a molecule has polar covalent bonds, it does not mean that the molecule is necessarily polar overall. Carbon dioxide (CO$_2$) and tetrachloromethane (CCl$_4$) molecules, for instance, have no net polarity because their symmetrical shapes cause the individual C=O and C—Cl bond polarities to cancel.

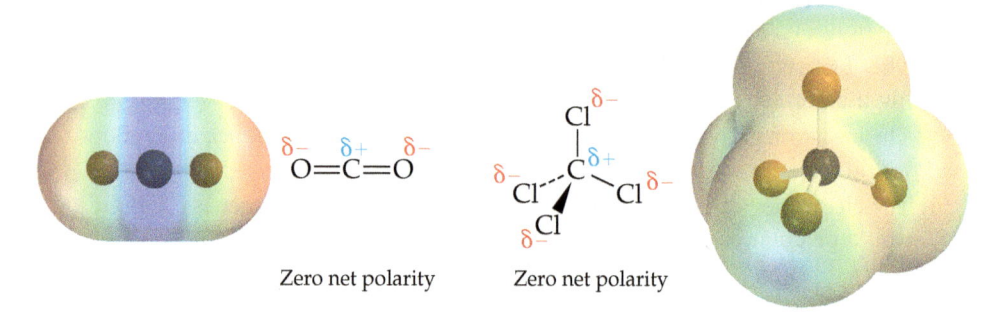

Zero net polarity   Zero net polarity

Polarity has a dramatic effect on the physical properties of molecules, particularly on melting points, boiling points, and solubilities. We will see numerous examples of such effects in subsequent chapters.

Worked Example 4.12 Electronegativity: Polar Bonds and Polar Molecules

Look at the structures of (a) hydrogen cyanide (HCN) and (b) vinyl chloride (H$_2$C=CHCl), described in Worked Examples 4.6 and 4.7, decide whether or not the molecules are polar, and show the direction of net polarity in each.

ANALYSIS Draw a Lewis structure for each molecule to find its shape, and identify any polar bonds using the electronegativity values in Figure 4.6. Then, decide on net polarity by adding the individual contributions.

SOLUTION

(a) The carbon atom in hydrogen cyanide has two charge clouds, making HCN a linear molecule. The C—H bond is relatively nonpolar, but the C≡N bonding electrons are pulled toward the electronegative nitrogen atom. In addition, a lone pair protrudes from nitrogen. Thus, the molecule has a net polarity:

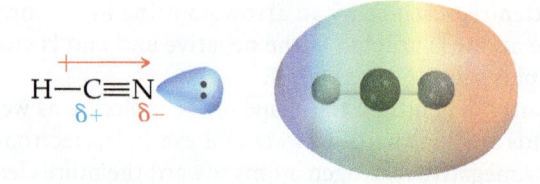

H—C≡N:

(b) Vinyl chloride, like ethylene, is a planar molecule. The C—H and C=C bonds are nonpolar, but the C—Cl bonding electrons are displaced toward the electronegative chlorine. Thus, the molecule has a net polarity:

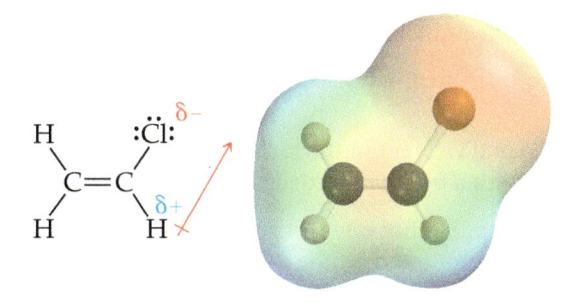

PROBLEM 4.21
Look at the molecular shape of formaldehyde (CH_2O) described on page 115, decide whether or not the molecule is polar, and show the direction of net polarity.

PROBLEM 4.22
Draw a Lewis structure for dimethyl ether (CH_3OCH_3), predict its shape, and tell whether or not the molecule is polar.

KEY CONCEPT PROBLEM 4.23
From this electrostatic potential map of methyllithium, identify the direction of net polarity in the molecule. Explain this polarity based on electronegativity values.

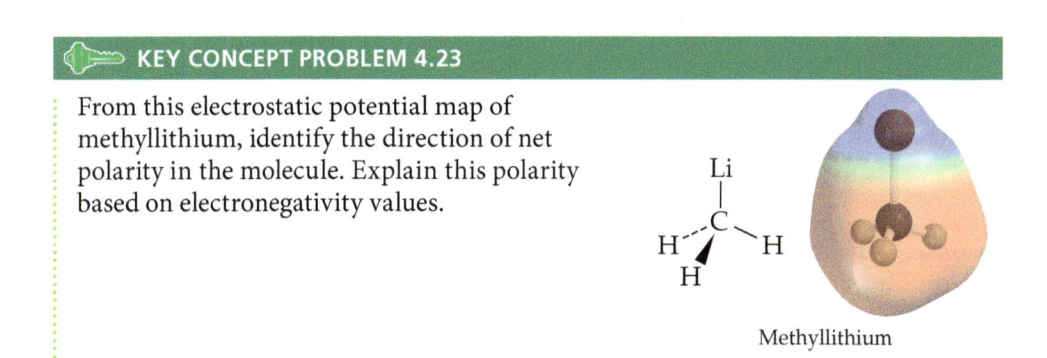

Methyllithium

4.11 Naming Binary Molecular Compounds

When two different elements combine, they form what is called a **binary compound**. The formulas of binary molecular compounds are usually written with the less electronegative element first. Thus, metals are always written before nonmetals, and a nonmetal farther left on the periodic table generally comes before a nonmetal farther right. For example,

Binary compound A compound formed by combination of two different elements.

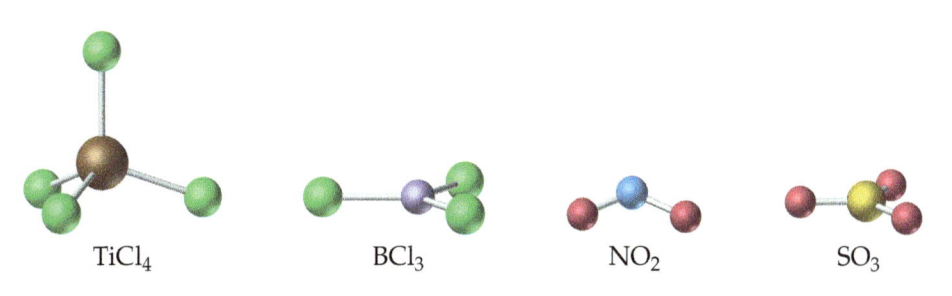

$TiCl_4$ BCl_3 NO_2 SO_3

As we learned in Section 3.9, the formulas of ionic compounds indicate the number of anions and cations necessary for a neutral formula unit, which depends on the charge on each of the ions. With molecular compounds, however, many combinations of atoms are possible, since nonmetals are capable of forming multiple covalent bonds.

124   CHAPTER 4   Molecular Compounds

TABLE 4.3 Numerical Prefixes Used in Chemical Names	
Number	Prefix
1	mono-
2	di-
3	tri-
4	tetra-
5	penta-
6	hexa-
7	hepta-
8	octa-
9	nona-
10	deca-

When naming binary molecular compounds, therefore, we must identify exactly how many atoms of each element are included in the molecular formula. The names of binary molecular compounds are assigned in two steps, using the prefixes listed in Table 4.3 to indicate the number of atoms of each element combined.

STEP 1: Name the first element in the formula, using a prefix if needed to indicate the number of atoms.

STEP 2: Name the second element in the formula, using an *-ide* ending like for anions (Section 3.7), along with a prefix if needed.

The prefix *mono-*, meaning one, is omitted except where needed to distinguish between two different compounds with the same elements. For example, the two oxides of carbon are named carbon *mon*oxide for CO and carbon *di*oxide for CO_2. (Note that we say *mon*oxide rather than *mono*oxide.) Some other examples are:

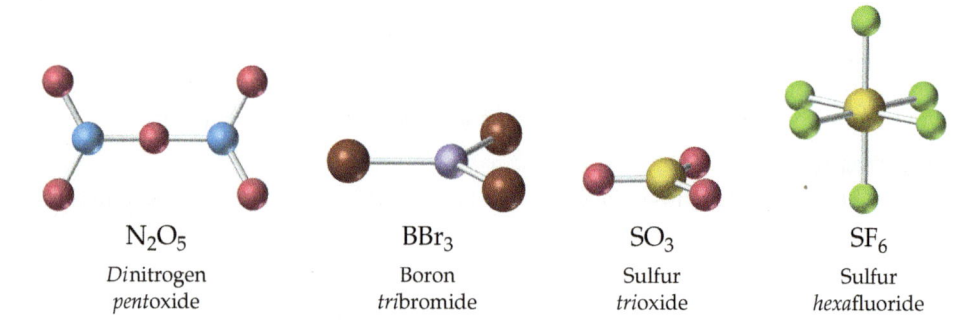

N_2O_5 — *Di*nitrogen *pent*oxide

BBr_3 — Boron *tri*bromide

SO_3 — Sulfur *tri*oxide

SF_6 — Sulfur *hexa*fluoride

Naming of molecular compounds can get complicated when more than two elements are present. This is particularly true for *organic compounds*, a class of molecular compounds composed largely of carbon (see examples in the Chemistry in Action on p. 125). The rules for naming these compounds will be discussed in later chapters.

Worked Example 4.13 Naming Molecular Compounds

Name the following compounds:

(a) N_2O_3 (b) $GeCl_4$ (c) PCl_5

SOLUTION

(a) Dinitrogen trioxide   (b) Germanium tetrachloride
(c) Phosphorus pentachloride

Worked Example 4.14 Writing Formulas for Molecular Compounds

Write molecular formulas for the following compounds:

(a) Nitrogen triiodide   (b) Silicon tetrachloride
(c) Carbon disulfide

SOLUTION

(a) NI_3 (b) $SiCl_4$ (c) CS_2

PROBLEM 4.24
Name the following compounds:
(a) S_2Cl_2 (b) ICl (c) ICl_3

PROBLEM 4.25
Write formulas for the following compounds:
(a) Selenium tetrafluoride   (b) Diphosphorus pentoxide   (c) Bromine trifluoride

PROBLEM 4.26

Geraniol, one of the components of rose oil (see the following Chemistry in Action), has the basic structure represented below. Draw the structural formula for geraniol to include any multiple bonds, and then write the condensed structure for geraniol.

$$CH_3-\underset{\underset{CH_3}{|}}{C}-\underset{\underset{H}{|}}{\overset{\overset{H}{|}}{C}}-\underset{\underset{H}{|}}{\overset{\overset{H}{|}}{C}}-\underset{\underset{H}{|}}{\overset{\overset{CH_3}{|}}{C}}-\underset{\underset{H}{|}}{\overset{\overset{H}{|}}{C}}-\underset{\underset{H}{|}}{\overset{\overset{H}{|}}{C}}-OH$$

CHEMISTRY IN ACTION

Damascenone by Any Other Name Would Smell as Sweet

What's in a name? According to Shakespeare's *Romeo and Juliet*, a rose by any other name would smell as sweet. Chemical names, however, often provoke less favorable responses: "It's unpronounceable." "It's too complicated." "It must be something bad."

But why are chemical names so complicated? The reason is obvious once you realize that there are more than 19 *million* known chemical compounds. The full name of a chemical compound has to include enough information to tell chemists the composition and structure of the compound. It is as if every person on earth had to have his or her own unique name that described height, hair color, and other identifying characteristics in sufficient detail to distinguish him or her from every other person. Consider, also, that subtle differences in structure can result in significant differences in chemical or physical properties. Geraniol, for example, is used as a flavor additive in the food industry, while citronellol is used in perfumes and insect repellants, such as citronella candles. The common names for these substances are easier to remember, but their *chemical* names give us precise information about their structural differences and similarities. Geraniol ($C_{10}H_{18}O$), also known as *3,7-dimethylocta-2, 6-dien-1-ol* differs from citronellol ($C_{10}H_{20}O$ or *3,7-dimethyloct-6-en-1-ol*) by only one C—C double bond.

Unfortunately, people sometimes conclude that everything with a chemical name is unnatural and dangerous. Neither is true, of course. Acetaldehyde, for instance, is present naturally in most tart, ripe fruits and is often added in small amounts to artificial flavorings. When *pure*, however, acetaldehyde

▲ The scent of these roses contains β-damascenone, β-ionone, citronellol, geraniol, nerol, eugenol, methyl eugenol, β-phenylethyl, alcohol, farnesol, linalool, terpineol, rose oxide, carvone, and many other natural substances.

is also a flammable gas that is toxic and explosive in high concentrations.

Similar comparisons of desirable and harmful properties can be made for almost all chemicals, including water, sugar, and salt. The properties of a substance and the conditions surrounding its use must be evaluated before judgments are made. Damascenone, geraniol, and citronellol, by the way, are chemicals that contribute to the wonderful aroma of roses.

See Chemistry in Action Problems 4.93 and 4.94 at the end of the chapter.

SUMMARY: REVISITING THE CHAPTER GOALS

1. What is a covalent bond? A *covalent bond* is formed by the sharing of electrons between atoms rather than by the complete transfer of electrons from one atom to another. Atoms that share 2 electrons are joined by a *single bond* (such as C—C), atoms that share 4 electrons are joined by a *double bond* (such as C=C), and atoms that share 6 electrons are joined by a *triple* bond (such as C≡C). The group of atoms held together by covalent bonds is called a *molecule*.

Electron sharing typically occurs when a singly occupied valence orbital on one atom *overlaps* a singly occupied valence orbital on another atom. The 2 electrons occupy both overlapping orbitals and belong to both atoms, thereby bonding the atoms

126 CHAPTER 4 Molecular Compounds

together. Alternatively, electron sharing can occur when a filled orbital containing an unshared, *lone pair* of electrons on one atom overlaps a vacant orbital on another atom to form a *coordinate covalent bond* (see Problems 33–35, 40, 41, 44, 45, 89, 92).

2. How does the octet rule apply to covalent bond formation? Depending on the number of valence electrons, different atoms form different numbers of covalent bonds. In general, an atom shares enough electrons to reach a noble gas configuration. Hydrogen, for instance, forms 1 covalent bond because it needs to share 1 more electron to achieve the helium configuration ($1s^2$). Carbon and other group 4A elements form 4 covalent bonds because they need to share 4 more electrons to reach an octet. In the same way, nitrogen and other group 5A elements form 3 covalent bonds, oxygen and other group 6A elements form 2 covalent bonds, and halogens (group 7A elements) form 1 covalent bond (see Problems 38, 39, 50, 51, 95).

3. What are the major differences between ionic and molecular compounds? *Molecular compounds* can be gases, liquids, or low-melting solids. They usually have lower melting points and boiling points than ionic compounds, many are water insoluble, and they do not conduct electricity when melted or dissolved (see Problems 33, 35–37, 42, 43, 47, 99, 102, 103).

4. How are molecular compounds represented? Formulas such as H_2O, NH_3, and CH_4, which show the numbers and kinds of atoms in a molecule, are called *molecular formulas*. More useful are *Lewis structures*, which show how atoms are connected in molecules. Covalent bonds are indicated as lines between atoms, and valence electron lone pairs are shown as dots. Lewis structures are drawn by counting the total number of valence electrons in a molecule or polyatomic ion and then placing shared pairs (bonding) and lone pairs (nonbonding) so that all electrons are accounted for (see Problems 30, 46–66, 94–100, 104–109).

5. What is the influence of valence-shell electrons on molecular shape? Molecules have specific shapes that depend on the number of electron charge clouds (bonds and lone pairs) surrounding the various atoms. These shapes can often be predicted using the *valence-shell electron-pair repulsion* (*VSEPR*) model. Atoms with two electron charge clouds adopt linear geometry, atoms with three charge clouds adopt trigonal planar geometry, and atoms with four charge clouds adopt tetrahedral geometry (see Problems 27, 28–31, 67–72, 81, 96, 100, 109).

6. When are bonds and molecules polar? Bonds between atoms are *polar covalent* if the bonding electrons are not shared equally between the atoms. The ability of an atom to attract electrons in a covalent bond is the atom's *electronegativity* and is highest for reactive nonmetal elements on the upper right of the periodic table and lowest for metals on the lower left. Comparing electronegativities allows prediction of whether a given bond is covalent, polar covalent, or ionic. Just as individual bonds can be polar, entire molecules can be polar if electrons are attracted more strongly to one part of the molecule than to another. Molecular polarity is due to the sum of all individual bond polarities and lone-pair contributions in the molecule (see Problems 32, 73–84, 96, 97, 101).

CONCEPT MAP: ELECTROSTATIC FORCES

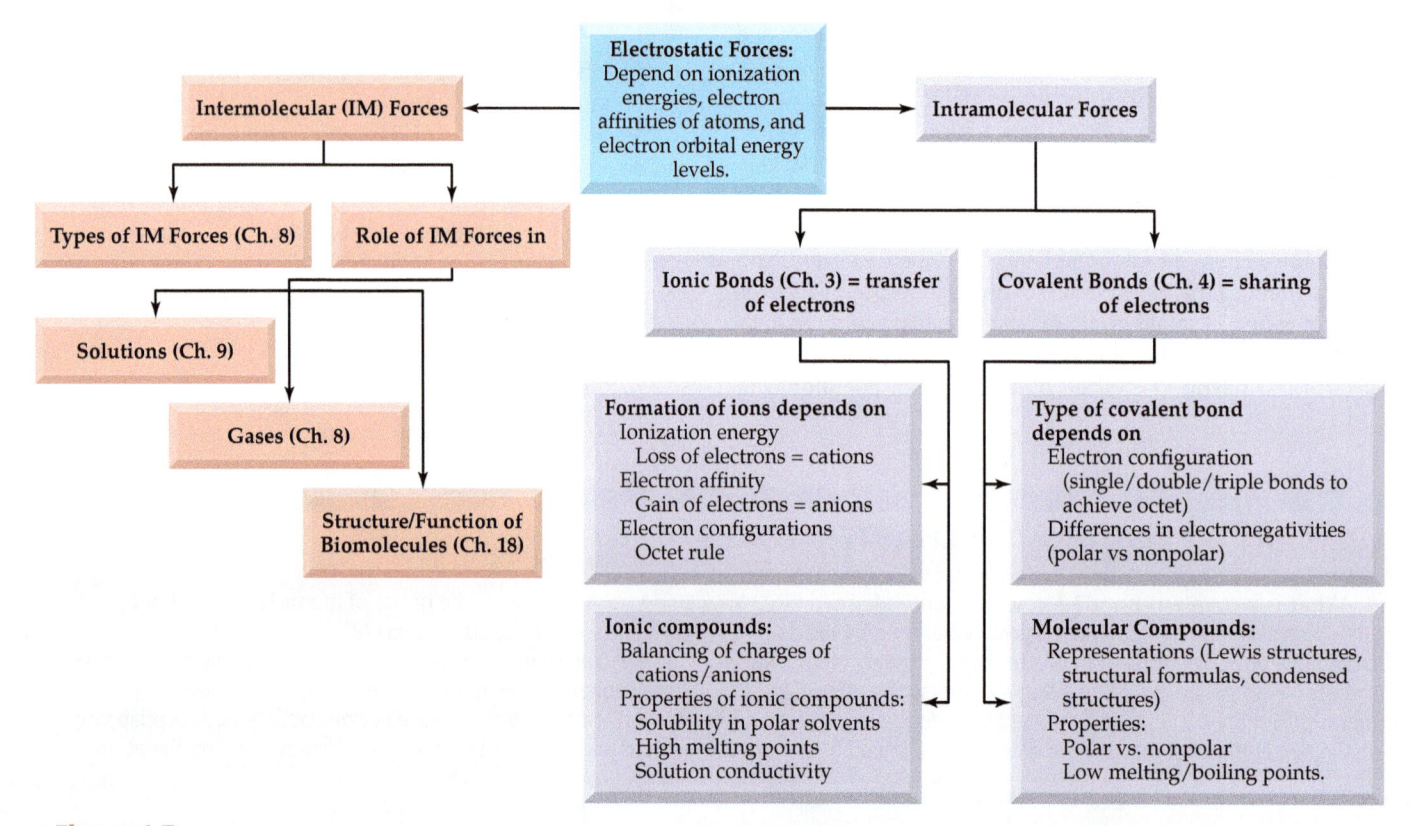

▲ Figure 4.7

Concept Maps. Chemistry, like most subjects, makes more sense when presented in context. When we understand the connections between concepts, or how one idea leads to another, it becomes easier to see the "big picture" and to appreciate why a certain concept is important. A concept map is one way of illustrating those connections and providing a context for what we have learned and what we will be learning in later chapters.

As you can see from the concept map in Figure 4.7, the electronic structure of atoms discussed in Chapter 2 plays a critical role in the chemical behavior of an element, specifically in terms of its tendency to form ionic compounds (Chapter 3) or molecular compounds (Chapter 4). Furthermore, the nature of the attractive forces between particles (intermolecular versus intramolecular) plays a role in the physical and chemical behavior of substances discussed in later chapters.

As we continue exploring new topics, we will expand certain areas of this concept map or add new branches as needed.

KEY WORDS

Binary compound, *p. 123*
Bond angle, *p. 115*
Bond length, *p. 100*
Condensed structure, *p. 109*
Coordinate covalent bond, *p. 106*
Covalent bond, *p. 99*
Double bond, *p. 104*

Electronegativity, *p. 120*
Lewis structure, *p. 108*
Lone pair, *p. 108*
Molecular compound, *p. 101*
Molecular formula, *p. 108*
Molecule, *p. 99*
Polar covalent bond, *p. 119*

Regular tetrahedron, *p. 116*
Single bond, *p. 104*
Structural formula, *p. 108*
Triple bond, *p. 104*
Valence-shell electron-pair repulsion (VSEPR) model, *p. 114*

UNDERSTANDING KEY CONCEPTS

4.27 What is the geometry around the central atom in the following molecular models? (There are no "hidden" atoms; all atoms in each model are visible.)

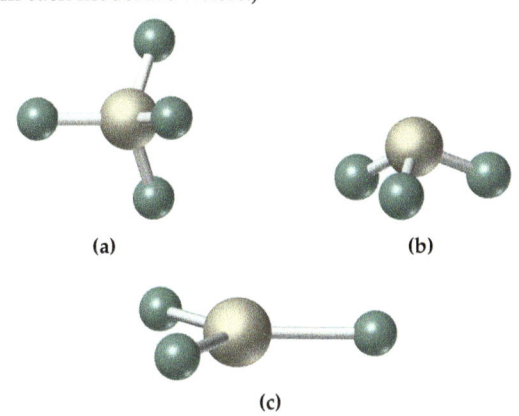

(a)   (b)

(c)

4.28 Three of the following molecular models have a tetrahedral central atom, and one does not. Which is the odd one?

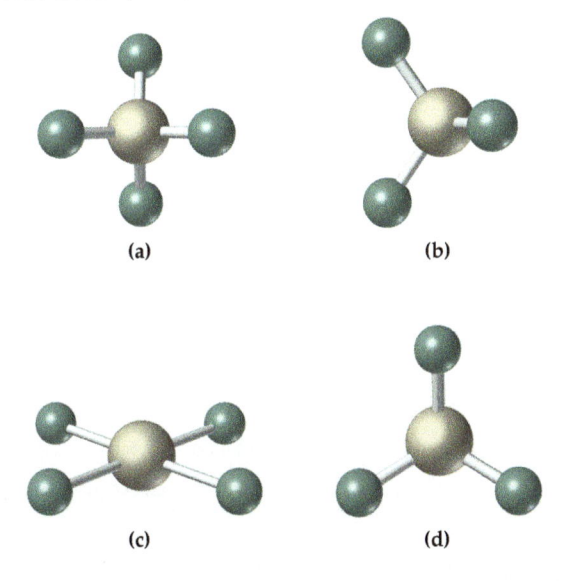

(a)   (b)

(c)   (d)

4.29 The ball-and-stick molecular model shown here is a representation of acetaminophen, the active ingredient in over-the-counter headache remedies such as Tylenol. The lines indicate only the connections between atoms, not whether the bonds are single, double, or triple (red = O, gray = C, blue = N, ivory = H).

(a) What is the molecular formula of acetaminophen?
(b) Indicate the positions of the multiple bonds in acetaminophen.
(c) What is the geometry around each carbon and each nitrogen?

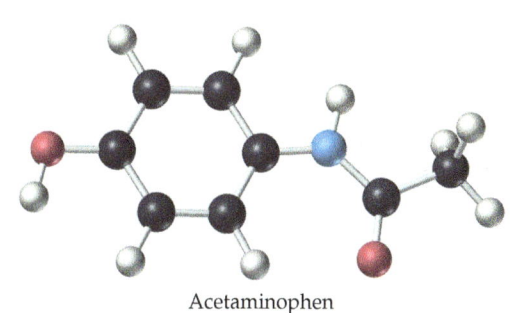

Acetaminophen

4.30 The atom-to-atom connections in vitamin C (ascorbic acid) are as shown here. Convert this skeletal drawing to a Lewis electron-dot structure for vitamin C by showing the positions of any multiple bonds and lone pairs of electrons.

Vitamin C

4.31 The ball-and-stick molecular model shown here is a representation of thalidomide, a drug that has been approved for treating leprosy, but causes severe birth defects when taken by expectant mothers. The lines indicate only the connections between atoms, not whether the bonds are single, double, or triple (red = O, gray = C, blue = N, ivory = H).

(a) What is the molecular formula of thalidomide?
(b) Indicate the positions of the multiple bonds in thalidomide.
(c) What is the geometry around each carbon and each nitrogen?

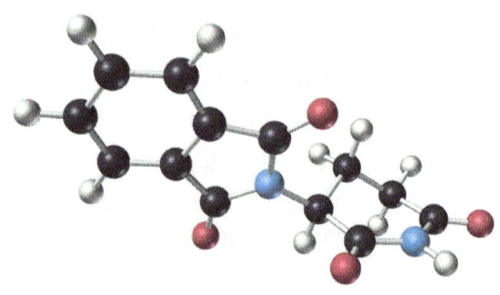

Thalidomide

4.32 Show the position of any electron lone pairs in this structure of acetamide, and indicate the electron-rich and electron-poor regions.

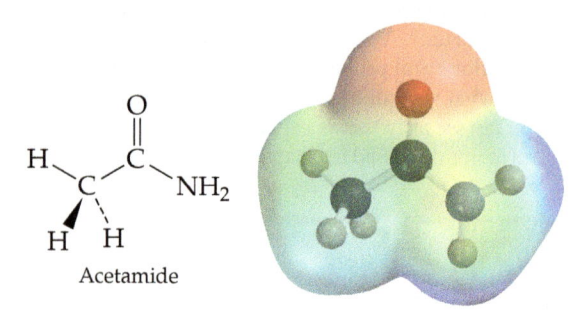

Acetamide

ADDITIONAL PROBLEMS

COVALENT BONDS

4.33 What is a covalent bond, and how does it differ from an ionic bond?

4.34 What is a coordinate covalent bond, and how does it differ from a covalent bond?

4.35 Which of the following elements would you expect to form (i) diatomic molecules, (ii) mainly covalent bonds, (iii) mainly ionic bonds, (iv) both covalent and ionic bonds? (More than one answer may apply; remember that some nonmetals can form ionic bonds with metals.)

(a) Oxygen   (b) Potassium   (c) Phosphorus
(d) Iodine   (e) Hydrogen   (f) Cesium

4.36 Identify the bonds formed between the following pairs of atoms as either covalent or ionic.

(a) Aluminum and bromine (b) Carbon and fluorine
(c) Cesium and iodine   (d) Zinc and fluorine
(e) Lithium and chlorine

4.37 Write electron-dot symbols to show the number of covalent bonds and the lone pairs of electrons in the molecules that are formed by reactions between the atoms in Problem 4.36.

4.38 Look up tellurium ($Z = 52$) in the periodic table and predict how many covalent bonds it is likely to form. Explain.

4.39 Look up antimony in the periodic table ($Z = 51$). How many covalent bonds would you expect it to form? Based on this information, which of the following antimony compounds is covalent and which is ionic: $SbCl_3$ or $SbCl_5$?

4.40 Which of the following contains a coordinate covalent bond? (Hint: how many covalent bonds would you expect the central atom (underlined) to form?)

(a) $\underline{Pb}Cl_2$ (b) $\underline{Cu}(NH_3)_4^{2+}$ (c) $\underline{N}H_4^+$

4.41 Which of the following contains a coordinate covalent bond? (Hint: how many covalent bonds would you expect the central atom (underlined) to form?)

(a) $H_2\underline{O}$ (b) $\underline{B}F_4^-$ (c) $H_3\underline{O}^+$

4.42 Tin forms both an ionic compound and a covalent compound with chlorine. The ionic compound is $SnCl_2$. Is the covalent compound more likely to be $SnCl_3$, $SnCl_4$, or $SnCl_5$? Explain.

4.43 A compound of gallium with chlorine has a melting point of 77 °C and a boiling point of 201 °C. Is the compound ionic or covalent? What is a likely formula?

4.44 Nitrous oxide, N_2O, has the following structure. Which bond in N_2O is a coordinate covalent bond?

:N≡N—Ö:
Nitrous oxide

4.45 Thionyl chloride, $SOCl_2$ has the following structure. Which bond in $SOCl_2$ is a coordinate covalent bond?

Thionyl chloride

STRUCTURAL FORMULAS

4.46 Distinguish between the following:

(a) A molecular formula and a structural formula
(b) A structural formula and a condensed structure
(c) A lone pair and a shared pair of electrons

4.47 Assume that you are given samples of two white crystalline compounds, one of them ionic and one covalent. Describe how you might tell which is which.

4.48 Determine the total number of valence electrons in the following molecules. If the molecule contains multiple bonds, indicate where the multiple bonds are located and whether they are double or triple bonds.

(a) N_2 (b) NOCl
(c) CH_3CH_2CHO (d) OF_2

4.49 Add lone pairs where appropriate to the following structures:

(a) C≡O (b) CH_3SH

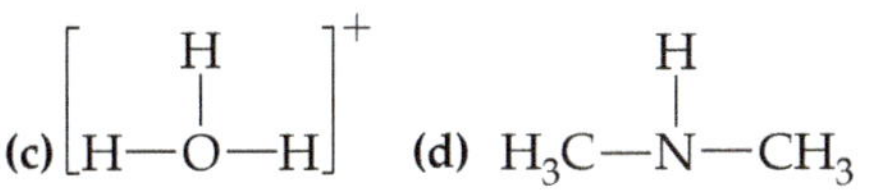

(c) $[H-O-H]^+$ (d) $H_3C-\overset{H}{\underset{}{N}}-CH_3$

4.50 If a research paper appeared reporting the structure of a new molecule with formula C_2H_8 most chemists would be highly skeptical. Why?

4.51 Consider the following possible structural formulas for $C_3H_6O_2$. If a structure is not reasonable, explain what changes could be made to convert it to a reasonable structure.

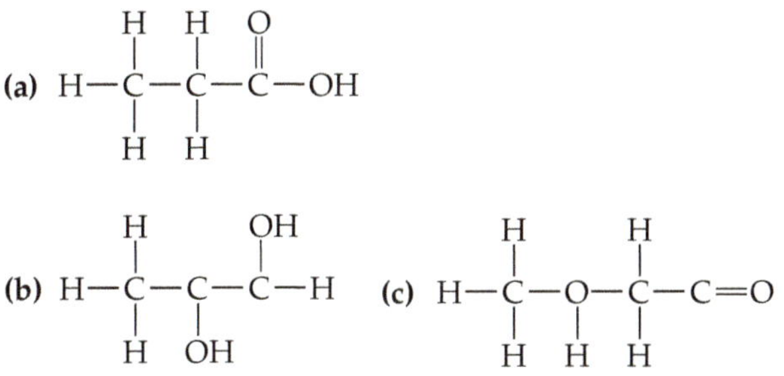

(a) H—C(H)(H)—C(H)(H)—C(=O)—OH

(b) H—C(H)(H)—C(OH)(OH)—C(H)(H)—H

(c) H—C(H)(H)—O—C(H)(H)—C=O with H

4.52 Convert the following Lewis structures into structural formulas in which lines replace the bonding electrons. Include the lone pairs.

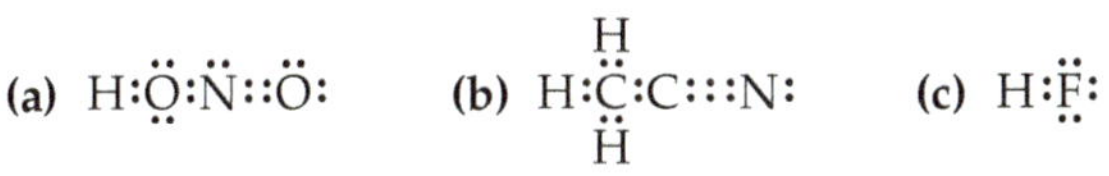

(a) H:Ö:N::Ö: (b) H:C̈:C:::N: with H (c) H:F̈:

4.53 Convert the following Lewis structure for the nitrate ion into a line structure that includes the lone pairs. Why does the nitrate ion have a −1 charge? $[:\ddot{O}:N:\ddot{O}:]$ with :Ö: on top

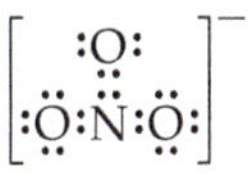

4.54 Convert the following structural formulas into condensed structures.

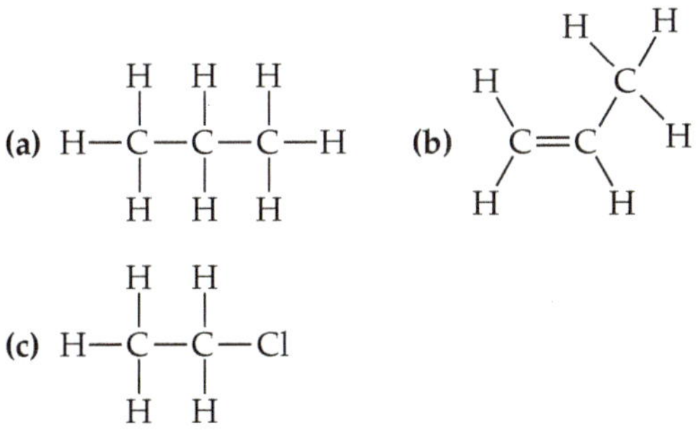

(a) H—C(H)(H)—C(H)(H)—C(H)(H)—H

(b) H,H\C=C/H,H

(c) H—C(H)(H)—C(H)(H)—Cl

4.55 Expand the following condensed structures into the correct structural formulas.

(a) $CH_3CH_2COCH(CH_3)_2$ (b) $CH_3CH_2COOCH_3$
(c) $CH_3CH_2OCH_2Cl$

4.56 Acetic acid is the major organic constituent of vinegar. Convert the following structural formula of acetic acid into a condensed structure similar to those shown in Problem 4.55.

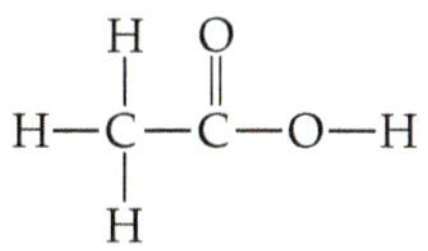

H—C(H)(H)—C(=O)—O—H

DRAWING LEWIS STRUCTURES

4.57 Draw a Lewis structure for the following molecules:

(a) SF_6 (b) $AlCl_3$ (c) CS_2 (d) SeF_4
(e) $BeCl_2$ (Note: this molecule does not follow the octet rule.)
(f) N_2O_4

4.58 Draw a Lewis structure for the following molecules:

(a) Nitrous acid, HNO_2 (H is bonded to an O atom)
(b) Ozone, O_3
(c) Acetaldehyde, CH_3CHO

4.59 Ethanol, or "grain alcohol," has the formula C_2H_6O and contains an O—H bond. Propose a structure for ethanol that is consistent with common bonding patterns.

4.60 Dimethyl ether has the same molecular formula as ethanol (Problem 4.59) but very different properties. Propose a structure for dimethyl ether in which the oxygen is bonded to two carbons.

4.61 Hydrazine, a substance used to make rocket fuel, has the formula N_2H_4. Propose a structure for hydrazine.

4.62 Tetrachloroethylene, C_2Cl_4, is used commercially as a dry-cleaning solvent. Propose a structure for tetrachloroethylene based on the common bonding patterns expected in organic molecules. What kind of carbon-carbon bond is present?

4.63 Dimethyl sulfoxide, also known as DMSO, is an important organic solvent often used for drug delivery since it readily penetrates the skin. The formula for DMSO is $(CH_3)_2SO$. Draw a Lewis structure of DMSO; both C atoms are attached to the S atom.

4.64 Draw a Lewis structure for hydroxylamine, NH_2OH.

4.65 The carbonate ion, CO_3^{2-}, contains a double bond. Draw a Lewis structure for the ion and show why it has a charge of −2.

4.66 Draw a Lewis structure for the following polyatomic ions:

(a) Formate, HCO_2^- (b) Sulfite, SO_3^{2-}
(c) Thiocyanate, SCN^- (d) Phosphate, PO_4^{3+}
(e) Chlorite, ClO_2^- (chlorine is the central atom)

MOLECULAR GEOMETRY

4.67 Predict the geometry and bond angles around atom A for molecules with the general formulas AB_3 and AB_2E, where B represents another atom and E represents an electron pair.

4.68 Predict the geometry and bond angles around atom A for molecules with the general formulas AB_4, AB_3E, and AB_2E_2, where B represents another atom and E represents an electron pair.

4.69 Sketch the three-dimensional shape of the following molecules:

(a) Methylamine, CH_3NH_2 (b) Iodoform, CHI_3
(c) Ozone, O_3
(d) Phosphorus pentachloride, PCl_5
(e) Chloric acid, $HClO_3$

4.70 Predict the three-dimensional shape of the following molecules:

(a) SiF_4 (b) CF_2Cl_2 (c) SO_3
(d) BBr_3 (e) NF_3

4.71 Predict the geometry around each carbon atom in the amino acid alanine.

$$CH_3CHCOH$$ with =O above C and NH_2 below (Alanine)

4.72 Predict the geometry around each carbon atom in vinyl acetate, a precursor of the polyvinyl alcohol polymer used in automobile safety glass.

$$H_2C=CH-O-C(=O)-CH_3$$ Vinyl acetate

POLARITY OF BONDS AND MOLECULES

4.73 Where in the periodic table are the most electronegative elements found, and where are the least electronegative elements found?

4.74 Predict the electronegativity of the yet-undiscovered element with $Z = 119$.

4.75 Look at the periodic table, and then order the following elements according to increasing electronegativity: K, Si, Be, O, B.

4.76 Look at the periodic table, and then order the following elements according to decreasing electronegativity: C, Ca, Cs, Cl, Cu.

4.77 Which of the following bonds are polar? If a bond is polar, identify the negative and positive ends of each bond by using $\delta+$ and $\delta-$.

(a) I—Br (b) O—H
(c) C—F (d) N—C
(e) C—C

4.78 Which of the following bonds are polar? If a bond is polar, identify the negative and positive ends of each bond by using $\delta+$ and $\delta-$.

(a) O—Cl (b) N—Cl
(c) P—H (d) C—I
(e) C—O

4.79 Based on electronegativity differences, would you expect bonds between the following pairs of atoms to be largely ionic or largely covalent?

(a) Be and F (b) Ca and Cl
(c) O and H (d) Be and Br

4.80 Arrange the following molecules in order of the increasing polarity of their bonds:

(a) HCl (b) PH_3
(c) H_2O (d) CF_4

4.81 Ammonia, NH_3, and phosphorus trihydride, PH_3, both have a trigonal pyramid geometry. Which one is more polar? Explain.

4.82 Decide whether each of the compounds listed in Problem 4.80 is polar, and show the direction of polarity.

4.83 Carbon dioxide is a nonpolar molecule, whereas sulfur dioxide is polar. Draw Lewis structures for each of these molecules to explain this observation.

4.84 Water (H_2O) is more polar than hydrogen sulfide (H_2S). Explain.

NAMES AND FORMULAS OF MOLECULAR COMPOUNDS

4.85 Name the following binary compounds:

(a) PI_3 (b) $AsCl_3$ (c) P_4S_3
(d) Al_2F_6 (e) N_2O_5 (f) $AsCl_5$

4.86 Name the following compounds:

(a) SeO_2 (b) XeO_4
(c) N_2S_5 (d) P_3Se_4

4.87 Write formulas for the following compounds:

(a) Nitrogen dioxide (b) Sulfur hexafluoride
(c) Bromine triiodide (d) Dinitrogen trioxide
(e) Nitrogen triiodide (f) Iodine heptafluoride

4.88 Write formulas for the following compounds:

(a) Silicon tetrachloride (b) Sodium hydride
(c) Antimony pentafluoride (d) Osmium tetroxide

CHEMISTRY IN ACTION

4.89 The CO molecule is highly reactive and will bind to the Fe^{2+} ion in hemoglobin and interfere with O_2 transport. What type of bond is formed between the CO molecule and the Fe^{2+} ion? [CO and NO: Pollutants or Miracle Molecules?, p. 113]

4.90 What is a vasodilator, and why would it be useful in treating hypertension (high blood pressure)? [*CO and NO: Pollutants or Miracle Molecules?, p. 113*]

4.91 How is a polymer formed? [*VERY Big Molecules, p. 118*]

4.92 Do any polymers exist in nature? Explain. [*VERY Big Molecules, p. 118*]

4.93 Why are many chemical names so complex? [*Damascenone by Any Other Name, p. 125*]

4.94 Citronellol, one of the compounds found in the scent of roses, is also used in perfumes and in insect repellent products. Write the condensed formula from the structural formula of citronellol shown below. [*Damascenone by Any Other Name, p. 125*]

$$CH_3-C(CH_3)=C(H)-C(H_2)-C(H_2)-C(CH_3)(H)-C(H_2)-OH$$

GENERAL QUESTIONS AND PROBLEMS

4.95 The discovery in the 1960s that xenon and fluorine react to form a molecular compound was a surprise to most chemists, because it had been thought that noble gases could not form bonds.

(a) Why was it thought that noble gases could not form bonds?

(b) Draw a Lewis structure of XeF_4 in which Xe is the central atom. How many electron clouds are there on the central atom?

(c) What type of bonds are the Xe—F bonds? Explain.

4.96 Acetone, a common solvent used in some nail polish removers, has the molecular formula C_3H_6O and contains a carbon-oxygen double bond.

(a) Propose two Lewis structures for acetone.

(b) What is the geometry around the carbon atoms in each of the structures?

(c) Which of the bonds in each structure are polar?

4.97 Draw the structural formulas for two compounds having the molecular formula C_2H_4O. What is the molecular geometry around the carbon atoms in each of these molecules? Would these molecules be polar or nonpolar? (Hint: there is one double bond.)

4.98 The following formulas are unlikely to be correct. What is wrong with each?

(a) CCl_3 (b) N_2H_5
(c) H_3S (d) C_2OS

4.99 Which of the compounds (a) through (d) contain ionic bonds? Which contain covalent bonds? Which contain coordinate covalent bonds? (A compound may contain more than one type of bond.)

(a) $BaCl_2$ (b) $Ca(NO_3)_2$
(c) BCl_4^- (d) $TiBr_4$

4.100 The phosphonium ion, PH_4^+, is formed by reaction of phosphine, PH_3, with an acid.

(a) Draw the Lewis structure of the phosphonium ion.

(b) Predict its molecular geometry.

(c) Describe how a fourth hydrogen can be added to PH_3.

(d) Explain why the ion has a +1 charge.

4.101 Compare the trend in electronegativity seen in Figure 4.6 (p. 120) with the trend in electron affinity shown in Figure 3.1 (p. 75). What similarities do you see? What differences? Explain.

4.102 Name the following compounds. Be sure to determine whether the compound is ionic or covalent so that you use the proper rules.

(a) $CaCl_2$ (b) $TeCl_2$ (c) BF_3
(d) $MgSO_4$ (e) K_2O (f) FeF_3
(g) PF_3

4.103 Titanium forms both molecular and ionic compounds with nonmetals, as, for example, $TiBr_4$ and TiO_2. One of these compounds has a melting point of 39 °C; and the other has a melting point of 1825 °C. Which is ionic and which is molecular? Explain your answer in terms of electronegativities of the atoms involved in each compound.

4.104 Draw a Lewis structure for chloral hydrate, known in detective novels as "knockout drops." Indicate all lone pairs.

$$Cl_3C-C(H)(OH)-O-H \quad \text{Chloral hydrate}$$

4.105 The dichromate ion, $Cr_2O_7^{2-}$, has neither Cr—Cr nor O—O bonds. Draw a Lewis structure.

4.106 Oxalic acid, $H_2C_2O_4$, is a substance found in uncooked spinach leaves and other greens that can be poisonous at high concentrations (for example, in raw rhubarb leaves). If oxalic acid has a C—C single bond and the H atoms are both connected to O atoms, draw its Lewis structure.

4.107 Identify the fourth row elements represented by "X" in the following compounds.

(a) $\ddot{O}=\ddot{X}=\ddot{O}$ (b) $:\ddot{F}-\ddot{X}-\ddot{F}:$

4.108 Write Lewis structures for molecules with the following connections, showing the positions of any multiple bonds and lone pairs of electrons.

(a) $Cl-C(=O)-O-C(H)-H$ (b) $H-C(H)(H)-C(H)-C(H)(H)-H$

4.109 Electron-pair repulsion influences the shapes of polyatomic ions in the same way it influences neutral molecules. Draw electron-dot symbols and predict the shape of the ammonium ion, NH_4^+, the sulfate ion, SO_4^{2-}, and the phosphite ion, PO_3^{3-}.

CHAPTER 5

Classification and Balancing of Chemical Reactions

CONTENTS

- **5.1** Chemical Equations
- **5.2** Balancing Chemical Equations
- **5.3** Classes of Chemical Reactions
- **5.4** Precipitation Reactions and Solubility Guidelines
- **5.5** Acids, Bases, and Neutralization Reactions
- **5.6** Redox Reactions
- **5.7** Recognizing Redox Reactions
- **5.8** Net Ionic Equations

◀ Water reclamation and purification plants utilize the many different types of chemical reactions discussed in this chapter.

CHAPTER GOALS

1. **How are chemical reactions written?**
   THE GOAL: Given the identities of reactants and products, be able to write a balanced chemical equation or net ionic equation.

2. **How are chemical reactions of ionic compounds classified?**
   THE GOAL: Be able to recognize precipitation, acid–base neutralization, and redox reactions.

3. **What are oxidation numbers, and how are they used?**
   THE GOAL: Be able to assign oxidation numbers to atoms in compounds and identify the substances oxidized and reduced in a given reaction. (◂◂◂ A.)

4. **What is a net ionic equation?**
   THE GOAL: Be able to recognize spectator ions and write the net ionic equation for reactions involving ionic compounds. (◂◂◂ A, B.)

CONCEPTS TO REVIEW

A. Periodic Properties and Ion Formation
(Section 3.2)

B. H^+ and OH^- Ions: An Introduction to Acids and Bases
(Section 3.11)

A log burns in the fireplace, an oyster makes a pearl, a seed grows into a plant—these and almost all the other changes you see taking place around you are the result of chemical reactions. The study of how and why chemical reactions happen is a major part of chemistry, providing information that is both fascinating and practical. In this chapter, we will begin to look at chemical reactions, starting with a discussion of how to represent them in writing. We will then examine how to balance reactions and how to recognize different types or classes of chemical reactions.

5.1 Chemical Equations

One way to view chemical reactions is to think of them as "recipes." Like recipes, all the "ingredients" in a chemical equation and their relative amounts are given, as well as the amount of product that would be obtained. Take, for example, a recipe for making S'mores, a concoction of chocolate, marshmallows, and graham crackers, which could be written as:

Graham crackers + Roasted marshmallows + Chocolate bars ⟶ S'mores

This recipe, however, is simply a list of ingredients and gives no indication of the relative amounts of each ingredient, or how many s'mores we would obtain. A more detailed recipe would be:

2 Graham crackers + 1 Roasted marshmallow + $\frac{1}{4}$ Chocolate bar ⟶ 1 S'more

In this case, the relative amounts of each ingredient are given, as well as the amount of the final product.

Let us extend this analogy to a typical chemical reaction. When sodium bicarbonate is heated in the range 50–100 °C, sodium carbonate, water, and carbon dioxide are produced. In words, we might write the reaction as:

Sodium bicarbonate $\xrightarrow{\text{Heat}}$ Sodium carbonate + Water + Carbon dioxide

Just as in the recipe, the starting materials and final products are listed. Replacing the chemical names with formulas converts the word description of this reaction into a **chemical equation:**

$$2\,\underbrace{NaHCO_3}_{\text{Reactant}} \xrightarrow{\text{Heat}} \underbrace{Na_2CO_3 + H_2O + CO_2}_{\text{Products}}$$

Look at how this equation is written. The **reactants** are written on the left, the **products** are written on the right, and an arrow is placed between them to indicate a chemical change. Conditions necessary for the reaction to occur—heat in this particular instance—are often specified above the arrow.

Why is the number 2 placed before $NaHCO_3$ in the equation? The 2 is necessary because of a fundamental law of nature called the **law of conservation of mass,** which states that matter can neither be created nor destroyed in a chemical reaction.

Chemical equation An expression in which symbols and formulas are used to represent a chemical reaction.

Reactant A substance that undergoes change in a chemical reaction and is written on the left side of the reaction arrow in a chemical equation.

Product A substance that is formed in a chemical reaction and is written on the right side of the reaction arrow in a chemical equation.

Balanced equation A chemical equation in which the numbers and kinds of atoms are the same on both sides of the reaction arrow.

Coefficient A number placed in front of a formula to balance a chemical equation.

The bonds between atoms in the reactants are rearranged to form new compounds in chemical reactions, but none of the atoms disappear and no new ones are formed. As a consequence, chemical equations must be **balanced**, meaning that *the numbers and kinds of atoms must be the same on both sides of the reaction arrow.*

Law of conservation of mass Matter is neither created nor destroyed in chemical reactions.

The numbers placed in front of formulas to balance equations are called **coefficients**, and they multiply all the atoms in a formula. Thus, the symbol "2 NaHCO$_3$" indicates two units of sodium bicarbonate, which contain 2 Na atoms, 2 H atoms, 2 C atoms, and 6 O atoms (2 × 3 = 6, the coefficient times the subscript for O). Count the numbers of atoms on the right side of the equation to convince yourself that it is indeed balanced.

The substances that take part in chemical reactions may be solids, liquids, or gases, or they may be dissolved in a solvent. Ionic compounds, in particular, frequently undergo reactions in *aqueous solution*—that is, when they are dissolved in water. Sometimes this information is added to an equation by placing the appropriate abbreviations after the formulas:

$$\begin{array}{cccc} (s) & (l) & (g) & (aq) \\ \text{Solid} & \text{Liquid} & \text{Gas} & \text{Aqueous solution} \end{array}$$

Thus, the decomposition of solid sodium bicarbonate can be written as

$$2\,NaHCO_3(s) \xrightarrow{\text{Heat}} Na_2CO_3(s) + H_2O(l) + CO_2(g)$$

Worked Example 5.1 Balancing Chemical Reactions

Use words to explain the following equation for the reaction used in extracting lead metal from its ores. Show that the equation is balanced.

$$2\,PbS(s) + 3\,O_2(g) \longrightarrow 2\,PbO(s) + 2\,SO_2(g)$$

SOLUTION
The equation can be read as, "Solid lead(II) sulfide plus gaseous oxygen yields solid lead(II) oxide plus gaseous sulfur dioxide."

To show that the equation is balanced, count the atoms of each element on each side of the arrow:

On the left:	2 Pb	2 S	(3 × 2) O = 6 O
On the right:	2 Pb	2 S	2 O + (2 × 2) O = 6 O

From 2 PbO — From 2 SO$_2$

The numbers of atoms of each element are the same in the reactants and products, so the equation is balanced.

PROBLEM 5.1
Interpret the following equations using words:
(a) $CoCl_2(s) + 2\,HF(g) \longrightarrow CoF_2(s) + 2\,HCl(g)$
(b) $Pb(NO_3)_2(aq) + 2\,KI(aq) \longrightarrow PbI_2(s) + 2\,KNO_3(aq)$

PROBLEM 5.2
Which of the following equations are balanced?
(a) $HCl + KOH \longrightarrow H_2O + KCl$
(b) $CH_4 + Cl_2 \longrightarrow CH_2Cl_2 + HCl$
(c) $H_2O + MgO \longrightarrow Mg(OH)_2$
(d) $Al(OH)_3 + H_3PO_4 \longrightarrow AlPO_4 + 2\,H_2O$

5.2 Balancing Chemical Equations

Just as a recipe indicates the appropriate amounts of each ingredient needed to make a given dish, a balanced chemical equation indicates the appropriate amounts of reactants needed to generate a given amount of product. Although balancing chemical equations often involves some trial and error, most reactions can be balanced by the following four-step approach:

STEP 1: Write an unbalanced equation, using the correct formulas for all given reactants and products. For example, hydrogen and oxygen must be written as H_2 and O_2, rather than as H and O, since we know that both elements exist as diatomic molecules. Remember that *the subscripts in chemical formulas cannot be changed in balancing an equation because doing so would change the identity of the substances in the reaction.*

STEP 2: Add appropriate coefficients to balance the numbers of atoms of each element. It helps to begin with elements that appear in only one compound or formula on each side of the equation, leaving elements that exist in elemental forms, such as oxygen and hydrogen, until last. For example, in the reaction of sulfuric acid with sodium hydroxide to give sodium sulfate and water, we might balance first for sodium. We could do this by adding a coefficient of 2 for NaOH:

$$H_2SO_4 + NaOH \longrightarrow Na_2SO_4 + H_2O \quad \text{(Unbalanced)}$$

$$H_2SO_4 + 2\,NaOH \longrightarrow Na_2SO_4 + H_2O \quad \text{(Balanced for Na)}$$

Add this coefficient . . .   . . . to balance these 2 Na.

If a polyatomic ion appears on both sides of an equation, it is treated as a single unit. For example, the sulfate ion (SO_4^{2-}) in our example is balanced because there is one on the left and one on the right:

$$H_2SO_4 + 2\,NaOH \longrightarrow Na_2SO_4 + H_2O \quad \text{(Balanced for Na and sulfate)}$$

One sulfate here . . .   . . . and one here.

At this point, the equation can be balanced for H and O by adding a coefficient of 2 for H_2O:

$$H_2SO_4 + 2\,NaOH \longrightarrow Na_2SO_4 + 2\,H_2O \quad \text{(Completely balanced)}$$

4 H and 2 O here.   4 H and 2 O here.

STEP 3: Check the equation to make sure the numbers and kinds of atoms on both sides of the equation are the same.

STEP 4: Make sure the coefficients are reduced to their lowest whole-number values. For example, the equation:

$$2\,H_2SO_4 + 4\,NaOH \longrightarrow 2\,Na_2SO_4 + 4\,H_2O$$

is balanced but can be simplified by dividing all coefficients by 2:

$$H_2SO_4 + 2\,NaOH \longrightarrow Na_2SO_4 + 2\,H_2O$$

Worked Example 5.2 Balancing Chemical Equations

Write a balanced chemical equation for the Haber process, an important industrial reaction in which elemental nitrogen and hydrogen combine to form ammonia.

SOLUTION

STEP 1: Write an unbalanced equation, using the correct formulas for all reactants and products.

$$N_2(g) + H_2(g) \longrightarrow NH_3(g)$$

By examination, we see that only two elements, N and H, need to be balanced. Both these elements exist in nature as diatomic gases, as indicated on the reactant side of the unbalanced equation.

STEP 2: Add appropriate coefficients to balance the numbers of atoms of each element. Remember that the subscript 2 in N_2 and H_2 indicates that these are diatomic molecules (that is, 2 N atoms or 2 H atoms per molecule). Since there are 2 nitrogen atoms on the left, we must add a coefficient of 2 in front of the NH_3 on the right side of the equation to balance the equation with respect to N:

$$N_2(g) + H_2(g) \longrightarrow 2\,NH_3(g)$$

Now we see that there are 2 H atoms on the left, but 6 H atoms on the right. We can balance the equation with respect to hydrogen by adding a coefficient of 3 in front of the $H_2(g)$ on the left side:

$$N_2(g) + 3\,H_2(g) \longrightarrow 2\,NH_3(g)$$

STEP 3: Check the equation to make sure the numbers and kinds of atoms on both sides of the equation are the same.

On the left: $(1 \times 2)\,N = 2\,N$ $(3 \times 2)\,H = 6\,H$

On the right: $(2 \times 1)\,N = 2\,N$ $(2 \times 3)\,H = 6\,H$

STEP 4: Make sure the coefficients are reduced to their lowest whole-number values. In this case, the coefficients already represent the lowest whole-number values.

Worked Example 5.3 Balancing Chemical Equations

Natural gas (methane, CH_4) burns in oxygen to yield water and carbon dioxide (CO_2). Write a balanced equation for the reaction.

SOLUTION

STEP 1: Write the unbalanced equation, using correct formulas for all substances:

$$CH_4 + O_2 \longrightarrow CO_2 + H_2O \quad \text{(Unbalanced)}$$

STEP 2: Since carbon appears in one formula on each side of the arrow, let us begin with that element. In fact, there is only 1 carbon atom in each formula, so the equation is already balanced for that element. Next, note that there are 4 hydrogen atoms on the left (in CH_4) and only 2 on the right (in H_2O). Placing a coefficient of 2 before H_2O gives the same number of hydrogen atoms on both sides:

$$CH_4 + O_2 \longrightarrow CO_2 + 2\,H_2O \quad \text{(Balanced for C and H)}$$

Finally, look at the number of oxygen atoms. There are 2 on the left (in O_2) but 4 on the right (2 in CO_2 and 1 in each H_2O). If we place a 2 before the O_2, the number of oxygen atoms will be the same on both sides, but the numbers of other elements will not change:

$$CH_4 + 2\,O_2 \longrightarrow CO_2 + 2\,H_2O \quad \text{(Balanced for C, H, and O)}$$

STEP 3: Check to be sure the numbers of atoms on both sides are the same.

On the left: 1 C   4 H $(2 \times 2)\,O = 4\,O$

On the right: 1 C $(2 \times 2)\,H = 4\,H$ $2\,O + 2\,O = 4\,O$
   From CO_2 From $2\,H_2O$

STEP 4: Make sure the coefficients are reduced to their lowest whole-number values. In this case, the answer is already correct.

Worked Example 5.4 Balancing Chemical Equations

Sodium chlorate ($NaClO_3$) decomposes when heated to yield sodium chloride and oxygen, a reaction used to provide oxygen for the emergency breathing masks in airliners. Write a balanced equation for this reaction.

SOLUTION

STEP 1: The unbalanced equation is:

$$NaClO_3 \longrightarrow NaCl + O_2$$

STEP 2: Both the Na and the Cl are already balanced, with only one atom of each on the left and right sides of the equation. There are 3 O atoms on the left, but only 2 on the right. The O atoms can be balanced by placing a coefficient of 1½ in front of O_2 on the right side of the equation:

$$NaClO_3 \longrightarrow NaCl + 1½\, O_2$$

STEP 3: Checking to make sure the same number of atoms of each type occurs on both sides of the equation, we see 1 atom of Na and Cl on both sides, and 3 O atoms on both sides.

STEP 4: In this case, obtaining all coefficients in their smallest whole-number values requires that we multiply all coefficients by 2 to obtain:

$$2\, NaClO_3 \longrightarrow 2\, NaCl + 3\, O_2$$

Checking gives

> *On the left:*   2 Na 2 Cl (2×3) O = 6 O
>
> *On the right:*   2 Na 2 Cl (3×2) O = 6 O

The oxygen in emergency breathing masks comes from heating sodium chlorate.

PROBLEM 5.3
Ozone (O_3) is formed in the earth's upper atmosphere by the action of solar radiation on oxygen molecules (O_2). Write a balanced equation for the formation of ozone from oxygen.

PROBLEM 5.4
Balance the following equations:
(a) $Ca(OH)_2 + HCl \longrightarrow CaCl_2 + H_2O$
(b) $Al + O_2 \longrightarrow Al_2O_3$
(c) $CH_3CH_3 + O_2 \longrightarrow CO_2 + H_2O$
(d) $AgNO_3 + MgCl_2 \longrightarrow AgCl + Mg(NO_3)_2$

KEY CONCEPT PROBLEM 5.5

The following diagram represents the reaction of A (red spheres) with B_2 (blue spheres). Write a balanced equation for the reaction.

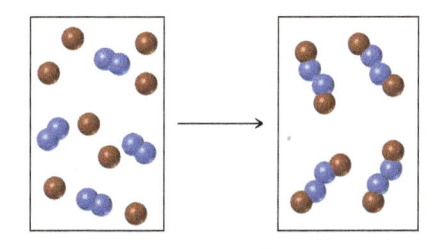

138 CHAPTER 5 Classification and Balancing of Chemical Reactions

5.3 Classes of Chemical Reactions

One of the best ways to understand any subject is to look for patterns that help us categorize large amounts of information. When learning about chemical reactions, for instance, it is helpful to group the reactions of ionic compounds into three general classes: *precipitation reactions, acid–base neutralization reactions*, and *oxidation-reduction reactions*. This is not the only possible way of categorizing reactions but it is useful nonetheless. Let us look briefly at examples of each of these three reaction classes before studying them in more detail in subsequent sections.

- **Precipitation reactions** are processes in which an insoluble solid called a **precipitate** forms when reactants are combined in aqueous solution. Most precipitations take place when the anions and cations of two ionic compounds change partners. For example, an aqueous solution of lead(II) nitrate reacts with an aqueous solution of potassium iodide to yield an aqueous solution of potassium nitrate plus an insoluble yellow precipitate of lead iodide:

$$Pb(NO_3)_2\,(aq) + 2\,KI(aq) \longrightarrow 2\,KNO_3(aq) + PbI_2(s)$$

Precipitate An insoluble solid that forms in solution during a chemical reaction.

- **Acid–base neutralization reactions** are processes in which an acid reacts with a base to yield water plus an ionic compound called a **salt**. We will look at both acids and bases in more detail in Chapter 10, but you might recall for the moment that we previously defined acids as compounds that produce H^+ ions and bases as compounds that produce OH^- ions when dissolved in water. Thus, a neutralization reaction removes H^+ and OH^- ions from solution and yields neutral H_2O. The reaction between hydrochloric acid and sodium hydroxide is a typical example:

$$HCl(aq) + NaOH(aq) \longrightarrow H_2O(l) + NaCl(aq)$$

Salt An ionic compound formed from reaction of an acid with a base.

▶▶ See Section 3.11 for more discussion of acids and bases.

Note that in this reaction, the "salt" produced is sodium chloride, or common table salt. In a general sense, however, *any* ionic compound produced in an acid–base reaction is also called a salt. Other examples include potassium nitrate (KNO_3), magnesium bromide ($MgBr_2$), and sodium sulfate (Na_2SO_4).

- **Oxidation–reduction reactions**, or **redox reactions**, are processes in which one or more electrons are transferred between reaction partners (atoms, molecules, or ions). As a result of this transfer, the number of electrons assigned to individual atoms in the various reactants change. When metallic magnesium reacts with iodine vapor, for instance, a magnesium atom gives an electron to each of 2 iodine atoms, forming a Mg^{2+} ion and 2 I^- ions. The charge on the magnesium changes from 0 to +2, and the charge on each iodine changes from 0 to −1:

$$Mg(s) + I_2(g) \longrightarrow MgI_2(s)$$

Oxidation–reduction (redox) reaction A reaction in which electrons are transferred from one atom to another.

Fundamentally, all reactions involving covalent compounds are classified as redox reactions, because electrons are rearranged as bonds are broken and new bonds are formed. The discussion here, however, will focus mainly on reactions involving ionic substances.

Worked Example 5.5 Classifying Chemical Reactions

Classify the following as a precipitation, an acid–base neutralization, or a redox reaction.

(a) $Ca(OH)_2(aq) + 2\,HBr(aq) \longrightarrow 2\,H_2O(l) + CaBr_2(aq)$
(b) $Pb(ClO_4)_2(aq) + 2\,NaCl(aq) \longrightarrow PbCl_2(s) + 2\,NaClO_4(aq)$
(c) $2\,AgNO_3(aq) + Cu(s) \longrightarrow 2\,Ag(s) + Cu(NO_3)_2(aq)$

ANALYSIS One way to identify the class of reaction is to examine the products that form and match them with the descriptions for the types of reactions provided in this section. By a process of elimination, we can readily identify the appropriate reaction classification.

▲ Reaction of aqueous $Pb(NO_3)_2$ with aqueous KI gives a yellow precipitate of PbI_2.

SOLUTION

(a) The products of this reaction are water and an ionic compound, or salt ($CaBr_2$). This is consistent with the description of an acid–base neutralization reaction.

(b) This reaction involves two aqueous reactants, $Pb(ClO_4)_2$ and NaCl, which combine to form a solid product, $PbCl_2$. This is consistent with a precipitation reaction.

(c) The products of this reaction are a solid, Ag(s), and an aqueous ionic compound, $Cu(NO_3)_2$. This does not match the description of a neutralization reaction, which would form *water* and an ionic compound. One of the products *is* a solid, but the reactants are not both aqueous compound; one of the reactants is *also* a solid (Cu). Therefore, this reaction would not be classified as a precipitation reaction. By the process of elimination, then, it must be a redox reaction.

PROBLEM 5.6
Classify each of the following as a precipitation, an acid–base neutralization, or a redox reaction.
(a) $AgNO_3(aq) + KCl(aq) \longrightarrow AgCl(s) + KNO_3(aq)$
(b) $2\,Al(s) + 3\,Br_2(l) \longrightarrow 2\,AlBr_3(s)$
(c) $Ca(OH)_2(aq) + 2\,HNO_3(aq) \longrightarrow 2\,H_2O(l) + Ca(NO_3)_2(aq)$

PROBLEM 5.7
The reaction involved in photosynthesis combines carbon dioxide and water to create simple sugars:

$$CO_2(g) + H_2O(l) \xrightarrow{\text{Sunlight}} C_6H_{12}O_6(s)$$

Balance the equation and classify the reaction.

5.4 Precipitation Reactions and Solubility Guidelines

Now let us look at precipitation reactions in more detail. To predict whether a precipitation reaction will occur upon mixing aqueous solutions of two ionic compounds, you must know the **solubilities** of the potential products—how much of each compound will dissolve in a given amount of solvent at a given temperature. If a substance has a low solubility in water, then it is likely to precipitate from an aqueous solution. If a substance has a high solubility in water, then no precipitate will form.

Solubility is a complex matter, and it is not always possible to make correct predictions. As a rule of thumb, though, the following solubility guidelines for ionic compounds are useful.

Solubility The amount of a compound that will dissolve in a given amount of solvent at a given temperature.

General Rules on Solubility

RULE 1: **A compound is probably soluble if it contains one of the following cations:**
- Group 1A cation: Li^+, Na^+, K^+, Rb^+, Cs^+
- Ammonium ion: NH_4^+

RULE 2: **A compound is probably soluble if it contains one of the following anions:**
- Halide: Cl^-, Br^-, I^- except Ag^+, Hg_2^{2+}, and Pb^{2+} compounds
- Nitrate (NO_3^-), perchlorate (ClO_4^-), acetate ($CH_3CO_2^-$), sulfate (SO_4^{2-}) except Ba^{2+}, Hg_2^{2+}, and Pb^{2+} sulfates

If a compound does *not* contain at least one of the ions listed above, it is probably *not* soluble. Thus, Na_2CO_3 is soluble because it contains a group 1A cation, and $CaCl_2$ is soluble because it contains a halide anion. The compound $CaCO_3$, however, is probably *insoluble* because it contains none of the ions listed above. These same guidelines are presented in table form in Table 5.1.

TABLE 5.1 General Solubility Guidelines for Ionic Compounds in Water

Soluble	Exceptions
Ammonium compounds (NH_4^+)	None
Lithium compounds (Li^+)	None
Sodium compounds (Na^+)	None
Potassium compounds (K^+)	None
Nitrates (NO_3^-)	None
Perchlorates (ClO_4^-)	None
Acetates ($CH_3CO_2^-$)	None
Chlorides (Cl^-)	
Bromides (Br^-)	Ag^+, Hg_2^{2+}, and Pb^{2+} compounds
Iodides (I^-)	
Sulfates (SO_4^{2-})	Ba^{2+}, Hg_2^{2+}, and Pb^{2+} compounds

CHEMISTRY IN ACTION

Gout and Kidney Stones: Problems in Solubility

One of the major pathways in the body for the breakdown of the nucleic acids DNA and RNA is by conversion to a substance called uric acid, $C_5H_4N_4O_3$, so named because it was first isolated from urine in 1776. Most people excrete about 0.5 g of uric acid every day in the form of sodium urate, the salt that results from an acid–base reaction of uric acid. Unfortunately, the amount of sodium urate that dissolves in water (or urine) is fairly low—only about 0.07 mg/mL at the normal body temperature of 37 °C. When too much sodium urate is produced or mechanisms for its elimination fail, its concentration in blood and urine rises, and the excess sometimes precipitates in the joints and kidneys.

Gout is a disorder of nucleic acid metabolism that primarily affects middle-aged men (only 5% of gout patients are women). It is characterized by an increased sodium urate concentration in blood, leading to the deposit of sodium urate crystals in soft tissue around the joints, particularly in the hands and at the base of the big toe. Deposits of the sharp, needlelike crystals cause an extremely painful inflammation that can lead ultimately to arthritis and even to bone destruction.

Just as increased sodium urate concentration in blood can lead to gout, increased concentration in urine can result in the formation of one kind of *kidney stones*, small crystals that precipitate in the kidney. Although often quite small, kidney stones cause excruciating pain when they pass through the ureter, the duct that carries urine from the kidney to the bladder. In some cases, complete blockage of the ureter occurs.

Treatment of excessive sodium urate production involves both dietary modification and drug therapy. Foods such as liver, sardines, and asparagus should be avoided, and drugs such as allopurinol can be taken to lower production of sodium urate. Allopurinol functions by inhibiting the action of an enzyme called *xanthine oxidase*, thereby blocking a step in nucleic acid metabolism.

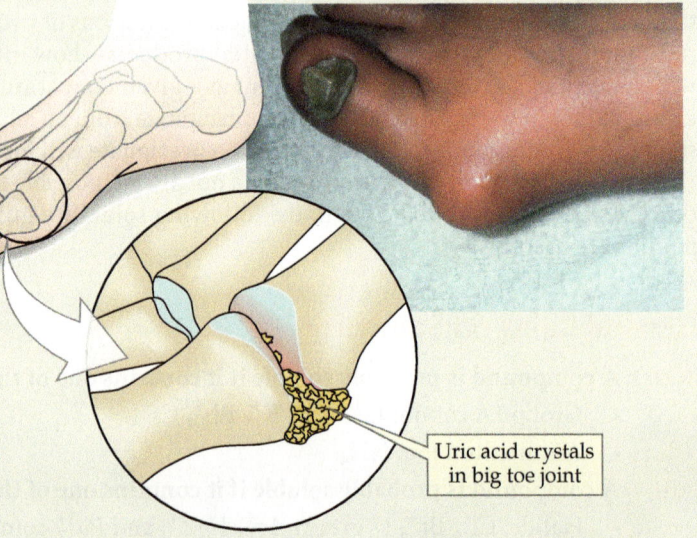

▲ Excess production of uric acid can cause gout, a painful condition characterized by the accumulation of sodium urate crystals in joints.

See Chemistry in Action Problems 5.63 and 5.64 at the end of the chapter.

Let us try a problem. What will happen if aqueous solutions of sodium nitrate ($NaNO_3$) and potassium sulfate (K_2SO_4) are mixed? To answer this question, look at the guidelines to find the solubilities of the two possible products, Na_2SO_4 and KNO_3. Because both have group 1A cations (Na^+ and K^+), both are water-soluble and no precipitation will occur. If aqueous solutions of silver nitrate ($AgNO_3$) and sodium carbonate (Na_2CO_3) are mixed, however, the guidelines predict that a precipitate of insoluble silver carbonate (Ag_2CO_3) will form.

$$2\,AgNO_3(aq) + Na_2CO_3(aq) \longrightarrow Ag_2CO_3(s) + 2\,NaNO_3(aq)$$

Worked Example 5.6 Chemical Reactions: Solubility Rules

Will a precipitation reaction occur when aqueous solutions of $CdCl_2$ and $(NH_4)_2S$ are mixed?

SOLUTION
Identify the two potential products, and predict the solubility of each using the guidelines in the text. In this instance, $CdCl_2$ and $(NH_4)_2S$ might give CdS and NH_4Cl. Since the guidelines predict that CdS is insoluble, a precipitation reaction will occur:

$$CdCl_2(aq) + (NH_4)_2S(aq) \longrightarrow CdS(s) + 2\,NH_4Cl(aq)$$

PROBLEM 5.8
Predict the solubility of the following compounds:
(a) $CdCO_3$
(b) Na_2S
(c) $PbSO_4$
(d) $(NH_4)_3PO_4$
(e) Hg_2Cl_2

PROBLEM 5.9
Predict whether a precipitation reaction will occur in the following situations. If a precipitation reaction occurs, write the balanced chemical equation for the reaction.
(a) $NiCl_2(aq) + (NH_4)_2S(aq) \longrightarrow$
(b) $AgNO_3(aq) + CaBr_2(aq) \longrightarrow$

PROBLEM 5.10
In addition to kidney stone formation by sodium urate (See Chemistry in Action on p. 140), many kidney stones are formed by precipitation of oxalate by calcium. Oxalates are found in many foods, including spinach, blueberries, and chocolate. Show the balanced chemical equation for the precipitation of calcium oxalate, starting with calcium chloride ($CaCl_2$) and sodium oxalate ($Na_2C_2O_4$).

5.5 Acids, Bases, and Neutralization Reactions

When acids and bases are mixed in the correct proportion, both acidic and basic properties disappear because of a **neutralization reaction**. The most common kind of neutralization reaction occurs between an acid (generalized as HA), and a metal hydroxide (generalized as MOH), to yield water and a salt. The H^+ ion from the acid combines with the OH^- ion from the base to give neutral H_2O, whereas the anion from the acid (A^-) combines with the cation from the base (M^+) to give the salt:

A neutralization reaction: $\underset{\text{Acid}}{HA(aq)} + \underset{\text{Base}}{MOH(aq)} \longrightarrow \underset{\text{Water}}{H_2O(l)} + \underset{\text{A salt}}{MA(aq)}$

Neutralization reaction The reaction of an acid with a base.

The reaction of hydrochloric acid with potassium hydroxide to produce potassium chloride is an example:

$$HCl(aq) + KOH(aq) \longrightarrow H_2O(l) + KCl(aq)$$

Another kind of neutralization reaction occurs between an acid and a carbonate (or bicarbonate) to yield water, a salt, and carbon dioxide. Hydrochloric acid reacts with potassium carbonate, for example, to give H_2O, KCl, and CO_2:

$$2\,HCl(aq) + K_2CO_3(aq) \longrightarrow H_2O(l) + 2\,KCl(aq) + CO_2(g)$$

The reaction occurs because the carbonate ion (CO_3^{2-}) reacts initially with H^+ to yield H_2CO_3, which is unstable and immediately decomposes to give CO_2 plus H_2O.

We will defer a more complete discussion of carbonates as bases until Chapter 10, but note for now that they yield OH^- ions when dissolved in water just as KOH and other bases do.

$$K_2CO_3(s) + H_2O(l) \xrightarrow{\text{Dissolve in water}} 2K^+(aq) + HCO_3^-(aq) + OH^-(aq)$$

LOOKING AHEAD ▶▶▶ Acids and bases are enormously important in biological chemistry. We will see in Chapter 18, for instance, how acids and bases affect the structure and properties of proteins.

Worked Example 5.7 Chemical Reactions: Acid–Base Neutralization

Write an equation for the neutralization reaction of aqueous HBr and aqueous $Ba(OH)_2$.

SOLUTION
The reaction of HBr with $Ba(OH)_2$ involves the combination of a proton (H^+) from the acid with OH^- from the base to yield water and a salt ($BaBr_2$).

$$2\,HBr(aq) + Ba(OH)_2(aq) \longrightarrow 2\,H_2O(l) + BaBr_2(aq)$$

PROBLEM 5.11
Write and balance equations for the following acid–base neutralization reactions:
(a) $CsOH(aq) + H_2SO_4(aq) \longrightarrow$
(b) $Ca(OH)_2(aq) + CH_3CO_2H(aq) \longrightarrow$
(c) $NaHCO_3(aq) + HBr(aq) \longrightarrow$

5.6 Redox Reactions

Oxidation–reduction (redox) reactions, the third and final category of reactions that we will discuss here, are more complex than precipitation and neutralization reactions. Look, for instance, at the following examples and see if you can tell what they have in common. Copper metal reacts with aqueous silver nitrate to form silver metal and aqueous copper(II) nitrate; iron rusts in air to form iron(III) oxide; the zinc metal container on the outside of a battery reacts with manganese dioxide and ammonium chloride inside the battery to generate electricity and give aqueous zinc chloride plus manganese(III) oxide. Although these and many thousands of other reactions appear unrelated, all are examples of redox reactions.

$$Cu(s) + 2\,AgNO_3(aq) \longrightarrow 2\,Ag(s) + Cu(NO_3)_2(aq)$$
$$2\,Fe(s) + 3\,O_2(g) \longrightarrow Fe_2O_3(s)$$
$$Zn(s) + 2\,MnO_2(s) + 2\,NH_4Cl(s) \longrightarrow$$
$$ZnCl_2(aq) + Mn_2O_3(s) + 2\,NH_3(aq) + H_2O(l)$$

Historically, the word *oxidation* referred to the combination of an element with oxygen to yield an oxide, and the word *reduction* referred to the removal of oxygen from an oxide to yield the element. Today, though, the words have taken on a much broader meaning. An **oxidation** is now defined as the loss of one or more electrons by an atom,

Oxidation The loss of one or more electrons by an atom.

and a **reduction** is the gain of one or more electrons. Thus, an oxidation–reduction reaction, or redox reaction, is one in which *electrons are transferred from one atom to another*.

Reduction The gain of one or more electrons by an atom.

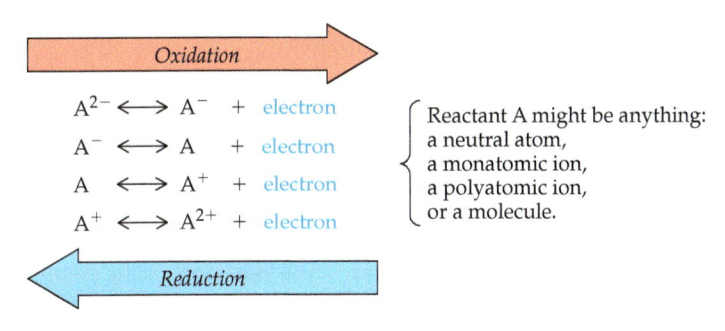

Oxidation →

$A^{2-} \longleftrightarrow A^- + $ electron
$A^- \longleftrightarrow A + $ electron
$A \longleftrightarrow A^+ + $ electron
$A^+ \longleftrightarrow A^{2+} + $ electron

Reactant A might be anything: a neutral atom, a monatomic ion, a polyatomic ion, or a molecule.

← **Reduction**

Take the reaction of copper with aqueous Ag^+ as an example, as shown in Figure 5.1. Copper metal gives an electron to each of 2 Ag^+ ions, forming Cu^{2+} and silver metal. Copper is oxidized in the process, and Ag^+ is reduced. You can follow the transfer of the electrons by noting that the charge on the copper increases from 0 to +2 when it loses 2 electrons, whereas the charge on Ag^+ decreases from +1 to 0 when it gains an electron.

+2 electrons = reduced!

$$Cu(s) + 2\,Ag^+(aq) \longrightarrow Cu^{2+}(aq) + 2\,Ag(s)$$

0 charge   +1 charge   +2 charge   0 charge

−2 electrons = oxidized!

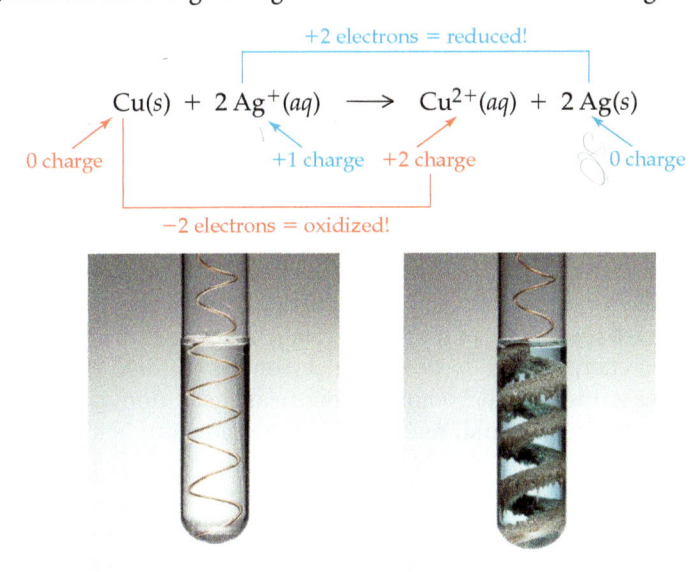

◄ **Figure 5.1**
The copper wire reacts with aqueous Ag^+ ion and becomes coated with metallic silver. At the same time, copper(II) ions go into solution, producing the blue color.

Similarly, in the reaction of aqueous iodide ion with bromine, iodide ion gives an electron to bromine, forming iodine and bromide ion. Iodide ion is oxidized as its charge increases from −1 to 0, and bromine is reduced as its charge decreases from 0 to −1.

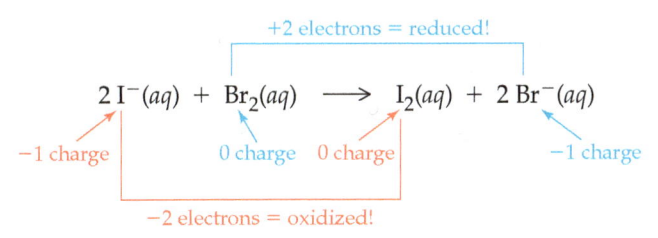

+2 electrons = reduced!

$$2\,I^-(aq) + Br_2(aq) \longrightarrow I_2(aq) + 2\,Br^-(aq)$$

−1 charge   0 charge   0 charge   −1 charge

−2 electrons = oxidized!

As these examples show, oxidation and reduction always occur together. Whenever one substance loses an electron (is oxidized), another substance must gain that electron (be reduced). The substance that gives up an electron and causes the reduction—the copper atom in the reaction of Cu with Ag^+ and the iodide ion in the reaction of I^- with Br_2—is called a **reducing agent**. The substance that gains an electron and causes the oxidation—the silver ion in the reaction of Cu with Ag^+ and the bromine molecule in the reaction of I^- with Br_2—is called an **oxidizing agent**. The charge on the reducing agent increases during the reaction, and the charge on the oxidizing agent decreases.

Reducing agent A reactant that causes a reduction in another reactant by giving up electron to it.

Oxidizing agent A reactant that causes an oxidation by taking electrons from another reactant.

Reducing agent   Loses one or more electrons
Causes reduction
Undergoes oxidation
Becomes more positive (less negative)
(May gain oxygen atoms)

Oxidizing agent   Gains one or more electrons
Causes oxidation
Undergoes reduction
Becomes more negative (less positive)
(May lose oxygen atoms)

Among the simplest of redox processes is the reaction of an element, usually a metal, with an aqueous cation to yield a different element and a different ion. Iron metal reacts with aqueous copper(II) ion, for example, to give iron(II) ion and copper metal. Similarly, magnesium metal reacts with aqueous acid to yield magnesium ion and hydrogen gas. In both cases, the reactant element (Fe or Mg) is oxidized, and the reactant ion (Cu^{2+} or H^+) is reduced.

$$Fe(s) + Cu^{2+}(aq) \longrightarrow Fe^{2+}(aq) + Cu(s)$$
$$Mg(s) + 2\,H^+(aq) \longrightarrow Mg^{2+}(aq) + H_2(g)$$

The reaction of a metal with water or aqueous acid (H^+) to release H_2 gas is a particularly important process. As you might expect based on the periodic properties discussed in Section 3.2, the alkali metals and alkaline earth metals (on the left side of the periodic table) are the most powerful reducing agents (electron donors), so powerful that they even react with pure water, in which the concentration of H^+ is very low. This is due in part to the fact that alkali metals and alkaline earth metals have low ionization energies. Ionization energy, which is a measure of how easily an element will lose an electron, tends to decrease as we move to the left and down in the periodic table. Thus, metals toward the middle of the periodic table, such as iron and chromium, have higher ionization energies and do not lose electrons as readily; they react only with aqueous acids but not with water. Those metals near the bottom right of the periodic table, such as platinum and gold, react with neither aqueous acid nor water. At the other extreme from the alkali metals, the reactive nonmetals at the top right of the periodic table have the highest ionization energies and are extremely weak reducing agents but powerful oxidizing agents (electron acceptors). This is, again, predictable based on the periodic property of electron affinity (Section 3.2), which becomes more energetically favored as we move up and to the right in the periodic table.

▶▶ The relationship between formation of ions and ionization energy/electronegativity was discussed in Chapter 3.

We can make a few generalizations about the redox behavior of metals and nonmetals.

1. In reactions involving metals and nonmetals, metals tend to lose electrons while nonmetals tend to gain electrons. The number of electrons lost or gained can often be predicted based on the position of the element in the periodic table. (Section 3.5)
2. In reactions involving nonmetals, the "more metallic" element (farther down and/or to the left in the periodic table) tends to lose electrons, and the "less metallic" element (up and/or to the right) tends to gain electrons.

    Redox reactions involve almost every element in the periodic table, and they occur in a vast number of processes throughout nature, biology, and industry. Here are just a few examples:

    - **Corrosion** is the deterioration of a metal by oxidation, such as the rusting of iron in moist air. The economic consequences of rusting are enormous: it has been estimated that up to one-fourth of the iron produced in the United States is used to replace bridges, buildings, and other structures that have been destroyed by corrosion. (The raised dot in the formula $Fe_2O_3 \cdot H_2O$ for rust indicates that one water molecule is associated with each Fe_2O_3 in an undefined way.)

- **Combustion** is the burning of a fuel by rapid oxidation with oxygen in air. Gasoline, fuel oil, natural gas, wood, paper, and other organic substances of carbon and hydrogen are the most common fuels that burn in air. Even some metals, though, will burn in air. Magnesium and calcium are examples.

$$CH_4(g) + 2\,O_2(g) \longrightarrow CO_2(g) + 2\,H_2O(l)$$
Methane
(natural gas)

$$2\,Mg(s) + O_2(g) \longrightarrow 2\,MgO(s)$$

- **Respiration** is the process of breathing and using oxygen for the many biological redox reactions that provide the energy required by living organisms. We will see in Chapters 21–22 that in the respiration process, energy is released from food molecules slowly and in complex, multistep pathways, but that the overall result is similar to that of the simpler combustion reactions. For example, the simple sugar glucose ($C_6H_{12}O_6$) reacts with O_2 to give CO_2 and H_2O according to the following equation:

$$C_6H_{12}O_6 + 6\,O_2 \longrightarrow 6\,CO_2 + 6\,H_2O + Energy$$
Glucose
(a carbohydrate)

- **Bleaching** makes use of redox reactions to decolorize or lighten colored materials. Dark hair is bleached to turn it blond, clothes are bleached to remove stains, wood pulp is bleached to make white paper, and so on. The oxidizing agent used depends on the situation: hydrogen peroxide (H_2O_2) is used for hair, sodium hypochlorite (NaOCl) for clothes, and elemental chlorine for wood pulp, but the principle is always the same. In all cases, colored organic materials are destroyed by reaction with strong oxidizing agents.
- **Metallurgy,** the science of extracting and purifying metals from their ores, makes use of numerous redox processes. Worldwide, approximately 800 million tons of iron are produced each year by reduction of the mineral hematite, Fe_2O_3, with carbon monoxide.

$$Fe_2O_3(s) + 3\,CO(g) \longrightarrow 2\,Fe(s) + 3\,CO_2(g)$$

Worked Example 5.8 Chemical Reactions: Redox Reactions

For the following reactions, indicate which atom is oxidized and which is reduced, based on the definitions provided in this section. Identify the oxidizing and reducing agents.

(a) $Cu(s) + Pt^{2+}(aq) \longrightarrow Cu^{2+}(aq) + Pt(s)$
(b) $2\,Mg(s) + CO_2(g) \longrightarrow 2\,MgO(s) + C(s)$

ANALYSIS The definitions for oxidation include a loss of electrons, an increase in charge, and a gain of oxygen atoms; reduction is defined as a gain of electrons, a decrease in charge, and a loss of oxygen atoms.

SOLUTION

(a) In this reaction, the charge on the Cu atom increases from 0 to 2+. This corresponds to a loss of 2 electrons. The Cu is therefore oxidized and acts as the reducing agent. Conversely, the Pt^{2+} ion undergoes a decrease in charge from 2+ to 0, corresponding to a gain of 2 electrons for the Pt^{2+} ion. The Pt^{2+} is reduced, and acts as the oxidizing agent.

(b) In this case, the gain or loss of oxygen atoms is the easiest way to identify which atoms are oxidized and reduced. The Mg atom is gaining oxygen to form MgO; therefore, the Mg is being oxidized and acts as the reducing agent. The C atom in CO_2 is losing oxygen. Therefore, the C atom in CO_2 is being reduced, and so CO_2 acts as the oxidizing agent.

Worked Example 5.9 Chemical Reactions: Identifying Oxidizing/Reducing Agents

For the respiration and metallurgy examples discussed previously, identify the atoms being oxidized and reduced, and label the oxidizing and reducing agents.

ANALYSIS Again, using the definitions of oxidation and reduction provided in this section, we can determine which atom(s) are gaining/losing electrons or gaining/losing oxygen atoms.

SOLUTION

$$\text{Respiration:} \quad C_6H_{12}O_6 + 6\,O_2 \longrightarrow 6\,CO_2 + 6\,H_2O$$

Because the charge associated with the individual atoms is not evident, we will use the definition of oxidation/reduction as the gaining/losing of oxygen atoms. In this reaction, there is only one reactant besides oxygen ($C_6H_{12}O_6$), so we must determine *which* atom in the compound is changing. The ratio of carbon to oxygen in $C_6H_{12}O_4$ is 1:1, while the ratio in CO_2 is 1:2. Therefore, the C atoms are gaining oxygen and are oxidized; the $C_6H_{12}O_{16}$ is the reducing agent and O_2 is the oxidizing agent. Note that the ratio of hydrogen to oxygen in $C_6H_{12}O_6$ and in H_2O is 2:1. The H atoms are neither oxidized nor reduced.

$$\text{Metallurgy:} \quad Fe_2O_3(s) + 3\,CO(g) \longrightarrow 2\,Fe(s) + 3\,CO_2(g)$$

The Fe_2O_3 is losing oxygen to form Fe(s); it is being reduced and acts as the oxidizing agent. In contrast, the CO is gaining oxygen to form CO_2; it is being oxidized and acts as the reducing agent.

Worked Example 5.10 Chemical Reactions: Identifying Redox Reactions

For the following reactions, identify the atom(s) being oxidized and reduced:

(a) $2\,Al(s) + 3\,Cl_2(g) \longrightarrow 2\,AlCl_3(s)$ (b) $C(s) + 2\,Cl_2(g) \longrightarrow CCl_4(l)$

ANALYSIS Again, there is no obvious increase or decrease in charge to indicate a gain or loss of electrons. Also, the reactions do not involve a gain or loss of oxygen. We can, however, evaluate the reactions in terms of the typical behavior of metals and nonmetals in reactions.

SOLUTION

(a) In this case, we have the reaction of a metal (Al) with a nonmetal (Cl_2). Because metals tend to lose electrons and nonmetals tend to gain electrons, we can assume that the Al atom is oxidized (loses electrons) and the Cl_2 is reduced (gains electrons).

(b) The carbon atom is the less electronegative element (farther to the left) and is less likely to gain an electron. The more electronegative element (Cl) will tend to gain electrons (be reduced).

PROBLEM 5.12

Identify the oxidized reactant, the reduced reactant, the oxidizing agent, and the reducing agent in the following reactions:

(a) $Fe(s) + Cu^{2+}(aq) \longrightarrow Fe^{2+}(aq) + Cu(s)$
(b) $Mg(s) + Cl_2(g) \longrightarrow MgCl_2(s)$
(c) $2\,Al(s) + Cr_2O_3(s) \longrightarrow 2\,Cr(s) + Al_2O_3(s)$

PROBLEM 5.13

Potassium, a silvery metal, reacts with bromine, a corrosive, reddish liquid, to yield potassium bromide, a white solid. Write the balanced equation, and identify the oxidizing and reducing agents.

PROBLEM 5.14

The redox reaction that provides energy for the lithium battery described in the Chemistry in Action on p. 147 is $2\,Li(s) + I_2(s) \rightarrow 2\,LiI(aq)$. Identify which reactant is being oxidized and which is being reduced in this reaction.

CHEMISTRY IN ACTION

Batteries

Imagine life without batteries: no cars (they do not start very easily without their batteries!), no heart pacemakers, no flashlights, no hearing aids, no laptops, no radios, no cell phones, nor thousands of other things. Modern society could not exist without batteries.

Although they come in many types and sizes, all batteries work using redox reactions. In a typical redox reaction carried out in the laboratory—say, the reaction of zinc metal with Ag^+ to yield Zn^{2+} and silver metal—the reactants are simply mixed in a flask and electrons are transferred by direct contact between the reactants. In a battery, however, the two reactants are kept in separate compartments and the electrons are transferred through a wire running between them.

The common household battery used for flashlights and radios is the *dry cell*, developed in 1866. One reactant is a can of zinc metal, and the other is a paste of solid manganese dioxide. A graphite rod sticks into the MnO_2 paste to provide electrical contact, and a moist paste of ammonium chloride separates the two reactants. If the zinc can and the graphite rod are connected by a wire, zinc sends electrons flowing through the wire toward the MnO_2 in a redox reaction. The resultant electrical current can then be used to power a lightbulb or a radio. The accompanying figure shows a cutaway view of a dry-cell battery.

$$Zn(s) + 2\,MnO_2(s) + 2\,NH_4Cl(s) \longrightarrow$$
$$ZnCl_2(aq) + Mn_2O_3(s) + 2\,NH_3(aq) + H_2O(l)$$

▲ Think of all the devices we use every day—laptop computers, cell phones, iPods—that depend on batteries.

Closely related to the dry-cell battery is the familiar *alkaline* battery, in which the ammonium chloride paste is replaced by an alkaline, or basic, paste of NaOH or KOH. The alkaline battery has a longer life than the standard dry-cell battery because the zinc container corrodes less easily under basic conditions. The redox reaction is:

$$Zn(s) + 2\,MnO_2(s) \longrightarrow ZnO(aq) + Mn_2O_3(s)$$

The batteries used in implanted medical devices such as pacemakers must be small, corrosion-resistant, reliable, and able to last up to 10 years. Nearly all pacemakers being implanted today—about 750,000 each year—use titanium-encased, lithium–iodine batteries, whose redox reaction is:

$$2\,Li(s) + I_2(s) \longrightarrow 2\,LiI(aq)$$

See Chemistry in Action Problems 5.65 and 5.66 at the end of the chapter.

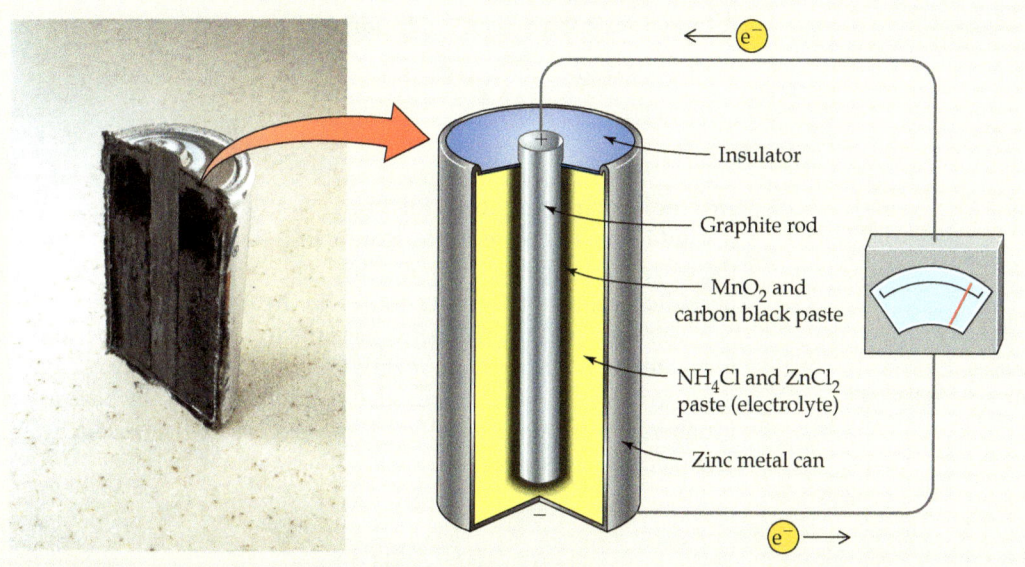

▲ A dry-cell battery. The cutaway view shows the two reactants that make up the redox reaction.

148 CHAPTER 5 Classification and Balancing of Chemical Reactions

5.7 Recognizing Redox Reactions

How can you tell when a redox reaction is taking place? When ions are involved, it is simply a matter of determining whether there is a change in the charges. For reactions involving metals and nonmetals, we can predict the gain or loss of electrons as discussed previously. When molecular substances are involved, though, it is not as obvious. Is the combining of sulfur with oxygen a redox reaction? If so, which partner is the oxidizing agent and which is the reducing agent?

$$S(s) + O_2(g) \longrightarrow SO_2(g)$$

One way to evaluate this reaction is in terms of the oxygen gain by sulfur, indicating that S atoms are oxidized and O atoms are reduced. But can we also look at this reaction in terms of the gain or loss of electrons by the S and O atoms? Because oxygen is more electronegative than sulfur, the oxygen atoms in SO_2 attract the electrons in the S—O bonds more strongly than sulfur does, giving the oxygen atoms a larger share of the electrons than sulfur. By extending the ideas of oxidation and reduction to an increase or decrease in electron *sharing* instead of complete electron *transfer*, we can say that the sulfur atom is oxidized in its reaction with oxygen because it loses a share in some electrons, whereas the oxygen atoms are reduced because they gain a share in some electrons.

▶▶ Electronegativity, or the propensity of an atom in a covalent bond to attract electrons, was introduced in Section 4.9.

A formal system has been devised for keeping track of changes in electron sharing, and thus for determining whether atoms are oxidized or reduced in reactions. To each atom in a substance, we assign a value called an **oxidation number** (or *oxidation state*), which indicates whether the atom is neutral, electron-rich, or electron-poor. By comparing the oxidation number of an atom before and after a reaction, we can tell whether the atom has gained or lost shares in electrons. Note that *oxidation numbers do not necessarily imply ionic charges*. They are simply a convenient device for keeping track of electrons in redox reactions.

Oxidation number A number that indicates whether an atom is neutral, electron-rich, or electron-poor.

The rules for assigning oxidation numbers are straightforward:

- **An atom in its elemental state has an oxidation number of 0.**

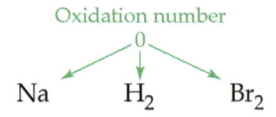

Oxidation number
0
Na   H₂   Br₂

- **A monatomic ion has an oxidation number equal to its charge.**

Oxidation number +1: Na⁺
Oxidation number +2: Ca²⁺
Oxidation number −1: Cl⁻
Oxidation number −2: O²⁻

- **In a molecular compound, an atom usually has the same oxidation number it would have if it were a monatomic ion.** Recall from Chapters 3 and 4 that the less electronegative elements (hydrogen and metals) on the left side of the periodic table tend to form cations, and the more electronegative elements (oxygen, nitrogen, and the halogens) near the top right of the periodic table tend to form anions. Hydrogen and metals therefore have positive oxidation numbers in most compounds, whereas reactive nonmetals generally have negative oxidation numbers. Hydrogen is usually +1, oxygen is usually −2, nitrogen is usually −3, and halogens are usually −1:

▶▶ Review the Important Points about Ion Formation and the Periodic Table listed in Section 3.6.

+1 −1   +1 −2 +1   +1 −3 +1
H—Cl   H—O—H   H—N—H
                              |
                              H ← +1

For compounds with more than one nonmetal element, such as SO_2, NO, or CO_2, the more electronegative element—oxygen in these examples—has a negative oxidation number and the less electronegative element has a positive oxidation number. Thus, in answer to the question posed at the beginning of this section, combining sulfur with oxygen to form SO_2 is a redox reaction because the oxidation number of sulfur increases from 0 to +4 and that of oxygen decreases from 0 to −2.

$$\overset{-2}{O}-\overset{+4}{S}=\overset{-2}{O} \qquad \overset{+2}{N}=\overset{-2}{O} \qquad \overset{-2}{O}=\overset{+4}{C}=\overset{-2}{O}$$

- **The sum of the oxidation numbers in a neutral compound is 0.** Using this rule, the oxidation number of any atom in a compound can be found if the oxidation numbers of the other atoms are known. In the SO_2 example just mentioned, each of the 2 O atoms has an oxidation number of −2, so the S atom must have an oxidation number of +4. In HNO_3, the H atom has an oxidation number of +1 and the strongly electronegative O atom has an oxidation number of −2, so the N atom must have an oxidation number of +5. In a polyatomic ion, the sum of the oxidation numbers equals the charge on the ion.

$$\overset{+1}{H}-\overset{-2}{O}-\overset{+5}{N}=\overset{-2}{O} \qquad \text{Total} = 1 + 5 + 3(-2) = 0$$
$$\underset{\overset{|}{O}_{-2}}{}$$

Worked Examples 5.11 and 5.12 show further instances of assigning and using oxidation numbers.

Worked Example 5.11 Redox Reactions: Oxidation Numbers

What is the oxidation number of the titanium atom in $TiCl_4$? Name the compound using a Roman numeral (Section 3.10).

SOLUTION
Chlorine, a reactive nonmetal, is more electronegative than titanium and has an oxidation number of −1. Because there are 4 chlorine atoms in $TiCl_4$, the oxidation number of titanium must be +4. The compound is named titanium(IV) chloride. Note that the Roman numeral IV in the name of this molecular compound refers to the oxidation number +4 rather than to a true ionic charge.

Worked Example 5.12 Redox Reactions: Identifying Redox Reactions

Use oxidation numbers to show that the production of iron metal from its ore (Fe_2O_3) by reaction with charcoal (C) is a redox reaction. Which reactant has been oxidized, and which has been reduced? Which reactant is the oxidizing agent, and which is the reducing agent?

$$2\,Fe_2O_3(s) + 3\,C(s) \longrightarrow 4\,Fe(s) + 3\,CO_2(g)$$

SOLUTION
The idea is to assign oxidation numbers to both reactants and products and see if there has been a change. In the production of iron from Fe_2O_3, the oxidation number of Fe changes from +3 to 0, and the oxidation number of C changes from 0 to +4. Iron has thus been reduced (decrease in oxidation number), and carbon has been oxidized (increase in oxidation number). Oxygen is neither oxidized nor reduced because its oxidation number does not change. Carbon is the reducing agent, and Fe_2O_3 is the oxidizing agent.

$$2\,\overset{+3\ -2}{Fe_2O_3} + 3\,\overset{0}{C} \longrightarrow 4\,\overset{0}{Fe} + 3\,\overset{+4\ -2}{CO_2}$$

PROBLEM 5.15
What are the oxidation numbers of the metal atoms in the following compounds? Name each, using the oxidation number as a Roman numeral.
(a) VCl_3 (b) $SnCl_4$ (c) CrO_3 (d) $Cu(NO_3)_2$ (e) $NiSO_4$

PROBLEM 5.16
Assign an oxidation number to each atom in the reactants and products shown here to determine which of the following reactions are redox reactions:
(a) $Na_2S(aq) + NiCl_2(aq) \longrightarrow 2\,NaCl(aq) + NiS(s)$
(b) $2\,Na(s) + 2\,H_2O(l) \longrightarrow 2\,NaOH(aq) + H_2(g)$
(c) $C(s) + O_2(g) \longrightarrow CO_2(g)$
(d) $CuO(s) + 2\,HCl(aq) \longrightarrow CuCl_2(aq) + H_2O(l)$
(e) $2\,MnO_4^-(aq) + 5\,SO_2(g) + 2\,H_2O(l) \longrightarrow$
$2\,Mn^{2+}(aq) + 5\,SO_4^{2-}(aq) + 4\,H^+(aq)$

PROBLEM 5.17
For each of the reactions you identified as redox reactions in Problem 5.16, identify the oxidizing agent and the reducing agent.

5.8 Net Ionic Equations

In the equations we have been writing up to this point, all the substances involved in reactions have been written using their full formulas. In the precipitation reaction of lead(II) nitrate with potassium iodide mentioned in Section 5.3, for example, only the parenthetical *aq* indicated that the reaction actually takes place in aqueous solution, and nowhere was it explicitly indicated that ions are involved:

$$Pb(NO_3)_2(aq) + 2\,KI(aq) \longrightarrow 2\,KNO_3(aq) + PbI_2(s)$$

In fact, lead(II) nitrate, potassium iodide, and potassium nitrate dissolve in water to yield solutions of ions. Thus, it is more accurate to write the reaction as an **ionic equation**, in which all the ions are explicitly shown:

An ionic equation: $\quad Pb^{2+}(aq) + 2\,NO_3^-(aq) + 2\,K^+(aq) + 2\,I^-(aq) \longrightarrow$
$2\,K^+(aq) + 2\,NO_3^-(aq) + PbI_2(s)$

Ionic equation An equation in which ions are explicitly shown.

A look at this ionic equation shows that the NO_3^- and K^+ ions undergo no change during the reaction. They appear on both sides of the reaction arrow and act merely as **spectator ions**, that is, they are present but play no role. The actual reaction, when stripped to its essentials, can be described more simply by writing a **net ionic equation**, which includes only the ions that undergo change and ignores all spectator ions:

Spectator ion An ion that appears unchanged on both sides of a reaction arrow.

Net ionic equation An equation that does not include spectator ions.

Ionic equation: $\quad Pb^{2+}(aq) + 2\,\cancel{NO_3^-}(aq) + 2\,\cancel{K^+}(aq) + 2\,I^-(aq) \longrightarrow$
$2\,\cancel{K^+}(aq) + 2\,\cancel{NO_3^-}(aq) + PbI_2(s)$

Net ionic equation: $\quad Pb^{2+}(aq) + 2\,I^-(aq) \longrightarrow PbI_2(s)$

Note that a net ionic equation, like all chemical equations, must be balanced both for atoms and for charge, with all coefficients reduced to their lowest whole numbers. Note also that all compounds that do *not* give ions in solution—all insoluble compounds and all molecular compounds—are represented by their full formulas.

We can apply the concept of ionic equations to acid–base neutralization reactions and redox reactions as well. Consider the neutralization reaction between KOH and HNO_3:

$$KOH(aq) + HNO_3(aq) \longrightarrow H_2O(l) + KNO_3(aq)$$

Since acids and bases are identified based on the ions they form when dissolved in aqueous solutions, we can write an ionic equation for this reaction:

Ionic equation: $\quad \cancel{K^+}(aq) + OH^-(aq) + H^+(aq) + \cancel{NO_3^-}(aq) \longrightarrow$
$H_2O(l) + \cancel{K^+}(aq) + \cancel{NO_3^-}(aq)$

Eliminating the spectator ions (K^+ and NO_3^-), we obtain the net ionic equation for the neutralization reaction:

Net ionic equation: $OH^-(aq) + H^+(aq) \longrightarrow H_2O(l)$

The net ionic equation confirms the basis of the acid–base neutralization; the OH^- from the base and the H^+ from the acid neutralize each other to form water.

Similarly, many redox reactions can be viewed in terms of ionic equations. Consider the reaction between $Cu(s)$ and $AgNO_3$ from Section 5.6:

$$Cu(s) + 2\,AgNO_3(aq) \longrightarrow 2\,Ag^+(aq) + Cu(NO_3)_2(aq)$$

The aqueous products and reactants can be written as dissolved ions:

Ionic equation: $Cu(s) + 2\,Ag^+(aq) + 2\,NO_3^-(aq) \longrightarrow$
$2\,Ag(s) + Cu^{2+}(aq) + 2\,NO_3^-(aq)$

Again, eliminating the spectator ions (NO_3^-), we obtain the net ionic equation for this redox reaction:

Net ionic equation: $Cu(s) + 2\,Ag^+(aq) \longrightarrow 2\,Ag(s) + Cu^{2+}(aq)$

It is now clear that the $Cu(s)$ loses 2 electrons and is oxidized, whereas each Ag^+ ion gains an electron and is reduced.

Worked Example 5.13 Chemical Reactions: Net Ionic Reactions

Write balanced net ionic equations for the following reactions:
(a) $AgNO_3(aq) + ZnCl_2(aq) \longrightarrow$
(b) $HCl(aq) + Ca(OH)_2(aq) \longrightarrow$
(c) $6\,HCl(aq) + 2\,Al(s) \longrightarrow 2\,AlCl_3(aq) + 3\,H_2(g)$

SOLUTION

(a) The solubility guidelines discussed in Section 5.4 predict that a precipitate of insoluble AgCl forms when aqueous solutions of Ag^+ and Cl^- are mixed. Writing all the ions separately gives an ionic equation, and eliminating spectator ions Zn^{2+} and NO_3^- gives the net ionic equation.

Ionic equation: $2\,Ag^+(aq) + 2NO_3^-(aq) + Zn^{2+}(aq) + 2\,Cl^-(aq) \longrightarrow$
$2\,AgCl(s) + Zn^{2+}(aq) + 2\,NO_3^-(aq)$

Net ionic equation: $2\,Ag^+(aq) + 2\,Cl^-(aq) \longrightarrow 2\,AgCl(s)$

The coefficients can all be divided by 2 to give:

Net ionic equation: $Ag^+(aq) + Cl^+(aq) \longrightarrow AgCl(s)$

A check shows that the equation is balanced for atoms and charge (zero on each side).

(b) Allowing the acid HCl to react with the base $Ca(OH)_2$ leads to a neutralization reaction. Writing the ions separately, and remembering to write a complete formula for water, gives an ionic equation. Then eliminating the spectator ions and dividing the coefficients by 2 gives the net ionic equation.

Ionic equation: $2\,H^+(aq) + 2Cl^-(aq) + Ca^{2+}(aq) + 2\,OH^-(aq) \longrightarrow$
$2\,H_2O(l) + Ca^{2+}(aq) + 2\,Cl^-(aq)$

Net ionic equation: $H^+(aq) + OH^-(aq) \longrightarrow H_2O(l)$

A check shows that atoms and charges are the same on both sides of the equation.

(c) The reaction of Al metal with acid (HCl) is a redox reaction. The Al is oxidized, since the oxidation number increases from $0 \rightarrow +3$, whereas the H in HCl is reduced from $+1 \rightarrow 0$. We write the ionic equation by showing the ions that are formed for each aqueous ionic species. Eliminating the spectator ions yields the net ionic equation.

Ionic equation: $6\,H^+(aq) + 6\,Cl^-(aq) + 2\,Al(s) \longrightarrow 2\,Al^{3+}(aq) + 6\,Cl^-(aq) + 3\,H_2(g)$

Net ionic equation: $6\,H^+(aq) + 2\,Al(s) \longrightarrow 2\,Al^{3+}(aq) + 3\,H_2(g)$

A check shows that atoms and charges are the same on both sides of the equation.

PROBLEM 5.18
Write net ionic equations for the following reactions:
(a) $Zn(s) + Pb(NO_3)_2(aq) \longrightarrow Zn(NO_3)_2(aq) + Pb(s)$
(b) $2\,KOH(aq) + H_2SO_4(aq) \longrightarrow K_2SO_4(aq) + 2\,H_2O(l)$
(c) $2\,FeCl_3(aq) + SnCl_2(aq) \longrightarrow 2\,FeCl_2(aq) + SnCl_4(aq)$

PROBLEM 5.19
Identify each of the reactions in Problem 5.18 as an acid–base neutralization, a precipitation, or a redox reaction.

SUMMARY: REVISITING THE CHAPTER GOALS

1. How are chemical reactions written? Chemical equations must be *balanced*; that is, the numbers and kinds of atoms must be the same in both the reactants and the products. To balance an equation, *coefficients* are placed before formulas but the formulas themselves cannot be changed (see *Problems 21–23, 26–37, 59, 60, 64, 67, 68, 71, 72, 75, 76, 79, 80*).

2. How are chemical reactions of ionic compounds classified? There are three common types of reactions of ionic compounds (see *Problems 38–50, 65, 70, 79, 81*).

Precipitation reactions are processes in which an insoluble solid called a *precipitate* is formed. Most precipitations take place when the anions and cations of two ionic compounds change partners. Solubility guidelines for ionic compounds are used to predict when precipitation will occur (see *Problems 24, 25, 43–46, 49, 69, 76–78*).

Acid–base neutralization reactions are processes in which an acid reacts with a base to yield water plus an ionic compound called a *salt*. Since acids produce H^+ ions and bases produce OH^- ions when dissolved in water, a neutralization reaction removes H^+ and OH^- ions from solution and yields neutral H_2O (see *Problems 37, 39, 75, 81*).

Oxidation–reduction (redox) reactions are processes in which one or more electrons are transferred between reaction partners. An *oxidation* is defined as the loss of one or more electrons by an atom, and a *reduction* is the gain of one or more electrons. An *oxidizing agent* causes the oxidation of another reactant by accepting electrons, and a *reducing agent* causes the reduction of another reactant by donating electrons (see *Problems 51–54, 57–62, 65, 66, 68, 82*).

3. What are oxidation numbers, and how are they used? *Oxidation numbers* are assigned to atoms in reactants and products to provide a measure of whether an atom is neutral, electron-rich, or electron-poor. By comparing the oxidation number of an atom before and after reaction, we can tell whether the atom has gained or lost shares in electrons and thus whether a redox reaction has occurred (see *Problems 51–62, 65, 66, 70–74, 82*).

4. What is a net ionic equation? The *net ionic equation* only includes those ions that are directly involved in the ionic reaction. These ions can be identified because they are found in different phases or compounds on the reactant and product sides of the chemical equation. The net ionic equation does not include *spectator ions*, which appear in the same state on both sides of the chemical equation (see *Problems 39, 47, 48, 50, 69, 76–78, 81*).

KEY WORDS

Balanced equation, *p. 134*
Chemical equation, *p. 133*
Coefficient, *p. 134*
Ionic equation, *p. 150*
Law of conservation of mass, *p. 133*
Net ionic equation, *p. 150*
Neutralization reaction, *p. 141*

Oxidation, *p. 143*
Oxidation number, *p. 148*
Oxidation–reduction (redox) reaction, *p. 138*
Oxidizing agent, *p. 143*
Precipitate, *p. 138*
Product, *p. 133*

Reactant, *p. 133*
Reducing agent, *p. 143*
Reduction, *p. 143*
Salt, *p. 138*
Solubility, *p. 139*
Spectator ion, *p. 150*

UNDERSTANDING KEY CONCEPTS

5.20 Assume that the mixture of substances in drawing (a) undergoes a reaction. Which of the drawings (b)–(d) represents a product mixture consistent with the law of conservation of mass?

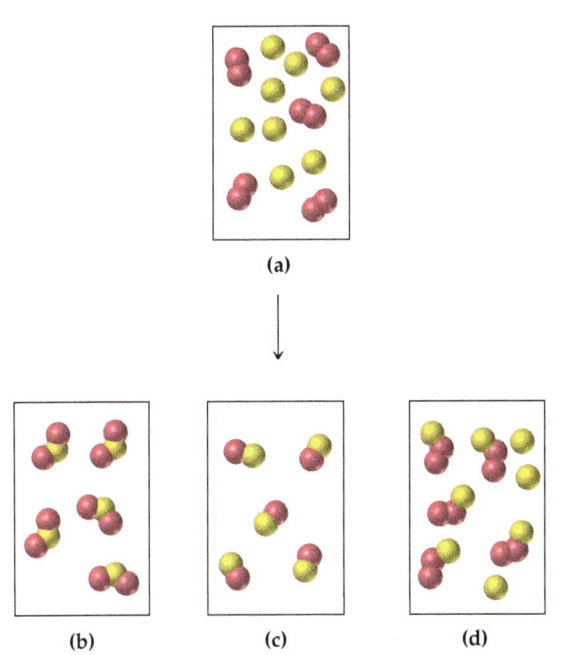

5.21 Reaction of A (green spheres) with B (blue spheres) is shown in the following diagram:

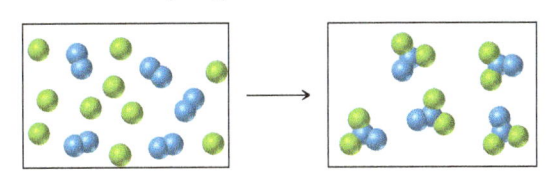

Which equation best describes the reaction?

(a) $A_2 + 2B \longrightarrow A_2B_2$
(b) $10A + 5B_2 \longrightarrow 5A_2B_2$
(c) $2A + B_2 \longrightarrow A_2B_2$
(d) $5A + 5B_2 \longrightarrow 5A_2B_2$

5.22 If blue spheres represent nitrogen atoms and red spheres represent oxygen atoms in the following diagrams, which box represents reactants and which represents products for the reaction $2\,NO(g) + O_2(g) \longrightarrow 2\,NO_2(g)$?

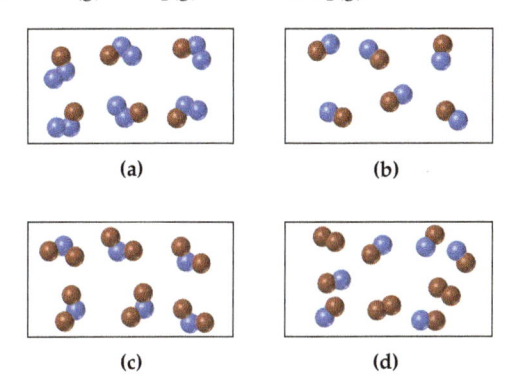

5.23 Assume that an aqueous solution of a cation (represented as red spheres in the diagram) is allowed to mix with a solution of an anion (represented as yellow spheres). Three possible outcomes are represented by boxes (1)–(3):

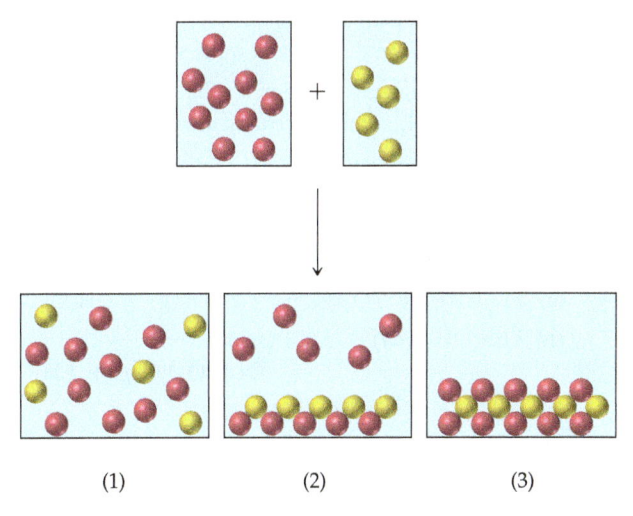

Which outcome corresponds to each of the following reactions?

(a) $2\,Na^+(aq) + CO_3^{2-}(aq) \longrightarrow$
(b) $Ba^{2+}(aq) + CrO_4^{2-}(aq) \longrightarrow$
(c) $2\,Ag^+(aq) + SO_3^{2-}(aq) \longrightarrow$

154 CHAPTER 5 Classification and Balancing of Chemical Reactions

5.24 An aqueous solution of a cation (represented as blue spheres in the diagram) is allowed to mix with a solution of an anion (represented as green spheres) and the following result is obtained:

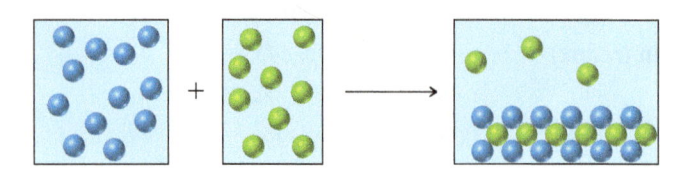

Which combinations of cation and anion, chosen from the following lists, are compatible with the observed results? Explain.

Cations: Na^+, Ca^{2+}, Ag^+, Ni^{2+}

Anions: Cl^-, CO_3^{2-}, CrO_4^{2-}, NO_3^-

5.25 A molecular view of two ionic solutions is presented right:

(a) Which compound is most likely dissolved in beaker A: KBr, $CaCl_2$, PbI_2, Na_2SO_4?

(b) Which compound is most likely dissolved in beaker B: Na_2CO_3, $BaSO_4$, $Cu(NO_3)_2$, $FeCl_3$?

(c) Identify the precipitate and spectator ions for any reaction that will result when beakers A and B are mixed.

Beaker A   Beaker B

2+   1−   2−   1+

ADDITIONAL PROBLEMS

BALANCING CHEMICAL EQUATIONS

5.26 What is meant by the term "balanced equation"?

5.27 Why is it not possible to balance an equation by changing the subscript on a substance, say from H_2O to H_2O_2?

5.28 Write balanced equations for the following reactions:

(a) Gaseous sulfur dioxide reacts with water to form aqueous sulfurous acid (H_2SO_3).

(b) Liquid bromine reacts with solid potassium metal to form solid potassium bromide.

(c) Gaseous propane (C_3H_8) burns in oxygen to form gaseous carbon dioxide and water vapor.

5.29 Balance the following equation for the synthesis of hydrazine, N_2H_4, a substance used as rocket fuel.

$$NH_3(g) + Cl_2(g) \longrightarrow N_2H_4(l) + NH_4Cl(s)$$

5.30 Which of the following equations are balanced? Balance those that need it.

(a) $2\,C_2H_6(g) + 5\,O_2(g) \longrightarrow 2\,CO_2(g) + 6\,H_2O(l)$

(b) $3\,Ca(OH)_2(aq) + 2\,H_3PO_4(aq) \longrightarrow Ca_3(PO_4)_2(aq) + 6\,H_2O(l)$

(c) $Mg(s) + O_2(g) \longrightarrow 2\,MgO(s)$

(d) $K(s) + H_2O(l) \longrightarrow KOH(aq) + H_2(g)$

5.31 Which of the following equations are balanced? Balance those that need it.

(a) $CaC_2 + 2\,H_2O \longrightarrow Ca(OH)_2 + C_2H_2$

(b) $C_2H_8N_2 + 2\,N_2O_4 \longrightarrow 2\,N_2 + 2\,CO_2 + 4\,H_2O$

(c) $3\,MgO + 2\,Fe \longrightarrow Fe_2O_3 + 3\,Mg$

(d) $N_2O \longrightarrow N_2 + O_2$

5.32 Balance the following equations:

(a) $Hg(NO_3)_2(aq) + LiI(aq) \longrightarrow LiNO_3(aq) + HgI_2(s)$

(b) $I_2(s) + Cl_2(g) \longrightarrow ICl_5(s)$

(c) $Al(s) + O_2(g) \longrightarrow Al_2O_3(s)$

(d) $CuSO_4(aq) + AgNO_3(aq) \longrightarrow Ag_2SO_4(s) + Cu(NO_3)_2(aq)$

(e) $Mn(NO_3)_3(aq) + Na_2S(aq) \longrightarrow Mn_2S_3(s) + NaNO_3(aq)$

5.33 Balance the following equations:

(a) $NO_2(g) + O_2(g) \longrightarrow N_2O_5(g)$

(b) $P_4O_{10}(s) + H_2O(l) \longrightarrow H_3PO_4(aq)$

(c) $B_2H_6(l) + O_2(g) \longrightarrow B_2O_3(s) + H_2O(l)$

(d) $Cr_2O_3(s) + CCl_4(l) \longrightarrow CrCl_3(s) + COCl_2(aq)$

(e) $Fe_3O_4(s) + O_2(g) \longrightarrow Fe_2O_3(s)$.

5.34 When organic compounds are burned, they react with oxygen to form CO_2 and H_2O. Write balanced equations for the combustion of the following:

(a) C_4H_{10} (butane, used in lighters)

(b) C_2H_6O (ethyl alcohol, used in gasohol and as race car fuel)

(c) C_8H_{18} (octane, a component of gasoline)

5.35 When organic compounds are burned without enough oxygen, carbon monoxide is formed as a product instead of carbon dioxide. Write and balance the combustion reactions from Problem 5.34 using CO as a product instead of CO_2.

5.36 Hydrofluoric acid (HF) is used to etch glass (SiO_2). The products of the reaction are silicon tetrafluoride and water. Write the balanced chemical equation.

5.37 Write a balanced equation for the reaction of aqueous sodium carbonate (Na_2CO_3) with aqueous nitric acid (HNO_3) to yield CO_2, $NaNO_3$, and H_2O.

TYPES OF CHEMICAL REACTIONS

5.38 Identify each of the following reactions as a precipitation, neutralization, or redox reaction:

(a) $Mg(s) + 2 HCl(aq) \longrightarrow MgCl_2(aq) + H_2(g)$
(b) $KOH(aq) + HNO_3(aq) \longrightarrow KNO_3(aq) + H_2O(l)$
(c) $Pb(NO_3)_2(aq) + 2 HBr(aq) \longrightarrow PbBr_2(s) + 2 HNO_3(aq)$
(d) $Ca(OH)_2(aq) + 2 HCl(aq) \longrightarrow 2 H_2O(l) + CaCl_2(aq)$

5.39 Write balanced ionic equations and net ionic equations for the following reactions:

(a) Aqueous sulfuric acid is neutralized by aqueous potassium hydroxide.
(b) Aqueous magnesium hydroxide is neutralized by aqueous hydrochloric acid.

5.40 Write balanced ionic equations and net ionic equations for the following reactions:

(a) A precipitate of barium sulfate forms when aqueous solutions of barium nitrate and potassium sulfate are mixed.
(b) Zinc ion and hydrogen gas form when zinc metal reacts with aqueous sulfuric acid.

5.41 Identify each of the reactions in Problem 5.30 as a precipitation, neutralization, or redox reaction.

5.42 Identify each of the reactions in Problem 5.32 as a precipitation, neutralization, or redox reaction.

5.43 Which of the following substances are likely to be soluble in water?

(a) $ZnSO_4$ (b) $NiCO_3$
(c) $PbCl_2$ (d) $Ca_3(PO_4)_2$

5.44 Which of the following substances are likely to be soluble in water?

(a) Ag_2O (b) $Ba(NO_3)_2$
(c) $SnCO_3$ (d) Al_2S_3

5.45 Use the solubility guidelines in Section 5.4 to predict whether a precipitation reaction will occur when aqueous solutions of the following substances are mixed.

(a) $NaOH + HClO_4$
(b) $FeCl_2 + KOH$
(c) $(NH_4)_2SO_4 + NiCl_2$

5.46 Use the solubility guidelines in Section 5.4 to predict whether precipitation reactions will occur between the listed pairs of reactants. Write balanced equations for those reactions that should occur.

(a) $NaBr$ and $Hg_2(NO_3)_2$
(b) $CuCl_2$ and K_2SO_4
(c) $LiNO_3$ and $Ca(CH_3CO_2)_2$
(d) $(NH_4)_2CO_3$ and $CaCl_2$
(e) KOH and $MnBr_2$
(f) Na_2S and $Al(NO_3)_3$

5.47 Write net ionic equations for the following reactions:

(a) $Mg(s) + CuCl_2(aq) \longrightarrow MgCl_2(aq) + Cu(s)$
(b) $2 KCl(aq) + Pb(NO_3)_2(aq) \longrightarrow PbCl_2(s) + 2 KNO_3(aq)$
(c) $2 Cr(NO_3)_3(aq) + 3 Na_2S(aq) \longrightarrow Cr_2S_3(s) + 6 NaNO_3(aq)$

5.48 Write net ionic equations for the following reactions:

(a) $2 AuCl_3(aq) + 3 Sn(s) \longrightarrow 3 SnCl_2(aq) + 2 Au(s)$
(b) $2 NaI(aq) + Br_2(l) \longrightarrow 2 NaBr(aq) + I_2(s)$
(c) $2 AgNO_3(aq) + Fe(s) \longrightarrow Fe(NO_3)_2(aq) + 2 Ag(s)$

5.49 Complete the following precipitation reactions using balanced chemical equations:

(a) $FeSO_4(aq) + Sr(OH)_2(aq) \longrightarrow$
(b) $Na_2S(aq) + ZnSO_4(aq) \longrightarrow$

5.50 Write net ionic equations for each of the reactions in Problem 5.49.

REDOX REACTIONS AND OXIDATION NUMBERS

5.51 Where in the periodic table are the best reducing agents found? The best oxidizing agents?

5.52 Where in the periodic table are the most easily reduced elements found? The most easily oxidized?

5.53 In each of the following, tell whether the substance gains electrons or loses electrons in a redox reaction:

(a) An oxidizing agent
(b) A reducing agent
(c) A substance undergoing oxidation
(d) A substance undergoing reduction

5.54 For the following substances, tell whether the oxidation number increases or decreases in a redox reaction:

(a) An oxidizing agent
(b) A reducing agent
(c) A substance undergoing oxidation
(d) A substance undergoing reduction

5.55 Assign an oxidation number to each element in the following compounds or ions:

(a) N_2O_5 (b) SO_3^{2-}
(c) CH_2O (d) $HClO_3$

5.56 Assign an oxidation number to the metal in the following compounds:

(a) $CoCl_3$ (b) $FeSO_4$
(c) UO_3 (d) CuF_2
(e) TiO_2 (f) SnS

5.57 Which element is oxidized and which is reduced in the following reactions?

(a) $Si(s) + 2\,Cl_2(g) \longrightarrow SiCl_4(l)$
(b) $Cl_2(g) + 2\,NaBr(aq) \longrightarrow Br_2(aq) + 2\,NaCl(aq)$
(c) $SbCl_3(s) + Cl_2(g) \longrightarrow SbCl_5(s)$

5.58 Which element is oxidized and which is reduced in the following reactions?

(a) $2\,SO_2(g) + O_2(g) \longrightarrow 2\,SO_3(g)$
(b) $2\,Na(s) + Cl_2(g) \longrightarrow 2\,NaCl(s)$
(c) $CuCl_2(aq) + Zn(s) \longrightarrow ZnCl_2(aq) + Cu(s)$
(d) $2\,NaCl(aq) + F_2(g) \longrightarrow 2\,NaF(aq) + Cl_2(g)$

5.59 Balance each of the following redox reactions:

(a) $Al(s) + H_2SO_4(aq) \longrightarrow Al_2(SO_4)_3(aq) + H_2(g)$
(b) $Fe(s) + Cl_2(g) \longrightarrow FeCl_3(s)$
(c) $CO(g) + I_2O_5(s) \longrightarrow I_2(s) + CO_2(g)$

5.60 Balance each of the following redox reactions:

(a) $N_2O_4(l) + N_2H_4(l) \longrightarrow N_2(g) + H_2O(g)$
(b) $CaH_2(s) + H_2O(l) \longrightarrow Ca(OH)_2(aq) + H_2(g)$
(c) $Al(s) + H_2O(l) \longrightarrow Al(OH)_3(s) + H_2(g)$

5.61 Identify the oxidizing agent and the reducing agent in Problem 5.59.

5.62 Identify the oxidizing agent and the reducing agent in Problem 5.60.

CHEMISTRY IN ACTION

5.63 Sodium urate, the principal constituent of some kidney stones and the substance responsible for gout, has the formula $NaC_5H_3N_4O_3$. In aqueous solution, the solubility of sodium urate is only 0.067 g/L. How many grams of sodium urate could be dissolved in the blood before precipitation might occur? (The average adult has a blood capacity of about 5 L.) [*Gout and Kidney Stones, p. 140*]

5.64 Uric acid is formed in the body by the metabolism of purines. The reaction can be represented as $C_5H_4N_4$ (purine) $+ O_2 \rightarrow C_5H_4N_4O_3$ (uric acid).

(a) Balance the reaction.
(b) What type of reaction is this? [*Gout and Kidney Stones, p. 140*]

5.65 The rechargeable NiCd battery uses the following reaction:

$2\,NiO(OH) + Cd + 2\,H_2O \longrightarrow 2\,Ni(OH)_2 + Cd(OH)_2$.

Which reactant is being oxidized and which is being reduced in this reaction? [*Batteries, p. 147*]

5.66 Identify the oxidizing and reducing agents in a typical dry-cell battery. [*Batteries, p. 147*]

GENERAL QUESTIONS AND PROBLEMS

5.67 Balance the following equations.

(a) The thermite reaction, used in welding:
$Al(s) + Fe_2O_3(s) \longrightarrow Al_2O_3(l) + Fe(l)$

(b) The explosion of ammonium nitrate:
$NH_4NO_3(s) \longrightarrow N_2(g) + O_2(g) + H_2O(g)$

5.68 Lithium oxide is used aboard the space shuttle to remove water from the atmosphere according to the equation:

$Li_2O(s) + H_2O(g) \longrightarrow LiOH(s)$

(a) Balance the chemical equation.
(b) Is this a redox reaction? Why or why not?

5.69 Look at the solubility guidelines in Section 5.4 and predict whether a precipitate forms when $CuCl_2(aq)$ and $Na_2CO_3(aq)$ are mixed. If so, write both the balanced equation and the net ionic equation for the process.

5.70 Balance the following equations and classify each as a precipitation, neutralization, or redox reaction:

(a) $Al(OH)_3(aq) + HNO_3(aq) \longrightarrow Al(NO_3)_3(aq) + H_2O(l)$
(b) $AgNO_3(aq) + FeCl_3(aq) \longrightarrow AgCl(s) + Fe(NO_3)_3(aq)$
(c) $(NH_4)_2Cr_2O_7(s) \longrightarrow Cr_2O_3(s) + H_2O(g) + N_2(g)$
(d) $Mn_2(CO_3)_3(s) \longrightarrow Mn_2O_3(s) + CO_2(g)$

5.71 White phosphorus (P_4) is a highly reactive form of elemental phosphorus that reacts with oxygen to form a variety of molecular compounds, including diphosphorus pentoxide.

(a) Write the balanced chemical equation for this reaction.
(b) Calculate the oxidation number for P and O on both sides of the reaction, and identify the oxidizing and reducing agents.

5.72 The combustion of fossil fuels containing sulfur contributes to the phenomenon known as acid rain. The combustion process releases sulfur in the form of sulfur dioxide, which is converted to sulfuric acid in a process involving two reactions.

(a) In the first reaction, sulfur dioxide reacts with molecular oxygen to form sulfur trioxide. Write the balanced chemical equation for this reaction.
(b) In the second reaction, sulfur trioxide reacts with water in the atmosphere to form sulfuric acid. Write the balanced chemical equation for this reaction.
(c) Calculate the oxidation number for the S atom in each compound in these reactions.

5.73 The transition metals form compounds with oxygen in which the metals have different oxidation states. Calculate the oxidation number for the transition metal in the following sets of compounds:

(a) Mn in MnO_2, Mn_2O_3, and $KMnO_4$
(b) Cr in CrO_2, CrO_3, and Cr_2O_3.

5.74 In the Breathalyzer test, blood alcohol is determined by reaction of the alcohol with potassium dichromate:

$$16H^+(aq) + 2Cr_2O_7^{2-}(aq) + C_2H_5OH(aq) \longrightarrow 4Cr^{3+}(aq) + 2CO_2(g) + 11H_2O(l)$$

(a) Calculate the oxidation number of Cr in $Cr_2O_7^{2-}$.
(b) Calculate the oxidation number of C in C_2H_5OH and in CO_2.
(c) Identify the oxidizing agent and the reducing agent in this reaction.

5.75 Milk of magnesia is a suspension of magnesium hydroxide in water that is used to neutralize excess stomach acid. Write the balanced chemical equation for this neutralization reaction.

5.76 Iron in drinking water is removed by precipitation of the Fe^{3+} ion by reaction with NaOH to produce iron(III) hydroxide. Write the balanced chemical equation and the net ionic equation for this reaction.

5.77 Hard water contains magnesium and calcium ions (Mg^{2+}, Ca^{2+}), which can precipitate out in hot water pipes and water heaters as carbonates. Write the net ionic equation for this reaction.

5.78 Pepto-Bismol™, an antacid and antidiarrheal, contains bismuth subsalicylate, $C_7H_5BiO_4$. Some users of this product can experience a condition known as "black tongue," which is caused by the reaction of bismuth(III) ions with trace amounts of S^{2-} in saliva to form a black precipitate. Write the balanced net ionic equation for this precipitation reaction.

5.79 Iron is produced from iron ore by reaction with carbon monoxide:

$$Fe_2O_3(s) + CO(g) \longrightarrow Fe(s) + CO_2(g)$$

(a) Balance the chemical equation.
(b) Classify the reaction as a precipitation, neutralization, or redox reaction.

5.80 Balance the reaction for the synthesis of urea, commonly used as a fertilizer:

$$CO_2(g) + NH_3(g) \longrightarrow NH_2CONH_2(s) + H_2O(l)$$

5.81 Geologists identify carbonate minerals by reaction with acids. Dolomite, for example, contains magnesium carbonate, which reacts with hydrochloric acid by the following reaction:

$$MgCO_3(s) + HCl(aq) \longrightarrow MgCl_2(aq) + CO_2(g) + H_2O(l)$$

(a) Balance the reaction and write the net ionic equation.
(b) Classify the reaction as a precipitation, neutralization, or redox reaction.

5.82 Iodine, used as an antiseptic agent, can be prepared in the laboratory by the following reaction:

$$2NaI(s) + 2H_2SO_4(aq) + MnO_2(s) \longrightarrow Na_2SO_4(aq) + MnSO_4(aq) + I_2(g) + 2H_2O(l)$$

(a) Determine the oxidation number for the Mn and I on both sides of the equation.
(b) Identify the oxidizing and reducing agents.

CHAPTER 6

Chemical Reactions: Mole and Mass Relationships

CONTENTS

6.1 The Mole and Avogadro's Number
6.2 Gram–Mole Conversions
6.3 Mole Relationships and Chemical Equations
6.4 Mass Relationships and Chemical Equations
6.5 Limiting Reagent and Percent Yield

◀ The amount of CO_2 and H_2O produced by the fuel combustion of airplanes and automobiles can be calculated using mole ratios and mole-to-mass conversions.

CHAPTER GOALS

1. **What is the mole, and why is it useful in chemistry?**
   THE GOAL: Be able to explain the meaning and uses of the mole and Avogadro's number.

2. **How are molar quantities and mass quantities related?**
   THE GOAL: Be able to convert between molar and mass quantities of an element or compound. (◀◀ A.)

3. **What are the limiting reagent, theoretical yield, and percent yield of a reaction?**
   THE GOAL: Be able to take the amount of product actually formed in a reaction, calculate the amount that could form theoretically, and express the results as a percent yield. (◀◀ A, B.)

CONCEPTS TO REVIEW

A. Problem Solving: Unit Conversions and Estimating Answers
(Section 1.12)

B. Balancing Chemical Equations
(Section 5.2)

When chefs prepare to cook a rice pudding, they don't count out individual grains of rice, or individual raisins, or individual sugar crystals. Rather, they measure out appropriate amounts of the necessary ingredients using more convenient units—such as cups, or tablespoons. When chemists prepare chemical reactions, they use the same approach. In this chapter we introduce the concept of the mole and how chemists use it when studying the quantitative relationships between reactants and products.

6.1 The Mole and Avogadro's Number

In the previous chapter, we learned how to use the balanced chemical equation to indicate what is happening at the molecular level during a reaction. Now, let us imagine a laboratory experiment: the reaction of ethylene (C_2H_4) with hydrogen chloride (HCl) to prepare ethyl chloride (C_2H_5Cl), a colorless, low-boiling liquid used by doctors and athletic trainers as a spray-on anesthetic. The reaction is represented as

$$C_2H_4(g) + HCl(g) \rightarrow C_2H_5Cl(g)$$

In this reaction, 1 molecule of ethylene reacts with 1 molecule of hydrogen chloride to produce 1 molecule of ethyl chloride.

How, though, can you be sure you have a 1 to 1 ratio of reactant molecules in your reaction flask? Since it is impossible to hand-count the number of molecules correctly, you must weigh them instead. (This is a common method for dealing with all kinds of small objects: Nails, nuts, and grains of rice are all weighed rather than counted.) But the weighing approach leads to another problem. How many molecules are there in 1 gram of ethylene, hydrogen chloride, or any other substance? The answer depends on the identity of the substance, because different molecules have different masses.

To determine how many molecules of a given substance are in a certain mass, it is helpful to define a quantity called *molecular weight*. Just as the *atomic weight* of an element is the average mass of the element's *atoms*, the **molecular weight (MW)** of a molecule is the average mass of a substance's *molecules*. Numerically, a substance's molecular weight (or **formula weight** for an ionic compound) is equal to the sum of the atomic weights for all the atoms in the molecule or formula unit.

For example, the molecular weight of ethylene (C_2H_4) is 28.0 amu, the molecular weight of HCl is 36.5 amu, and the molecular weight of ethyl chloride (C_2H_5Cl) is 64.5 amu. (The actual values are known more precisely but are rounded off here for convenience.)

For ethylene, C_2H_4:

Atomic weight of 2 C	= 2 × 12.0 amu	= 24.0 amu
Atomic weight of 4 H	= 4 × 1.0 amu	= 4.0 amu
MW of C_2H_4		= 28.0 amu

Molecular weight The sum of atomic weights of all atoms in a molecule.

◀◀ See Section 2.3 for discussion of atomic weight.

Formula weight The sum of atomic weights of all atoms in one formula unit of any compound, whether molecular or ionic.

160 CHAPTER 6 Chemical Reactions: Mole and Mass Relationships

These samples of sulfur, copper, mercury, and helium each contain 1 mol. Do they all have the same mass?

For hydrogen chloride, **HCl**:

$$\text{Atomic weight of H} = 1.0 \text{ amu}$$
$$\underline{\text{Atomic weight of Cl} = 35.5 \text{ amu}}$$
$$\text{MW of HCl} = 36.5 \text{ amu}$$

For ethyl chloride, **C$_2$H$_5$Cl**:

$$\text{Atomic weight of 2 C} = 2 \times 12.0 \text{ amu} = 24.0 \text{ amu}$$
$$\text{Atomic weight of 5 H} = 5 \times 1.0 \text{ amu} = 5.0 \text{ amu}$$
$$\underline{\text{Atomic weight of Cl} = 35.5 \text{ amu}}$$
$$\text{MW of C}_2\text{H}_5\text{Cl} = 64.5 \text{ amu}$$

How are molecular weights used? Since the mass ratio of 1 ethylene molecule to 1 HCl molecule is 28.0 to 36.5, the mass ratio of *any* given number of ethylene molecules to the same number of HCl molecules is also 28.0 to 36.5. In other words, a 28.0 to 36.5 *mass* ratio of ethylene and HCl always guarantees a 1 to 1 *number* ratio. *Samples of different substances always contain the same number of molecules or formula units whenever their mass ratio is the same as their molecular or formula weight ratio* (Figure 6.1).

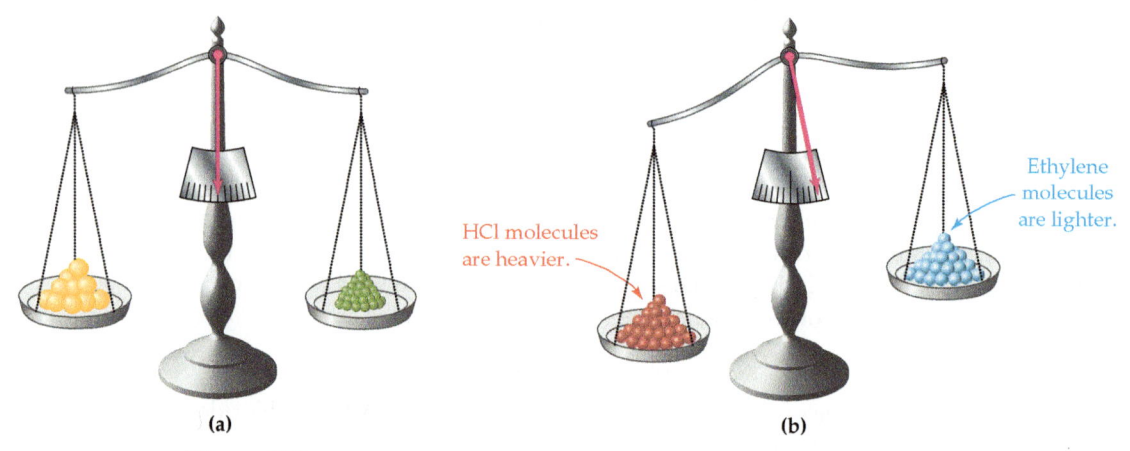

▲ **Figure 6.1**
(a) Because the yellow balls (left pan) are bigger than the green balls (right pan), you cannot get an equal number by taking equal weights. The same is true for atoms or molecules of different substances. (b) Equal numbers of ethylene and HCl molecules always have a mass ratio equal to the ratio of their molecular weights, 28.0 to 36.5.

A particularly convenient way to use this mass/number relationship for molecules is to measure amounts in grams that are numerically equal to molecular weights. If, for instance, you were to carry out your experiment with 28.0 g of ethylene and 36.5 g of HCl, you could be certain that you would have a 1 to 1 ratio of reactant molecules.

When referring to the vast numbers of molecules or formula units that take part in a visible chemical reaction, it is convenient to use a counting unit called a **mole**, abbreviated *mol*. One mole of any substance is the amount whose mass in grams—its **molar mass**—is numerically equal to its molecular or formula weight in amu. One mole of ethylene has a mass of 28.0 g, one mole of HCl has a mass of 36.5 g, and one mole of ethyl chloride has a mass of 64.5 g.

Just how many molecules are there in a mole? Think back to Chapter 2 where we learned to calculate the number of atoms in a sample of an element given its weight in grams, the atomic mass of the atom, and a gram/amu conversion factor. In Problem 2.2, you (hopefully!) found that a 1 gram sample of hydrogen (atomic mass 1 amu) and a 12 gram sample of carbon (atomic mass 12 amu) each contain 6.022×10^{23} atoms. One mole of any substance, therefore, contains 6.022×10^{23} formula units, a value called **Avogadro's number** (abbreviated N_A) after the Italian scientist who first recognized the importance of the mass/number relationship in molecules. Avogadro's

Mole The amount of a substance whose mass in grams is numerically equal to its molecular or formula weight.

Molar mass The mass in grams of 1 mole of a substance, numerically equal to molecular weight.

number of formula units of any substance—that is, one mole—has a mass in grams numerically equal to the molecular weight of the substance.

Avogadro's number (N_A) The number of formula units in 1 mole of anything; 6.022×10^{23}.

1 mol HCl = 6.022×10^{23} HCl molecules = 36.5 g HCl

1 mol C_2H_4 = 6.022×10^{23} C_2H_4 molecules = 28.0 g C_2H_4

1 mol C_2H_5Cl = 6.022×10^{23} C_2H_5Cl molecules = 64.5 g C_2H_5Cl

How big is Avogadro's number? Our minds cannot really conceive of the magnitude of a number like 6.022×10^{23}, but the following comparisons will give you a sense of the scale:

Avogadro's number: 602,200,000,000,000,000,000,000

- Amount of water in world's oceans (liters)
- Age of earth in (seconds)
- Population of earth
- Distance from earth to sun (centimeters)
- Average college tuition (U.S. dollars)

Worked Example 6.1 Molar Mass and Avogadro's Number: Number of Molecules

Pseudoephedrine hydrochloride ($C_{10}H_{16}ClNO$) is a nasal decongestant commonly found in cold medication. (a) What is the molar mass of pseudoephedrine hydrochloride? (b) How many molecules of pseudoephedrine hydrochloride are in a tablet that contains 30.0 mg of this decongestant?

ANALYSIS We are given a mass and need to convert to a number of molecules. This is most easily accomplished by using the molar mass of pseudoephedrine hydrochloride calculated in part (a) as the conversion factor from mass to moles and realizing that this mass (in grams) contains Avogadro's number of molecules (6.022×10^{23}).

BALLPARK ESTIMATE The formula for pseudoephedrine contains 10 carbon atoms (each one of atomic weight 12.0 amu), so the molecular weight is greater than 120 amu, probably near 200 amu. Thus, the molecular weight should be near 200 g/mol. The mass of 30 mg of pseudoepinephrine HCl is less than the mass of 1 mol of this compound by a factor of roughly 10^4 (0.03 g versus 200 g), which means that the number of molecules should also be smaller by a factor of 10^4 (on the order of 10^{19} in the tablet versus 10^{23} in 1 mol).

SOLUTION

(a) The molecular weight of pseudoephedrine is found by summing the atomic weights of all atoms in the molecule:

Atomic Weight of 10 atoms of C:	10×12.011 amu =	120.11 amu
16 atoms of H:	16×1.00794 amu =	16.127 amu
1 atom of Cl:	1×35.4527 amu =	35.4527 amu
1 atom of N:	1×14.0067 amu =	14.0067 amu
1 atom of O:	1×15.9994 amu =	15.9994 amu

MW of $C_{10}H_{16}ClNO$ = 201.6958 amu ⟶ 201.70 g/mol

Remember that atomic mass in amu converts directly to molar mass in g/mol. Also, following the rules for significant figures from Sections 1.9 and 1.11, our final answer is rounded to the second decimal place.

(b) Since this problem involves unit conversions, we can use the step-wise solution introduced in Chapter 1.

STEP 1: Identify known information. We are given the mass of pseudoephedrine hydrochloride (in mg).

30.0 mg pseudoephedrine hydrochloride

STEP 2: Identify answer and units. We are looking for the number of molecules of pseudoephedrine hydrochloride in a 30 mg tablet.

?? = molecules

STEP 3: Identify conversion factors. Since the molecular weight of pseudoephedrine hydrochloride is 201.70 amu, 201.70 g contains 6.022×10^{23} molecules. We can use this ratio as a conversion factor to convert from mass to molecules. We will also need to convert 30 mg to g.

$$\frac{6.022 \times 10^{23} \text{ molecules}}{201.70 \text{ g}}$$

$$\frac{.001 \text{ g}}{1 \text{ mg}}$$

STEP 4: Solve. Set up an equation so that unwanted units cancel.	$(30.0 \text{ mg pseudoephedrine hydrochloride}) \times \left(\dfrac{.001 \text{ g}}{1 \text{ mg}}\right) \times \left(\dfrac{6.022 \times 10^{23} \text{ molecules}}{201.70 \text{ g}}\right)$ $= 8.96 \times 10^{19}$ molecules of pseudoephedrine hydrochloride **BALLPARK CHECK** Our estimate for the number of molecules was on the order of 10^{19}, which is consistent with the calculated answer.

Worked Example 6.2 Avogadro's Number: Atom to Mass Conversions

A tiny pencil mark just visible to the naked eye contains about 3×10^{17} atoms of carbon. What is the mass of this pencil mark in grams?

ANALYSIS We are given a number of atoms and need to convert to mass. The conversion factor can be obtained by realizing that the atomic weight of carbon in grams contains Avogadro's number of atoms (6.022×10^{23}).

BALLPARK ESTIMATE Since we are given a number of atoms that is six orders of magnitude less than Avogadro's number, we should get a corresponding mass that is six orders of magnitude less than the molar mass of carbon, which means a mass for the pencil mark of about 10^{-6} g.

SOLUTION

STEP 1: Identify known information. We know the number of carbon atoms in the pencil mark.	3×10^{17} atoms of carbon
STEP 2: Identify answer and units.	Mass of carbon = ?? g
STEP 3: Identify conversion factors. The atomic weight of carbon is 12.01 amu, so 12.01 g of carbon contains 6.022×10^{23} atoms.	$\dfrac{12.01 \text{ g carbon}}{6.022 \times 10^{23} \text{ atoms}}$
STEP 4: Solve. Set up an equation using the conversion factors so that unwanted units cancel.	$(3 \times 10^{17} \text{ atoms})\left(\dfrac{12.01 \text{ g carbon}}{6.022 \times 10^{23} \text{ atoms}}\right) = 6 \times 10^{-6}$ g carbon

BALLPARK CHECK The answer is of the same magnitude as our estimate and makes physical sense.

PROBLEM 6.1
Calculate the molecular weight of the following substances:
(a) Ibuprofen, $C_{13}H_{18}O_2$ (b) Phenobarbital, $C_{12}H_{12}N_2O_3$

PROBLEM 6.2
How many molecules of ascorbic acid (vitamin C, $C_6H_8O_6$) are in a 500 mg tablet?

PROBLEM 6.3
What is the mass in grams of 5.0×10^{20} molecules of aspirin ($C_9H_8O_4$)?

KEY CONCEPT PROBLEM 6.4

What is the molecular weight of cytosine, a component of DNA (deoxyribonucleic acid)? (black = C, blue = N, red = O, white = H.)

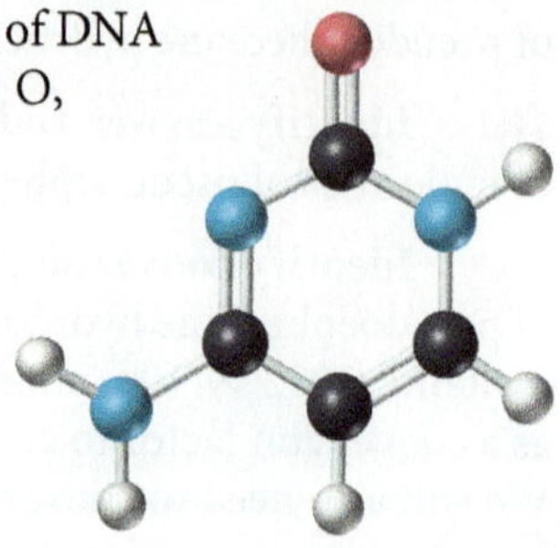

Cytosine

6.2 Gram–Mole Conversions

To ensure that we have the correct molecule to molecule (or mole to mole) relationship between reactants as specified by the balanced chemical equation, we can take advantage of the constant mass ratio between reactants. The mass in grams of 1 mol of any substance (that is, Avogadro's number of molecules or formula units) is called the molar mass of the substance.

$$\begin{aligned}\text{Molar mass} &= \text{Mass of 1 mol of substance} \\ &= \text{Mass of } 6.022 \times 10^{23} \text{ molecules (formula units) of substance} \\ &= \text{Molecular (formula) weight of substance in grams}\end{aligned}$$

In effect, molar mass serves as a conversion factor between numbers of moles and mass. If you know how many moles you have, you can calculate their mass; if you know the mass of a sample, you can calculate the number of moles. Suppose, for example, we need to know how much 0.25 mol of water weighs. The molecular weight of H_2O is $(2 \times 1.0 \text{ amu}) + 16.0 \text{ amu} = 18.0$ amu, so the molar mass of water is 18.0 g/mol. Thus, the conversion factor between moles of water and mass of water is 18.0 g/mol:

$$0.25 \text{ mol } H_2O \times \underbrace{\frac{18.0 \text{ g } H_2O}{1 \text{ mol } H_2O}}_{\text{Molar mass used as conversion factor}} = 4.5 \text{ g } H_2O$$

Alternatively, suppose we need to know how many moles of water are in 27 g of water. The conversion factor is 1 mol/18.0 g:

$$27 \text{ g } H_2O \times \underbrace{\frac{1 \text{ mol } H_2O}{18.0 \text{ g } H_2O}}_{\text{Molar mass used as conversion factor}} = 1.5 \text{ mol } H_2O$$

Note that the 1 mol in the numerator is an exact number, so the number of significant figures in the final answer is based on the 27 g H_2O (2 sig figs.). Worked Examples 6.3 and 6.4 give more practice in gram–mole conversions.

Worked Example 6.3 Molar Mass: Mole to Gram Conversion

The nonprescription pain relievers Advil and Nuprin contain ibuprofen ($C_{13}H_{18}O_2$), whose molecular weight is 206.3 amu (Problem 6.1a). If all the tablets in a bottle of pain reliever together contain 0.082 mol of ibuprofen, what is the number of grams of ibuprofen in the bottle?

ANALYSIS We are given a number of moles and asked to find the mass. Molar mass is the conversion factor between the two.

BALLPARK ESTIMATE Since 1 mol of ibuprofen has a mass of about 200 g, 0.08 mol has a mass of about $0.08 \times 200 \text{ g} = 16 \text{ g}$.

SOLUTION

STEP 1: Identify known information.	0.082 mol ibuprofen in bottle
STEP 2: Identify answer and units.	mass ibuprofen in bottle = ?? g
STEP 3: Identify conversion factor. We use the molecular weight of ibuprofen to convert from moles to grams.	1 mol ibuprofen = 206.3 g $$\frac{206.3 \text{ g ibuprofen}}{1 \text{ mol ibuprofen}}$$
STEP 4: Solve. Set up an equation using the known information and conversion factor so that unwanted units cancel.	$$0.082 \text{ mol } C_{13}H_{18}O_2 \times \frac{206.3 \text{ g ibuprofen}}{1 \text{ mol ibuprofen}} = 17 \text{ g } C_{13}H_{18}O_2$$

BALLPARK CHECK The calculated answer is consistent with our estimate of 16 g.

CHEMISTRY IN ACTION

Did Ben Franklin Have Avogadro's Number? A Ballpark Calculation

"At" length being at Clapham, where there is on the common a large pond . . . I fetched out a cruet of oil and dropped a little of it on the water. I saw it spread itself with surprising swiftness upon the surface. The oil, though not more than a teaspoonful, produced an instant calm over a space several yards square which spread amazingly and extended itself gradually . . . making all that quarter of the pond, perhaps half an acre, as smooth as a looking glass. *Excerpt from a letter of Benjamin Franklin to William Brownrigg, 1773.*

▲ What did these two have in common? [Benjamin Franklin (left), Amedeo Avogadro (right)]

Benjamin Franklin, author and renowned statesman, was also an inventor and a scientist. Every school-child knows of Franklin's experiment with a kite and a key, demonstrating that lightning is electricity. Less well known is that his measurement of the extent to which oil spreads on water makes possible a simple estimate of molecular size and Avogadro's number.

The calculation goes like this: Avogadro's number is the number of molecules in 1 mole of any substance. So, if we can estimate both the number of molecules and the number of moles in Franklin's teaspoon of oil, we can calculate Avogadro's number. Let us start by calculating the number of molecules in the oil.

1. The volume (V) of oil Franklin used was 1 tsp = 4.9 cm^3, and the area (A) covered by the oil was 1/2 acre = 2.0×10^7 cm^2. We will assume that the oil molecules are tiny cubes that pack closely together and form a layer only one molecule thick. As shown in the accompanying figure, the volume of the oil is equal to the surface area of the layer times the length (l) of the side of one molecule: $V = A \times l$. Rearranging this equation to find the length then gives us an estimate of molecular size:

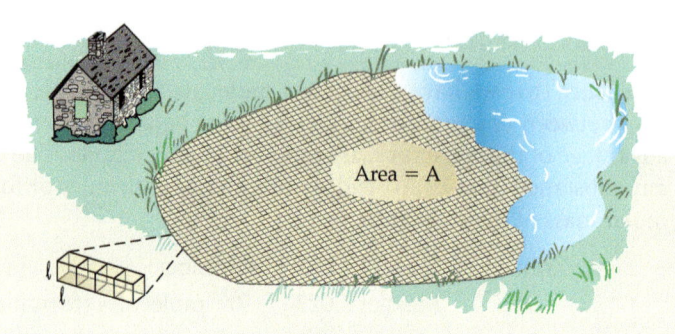

$$l = \frac{V}{A} = \frac{4.9 \text{ cm}^3}{2.0 \times 10^7 \text{ cm}^2} = 2.5 \times 10^{-7} \text{ cm}$$

2. The area of the oil layer is the area of the side of one molecule (l^2) times the number of molecules (N) of oil: $A = l^2 \times N$. Rearranging this equation gives us the number of molecules:

$$N = \frac{A}{l^2} = \frac{2 \times 10^7 \text{ cm}^2}{(2.5 \times 10^{-7} \text{ cm})^2} = 3.2 \times 10^{20} \text{ molecules}$$

3. To calculate the number of moles, we first need to know the mass (M) of the oil. This could have been determined by weighing the oil, but Franklin neglected to do so. Let us therefore estimate the mass by multiplying the volume (V) of the oil by the density (D) of a typical oil, 0.95 g/cm^3. (Since oil floats on water, it is not surprising that the density of oil is a bit less than the density of water, which is 1.00 g/cm^3.)

$$M = V \times D = 4.9 \text{ cm}^3 \times 0.95 \frac{\text{g}}{\text{cm}^3} = 4.7 \text{ g}$$

4. We now have to make one final assumption about the molecular weight of the oil before we complete the calculation. Assuming that a typical oil has MW = 200 amu, then the mass of 1 mol of oil is 200 g. Dividing the mass of the oil (M) by the mass of 1 mol gives the number of moles of oil:

$$\text{Moles of oil} = \frac{4.7 \text{ g}}{200 \text{ g/mol}} = 0.024 \text{ mol}$$

5. Finally, the number of molecules per mole—Avogadro's number—can be obtained by dividing the estimated number of molecules (step 2) by the estimated moles (step 4):

$$\text{Avogadro's number} = \frac{3.2 \times 10^{20} \text{ molecules}}{0.024 \text{ mol}} = 1.3 \times 10^{22}$$

The calculation is not very accurate, of course, but Ben was not really intending for us to calculate Avogadro's number when he made a rough estimate of how much his oil spread out. Nevertheless, the result is not too bad for such a simple experiment.

See Chemistry in Action Problem 6.58 at the end of the chapter.

Worked Example 6.4 Molar Mass: Gram to Mole Conversion

The maximum dose of sodium hydrogen phosphate (Na_2HPO_4, MW = 142.0 molar mass) that should be taken in one day for use as a laxative is 3.8 g. How many moles of sodium hydrogen phosphate, how many moles of Na^+ ions, and how many total moles of ions are in this dose?

ANALYSIS Molar mass is the conversion factor between mass and number of moles. The chemical formula Na_2HPO_4 shows that each formula unit contains 2 Na^+ ions and 1 HPO_4^{2-} ion.

BALLPARK ESTIMATE The maximum dose is about two orders of magnitude smaller than the molecular weight (approximately 4 g compared to 142 g). Thus, the number of moles of sodium hydrogen phosphate in 3.8 g should be about two orders of magnitude less than one mole. The number of moles of Na_2HPO_4 and total moles of ions, then, should be on the order of 10^{-2}.

SOLUTION

STEP 1: Identify known information. We are given the mass and molecular weight of Na_2HPO_4.

3.8 g Na_2HPO_4; MW = 142.0 amu

STEP 2: Identify answer and units. We need to find the number of moles of Na_2HPO_4, and the total number of moles of ions.

Moles of Na_2HPO_4 = ?? mol
Moles of Na^+ ions = ?? mol
Total moles of ions = ?? mol

STEP 3: Identify conversion factor. We can use the molecular weight of Na_2HPO_4 to convert from grams to moles.

$$\frac{1 \text{ mol } Na_2HPO_4}{142.0 \text{ g } Na_2HPO_4}$$

STEP 4: Solve. We use the known information and conversion factor to obtain moles of Na_2HPO_4; since 1 mol of Na_2HPO_4 contains 2 mol of Na^+ ions and 1 mol of HPO_4^{2-} ions, we multiply these values by the number of moles in the sample.

$$3.8 \text{ g } Na_2HPO_4 \times \frac{1 \text{ mol } Na_2HPO_4}{142.0 \text{ g } Na_2HPO_4} = 0.027 \text{ mol } Na_2HPO_4$$

$$\frac{2 \text{ mol } Na^+}{1 \text{ mol } Na_2HPO_4} \times 0.027 \text{ mol } Na_2HPO_4 = 0.054 \text{ mol } Na^+$$

$$\frac{3 \text{ mol ions}}{1 \text{ mol } Na_2HPO_4} \times 0.027 \text{ mol } Na_2HPO_4 = 0.081 \text{ mol ions}$$

BALLPARK CHECK: The calculated answers (0.027 mol Na_2HPO_4, 0.081 mol ions) are on the order of 10^{-2}, consistent with our estimate.

PROBLEM 6.5
How many moles of ethyl alcohol, C_2H_6O, are in a 10.0 g sample? How many grams are in a 0.10 mol sample of ethyl alcohol?

PROBLEM 6.6
Which weighs more, 5.00 g or 0.0225 mol of acetaminophen ($C_8H_9NO_2$)?

PROBLEM 6.7
How would our estimate of Avogadro's number be affected if we were to assume that Benjamin Franklin's oil molecules were spherical rather than cubes (see Chemistry in Action on p. 164)? If the density of the oil was 0.90 g/mL? If the molar mass was 150 g/mol rather than 200 g/mol?

6.3 Mole Relationships and Chemical Equations

In a typical recipe, the amounts of ingredients needed are specified using a variety of units: the amount of flour, for example, is usually specified in cups, whereas the amount of salt or vanilla flavoring might be indicated in teaspoons. In chemical reactions, the appropriate unit to specify the relationship between reactants and products is the mole.

The coefficients in a balanced chemical equation tell how many *molecules*, and thus, how many *moles*, of each reactant are needed and how many molecules, and thus, moles, of each product are formed. You can then use molar mass to calculate reactant and product masses. If, for example, you saw the following balanced equation for the industrial synthesis of ammonia, you would know that 3 mol of H$_2$ (3 mol × 2.0 g/mol = 6.0 g) are required for reaction with 1 mol of N$_2$ (28.0 g) to yield 2 mol of NH$_3$ (2 mol × 17.0 g/mol = 34.0 g).

$$\underset{\text{This number of moles of hydrogen...}}{3\ H_2} + \underset{\text{...reacts with this number of moles of nitrogen...}}{1\ N_2} \longrightarrow \underset{\text{to yield this number of moles of ammonia.}}{2\ NH_3}$$

The coefficients can be put in the form of *mole ratios*, which act as conversion factors when setting up factor-label calculations. In the ammonia synthesis, for example, the mole ratio of H$_2$ to N$_2$ is 3:1, the mole ratio of H$_2$ to NH$_3$ is 3:2, and the mole ratio of N$_2$ to NH$_3$ is 1:2:

$$\frac{3\ mol\ H_2}{1\ mol\ N_2} \quad \frac{3\ mol\ H_2}{2\ mol\ NH_3} \quad \frac{1\ mol\ N_2}{2\ mol\ NH_3}$$

Worked Example 6.5 shows how to set up and use mole ratios.

Worked Example 6.5 Balanced Chemical Equations: Mole Ratios

Rusting involves the reaction of iron with oxygen to form iron(III) oxide, Fe$_2$O$_3$:

$$4\ Fe(s) + 3\ O_2(g) \longrightarrow 2\ Fe_2O_3(s)$$

(a) What are the mole ratios of the product to each reactant and of the reactants to each other?

(b) How many moles of iron(III) oxide are formed by the complete oxidation of 6.2 mol of iron?

ANALYSIS AND SOLUTION

(a) The coefficients of a balanced equation represent the mole ratios:

$$\frac{2\ mol\ Fe_2O_3}{4\ mol\ Fe} \quad \frac{2\ mol\ Fe_2O_3}{3\ mol\ O_2} \quad \frac{4\ mol\ Fe}{3\ mol\ O_2}$$

(b) To find how many moles of Fe$_2$O$_3$ are formed, write down the known information—6.2 mol of iron—and select the mole ratio that allows the quantities to cancel, leaving the desired quantity:

$$6.2\ mol\ Fe \times \frac{2\ mol\ Fe_2O_3}{4\ mol\ Fe} = 3.1\ mol\ Fe_2O_3$$

Note that mole ratios are exact numbers and therefore do not limit the number of significant figures in the result of a calculation.

PROBLEM 6.8

(a) Balance the following equation, and tell how many moles of nickel will react with 9.81 mol of hydrochloric acid.

$$Ni(s) + HCl(aq) \longrightarrow NiCl_2(aq) + H_2(g)$$

(b) How many moles of NiCl$_2$ can be formed in the reaction of 6.00 mol of Ni and 12.0 mol of HCl?

PROBLEM 6.9

Plants convert carbon dioxide and water to glucose (C$_6$H$_{12}$O$_6$) and oxygen in the process of photosynthesis. Write a balanced equation for this reaction, and determine how many moles of CO$_2$ are required to produce 15.0 mol of glucose.

6.4 Mass Relationships and Chemical Equations

It is important to remember that the coefficients in a balanced chemical equation represent molecule to molecule (or mole to mole) relationships between reactants and products. Mole ratios make it possible to calculate the molar amounts of reactants and products, but actual amounts of substances used in the laboratory are weighed out in grams. Regardless of what units we use to specify the amount of reactants and/or products (mass, volume, number of molecules, and so on), the reaction always takes place on a mole to mole basis. Thus, we need to be able to carry out three kinds of conversions when doing chemical arithmetic:

- **Mole to mole conversions** are carried out using *mole ratios* as conversion factors. Worked Example 6.5 at the end of the preceding section is an example of this kind of calculation.

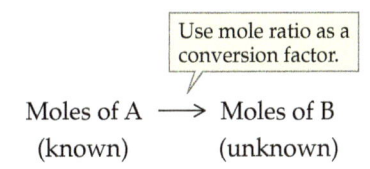

$$\text{Moles of A} \xrightarrow{\text{Use mole ratio as a conversion factor.}} \text{Moles of B}$$
$$\text{(known)} \qquad\qquad \text{(unknown)}$$

- **Mole to mass and mass to mole conversions** are carried out using *molar mass* as a conversion factor. Worked Examples 6.3 and 6.4 at the end of Section 6.2 are examples of this kind of calculation.

$$\text{Moles of A} \xleftrightarrow{\text{Use molar mass as a conversion factor.}} \text{Mass of A (in grams)}$$

- **Mass to mass conversions** are frequently needed but cannot be carried out directly. If you know the mass of substance A and need to find the mass of substance B, you must first convert the mass of A into moles of A, then carry out a mole to mole conversion to find moles of B, and then convert moles of B into the mass of B (Figure 6.2).

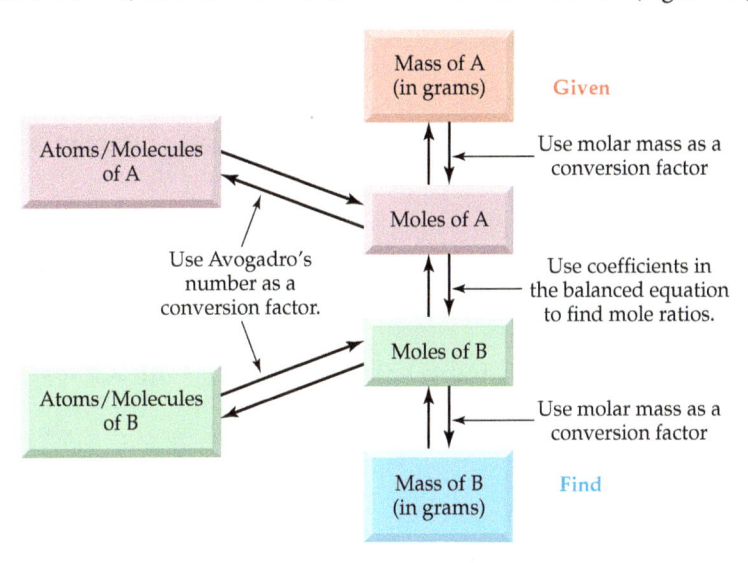

◀ **Figure 6.2**
A summary of conversions between moles, grams, and number of atoms or molecules for substances in a chemical reaction.
The numbers of moles tell how many molecules of each substance are needed, as given by the coefficients in the balanced equation; the numbers of grams tell what mass of each substance is needed.

Overall, there are four steps for determining mass relationships among reactants and products:

STEP 1: Write the balanced chemical equation.

STEP 2: Choose molar masses and mole ratios to convert the known information into the needed information.

STEP 3: Set up the factor-label expressions.

STEP 4: Calculate the answer and check the answer against the ballpark estimate you made before you began your calculations.

168 CHAPTER 6 Chemical Reactions: Mole and Mass Relationships

Worked Example 6.6 Mole Ratios: Mole to Mass Conversions

In the atmosphere, nitrogen dioxide reacts with water to produce NO and nitric acid, which contributes to pollution by acid rain:

$$3\ NO_2(g) + H_2O(l) \longrightarrow 2\ HNO_3(aq) + NO(g)$$

How many grams of HNO_3 are produced for every 1.0 mol of NO_2 that reacts? The molecular weight of HNO_3 is 63.0 amu.

ANALYSIS We are given the number of moles of a reactant and are asked to find the mass of a product. Problems of this sort always require working in moles and then converting to mass, as outlined in Figure 6.2.

BALLPARK ESTIMATE The molar mass of nitric acid is approximately 60 g/mol, and the coefficients in the balanced equation say that 2 mol of HNO_3 are formed for each 3 mol of NO_2 that undergo reaction. Thus, 1 mol of NO_2 should give about 2/3 mol HNO_3, or 2/3 mol × 60 g/mol = 40 g.

SOLUTION

STEP 1: Write balanced equation.

$$3\ NO_2(g) + H_2O(l) \longrightarrow 2\ HNO_3(aq) + NO(g)$$

STEP 2: Identify conversion factors. We need a mole to mole conversion to find the number of moles of product, and then a mole to mass conversion to find the mass of product. For the first conversion we use the mole ratio of HNO_3 to NO_2 as a conversion factor, and for the mole to mass calculation, we use the molar mass of HNO_3 (63.0 g/mol) as a conversion factor.

$$\frac{2\ mol\ HNO_3}{3\ mol\ NO_2}$$

$$\frac{63.0\ g\ HNO_3}{1\ mol\ HNO_3}$$

STEP 3: Set up factor labels. Identify appropriate mole ratio factor labels to convert moles NO_2 to moles HNO_3, and moles HNO_3 to grams.

$$1.0\ mol\ NO_2 \times \frac{2\ mol\ HNO_3}{3\ mol\ NO_2} \times \frac{63.0\ g\ HNO_3}{1\ mol\ HNO_3} = 42\ g\ HNO_3$$

STEP 4: Solve.

BALLPARK CHECK Our estimate was 40 g!

Worked Example 6.7 Mole Ratios: Mass to Mole/Mole to Mass Conversions

The following reaction produced 0.022 g of calcium oxalate (CaC_2O_4). What mass of calcium chloride was used as reactant? (The molar mass of CaC_2O_4 is 128.1 g/mol, and the molar mass of $CaCl_2$ is 111.0 g/mol.)

$$CaCl_2(aq) + Na_2C_2O_4(aq) \longrightarrow CaC_2O_4(s) + 2\ NaCl(aq)$$

ANALYSIS Both the known information and that to be found are masses, so this is a mass to mass conversion problem. The mass of CaC_2O_4 is first converted into moles, a mole ratio is used to find moles of $CaCl_2$, and the number of moles of $CaCl_2$ is converted into mass.

BALLPARK ESTIMATE The balanced equation says that 1 mol of CaC_2O_4 is formed for each mole of $CaCl_2$ that reacts. Because the formula weights of the two substances are similar, it should take about 0.02 g of $CaCl_2$ to form 0.02 g of CaC_2O_4.

SOLUTION

STEP 1: Write the balanced equation.

$$CaCl_2(aq) + Na_2C_2O_4(aq) \longrightarrow CaC_2O_4(s) + 2\ NaCl(aq)$$

STEP 2: Identify conversion factors. Convert the mass of CaC_2O_4 into moles, use a mole ratio to find moles of $CaCl_2$, and convert the number of moles of $CaCl_2$ to mass. We will need three conversion factors.

mass CaC_2O_4 to moles: $\dfrac{1\ mol\ CaC_2O_4}{128.1\ g}$

moles CaC_2O_4 to moles $CaCl_2$: $\dfrac{1\ mol\ CaCl_2}{1\ mol\ CaC_2O_4}$

moles $CaCl_2$ to mass: $\dfrac{111.0\ g\ CaCl_2}{1\ mol\ CaCl_2}$

STEP 3: Set up factor-labels. We will need to perform gram to mole and mole to mole conversions to get from grams CaC_2O_4 to grams $CaCl_2$.

$$0.022\ g\ CaC_2O_4 \times \frac{1\ mol\ CaC_2O_4}{128.1\ g\ CaC_2O_4} \times \frac{1\ mol\ CaCl_2}{1\ mol\ CaC_2O_4} \times \frac{111.0\ g\ CaCl_2}{1\ mol\ CaCl_2} = 0.019\ g\ CaCl_2$$

STEP 4: Solve.

BALLPARK CHECK The calculated answer (0.019 g) is consistent with our estimate (0.02 g).

PROBLEM 6.10
Hydrogen fluoride is one of the few substances that react with glass (which is made of silicon dioxide, SiO_2).

$$4\,HF(g) + SiO_2(s) \longrightarrow SiF_4(g) + 2\,H_2O(l)$$

(a) How many moles of HF will react completely with 9.90 mol of SiO_2?
(b) What mass of water (in grams) is produced by the reaction of 23.0 g of SiO_2?

PROBLEM 6.11
The tungsten metal used for filaments in light bulbs is made by reaction of tungsten trioxide with hydrogen:

$$WO_3(s) + 3\,H_2(g) \longrightarrow W(s) + 3\,H_2O(g)$$

How many grams of tungsten trioxide, and how many grams of hydrogen must you start with to prepare 5.00 g of tungsten? (For WO_3, MW = 231.8 amu.)

6.5 Limiting Reagent and Percent Yield

All the calculations we have done in the last several sections have assumed that 100% of the reactants are converted to products. Only rarely is this the case in practice, though. Let us return to the recipe for s'mores presented in the previous chapter:

2 Graham crackers + 1 Roasted marshmallow + $\frac{1}{4}$ Chocolate bar $\longrightarrow$ 1 S'more

When you check your supplies, you find that you have 20 graham crackers, 8 marshmallows, and 3 chocolate bars. How many s'mores can you make? (Answer = 8!) You have enough graham crackers and chocolate bars to make more, but you will run out of marshmallows after you have made eight s'mores. In a similar way, when running a chemical reaction we don't always have the exact amounts of reagents to allow all of them to react completely. The reactant that is exhausted first in such a reaction is called the **limiting reagent**. The amount of product you obtain if the limiting reagent is completely consumed is called the **theoretical yield** of the reaction.

Suppose that, while you are making s'mores, one of your eight marshmallows gets burned to a crisp. If this happens, the actual number of s'mores produced will be less than what you predicted based on the amount of starting materials. Similarly, chemical reactions do not always yield the exact amount of product predicted by the initial amount of reactants. More frequently, a majority of the reactant molecules behave as written, but other processes, called *side reactions*, also occur. In addition, some of the product may be lost in handling. As a result, the amount of product actually formed—the reaction's **actual yield**—is somewhat less than the theoretical yield. The amount of product actually obtained in a reaction is usually expressed as a **percent yield**:

$$\text{Percent yield} = \frac{\text{Actual yield}}{\text{Theoretical yield}} \times 100\%$$

A reaction's actual yield is found by weighing the amount of product obtained. The theoretical yield is found by using the amount of limiting reagent in a mass to mass calculation like those illustrated in the preceding section (see Worked Example 6.7). Worked Examples 6.8–6.10 involve limiting reagent, percent yield, actual yield, and theoretical yield calculations.

Limiting reagent The reactant that runs out first in any given reaction.

Theoretical yield The amount of product formed, assuming complete reaction of the limiting reagent.

Actual yield The amount of product actually formed in a reaction.

Percent yield The percent of the theoretical yield actually obtained from a chemical reaction.

Worked Example 6.8 Percent Yield

The combustion of acetylene gas (C_2H_2) produces carbon dioxide and water, as indicated in the following reaction:

$$2\ C_2H_2(g) + 5\ O_2(g) \longrightarrow 4\ CO_2(g) + 2\ H_2O(g)$$

When 26.0 g of acetylene is burned in sufficient oxygen for complete reaction, the theoretical yield of CO_2 is 88.0 g. Calculate the percent yield for this reaction if the actual yield is only 72.4 g CO_2.

ANALYSIS The percent yield is calculated by dividing the actual yield by the theoretical yield and multiplying by 100.

BALLPARK ESTIMATE The theoretical yield (88.0 g) is close to 100 g. The actual yield (72.4 g) is about 15 g less than the theoretical yield. The actual yield is thus about 15% less than the theoretical yield, so the percent yield is about 85%.

SOLUTION

$$\text{Percent yield} = \frac{\text{Actual yield}}{\text{Theoretical yield}} \times 100 = \frac{72.4 \text{ g } CO_2}{88.0 \text{ g } CO_2} \times 100 = 82.3\%$$

BALLPARK CHECK The calculated percent yield agrees very well with our estimate of 85%.

Worked Example 6.9 Mass to Mole Conversions: Limiting Reagent and Theoretical Yield

The element boron is produced commercially by the reaction of boric oxide with magnesium at high temperature:

$$B_2O_3(l) + 3\ Mg(s) \longrightarrow 2\ B(s) + 3\ MgO(s)$$

What is the theoretical yield of boron when 2350 g of boric oxide is reacted with 3580 g of magnesium? The molar masses of boric oxide and magnesium are 69.6 g/mol and 24.3 g/mol, respectively.

ANALYSIS To calculate theoretical yield, we first have to identify the limiting reagent. The theoretical yield in grams is then calculated from the amount of limiting reagent used in the reaction. The calculation involves the mass to mole and mole to mass conversions discussed in the preceding section.

SOLUTION

STEP 1: Identify known information. We have the masses and molar masses of the reagents.

2350 g B_2O_3, molar mass 69.6 g/mol
3580 g Mg, molar mass 24.3 g/mol

STEP 2: Identify answer and units. We are solving for the theoretical yield of boron.

Theoretical mass of B = ?? g

STEP 3: Identify conversion factors. We can use the molar masses to convert from masses to moles of reactants (B_2O_3, Mg). From moles of reactants, we can use mole ratios from the balanced chemical equation to find the number of moles of B produced, assuming complete conversion of a given reactant. B_2O_3 is the limiting reagent, since complete conversion of this reagent yields less product (67.6 mol B formed) than does complete conversion of Mg (98.0 mol B formed).

$$(2350 \text{ g } B_2O_3) \times \frac{1 \text{ mol } B_2O_3}{69.6 \text{ g } B_2O_3} = 33.8 \text{ mol } B_2O_3$$

$$(3580 \text{ g Mg}) \times \frac{1 \text{ mol Mg}}{24.3 \text{ g Mg}} = 147 \text{ mol Mg}$$

$$33.8 \text{ mol } B_2O_3 \times \frac{2 \text{ mol B}}{1 \text{ mol } B_2O_3} = 67.6 \text{ mol B*}$$

$$147 \text{ mol Mg} \times \frac{2 \text{ mol B}}{3 \text{ mol Mg}} = 98.0 \text{ mol B}$$

(*B_2O_3 is the limiting reagent because it yields fewer moles of B!)

STEP 4: Solve. Once the limiting reagent has been identified (B_2O_3), the theoretical amount of B that should be formed can be calculated using a mole to mass conversion.

$$67.6 \text{ mol B} \times \frac{10.8 \text{ g B}}{1 \text{ mol B}} = 730 \text{ g B}$$

Worked Example 6.10 Mass to Mole Conversion: Percent Yield

The reaction of ethylene with water to give ethyl alcohol (CH_3CH_2OH) occurs with 78.5% actual yield. How many grams of ethyl alcohol are formed by reaction of 25.0 g of ethylene? (For ethylene, MW = 28.0 amu; for ethyl alcohol, MW = 46.0 amu.)

$$H_2C=CH_2 + H_2O \longrightarrow CH_3CH_2OH$$

ANALYSIS Treat this as a typical mass relationship problem to find the amount of ethyl alcohol that can theoretically be formed from 25.0 g of ethylene, and then multiply the answer by 0.785 (the fraction of the theoretical yield actually obtained) to find the amount actually formed.

BALLPARK ESTIMATE The 25.0 g of ethylene is a bit less than 1 mol; since the percent yield is about 78%, a bit less than 0.78 mol of ethyl alcohol will form—perhaps about 3/4 mol, or 3/4 × 46 g = 34 g.

SOLUTION
The theoretical yield of ethyl alcohol is:

$$25.0 \text{ g ethylene} \times \frac{1 \text{ mol ethylene}}{28.0 \text{ g ethylene}} \times \frac{1 \text{ mol ethyl alc.}}{1 \text{ mol ethylene}} \times \frac{46.0 \text{ g ethyl alc.}}{1 \text{ mol ethyl alc.}}$$
$$= 41.1 \text{ g ethyl alcohol}$$

and so the actual yield is:

$$41.1 \text{ g ethyl alc.} \times 0.785 = 32.3 \text{ g ethyl alcohol}$$

BALLPARK CHECK The calculated result (32.3 g) is close to our estimate (34 g).

PROBLEM 6.12
What is the theoretical yield of ethyl chloride in the reaction of 19.4 g of ethylene with 50 g of hydrogen chloride? What is the percent yield if 25.5 g of ethyl chloride is actually formed? (For ethylene, MW = 28.0 amu; for hydrogen chloride, MW = 36.5 amu; for ethyl chloride, MW = 64.5 amu.)

$$H_2C=CH_2 + HCl \longrightarrow CH_3CH_2Cl$$

PROBLEM 6.13
The reaction of ethylene oxide with water to give ethylene glycol (automobile antifreeze) occurs in 96.0% actual yield. How many grams of ethylene glycol are formed by reaction of 35.0 g of ethylene oxide? (For ethylene oxide, MW = 44.0 amu; for ethylene glycol, MW = 62.0 amu.)

$$\underset{\text{Ethylene oxide}}{H_2C\overset{O}{\overset{\triangle}{-}}CH_2} + H_2O \longrightarrow \underset{\text{Ethylene glycol}}{HOCH_2CH_2OH}$$

PROBLEM 6.14
The recommended daily intake of iron is 8 mg for adult men and 18 mg for premenopausal women (see Chemistry in Action on p. 172). Convert these masses of iron into moles.

172 CHAPTER 6 Chemical Reactions: Mole and Mass Relationships

> **KEY CONCEPT PROBLEM 6.15**
>
> Identify the limiting reagent in the reaction mixture shown below. The balanced reaction is:
>
> $$A_2 + 2\,B_2 \longrightarrow 2\,AB_2$$
>
> $A_2 =$ (red-red) $B_2 =$ (blue-blue)

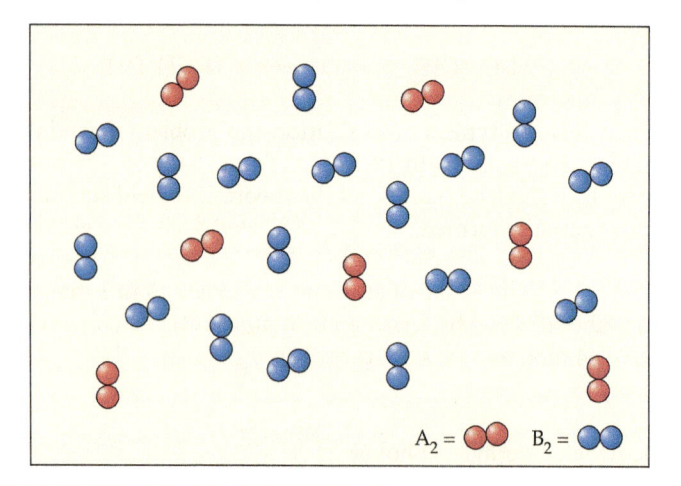

CHEMISTRY IN ACTION

Anemia – A Limiting Reagent Problem?

Anemia is the most commonly diagnosed blood disorder, with symptoms typically including lethargy, fatigue, poor concentration, and sensitivity to cold. Although anemia has many causes, including genetic factors, the most common cause is insufficient dietary intake or absorption of iron.

Hemoglobin (abbreviated Hb), the iron-containing protein found in red blood cells, is responsible for oxygen transport throughout the body. Low iron levels in the body result in decreased production and incorporation of Hb into red blood cells. In addition, blood loss due to injury or to menstruation in women increases the body's demand for iron in order to replace lost Hb. In the United States, nearly 20% of women of childbearing age suffer from iron-deficiency anemia compared to only 2% of adult men.

The recommended minimum daily iron intake is 8 mg for adult men and 18 mg for premenopausal women. One way to ensure sufficient iron intake is a well-balanced diet that includes iron-fortified grains and cereals, red meat, egg yolks, leafy green vegetables, tomatoes, and raisins. Vegetarians should pay extra attention to their diet, because the iron in fruits and vegetables is not as readily absorbed by the body as the iron in meat, poultry, and fish. Vitamin supplements containing folic acid and either ferrous sulfate or ferrous gluconate can decrease iron deficiencies, and vitamin C increases the absorption of iron by the body.

▲ Can cooking in cast iron pots decrease anemia?

However, the simplest way to increase dietary iron may be to use cast iron cookware. Studies have demonstrated that the iron content of many foods increases when cooked in an iron pot. Other studies involving Ethiopian children showed that those who ate food cooked in iron cookware were less likely to suffer from iron-deficiency anemia than their playmates who ate similar foods prepared in aluminum cookware.

See Chemistry in Action Problems 6.59 and 6.60 at the end of the chapter.

▶▶ We'll explore the role of hemoglobin in oxygen transport in greater detail in Chapter 9.

SUMMARY   Revisiting the Chapter Goals   173

CONCEPT MAP: CHEMICAL REACTIONS (CHAPTERS 5, 6)

```
                    ┌─────────────────────┐
                    │ Intramolecular Forces│
                    └─────────────────────┘
                       ↙              ↘
        ┌──────────────────────┐   ┌──────────────────────┐
        │ Ionic Bonds (Ch. 3) =│   │ Covalent Bonds (Ch. 4)│
        │ transfer of electrons│   │ = sharing of electrons│
        └──────────────────────┘   └──────────────────────┘
                       ↘              ↙
                ┌────────────────────────────────┐
                │ Chemical Reactions = rearrangement of │
                │ atoms and ions to form new compounds. │
                └────────────────────────────────┘
                       ↙              ↘
```

Types of reactions (Chapter 5):
Precipitation: depends on solubility rules
Neutralization:
    Acids/Bases (Chapter 10)
Redox: change in number of electrons associated with atoms in a compound.

Quantitative Relationships in Chemical Reactions (Chapter 6):
Conservation of Mass– reactants and products must be balanced! (Chapter 5)
Molar relationships between reactants and products
Avogadro's number = particle to mole conversions
Molar masses = gram to mole conversions
Limiting reagents, theoretical and percent yields.

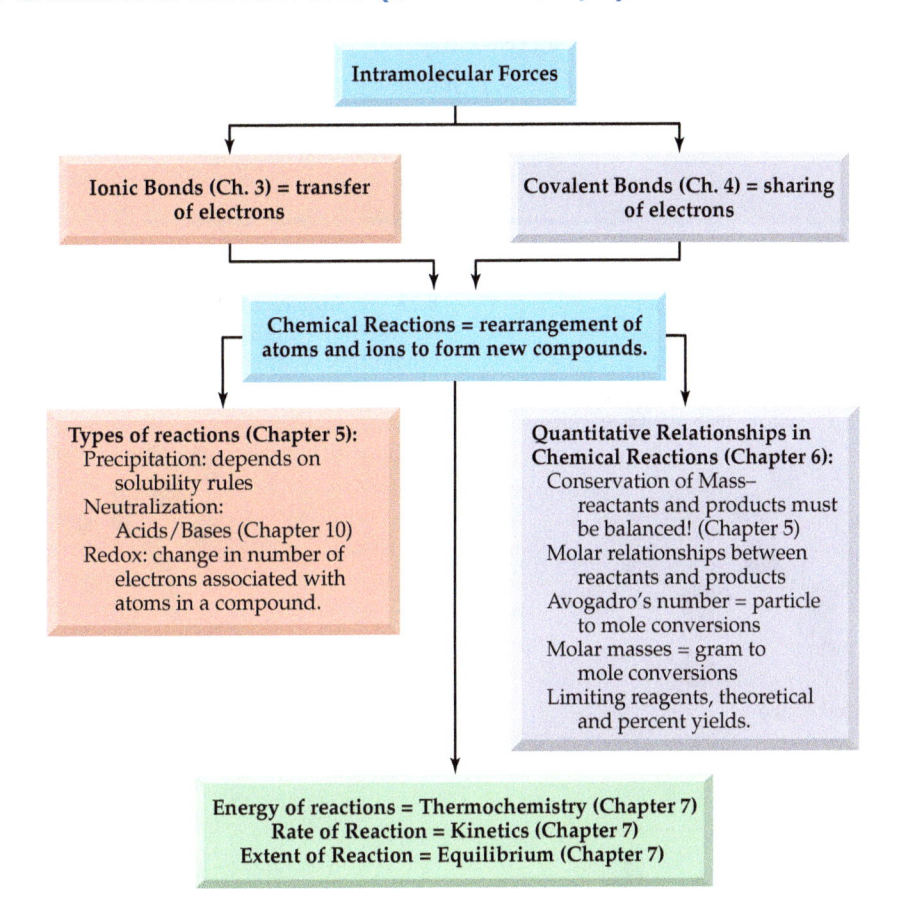

Energy of reactions = Thermochemistry (Chapter 7)
Rate of Reaction = Kinetics (Chapter 7)
Extent of Reaction = Equilibrium (Chapter 7)

▶ Figure 6.3

Concept Maps. In Chapters 5 and 6, we examined chemical reactions. As shown in the concept map above, chemical reactions represent a rearrangement of the intermolecular forces within compounds as bonds in the reactants are broken and new bonds are formed to generate products. Ionic reactions can be classified as precipitation reactions, neutralization reactions, or redox reactions. One class of reactions (acid–base neutralization) will be examined in further detail in Chapter 10. Other characteristics of reactions will be examined in Chapter 7.

SUMMARY: REVISITING THE CHAPTER GOALS

1. What is the mole, and why is it useful in chemistry?
A *mole* refers to *Avogadro's number* (6.022×10^{23}) of formula units of a substance. One mole of any substance has a mass (*molar mass*) equal to the molecular or formula weight of the substance in grams. Because equal numbers of moles contain equal numbers of formula units, molar masses act as conversion factors between numbers of moles and masses in grams (see Problems 16, 21–25, 27, 28, 32, 33, 38, 41, 62, 63).

2. How are molar quantities and mass quantities related?
The coefficients in a balanced chemical equation represent the numbers of moles of reactants and products in a reaction. Thus, the ratios of coefficients act as *mole ratios* that relate amounts of reactants and/or products. By using molar masses and mole ratios in factor-label calculations, unknown masses or molar amounts can be found from known masses or molar amounts (see Problems 17, 20, 25, 26, 29–31, 34–57, 59–61, 63–76).

3. What are the limiting reagent, theoretical yield, and percent yield of a reaction? The *limiting reagent* is the reactant that runs out first. The *theoretical yield* is the amount of product that would be formed based on the amount of the limiting reagent. The *actual yield* of a reaction is the amount of product obtained. The *percent yield* is the amount of product obtained divided by the amount theoretically possible and multiplied by 100% (see Problems 18, 19, 52–57, 71, 72).

174 CHAPTER 6 Chemical Reactions: Mole and Mass Relationships

KEY WORDS

Actual yield, p. 169
Avogadro's number (N_A), p. 161
Formula weight, p. 159
Limiting reagent, p. 169
Molar mass, p. 160
Mole, p. 160
Molecular weight (MW), p. 159
Percent yield, p. 169
Theoretical yield, p. 169

UNDERSTANDING KEY CONCEPTS

6.16 Methionine, an amino acid used by organisms to make proteins, can be represented by the following ball-and-stick molecular model. Write the formula for methionine, and give its molecular weight (red = O, black = C, blue = N, yellow = S, white = H).

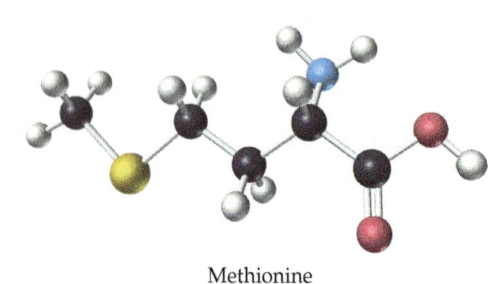

Methionine

6.17 The following diagram represents the reaction of A_2 (red spheres) with B_2 (blue spheres):

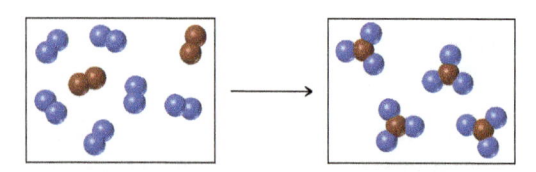

(a) Write a balanced equation for the reaction.
(b) How many moles of product can be made from 1.0 mol of A_2? From 1.0 mol of B_2?

6.18 Consider the balanced chemical equation: $2A + B_2 \rightarrow 2AB$. Given the reaction vessel below, determine the theoretical yield of product.

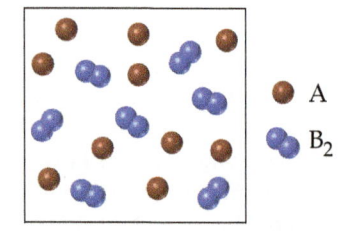

A
B_2

6.19 Consider the balanced chemical equation: $A_2 + 2B_2 \rightarrow 2AB_2$. A reaction is performed with the initial amounts of A_2 and B_2 shown in part (a). The amount of product obtained is shown in part (b). Calculate the percent yield.

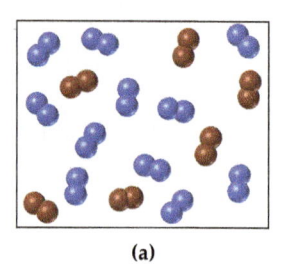

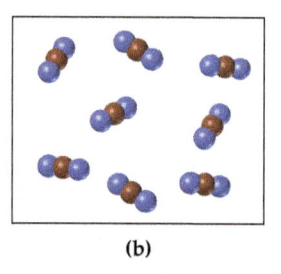

(a)

(b)

6.20 The following drawing represents the reaction of ethylene oxide with water to give ethylene glycol, a compound used as automobile antifreeze. What mass in grams of ethylene oxide is needed to react with 9.0 g of water, and what mass in grams of ethylene glycol is formed?

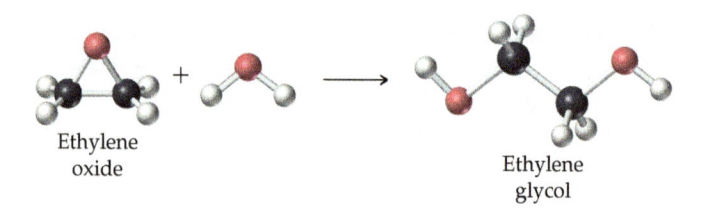

Ethylene oxide + → Ethylene glycol

ADDITIONAL PROBLEMS

MOLAR MASSES AND MOLES

6.21 What is a mole of a substance? How many molecules are in 1 mol of a molecular compound?

6.22 What is the difference between molecular weight and formula weight? Between molecular weight and molar mass?

6.23 How many Na^+ ions are in a mole of Na_2SO_4? How many SO_4^{2-} ions?

6.24 How many moles of ions are in 1.75 mol of K_2SO_4?

6.25 How many calcium atoms are in 16.2 g of calcium?

6.26 What is the mass in grams of 2.68×10^{22} atoms of uranium?

6.27 Calculate the molar mass of each of the following compounds:

(a) Calcium carbonate, $CaCO_3$
(b) Urea, $CO(NH_2)_2$
(c) Ethylene glycol, $C_2H_6O_2$

6.28 How many moles of carbon atoms are there in 1 mol of each compound in Problem 6.27?

6.29 How many atoms of carbon, and how many grams of carbon are there in 1 mol of each compound in Problem 6.27?

6.30 Caffeine has the formula $C_8H_{10}N_4O_2$. If an average cup of coffee contains approximately 125 mg of caffeine, how many moles of caffeine are in one cup?

6.31 How many moles of aspirin, $C_9H_8O_4$, are in a 500 mg tablet?

6.32 What is the molar mass of diazepam (Valium), $C_{16}H_{13}ClN_2O$?

6.33 Calculate the molar masses of the following substances:

(a) Aluminum sulfate, $Al_2(SO_4)_3$
(b) Sodium bicarbonate, $NaHCO_3$
(c) Diethyl ether, $(C_2H_5)_2O$
(d) Penicillin V, $C_{16}H_{18}N_2O_5S$

6.34 How many moles are present in a 4.50 g sample of each compound listed in Problem 6.33?

6.35 The recommended daily dietary intake of calcium for adult men and premenopausal women is 1000 mg/day. Calcium citrate, $Ca_3(C_6H_5O_7)_2$ (MW = 498.5 amu), is a common dietary supplement. What mass of calcium citrate would be needed to provide the recommended daily intake of calcium?

6.36 What is the mass in grams of 0.0015 mol of aspirin, $C_9H_8O_4$? How many aspirin molecules are there in this 0.0015 mol sample?

6.37 How many grams are present in a 0.075 mol sample of each compound listed in Problem 6.33?

6.38 The principal component of many kidney stones is calcium oxalate, CaC_2O_4. A kidney stone recovered from a typical patient contains 8.5×10^{20} formula units of calcium oxalate. How many moles of CaC_2O_4 are present in this kidney stone? What is the mass of the kidney stone in grams?

MOLE AND MASS RELATIONSHIPS FROM CHEMICAL EQUATIONS

6.39 At elevated temperatures in an automobile engine, N_2 and O_2 can react to yield NO, an important cause of air pollution.

(a) Write a balanced equation for the reaction.
(b) How many moles of N_2 are needed to react with 7.50 mol of O_2?
(c) How many moles of NO can be formed when 3.81 mol of N_2 reacts?
(d) How many moles of O_2 must react to produce 0.250 mol of NO?

6.40 Ethyl acetate reacts with H_2 in the presence of a catalyst to yield ethyl alcohol:

$$C_4H_8O_2(l) + H_2(g) \longrightarrow C_2H_6O(l)$$

(a) Write a balanced equation for the reaction.
(b) How many moles of ethyl alcohol are produced by reaction of 1.5 mol of ethyl acetate?
(c) How many grams of ethyl alcohol are produced by reaction of 1.5 mol of ethyl acetate with H_2?
(d) How many grams of ethyl alcohol are produced by reaction of 12.0 g of ethyl acetate with H_2?
(e) How many grams of H_2 are needed to react with 12.0 g of ethyl acetate?

6.41 The active ingredient in Milk of Magnesia (an antacid) is magnesium hydroxide, $Mg(OH)_2$. A typical dose (one tablespoon) contains 1.2 g of $Mg(OH)_2$. Calculate (a) the molar mass of magnesium hydroxide, and (b) the amount of magnesium hydroxide (in moles) in one teaspoon.

6.42 Ammonia, NH_3, is prepared for use as a fertilizer by reacting N_2 with H_2.

(a) Write a balanced equation for the reaction.
(b) How many moles of N_2 are needed for reaction to make 16.0 g of NH_3?
(c) How many grams of H_2 are needed to react with 75.0 g of N_2?

6.43 Hydrazine, N_2H_4, a substance used as rocket fuel, reacts with oxygen as follows:

$$N_2H_4(l) + O_2(g) \longrightarrow NO_2(g) + H_2O(g)$$

(a) Balance the equation.
(b) How many moles of oxygen are needed to react with 165 g of hydrazine?
(c) How many grams of oxygen are needed to react with 165 g of hydrazine?

6.44 One method for preparing pure iron from Fe_2O_3 is by reaction with carbon monoxide:

$$Fe_2O_3(s) + CO(g) \longrightarrow Fe(s) + CO_2(g)$$

(a) Balance the equation.
(b) How many grams of CO are needed to react with 3.02 g of Fe_2O_3?
(c) How many grams of CO are needed to react with 1.68 mol of Fe_2O_3?

6.45 Magnesium metal burns in oxygen to form magnesium oxide, MgO.

(a) Write a balanced equation for the reaction.
(b) How many grams of oxygen are needed to react with 25.0 g of Mg? How many grams of MgO will result?

(c) How many grams of Mg are needed to react with 25.0 g of O_2? How many grams of MgO will result?

6.46 Titanium metal is obtained from the mineral rutile, TiO_2. How many kilograms of rutile are needed to produce 95 kg of Ti?

6.47 In the preparation of iron from hematite (Problem 6.44), how many moles of carbon monoxide are needed to react completely with 105 kg of Fe_2O_3?

6.48 The eruption of Mount St. Helens volcano in 1980 injected 4×10^8 kg of SO_2 into the atmosphere. If all this SO_2 was converted to sulfuric acid, how many moles of H_2SO_4 would be produced? How many kg?

6.49 The thermite reaction was used to produce molten iron for welding applications before arc welding was available. The thermite reaction is:

$$Fe_2O_3(s) + 2\,Al(s) \longrightarrow Al_2O_3(s) + 2\,Fe(l)$$

How many moles of molten iron can be produced from 1.5 kg of iron(III) oxide?

6.50 Pyrite, also known as fool's gold, is composed of iron disulfide, FeS_2. It is used commercially to produce SO_2 used in the production of paper products. How many moles of SO_2 can be produced from 1.0 kg of pyrite?

6.51 Diborane (B_2H_6) is a gas at room temperature that forms explosive mixtures with air. It reacts with oxygen according to the following equation:

$$B_2H_6(g) + 3\,O_2(g) \longrightarrow B_2O_3(s) + 3\,H_2O(l)$$

How many grams of diborane will react with 7.5 mol of O_2?

LIMITING REAGENT AND PERCENT YIELD

6.52 Once made by heating wood in the absence of air, methanol (CH_3OH) is now made by reacting carbon monoxide and hydrogen at high pressure:

$$CO(g) + 2\,H_2(g) \longrightarrow CH_3OH(l)$$

(a) If 25.0 g of CO is reacted with 6.00 g of H_2, which is the limiting reagent?
(b) How many grams of CH_3OH can be made from 10.0 g of CO if it all reacts?
(c) If 9.55 g of CH_3OH is recovered when the amounts in part (b) are used, what is the percent yield?

6.53 In Problem 6.43, hydrazine reacted with oxygen according to the (unbalanced) equation:

$$N_2H_4(l) + O_2(g) \longrightarrow NO_2(g) + H_2O(g)$$

(a) If 75.0 kg of hydrazine are reacted with 75.0 kg of oxygen, which is the limiting reagent?
(b) How many kilograms of NO_2 are produced from the reaction of 75.0 kg of the limiting reagent?
(c) If 59.3 kg of NO_2 are obtained from the reaction in part (a), what is the percent yield?

6.54 Dichloromethane, CH_2Cl_2, the solvent used to decaffeinate coffee beans, is prepared by reaction of CH_4 with Cl_2.
(a) Write the balanced equation. (HCl is also formed.)
(b) How many grams of Cl_2 are needed to react with 50.0 g of CH_4?
(c) How many grams of dichloromethane are formed from 50.0 g of CH_4 if the percent yield for the reaction is 76%?

6.55 Cisplatin $[Pt(NH_3)_2Cl_2]$, a compound used in cancer treatment, is prepared by reaction of ammonia with potassium tetrachloroplatinate:

$$K_2PtCl_4 + 2\,NH_3 \longrightarrow 2\,KCl + Pt(NH_3)_2Cl_2$$

(a) How many grams of NH_3 are needed to react with 55.8 g of K_2PtCl_4?
(b) How many grams of cisplatin are formed from 55.8 g of K_2PtCl_4 if the percent yield for the reaction is 95%?

6.56 Nitrobenzene $(C_6H_5NO_2)$ is used in small quantities as a flavoring agent or in perfumes, but can be toxic in large amounts. It is produced by reaction of benzene (C_6H_6) with nitric acid:

$$C_6H_6(l) + HNO_3(aq) \longrightarrow C_6H_5NO_2(l) + H_2O(l).$$

(a) Identify the limiting reagent in the reaction of 27.5 g of nitric acid with 75 g of benzene.
(b) Calculate the theoretical yield for this reaction.

6.57 Calculate the percent yield if 48.2 g of nitrobenzene is obtained from the reaction described in Problem 6.56.

CHEMISTRY IN ACTION

6.58 What do you think might be some of the errors involved in calculating Avogadro's number by spreading oil on a pond? [*Did Ben Franklin Have Avogadro's Number?* p. 164]

6.59 Dietary iron forms a 1:1 complex with hemoglobin (Hb), which is responsible for O_2 transport in the body based on the following equation:

$$Hb + 4\,O_2 \longrightarrow Hb(O_2)_4$$

How many moles of oxygen could be transported by the hemoglobin complex formed from 8 mg of dietary iron? [*Anemia—A Limiting Reagent Problem?* p. 172]

6.60 Ferrous sulfate is one dietary supplement used to treat iron-deficiency anemia. What are the molecular formula and molecular weight of this compound? How many milligrams of iron are in 250 mg of ferrous sulfate?

GENERAL QUESTIONS AND PROBLEMS

6.61 Zinc metal reacts with hydrochloric acid (HCl) according to the equation:

$$Zn(s) + 2\,HCl(aq) \longrightarrow ZnCl_2(aq) + H_2(g)$$

(a) How many grams of hydrogen are produced if 15.0 g of zinc reacts?
(b) Is this a redox reaction? If so, tell what is reduced, what is oxidized, and identify the reducing and oxidizing agents.

6.62 Batrachotoxin, $C_{31}H_{42}N_2O_6$, an active component of South American arrow poison, is so toxic that 0.05 μg can kill a person. How many molecules is this?

6.63 Lovastatin, a drug used to lower serum cholesterol, has the molecular formula of $C_{24}H_{36}O_5$.

(a) Calculate the molar mass of lovastatin.

(b) How many moles of lovastatin are present in a typical dose of one 10 mg tablet?

6.64 When table sugar (sucrose, $C_{12}H_{22}O_{11}$) is heated, it decomposes to form C and H_2O.

(a) Write a balanced equation for the process.

(b) How many grams of carbon are formed by the breakdown of 60.0 g of sucrose?

(c) How many grams of water are formed when 6.50 g of carbon are formed?

6.65 Although Cu is not sufficiently active to react with acids, it can be dissolved by concentrated nitric acid, which functions as an oxidizing agent according to the following equation:

$$Cu(s) + 4\,HNO_3(aq) \longrightarrow Cu(NO_3)_2(aq) + 2\,NO_2(g) + 2\,H_2O(l)$$

(a) Write the net ionic equation for this process.

(b) Is 35.0 g of HNO_3 sufficient to dissolve 5.00 g of copper?

6.66 The net ionic equation for the Breathalyzer test used to indicate alcohol concentration in the body is

$$16\,H^+(aq) + 2\,Cr_2O_7^{2-}(aq) + 3\,C_2H_6O(aq) \longrightarrow 3\,C_2H_4O_2(aq) + 4\,Cr^{3+}(aq) + 11\,H_2O(l)$$

(a) How many grams of $K_2Cr_2O_7$ must be used to consume 1.50 g of C_2H_6O?

(b) How many grams of $C_2H_4O_2$ can be produced from 80.0 g of C_2H_6O?

6.67 Ethyl alcohol is formed by enzyme action on sugars and starches during fermentation:

$$C_6H_{12}O_6 \longrightarrow 2\,CO_2 + 2\,C_2H_6O$$

If the density of ethyl alcohol is 0.789 g/mL, how many quarts can be produced by the fermentation of 100.0 lb of sugar?

6.68 Gaseous ammonia reacts with oxygen in the presence of a platinum catalyst to produce nitrogen monoxide and water vapor.

(a) Write a balanced chemical equation for this reaction.

(b) What mass of nitrogen monoxide would be produced by complete reaction of 17.0 g of ammonia?

6.69 Sodium hypochlorite, the primary component in commercial bleach, is prepared by bubbling chlorine gas through solutions of sodium hydroxide:

$$NaOH(aq) + Cl_2(g) \longrightarrow NaOCl(aq) + H_2O(l)$$

How many moles of sodium hypochlorite can be prepared from 32.5 g of NaOH?

6.70 Barium sulfate is an insoluble ionic compound swallowed by patients before having an X-ray of their gastrointestinal tract.

(a) Write the balanced chemical equation for the precipitation reaction between barium chloride and sodium sulfate.

(b) What mass of barium sulfate can be produced by complete reaction of 27.4 g of Na_2SO_4?

6.71 The last step in the production of nitric acid is the reaction of nitrogen dioxide with water:

$$NO_2(g) + H_2O(l) \longrightarrow HNO_3(aq) + NO(g)$$

(a) Balance the chemical equation.

(b) If 65.0 g of nitrogen dioxide is reacted with excess water, calculate the theoretical yield.

(c) If only 43.8 g of nitric acid is obtained, calculate the percent yield.

6.72 Acetylsalicylic acid, the active ingredient in aspirin, is prepared from salicylic acid by reaction with acetic anhydride:

$$C_7H_6O_3 + C_4H_6O_3 \longrightarrow C_9H_8O_4 + C_2H_4O_2$$
(salicylic acid) (acetic anhydride) (acetylsalicylic acid) (acetic acid)

(a) Calculate the theoretical yield if 47 g of salicylic acid is reacted with 25 g of acetic anhydride.

(b) What is the percent yield if only 35 g is obtained?

6.73 Jewelry and tableware can be silver-plated by reduction of silver ions from a solution of silver nitrate. The net ionic equation is $Ag^+(aq) + e^- \longrightarrow Ag(s)$. How many grams of silver nitrate would be needed to plate 15.2 g of silver on a piece of jewelry?

6.74 Elemental phosphorus exists as molecules of P_4. It reacts with $Cl_2(g)$ to produce phosphorus pentachloride.

(a) Write the balanced chemical equation for this reaction.

(b) What mass of phosphorus pentachloride would be produced by the complete reaction of 15.2 g of P_4?

6.75 Lithium oxide is used aboard the space shuttle to remove water from the atmosphere according to the equation

$$Li_2O(s) + H_2O(g) \longrightarrow 2\,LiOH(s)$$

How many grams of Li_2O must be carried on board to remove 80.0 kg of water?

6.76 One of the reactions used to provide thrust for space shuttle launch involves the reaction of ammonium perchlorate with aluminum to produce $AlCl_3(s)$, $H_2O(g)$, and $NO(g)$.

(a) Write the balanced chemical equation for this reaction.

(b) How many moles of gas are produced by the reaction of 14.5 kg of ammonium perchlorate?

CHAPTER 7
Chemical Reactions: Energy, Rates, and Equilibrium

▲ Many spontaneous chemical reactions are accompanied by the release of energy, in some cases explosively.

CONTENTS

- 7.1 Energy and Chemical Bonds
- 7.2 Heat Changes during Chemical Reactions
- 7.3 Exothermic and Endothermic Reactions
- 7.4 Why Do Chemical Reactions Occur? Free Energy
- 7.5 How Do Chemical Reactions Occur? Reaction Rates
- 7.6 Effects of Temperature, Concentration, and Catalysts on Reaction Rates
- 7.7 Reversible Reactions and Chemical Equilibrium
- 7.8 Equilibrium Equations and Equilibrium Constants
- 7.9 Le Châtelier's Principle: The Effect of Changing Conditions on Equilibria

CHAPTER GOALS

1. **What energy changes take place during reactions?**
   THE GOAL: Be able to explain the factors that influence energy changes in chemical reactions. (◀◀ A, B, and C)

2. **What is "free energy," and what is the criterion for spontaneity in chemistry?**
   THE GOAL: Be able to define enthalpy, entropy, and free-energy changes, and explain how the values of these quantities affect chemical reactions.

3. **What determines the rate of a chemical reaction?**
   THE GOAL: Be able to explain activation energy and other factors that determine reaction rate. (◀◀ D)

4. **What is chemical equilibrium?**
   THE GOAL: Be able to describe what occurs in a reaction at equilibrium, and write the equilibrium equation for a given reaction. (◀◀ D)

5. **What is Le Châtelier's principle?**
   THE GOAL: Be able to state Le Châtelier's principle, and use it to predict the effect of changes in temperature, pressure, and concentration on reactions.

CONCEPTS TO REVIEW

◀◀

A. Energy and Heat
(Section 1.13)

B. Ionic Bonds
(Section 3.3)

C. Covalent Bonds
(Section 4.1)

D. Chemical Equations
(Section 5.1)

We have yet to answer many questions about reactions. Why, for instance, do reactions occur? Just because a balanced equation can be written it does not mean it will take place. We can write a balanced equation for the reaction of gold with water, for example, but the reaction does not occur in practice—so your gold jewelry is safe in the shower.

Balanced, but does not occur $2\,Au(s) + 3\,H_2O(l) \longrightarrow Au_2O_3(s) + 3\,H_2(g)$

To describe reactions more completely, several fundamental questions are commonly asked: Is energy released or absorbed when a reaction occurs? Is a given reaction fast or slow? Does a reaction continue until all reactants are converted to products, or is there a point beyond which no additional product forms?

7.1 Energy and Chemical Bonds

There are two fundamental and interconvertible kinds of energy: *potential* and *kinetic*. **Potential energy** is stored energy. The water in a reservoir behind a dam, an automobile poised to coast downhill, and a coiled spring have potential energy waiting to be released. **Kinetic energy**, by contrast, is the energy of motion. When the water falls over the dam and turns a turbine, when the car rolls downhill, or when the spring uncoils and makes the hands on a clock move, the potential energy in each is converted to kinetic energy. Of course, once all the potential energy is converted, nothing further occurs. The water at the bottom of the dam, the car at the bottom of the hill, and the uncoiled spring no longer have potential energy and thus, undergo no further change.

In chemical compounds, the attractive forces between ions or atoms are a form of potential energy, similar to the attractive forces between the poles of a magnet. When these attractive forces result in the formation of ionic or covalent bonds between ions or atoms, the potential energy is often converted into **heat**—a measure of the kinetic energy of the particles that make up the molecule. Breaking these bonds requires an input of energy.

In chemical reactions, some of the chemical bonds in the reactants must break (energy in) so that new bonds can form in the products (energy out). If the reaction products have less potential energy than the reactants, we say that the products are *more stable* than the reactants. The term "stable" is used in chemistry to describe a substance that has little remaining potential energy and consequently little tendency to undergo further change. Whether a reaction occurs, and how much energy or heat

Potential energy Stored energy.

Kinetic energy The energy of an object in motion.

Heat A measure of the transfer of thermal energy.

is associated with the reaction, depends on the difference in the amount of potential energy contained in the reactants and products.

7.2 Heat Changes during Chemical Reactions

Why does chlorine react so easily with many elements and compounds, but nitrogen does not? What difference between Cl_2 molecules and N_2 molecules accounts for their different reactivities? The answer is that the nitrogen–nitrogen triple bond is much *stronger* than the chlorine–chlorine single bond and cannot be broken as easily in chemical reactions.

Bond dissociation energy The amount of energy that must be supplied to break a bond and separate the atoms in an isolated gaseous molecule.

The strength of a covalent bond is measured by its **bond dissociation energy**, defined as the amount of energy that must be supplied to break the bond and separate the atoms in an isolated gaseous molecule. The greater the bond dissociation energy, the more stable the chemical bond between the atoms or ions. The triple bond in N_2, for example, has a bond dissociation energy of 226 kcal/mol (946 kJ/mol), whereas the single bond in chlorine has a bond dissociation energy of only 58 kcal/mol (243 kJ/mol):

:N:::N: $\xrightarrow{226 \text{ kcal/mol}}$:N· + ·N: N_2 bond dissociation energy = 226 kcal/mol (946 kJ/mol)

:Cl:Cl: $\xrightarrow{58 \text{ kcal/mol}}$:Cl· + ·Cl: Cl_2 bond dissociation energy = 58 kcal/mol (243 kJ/mol)

The greater stability of the triple bond in N_2 explains why nitrogen molecules are less reactive than Cl_2 molecules. Some typical bond dissociation energies are given in Table 7.1.

TABLE 7.1 Average Bond Dissociation Energies

Bond	Bond Dissociation Energy (kcal/mol, kJ/mol)	Bond	Bond Dissociation Energy (kcal/mol, kJ/mol)	Bond	Bond Dissociation Energy (kcal/mol, kJ/mol)
C—H	99, 413	N—H	93, 391	C=C	147, 614
C—C	83, 347	N—N	38, 160	C≡C	201, 839
C—N	73, 305	N—Cl	48, 200	C=O*	178, 745
C—O	86, 358	N—O	48, 201	O=O	119, 498
C—Cl	81, 339	H—H	103, 432	N=O	145, 607
Cl—Cl	58, 243	O—H	112, 467	C≡N	213, 891
H—Cl	102, 427	O—Cl	49, 203	N≡N	226, 946

*The C=O bond dissociation energies in CO_2 are 191 kcal/mol (799 kJ/mol).

Endothermic A process or reaction that absorbs heat.

Exothermic A process or reaction that releases heat.

A chemical change that absorbs heat, like the breaking of bonds, is said to be **endothermic**, from the Greek words *endon* (within) and *therme* (heat), meaning that *heat is put in*. The reverse of bond breaking is bond formation, a process that *releases* heat and is described as **exothermic**, from the Greek *exo* (outside), meaning that heat goes *out*. The amount of energy released in forming a bond is numerically the same as that absorbed in breaking it. When nitrogen atoms combine to give N_2, 226 kcal/mol (946 kJ/mol) of heat is released. Similarly, when chlorine atoms combine to give Cl_2, 58 kcal/mol (243 kJ/mol) of heat is released. We indicate the direction of energy flow in a chemical change by the sign associated with the number. If heat is absorbed (endothermic) then the sign is positive to indicate energy is *gained* by the substance. If heat is released (exothermic) then the sign is negative to indicate energy is *lost* by the substance during the change.

:N· + ·N: ⟶ :N:::N: + 226 kcal/mol (946 kJ/mol) heat released

:Cl· + ·Cl: ⟶ :Cl:Cl: + 58 kcal/mol (243 kJ/mol) heat released

The same energy relationships that govern bond breaking and bond formation apply to every physical or chemical change. That is, the amount of heat transferred during a change in one direction is numerically equal to the amount of heat transferred during the change in the opposite direction. Only the *direction* of the heat transfer is different. This relationship reflects a fundamental law of nature called the *law of conservation of energy*:

Law of conservation of energy Energy can be neither created nor destroyed in any physical or chemical change.

If more energy could be released by an exothermic reaction than was consumed in its reverse, the law would be violated, and we could "manufacture" energy out of nowhere by cycling back and forth between forward and reverse reactions—a clear impossibility.

In every chemical reaction, some bonds in the reactants are broken, and new bonds are formed in the products. The difference between the heat energy absorbed in breaking bonds and the heat energy released in forming bonds is called the **heat of reaction** and is a quantity that we can measure. Heats of reaction that are measured when a reaction is held at constant pressure are represented by the abbreviation ΔH, where Δ (the Greek capital letter delta) is a general symbol used to indicate "a change in," and H is a quantity called **enthalpy**. Thus, the value of ΔH represents the **enthalpy change** that occurs during a reaction. The terms *enthalpy change* and *heat of reaction* are often used interchangeably, but we will generally use the latter term in this book.

Heat of reaction, or **Enthalpy change** (ΔH) The difference between the energy of bonds broken in reactants and the energy of bonds formed in products.

Enthalpy (H) A measure of the amount of energy associated with substances involved in a reaction.

7.3 Exothermic and Endothermic Reactions

When the total strength of the bonds formed in the products is *greater* than the total strength of the bonds broken in the reactants, the net result is that energy is released and a reaction is exothermic. All combustion reactions are exothermic; for example, burning 1 mol of methane releases 213 kcal (891 kJ) of energy in the form of heat. The heat released in an exothermic reaction can be thought of as a reaction product, and the heat of reaction ΔH is assigned a *negative* value, because overall, heat is *lost* during the reaction.

An exothermic reaction—negative ΔH

$$CH_4(g) + 2\,O_2(g) \longrightarrow CO_2(g) + 2\,H_2O(l) + 213\text{ kcal }(891\text{ kJ})$$

(Heat is a product.)

or

$$CH_4(g) + 2\,O_2(g) \longrightarrow CO_2(g) + 2\,H_2O(l) \qquad \Delta H = -213\text{ kcal/mol }(-891\text{ kJ/mol})$$

The heat of reaction can be calculated as the difference between the bond dissociation energies in the products and the bond dissociation energies of the reactants:

$$\Delta H = \Sigma(\text{Bond dissociation energies})_{\text{reactants}} - \Sigma(\text{Bond dissociation energies})_{\text{products}}$$

Look again at the reaction involving the combustion of methane. By counting the number of bonds on each side of the chemical equation, we can use the average bond dissociation energies from Table 7.1 to estimate ΔH for the reaction.

Reactants	Bond Dissociation Energies (kcal/mol)	Products	Bond Dissociation Energies (kcal/mol)
(C—H) × 4	99 × 4 = 396 kcal	(C=O) × 2	191 × 2 = 382 kcal
(O=O) × 2	119 × 2 = 238 kcal	(H—O) × 4	112 × 4 = 448 kcal
Total:	= 634 kcal		= 830 kcal

$$\Delta H = (634\text{ kcal})_{\text{reactants}} - (830\text{ kcal})_{\text{products}} = -196\text{ kcal}\,(-820\text{ kJ})$$

▲ The reaction between aluminum metal and iron(III) oxide, called the *thermite reaction*, is so strongly exothermic that it melts iron.

In this reaction, the input of energy needed to break the bonds in the reactants is less than the amount of energy released when forming bonds in the products. The excess energy is released as heat, and the reaction is exothermic (ΔH = negative).

It should be noted that the bond energies in Table 7.1 are average values, and that actual bond energies may vary depending on the chemical environment in which the bond is found. The average C=O bond energy, for example, is 178 kcal/mol, but the actual value for the C=O bonds in the CO_2 molecule is 191 kcal/mol. The average C—H bond energy is 99 kca/mol (413 kJ/mol), but in CH_3CH_3 the C—H bond dissociation energy is actually 101 kcal/mol (423 kJ/mol). Thus, the calculated ΔH for a reaction using average bond energies may differ slightly from the value obtained by experiment. For the combustion of methane, for example, the ΔH estimated from bond energies is −196 kcal/mol (−820 kJ/mol), while the value measured experimentally is −213 kcal/mol (−891 kJ/mol), a difference of about 9%.

Note that ΔH is given in units of kilocalories or kilojoules per mole, where "per mole" means the reaction of *molar amounts of products and reactants as represented by the coefficients of the balanced equation.* Thus, the experimental value ΔH = −213 kcal/mol (−891 kJ/mol) refers to the amount of heat released when 1 mol (16.0 g) of methane reacts with 2 mol of O_2 to give 1 mol of CO_2 gas and 2 mol of liquid H_2O. If we were to double the amount of methane from 1 mol to 2 mol, the amount of heat released would also double.

The quantities of heat released in the combustion of several fuels, including natural gas (which is primarily methane), are compared in Table 7.2. The values are given in kilocalories and kilojoules per gram to make comparisons easier. You can see from the table why there is interest in the potential of hydrogen as a fuel.

TABLE 7.2 Energy Values of Some Common Fuels

Fuel	Energy Value (kcal/g, kJ/g)
Wood (pine)	4.3, 18.0
Ethyl alcohol	7.1, 29.7
Coal (anthracite)	7.4, 31.0
Crude oil (Texas)	10.5, 43.9
Gasoline	11.5, 48.1
Natural gas	11.7, 49.0
Hydrogen	34.0, 142

When the total energy released upon bond formation in the products is *less* than the total energy added to break the bonds in the reactants, the net result is that energy is absorbed and a reaction is endothermic. The combination of nitrogen and oxygen to give nitrogen oxide (also known as nitric oxide), a gas present in automobile exhaust, is such a reaction. The heat added in an endothermic reaction is like a reactant, and ΔH is assigned a *positive* value because heat is *added*.

An endothermic reaction—positive ΔH

Heat is a reactant.

$$N_2(g) + O_2(g) + 43 \text{ kcal } (180 \text{ kJ}) \longrightarrow 2 \text{ NO}(g)$$

or

$$N_2(g) + O_2(g) \longrightarrow 2 \text{ NO}(g) \quad \Delta H = +43 \text{ kcal/mol } (+180 \text{ kJ/mol})$$

SECTION 7.3 Exothermic and Endothermic Reactions

Important Points about Heat Transfers and Chemical Reactions

- An exothermic reaction releases heat to the surroundings; ΔH is negative.
- An endothermic reaction absorbs heat from the surroundings; ΔH is positive.
- The reverse of an exothermic reaction is endothermic.
- The reverse of an endothermic reaction is exothermic.
- The amount of heat absorbed or released in the reverse of a reaction is equal to that released or absorbed in the forward reaction, but ΔH has the opposite sign.

Worked Examples 7.1–7.4 show how to calculate the amount of heat absorbed or released for reaction of a given amount of reactant. All that is needed is the balanced equation and its accompanying ΔH or the bond dissociation energies to permit calculation of ΔH. Mole ratios and molar masses are used to convert between masses and moles of reactants or products, as discussed in Sections 6.3 and 6.4.

Worked Example 7.1 Heat of Reaction from Bond Energies

Estimate the ΔH (in kcal/mol) for the reaction of hydrogen and oxygen to form water:

$$2\,H_2 + O_2 \longrightarrow 2\,H_2O \quad \Delta H = ?$$

ANALYSIS The individual bond energies from Table 7.1 can be used to calculate the total bond energies of reactants and products. ΔH can then be calculated as

$$\Delta H = \Sigma(\text{Bond dissociation energies})_{\text{reactants}} - \Sigma(\text{Bond dissociation energies})_{\text{products}}$$

BALLPARK ESTIMATE The average H—H bond energy is ~100 kcal/mol and the O=O bond energy is ~120 kcal/mol. Thus, the total energy needed to break reactant bonds is ~ (200 + 120) = 320 kcal/mol. The O—H bonds are ~110 kcal/mol, so the total energy released when product bonds are formed is ~440 kcal/mol. Based on these estimates, $\Delta H \sim -120$ kcal/mol.

SOLUTION

$$\Delta H = \Sigma(\text{Bond dissociation energies})_{\text{reactants}} - \Sigma(\text{Bond dissociation energies})_{\text{products}}$$
$$= (2(\text{H—H}) + (\text{O=O})) - (4(\text{O—H}))$$
$$= (2(103\,\text{kcal/mol}) + (119\,\text{kcal/mol})) - (4(112\,\text{kcal/mol})) = -123\,\text{kcal/mol}$$

BALLPARK CHECK Our estimate was -120 kcal/mol, within 3% of the calculated answer.

Worked Example 7.2 Heat of Reaction: Moles

Methane undergoes combustion with O_2 according to the following equation:

$$CH_4(g) + 2\,O_2(g) \longrightarrow CO_2(g) + 2\,H_2O(l) \quad \Delta H = -213\,\frac{\text{kcal}}{\text{mol CH}_4}$$

How much heat (in kcal and kJ) is released during the combustion of 0.35 mol of methane?

ANALYSIS Since the value of ΔH for the reaction (213 kcal/mol) is negative, it indicates the amount of heat released when 1 mol of methane reacts with O_2. We need to find the amount of heat released when an amount other than 1 mol reacts, using appropriate factor-label calculations to convert from our known or given units to kilocalories, and then to kilojoules.

BALLPARK ESTIMATE Since 213 kcal is released for each mole of methane that reacts, 0.35 mol of methane should release about one-third of 213 kcal, or about 70 kcal. There are about 4 kJ per kcal, so 70 kcal is about 280 kJ.

184 CHAPTER 7 Chemical Reactions: Energy, Rates, and Equilibrium

SOLUTION
To find the amount of heat released (in kilocalories) by combustion of 0.35 mol of methane, we use a conversion factor of kcal/mol, and then we can convert to kilojoules using a kJ/kcal conversion factor (see Section 1.13):

$$0.35 \text{ mol CH}_4 \times \frac{-213 \text{ kcal}}{1 \text{ mol CH}_4} = -75 \text{ kcal}$$

$$-75 \text{ kcal} \times \left(\frac{4.184 \text{ kJ}}{\text{kcal}}\right) = -314 \text{ kJ}$$

The negative sign indicates that the 75 kcal (314 kJ) of heat is released.

BALLPARK CHECK The calculated answer is consistent with our estimate (70 kcal or 280 kJ).

Worked Example 7.3 Heat of Reaction: Mass to Mole Conversion

How much heat is released during the combustion of 7.50 g of methane (molar mass = 16.0 g/mol)?

$$CH_4(g) + 2\,O_2(g) \longrightarrow CO_2(g) + 2\,H_2O(l) \quad \Delta H = -213\frac{\text{kcal}}{\text{mol CH}_4} = -891\frac{\text{kJ}}{\text{mol CH}_4}$$

ANALYSIS We can find the moles of methane involved in the reaction by using the molecular weight in a mass to mole conversion, and then use ΔH to find the heat released.

BALLPARK ESTIMATE Since 1 mol of methane (molar mass = 16.0 g/mol) has a mass of 16.0 g, 7.50 g of methane is a little less than 0.5 mol. Thus, less than half of 213 kcal, or about 100 kcal (418 kJ), is released from combustion of 7.50 g.

SOLUTION
Going from a given mass of methane to the amount of heat released in a reaction requires that we first find the number of moles of methane by including molar mass (in mol/g) in the calculation and then converting moles to kilocalories or kilojoules:

$$7.50 \text{ g CH}_4 \times \frac{1 \text{ mol CH}_4}{16.0 \text{ g CH}_4} \times \frac{-213 \text{ kcal}}{1 \text{ mol CH}_4} = -99.8 \text{ kcal}$$

or

$$7.50 \text{ g CH}_4 \times \frac{1 \text{ mol CH}_4}{16.0 \text{ g CH}_4} \times \frac{-891 \text{ kJ}}{1 \text{ mol CH}_4} = -418 \text{ kJ}$$

The negative sign indicates that the 99.8 kcal (418 kJ) of heat is released.

BALLPARK CHECK Our estimate was -100 kcal (-418 kJ)!

Worked Example 7.4 Heat of Reaction: Mole Ratio Calculations

How much heat is released in kcal and kJ when 2.50 mol of O_2 reacts completely with methane?

$$CH_4(g) + 2\,O_2(g) \longrightarrow CO_2(g) + 2\,H_2O(l) \quad \Delta H = -213\frac{\text{kcal}}{\text{mol CH}_4} = -891\frac{\text{kJ}}{\text{mol CH}_4}$$

ANALYSIS Since the ΔH for the reaction is based on the combustion of 1 mol of methane, we will need to perform a mole ratio calculation.

BALLPARK ESTIMATE The balanced equation shows that 213 kcal (891 kJ) is released for each 2 mol of oxygen that reacts. Thus, 2.50 mol of oxygen should release a bit more than 213 kcal, perhaps about 250 kcal (1050 kJ).

SOLUTION
To find the amount of heat released by combustion of 2.50 mol of oxygen, we include in our calculation a mole ratio based on the balanced chemical equation:

$$2.50 \text{ mol } O_2 \times \frac{1 \text{ mol } CH_4}{2 \text{ mol } O_2} \times \frac{-213 \text{ kcal}}{1 \text{ mol } CH_4} = -266 \text{ kcal}$$

or

$$2.50 \text{ mol } O_2 \times \frac{1 \text{ mol } CH_4}{2 \text{ mol } O_2} \times \frac{-891 \text{ kJ}}{1 \text{ mol } CH_4} = -1110 \text{ kJ}$$

The negative sign indicates that the 266 kcal (1110 kJ) of heat is released.

BALLPARK CHECK The calculated answer is close to our estimate (-250 kcal or -1050 kJ).

CHEMISTRY IN ACTION

Energy from Food

Any serious effort to lose weight usually leads to studying the caloric values of foods. Have you ever wondered how the numbers quoted on food labels are obtained?

Food is "burned" in the body to yield H_2O, CO_2, and energy, just as natural gas is burned in furnaces to yield the same products. In fact, the "caloric value" of a food is just the heat of reaction for complete combustion of the food (minus a small correction factor). The value is the same whether the food is burned in the body or in the laboratory. One gram of protein releases 4 kcal, 1 g of table sugar (a carbohydrate) releases 4 kcal, and 1 g of fat releases 9 kcal (see Table).

Caloric Values of Some Foods

Substance, Sample Size	Caloric Value (kcal, kJ)
Protein, 1 g	4, 17
Carbohydrate, 1 g	4, 17
Fat, 1 g	9, 38
Alcohol, 1 g	7.1, 29.7
Cola drink, 12 fl oz (369 g)	160, 670
Apple, one medium (138 g)	80, 330
Iceberg lettuce, 1 cup shredded (55 g)	5, 21
White bread, 1 slice (25 g)	65, 270
Hamburger patty, 3 oz (85 g)	245, 1030
Pizza, 1 slice (120 g)	290, 1200
Vanilla ice cream, 1 cup (133 g)	270, 1130

The caloric value of a food is usually given in "Calories" (note the capital C), where 1 Cal = 1000 cal = 1 kcal = 4.184 kJ. To determine these values experimentally, a carefully dried and weighed food sample is placed together with oxygen in an instrument called a *calorimeter*, the food is ignited, the temperature change is measured, and the amount of heat given off is calculated from the temperature change. In the calorimeter, the heat from the food is released very quickly and the temperature rises dramatically. Clearly, though, something a bit different goes on when food is burned in the body, otherwise we would burst into flames after a meal!

▲ This frosted donut provides your body with 330 Calories. Burning this donut in a calorimeter releases 330 kcal (1380 kJ) as heat.

It is a fundamental principle of chemistry that the total heat released or absorbed in going from reactants to products is the same, no matter how many reactions are involved. The body applies this principle by withdrawing energy from food a bit at a time in a long series of interconnected reactions rather than all at once in a single reaction. These and other reactions that are continually taking place in the body—called the body's *metabolism*—will be examined in later chapters.

See Chemistry in Action Problems 7.70 and 7.71 at the end of the chapter.

PROBLEM 7.1

In photosynthesis, green plants convert carbon dioxide and water into glucose ($C_6H_{12}O_6$) according to the following equation:

$$6\,CO_2(g) + 6\,H_2O(l) \longrightarrow C_6H_{12}O_6(aq) + 6\,O_2(g)$$

(a) Estimate ΔH for the reaction using bond dissociation energies from Table 7.1. Give your answer in kcal/mol and kJ/mol. $C_6H_{12}O_6$ has five C—C bonds, seven C—H bonds, seven C—O bonds, and five O—H bonds).

(b) Is the reaction endothermic or exothermic?

PROBLEM 7.2

The following equation shows the conversion of aluminum oxide (from the ore bauxite) to aluminum:

$$2\,Al_2O_3(s) \longrightarrow 4\,Al(s) + 3\,O_2(g) \quad \Delta H = +801 \text{ kcal/mol } (+3350 \text{ kJ/mol})$$

(a) Is the reaction exothermic or endothermic?

(b) How many kilocalories are required to produce 1.00 mol of aluminum? How many kilojoules?

(c) How many kilocalories are required to produce 10.0 g of aluminum? How many kilojoules?

PROBLEM 7.3

How much heat is absorbed (in kilocalories and kilojoules) during production of 127 g of NO by the combination of nitrogen and oxygen?

$$N_2(g) + O_2(g) \longrightarrow 2\,NO(g) \quad \Delta H = +43 \text{ kcal/mol } (+180 \text{ kJ/mol})$$

PROBLEM 7.4

Once consumed, the body metabolizes alcohol (ethanol, CH_3CH_2OH; MW = 46 g/mol) to carbon dioxide and water. The balanced reaction is: $CH_3CH_2OH + 3\,O_2 \longrightarrow 2\,CO_2 + 3\,H_2O$. Using the bond energies in Table 7.1, estimate the ΔH for this reaction in kcal/mol. How does it compare to the caloric value of alcohol (in Cal/g) given in Chemistry in Action: Energy from Food on p. 185)?

7.4 Why Do Chemical Reactions Occur? Free Energy

Events that lead to lower energy states tend to occur spontaneously. Water falls downhill, for instance, releasing its stored (potential) energy and reaching a lower-energy, more stable position. Similarly, a wound-up spring uncoils when set free. Applying this lesson to chemistry, the obvious conclusion is that exothermic processes—those that release heat energy—should be spontaneous. A log burning in a fireplace is just one example of a spontaneous reaction that releases heat. At the same time, endothermic processes, which absorb heat energy, should not be spontaneous. Often, these conclusions are correct, but not always. Many, but not all, exothermic processes take place spontaneously, and many, but not all, endothermic processes are nonspontaneous.

Before exploring the situation further, it is important to understand what the word "spontaneous" means in chemistry, which is not quite the same as in everyday language. A **spontaneous process** is one that, once started, proceeds on its own without any external influence. The change does not necessarily happen quickly, like a spring suddenly uncoiling or a car coasting downhill. It can also happen slowly, like the gradual rusting away of an abandoned bicycle. A *nonspontaneous process*, by contrast, takes place only in the presence of a continuous external influence. Energy must be continually expended to rewind a spring or push a car uphill. The reverse of a spontaneous process is always nonspontaneous.

As an example of a process that takes place spontaneously yet absorbs heat, think about what happens when you take an ice cube out of the freezer. The ice spontaneously

▲ Events that lead to lower energy tend to occur spontaneously. Thus, water always flows *down* a waterfall, not up.

Spontaneous process A process or reaction that, once started, proceeds on its own without any external influence.

melts to give liquid water above 0 °C, even though it *absorbs* heat energy from the surroundings. What this and other spontaneous endothermic processes have in common is *an increase in molecular disorder, or randomness*. When the solid ice melts, the H_2O molecules are no longer locked in position but are now free to move around randomly in the liquid water.

The amount of disorder in a system is called the system's **entropy**, symbolized by S and expressed in units of calories (or Joules) per mole-kelvin [$cal/(mol \cdot K)$ or $J/(mol \cdot K)$]. The greater the disorder, or randomness, of the particles in a substance or mixture, the larger the value of S (Figure 7.1). Gases have more disorder and therefore higher entropy than liquids because particles in the gas move around more freely than particles in the liquid. Similarly, liquids have higher entropy than solids. In chemical reactions, entropy increases when, for example, a gas is produced from a solid or when 2 mol of reactants split into 4 mol of products.

Entropy (S) A measure of the amount of molecular disorder in a system.

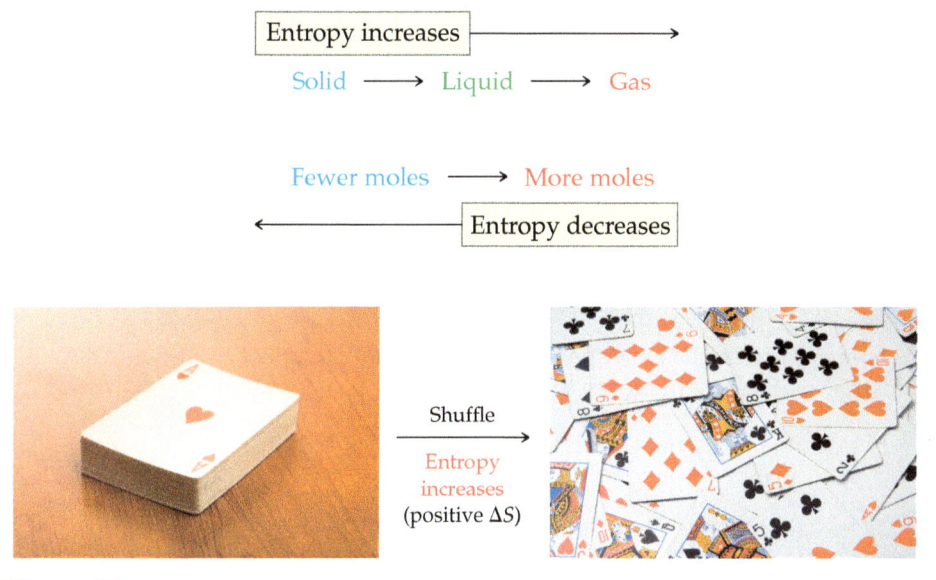

Entropy increases ⟶
Solid ⟶ Liquid ⟶ Gas

Fewer moles ⟶ More moles
⟵ Entropy decreases

Shuffle
Entropy increases
(positive ΔS)

▲ **Figure 7.1**
Entropy and values of S.
A new deck of cards, neatly stacked, has more order and lower entropy than the randomly shuffled and strewn cards on the right. The value of the entropy change, ΔS, for converting the system on the left to that on the right is positive because entropy increases.

The **entropy change** for a process, ΔS, has a *positive* value if disorder increases because the process adds disorder to the system. The melting of ice to give water is an example. Conversely, ΔS has a *negative* value if the disorder of a system decreases. The freezing of water to give ice is an example.

Entropy change ΔS A measure of the increase in disorder ($\Delta S = +$) or decrease in disorder ($\Delta S = -$) as a chemical reaction or physical change occurs.

It thus appears that two factors determine the spontaneity of a chemical or physical change: the release or absorption of heat, ΔH, and the increase or decrease in entropy, ΔS. *To decide whether a process is spontaneous, both the enthalpy change and the entropy change must be taken into account.* We have already seen that a negative ΔH favors spontaneity, but what about ΔS? The answer is that an increase in molecular disorder (ΔS positive) favors spontaneity. A good analogy is the bedroom or office that seems to spontaneously become more messy over time (an increase in disorder, ΔS positive); to clean it up (a decrease in disorder, ΔS negative) requires an input of energy, a nonspontaneous process. Using our chemical example, the combustion of a log spontaneously converts large, complex molecules like lignin and cellulose (high molecular order, low entropy) into CO_2 and H_2O (a large number of small molecules with higher entropy). For this process, the level of disorder increases, and so ΔS is positive. The reverse process—turning CO_2 and H_2O back into cellulose—does occur in photosynthesis, but it requires a significant input of energy in the form of sunlight.

When enthalpy and entropy are both favorable (ΔH negative, ΔS positive), a process is spontaneous; when both are unfavorable, a process is nonspontaneous. Clearly,

Free-energy change ΔG A measure of the change in free energy as a chemical reaction or physical change occurs.

however, the two factors do not have to operate in the same direction. It is possible for a process to be *unfavored* by enthalpy (the process absorbs heat, and so, has a positive ΔH) and yet be *favored* by entropy (there is an increase in disorder, and so, ΔS is positive). The melting of an ice cube above 0 °C, for which $\Delta H = +1.44$ kcal/mol ($+6.02$ kJ/mol) and $\Delta S = +5.26$ cal/(mol·K)($+22.0$ J/(mol·K)) is such a process. To take both heat of reaction (ΔH) and change in disorder (ΔS) into account when determining the spontaneity of a process, a quantity called the **free-energy change** (ΔG) is needed:

Free-energy change

$$\Delta G = \Delta H - T\Delta S$$

where ΔH is the Heat of reaction, T is the Temperature (in kelvins), and ΔS is the Entropy change.

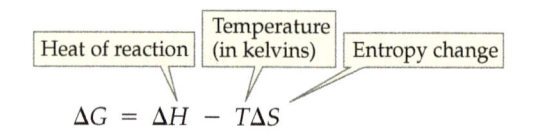

The value of the free-energy change, ΔG, determines spontaneity. A negative value for ΔG means that free energy is released and the reaction or process is spontaneous. Such events are said to be **exergonic**. A positive value for ΔG means that free energy must be added and the process is nonspontaneous. Such events are said to be **endergonic**.

Exergonic A spontaneous reaction or process that releases free energy and has a negative ΔG.

Endergonic A nonspontaneous reaction or process that absorbs free energy and has a positive ΔG.

The equation for the free-energy change shows that spontaneity also depends on temperature (T). At low temperatures, the value of $T\Delta S$ is often small so that ΔH is the dominant factor. At a high enough temperature, however, the value of $T\Delta S$ can become larger than ΔH. Thus, an endothermic process that is nonspontaneous at a low temperature can become spontaneous at a higher temperature. An example is the industrial synthesis of hydrogen by reaction of carbon with water:

$$C(s) + H_2O(l) \longrightarrow CO(g) + H_2(g)$$

$\Delta H = +31.3$ kcal/mol ($+131.0$ kJ/mol)   (Unfavorable)
$\Delta S = +32$ cal/(mol·K)($+134$ J/(mol·K))   (Favorable)

The reaction has an unfavorable (positive) ΔH term but a favorable (positive) ΔS term because disorder increases when a solid and a liquid are converted into two gases. No reaction occurs if carbon and water are mixed together at 25 °C (298 K) because the unfavorable ΔH is larger than the favorable $T\Delta S$. Above about 700 °C (973 K), however, the favorable $T\Delta S$ becomes larger than the unfavorable ΔH, so the reaction becomes spontaneous.

Important Points about Spontaneity and Free Energy

- A spontaneous process, once begun, proceeds without any external assistance and is exergonic; that is, free energy is released and it has a negative value of ΔG.
- A nonspontaneous process requires continuous external influence and is endergonic; that is, free energy is added and it has a positive value of ΔG.
- The value of ΔG for the reverse of a reaction is numerically equal to the value of ΔG for the forward reaction, but has the opposite sign.
- Some nonspontaneous processes become spontaneous with a change in temperature.

LOOKING AHEAD ▶▶ In later chapters, we will see that a knowledge of free-energy changes is especially important for understanding how metabolic reactions work. Living organisms cannot raise their temperatures to convert nonspontaneous reactions into spontaneous reactions, so they must resort to other strategies, which we will explore in Chapter 20.

Worked Example 7.5 Entropy Change of Processes

Does entropy increase or decrease in the following processes?
(a) Smoke from a cigarette disperses throughout a room rather than remaining in a cloud over the smoker's head.
(b) Water boils, changing from liquid to vapor.
(c) A chemical reaction occurs: $3 H_2(g) + N_2(g) \longrightarrow 2 NH_3(g)$

ANALYSIS Entropy is a measure of molecular disorder. Entropy increases when the products are more disordered than the reactants; entropy decreases when the products are less disordered than the reactants.

SOLUTION
(a) Entropy increases because smoke particles are more disordered when they are randomly distributed in the larger volume.
(b) Entropy increases because H_2O molecules have more freedom and disorder in the gas phase than in the liquid phase.
(c) Entropy decreases because 4 mol of reactant gas particles become 2 mol of product gas particles, with a consequent decrease in freedom and disorder.

Worked Example 7.6 Spontaneity of Reactions: Enthalpy, Entropy, and Free Energy

The industrial method for synthesizing hydrogen by reaction of carbon with water has $\Delta H = +31.3$ kcal/mol ($+131$ kJ/mol) and $\Delta S = +32$ cal/[mol·K]($+134$ J/[mol·K]). What is the value of ΔG (in kcal and kJ) for the reaction at 27 °C (300 K)? Is the reaction spontaneous or nonspontaneous at this temperature?

$$C(s) + H_2O(l) \longrightarrow CO(g) + H_2(g)$$

ANALYSIS The reaction is endothermic (ΔH positive) and does not favor spontaneity, whereas the ΔS indicates an increase in disorder (ΔS positive), which *does* favor spontaneity. Calculate ΔG to determine spontaneity.

BALLPARK ESTIMATE The unfavorable ΔH ($+31.3$ kcal/mol) is 1000 times greater than the favorable ΔS ($+32$ cal/mol·K), so the reaction will be spontaneous (ΔG negative) only when the temperature is high enough to make the $T\Delta S$ term in the equation for ΔG larger than the ΔH term. This happens at $T \geq 1000$ K. Since $T = 300$ K, expect ΔG to be positive and the reaction to be nonspontaneous.

SOLUTION
Use the free-energy equation to determine the value of ΔG at this temperature. (Remember that ΔS has units of *calories* per mole-kelvin or *joules* per mole-kelvin, not kilocalories per mole-kelvin or kilojoules per mole-kelvin.)

$$\Delta G = \Delta H - T\Delta S$$

$$\Delta G = +31.3 \frac{\text{kcal}}{\text{mol}} - (300 \text{ K})\left(+32 \frac{\text{cal}}{\text{mol} \cdot \text{K}}\right)\left(\frac{1 \text{ kcal}}{1000 \text{ cal}}\right) = +21.7 \frac{\text{kcal}}{\text{mol}}$$

$$\Delta G = +131 \frac{\text{kJ}}{\text{mol}} - (300 \text{ K})\left(+134 \frac{\text{J}}{\text{mol} \cdot \text{K}}\right)\left(\frac{1 \text{ kJ}}{1000 \text{ J}}\right) = +90.8 \frac{\text{kJ}}{\text{mol}}$$

BALLPARK CHECK Because ΔG is positive, the reaction is nonspontaneous at 300 K, consistent with our estimate.

PROBLEM 7.5
Does entropy increase or decrease in the following processes?
(a) Complex carbohydrates are metabolized by the body, converted into simple sugars.
(b) Steam condenses on a glass surface.
(c) $2 SO_2(g) + O_2(g) \longrightarrow 2 SO_3(g)$

PROBLEM 7.6
Lime (CaO) is prepared by the decomposition of limestone ($CaCO_3$).

$$CaCO_3(s) \longrightarrow CaO(s) + CO_2(g) \quad \Delta H = +42.6 \text{ kcal/mol } (+178.3 \text{ kJ/mol});$$
$$\Delta S = +38.0 \text{ cal/(mol} \cdot \text{K)}(+159 \text{ J/(mol} \cdot \text{K)}) \text{ at } 25\,°C$$

(a) Calculate ΔG at 25 °C. Give your answer in kcal/mol and kJ/mol. Does the reaction occur spontaneously?

(b) Would you expect the reaction to be spontaneous at higher or lower temperatures?

PROBLEM 7.7
The melting of solid ice to give liquid water has $\Delta H = +1.44 \text{ kcal/mol } (+6.02 \text{ kJ/mol})$ and $\Delta S = +5.26 \text{ cal/(mol} \cdot \text{K)}(+22.0 \text{ J/(mol} \cdot \text{K)})$. What is the value of ΔG for the melting process at the following temperatures? Give your answer in kcal/mol and kJ/mol. Is the melting spontaneous or nonspontaneous at these temperatures?

(a) −10 °C (263 K) (b) 0 °C (273 K) (c) +10 °C (283 K)

KEY CONCEPT PROBLEM 7.8

The following diagram portrays a reaction of the type $A(s) \longrightarrow B(s) + C(g)$, where the different colored spheres represent different molecular structures. Assume that the reaction has $\Delta H = -23.5 \text{ kcal/mol } (-98.3 \text{ kJ/mol})$.

(a) What is the sign of ΔS for the reaction?

(b) Is the reaction likely to be spontaneous at all temperatures, nonspontaneous at all temperatures, or spontaneous at some but nonspontaneous at others?

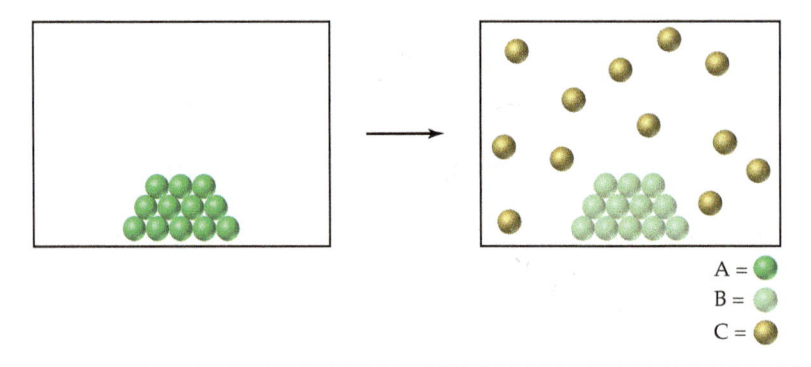

7.5 How Do Chemical Reactions Occur? Reaction Rates

Just because a chemical reaction has a favorable free-energy change does not mean that it occurs rapidly. The value of ΔG tells us only whether a reaction *can* occur; it says nothing about how *fast* the reaction will occur or about the details of the molecular changes that take place during the reaction. It is now time to look into these other matters.

For a chemical reaction to occur, reactant particles must collide, some chemical bonds have to break, and new bonds have to form. Not all collisions lead to products, however. One requirement for a productive collision is that the colliding molecules must approach with the correct orientation so that the atoms about to form new bonds can connect. In the reaction of ozone (O_3) with nitric oxide (NO) to give oxygen (O_2) and nitrogen dioxide (NO_2), for example, the two reactants must collide so that the nitrogen atom of NO strikes a terminal oxygen atom of O_3 (Figure 7.2).

Another requirement for a reaction to occur is that the collision must take place with enough energy to break the appropriate bonds in the reactant. If the reactant particles are moving slowly, collisions might be too gentle to overcome the repulsion between electrons in the different reactants, and the particles will simply bounce apart. A reaction will only occur if the collisions between reactant molecules are sufficiently energetic.

SECTION 7.5 How Do Chemical Reactions Occur? Reaction Rates 191

Effective collision:

Bond can form

Ineffective collision:

No bond can form

◀ **Figure 7.2**
How do chemical reactions occur? For a collision between NO and O_3 molecules to give O_2 and NO_2, the molecules must collide so that the correct atoms come into contact. No bond forms if the molecules collide with the wrong orientation.

For this reason, many reactions with a favorable free-energy change do not occur at room temperature. To get such a reaction started, energy (heat) must be added. The heat causes the reactant particles to move faster, thereby increasing both the frequency and the force of the collisions. We all know that matches burn, for instance, but we also know that they do not burst into flame until struck. The heat of friction provides enough energy for a few molecules to react. Once started, the reaction sustains itself as the energy released by reacting molecules gives other molecules enough energy to react.

The energy change that occurs during the course of a chemical reaction can be visualized in an energy diagram like that in Figure 7.3. At the beginning of the reaction (left side of the diagram), the reactants are at the energy level indicated. At the end of the reaction (right side of the diagram), the products are at a lower energy level than the reactants if the reaction is exergonic (Figure 7.3a) but higher than the reactants if the reaction is endergonic (Figure 7.3b).

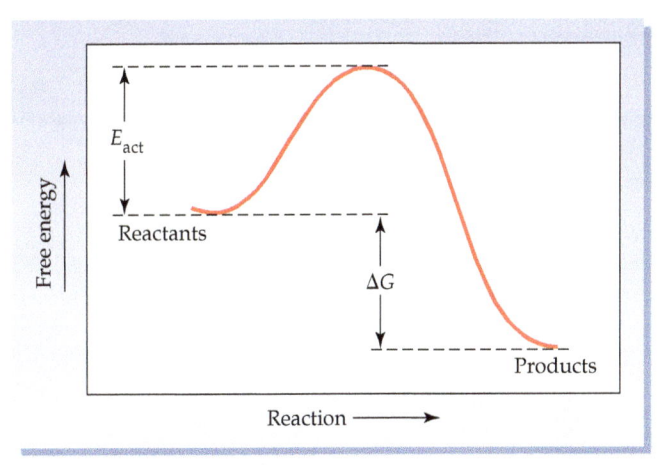

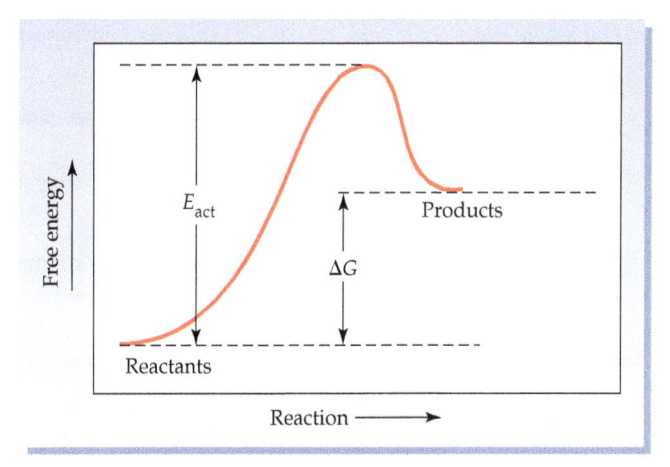

(a) An exergonic reaction

(b) An endergonic reaction

▲ **Figure 7.3**
Reaction energy diagrams show energy changes during a chemical reaction.
A reaction begins on the left and proceeds to the right. (a) In an exergonic reaction, the product energy level is lower than that of reactants. (b) In an endergonic reaction, the situation is reversed. The height of the barrier between reactant and product energy levels is the activation energy, E_{act}. The difference between reactant and product energy levels is the free-energy change, ΔG.

Lying between the reactants and the products is an energy "barrier" that must be surmounted. The height of this barrier represents the amount of energy the colliding particles must have for productive collisions to occur, an amount called the **activation energy** (E_{act}) of the reaction. The size of the activation energy determines the **reaction rate**, or how fast the reaction occurs. The lower the activation energy, the greater the number of productive collisions in a given amount of time, and the faster the reaction. Conversely, the higher the activation energy, the lower the number of productive collisions, and the slower the reaction.

Activation energy (E_{act}) The amount of energy necessary for reactants to surmount the energy barrier to reaction; determines reaction rate.

Reaction rate A measure of how rapidly a reaction occurs; determined by E_{act}.

192 CHAPTER 7 Chemical Reactions: Energy, Rates, and Equilibrium

Note that the size of the activation energy and the size of the free-energy change are unrelated. A reaction with a large E_{act} takes place very slowly even if it has a large negative ΔG. Every reaction is different; each has its own characteristic activation energy and free-energy change.

Worked Example 7.7 Energy of Reactions: Energy Diagrams

Draw an energy diagram for a reaction that is very fast but has a small negative free-energy change.

ANALYSIS A very fast reaction has a small E_{act}. A reaction with a small negative free-energy change is a favorable reaction with a small energy difference between starting materials and products.

SOLUTION

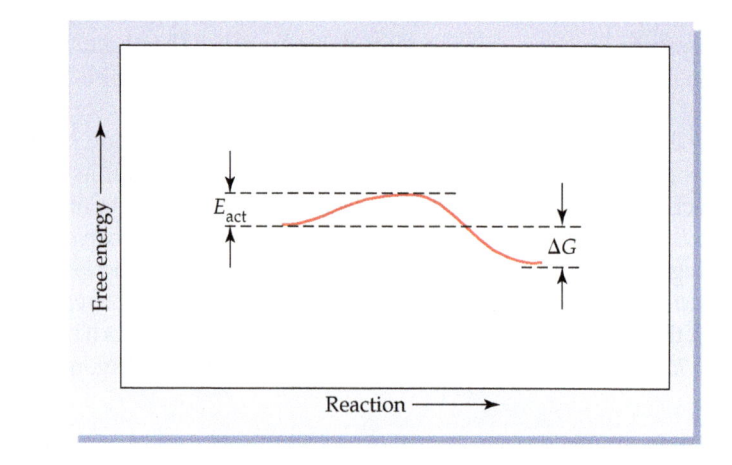

PROBLEM 7.9
Draw an energy diagram for a reaction that is very slow but highly favorable.

PROBLEM 7.10
Draw an energy diagram for a reaction that is slightly unfavorable.

7.6 Effects of Temperature, Concentration, and Catalysts on Reaction Rates

Several things can be done to help reactants over an activation energy barrier and thereby speed up a reaction. Let us look at some possibilities.

Temperature

One way to increase reaction rate is to add energy to the reactants by raising the temperature. With more energy in the system, the reactants move faster, so the frequency of collisions increases. Furthermore, the force with which collisions occur increases, making them more likely to overcome the activation barrier. As a rule of thumb, a 10 °C rise in temperature causes a reaction rate to double.

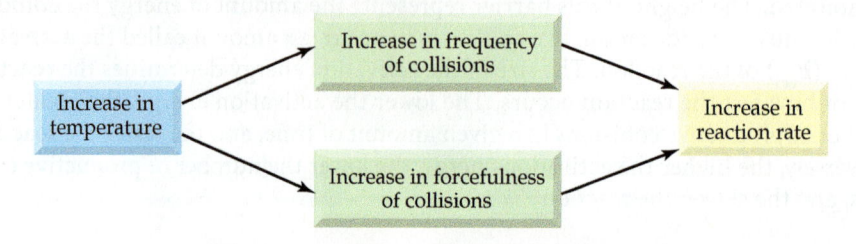

Concentration

A second way to speed up a reaction is to increase the **concentrations** of the reactants. As the concentration increases, reactants are crowded together, and collisions between reactant molecules become more frequent. As the frequency of collisions increases, reactions between molecules become more likely. Flammable materials burn more rapidly in pure oxygen than in air, for instance, because the concentration of O_2 molecules is higher (air is approximately 21% oxygen). Hospitals must therefore take extraordinary precautions to ensure that no flames are used near patients receiving oxygen. Although different reactions respond differently to concentration changes, doubling or tripling a reactant concentration often doubles or triples the reaction rate.

Concentration A measure of the amount of a given substance in a mixture.

Increase in concentration → Increase in frequency of collisions → Increase in reaction rate

Catalysts

A third way to speed up a reaction is to add a **catalyst**—a substance that accelerates a chemical reaction but is itself unchanged in the process. For example, metals such as nickel, palladium, and platinum catalyze the addition of hydrogen to the carbon–carbon double bonds in vegetable oils to yield semisolid margarine. Without the metal catalyst, the reaction does not occur.

Catalyst A substance that speeds up the rate of a chemical reaction but is itself unchanged.

A double bond in vegetable oil + H_2 →(Ni, Pd, or Pt catalyst)→ A single bond in margarine

A catalyst does not affect the energy level of either reactants or products. Rather, it increases reaction rate either by letting a reaction take place through an alternative pathway with a lower energy barrier, or by orienting the reacting molecules appropriately. In a reaction energy diagram, the catalyzed reaction has a lower activation energy (Figure 7.4). It is worth noting that the free-energy change for a reaction depends *only* on the difference in the energy levels of the reactants and products, and *not* on the pathway of the reaction. Therefore, a catalyzed reaction releases (or absorbs) the same amount of energy as an uncatalyzed reaction. It simply occurs more rapidly.

◄ **Figure 7.4**
A reaction energy diagram for a reaction in the presence (green curve) and absence (blue curve) of a catalyst. The catalyzed reaction has a lower (E_{act}) because it uses an alternative pathway (represented by the multiple bumps in the green line) with a lower energy barrier. The free-energy change, ΔG, is unaffected by the presence of a catalyst.

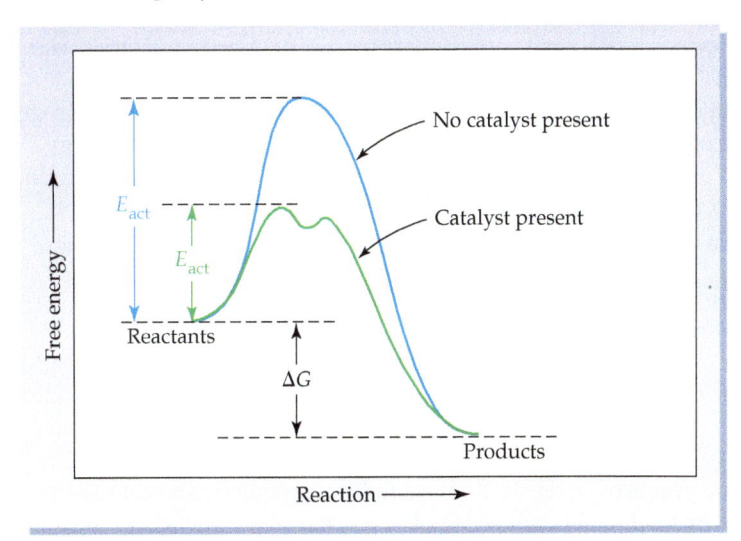

In addition to their widespread use in industry, we also rely on catalysts to reduce the air pollution created by exhaust from automobile engines. The catalytic converters in

194 CHAPTER 7 Chemical Reactions: Energy, Rates, and Equilibrium

most automobiles are tubes packed with catalysts of two types (Figure 7.5). One catalyst accelerates the complete combustion of hydrocarbons and CO in the exhaust to give CO_2 and H_2O, and the other decomposes NO to N_2 and O_2.

▶ **Figure 7.5**
A catalytic converter.
The exhaust gases from an automobile pass through a two-stage catalytic converter. In one stage, carbon monoxide and unburned hydrocarbons are converted to CO_2 and H_2O. In the second stage, NO is converted to N_2 and O_2.

C_xH_y, CO, NO, O_2 → CO_2, H_2O, N_2, O_2

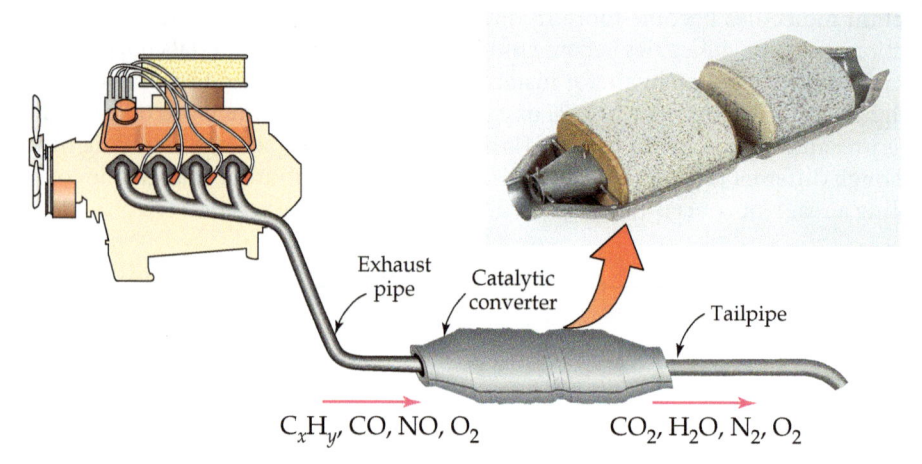

Table 7.3 summarizes the effects of changing conditions on reaction rates.

TABLE 7.3 Effects of Changes in Reaction Conditions on Reaction Rates

Change	Effect
Concentration	Increase in reactant concentration increases rate. Decrease in reactant concentration decreases rate.
Temperature	Increase in temperature increases rate. Decrease in temperature decreases rate.
Catalyst added	Increases reaction rate.

LOOKING AHEAD ▶▶▶ The thousands of biochemical reactions continually taking place in our bodies are catalyzed by large protein molecules called *enzymes*, which promote reactions by controlling the orientation of the reacting molecules. Since almost every reaction is catalyzed by its own specific enzyme, the study of enzyme structure, activity, and control is a central part of biochemistry. We will look more closely at enzymes and how they work in Chapter 19.

PROBLEM 7.11
Ammonia is synthesized industrially by reaction of nitrogen and hydrogen according to the equation $3 H_2(g) + N_2(g) \longrightarrow 2 NH_3(g)$. The free-energy change for this reaction is $\Delta G = -3.8$ kcal/mol (-16 kJ/mol), yet this reaction does not readily occur at room temperature.
(a) Draw a reaction energy diagram for this reaction, indicating E_{act} and ΔG.
(b) List three ways to increase the rate of this reaction.

PROBLEM 7.12
As we exercise, our bodies metabolize glucose, converting it to CO_2 and H_2O, to supply the energy necessary for physical activity. The simplified reaction is:

$$C_6H_{12}O_6(aq) + 6 O_2(g) \longrightarrow 6 CO_2(g) + 6 H_2O(l) + 678 \text{ kcal} (2840 \text{ kJ})$$

How many grams of water would have to be evaporated as sweat to remove the heat generated by the metabolism of 1 mol of glucose? (See Chemistry in Action: Regulation of Body Temperature on p. 195).

CHEMISTRY IN ACTION

Regulation of Body Temperature

Maintaining normal body temperature is crucial. If the body's thermostat is unable to maintain a temperature of 37 °C, the rates of the many thousands of chemical reactions that take place constantly in the body will change accordingly, with potentially disastrous consequences.

If, for example, a skater fell through the ice of a frozen lake, *hypothermia* could soon result. Hypothermia is a dangerous state that occurs when the body is unable to generate enough heat to maintain normal temperature. All chemical reactions in the body slow down because of the lower temperature, energy production drops, and death can result. Slowing the body's reactions can also be used to advantage, however. During open-heart surgery, the heart is stopped and maintained at about 15 °C, while the body, which receives oxygenated blood from an external pump, is cooled to 25–32 °C.

In this case, the body is receiving oxygenated blood from an external pump in an operating chamber under medical supervision. If hypothermia occurred due to some other environmental condition, the heart would slow down, respiration would decrease, and the body would not receive sufficient oxygen and death would result.

Conversely, a marathon runner on a hot, humid day might become overheated, and *hyperthermia* could result. Hyperthermia, also called *heat stroke*, is an uncontrolled rise in temperature as the result of the body's inability to lose sufficient heat. Chemical reactions in the body are accelerated at higher temperatures, the heart struggles to pump blood faster to supply increased oxygen, and brain damage can result if the body temperature rises above 41 °C.

Body temperature is maintained both by the thyroid gland and by the hypothalamus region of the brain, which act together to regulate metabolic rate. When the body's environment changes, temperature receptors in the skin, spinal cord, and abdomen send signals to the hypothalamus, which contains both heat-sensitive and cold-sensitive neurons.

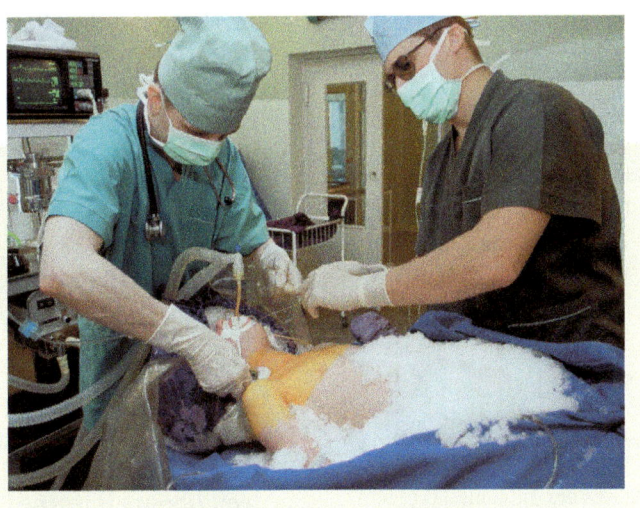

▲ The body is cooled to 25–32°C by immersion in ice prior to open-heart surgery to slow down metabolism.

Stimulation of the heat-sensitive neurons on a hot day causes a variety of effects: Impulses are sent to stimulate the sweat glands, dilate the blood vessels of the skin, decrease muscular activity, and reduce metabolic rate. Sweating cools the body through evaporation; approximately 540 cal (2260 J) is removed by evaporation of 1.0 g of sweat. Dilated blood vessels cool the body by allowing more blood to flow close to the surface of the skin, where heat is removed by contact with air. Decreased muscular activity and a reduced metabolic rate cool the body by lowering internal heat production.

Stimulation of the cold-sensitive neurons on a cold day also causes a variety of effects: The hormone epinephrine is released to stimulate metabolic rate; peripheral blood vessels contract to decrease blood flow to the skin and prevent heat loss; and muscular contractions increase to produce more heat, resulting in shivering and "goosebumps."

One further comment: Drinking alcohol to warm up on a cold day actually has the opposite effect. Alcohol causes blood vessels to dilate, resulting in a warm feeling as blood flow to the skin increases. Although the warmth feels good temporarily, body temperature ultimately drops as heat is lost through the skin at an increased rate.

See Chemistry in Action Problems 7.72 and 7.73 at the end of the chapter.

7.7 Reversible Reactions and Chemical Equilibrium

Many chemical reactions result in the virtually complete conversion of reactants into products. When sodium metal reacts with chlorine gas, for example, both are entirely consumed. The sodium chloride product is so much more stable than the reactants that, once started, the reaction keeps going until it is complete.

What happens, though, when the reactants and products are of approximately equal stability? This is the case, for example, in the reaction of acetic acid (the main organic constituent of vinegar) with ethyl alcohol to yield ethyl acetate, a solvent used in nail-polish remover and glue.

$$\underset{\text{Acetic acid}}{CH_3COH} + \underset{\text{Ethyl alcohol}}{HOCH_2CH_3} \underset{\text{Or this direction?}}{\overset{\text{This direction?}}{\rightleftarrows}} \underset{\text{Ethyl acetate}}{CH_3COCH_2CH_3} + \underset{\text{Water}}{H_2O}$$

196 CHAPTER 7 Chemical Reactions: Energy, Rates, and Equilibrium

Imagine the situation if you mix acetic acid and ethyl alcohol. The two begin to form ethyl acetate and water. But as soon as ethyl acetate and water form, they begin to go back to acetic acid and ethyl alcohol. Such a reaction, which easily goes in either direction, is said to be **reversible** and is indicated by a double arrow ($\rightleftharpoons$) in equations. The reaction read from left to right as written is referred to as the *forward reaction*, and the reaction from right to left is referred to as the *reverse reaction*.

Reversible reaction A reaction that can go in either direction, from products to reactants or reactants to products.

Now suppose you mix some ethyl acetate and water. The same thing occurs: As soon as small quantities of acetic acid and ethyl alcohol form, the reaction in the other direction begins to take place. No matter which pair of reactants is mixed together, both reactions occur until ultimately the concentrations of reactants and products reach constant values and undergo no further change. At this point, the reaction vessel contains all four substances—acetic acid, ethyl acetate, ethyl alcohol, and water—and the reaction is said to be in a state of **chemical equilibrium**.

Chemical equilibrium A state in which the rates of forward and reverse reactions are the same.

Since the reactant and product concentrations undergo no further change once equilibrium is reached, you might conclude that the forward and reverse reactions have stopped. That is not the case, however. The forward reaction takes place rapidly at the beginning of the reaction but then slows down as reactant concentrations decrease. At the same time, the reverse reaction takes place slowly at the beginning but then speeds up as product concentrations increase (Figure 7.6). Ultimately, the forward and reverse rates become equal and change no further.

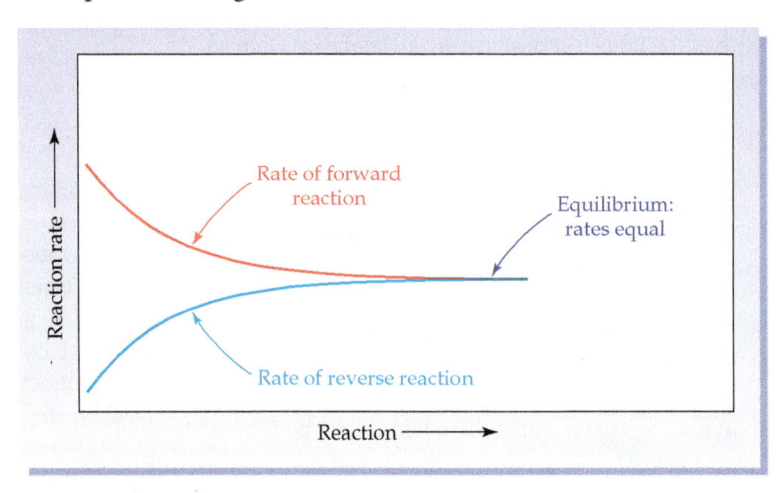

▶ **Figure 7.6**
Reaction rates in an equilibrium reaction.
The forward rate is large initially but decreases as the concentrations of reactants drop. The reverse rate is small initially but increases as the concentrations of products increase. At equilibrium, the forward and reverse reaction rates are equal.

Chemical equilibrium is an active, dynamic condition. All substances present are continuously being made and unmade at the same rate, so their concentrations are constant at equilibrium. As an analogy, think of two floors of a building connected by up and down escalators. If the number of people moving up is the same as the number of people moving down, the numbers of people on each floor remain constant. *Individual people* are continuously changing from one floor to the other, but the *total populations* of the two floors are in equilibrium.

Note that it is not necessary for the concentrations of reactants and products at equilibrium to be equal (just as it is not necessary for the numbers of people on two floors connected by escalators to be equal). Equilibrium can be reached at any point between pure products and pure reactants. The extent to which the forward or reverse reaction is favored over the other is a characteristic property of a given reaction under given conditions.

▲ When the number of people moving up is the same as the number of people moving down, the number of people on each floor remains constant, and the two populations are in equilibrium.

7.8 Equilibrium Equations and Equilibrium Constants

Remember that the rate of a reaction depends on the number of collisions between molecules (Section 7.5), and that the number of collisions in turn depends on concentration, i.e., the number of molecules in a given volume (Section 7.6). For a reversible reaction, then, the rates of both the forward *and* the reverse reactions must depend on the concentration of reactants and products, respectively. When a reaction reaches equilibrium, the rates of the forward and reverse reactions are equal, and the concentrations of reactants and products remain constant. We can use this fact to obtain useful information about a reaction.

Let us look at the details of a specific equilibrium reaction. Suppose that you allow various mixtures of sulfur dioxide and oxygen to come to equilibrium with sulfur trioxide at a temperature of 727 °C and then measure the concentrations of all three gases in the mixtures.

$$2\,SO_2(g) + O_2(g) \rightleftharpoons 2\,SO_3(g)$$

In one experiment, we start with only 1.00 mol of SO_2 and 1.00 mol of O_2 in a 1.00 L container. In other words, the initial concentrations of reactants are 1.00 mol/L. When the reaction reaches equilibrium, we have 0.0620 mol/L of SO_2, 0.538 mol/L of O_2, and 0.938 mol/L of SO_3. In another experiment, we start with 1.00 mol/L of SO_3. When this reaction reaches equilibrium, we have 0.150 mol/L of SO_2, 0.0751 mol/L of O_2, and 0.850 mol/L of SO_3. In both cases, we see that there is substantially more product (SO_3) than reactants when the reaction reaches equilibrium, regardless of the starting conditions. Is it possible to predict what the equilibrium conditions will be for any given reaction?

As it turns out, the answer is YES! No matter what the original concentrations were, and no matter what concentrations remain at equilibrium, we find that a constant numerical value is obtained if the equilibrium concentrations are substituted into the expression

$$\frac{[SO_3]^2}{[SO_2]^2[O_2]} = \text{constant at a given T}$$

The square brackets in this expression indicate the concentration of each substance expressed as moles per liter. Using the equilibrium concentrations for each of the experiments described above, we can calculate the value and verify that it is constant:

Experiment 1. $\quad \dfrac{[SO_3]^2}{[SO_2]^2[O_2]} = \dfrac{(0.938\text{ mol/L})^2}{(0.0620\text{ mol/L})^2(0.538\text{ mol/L})} = 425$

Experiment 2. $\quad \dfrac{[SO_3]^2}{[SO_2]^2[O_2]} = \dfrac{(0.850\text{ mol/L})^2}{(0.150\text{ mol/L})^2(0.0751\text{ mol/L})} = 428$

At a temperature of 727 °C, the actual value of the constant is 429. Within experimental error, the ratios of product and reactant concentrations for the two experiments at equilibrium yield the same result. Numerous experiments like those just described have led to a general equation that is valid for any reaction. Consider a general reversible reaction:

$$aA + bB + \ldots \rightleftharpoons mM + nN + \ldots$$

where A, B, ... are reactants; M, N, ... are products; and $a, b, \ldots, m, n, \ldots$ are coefficients in the balanced equation. At equilibrium, the composition of the reaction mixture obeys the following *equilibrium equation*, where K is the **equilibrium constant**.

Equilibrium equation $\quad K = \dfrac{[M]^m[N]^n \cdots}{[A]^a[B]^b \cdots}$ ← Product concentrations / Reactant concentrations

Equilibrium constant

Equilibrium constant (K) Value obtained at a given temperature from the ratio of the concentrations of products and reactants, each raised to a power equal to its coefficient in the balanced equation.

The equilibrium constant K is the number obtained by multiplying the equilibrium concentrations of the products and dividing by the equilibrium concentrations of the reactants, with the concentration of each substance raised to a power equal to its coefficient in the balanced equation. If we take another look at the reaction between sulfur dioxide and oxygen, we can now see how the equilibrium constant was obtained:

$$2\,SO_2(g) + O_2(g) \rightleftharpoons 2\,SO_3(g)$$

$$K = \frac{[SO_3]^2}{[SO_2]^2[O_2]}$$

Note that if there is no coefficient for a reactant or product in the reaction equation, it is assumed to be 1. The value of K varies with temperature (25 °C) is assumed unless otherwise specified—and units are usually omitted.

For reactions that involve pure solids or liquids, these pure substances are omitted when writing the equilibrium constant expression. To explain why, consider the decomposition of limestone from Problem 7.6:

$$CaCO_3(s) \longrightarrow CaO(s) + CO_2(g)$$

▶▶▶ The practice of omitting pure substances in the equilibrium constant expression will be utilized in Chapter 10 when we discuss equilibria involving acids and bases.

Writing the equilibrium constant expression for this reaction as the concentration of products over the concentration of reactions would yield

$$K = \frac{[CaO][CO_2]}{[CaCO_3]}$$

Consider the solids CaO and $CaCO_3$. Their concentrations (in moles/L) can be calculated from their molar masses and densities at a given temperature. For example, the concentration of CaO at 25 °C can be calculated as

$$\frac{\left(3.25\,\frac{g\,CaO}{cm^3}\right) \cdot \left(\frac{1000\,cm^3}{L}\right)}{56.08\,\frac{g\,CaO}{mol\,CaO}} = 58.0\,\frac{mol\,CaO}{L}$$

The ratio of products over reactants would change if CO_2 was added to or removed from the reaction. The concentration of CaO, however, is the same whether we have 10 grams or 500 grams. Adding solid CaO will not change the ratio of products over reactants. Since the concentration of solids is independent of the amount of solid present, these concentrations are omitted and the expression for K becomes

$$K = \frac{[\cancel{CaO}][CO_2]}{[\cancel{CaCO_3}]} = [CO_2]$$

The value of the equilibrium constant indicates the position of a reaction at equilibrium. If the forward reaction is favored, the product term $[M]^m[N]^n$ is larger than the reactant term $[A]^a[B]^b$, and the value of K is larger than 1. If instead the reverse reaction is favored, $[M]^m[N]^n$ is smaller than $[A]^a[B]^b$ at equilibrium, and the value of K is smaller than 1.

For a reaction such as the combination of hydrogen and oxygen to form water vapor, the equilibrium constant is enormous (3.1×10^{81}), showing how greatly the formation of water is favored. Equilibrium is effectively nonexistent for such reactions, and the reaction is described as *going to completion*.

On the other hand, the equilibrium constant is very small for a reaction such as the combination of nitrogen and oxygen at 25 °C to give NO (4.7×10^{-31}), showing what we know from observation—that N_2 and O_2 in the air do not combine noticeably at room temperature:

$$N_2(g) + O_2(g) \rightleftharpoons 2\,NO(g) \quad K = \frac{[NO]^2}{[N_2][O_2]} = 4.7 \times 10^{-31}$$

SECTION 7.8 Equilibrium Equations and Equilibrium Constants

When K is close to 1, say between 10^3 and 10^{-3}, significant amounts of both reactants and products are present at equilibrium. An example is the reaction of acetic acid with ethyl alcohol to give ethyl acetate (Section 7.7). For this reaction, $K = 3.4$.

$$CH_3CO_2H + CH_3CH_2OH \rightleftharpoons CH_3CO_2CH_2CH_3 + H_2O$$

$$K = \frac{[CH_3CO_2CH_2CH_3][H_2O]}{[CH_3CO_2H][CH_3CH_2OH]} = 3.4$$

We can summarize the meaning of equilibrium constants in the following way:

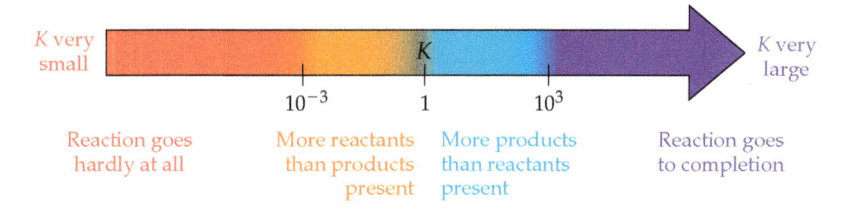

```
K very                                                K very
small              K                                  large
   |───────────────|───────────────|
 10⁻³              1              10³

Reaction goes   More reactants  More products   Reaction goes
hardly at all   than products   than reactants  to completion
                present         present
```

K much smaller than 0.001 Only reactants are present at equilibrium; essentially no reaction occurs.

K between 0.001 and 1 More reactants than products are present at equilibrium.

K between 1 and 1000 More products than reactants are present at equilibrium.

K much larger than 1000 Only products are present at equilibrium; reaction goes essentially to completion.

Worked Example 7.8 Writing Equilibrium Equations

The first step in the industrial synthesis of hydrogen is the reaction of steam with methane to give carbon monoxide and hydrogen. Write the equilibrium equation for the reaction.

$$H_2O(g) + CH_4(g) \rightleftharpoons CO(g) + 3\,H_2(g)$$

ANALYSIS The equilibrium constant K is the number obtained by multiplying the equilibrium concentrations of the products (CO and H_2) and dividing by the equilibrium concentrations of the reactants (H_2O and CH_4), with the concentration of each substance raised to the power of its coefficient in the balanced equation.

SOLUTION

$$K = \frac{[CO][H_2]^3}{[H_2O][CH_4]}$$

Worked Example 7.9 Equilibrium Equations: Calculating K

In the reaction of Cl_2 with PCl_3, the concentrations of reactants and products were determined experimentally at equilibrium and found to be 7.2 mol/L for PCl_3, 7.2 mol/L for Cl_2, and 0.050 mol/L for PCl_5.

$$PCl_3(g) + Cl_2(g) \rightleftharpoons PCl_5(g)$$

Write the equilibrium equation, and calculate the equilibrium constant for the reaction. Which reaction is favored, the forward one or the reverse one?

ANALYSIS All the coefficients in the balanced equation are 1, so the equilibrium constant equals the concentration of the product, PCl_5, divided by the product of the concentrations of the two reactants, PCl_3 and Cl_2. Insert the values given for each concentration, and calculate the value of K.

BALLPARK ESTIMATE At equilibrium, the concentration of the reactants (7.2 mol/L for each reactant) is higher than the concentration of the product (0.05 mol/L), so we expect a value of K less than 1.

SOLUTION

$$K = \frac{[PCl_5]}{[PCl_3][Cl_2]} = \frac{0.050 \text{ mol/L}}{(7.2 \text{ mol/L})(7.2 \text{ mol/L})} = 9.6 \times 10^{-4}$$

The value of K is less than 1, so the reverse reaction is favored. Note that units for K are omitted.

BALLPARK CHECK Our calculated value of K is just as we predicted: $K < 1$.

PROBLEM 7.13

Write equilibrium equations for the following reactions:
(a) $N_2O_4(g) \rightleftharpoons 2\,NO_2(g)$
(b) $2\,H_2S(g) + O_2(g) \rightleftharpoons 2\,S(s) + 2\,H_2O(g)$
(c) $2\,BrF_5(g) \rightleftharpoons Br_2(g) + 5\,F_2(g)$

PROBLEM 7.14

Do the following reactions favor reactants or products at equilibrium? Give relative concentrations at equilibrium.
(a) $Sucrose(aq) + H_2O(l) \rightleftharpoons Glucose(aq) + Fructose(aq) \quad K = 1.4 \times 10^5$
(b) $NH_3(aq) + H_2O(l) \rightleftharpoons NH_4^+(aq) + OH^-(aq) \quad K = 1.6 \times 10^{-5}$
(c) $Fe_2O_3(s) + 3\,CO(g) \rightleftharpoons 2\,Fe(s) + 3\,CO_2(g) \quad K \text{ (at 727 °C)} = 24.2$

PROBLEM 7.15

For the reaction $H_2(g) + I_2(g) \rightleftharpoons 2\,HI(g)$, equilibrium concentrations at 25 °C are $[H_2] = 0.0510$ mol/L, $[I_2] = 0.174$ mol/L, and $[HI] = 0.507$ mol/L. What is the value of K at 25 °C?

KEY CONCEPT PROBLEM 7.16

The following diagrams represent two similar reactions that have achieved equilibrium:

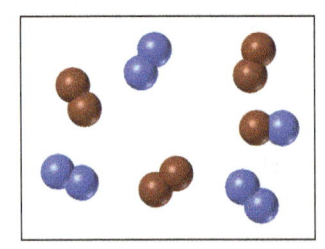

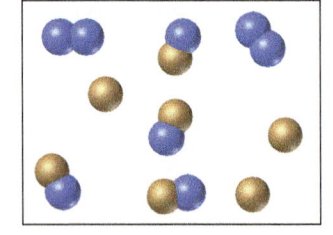

$A_2 + B_2 \longrightarrow 2\,AB$ \qquad $A_2 + 2B \longrightarrow 2\,AB$

(a) Write the expression for the equilibrium constant for each reaction.
(b) Calculate the value for the equilibrium constant for each reaction.

7.9 Le Châtelier's Principle: The Effect of Changing Conditions on Equilibria

The effect of a change in reaction conditions on chemical equilibrium is predicted by a general rule called *Le Châtelier's principle*:

Le Châtelier's principle When a stress is applied to a system at equilibrium, the equilibrium shifts to relieve the stress.

SECTION 7.9 Le Châtelier's Principle: The Effect of Changing Conditions on Equilibria

The word "stress" in this context means any change in concentration, pressure, volume, or temperature that disturbs the original equilibrium and causes the rates of the forward and reverse reactions to become temporarily unequal.

We saw in Section 7.6 that reaction rates are affected by changes in temperature and concentration, and by addition of a catalyst. But what about equilibria? Are they similarly affected? The answer is that changes in concentration, temperature, and pressure *do* affect equilibria, but that addition of a catalyst does not (except to reduce the time it takes to reach equilibrium). The change caused by a catalyst affects forward and reverse reactions equally so that equilibrium concentrations are the same in both the presence and the absence of the catalyst.

Effect of Changes in Concentration

Let us look at the effect of a concentration change by considering the reaction of CO with H_2 to form CH_3OH (methanol). Once equilibrium is reached, the concentrations of the reactants and product are constant, and the forward and reverse reaction rates are equal.

$$CO(g) + 2\,H_2(g) \rightleftharpoons CH_3OH(g)$$

What happens if the concentration of CO is increased? To relieve the "stress" of added CO, according to Le Châtelier's principle, the extra CO must be used up. In other words, the rate of the forward reaction must increase to consume CO. Think of the CO added on the left as "pushing" the equilibrium to the right:

$$[CO \longrightarrow]$$
$$CO(g) + 2\,H_2(g) \rightleftharpoons CH_3OH(g)$$

Of course, as soon as more CH_3OH forms, the reverse reaction also speeds up, some CH_3OH converts back to CO and H_2. Ultimately, the forward and reverse reaction rates adjust until they are again equal, and equilibrium is reestablished. At this new equilibrium state, the value of $[H_2]$ is lower because some of the H_2 reacted with the added CO and the value of $[CH_3OH]$ is higher because CH_3OH formed as the reaction was driven to the right by the addition of CO. The changes offset each other, however, so that the value of the equilibrium constant K remains constant.

$$CO(g) + 2\,H_2(g) \rightleftharpoons CH_3OH(g)$$

If this increases . . .   . . . then this decreases . . .   . . . and this increases . . .

. . . but this remains constant. $K = \dfrac{[CH_3OH]}{[CO][H_2]^2}$

What happens if CH_3OH is added to the reaction at equilibrium? Some of the methanol reacts to yield CO and H_2, making the values of $[CO]$, $[H_2]$, and $[CH_3OH]$ higher when equilibrium is reestablished. As before, the value of K does not change.

If this increases . . .

$$CO(g) + 2\,H_2(g) \rightleftharpoons CH_3OH(g)$$

. . . then this increases . . .   . . . and this increases . . .

. . . but this remains constant. $K = \dfrac{[CH_3OH]}{[CO][H_2]^2}$

Alternatively, we can view chemical equilibrium as a *balance* between the free energy of the reactants (on the left) and the free energy of the products (on the right). Adding more reactants tips the balance in favor of the reactants. In order to restore the balance, reactants must be converted to products, or the reaction must shift to the right. If, instead, we remove reactants, then the balance is too heavy on the product side and the reaction must shift left, generating more reactants to restore balance.

▶ **Equilibrium represents a balance between the free energy of reactants and products. Adding reactants (or products) to one side upsets the balance, and the reaction will proceed in a direction to restore the balance.**

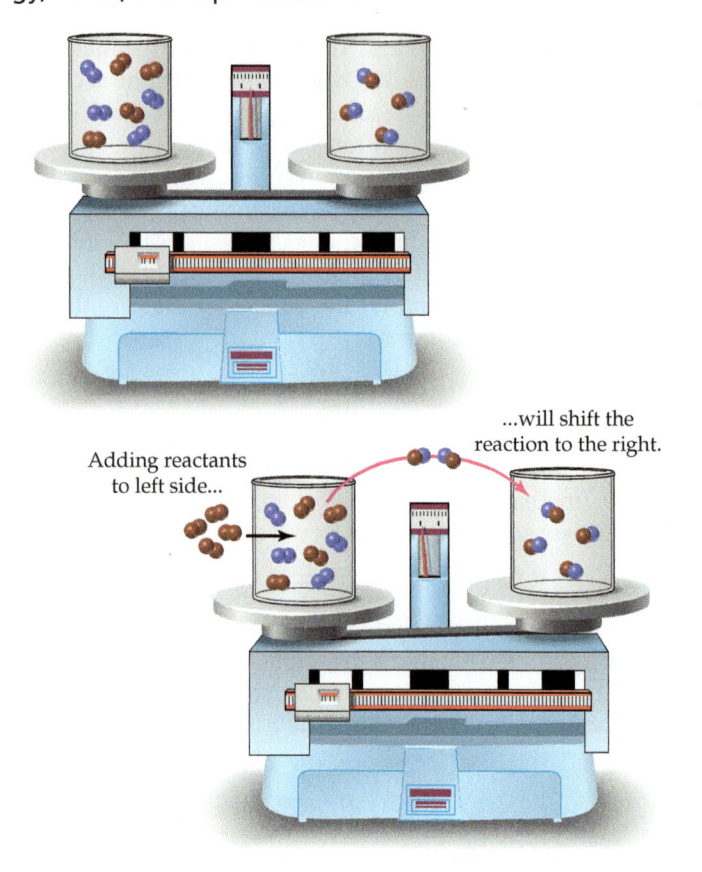

...will shift the reaction to the right.

Adding reactants to left side...

Finally, what happens if a reactant is continuously supplied or a product is continuously removed? Because the concentrations are continuously changing, equilibrium can never be reached. As a result, it is sometimes possible to force a reaction to produce large quantities of a desirable product even when the equilibrium constant is unfavorable. Take the reaction of acetic acid with ethanol to yield ethyl acetate, for example. As discussed in the preceding section, the equilibrium constant K for this reaction is 3.4, meaning that substantial amounts of reactants and products are both present at equilibrium. If, however, the ethyl acetate is removed as soon as it is formed, the production of more and more product is forced to occur, in accord with Le Châtelier's principle.

Continuously removing this product from the reaction forces more of it to be produced.

$$CH_3\overset{\overset{O}{\|}}{C}OH + CH_3CH_2OH \rightleftarrows CH_3\overset{\overset{O}{\|}}{C}OCH_2CH_3 + H_2O$$

Acetic acid   Ethyl alcohol   Ethyl acetate

Metabolic reactions sometimes take advantage of this effect, with one reaction prevented from reaching equilibrium by the continuous consumption of its product in a further reaction.

Effect of Changes in Temperature and Pressure

We noted in Section 7.2 that the reverse of an exothermic reaction is always endothermic. Equilibrium reactions are therefore exothermic in one direction and endothermic in the other. Le Châtelier's principle predicts that an increase in temperature will cause an equilibrium to shift in favor of the endothermic reaction so the additional heat is absorbed. Conversely, a decrease in temperature will cause an equilibrium to shift in favor of the exothermic reaction so additional heat is released. In other words, you can think of heat as a reactant or product whose increase or decrease stresses an equilibrium just as a change in reactant or product concentration does.

Endothermic reaction   Favored by increase in temperature
(Heat is absorbed)

Exothermic reaction   Favored by decrease in temperature
(Heat is released)

In the exothermic reaction of N_2 with H_2 to form NH_3, for example, raising the temperature favors the reverse reaction, which absorbs the heat:

$$[\longleftarrow \text{Heat}]$$
$$N_2(g) + 3H_2(g) \rightleftharpoons 2NH_3(g) + \text{Heat}$$

We can also use the balance analogy to predict the effect of temperature on an equilibrium mixture; this time, we think of heat as a reactant or product. Increasing the temperature of the reaction is the same as adding heat to the left side (for an endothermic reaction) or to the right side (for an exothermic reaction). The reaction then proceeds in the appropriate direction to restore "balance" to the system.

What about changing the pressure? Pressure influences an equilibrium only if one or more of the substances involved is a gas. As predicted by Le Châtelier's principle, decreasing the volume to increase the pressure in such a reaction shifts the equilibrium in the direction that decreases the number of molecules in the gas phase and thus, decreases the pressure. For the ammonia synthesis, decreasing the volume *increases* the concentration of reactants and products, but has a greater effect on the reactant side of the equilibrium since there are more moles of gas phase reactants. Increasing the pressure, therefore, favors the forward reaction because 4 mol of gas is converted to 2 mol of gas.

$$[\text{Pressure} \longrightarrow]$$
$$\underbrace{N_2(g) + 3H_2(g)}_{\text{4 mol of gas}} \rightleftharpoons \underbrace{2NH_3(g)}_{\text{2 mol of gas}}$$

The effects of changing reaction conditions on equilibria are summarized in Table 7.4

TABLE 7.4 Effects of Changes in Reaction Conditions on Equilibria

Change	Effect
Concentration	Increase in reactant concentration or decrease in product concentration favors forward reaction. Increase in product concentration or decrease in reactant concentration favors reverse reaction.
Temperature	Increase in temperature favors endothermic reaction. Decrease in temperature favors exothermic reaction.
Pressure	Increase in pressure favors side with fewer moles of gas. Decrease in pressure favors side with more moles of gas.
Catalyst added	Equilibrium reached more quickly; value of K unchanged.

LOOKING AHEAD   In Chapter 20, we will see how Le Châtelier's principle is exploited to keep chemical "traffic" moving through the body's metabolic pathways. It often happens that one reaction in a series is prevented from reaching equilibrium because its product is continuously consumed in another reaction.

CHEMISTRY IN ACTION

Coupled Reactions

Living organisms are highly complex systems that use chemical reactions to produce the energy needed for daily activity. Many of these reactions occur very slowly—if at all—at normal body temperature, so organisms use several different strategies discussed in this chapter to obtain the energy they need and to function optimally. For example, the rates of slow reactions are increased by using biocatalysts, otherwise known as enzymes (Chapter 19). Le Châtelier's principle is used for regulation of critical processes, including oxygen transport (Chemistry in Action: Breathing and O_2 Transport, p. 263) and blood pH (Chemistry in Action: Buffers in the Body, p. 312). But what about reactions that do not occur spontaneously? One useful strategy is to "couple" a nonspontaneous reaction with a spontaneous one.

Coupling of reactions is a common strategy in both biochemical and industrial applications. Consider the following reaction for the recovery of copper metal from the smelting of ore containing Cu_2S:

$$Cu_2S(s) \longrightarrow 2\,Cu(s) + S(s) \quad \Delta G = +86.2 \text{ kJ } (+21.6 \text{ kcal})$$

Since ΔG for this process is positive (endergonic), this reaction will not proceed spontaneously. But when the smelting process is performed at elevated temperatures in the presence of oxygen, this reaction can be "coupled" with another reaction:

$$Cu_2S(s) \longrightarrow 2\,Cu(s) + S(s) \quad \Delta G = +86.2 \text{ kJ } (+21.6 \text{ kcal})$$
$$S(s) + O_2(g) \longrightarrow SO_2(g) \quad \Delta G = -300.1 \text{ kJ } (-71.7 \text{ kcal})$$
$$\text{Net Reaction: } Cu_2S(s) + O_2(g) \longrightarrow 2\,Cu(s) + SO_2(g)$$
$$\Delta G = -213.9 \text{ kJ } (-51.1 \text{ kcal})$$

The overall reaction has a negative ΔG (exergonic) to produce pure copper spontaneously.

Coupled Reactions in Biochemistry

An important example of coupled reactions in biochemistry is the endergonic phosphorylation of glucose (Section 22.6), which is the essential first step in the metabolism of glucose. It is combined with the hydrolysis of adenosine triphosphate (ATP) to form adenosine diphosphate (ADP), an exergonic process:

$$\text{Glucose} + HOPO_3^{2-} \longrightarrow \text{Glucose-6-phosphate} + H_2O$$
$$\Delta G = +13.8 \text{ kJ/mol}$$
$$ATP + H_2O \longrightarrow ADP + HOPO_3^{2-} + H^+ \quad \Delta G = -30.5 \text{ kJ/mol}$$
$$\text{Net Reaction: Glucose} + ATP \longrightarrow ADP + \text{Glucose-6-phosphate}$$
$$\Delta G = -16.7 \text{ kJ/mol}$$

In addition to the production of glucose-6-phosphate, which is critical for metabolic activity, any heat that is generated by the coupled reactions can be used to maintain body temperature.

See Chemistry in Action Problems 7.74 and 7.75 at the end of the chapter.

Worked Example 7.10 Le Châtelier's Principle and Equilibrium Mixtures

Nitrogen reacts with oxygen to give NO:

$$N_2(g) + O_2(g) \rightleftharpoons 2\,NO(g) \quad \Delta H = +43 \text{ kcal/mol } (+180 \text{ kJ/mol})$$

Explain the effects of the following changes on reactant and product concentrations:

(a) Increasing temperature
(b) Increasing the concentration of NO
(c) Adding a catalyst

SOLUTION

(a) The reaction is endothermic (positive ΔH), so increasing the temperature favors the forward reaction. The concentration of NO will be higher at equilibrium.

(b) Increasing the concentration of NO, a product, favors the reverse reaction. At equilibrium, the concentrations of both N_2 and O_2, as well as that of NO, will be higher.

(c) A catalyst accelerates the rate at which equilibrium is reached, but the concentrations at equilibrium do not change.

PROBLEM 7.17
Is the yield of SO_3 at equilibrium favored by a higher or lower pressure? By a higher or lower temperature?

$$2\,SO_2(g) + O_2(g) \rightleftharpoons 2\,SO_3(g) \quad \Delta H = -47\text{ kcal/mol}$$

PROBLEM 7.18
What effect do the listed changes have on the position of the equilibrium in the reaction of carbon with hydrogen?

$$C(s) + 2\,H_2(g) \rightleftharpoons CH_4(g) \quad \Delta H = -18\text{ kcal/mol } (-75\text{ kJ/mol})$$

(a) Increasing temperature
(b) Increasing pressure by decreasing volume
(c) Allowing CH_4 to escape continuously from the reaction vessel

PROBLEM 7.19
Another example of a coupled reaction used in the smelting of copper ore (Chemistry in Action: Coupled Reactions, p. 204) involves the following two reactions performed at 375 °C:

(1) $Cu_2O(s) \longrightarrow 2\,Cu(s) + \frac{1}{2}O_2(g) \quad \Delta G \text{ (at 375 °C)} = +140.0\text{ kJ } (+33.5\text{ kcal})$
(2) $C(s) + \frac{1}{2}O_2(g) \longrightarrow CO(g) \quad \Delta G \text{ (at 375 °C)} = -143.8\text{ kJ } (-34.5\text{ kcal})$

Derive the overall reaction and calculate the net free-energy change for the coupled reaction.

SUMMARY: REVISITING THE CHAPTER GOALS

1. What energy changes take place during reactions? The strength of a covalent bond is measured by its *bond dissociation energy*, the amount of energy that must be supplied to break the bond in an isolated gaseous molecule. For any reaction, the heat released or absorbed by changes in bonding is called the *heat of reaction*, or *enthalpy change* (ΔH). If the total strength of the bonds formed in a reaction is greater than the total strength of the bonds broken, then heat is released (negative ΔH) and the reaction is said to be *exothermic*. If the total strength of the bonds formed in a reaction is less than the total strength of the bonds broken, then heat is absorbed (positive ΔH) and the reaction is said to be *endothermic* (see Problems 26–33, 40, 62, 63, 70, 71, 76–78, 80, 81, 83, 85).

2. What is "free-energy," and what is the criterion for spontaneity in chemistry? *Spontaneous reactions* are those that, once started, continue without external influence; nonspontaneous reactions require a continuous external influence. Spontaneity depends on two factors, the amount of heat absorbed or released in a reaction (ΔH) and the *entropy change* (ΔS), which measures the change in molecular disorder in a reaction. Spontaneous reactions are favored by a release of heat (negative ΔH) and an increase in disorder (positive ΔS). The *free-energy change* (ΔG) takes both factors into account, according to the equation $\Delta G = \Delta H - T\Delta S$. A negative value for ΔG indicates spontaneity, and a positive value for ΔG indicates nonspontaneity (see Problems 20–22, 25, 34–43, 46, 50, 51, 73, 84).

3. What determines the rate of a chemical reaction? A chemical reaction occurs when reactant particles collide with proper orientation and sufficient energy. The exact amount of collision energy necessary is called the *activation energy* (E_{act}). A high activation energy results in a slow reaction because few collisions occur with sufficient force, whereas a low activation energy results in a fast reaction. Reaction rates can be increased by raising the temperature, by raising the concentrations of reactants, or by adding a *catalyst*, which accelerates a reaction without itself undergoing any change (see Problems 23, 24, 44–51, 75).

4. What is chemical equilibrium? A reaction that can occur in either the forward or reverse direction is *reversible* and will ultimately reach a state of *chemical equilibrium*. At equilibrium, the forward and reverse reactions occur at the same rate, and the concentrations of reactants and products are constant. Every reversible reaction has a characteristic *equilibrium constant* (K), given by an *equilibrium equation* (see Problems 52–63, 78, 82).

For the reaction: $\quad a\text{A} + b\text{B} + \cdots \rightleftharpoons m\text{M} + n\text{N} + \cdots$

$$K = \frac{[M]^m[N]^n \cdots}{[A]^a[B]^b \cdots}$$

Product concentrations raised to powers equal to coefficients
Reactant concentrations raised to powers equal to coefficients

5. What is Le Châtelier's principle? *Le Châtelier's principle* states that when a stress is applied to a system in equilibrium, the equilibrium shifts so that the stress is relieved. Applying this principle allows prediction of the effects of changes in temperature, pressure, and concentration (see Problems 62–69, 79, 82).

206 CHAPTER 7 Chemical Reactions: Energy, Rates, and Equilibrium

CONCEPT MAP: CHEMICAL REACTIONS: ENERGY, RATES, AND EQUILIBRIUM

Intramolecular Forces
- Ionic Bonds (Ch. 3) = transfer of electrons
- Covalent Bonds (Ch. 4) = sharing of electrons

Chemical reactions (Chapters 5 and 6)

Energy of reactions (Thermochemistry):
Heat of reaction (ΔH):
- Endothermic (ΔH = positive) or exothermic (ΔH = negative)
- Difference in bond energies of products and reactants

Rate of reactions (Kinetics):
Factors affecting rates:
- Collisions between molecules:
  Concentration of reactants
- Orientation of colliding molecules
- Energy of collisions:
  Must exceed **Activation Energy** (E_{act})
  Temperature; increases kinetic energy of colliding molecules
- Catalyst:
  Lowers E_{act} and/or provides favorable orientation of molecules.

Spontaneity of reactions (Thermodynamics):
Free energy (ΔG):
- $\Delta G = \Delta H - T\Delta S$
- Spontaneous = Exergonic (ΔG = negative)
- Nonspontaneous = Endergonic (ΔG = positive)

Extent of reaction:
Equilibrium:
- Rates of forward and reverse reactions are equal.
- Concentrations of products/reactants do not change.
Equilibrium constant:
- K = [products]/[reactants]
- Large K (>10^3) favors products; small K (<10^{-3}) favors reactants.
LeChatelier's Principle — position of equilibrium will be affected by:
  Changing concentration of reactants or products
  Changing temperature
  Changing volume

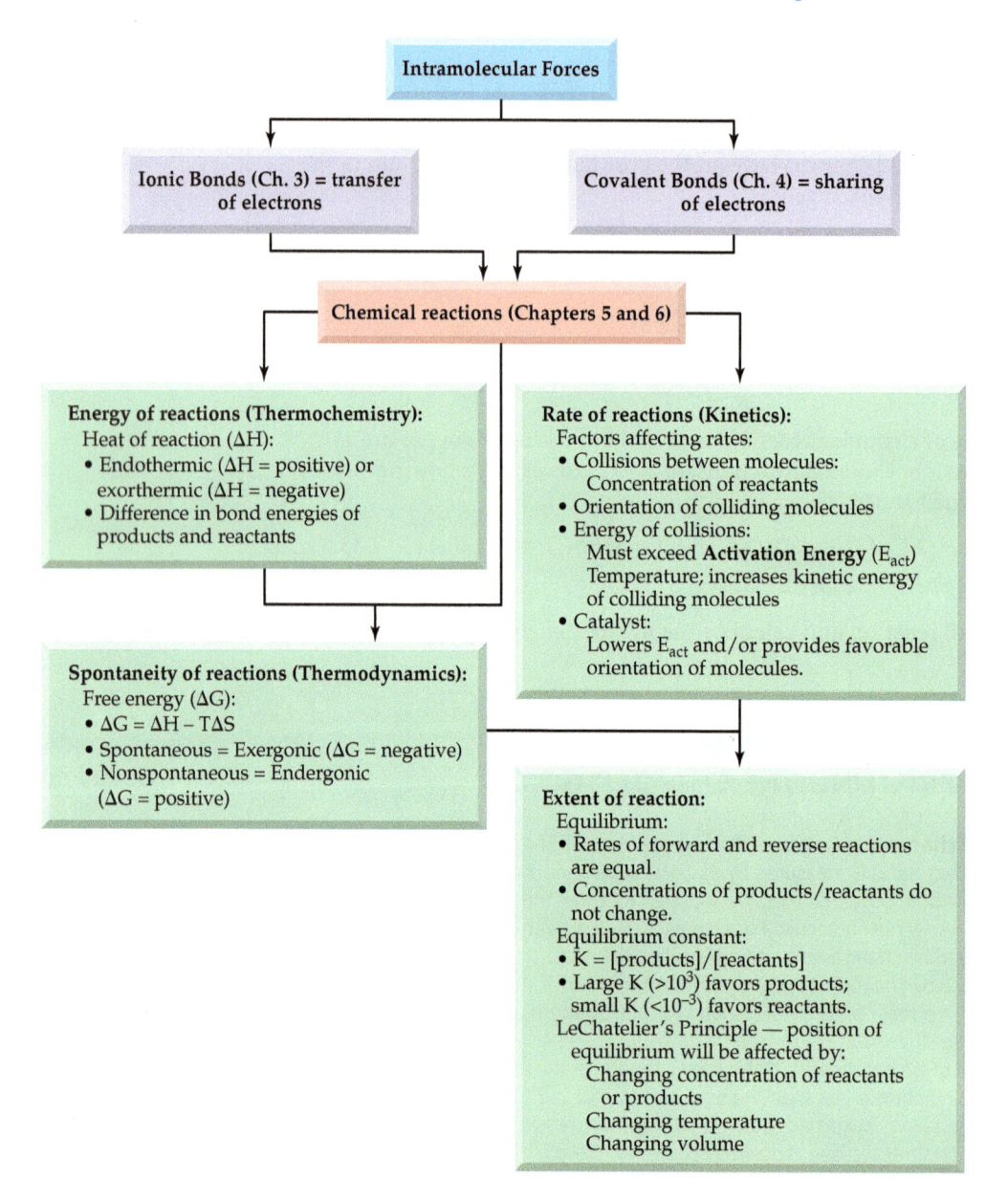

▶ **Figure 7.7**

Concept Map: We discussed the fundamentals of chemical reactions in Chapters 5 and 6. In Chapter 7 we looked at the heats of reaction, rates of reaction, spontaneity of reactions, and the extent of reaction as indicated by the equilibrium constant, *K*. These concepts, and the connections between them and previous concepts, are shown here in Figure 7.7.

KEY WORDS

Activation energy (E_{act}), p. 191
Bond dissociation energy, p. 180
Catalyst, p. 193
Chemical equilibrium, p. 196
Concentration, p. 193
Endergonic, p. 188
Endothermic, p. 180
Enthalpy (*H*), p. 181
Enthalpy change (ΔH), p. 181
Entropy (*S*), p. 187
Entropy change (ΔS), p. 187
Equilibrium constant (*K*), p. 197
Exergonic, p. 188
Exothermic, p. 180
Free-energy change (ΔG), p. 188

Heat, *p. 179*
Heat of reaction, *p. 181*
Kinetic energy, *p. 179*
Law of conservation of energy, *p. 181*
Le Châtelier's principle, *p. 200*
Potential energy, *p. 179*
Reaction rate, *p. 191*
Reversible reaction, *p. 196*
Spontaneous process, *p. 186*

UNDERSTANDING KEY CONCEPTS

7.20 What are the signs of ΔH, ΔS, and ΔG for the spontaneous conversion of a crystalline solid into a gas? Explain.

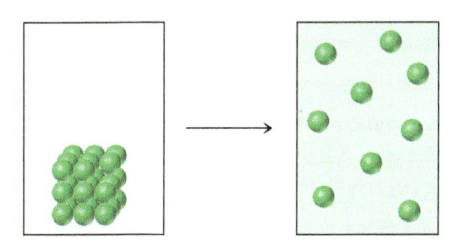

7.21 What are the signs of ΔH, ΔS, and ΔG for the spontaneous condensation of a vapor to a liquid? Explain.

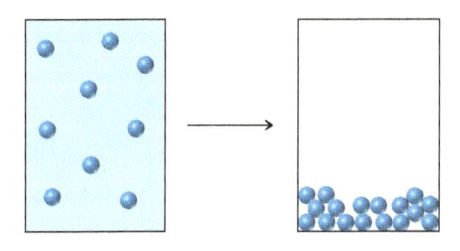

7.22 Consider the following spontaneous reaction of A_2 molecules (red) and B_2 molecules (blue):

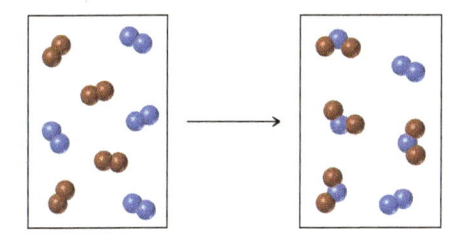

(a) Write a balanced equation for the reaction.
(b) What are the signs of ΔH, ΔS, and ΔG for the reaction? Explain.

7.23 Two curves are shown in the following energy diagram:

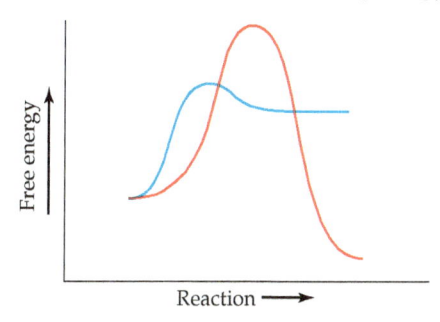

(a) Which curve represents the faster reaction, and which the slower?
(b) Which curve represents the spontaneous reaction, and which the nonspontaneous?

7.24 Draw energy diagrams for the following situations:

(a) A slow reaction with a large negative ΔG
(b) A fast reaction with a small positive ΔG

7.25 The following diagram portrays a reaction of the type $A(s) \longrightarrow B(g) + C(g)$, where the different colored spheres represent different molecular structures. Assume that the reaction has $\Delta H = +9.1$ kcal/mol ($+38.1$ kJ/mol).

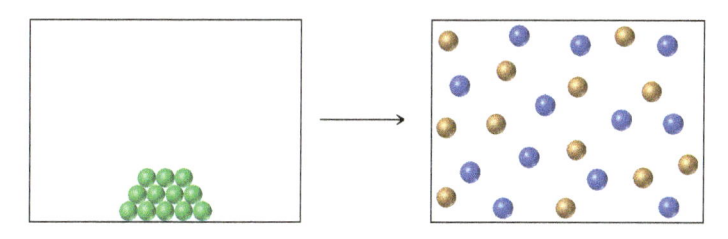

(a) What is the sign of ΔS for the reaction?
(b) Is the reaction likely to be spontaneous at all temperatures, nonspontaneous at all temperatures, or spontaneous at some but nonspontaneous at others?

ADDITIONAL PROBLEMS

ENTHALPY AND HEAT OF REACTION

7.26 Is the total enthalpy (*H*) of the reactants for an endothermic reaction greater than or less than the total enthalpy of the products?

7.27 What is meant by the term *heat of reaction*? What other name is a synonym for this term?

7.28 The vaporization of Br_2 from the liquid to the gas state requires 7.4 kcal/mol (31.0 kJ/mol).

(a) What is the sign of ΔH for this process? Write a reaction showing heat as a product or reactant.

(b) How many kilocalories are needed to vaporize 5.8 mol of Br_2?

(c) How many kilojoules are needed to evaporate 82 g of Br_2?

7.29 Converting liquid water to solid ice releases 1.44 kcal/mol (6.02 kJ/mol).

(a) What is the sign of ΔH for this process? Write a reaction showing heat as a product or reactant.

(b) How many kilojoules are released by freezing 2.5 mol of H_2O?

(c) How many kilocalories are released by freezing 32 g of H_2O?

(d) How many kilocalories are absorbed by melting 1 mol of ice?

7.30 Acetylene (H—C≡C—H) is the fuel used in welding torches.

(a) Write the balanced chemical equation for the combustion reaction of 1 mol of acetylene with $O_2(g)$ to produce $CO_2(g)$ and water vapor.

(b) Estimate ΔH for this reaction (in kJ/mol) using the bond energies listed in Table 7.1.

(c) Calculate the energy value (in kJ/g) for acetylene. How does it compare to the energy values for other fuels in Table 7.2?

7.31 Nitrogen in air reacts at high temperatures to form NO_2 according to the following reaction: $N_2 + 2 O_2 \longrightarrow 2 NO_2$

(a) Draw structures for the reactant and product molecules indicating single, double, and triple bonds.

(b) Estimate ΔH for this reaction (in kcal and kJ) using the bond energies from Table 7.1.

7.32 Glucose, also known as "blood sugar" when measured in blood, has the formula $C_6H_{12}O_6$.

(a) Write the equation for the combustion of glucose with O_2 to give CO_2 and H_2O.

(b) If 3.8 kcal (16 kJ) is released by combustion of each gram of glucose, how many kilocalories are released by the combustion of 1.50 mol of glucose? How many kilojoules?

(c) What is the minimum amount of energy (in kJ) a plant must absorb to produce 15.0 g of glucose?

7.33 During the combustion of 5.00 g of octane, C_8H_{18}, 239.5 kcal (1002 kJ) is released.

(a) Write a balanced equation for the combustion reaction.

(b) What is the sign of ΔH for this reaction?

(c) How much energy (in kJ) is released by the combustion of 1.00 mol of C_8H_{18}?

(d) How many grams and how many moles of octane must be burned to release 450.0 kcal?

(e) How many kilojoules are released by the combustion of 17.0 g of C_8H_{18}?

ENTROPY AND FREE ENERGY

7.34 Which of the following processes results in an increase in entropy of the system?

(a) A drop of ink spreading out when it is placed in water

(b) Steam condensing into drops on windows

(c) Constructing a building from loose bricks

7.35 For each of the following processes, specify whether entropy increases or decreases. Explain each of your answers.

(a) Assembling a jigsaw puzzle

(b) $I_2(s) + 3 F_2(g) \longrightarrow 2 IF_3(g)$

(c) A precipitate forming when two solutions are mixed

(d) $C_6H_{12}O_6(aq) + 6 O_2(g) \longrightarrow 6 CO_2(g) + 6 H_2O(g)$

(e) $CaCO_3(s) \longrightarrow CaO(s) + CO_2(g)$

(f) $Pb(NO_3)_2(aq) + 2 NaCl(aq) \longrightarrow PbCl_2(s) + 2 NaNO_3(aq)$

7.36 What two factors affect the spontaneity of a reaction?

7.37 What is the difference between an exothermic reaction and an exergonic reaction?

7.38 Why are most spontaneous reactions exothermic?

7.39 Under what conditions might a reaction be endothermic, but exergonic? Explain.

7.40 For the reaction

$$NaCl(s) \xrightarrow{Water} Na^+(aq) + Cl^-(aq),$$
$$\Delta H = +1.00 \text{ kcal/mol } (+4.184 \text{ kJ/mol})$$

(a) Is this process endothermic or exothermic?

(b) Does entropy increase or decrease in this process?

(c) Table salt (NaCl) readily dissolves in water. Explain, based on your answers to parts (a) and (b).

7.41 For the reaction $2 Hg(l) + O_2(g) \longrightarrow 2 HgO(s)$,

$$\Delta H = -43 \text{ kcal/mol } (-180 \text{ kJ/mol}).$$

(a) Does entropy increase or decrease in this process? Explain.

(b) Under what conditions would you expect this process to be spontaneous?

7.42 The reaction of gaseous H_2 and liquid Br_2 to give gaseous HBr has $\Delta H = -17.4$ kcal/mol (-72.8 kJ/mol) and $\Delta S = 27.2$ cal/(mol·K)(114 J/(mol·K)).

(a) Write the balanced equation for this reaction.

(b) Does entropy increase or decrease in this process?

(c) Is this process spontaneous at all temperatures? Explain.

(d) What is the value of ΔG (in kcal and kJ) for the reaction at 300 K?

7.43 The following reaction is used in the industrial synthesis of PVC polymer:

$$Cl_2(g) + H_2C=CH_2(g) \longrightarrow ClCH_2CH_2Cl(l) \quad \Delta H = -52 \text{ kcal/mol } (-218 \text{ kJ/mol})$$

(a) Is ΔS positive or negative for this process?

(b) Is this process spontaneous at all temperatures? Explain.

RATES OF CHEMICAL REACTIONS

7.44 What is the activation energy of a reaction?

7.45 Which reaction is faster, one with $E_{act} = +10$ kcal/mol (+41.8 kJ/mol) or one with $E_{act} = +5$ kcal/mol (+20.9 kJ/mol)? Explain.

7.46 Draw energy diagrams for exergonic reactions that meet the following descriptions:

(a) A slow reaction that has a small free-energy change

(b) A fast reaction that has a large free-energy change

7.47 Why does increasing concentration generally increase the rate of a reaction?

7.48 What is a catalyst, and what effect does it have on the activation energy of a reaction?

7.49 If a catalyst changes the activation energy of a forward reaction from 28.0 kcal/mol to 23.0 kcal/mol, what effect does it have on the reverse reaction?

7.50 For the reaction C(s, diamond) ⟶ C(s, graphite),

$$\Delta G = -0.693 \text{ kcal/mol} (-2.90 \text{ kJ/mol}) \text{ at } 25 \,°C.$$

(a) According to this information, do diamonds spontaneously turn into graphite?

(b) In light of your answer to part (a), why can diamonds be kept unchanged for thousands of years?

7.51 The reaction between hydrogen gas and carbon to produce the gas known as ethylene is

$$2 \text{ H}_2(g) + 2 \text{ C}(s) \longrightarrow \text{H}_2\text{C}=\text{CH}_2(g),$$
$$\Delta G = +16.3 \text{ kcal/mol} (+68.2 \text{ kJ/mol}) \text{ at } 25 \,°C.$$

(a) Is this reaction spontaneous at 25 °C?

(b) Would it be reasonable to try to develop a catalyst for the reaction run at 25 °C? Explain.

CHEMICAL EQUILIBRIA

7.52 What is meant by the term "chemical equilibrium"? Must amounts of reactants and products be equal at equilibrium?

7.53 Why do catalysts not alter the amounts of reactants and products present at equilibrium?

7.54 Write the equilibrium constant expressions for the following reactions:

(a) $2 \text{ CO}(g) + \text{O}_2(g) \rightleftharpoons 2 \text{ CO}_2(g)$

(b) $\text{Mg}(s) + \text{HCl}(aq) \rightleftharpoons \text{MgCl}_2(aq) + \text{H}_2(g)$

(c) $\text{HF}(aq) + \text{H}_2\text{O}(l) \rightleftharpoons \text{H}_3\text{O}^+(aq) + \text{F}^-(aq)$

(d) $\text{S}(s) + \text{O}_2(g) \rightleftharpoons \text{SO}_2(g)$

7.55 Write the equilibrium constant expressions for the following reactions.

(a) $\text{S}_2(g) + 2 \text{ H}_2(g) \rightleftharpoons 2 \text{ H}_2\text{S}(g)$

(b) $\text{H}_2\text{S}(aq) + \text{Cl}_2(aq) \rightleftharpoons \text{S}(s) + 2 \text{ HCl}(aq)$

(c) $\text{Br}_2(g) + \text{Cl}_2(g) \rightleftharpoons 2 \text{ BrCl}(g)$

(d) $\text{C}(s) + \text{H}_2\text{O}(g) \rightleftharpoons \text{CO}(g) + \text{H}_2(g)$

7.56 For the reaction $\text{N}_2\text{O}_4(g) \rightleftharpoons 2 \text{ NO}_2(g)$, the equilibrium concentrations at 25 °C are $[\text{NO}_2] = 0.0325$ mol/L and $[\text{N}_2\text{O}_4] = 0.147$ mol/L.

(a) What is the value of K at 25 °C? Are reactants or products favored?

7.57 For the reaction $2 \text{ CO}(g) + \text{O}_2(g) \rightleftharpoons 2 \text{ CO}_2(g)$, the equilibrium concentrations at a certain temperature are $[\text{CO}_2] = 0.11$ mol/L, $[\text{O}_2] = 0.015$ mol/L, $[\text{CO}] = 0.025$ mol/L.

(a) Write the equilibrium constant expression for the reaction.

(b) What is the value of K at this temperature? Are reactants or products favored?

7.58 Use your answer from Problem 7.56 to calculate the following:

(a) $[\text{N}_2\text{O}_4]$ at equilibrium when $[\text{NO}_2] = 0.0250$ mol/L

(b) $[\text{NO}_2]$ at equilibrium when $[\text{N}_2\text{O}_4] = 0.0750$ mol/L

7.59 Use your answer from Problem 7.57 to calculate the following:

(a) $[\text{O}_2]$ at equilibrium when $[\text{CO}_2] = 0.18$ mol/L and $[\text{CO}] = 0.0200$ mol/L

(b) $[\text{CO}_2]$ at equilibrium when $[\text{CO}] = 0.080$ mol/L and $[\text{O}_2] = 0.520$ mol/L

7.60 Would you expect to find relatively more reactants or more products for the reaction in Problem 7.56 if the pressure is raised? Explain.

7.61 Would you expect to find relatively more reactants or more products for the reaction in Problem 7.57 if the pressure is lowered?

LE CHÂTELIER'S PRINCIPLE

7.62 Oxygen can be converted into ozone by the action of lightning or electric sparks:

$$3 \text{ O}_2(g) \rightleftharpoons 2 \text{ O}_3(g)$$

For this reaction, $\Delta H = +68$ kcal/mol $(+285$ kJ/mol$)$ and $K = 2.68 \times 10^{-29}$ at 25 °C.

(a) Is the reaction exothermic or endothermic?

(b) Are the reactants or the products favored at equilibrium?

(c) Explain the effect on the equilibrium of

(1) increasing pressure by decreasing volume.
(2) increasing the concentration of $\text{O}_2(g)$.
(3) increasing the concentration of $\text{O}_3(g)$.
(4) adding a catalyst.
(5) increasing the temperature.

7.63 Hydrogen chloride can be made from the reaction of chlorine and hydrogen:

$$\text{Cl}_2(g) + \text{H}_2(g) \longrightarrow 2 \text{ HCl}(g)$$

For this reaction, $K = 26 \times 10^{33}$ and $\Delta H = -44$ kcal/mol $(-184$ kJ/mol$)$ at 25 °C.

(a) Is the reaction endothermic or exothermic?

(b) Are the reactants or the products favored at equilibrium?

(c) Explain the effect on the equilibrium of
 (1) Increasing pressure by decreasing volume
 (2) Increasing the concentration of $HCl(g)$
 (3) Decreasing the concentration of $Cl_2(g)$
 (4) Increasing the concentration of $H_2(g)$
 (5) Adding a catalyst

7.64 When the following equilibria are disturbed by increasing the pressure, does the concentration of reaction products increase, decrease, or remain the same?
 (a) $2\,CO_2(g) \rightleftharpoons 2\,CO(g) + O_2(g)$
 (b) $N_2(g) + O_2(g) \rightleftharpoons 2\,NO(g)$
 (c) $Si(s) + 2\,Cl_2(g) \rightleftharpoons SiCl_4(g)$

7.65 For the following equilibria, use Le Châtelier's principle to predict the direction of the reaction when the pressure is increased by decreasing the volume of the equilibrium mixture.
 (a) $C(s) + H_2O(g) \rightleftharpoons CO(g) + H_2(g)$
 (b) $2\,H_2(g) + O_2(g) \rightleftharpoons 2\,H_2O(g)$
 (c) $2\,Fe(s) + 3\,H_2O(g) \rightleftharpoons Fe_2O_3(s) + 3\,H_2(g)$

7.66 The reaction $CO(g) + H_2O(g) \rightleftharpoons CO_2(g) + H_2(g)$ has $\Delta H = -9.8$ kcal/mol (-41 kJ/mol). Does the amount of H_2 in an equilibrium mixture increase or decrease when the temperature is decreased?

7.67 The reaction $3\,O_2(g) \rightleftharpoons 2\,O_3(g)$ has $\Delta H = +68$ kcal/mol ($+285$ kJ/mol). Does the equilibrium constant for the reaction increase or decrease when the temperature increases?

7.68 The reaction $H_2(g) + I_2(g) \rightleftharpoons 2\,HI(g)$ has $\Delta H = -2.2$ kcal/mol (-9.2 kJ/mol). Will the equilibrium concentration of HI increase or decrease when
 (a) I_2 is added?
 (b) H_2 is removed?
 (c) a catalyst is added?
 (d) the temperature is increased?

7.69 The reaction $Fe^{3+}(aq) + Cl^-(aq) \rightleftharpoons FeCl^{2+}(aq)$ is endothermic. How will the equilibrium concentration of $FeCl^{2+}$ change when
 (a) $Fe(NO_3)_3$ is added?
 (b) Cl^- is precipitated by addition of $AgNO_3$?
 (c) the temperature is increased?
 (d) a catalyst is added?

CHEMISTRY IN ACTION

7.70 Which provides more energy, 1 g of carbohydrate or 1 g of fat? [*Energy from Food, p. 185*]

7.71 How many Calories (that is, kilocalories) are in a 45.0 g serving of potato chips if we assume that they are essentially 50% carbohydrate and 50% fats? [*Energy from Food, p. 185*]

7.72 Which body organs help to regulate body temperature? [*Regulation of Body Temperature, p. 195*]

7.73 What is the purpose of blood vessel dilation? [*Regulation of Body Temperature, p. 195*]

7.74 The ATP required for the production of glucose-6-phosphate is regenerated by another coupled reaction:

$ADP + HOPO_3^{2-} \longrightarrow ATP + H_2O \quad \Delta G = +30.5$ kJ/mol
Phosphoenolpyruvate $+ H_2O \longrightarrow$ pyruvate $+ HOPO_3^{2-}$
$\Delta G = -61.9$ kJ/mol

Derive the net reaction and calculate ΔG for the coupled reaction. [*Coupled Reactions, p. 204*]

7.75 The coupling of reactions in the smelting of copper at elevated temperatures yields an overall reaction that is energetically favorable. Why is the use of elevated temperature not feasible for most living organisms, and what other strategies do they use to make reactions occur at normal body temperatures? [*Coupled Reactions, p. 204*]

GENERAL QUESTIONS AND PROBLEMS

7.76 For the unbalanced combustion reaction shown below, 1 mol of ethanol, C_2H_5OH, releases 327 kcal (1370 kJ).

$$C_2H_5OH + O_2 \longrightarrow CO_2 + H_2O$$

 (a) Write a balanced equation for the combustion reaction.
 (b) What is the sign of ΔH for this reaction?
 (c) How much heat (in kilocalories) is released from the combustion of 5.00 g of ethanol?
 (d) How many grams of C_2H_5OH must be burned to raise the temperature of 500.0 mL of water from 20.0 °C to 100.0 °C? (The specific heat of water is 1.00 cal/g·°C or 4.184 J/g·°C. See Section 1.13.)
 (e) If the density of ethanol is 0.789 g/mL, calculate the combustion energy of ethanol in kilocalories/milliliter and kilojoules/milliliter

7.77 For the production of ammonia from its elements, $\Delta H = -22$ kcal/mol (-92 kJ/mol).
 (a) Is this process endothermic or exothermic?
 (b) How much energy (in kilocalories and kilojoules) is involved in the production of 0.700 mol of NH_3?

7.78 Magnetite, an iron ore with formula Fe_3O_4, can be reduced by treatment with hydrogen to yield iron metal and water vapor.
 (a) Write the balanced equation.
 (b) This process requires 36 kcal (151 kJ) for every 1.00 mol of Fe_3O_4 reduced. How much energy (in kilocalories and kilojoules) is required to produce 55 g of iron?
 (c) How many grams of hydrogen are needed to produce 75 g of iron?
 (d) This reaction has $K = 2.3 \times 10^{-18}$. Are the reactants or the products favored?

Additional Problems

7.79 Hemoglobin (Hb) reacts reversibly with O_2 to form HbO_2, a substance that transfers oxygen to tissues:

$$Hb(aq) + O_2(aq) \rightleftharpoons HbO_2(aq)$$

Carbon monoxide (CO) is attracted to Hb 140 times more strongly than O_2 and establishes another equilibrium.

(a) Explain, using Le Châtelier's principle, why inhalation of CO can cause weakening and eventual death.

(b) Still another equilibrium is established when both O_2 and CO are present:

$$Hb(CO)(aq) + O_2(aq) \rightleftharpoons HbO_2(aq) + CO(aq)$$

Explain, using Le Châtelier's principle, why pure oxygen is often administered to victims of CO poisoning.

7.80 Urea is a metabolic waste product that decomposes to ammonia and water according to the following reaction:

$$NH_2CONH_2 + H_2O \longrightarrow 2\,NH_3 + CO_2.$$

(a) Draw the Lewis structure for urea.

(b) Estimate ΔH (in kcal and kJ) for this reaction using the bond energies from Table 7.1.

7.81 For the evaporation of water, $H_2O(l) \longrightarrow H_2O(g)$, at 100 °C, $\Delta H = +9.72$ kcal/mol ($+40.7$ kJ/mol).

(a) How many kilocalories are needed to vaporize 10.0 g of $H_2O(l)$?

(b) How many kilojoules are released when 10.0 g of $H_2O(g)$ is condensed?

7.82 Ammonia reacts slowly in air to produce nitrogen monoxide and water vapor:

$$NH_3(g) + O_2(g) \rightleftharpoons NO(g) + H_2O(g) + Heat$$

(a) Balance the equation.

(b) Write the equilibrium equation.

(c) Explain the effect on the equilibrium of
  (1) raising the pressure.
  (2) adding NO(g).
  (3) decreasing the concentration of NH_3.
  (4) lowering the temperature.

7.83 Methanol, CH_3OH, is used as race car fuel.

(a) Write the balanced equation for the combustion reaction of methanol with O_2 to form CO_2 and H_2O.

(b) $\Delta H = -174$ kcal/mol (-728 kJ/mol) methanol for the process. How many kilocalories are released by burning 1.85 mol of methanol?

(c) How many kilojoules are released by burning 50.0 g of methanol?

7.84 Sketch an energy diagram for a system in which the forward reaction has $E_{act} = +25$ kcal/mol ($+105$ kJ/mol) and the reverse reaction has $E_{act} = +35$ kcal/mol ($+146$ kJ/mol).

(a) Is the forward process endergonic or exergonic?

(b) What is the value of ΔG for the reaction?

7.85 The thermite reaction (photograph, p. 181), in which aluminum metal reacts with iron(III) oxide to produce a spectacular display of sparks, is so exothermic that the product (iron) is in the molten state:

$$2\,Al(s) + Fe_2O_3(s) \longrightarrow 2\,Al_2O_3(s) + 2\,Fe(l)$$
$$\Delta H = -202.9 \text{ kcal/mol } (-848.9 \text{ kJ/mol})$$

(a) How much heat is released (in kilojoules) when 0.255 mol of Al is used in this reaction?

(b) How much heat (in kilocalories) is released when 5.00 g of Al is used in the reaction?

7.86 How much heat (in kilocalories) is evolved or absorbed in the reaction of 1.00 g of Na with H_2O? Is the reaction exothermic or endothermic?

$$2\,Na(s) + 2\,H_2O(l) \longrightarrow 2\,NaOH(aq) + H_2(g)$$
$$\Delta H = -88.0 \text{ kcal/mol } (-368 \text{ kJ/mol})$$

CHAPTER 8

Gases, Liquids, and Solids

CONTENTS

8.1 States of Matter and Their Changes
8.2 Intermolecular Forces
8.3 Gases and the Kinetic–Molecular Theory
8.4 Pressure
8.5 Boyle's Law: The Relation between Volume and Pressure
8.6 Charles's Law: The Relation between Volume and Temperature
8.7 Gay-Lussac's Law: The Relation between Pressure and Temperature
8.8 The Combined Gas Law
8.9 Avogadro's Law: The Relation between Volume and Molar Amount
8.10 The Ideal Gas Law
8.11 Partial Pressure and Dalton's Law
8.12 Liquids
8.13 Water: A Unique Liquid
8.14 Solids
8.15 Changes of State

◀ This winter scene in Yellowstone National Park shows the three states of matter for water—solid (snow/ice), liquid (water), and gas (steam/water vapor)—all present at the same time.

CHAPTER GOALS

1. **What are the major intermolecular forces, and how do they affect the states of matter?**
   THE GOAL: Be able to explain dipole–dipole forces, London dispersion forces, and hydrogen bonding, recognize which of these forces affect a given molecule, and understand how these forces are related to the physical properties of a substance. (◀◀ B.)

2. **How do scientists explain the behavior of gases?**
   THE GOAL: Be able to state the assumptions of the kinetic–molecular theory and use these assumptions to explain the behavior of gases. (◀◀ B.)

3. **How do gases respond to changes in temperature, pressure, and volume?**
   THE GOAL: Be able to use Boyle's law, Charles's law, Gay-Lussac's law, and Avogadro's law to explain the effect on gases of a change in pressure, volume, or temperature.

4. **What is the ideal gas law?**
   THE GOAL: Be able to use the ideal gas law to find the pressure, volume, temperature, or molar amount of a gas sample.

5. **What is partial pressure?**
   THE GOAL: Be able to define partial pressure and use Dalton's law of partial pressures.

6. **What are the various kinds of solids, and how do they differ?**
   THE GOAL: Be able to recognize the different kinds of solids and describe their characteristics. (◀◀ A., B.)

7. **What factors affect a change of state?**
   THE GOAL: Be able to apply the concepts of heat change, equilibrium, vapor pressure, and intermolecular forces to changes of state. (◀◀ A., B., C.)

CONCEPTS TO REVIEW

A. Ionic Bonds
(Section 3.3)

B. Polar Covalent Bonds and Polar Molecules
(Sections 4.9 and 4.10)

C. Enthalpy, Entropy, and Free Energy
(Sections 7.2–7.4)

The previous seven chapters dealt with matter at the atomic level. We have seen that all matter is composed of atoms, ions, or molecules; that these particles are in constant motion; that atoms combine to make compounds using chemical bonds; and that physical and chemical changes are accompanied by the release or absorption of energy. Now it is time to look at a different aspect of matter, concentrating not on the properties and small-scale behavior of individual atoms but on the properties and large-scale behavior of visible amounts of matter and the factors that affect those properties.

8.1 States of Matter and Their Changes

Matter exists in any of three phases, or *states*—solid, liquid, or gas. The state in which a compound exists under a given set of conditions depends on the relative strength of the attractive forces between particles compared to the kinetic energy of the particles. Kinetic energy (Section 7.1) is energy associated with motion and is related to the temperature of the substance. In gases, the attractive forces between particles are very weak compared to their kinetic energy, so the particles move about freely, are far apart, and have almost no influence on one another. In liquids, the attractive forces between particles are stronger, pulling the particles close together but still allowing them considerable freedom to move about. In solids, the attractive forces are much stronger than the kinetic energy of the particles, so the atoms, molecules, or ions are held in a specific arrangement and can only wiggle around in place (Figure 8.1).

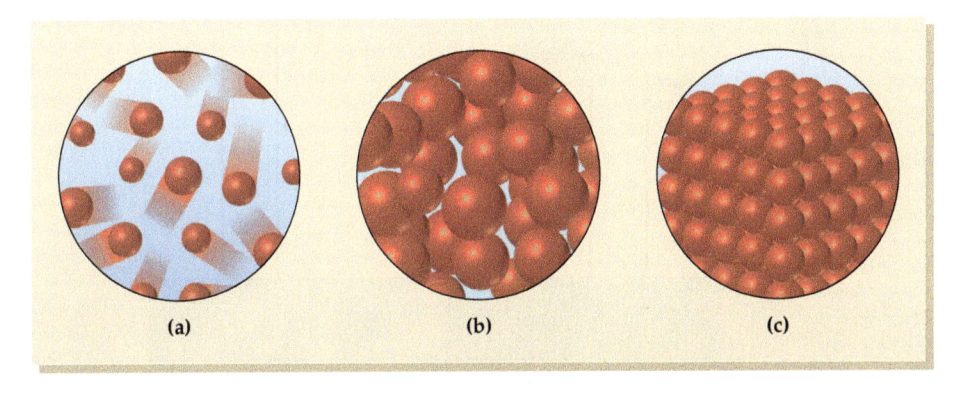

◀ **Figure 8.1**
A molecular comparison of gases, liquids, and solids.
(a) In gases, the particles feel little attraction for one another and are free to move about randomly. (b) In liquids, the particles are held close together by attractive forces but are free to slide over one another. (c) In solids, the particles are strongly attracted to one another. They can move slightly, but are held in a fairly rigid arrangement with respect to one another.

214 CHAPTER 8 Gases, Liquids, and Solids

Change of state The change of a substance from one state of matter (gas, liquid, or solid) to another.

▶▶▶ You might want to reread Section 7.4 to brush up on these concepts.

The transformation of a substance from one state to another is called a *phase change*, or a **change of state**. Every change of state is reversible and, like all chemical and physical processes, is characterized by a free-energy change, ΔG. A change of state that is spontaneous in one direction (exergonic, negative ΔG) is nonspontaneous in the other direction (endergonic, positive ΔG). As always, the free-energy change ΔG has both an enthalpy term ΔH and a temperature-dependent entropy term ΔS, according to the equation $\Delta G = \Delta H - T\Delta S$.

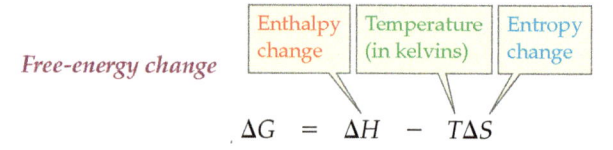

Free-energy change $\Delta G = \Delta H - T\Delta S$

(Enthalpy change, Temperature (in kelvins), Entropy change)

The enthalpy change ΔH is a measure of the heat absorbed or released during a given change of state. In the melting of a solid to a liquid, for example, heat is absorbed and ΔH is positive (endothermic). In the reverse process—the freezing of a liquid to a solid—heat is released and ΔH is negative (exothermic). Look at the change between ice and water for instance:

Melting: $H_2O(s) \longrightarrow H_2O(l)$ $\Delta H = +1.44$ kcal/mol or $+6.02$ kJ/mol
Freezing: $H_2O(l) \longrightarrow H_2O(s)$ $\Delta H = -1.44$ kcal/mol or -6.02 kJ/mol

The entropy change ΔS is a measure of the change in molecular disorder or freedom that occurs during a process. In the melting of a solid to a liquid, for example, disorder increases because particles gain freedom of motion, so ΔS is positive. In the reverse process—the freezing of a liquid to a solid—disorder decreases as particles are locked into position, so ΔS is negative. Look at the change between ice and water:

Melting: $H_2O(s) \longrightarrow H_2O(l)$ $\Delta S = +5.26$ cal/(mol·K) or $+22.0$ J/(mol·K)
Freezing: $H_2O(l) \longrightarrow H_2O(s)$ $\Delta S = -5.26$ cal/(mol·K) or -22.0 J/(mol·K)

As with all processes that are unfavored by one term in the free-energy equation but favored by the other, the sign of ΔG depends on the temperature (Section 7.4). The melting of ice, for instance, is unfavored by a positive ΔH but favored by a positive ΔS. Thus, at a low temperature, the unfavorable ΔH is larger than the favorable $T\Delta S$, so ΔG is positive and no melting occurs. At a higher temperature, however, $T\Delta S$ becomes larger than ΔH, so ΔG is negative and melting *does* occur. The exact temperature at

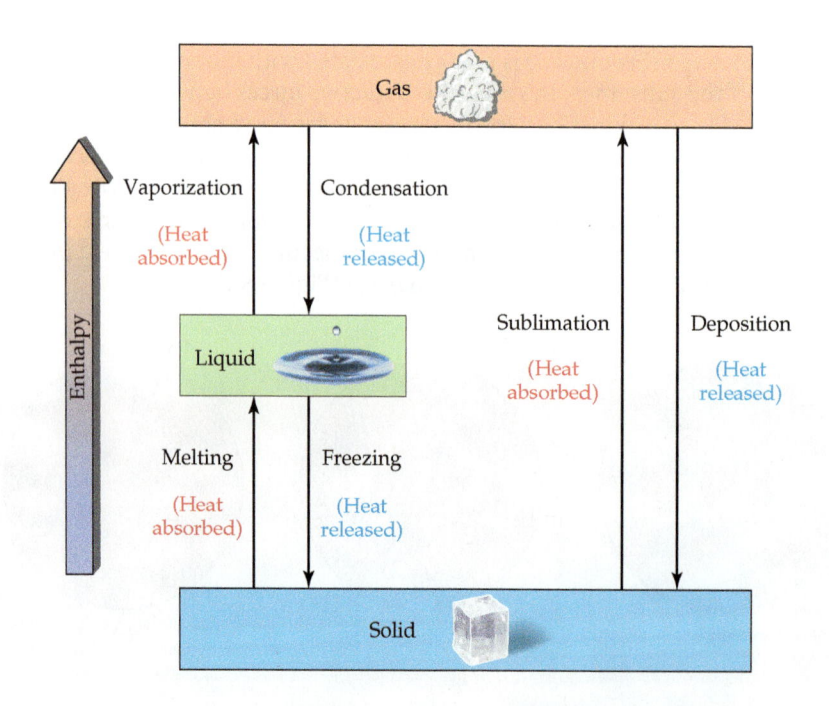

▶ **Figure 8.2**
Changes of state.
The changes are endothermic from bottom to top and exothermic from top to bottom. Solid and liquid states are in equilibrium at the melting point; liquid and gas states are in equilibrium at the boiling point.

Gas

Vaporization (Heat absorbed) | Condensation (Heat released)

Liquid

Sublimation (Heat absorbed) | Deposition (Heat released)

Melting (Heat absorbed) | Freezing (Heat released)

Solid

Enthalpy

which the changeover in behavior occurs is called the **melting point (mp)** and represents the temperature at which solid and liquid coexist in equilibrium. In the corresponding change from a liquid to a gas, the two states are in equilibrium at the **boiling point (bp)**.

The names and enthalpy changes associated with the different changes of state are summarized in Figure 8.2. Note that a solid can change directly to a gas without going through the liquid state—a process called *sublimation*. Dry ice (solid CO_2) at atmospheric pressure, for example, changes directly to a gas without melting.

Melting point (mp) The temperature at which solid and liquid are in equilibrium.

Boiling point (bp) The temperature at which liquid and gas are in equilibrium.

Worked Example 8.1 Change of State: Enthalpy, Entropy, and Free Energy

The change of state from liquid to gas for chloroform, formerly used as an anesthetic, has $\Delta H = +6.98$ kcal/mol ($+29.2$ kJ/mol) and a $\Delta S = +20.9$ cal/(mol·K) [$+87.4$ J/(mol·K)].

(a) Is the change of state from liquid to gas favored or unfavored by ΔH? by ΔS?
(b) Is the change of state from liquid to gas favored or unfavored at 35 °C?
(c) Is this change of state spontaneous at 65 °C?

ANALYSIS A process will be favored if energy is released ($\Delta H =$ negative) and if there is a decrease in disorder ($\Delta S =$ positive). In cases in which one factor is favorable and the other is unfavorable, then we can calculate the free-energy change to determine if the process is favored:

$$\Delta G = \Delta H - T\Delta S$$

When ΔG is negative, the process is favored.

SOLUTION

(a) The ΔH does NOT favor this change of state ($\Delta H =$ positive), but the ΔS does favor the process. Since the two factors are not in agreement, we must use the equation for free-energy change to determine if the process is favored at a given temperature.

(b) Substituting the values for ΔH and ΔS into the equation for free-energy change we can determine if ΔG is positive or negative at 35 °C (308 K). Note that we must first convert degrees celsius to kelvins and convert the ΔS from cal to kcal so the units can be added together.

$$\Delta G = \Delta H - T\Delta S = \left(\frac{6.98 \text{ kcal}}{\text{mol}}\right) - (308 \text{ K})\left(\frac{20.9 \text{ cal}}{\text{mol} \cdot \text{K}}\right)\left(\frac{1 \text{ kcal}}{1000 \text{ cal}}\right)$$

$$= 6.98 \frac{\text{kcal}}{\text{mol}} - 6.44 \frac{\text{kcal}}{\text{mol}} = +0.54 \frac{\text{kcal}}{\text{mol}}$$

$$\left(+0.54 \frac{\text{kcal}}{\text{mol}}\right)\left(\frac{4.184 \text{ kJ}}{\text{kcal}}\right) = +2.26 \frac{\text{kJ}}{\text{mol}}$$

Since the $\Delta G =$ positive, this change of state is not favored at 35 °C.

(c) Repeating the calculation using the equation for free-energy change at 65 °C (338 K):

$$\Delta G = \Delta H - T\Delta S = \left(\frac{6.98 \text{ kcal}}{\text{mol}}\right) - (338 \text{ K})\left(\frac{20.9 \text{ cal}}{\text{mol} \cdot \text{K}}\right)\left(\frac{1 \text{ kcal}}{1000 \text{ cal}}\right)$$

$$= 6.98 \frac{\text{kcal}}{\text{mol}} - 7.06 \frac{\text{kcal}}{\text{mol}} = -0.08 \frac{\text{kcal}}{\text{mol}} \left(\text{or } -0.33 \frac{\text{kJ}}{\text{mol}}\right)$$

Because ΔG is negative in this case, the change of state is favored at this temperature.

216 CHAPTER 8 Gases, Liquids, and Solids

> **PROBLEM 8.1**
> The change of state from liquid H_2O to gaseous H_2O has $\Delta H = +9.72$ kcal/mol $(+40.7$ kJ/mol$)$ and $\Delta S = -26.1$ cal/(mol·K)$[-109$ J/(mol·K)$]$.
> **(a)** Is the change from liquid to gaseous H_2O favored or unfavored by ΔH? By ΔS?
> **(b)** What is the value of ΔG (in kcal/mol and kJ/mol) for the change from liquid to gaseous H_2O at 373 K?
> **(c)** What are the values of ΔH and ΔS (in kcal/mol and kJ/mol) for the change from gaseous to liquid H_2O?

8.2 Intermolecular Forces

What determines whether a substance is a gas, a liquid, or a solid at a given temperature? Why does rubbing alcohol evaporate much more readily than water? Why do molecular compounds have lower melting points than ionic compounds? To answer these and a great many other such questions, we need to look into the nature of **intermolecular forces**—the forces that act *between different molecules* rather than within an individual molecule.

In gases, the intermolecular forces are negligible, so the gas molecules act independently of one another. In liquids and solids, however, intermolecular forces are strong enough to hold the molecules in close contact. As a general rule, the stronger the intermolecular forces in a substance, the more difficult it is to separate the molecules, and the higher the melting and boiling points of the substance.

There are three major types of intermolecular forces: *dipole–dipole, London dispersion,* and *hydrogen bonding*. We will discuss each in turn.

Dipole–Dipole Forces

Many molecules contain polar covalent bonds and may therefore have a net molecular polarity. In such cases, the positive and negative ends of different molecules are attracted to one another by what is called a **dipole–dipole force** (Figure 8.3).

Dipole–dipole forces are weak, with strengths on the order of 1 kcal/mol (4 kJ/mol) compared to the 70–100 kcal/mol (300–400 kJ/mol) typically found for the strength of a covalent bond (see Table 7.1). Nevertheless, the effects of dipole–dipole forces are important, as can be seen by looking at the difference in boiling points between polar and nonpolar molecules. Butane, for instance, is a nonpolar molecule with a molecular weight of 58 amu and a boiling point of −0.5 °C, whereas acetone has the same molecular weight yet boils 57 °C higher because it is polar.

Intermolecular force A force that acts between molecules and holds molecules close to one another.

▶▶ Recall from Sections 4.9 and 4.10 that a polar covalent bond is one in which the electrons are attracted more strongly by one atom than by the other.

Dipole–dipole force The attractive force between positive and negative ends of polar molecules.

▶▶ Recall from Section 4.9 how molecular polarities can be visualized using electrostatic potential maps.

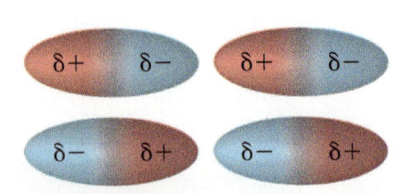

◀ **Figure 8.3**
Dipole–dipole forces.
The positive and negative ends of polar molecules are attracted to one another by dipole–dipole forces. As a result, polar molecules have higher boiling points than nonpolar molecules of similar size.

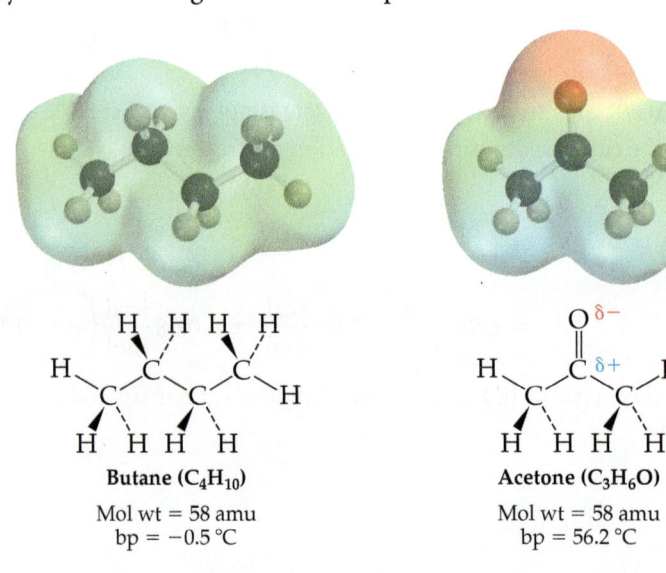

Butane (C_4H_{10})
Mol wt = 58 amu
bp = −0.5 °C

Acetone (C_3H_6O)
Mol wt = 58 amu
bp = 56.2 °C

London Dispersion Forces

Only polar molecules experience dipole–dipole forces, but all molecules, regardless of structure, experience *London dispersion forces*. **London dispersion forces** are caused by the constant motion of electrons within molecules. Take even a simple nonpolar molecule like Br_2, for example. Averaged over time, the distribution of electrons throughout the molecule is uniform, but at any given *instant* there may be more electrons at one end of the molecule than at the other (Figure 8.4). At that instant, the molecule has a short-lived polarity. Electrons in neighboring molecules are attracted to the positive end of the polarized molecule, resulting in a polarization of the neighbor and creation of an attractive London dispersion force that holds the molecules together. As a result, Br_2 is a liquid at room temperature rather than a gas.

London dispersion force The short-lived attractive force due to the constant motion of electrons within molecules.

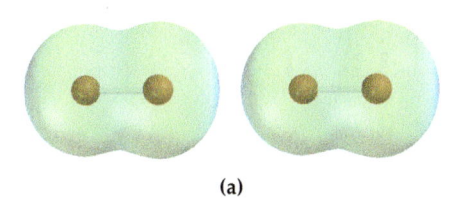

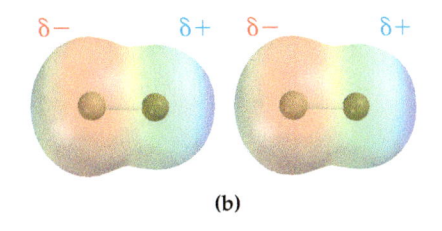

◀ **Figure 8.4**
(a) Averaged over time, the electron distribution in a Br_2 molecule is symmetrical. (b) At any given instant, however, the electron distribution may be unsymmetrical, resulting in a temporary polarity that induces a complementary polarity in neighboring molecules.

London dispersion forces are weak—in the range 0.5–2.5 kcal/mol (2–10 kJ/mol)—but they increase with molecular weight and amount of surface area available for interaction between molecules. The larger the molecular weight, the more electrons there are moving about and the greater the temporary polarization of a molecule. The larger the amount of surface contact, the greater the close interaction between different molecules.

The effect of surface area on the magnitude of London dispersion forces can be seen by comparing a roughly spherical molecule with a flatter, more linear one having the same molecular weight. Both 2,2-dimethylpropane and pentane, for instance, have the same formula (C_5H_{12}), but the nearly spherical shape of 2,2-dimethylpropane allows for less surface contact with neighboring molecules than does the more linear shape of pentane (Figure 8.5). As a result, London dispersion forces are smaller for 2,2-dimethylpropane, molecules are held together less tightly, and the boiling point is correspondingly lower: 9.5 °C for 2,2-dimethylpropane versus 36 °C for pentane.

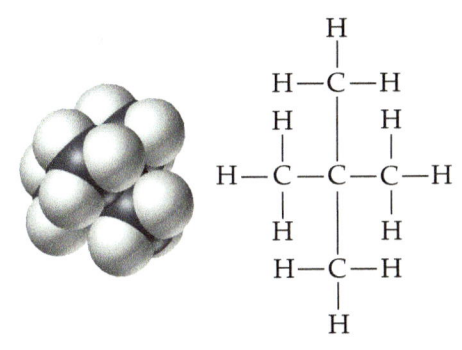

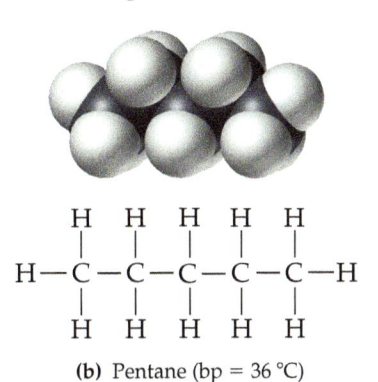

(a) 2,2-Dimethylpropane (bp = 9.5 °C)

(b) Pentane (bp = 36 °C)

◀ **Figure 8.5**
London dispersion forces. More compact molecules like 2,2-dimethylpropane have smaller surface areas, weaker London dispersion forces, and lower boiling points. By comparison, flatter, less compact molecules like pentane have larger surface areas, stronger London dispersion forces, and higher boiling points.

Hydrogen Bonds

In many ways, hydrogen bonding is responsible for life on earth. It causes water to be a liquid rather than a gas at ordinary temperatures, and it is the primary intermolecular force that holds huge biomolecules in the shapes needed to play their essential roles in biochemistry. Deoxyribonucleic acid (DNA) and keratin (Figure 8.6), for instance, are long molecular chains that form a α-helix, held in place largely due to hydrogen bonding.

A **hydrogen bond** is an attractive interaction between an unshared electron pair on an electronegative O, N, or F atom and a positively polarized hydrogen atom bonded to

Hydrogen bond The attraction between a hydrogen atom bonded to an electronegative O, N, or F atom and another nearby electronegative O, N, or F atom.

218 CHAPTER 8 Gases, Liquids, and Solids

▶ **Figure 8.6**
The α-helical structure of keratin results from hydrogen bonding along the amino acid backbone of the molecule. Hydrogen bonding is represented by gray dots in the ball and stick model on the left and red dots in the molecular structure on the right.

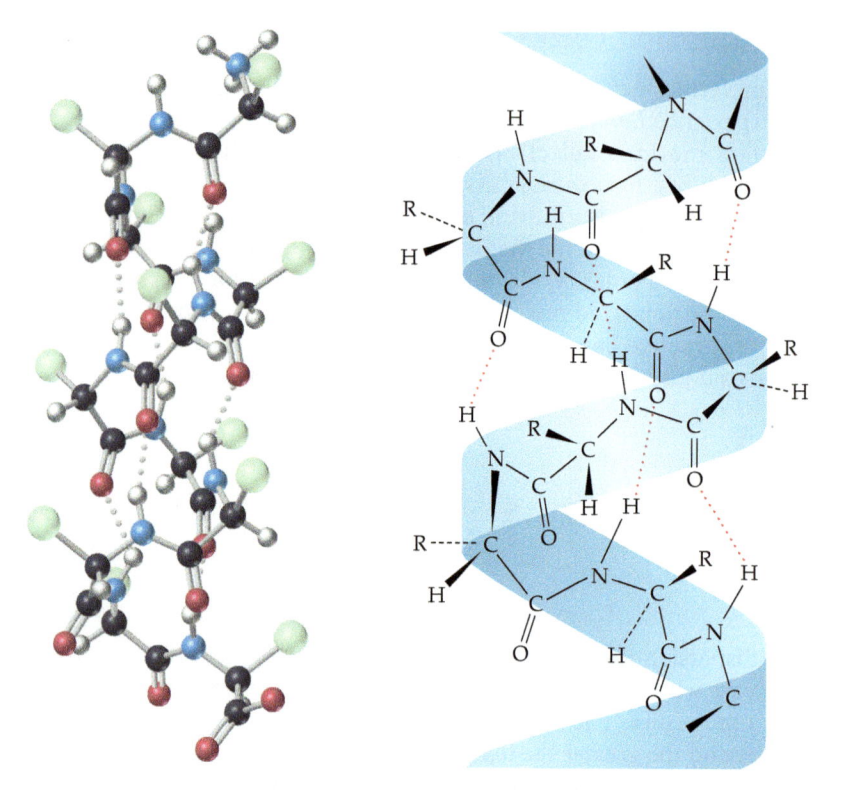

another electronegative O, N, or F. For example, hydrogen bonds occur in both water and ammonia:

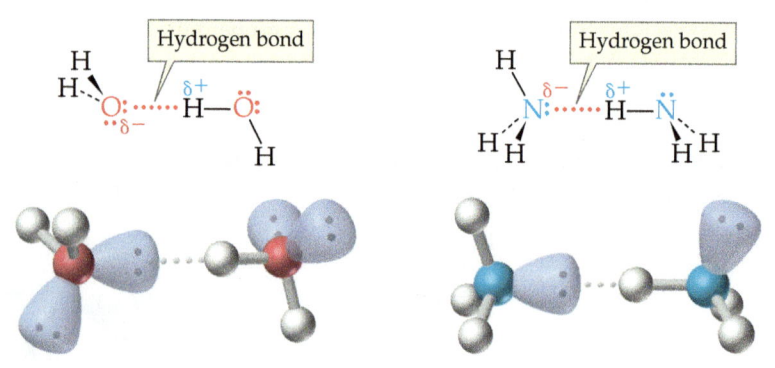

Hydrogen bonding is really just a special kind of dipole–dipole interaction. The O—H, N—H, and F—H bonds are highly polar, with a partial positive charge on the hydrogen and a partial negative charge on the electronegative atom. In addition, the hydrogen atom has no inner-shell electrons to act as a shield around its nucleus, and it is small, so it can be approached closely. As a result, the dipole–dipole attractions involving positively polarized hydrogens are unusually strong, and hydrogen bonds result. Water, in particular, is able to form a vast three-dimensional network of hydrogen bonds because each H_2O molecule has two hydrogens and two electron pairs (Figure 8.7).

▶ **Figure 8.7**
Hydrogen bonding in water.
The intermolecular attraction in water is especially strong because each oxygen atom has two lone pairs and two hydrogen atoms, allowing the formation of as many as four hydrogen bonds per molecule. Individual hydrogen bonds are constantly being formed and broken.

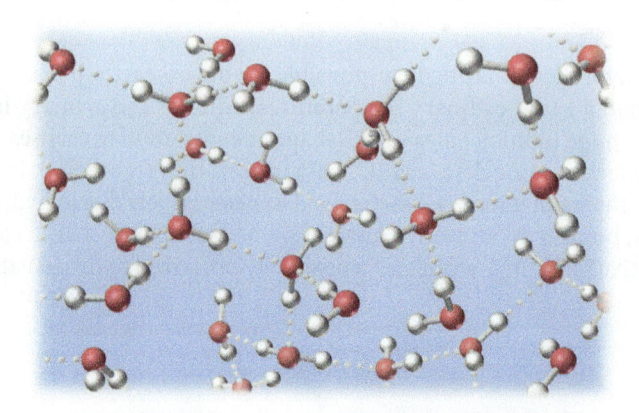

Hydrogen bonds can be quite strong, with energies up to 10 kcal/mol (40 kJ/mol). To see the effect of hydrogen bonding, look at Table 8.1, which compares the boiling points of binary hydrogen compounds of second-row elements with their third-row counterparts. Because NH_3, H_2O, and HF molecules are held tightly together by hydrogen bonds, an unusually large amount of energy must be added to separate them in the boiling process. As a result, the boiling points of NH_3, H_2O, and HF are much higher than the boiling points of their second-row neighbor CH_4 and of related third-row compounds.

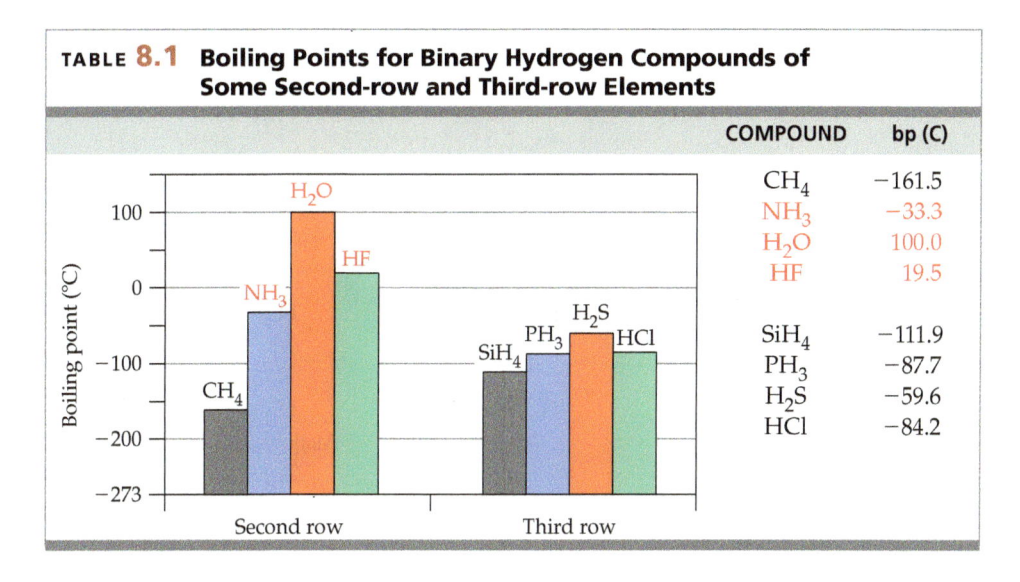

TABLE 8.1 Boiling Points for Binary Hydrogen Compounds of Some Second-row and Third-row Elements

COMPOUND	bp (C)
CH_4	−161.5
NH_3	−33.3
H_2O	100.0
HF	19.5
SiH_4	−111.9
PH_3	−87.7
H_2S	−59.6
HCl	−84.2

A summary and comparison of the various kinds of intermolecular forces is shown in Table 8.2.

TABLE 8.2 A Comparison of Intermolecular Forces

Force	Strength	Characteristics
Dipole–dipole	Weak (1 kcal/mol, 4 kJ/mol))	Occurs between polar molecules
London dispersion	Weak (0.5–2.5 kcal/mol, 2–10 kJ/mol)	Occurs between all molecules; strength depends on size
Hydrogen bond	Moderate (2–10 kcal/mol, 8–40 kJ/mol)	Occurs between molecules with O—H, N—H, and F—H bonds

LOOKING AHEAD ▶▶ Dipole–dipole forces, London dispersion forces, and hydrogen bonds are traditionally called "intermolecular forces" because of their influence on the properties of molecular compounds. But these same forces can also operate between different parts of a very large molecule. In this context, they are often referred to as "noncovalent interactions." In later chapters, we will see how noncovalent interactions determine the shapes of biologically important molecules such as proteins and nucleic acids.

Worked Example 8.2 Identifying Intermolecular Forces: Polar versus Nonpolar

Identify the intermolecular forces that influence the properties of the following compounds:
(a) Methane, CH_4 (b) HCl   (c) CH_3COOH

ANALYSIS The intermolecular forces will depend on the molecular structure, what type of bonds are in the molecule (polar or non-polar), and how the bonds are arranged.

SOLUTION

(a) Since methane contains only C—H bonds, it is a nonpolar molecule; it has only London dispersion forces.

(b) The H—Cl bond is polar, so this is a polar molecule; it has both dipole–dipole forces and London dispersion forces.

(c) Acetic acid is a polar molecule with an O—H bond. Thus, it has dipole–dipole forces, London dispersion forces, and hydrogen bonds.

PROBLEM 8.2
Would you expect the boiling points to increase or decrease in the following series? Explain.
(a) Kr, Ar, Ne   (b) Cl_2, Br_2, I_2

PROBLEM 8.3
Which of the following compounds form hydrogen bonds?

Methyl alcohol (a)   Ethylene (b)   Methylamine (c)

PROBLEM 8.4
Identify the intermolecular forces (dipole–dipole, London dispersion, hydrogen bonding) that influence the properties of the following compounds:
(a) Ethane, CH_3CH_3
(b) Ethyl alcohol, CH_3CH_2OH
(c) Ethyl chloride, CH_3CH_2Cl

8.3 Gases and the Kinetic–Molecular Theory

Gases behave quite differently from liquids and solids. Gases, for instance, have low densities and are easily compressed to a smaller volume when placed under pressure, a property that allows them to be stored in large tanks. Liquids and solids, by contrast, are much more dense and much less compressible. Furthermore, gases undergo a far larger expansion or contraction when their temperature is changed than do liquids and solids.

The behavior of gases can be explained by a group of assumptions known as the **kinetic–molecular theory of gases**. We will see in the next several sections how the following assumptions account for the observable properties of gases:

Kinetic–molecular theory of gases A group of assumptions that explain the behavior of gases.

- **A gas consists of many particles, either atoms or molecules, moving about at random with no attractive forces between them.** Because of this random motion, different gases mix together quickly.
- **The amount of space occupied by the gas particles themselves is much smaller than the amount of space between particles.** Most of the volume taken up by gases is empty space, accounting for the ease of compression and low densities of gases.
- **The average kinetic energy of gas particles is proportional to the Kelvin temperature.** Thus, gas particles have more kinetic energy and move faster as the temperature increases. (In fact, gas particles move much faster than you might suspect. The average speed of a helium atom at room temperature and atmospheric pressure is approximately 1.36 km/s, or 3000 mi/hr, nearly that of a rifle bullet.)

- **Collisions of gas particles, either with other particles or with the wall of their container, are elastic;** that is, the total kinetic energy of the particles is constant. The pressure of a gas against the walls of its container is the result of collisions of the gas particles with the walls. The more collisions and the more forceful each collision, the higher the pressure.

A gas that obeys all the assumptions of the kinetic–molecular theory is called an **ideal gas**. In practice, though, there is no such thing as a perfectly ideal gas. All gases behave somewhat differently than predicted when, at very high pressures or very low temperatures, their particles get closer together and interactions between particles become significant. As a rule, however, most real gases display nearly ideal behavior under normal conditions.

Ideal gas A gas that obeys all the assumptions of the kinetic–molecular theory.

Pressure (P) The force per unit area pushing against a surface.

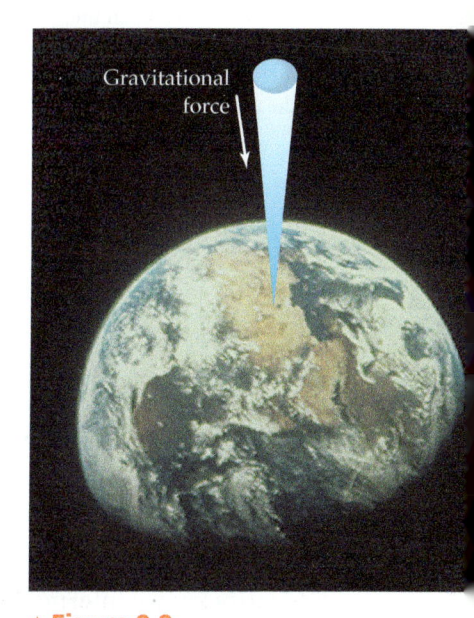

8.4 Pressure

We are all familiar with the effects of air pressure. When you fly in an airplane, the change in air pressure against your eardrums as the plane climbs or descends can cause a painful "popping." When you pump up a bicycle tire, you increase the pressure of air against the inside walls of the tire until the tire feels hard.

In scientific terms, **pressure** (P) is defined as a force (F) per unit area (A) pushing against a surface; that is, $P = F/A$. In the bicycle tire, for example, the pressure you feel is the force of air molecules colliding with the inside walls of the tire. The units you probably use for tire pressure are pounds per square inch (psi), where 1 psi is equal to the pressure exerted by a 1-pound object resting on a 1-square inch surface.

We on earth are under pressure from the atmosphere, the blanket of air pressing down on us (Figure 8.8). Atmospheric pressure is not constant, however; it varies slightly from day to day depending on the weather, and it also varies with altitude. Due to gravitational forces, the density of air is greatest at the earth's surface and decreases with increasing altitude. As a result, air pressure is greatest at the surface: it is about 14.7 psi at sea level but only about 4.7 psi on the summit of Mt. Everest.

One of the most commonly used units of pressure is the *millimeter of mercury*, abbreviated *mmHg* and often called a *torr* (after the Italian physicist Evangelista Torricelli). This unusual unit dates back to the early 1600s when Torricelli made the first mercury *barometer*. As shown in Figure 8.9, a barometer consists of a long, thin tube that is sealed at one end, filled with mercury, and then inverted into a dish of mercury. Some mercury runs from the tube into the dish until the downward pressure of the mercury in the column is exactly balanced by the outside atmospheric pressure, which presses down on the mercury in the dish and pushes it up into the column. The height of the mercury column varies depending on the altitude and weather conditions, but standard atmospheric pressure at sea level is defined to be exactly 760 mm.

Gas pressure inside a container is often measured using an open-ended *manometer*, a simple instrument similar in principle to the mercury barometer. As shown in Figure 8.10, an open-ended manometer consists of a U-tube filled with mercury, with one end connected to a gas-filled container and the other end open to the atmosphere. The difference between the heights of the mercury levels in the two arms of the U-tube indicates the difference between the pressure of the gas in the container and the pressure of the atmosphere. If the gas pressure inside the container is less than atmospheric, the mercury level is higher in the arm connected to the container (Figure 8.10a). If the gas pressure inside the container is greater than atmospheric, the mercury level is higher in the arm open to the atmosphere (Figure 8.10b).

Pressure is given in the SI system (Section 2.1) by a unit named the *pascal* (Pa), where 1 Pa = 0.007500 mmHg (or 1 mmHg = 133.32 Pa). Measurements in pascals are becoming more common, and many clinical laboratories have made the switchover. Higher pressures are often still given in *atmospheres* (atm), where 1 atm = 760 mmHg exactly.

Pressure units: 1 atm = 760 mmHg = 14.7 psi = 101,325 Pa

1 mmHg = 1 torr = 133.32 Pa

▲ **Figure 8.8**
Atmospheric pressure.
A column of air weighing 14.7 lb presses down on each square inch of the earth's surface at sea level, resulting in what we call atmospheric pressure.

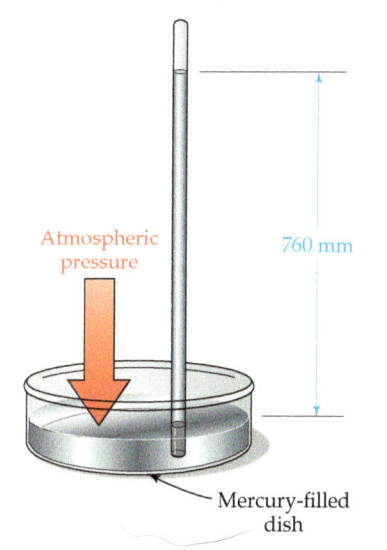

▲ **Figure 8.9**
Measuring atmospheric pressure.
A mercury barometer measures atmospheric pressure by determining the height of a mercury column in a sealed glass tube. The downward pressure of the mercury in the column is exactly balanced by the outside atmospheric pressure, which presses down on the mercury in the dish and pushes it up into the column.

222 CHAPTER 8 Gases, Liquids, and Solids

▶ **Figure 8.10**
Open-ended manometers for measuring pressure in a gas-filled bulb. (a) When the pressure in the gas-filled container is lower than atmospheric, the mercury level is higher in the arm open to the container. (b) When the pressure in the container is higher than atmospheric, the mercury level is higher in the arm open to the atmosphere.

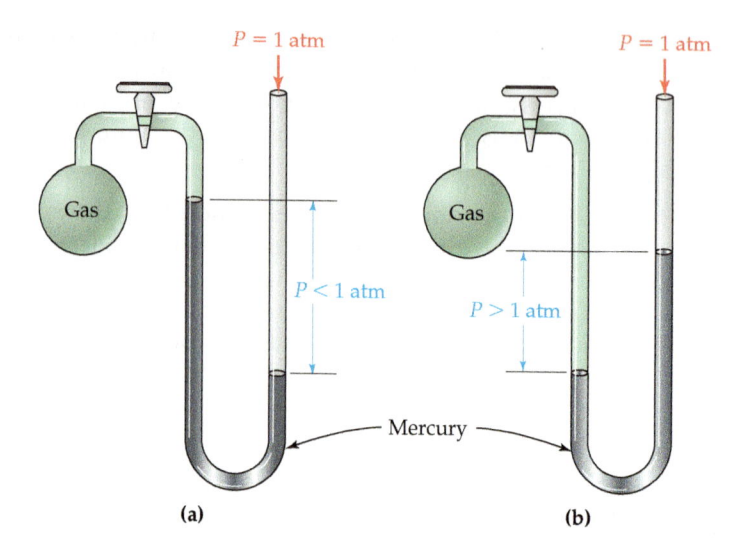

Worked Example 8.3 Unit Conversions (Pressure): psi, Atmospheres, and Pascals

A typical bicycle tire is inflated with air to a pressure of 55 psi. How many atmospheres is this? How many pascals?

ANALYSIS Using the starting pressure in psi, the pressure in atm and pascals can be calculated using the equivalent values in appropriate units as conversion factors.

SOLUTION

STEP 1: Identify known information. Pressure = 55 psi

STEP 2: Identify answer and units. Pressure = ?? atm = ?? pascals

STEP 3: Identify conversion factors. Using equivalent values in appropriate units, we can obtain conversion factors to convert to atm and pascals.

14.7 psi = 1 atm → $\frac{1 \text{ atm}}{14.7 \text{ psi}}$

14.7 psi = 101,325 Pa → $\frac{101,325 \text{ Pa}}{14.7 \text{ psi}}$

STEP 4: Solve. Use the appropriate conversion factors to set up an equation in which unwanted units cancel.

$(55 \text{ psi}) \times \left(\frac{1 \text{ atm}}{14.7 \text{ psi}}\right) = 3.7 \text{ atm}$

$(55 \text{ psi}) \times \left(\frac{101,325 \text{ Pa}}{14.7 \text{ psi}}\right) = 3.8 \times 10^5 \text{ Pa}$

Worked Example 8.4 Unit Conversions (Pressure): mmHg to Atmospheres

The pressure in a closed flask is measured using a manometer. If the mercury level in the arm open to the sealed vessel is 23.6 cm higher than the level of mercury in the arm open to the atmosphere, what is the gas pressure (in atm) in the closed flask?

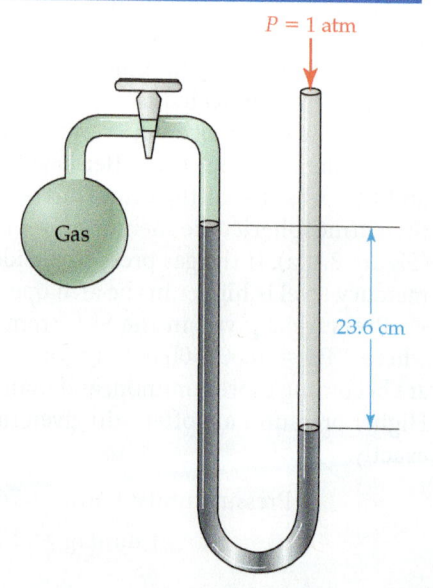

ANALYSIS Since the mercury level is higher in the arm open to the flask, the gas pressure in the flask is lower than atmospheric pressure (1 atm = 760 mmHg). We can convert the difference in the level of mercury in the two arms of the manometer from mmHg to atmospheres to determine the difference in pressure.

BALLPARK ESTIMATE The height difference (23.6 cm) is about one-third the height of a column of Hg that is equal to 1 atm (or 76 cm Hg). Therefore, the pressure in the flask should be about 0.33 atm lower than atmospheric pressure, or about 0.67 atm.

SOLUTION
Since the height difference is given in cm Hg, we must first convert to mmHg, and then to atm. The result is the difference in gas pressure between the flask and the open atmosphere (1 atm).

$$(23.6 \text{ cm Hg})\left(\frac{10 \text{ mmHg}}{\text{cm Hg}}\right)\left(\frac{1 \text{ atm}}{760 \text{ mmHg}}\right) = 0.311 \text{ atm}$$

The pressure in the flask is calculated by subtracting this difference from 1 atm:

$$1 \text{ atm} - 0.311 \text{ atm} = 0.689 \text{ atm}$$

BALLPARK CHECK This result agrees well with our estimate of 0.67 atm.

PROBLEM 8.5
The air pressure outside a jet airliner flying at 35,000 ft is about 0.289 atm. Convert this pressure to mmHg, psi, and pascals.

PROBLEM 8.6
The increase in atmospheric CO_2 levels has been correlated with the combustion of fossil fuels (see Chemistry in Action: Greenhouse Gases and Global Warming on p. 224). How would the atmospheric CO_2 levels be affected by a shift to corn-based ethanol or some other biomass-based fuel? Explain.

KEY CONCEPT PROBLEM 8.7
What is the pressure of the gas inside the following manometer (in mmHg) if outside pressure is 750 mmHg?

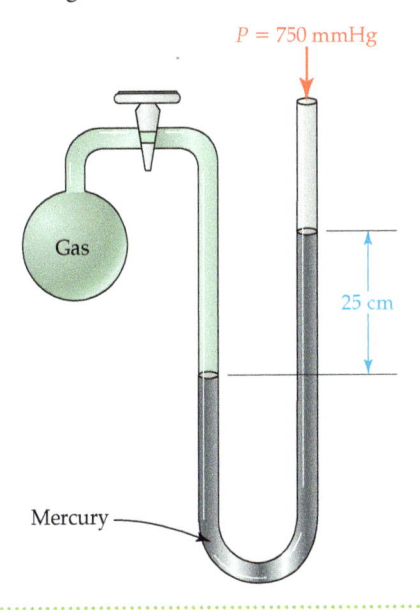

P = 750 mmHg

Gas

25 cm

Mercury

CHEMISTRY IN ACTION

Greenhouse Gases and Global Warming

The mantle of gases surrounding the earth is far from the uniform mixture you might expect, consisting of layers that vary in composition and properties at different altitudes. The ability of the gases in these layers to absorb radiation is responsible for life on earth as we know it.

The *stratosphere*—the layer extending from about 12 km up to 50 km altitude—contains the ozone layer that is responsible for absorbing harmful UV radiation. The *troposphere* is the layer extending from the surface up to about 12 km altitude. It should not surprise you to learn that the troposphere is the layer most easily disturbed by human activities and that this layer has the greatest impact on the earth's surface conditions. Among those impacts, a process called the *greenhouse effect* is much in the news today.

The greenhouse effect refers to the warming that occurs in the troposphere as gases absorb radiant energy. Much of the radiant energy reaching the Earth's surface from the sun is reflected back into space, but some is absorbed by atmospheric gases, particularly those referred to as *greenhouse gases* (GHGs)—water vapor, carbon dioxide, and methane. This absorbed radiation warms the atmosphere and acts to maintain a relatively stable temperature of 15 °C (59 °F) at the Earth's surface. Without the greenhouse effect, the average surface temperature would be about −18 °C (0 °F)—a temperature so low that Earth would be frozen and unable to sustain life.

The basis for concern about the greenhouse effect is the fear that human activities over the past century have disturbed the earth's delicate thermal balance. Should increasing amounts of radiation be absorbed, increased atmospheric heating will result, and global temperatures will continue to rise.

Measurements show that the concentration of atmospheric CO_2 has been rising in the last 150 years, from an estimated 290 parts per million (ppm) in 1850 to current levels approaching 400 ppm. The increase in CO_2 levels is largely because of the increased burning of fossil fuels and correlates with a concurrent increase in average global temperatures. The latest Assessment Report of the Intergovernmental Panel on Climate Change published in November 2007 concluded that "[W]arming of the climate system is unequivocal, as is now evident from observations of increases in global average air and ocean temperatures, widespread melting of snow and ice and rising global average sea level. . . . Continued GHG emissions at or above current rates would cause further warming and induce many changes in the global climate system during the 21st century that would *very likely* be larger than those observed during the 20th century."

Increased international concerns about the political and economic impacts of global climate change prompted development of the Kyoto Protocol to the United Nations Framework Convention on Climate Change (UNFCCC). Under the protocol, countries commit to a reduction in the production and emission of greenhouse gases, including CO_2, methane, and chlorofluorocarbons (CFCs). As of April 2010, 191 countries have signed and ratified the protocol. These concerns have also resulted in market pressures to develop sustainable and renewable energy sources, as well as more efficient technologies, such as hybrid electric vehicles.

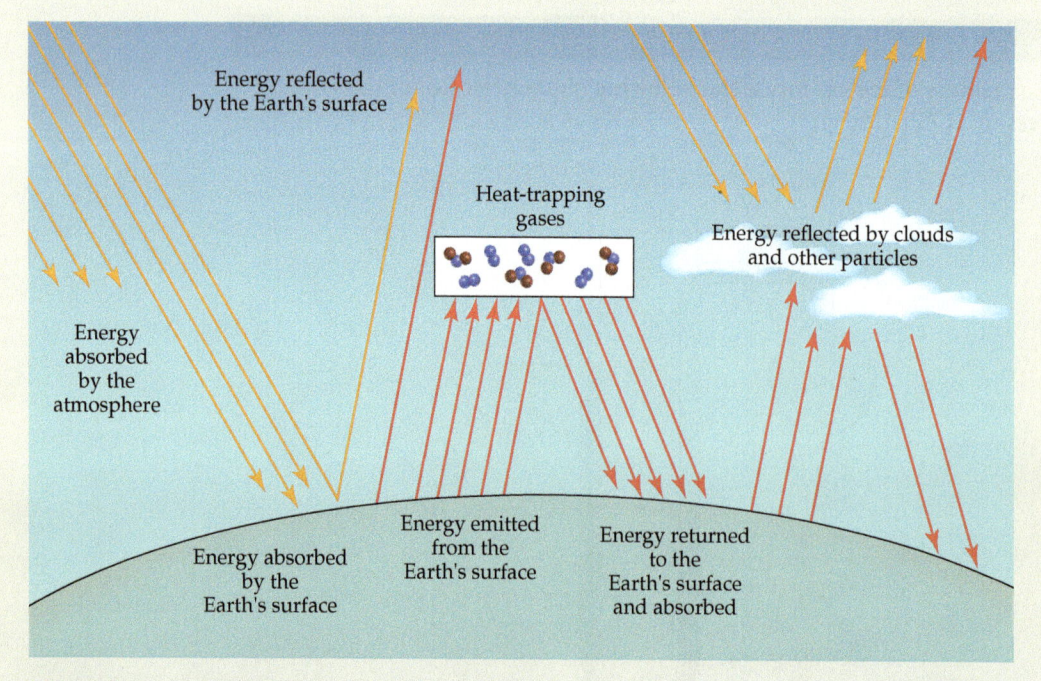

▲ Greenhouse gases (GHG) trap heat reflected from the earth's surface, resulting in the increase in surface temperatures known as global warming.

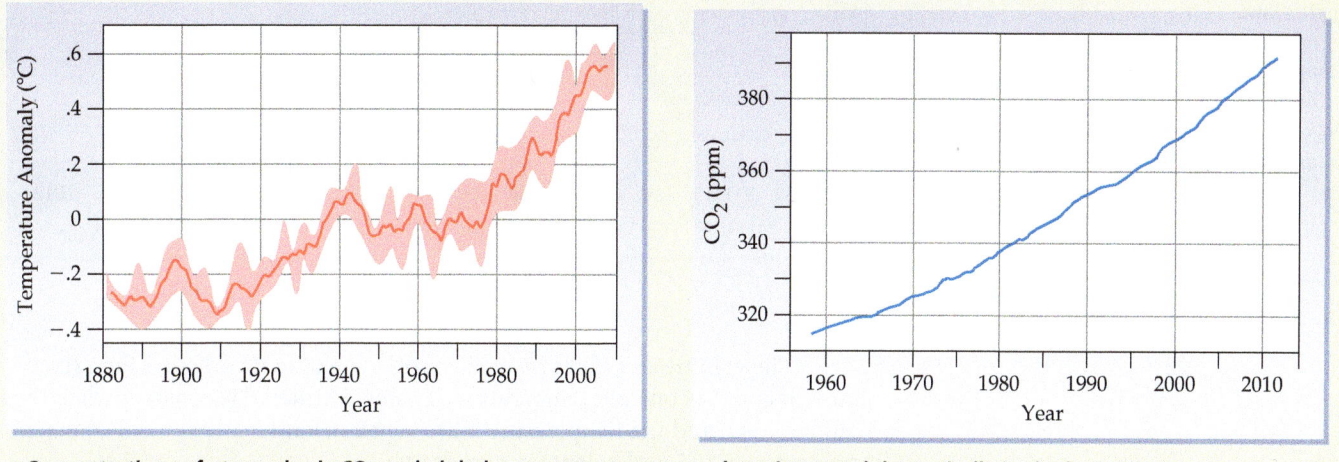

▲ Concentrations of atmospheric CO_2 and global average temperatures have increased dramatically in the last 150 years because of increased fossil fuel use, causing serious changes in earth's climate system.
© NASA, GISS Surface Temperature Analysis.

See Chemistry in Action Problems 8.100 and 8.101 at the end of the chapter.

8.5 Boyle's Law: The Relation between Volume and Pressure

The physical behavior of all gases is much the same, regardless of identity. Helium and chlorine, for example, are completely different in their *chemical* behavior, but are very similar in many of their physical properties. Observations of many different gases by scientists in the 1700s led to the formulation of what are now called the **gas laws**, which make it possible to predict the influence of pressure (P), volume (V), temperature (T), and molar amount (n) on any gas or mixture of gases. We will begin by looking at *Boyle's law*, which describes the relation between volume and pressure.

Imagine that you have a sample of gas inside a cylinder that has a movable plunger at one end (Figure 8.11). What happens if you double the pressure on the gas by pushing the plunger down, while keeping the temperature constant? Since the gas particles are forced closer together, the volume of the sample decreases.

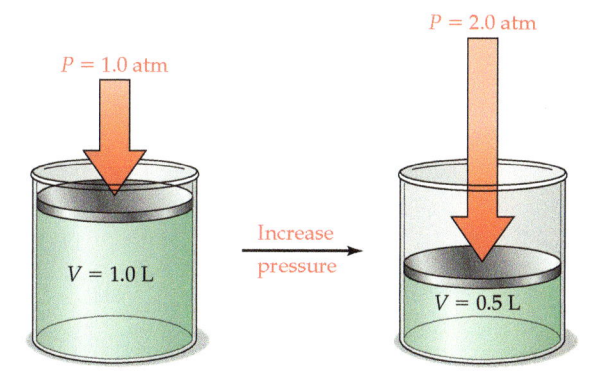

Gas laws A series of laws that predict the influence of pressure (P), volume (V), and temperature (T) on any gas or mixture of gases.

◀ **Figure 8.11**
Boyle's law.
The volume of a gas decreases proportionately as its pressure increases. For example, if the pressure of a gas sample is doubled, the volume is halved.

According to **Boyle's law**, the volume of a fixed amount of gas at a constant temperature is inversely proportional to its pressure, meaning that volume and pressure change in opposite directions. As pressure goes up, volume goes down; as pressure goes down, volume goes up (Figure 8.12). This observation is consistent with the kinetic–molecular theory. Since most of the volume occupied by gases is empty space, gases are easily compressed into smaller volumes. Since the average kinetic energy remains constant, the number of collisions must increase as the interior surface area of the container decreases, leading to an increase in pressure.

226 CHAPTER 8 Gases, Liquids, and Solids

▶ **Figure 8.12**
Boyle's law.
Pressure and volume are inversely related. Graph (a) demonstrates the decrease in volume as pressure increases, whereas graph (b) shows the linear relationship between V and 1/P.

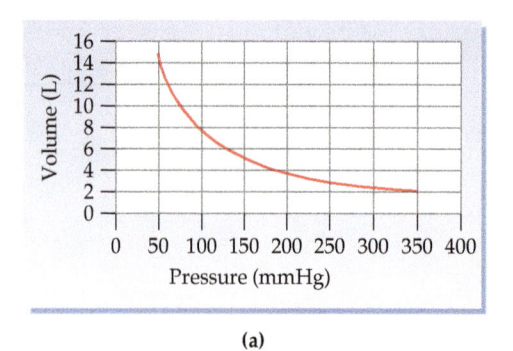

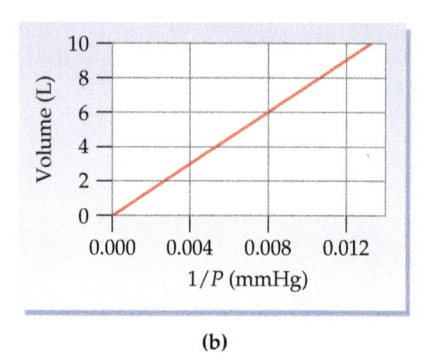

(a)

(b)

Boyle's law The volume of a gas is inversely proportional to its pressure for a fixed amount of gas at a constant temperature. That is, P times V is constant when the amount of gas n and the temperature T are kept constant. (The symbol ∝ means "is proportional to," and k denotes a constant value.)

$$\text{Volume } (V) \propto \frac{1}{\text{Pressure } (P)}$$

or $PV = k$ (A constant value)

Because $P \times V$ is a constant value for a fixed amount of gas at a constant temperature, the starting pressure (P_1) times the starting volume (V_1) must equal the final pressure (P_2) times the final volume (V_2). Thus, Boyle's law can be used to find the final pressure or volume when the starting pressure or volume is changed.

Since $P_1V_1 = k$ and $P_2V_2 = k$

then $P_1V_1 = P_2V_2$

so $P_2 = \dfrac{P_1V_1}{V_2}$ and $V_2 = \dfrac{P_1V_1}{P_2}$

As an example of Boyle's law behavior, think about what happens every time you breathe. Between breaths, the pressure inside your lungs is equal to atmospheric pressure. When inhalation takes place, your diaphragm lowers and the rib cage expands, increasing the volume of the lungs and thereby decreasing the pressure inside them (Figure 8.13). Air

▶ **Figure 8.13**
Boyle's law in breathing.
During inhalation, the diaphragm moves down and the rib cage moves up and out, thus increasing lung volume, decreasing pressure, and drawing in air. During exhalation, the diaphragm moves back up, lung volume decreases, pressure increases, and air moves out.

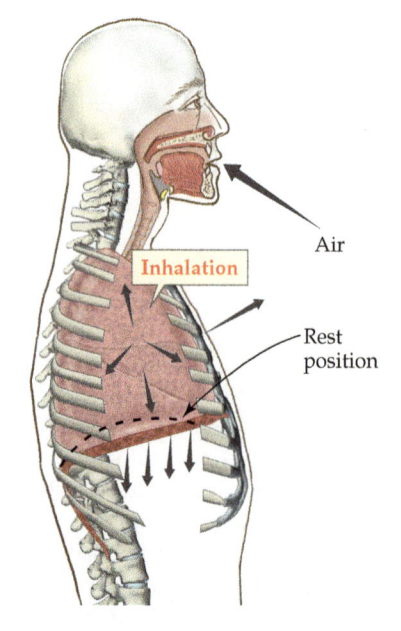

Lung volume increases, causing pressure in lungs to *decrease*. Air flows *in*.

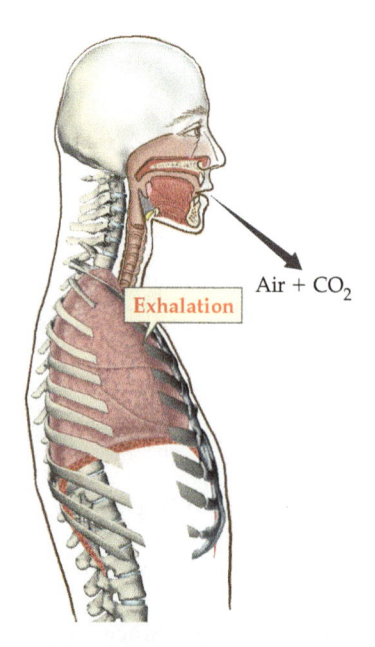

Lung volume decreases, causing pressure in lungs to *increase*. Air flows *out*.

must then move into the lungs to equalize their pressure with that of the atmosphere. When exhalation takes place, the diaphragm rises and the rib cage contracts, decreasing the volume of the lungs and increasing pressure inside them. Now gases move out of the lungs until pressure is again equalized with the atmosphere.

Worked Example 8.5 Using Boyle's Law: Finding Volume at a Given Pressure

In a typical automobile engine, the fuel/air mixture in a cylinder is compressed from 1.0 atm to 9.5 atm. If the uncompressed volume of the cylinder is 750 mL, what is the volume when fully compressed?

ANALYSIS This is a Boyle's law problem because the volume and pressure in the cylinder change but the amount of gas and the temperature remain constant. According to Boyle's law, the pressure of the gas times its volume is constant:

$$P_1 V_1 = P_2 V_2$$

Knowing three of the four variables in this equation, we can solve for the unknown.

◀ A cut-away diagram of an internal combustion engine shows movement of pistons during expansion and compression cycles.

BALLPARK ESTIMATE Since the pressure *increases* approximately 10-fold (from 1.0 atm to 9.5 atm), the volume must *decrease* to approximately 1/10, from 750 mL to about 75 mL.

SOLUTION

STEP 1: Identify known information. Of the four variables in Boyle's law, we know P_1, V_1, and P_2.

$P_1 = 1.0$ atm
$V_1 = 750$ mL
$P_2 = 9.5$ atm

STEP 2: Identify answer and units.

$V_2 = ??$ mL

STEP 3: Identify equation. In this case, we simply substitute the known variables into Boyle's law and rearrange to isolate the unknown.

$$P_1 V_1 = P_2 V_2 \implies V_2 = \frac{P_1 V_1}{P_2}$$

STEP 4: Solve. Substitute the known information into the equation. Make sure units cancel so that the answer is given in the units of the unknown variable.

$$V_2 = \frac{P_1 V_1}{P_2} = \frac{(1.0 \text{ atm})(750 \text{ mL})}{(9.5 \text{ atm})} = 79 \text{ mL}$$

BALLPARK CHECK Our estimate was 75 mL.

PROBLEM 8.8
An oxygen cylinder used for breathing has a volume of 5.0 L at 90 atm pressure. What is the volume of the same amount of oxygen at the same temperature if the pressure is 1.0 atm? (Hint: Would you expect the volume of gas at this pressure to be greater than or less than the volume at 90 atm?)

PROBLEM 8.9
A sample of hydrogen gas at 273 K has a volume of 3.2 L at 4.0 atm pressure. What is the volume if the pressure is increased to 10.0 atm? If the pressure is decreased to 0.70 atm?

PROBLEM 8.10
A typical blood pressure measured using a sphygmomanometer is reported as 112/75 (see Chemistry in Action: Blood Pressure on p. 228). How would this pressure be recorded if the sphygmomanometer used units of psi instead of mmHg?

CHEMISTRY IN ACTION

Blood Pressure

Having your blood pressure measured is a quick and easy way to get an indication of the state of your circulatory system. Although blood pressure varies with age, a normal adult male has a reading near 120/80 mmHg, and a normal adult female has a reading near 110/70 mmHg. Abnormally high values signal an increased risk of heart attack and stroke.

Pressure varies greatly in different types of blood vessels. Usually, though, measurements are carried out on arteries in the upper arm as the heart goes through a full cardiac cycle. *Systolic pressure* is the maximum pressure developed in the artery just after contraction, as the heart forces the maximum amount of blood into the artery. *Diastolic pressure* is the minimum pressure that occurs at the end of the heart cycle.

Blood pressure is most often measured by a *sphygmomanometer*, a device consisting of a squeeze bulb, a flexible cuff, and a mercury manometer. (1) The cuff is placed around the upper arm over the brachial artery and inflated by the squeeze bulb to about 200 mmHg pressure, an amount great enough to squeeze the artery shut and prevent blood flow. Air is then slowly released from the cuff, and pressure drops (2). As cuff pressure reaches the systolic pressure, blood spurts through the artery, creating a turbulent tapping sound that can be heard through a stethoscope. The pressure registered on the manometer at the moment the first sounds are heard is the systolic blood pressure.

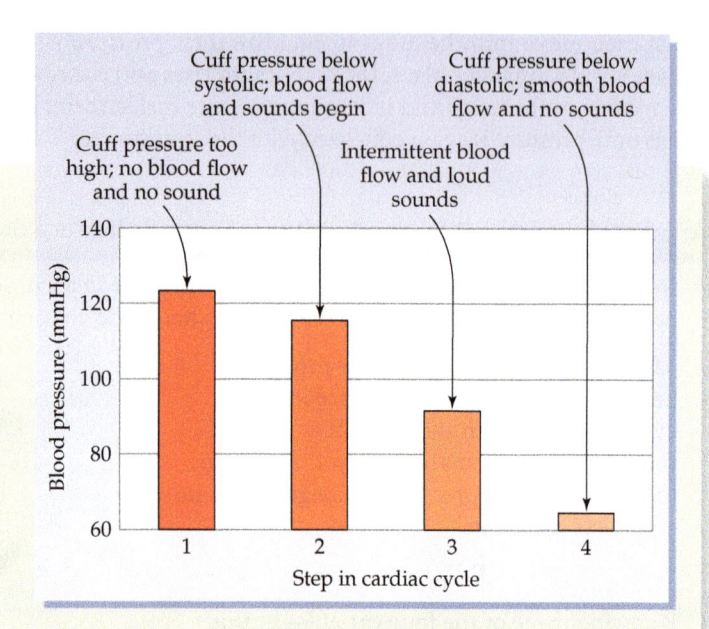

▲ The sequence of events during blood pressure measurement, including the sounds heard.

(3) Sounds continue until the pressure in the cuff becomes low enough to allow diastolic blood flow. (4) At this point, blood flow becomes smooth, no sounds are heard, and a diastolic blood pressure reading is recorded on the manometer. Readings are usually recorded as systolic/diastolic, for example, 120/80. The accompanying figure shows the sequence of events during measurement.

See Chemistry in Action Problems 8.102 and 103 at the end of the chapter.

8.6 Charles's Law: The Relation between Volume and Temperature

Imagine that you again have a sample of gas inside a cylinder with a plunger at one end. What happens if you double the sample's kelvin temperature while letting the plunger move freely to keep the pressure constant? The gas particles move with twice as much energy and collide twice as forcefully with the walls. To maintain a constant pressure, the volume of the gas in the cylinder must double (Figure 8.14).

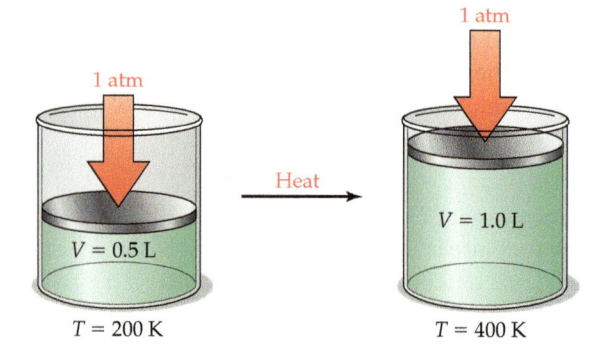

▲ The volume of the gas in the balloon increases as it is heated, causing a decrease in density and allowing the balloon to rise.

▲ **Figure 8.14**
Charles's law.
The volume of a gas is directly proportional to its kelvin temperature at constant *n* and *P*. If the kelvin temperature of the gas is doubled, its volume doubles.

SECTION 8.6 Charles's Law: The Relation between Volume and Temperature

According to **Charles's law**, the volume of a fixed amount of gas at constant pressure is directly proportional to its kelvin temperature. Note the difference between *directly* proportional in Charles's law and *inversely* proportional in Boyle's law. Directly proportional quantities change in the same direction: as temperature goes up or down, volume also goes up or down (Figure 8.15).

Charles's law The volume of a gas is directly proportional to its kelvin temperature for a fixed amount of gas at a constant pressure. That is, V divided by T is constant when n and P are held constant.

$$V \propto T \quad \text{(In kelvins)}$$

$$\text{or } \frac{V}{T} = k \quad \text{(A constant value)}$$

$$\text{or } \frac{V_1}{T_1} = \frac{V_2}{T_2}$$

▲ **Figure 8.15**
Charles's law.
Volume is directly proportional to the kelvin temperature for a fixed amount of gas at a constant pressure. As the temperature goes up, the volume also goes up.

This observation is consistent with the kinetic–molecular theory. As temperature increases, the average kinetic energy of the gas molecules increases, as does the energy of molecular collisions with the interior surface of the container. The volume of the container must increase to maintain a constant pressure. As an example of Charles's law, think about what happens when a hot-air balloon is inflated. Heating causes the air inside to expand and fill the balloon. The air inside the balloon is less dense than the air outside the balloon, creating the buoyancy effect.

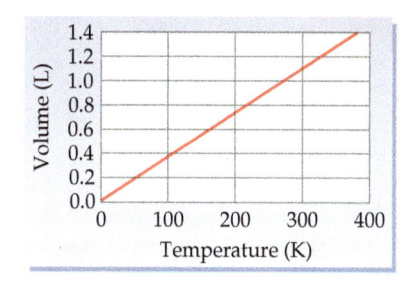

Worked Example 8.6 Using Charles's Law: Finding Volume at a Given Temperature

An average adult inhales a volume of 0.50 L of air with each breath. If the air is warmed from room temperature (20 °C = 293 K) to body temperature (37 °C = 310 K) while in the lungs, what is the volume of the air exhaled?

ANALYSIS This is a Charles's law problem because the volume and temperature of the air change while the amount and pressure remain constant. Knowing three of the four variables, we can rearrange Charles's law to solve for the unknown.

BALLPARK ESTIMATE Charles's law predicts an increase in volume directly proportional to the increase in temperature from 273 K to 310 K. The increase of less than 20 K represents a relatively small change compared to the initial temperature of 273 K. A 10% increase, for example, would be equal to a temperature change of 27 K; so a 20-K change would be less than 10%. We would therefore expect the volume to increase by less than 10%, from 0.50 L to a little less than 0.55 L.

SOLUTION

STEP 1: Identify known information. Of the four variables in Charles's law, we know T_1, V_1, and T_2.

$T_1 = 293 \text{ K}$
$V_1 = 0.50 \text{ L}$
$T_2 = 310 \text{ K}$

STEP 2: Identify answer and units.

$V_2 = \text{?? L}$

STEP 3: Identify equation. Substitute the known variables into Charles's law and rearrange to isolate the unknown.

$$\frac{V_1}{T_1} = \frac{V_2}{T_2} \Rightarrow V_2 = \frac{V_1 T_2}{T_1}$$

STEP 4: Solve. Substitute the known information into Charles's law; check to make sure units cancel.

$$V_2 = \frac{V_1 T_2}{T_1} = \frac{(0.50 \text{ L})(310 \text{ K})}{293 \text{ K}} = 0.53 \text{ L}$$

BALLPARK CHECK This is consistent with our estimate!

PROBLEM 8.11
A sample of chlorine gas has a volume of 0.30 L at 273 K and 1 atm pressure. What temperature (in °C) would be required to increase the volume to 1.0 L? To decrease the volume to 0.20 L?

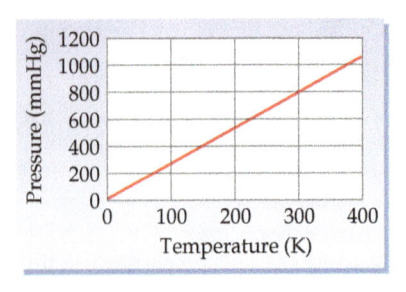

8.7 Gay-Lussac's Law: The Relation between Pressure and Temperature

Imagine next that you have a fixed amount of gas in a sealed container whose volume remains constant. What happens if you double the temperature (in kelvins)? The gas particles move with twice as much energy and collide with the walls of the container with twice as much force. Thus, the pressure in the container doubles. According to **Gay-Lussac's law**, the pressure of a fixed amount of gas at constant volume is directly proportional to its Kelvin temperature. As temperature goes up or down, pressure also goes up or down (Figure 8.16).

Gay-Lussac's law The pressure of a gas is directly proportional to its Kelvin temperature for a fixed amount of gas at a constant volume. That is, P divided by T is constant when n and V are held constant.

$$P \propto T \quad \text{(In kelvins)}$$

$$\text{or } \frac{P}{T} = k \quad \text{(A constant value)}$$

$$\text{or } \frac{P_1}{T_1} = \frac{P_2}{T_2}$$

▲ **Figure 8.16**
Gay-Lussac's law. Pressure is directly proportional to the temperature in kelvins for a fixed amount of gas at a constant volume. As the temperature goes up, the pressure also goes up.

According to the kinetic–molecular theory, the kinetic energy of molecules is directly proportional to absolute temperature. As the average kinetic energy of the molecules increases, the energy of collisions with the interior surface of the container increases, causing an increase in pressure. As an example of Gay-Lussac's law, think of what happens when an aerosol can is thrown into an incinerator. As the can gets hotter, pressure builds up inside and the can explodes (hence the warning statement on aerosol cans).

Worked Example 8.7 Using Gay-Lussac's Law: Finding Pressure at a Given Temperature

What does the inside pressure become if an aerosol can with an initial pressure of 4.5 atm is heated in a fire from room temperature (20 °C) to 600 °C?

ANALYSIS This is a Gay-Lussac's law problem because the pressure and temperature of the gas inside the can change while its amount and volume remain constant. We know three of the four variables in the equation for Gay-Lussac's law, and can find the unknown by substitution and rearrangement.

BALLPARK ESTIMATE Gay-Lussac's law states that pressure is directly proportional to temperature. Since the Kelvin temperature increases approximately threefold (from about 300 K to about 900 K), we expect the pressure to also increase by approximately threefold, from 4.5 atm to about 14 atm.

SOLUTION

STEP 1: Identify known information. Of the four variables in Gay-Lussac's law, we know P_1, T_1 and T_2. (Note that T must be in kelvins.)

$P_1 = 4.5$ atm
$T_1 = 20\,°C = 293$ K
$T_2 = 600\,°C = 873$ K

STEP 2: Identify answer and units.

$P_2 = ??$ atm

STEP 3: Identify equation. Substituting the known variables into Gay-Lussac's law, we rearrange to isolate the unknown.

$$\frac{P_1}{T_1} = \frac{P_2}{T_2} \quad \Rightarrow \quad P_2 = \frac{P_1 T_2}{T_1}$$

STEP 4: Solve. Substitute the known information into Gay-Lussac's law; check to make sure units cancel.

$$P_2 = \frac{P_1 T_2}{T_1} = \frac{(4.5 \text{ atm})(873 \text{ K})}{293 \text{ K}} = 13 \text{ atm}$$

BALLPARK CHECK Our estimate was 14 atm.

PROBLEM 8.12
Driving on a hot day causes tire temperature to rise. What is the pressure inside an automobile tire at 45 °C if the tire has a pressure of 30 psi at 15 °C? Assume that the volume and amount of air in the tire remain constant.

8.8 The Combined Gas Law

Since PV, V/T, and P/T all have constant values for a fixed amount of gas, these relationships can be merged into a **combined gas law**, which holds true whenever the amount of gas is fixed.

Combined gas law $\quad \dfrac{PV}{T} = k \quad$ (A constant value)

$$\text{or } \dfrac{P_1 V_1}{T_1} = \dfrac{P_2 V_2}{T_2}$$

If any five of the six quantities in this equation are known, the sixth quantity can be calculated. Furthermore, if any of the three variables T, P, or V is constant, that variable drops out of the equation, leaving behind Boyle's law, Charles's law, or Gay-Lussac's law. As a result, *the combined gas law is the only equation you need to remember for a fixed amount of gas.* Worked Example 8.8 gives a sample calculation.

$$\text{Since} \quad \dfrac{P_1 V_1}{T_1} = \dfrac{P_2 V_2}{T_2}$$

At constant T: $\quad \dfrac{P_1 V_1}{T} = \dfrac{P_2 V_2}{T} \quad$ gives $\quad P_1 V_1 = P_2 V_2 \quad$ (Boyle's law)

At constant P: $\quad \dfrac{P V_1}{T_1} = \dfrac{P V_2}{T_2} \quad$ gives $\quad \dfrac{V_1}{T_1} = \dfrac{V_2}{T_2} \quad$ (Charles's law)

At constant V: $\quad \dfrac{P_1 V}{T_1} = \dfrac{P_2 V}{T_2} \quad$ gives $\quad \dfrac{P_1}{T_1} = \dfrac{P_2}{T_2} \quad$ (Gay-Lussac's law)

Worked Example 8.8 Using the Combined Gas Law: Finding Temperature

A 6.3 L sample of helium gas stored at 25 °C and 1.0 atm pressure is transferred to a 2.0 L tank and maintained at a pressure of 2.8 atm. What temperature is needed to maintain this pressure?

ANALYSIS This is a combined gas law problem because pressure, volume, and temperature change while the amount of helium remains constant. Of the six variables in this equation, we know P_1, V_1, T_1, P_2, and V_2, and we need to find T_2.

BALLPARK ESTIMATE Since the volume goes down by a little more than a factor of about 3 (from 6.3 L to 2.0 L) and the pressure goes up by a little less than a factor of about 3 (from 1.0 atm to 2.8 atm), the two changes roughly offset each other, and so the temperature should not change much. Since the volume-decrease factor (3.2) is slightly greater than the pressure-increase factor (2.8), the temperature will drop slightly ($T \propto V$).

SOLUTION

STEP 1: Identify known information. Of the six variables in combined gas law we know P_1, V_1, T_1, P_2, and V_2. (As always, T must be converted from Celsius degrees to kelvins.)

$P_1 = 1.0$ atm, $P_2 = 2.8$ atm
$V_1 = 6.3$ L, $V_2 = 2.0$ L
$T_1 = 25 \,°C = 298$ K

STEP 2: Identify answer and units.

$T_2 = ??$ kelvin

STEP 3: Identify the equation. Substitute the known variables into the equation for the combined gas law and rearrange to isolate the unknown.

$$\dfrac{P_1 V_1}{T_1} = \dfrac{P_2 V_2}{T_2} \Rightarrow T_2 = \dfrac{P_2 V_2 T_1}{P_1 V_1}$$

STEP 4: Solve. Solve the combined gas law equation for T_2; check to make sure units cancel.

$$T_2 = \dfrac{P_2 V_2 T_1}{P_1 V_1} = \dfrac{(2.8 \text{ atm})(2.0 \text{ L})(298 \text{ K})}{(1.0 \text{ atm})(6.3 \text{ L})} = 260 \text{ K} \,(\Delta T = 2.38 \,°C)$$

BALLPARK CHECK The relatively small decrease in temperature (38 °C, or 13% compared to the original temperature) is consistent with our prediction.

232 CHAPTER 8 Gases, Liquids, and Solids

PROBLEM 8.13
A weather balloon is filled with helium to a volume of 275 L at 22 °C and 752 mmHg. The balloon ascends to an altitude where the pressure is 480 mmHg, and the temperature is −32 °C. What is the volume of the balloon at this altitude?

KEY CONCEPT PROBLEM 8.14
A balloon is filled under the initial conditions indicated below. If the pressure is then increased to 2 atm while the temperature is increased to 50 °C, which balloon on the right, (a) or (b), represents the new volume of the balloon?

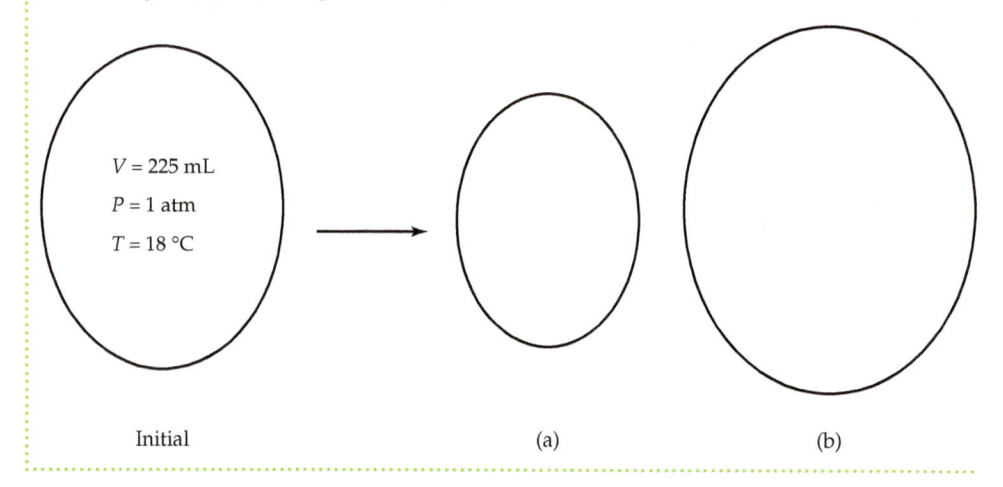

V = 225 mL
P = 1 atm
T = 18 °C

Initial   (a)   (b)

8.9 Avogadro's Law: The Relation between Volume and Molar Amount

Here we look at one final gas law, which takes changes in amount of gas into account. Imagine that you have two different volumes of a gas at the same temperature and pressure. How many moles does each sample contain? According to **Avogadro's law**, the volume of a gas is directly proportional to its molar amount at a constant pressure and temperature (Figure 8.17). A sample that contains twice the molar amount has twice the volume.

Avogadro's law The volume of a gas is directly proportional to its molar amount at a constant pressure and temperature. That is, V divided by n is constant when P and T are held constant.

Volume (V) ∝ Number of moles (n)

or $\dfrac{V}{n} = k$ (A constant value; the same for all gases)

or $\dfrac{V_1}{n_1} = \dfrac{V_2}{n_2}$

▲ **Figure 8.17**
Avogadro's law.
Volume is directly proportional to the molar amount, n, at a constant temperature and pressure. As the number of moles goes up, the volume also goes up.

Because the particles in a gas are so tiny compared to the empty space surrounding them, there is no interaction among gas particles as proposed by the kinetic–molecular theory. As a result, the chemical identity of the particles does not matter and the value of the constant k in the equation $V/n = k$ is the same for all gases. It is therefore possible to compare the molar amounts of *any* two gases simply by comparing their volumes at the same temperature and pressure.

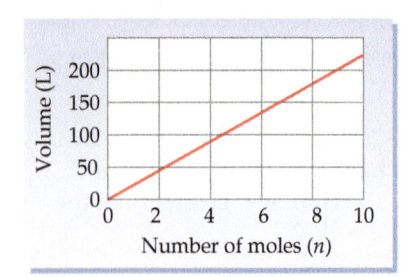

Notice that the *values* of temperature and pressure do not matter; it is only necessary that T and P be the same for both gases. To simplify comparisons of gas samples, however, it is convenient to define a set of conditions called **standard temperature and pressure (STP)**, which specifies a temperature of 0 °C (273 K) and a pressure of 1 atm (760 mmHg).

At standard temperature and pressure, 1 mol of any gas (6.02×10^{23} particles) has a volume of 22.4 L, a quantity called the **standard molar volume** (Figure 8.18).

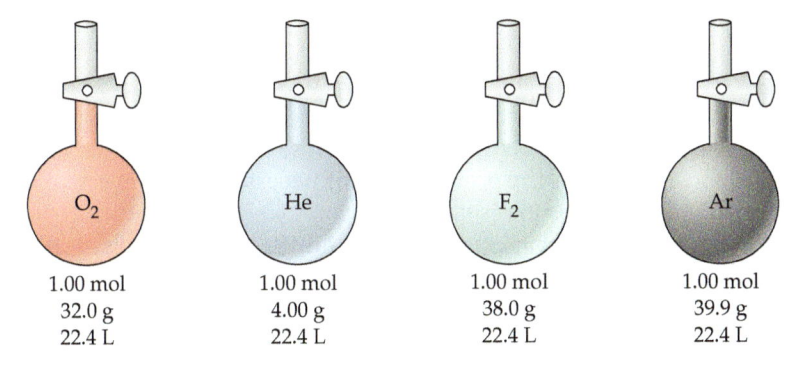

Standard temperature and pressure (STP) 0 °C (273.15 K); 1 atm (760 mmHg)

Standard molar volume of any ideal gas at STP 22.4 L/mol

◀ **Figure 8.18**
Avogadro's law.
Each of these 22.4 L bulbs contains 1.00 mol of gas at 0 °C and 1 atm pressure. Note that the volume occupied by 1 mol of gas is the same even though the mass (in grams) of 1 mol of each gas is different.

O_2 — 1.00 mol, 32.0 g, 22.4 L
He — 1.00 mol, 4.00 g, 22.4 L
F_2 — 1.00 mol, 38.0 g, 22.4 L
Ar — 1.00 mol, 39.9 g, 22.4 L

Worked Example 8.9 Using Avogadro's Law: Finding Moles in a Given Volume at STP

Use the standard molar volume of a gas at STP (22.4 L) to find how many moles of air at STP are in a room measuring 4.11 m wide by 5.36 m long by 2.58 m high.

ANALYSIS We first find the volume of the room and then use standard molar volume as a conversion factor to find the number of moles.

SOLUTION

STEP 1: Identify known information. We are given the room dimensions.

Length = 5.36 m
Width = 4.11 m
Height = 2.58 m

STEP 2: Identify answer and units.

Moles of air = ?? mol

STEP 3: Identify the equation. The volume of the room is the product of its three dimensions. Once we have the volume (in m³), we can convert to liters and use the molar volume at STP as a conversion factor to obtain moles of air.

Volume = (4.11 m)(5.36 m)(2.58 m) = 56.8 m³

$$= 56.8 \text{ m}^3 \times \frac{1000 \text{ L}}{1 \text{ m}^3} = 5.68 \times 10^4 \text{ L}$$

$$1 \text{ mol} = 22.4 \text{ L} \rightarrow \frac{1 \text{ mol}}{22.4 \text{ L}}$$

STEP 4: Solve. Use the room volume and the molar volume at STP to set up an equation, making sure unwanted units cancel.

$$5.68 \times 10^4 \text{ L} \times \frac{1 \text{ mol}}{22.4 \text{ L}} = 2.54 \times 10^3 \text{ mol}$$

PROBLEM 8.15
How many moles of methane gas, CH_4, are in a 1.00×10^5 L storage tank at STP? How many grams of methane is this? How many grams of carbon dioxide gas could the same tank hold?

8.10 The Ideal Gas Law

The relationships among the four variables *P, V, T,* and *n* for gases can be combined into a single expression called the **ideal gas law**. If you know the values of any three of the four quantities, you can calculate the value of the fourth.

Ideal gas law $\dfrac{PV}{nT} = R$ (A constant value)

or $PV = nRT$

234 CHAPTER 8 Gases, Liquids, and Solids

Gas constant (R) The constant R in the ideal gas law, $PV = nRT$.

The constant R in the ideal gas law (instead of the usual k) is called the **gas constant**. Its value depends on the units chosen for pressure, with the two most common values being

$$\text{For } P \text{ in atmospheres:} \quad R = 0.0821 \frac{\text{L} \cdot \text{atm}}{\text{mol} \cdot \text{K}}$$

$$\text{For } P \text{ in millimeters Hg:} \quad R = 62.4 \frac{\text{L} \cdot \text{mmHg}}{\text{mol} \cdot \text{K}}$$

In using the ideal gas law, it is important to choose the value of R having pressure units that are consistent with the problem and, if necessary, to convert volume into liters and temperature into kelvins.

Table 8.3 summarizes the various gas laws, and Worked Examples 8.10 and 8.11 show how to use the ideal gas law.

TABLE 8.3 A Summary of the Gas Laws

	Gas Law	Variables	Constant
Boyle's law	$P_1V_1 = P_2V_2$	P, V	n, T
Charles's law	$V_1/T_1 = V_2/T_2$	V, T	n, P
Gay-Lussac's law	$P_1/T_1 = P_2/T_2$	P, T	n, V
Combined gas law	$P_1V_1/T_1 = P_2V_2/T_2$	P, V, T	n
Avogadro's law	$V_1/n_1 = V_2/n_2$	V, n	P, T
Ideal gas law	$PV = nRT$	P, V, T, n	R

Worked Example 8.10 Using the Ideal Gas Law: Finding Moles

How many moles of air are in the lungs of an average person with a total lung capacity of 3.8 L? Assume that the person is at 1.0 atm pressure and has a normal body temperature of 37 °C.

ANALYSIS This is an ideal gas law problem because it asks for a value of n when P, V, and T are known: $n = PV/RT$. The volume is given in the correct unit of liters, but temperature must be converted to kelvins.

SOLUTION

STEP 1: Identify known information. We know three of the four variables in the ideal gas law.

$P = 1.0$ atm
$V = 3.8$ L
$T = 37\,°C = 310$ K

STEP 2: Identify answer and units.

Moles of air, $n = $?? mol

STEP 3: Identify the equation. Knowing three of the four variables in the ideal gas law, we can rearrange and solve for the unknown variable, n. Note: because pressure is given in atm, we use the value of R that is expressed in atm:

$$PV = nRT \Rightarrow n = \frac{PV}{RT}$$

$$R = 0.0821 \frac{\text{L} \cdot \text{atm}}{\text{mol} \cdot \text{K}}$$

STEP 4: Solve. Substitute the known information and the appropriate value of R into the ideal gas law equation and solve for n.

$$n = \frac{PV}{RT} = \frac{(1.0 \text{ atm})(3.8 \text{ L})}{\left(0.0821 \frac{\text{L} \cdot \text{atm}}{\text{mol} \cdot \text{K}}\right)(310 \text{ K})} = 0.15 \text{ mol}$$

Worked Example 8.11 Using the Ideal Gas Law: Finding Pressure

Methane gas is sold in steel cylinders with a volume of 43.8 L containing 5.54 kg. What is the pressure in atmospheres inside the cylinder at a temperature of 20.0 °C (293.15 K)? The molar mass of methane (CH_4) is 16.0 g/mol.

ANALYSIS This is an ideal gas law problem because it asks for a value of P when V, T, and n are given. Although not provided directly, enough information is given so that we can calculate the value of n ($n = g/MW$).

SOLUTION

STEP 1: Identify known information. We know two of the four variables in the ideal gas law; V, T, and can calculate the third, n, from the information provided.

$V = 43.8$ L
$T = 37 °C = 310$ K

STEP 2: Identify answer and units.

Pressure, $P = $?? atm

STEP 3: Identify equation. First, calculate the number of moles, n, of methane in the cylinder by using molar mass (16.0 g/mol) as a conversion factor. Then use the ideal gas law to calculate the pressure.

$$n = (5.54 \text{ kg methane})\left(\frac{1000 \text{ g}}{1 \text{ kg}}\right)\left(\frac{1 \text{ mol}}{16.0 \text{ g}}\right) = 346 \text{ mol methane}$$

$$PV = nRT \implies P = \frac{nRT}{V}$$

STEP 4: Solve. Substitute the known information and the appropriate value of R into the ideal gas law equation and solve for P.

$$P = \frac{nRT}{V} = \frac{(346 \text{ mol})\left(0.0821 \frac{\text{L} \cdot \text{atm}}{\text{mol} \cdot \text{K}}\right)(293 \text{ K})}{43.8 \text{ L}} = 190 \text{ atm}$$

PROBLEM 8.16
An aerosol spray can of deodorant with a volume of 350 mL contains 3.2 g of propane gas (C_3H_8) as propellant. What is the pressure in the can at 20 °C?

PROBLEM 8.17
A helium gas cylinder of the sort used to fill balloons has a volume of 180 L and a pressure of 2200 psi (150 atm) at 25 °C. How many moles of helium are in the tank? How many grams?

🔑 KEY CONCEPT PROBLEM 8.18

Show the approximate level of the movable piston in drawings (a) and (b) after the indicated changes have been made to the initial gas sample (assume a constant pressure of 1 atm).

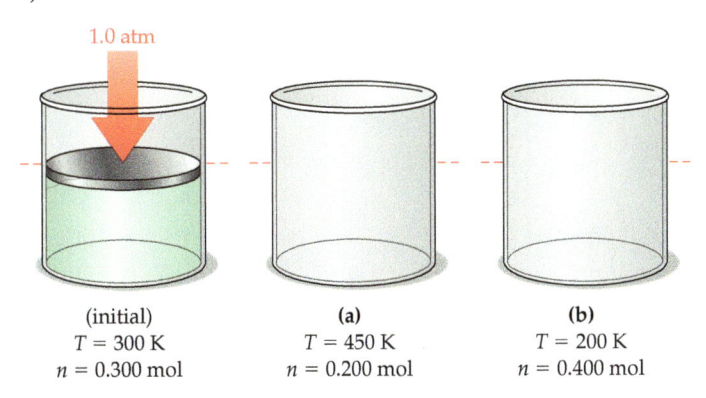

(initial)
$T = 300$ K
$n = 0.300$ mol

(a)
$T = 450$ K
$n = 0.200$ mol

(b)
$T = 200$ K
$n = 0.400$ mol

8.11 Partial Pressure and Dalton's Law

According to the kinetic–molecular theory, each particle in a gas acts independently of all others because there are no attractive forces between them and they are so far apart. To any individual particle, the chemical identity of its neighbors is irrelevant. Thus, *mixtures* of gases behave the same as pure gases and obey the same laws.

Dry air, for example, is a mixture of about 21% oxygen, 78% nitrogen, and 1% argon by volume, which means that 21% of atmospheric air pressure is caused by O_2 molecules, 78% by N_2 molecules, and 1% by Ar atoms. The contribution of each gas in a mixture to the total pressure of the mixture is called the **partial pressure** of that gas. According to **Dalton's law**, the total pressure exerted by a gas mixture (P_{total}) is the sum of the partial pressures of the components in the mixture:

Partial pressure The contribution of a given gas in a mixture to the total pressure.

Dalton's law $P_{total} = P_{gas\ 1} + P_{gas\ 2} + \cdots$

In dry air at a total air pressure of 760 mmHg, the partial pressure caused by the contribution of O_2 is 0.21×760 mmHg $= 160$ mmHg, the partial pressure of N_2 is 0.78×760 mmHg $= 593$ mmHg, and that of argon is 7 mmHg. *The partial pressure exerted by each gas in a mixture is the same pressure that the gas would exert if it were alone.* Put another way, the pressure exerted by each gas depends on the frequency of collisions of its molecules with the walls of the container. However, this frequency does not change when other gases are present, because the different molecules have no influence on one another.

To represent the partial pressure of a specific gas, we add the formula of the gas as a subscript to P, the symbol for pressure. You might see the partial pressure of oxygen represented as P_{O_2}, for instance. Moist air inside the lungs at 37 °C and atmospheric pressure has the following average composition at sea level. Note that P_{total} is equal to atmospheric pressure, 760 mmHg.

$$P_{total} = P_{N_2} + P_{O_2} + P_{CO_2} + P_{H_2O}$$
$$= 573\ \text{mmHg} + 100\ \text{mmHg} + 40\ \text{mmHg} + 47\ \text{mmHg}$$
$$= 760\ \text{mmHg}$$

The composition of air does not change appreciably with altitude, but the total pressure decreases rapidly. The partial pressure of oxygen in air therefore decreases with increasing altitude, and it is this change that leads to difficulty in breathing at high elevations.

Worked Example 8.12 Using Dalton's Law: Finding Partial Pressures

Humid air on a warm summer day is approximately 20% oxygen, 75% nitrogen, 4% water vapor, and 1% argon. What is the partial pressure of each component if the atmospheric pressure is 750 mmHg?

ANALYSIS According to Dalton's law, the partial pressure of any gas in a mixture is equal to the percent concentration of the gas times the total gas pressure (750 mmHg). In this case,

$$P_{total} = P_{O_2} + P_{N_2} + P_{H_2O} + P_{Ar}$$

SOLUTION

Oxygen partial pressure (P_{O_2}): $\quad 0.20 \times 750$ mmHg $= 150$ mmHg
Nitrogen partial pressure (P_{N_2}): $\quad 0.75 \times 750$ mmHg $= 560$ mmHg
Water vapor partial pressure (P_{H_2O}): $\quad 0.04 \times 750$ mmHg $= 30$ mmHg
Argon partial pressure (P_{Ar}): $\quad 0.01 \times 750$ mmHg $= 8$ mmHg

Total pressure $= 748$ mmHg $\rightarrow 750$ mmHg (rounding to 2 significant figures!)

Note that the sum of the partial pressures must equal the total pressure (within rounding error).

PROBLEM 8.19
Assuming a total pressure of 9.5 atm, what is the partial pressure of each component in the mixture of 98% helium and 2.0% oxygen breathed by deep-sea divers? How does the partial pressure of oxygen in diving gas compare with its partial pressure in normal air?

PROBLEM 8.20
Determine the percent composition of air in the lungs from the following composition in partial pressures: $P_{N_2} = 573$ mmHg, $P_{O_2} = 100$ mmHg, $P_{CO_2} = 40$ mmHg, $P_{H_2O} = 47$ mmHg; all at 37 °C and 1 atm pressure.

PROBLEM 8.21
The atmospheric pressure on the top of Mt. Everest, an altitude of 29,035 ft, is only 265 mmHg. What is the partial pressure of oxygen in the lungs at this altitude (assuming that the % O_2 is the same as in dry air)?

KEY CONCEPT PROBLEM 8.22
Assume that you have a mixture of He (MW = 4 amu) and Xe (MW = 131 amu) at 300 K. The total pressure of the mixture is 750 mmHg. What are the partial pressures of each of the gases? (blue = He; green = Xe)?

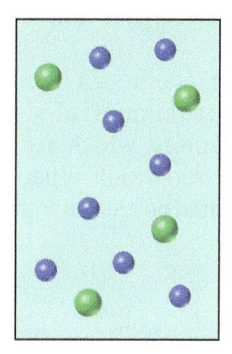

8.12 Liquids

Molecules are in constant motion in the liquid state, just as they are in gases. If a molecule happens to be near the surface of a liquid, and if it has enough energy, it can break free of the liquid and escape into the gas state, called **vapor**. In an open container, the now gaseous molecule will wander away from the liquid, and the process will continue until all the molecules escape from the container (Figure 8.19a). This, of course, is what happens during *evaporation*. We are all familiar with puddles of water evaporating after a rainstorm.

If the liquid is in a closed container, the situation is different because the gaseous molecules cannot escape. Thus, the random motion of the molecules occasionally brings them back into the liquid. After the concentration of molecules in the gas state has increased sufficiently, the number of molecules reentering the liquid becomes equal to the number escaping from the liquid (Figure 8.19b). At this point, a dynamic equilibrium exists, exactly as in a chemical reaction at equilibrium. Evaporation and condensation take place at the same rate, and the concentration of vapor in the container is constant as long as the temperature does not change.

Once molecules have escaped from the liquid into the gas state, they are subject to all the gas laws previously discussed. In a closed container at equilibrium, for example, the vapor molecules will make their own contribution to the total pressure of gases above the liquid according to Dalton's Law (Section 8.11). We call this contribution the **vapor pressure** of the liquid.

Vapor The gas molecules are in equilibrium with a liquid.

Vapor pressure The partial pressure of vapor molecules in equilibrium with a liquid.

Figure 8.19
The transfer of molecules between liquid and gas states.
(a) Molecules escape from an open container and drift away until the liquid has entirely evaporated.
(b) Molecules in a closed container cannot escape. Instead, they reach an equilibrium in which the rates of molecules leaving the liquid and returning to the liquid are equal, and the concentration of molecules in the gas state is constant.

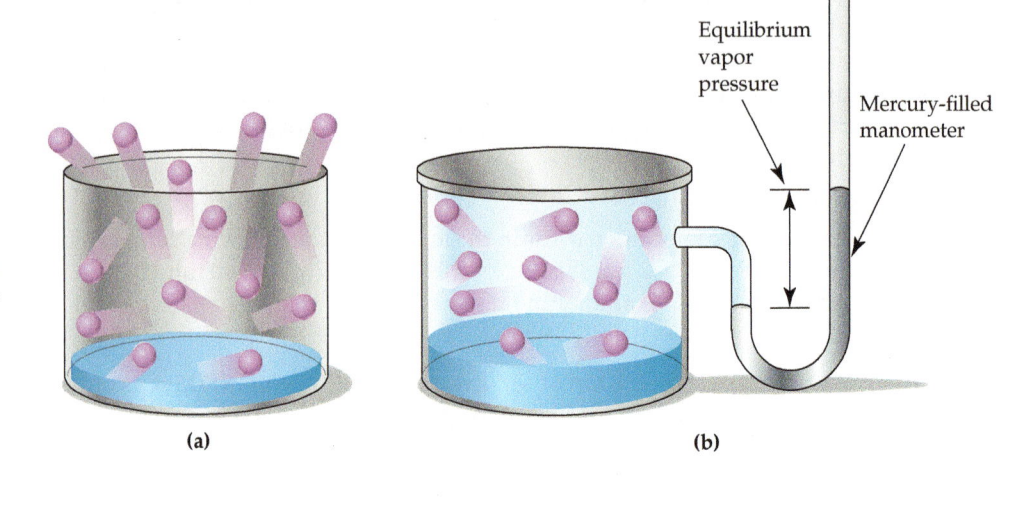

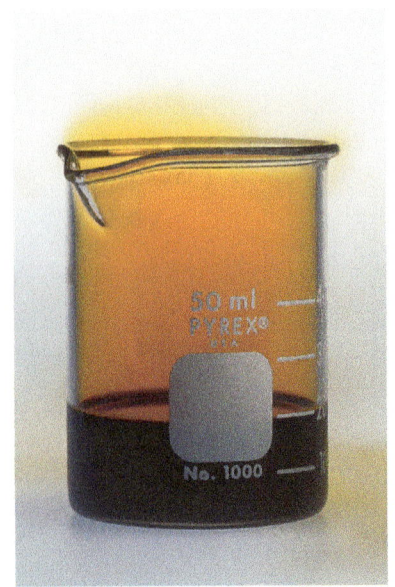

▲ Because bromine is colored, it is possible to see its gaseous reddish vapor above the liquid.

Normal boiling point The boiling point at a pressure of exactly 1 atmosphere.

Vapor pressure depends on both temperature and the chemical identity of a liquid. As the temperature rises, molecules become more energetic and more likely to escape into the gas state. Thus, vapor pressure rises with increasing temperature until ultimately it becomes equal to the pressure of the atmosphere. At this point, bubbles of vapor form under the surface and force their way to the top, giving rise to the violent action observed during a vigorous boil. At an atmospheric pressure of exactly 760 mmHg, boiling occurs at what is called the **normal boiling point**.

The vapor pressure and boiling point of a liquid will also depend on the intermolecular forces at work between liquid molecules. Ether molecules, for example, can engage in dipole–dipole interactions, which are weaker than the hydrogen bonds formed between water molecules. As a result, ether exhibits both lower vapor pressures and a lower boiling point than water, as seen in Figure 8.20.

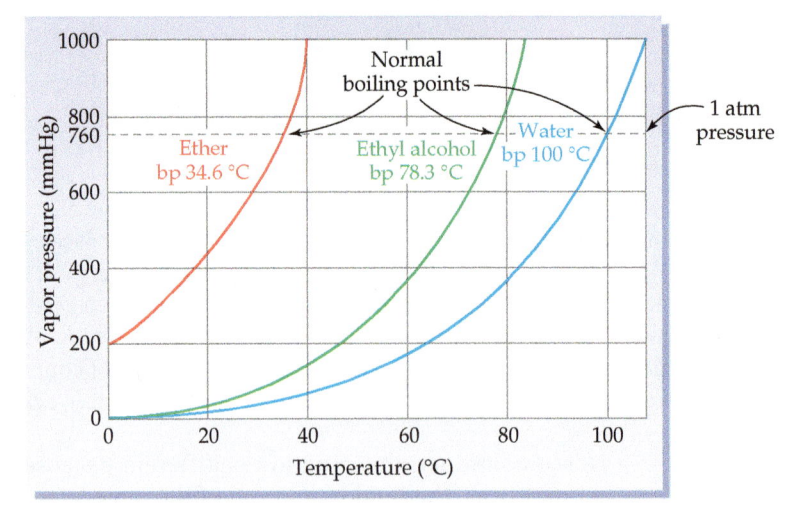

▲ **Figure 8.20**
A plot of the change of vapor pressure with temperature for ethyl ether, ethyl alcohol, and water. At a liquid's boiling point, its vapor pressure is equal to atmospheric pressure. Commonly reported boiling points are those at 760 mmHg.

If atmospheric pressure is higher or lower than normal, the boiling point of a liquid changes accordingly. At high altitudes, for example, atmospheric pressure is lower than at sea level, and boiling points are also lower. On top of Mt. Everest (29,035 ft; 8850 m), atmospheric pressure is about 245 mmHg and the boiling temperature of water is only 71 °C. If the atmospheric pressure is higher than normal, the boiling point is also higher. This principle is used in strong vessels known as *autoclaves*, in which water at high pressure is heated to the temperatures needed for sterilizing medical and dental instruments (170 °C).

Many familiar properties of liquids can be explained by the intermolecular forces just discussed. We all know, for instance, that some liquids, such as water or gasoline, flow easily when poured, whereas others, such as motor oil or maple syrup, flow sluggishly.

The measure of a liquid's resistance to flow is called its *viscosity*. Not surprisingly, viscosity is related to the ease with which individual molecules move around in the liquid and thus to the intermolecular forces present. Substances such as gasoline, which have small, nonpolar molecules, experience only weak intermolecular forces and have relatively low viscosities, whereas more polar substances such as glycerin $[C_3H_5(OH)_3]$ experience stronger intermolecular forces and so have higher viscosities.

Another familiar property of liquids is *surface tension*, the resistance of a liquid to spreading out and increasing its surface area. The beading-up of water on a newly waxed car and the ability of a water strider to walk on water are both due to surface tension.

Surface tension is caused by the difference between the intermolecular forces experienced by molecules at the surface of the liquid and those experienced by molecules in the interior. Molecules in the interior of a liquid are surrounded and experience maximum intermolecular forces, whereas molecules at the surface have fewer neighbors and feel weaker forces. Surface molecules are therefore less stable, and the liquid acts to minimize their number by minimizing the surface area (Figure 8.21).

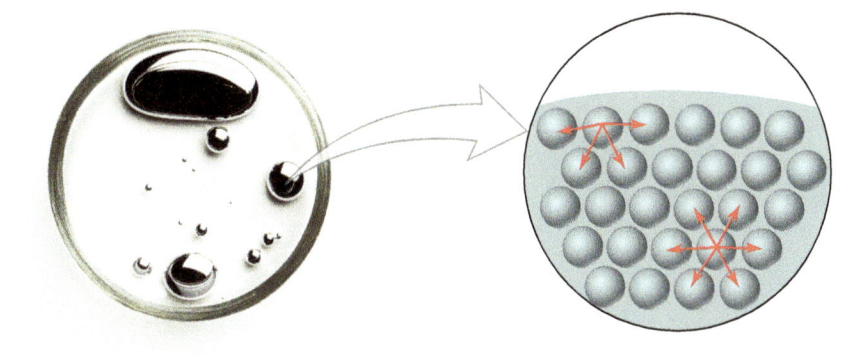

▲ Surface tension allows a water strider to walk on water without penetrating the surface.

◀ **Figure 8.21**
Surface tension.
Surface tension is caused by the different forces experienced by molecules in the interior of a liquid and those on the surface. Molecules on the surface are less stable because they feel fewer attractive forces, so the liquid acts to minimize their number by minimizing surface area.

▶▶ Recall from Section 1.13 that specific heat is the amount of heat required to raise the temperature of 1g of a substance by 1 °C.

8.13 Water: A Unique Liquid

Ours is a world based on water. Water covers nearly 71% of the earth's surface, it accounts for 66% of the mass of an adult human body, and it is needed by all living things. The water in our blood forms the transport system that circulates substances throughout our body, and water is the medium in which all biochemical reactions are carried out. Largely because of its strong hydrogen bonding, water has many properties that are quite different from those of other compounds.

Water has the highest specific heat of any liquid, giving it the capacity to absorb a large quantity of heat while changing only slightly in temperature. As a result, large lakes and other bodies of water tend to moderate the air temperature and climate of surrounding areas. Another consequence of the high specific heat of water is that the human body is better able to maintain a steady internal temperature under changing outside conditions.

In addition to a high specific heat, water has an unusually high *heat of vaporization* (540 cal/g or 2.3 kJ/g), meaning that it carries away a large amount of heat when it evaporates. You can feel the effect of water evaporation on your wet skin when the wind blows. Even when comfortable, your body is still relying for cooling on the heat carried away from the skin and lungs by evaporating water. The heat generated by the

▲ The moderate year-round temperatures in San Francisco are due to the large heat capacity of the surrounding waters.

chemical reactions of metabolism is carried by blood to the skin, where water moves through cell walls to the surface and evaporates. When metabolism, and therefore heat generation, speeds up, blood flow increases and capillaries dilate so that heat is brought to the surface faster.

Water is also unique in what happens as it changes from a liquid to a solid. Most substances are more dense as solids than as liquids because molecules are more closely packed in the solid than in the liquid. Water, however, is different. Liquid water has a maximum density of 1.000 g/mL at 3.98 °C but then becomes *less* dense as it cools. When it freezes, its density decreases still further to 0.917 g/mL.

As water freezes, each molecule is locked into position by hydrogen bonding to four other water molecules (Figure 8.22). The resulting structure has more open space than does liquid water, accounting for its lower density. As a result, ice floats on liquid water, and lakes and rivers freeze from the top down. If the reverse were true, fish would be killed in winter as they became trapped in ice at the bottom.

▶ **Figure 8.22**
Ice.
Ice consists of individual H_2O molecules held rigidly together in an ordered manner by hydrogen bonds. The open, cage-like crystal structure shows why ice is less dense than liquid water.

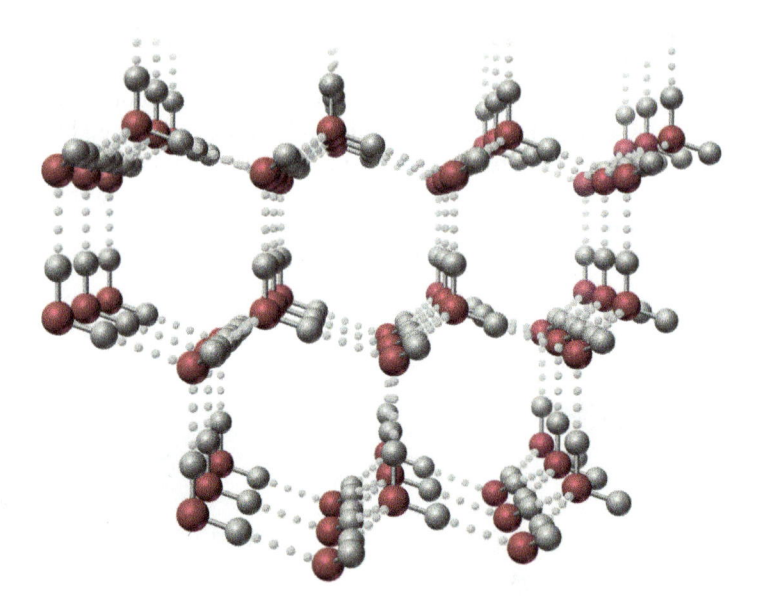

8.14 Solids

A brief look around us reveals that most substances are solids rather than liquids or gases. It is also obvious that there are many different kinds of solids. Some, such as iron and aluminum, are hard and metallic; others, such as sugar and table salt, are crystalline and easily broken; and still others, such as rubber and many plastics, are soft and amorphous.

▲ Crystalline solids, such as pyrite (left) and fluorite (right) have flat faces and distinct angles. The octahedral shape of pyrite and the cubic shape of fluorite reflect similarly ordered arrangements of particles at the atomic level.

The most fundamental distinction between solids is that some are crystalline and some are amorphous. A **crystalline solid** is one whose particles—whether atoms, ions, or molecules—have an ordered arrangement extending over a long range. This order on the atomic level is also seen on the visible level, because crystalline solids usually have flat faces and distinct angles.

Crystalline solids can be further categorized as ionic, molecular, covalent network, or metallic. *Ionic solids* are those like sodium chloride, whose constituent particles are ions. A crystal of sodium chloride is composed of alternating Na^+ and Cl^- ions ordered in a regular three-dimensional arrangement and held together by ionic bonds (see Figure 3.3). *Molecular solids* are those like sucrose or ice, whose constituent particles are molecules held together by the intermolecular forces discussed in Section 8.2. *Covalent network solids* are those like diamond (Figure 8.23) or quartz (SiO_2), whose atoms are linked together by covalent bonds into a giant three-dimensional array. In effect, a covalent network solid is one *very* large molecule.

Crystalline solid A solid whose atoms, molecules, or ions are rigidly held in an ordered arrangement.

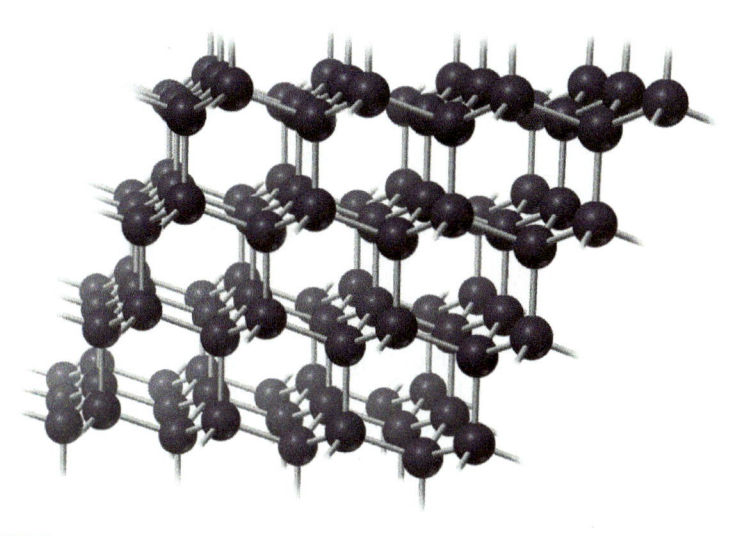

▲ **Figure 8.23**

Diamond. Diamond is a covalent network solid—one very large molecule of carbon atoms linked by covalent bonds.

Metallic solids, such as silver or iron, can be viewed as vast three-dimensional arrays of metal cations immersed in a sea of electrons that are free to move about. This continuous electron sea acts both as a glue to hold the cations together and as a mobile carrier of charge to conduct electricity. Furthermore, the fact that bonding attractions extend uniformly in all directions explains why metals are malleable rather than brittle. When a metal crystal receives a sharp blow, no spatially oriented bonds are broken; instead, the electron sea simply adjusts to the new distribution of cations.

An **amorphous solid**, by contrast with a crystalline solid, is one whose constituent particles are randomly arranged and have no ordered long-range structure. Amorphous solids often result when liquids cool before they can achieve internal order, or when their molecules are large and tangled together, as happens in many polymers. Glass is an amorphous solid, as are tar, the gemstone opal, and some hard candies. Amorphous solids differ from crystalline solids by softening over a wide temperature range rather than having sharp melting points and by shattering to give pieces with curved rather than planar faces.

Amorphous solid A solid whose particles do not have an orderly arrangement.

A summary of the different types of solids and their characteristics is given in Table 8.4.

TABLE 8.4 **Types of Solids**

Substance	Smallest Unit	Interparticle Forces	Properties	Examples
Ionic solid	Ions	Attraction between positive and negative ions	Brittle and hard; high mp; crystalline	NaCl, KI, $Ca_3(PO_4)_2$
Molecular solid	Molecules	Intermolecular forces	Soft; low to moderate mp; crystalline	Ice, wax, frozen CO_2, all solid organic compounds
Covalent network	Atoms	Covalent bonds	Very hard; very high mp; crystalline	Diamond, quartz (SiO_2), tungsten carbide (WC)
Metal or alloy	Metal atoms	Metallic bonding (attraction between metal ions and surrounding mobile electrons)	Lustrous; soft (Na) to hard (Ti); high melting; crystalline	Elements (Fe, Cu, Sn, ...), bronze (CuSn alloy), amalgams (Hg+ other metals)
Amorphous solid	Atoms, ions, or molecules (including polymer molecules)	Any of the above	Noncrystalline; no sharp mp; able to flow (may be very slow); curved edges when shattered	Glasses, tar, some plastics

8.15 Changes of State

What happens when a solid is heated? As more and more energy is added, molecules begin to stretch, bend, and vibrate more vigorously, and atoms or ions wiggle about with more energy. Finally, if enough energy is added and the motions become vigorous enough, particles start to break free from one another and the substance starts to melt. Addition of more heat continues the melting process until all particles have broken free and are in the liquid phase. The quantity of heat required to completely melt a substance once it reaches its melting point is called its **heat of fusion**. After melting is complete, further addition of heat causes the temperature of the liquid to rise.

The change of a liquid into a vapor proceeds in the same way as the change of a solid into a liquid. When you first put a pan of water on the stove, all the added heat goes into raising the temperature of the water. Once the boiling point is reached, further absorbed heat goes into freeing molecules from their neighbors as they escape into the gas state. The quantity of heat needed to completely vaporize a liquid once it reaches its boiling point is called its **heat of vaporization**. A liquid with a low heat of vaporization, like rubbing alcohol (isopropyl alcohol), evaporates rapidly and is said to be *volatile*. If you spill a volatile liquid on your skin, you will feel a cooling effect as it evaporates because it is absorbing heat from your body.

It is important to know the difference between heat that is added or removed to change the *temperature* of a substance and heat that is added or removed to change the *phase* of a substance. Remember that temperature is a measure of the kinetic energy in a substance (see Section 7.1). When a substance is above or below its phase-change temperature (i.e., melting point or boiling point), adding or removing heat will simply change the kinetic energy and, hence, the temperature of the substance. The amount of heat needed to produce a given temperature change was presented previously (Section 1.13), but is worth presenting again here:

$$\text{Heat (cal or J)} = \text{Mass (g)} \times \text{Temperature change (°C)} \times \text{Specific heat} \left(\frac{\text{cal or J}}{\text{g} \times \text{°C}}\right)$$

Heat of fusion The quantity of heat required to completely melt one gram of a substance once it has reached its melting point.

Heat of vaporization The quantity of heat needed to completely vaporize one gram of a liquid once it has reached its boiling point.

In contrast, when a substance is at its phase-change temperature, heat that is added is being used to overcome the intermolecular forces holding particles in that phase. The temperature remains constant until *all* particles have been converted to the next phase. The energy needed to complete the phase change depends only on the amount of the substance and the heat of fusion (for melting) or the heat of vaporization (for boiling).

$$\text{Heat (cal or J)} = \text{Mass (g)} \times \text{Heat of fusion} \left(\frac{\text{cal or J}}{g} \right)$$

$$\text{Heat (cal or J)} = \text{Mass (g)} \times \text{Heat of vaporization} \left(\frac{\text{cal or J}}{g} \right)$$

If the intermolecular forces are strong then large amounts of heat must be added to overcome these forces, and the heats of fusion and vaporization will be large. A list of heats of fusion and heats of vaporization for some common substances is given in Table 8.5. Butane, for example, has a small heat of vaporization since the predominant intermolecular forces in butane (dispersion) are relatively weak. Water, on the other hand, has a particularly high heat of vaporization because of its unusually strong hydrogen bonding interactions. Thus, water evaporates more slowly than many other liquids, takes a long time to boil away, and absorbs more heat in the process. A so-called *heating curve*, which indicates the temperature and state changes as heat is added, is shown in Figure 8.24.

TABLE 8.5 Melting Points, Boiling Points, Heats of Fusion, and Heats of Vaporization of Some Common Substances

Substance	Melting Point (°C)	Boiling Point (°C)	Heat of Fusion (cal/g; J/g)	Heat of Vaporization (cal/g; J/g)
Ammonia	−77.7	−33.4	84.0; 351	327; 1370
Butane	−138.4	−0.5	19.2; 80.3	92.5; 387
Ether	−116	34.6	23.5; 98.3	85.6; 358
Ethyl alcohol	−117.3	78.5	26.1; 109	200; 837
Isopropyl alcohol	−89.5	82.4	21.4; 89.5	159; 665
Sodium	97.8	883	14.3; 59.8	492; 2060
Water	0.0	100.0	79.7; 333	540; 2260

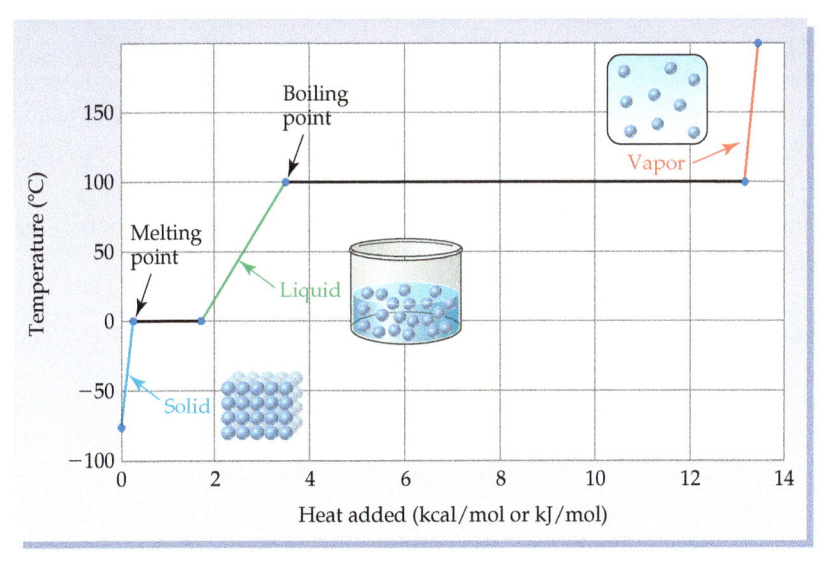

▲ **Figure 8.24**
A heating curve for water, showing the temperature and state changes that occur when heat is added.
The horizontal lines at 0 °C and 100 °C represent the heat of fusion and heat of vaporization, respectively.

244 CHAPTER 8 Gases, Liquids, and Solids

Worked Example 8.13 Heat of Fusion: Calculating Total Heat of Melting

Naphthalene, an organic substance often used in mothballs, has a heat of fusion of 35.7 cal/g (149 J/g) and a molar mass of 128.0 g/mol. How much heat in kilocalories is required to melt 0.300 mol of naphthalene?

ANALYSIS The heat of fusion tells how much heat is required to melt 1 g. To find the amount of heat needed to melt 0.300 mol, we need a mole-to-mass conversion.

BALLPARK ESTIMATE Naphthalene has a molar mass of 128.0 g/mol, so 0.300 mol has a mass of about one-third this amount, or about 40 g. Approximately 35 cal or 150 J is required to melt 1 g, so we need about 40 times this amount of heat, or ($35 \times 40 = 1400$ cal $= 1.4$ kcal, or $150 \times 40 = 6000$ J $= 6.0$ kJ).

SOLUTION

STEP 1: Identify known information. We know heat of fusion (cal/g), and the number of moles of naphthalene.

Heat of fusion = 35.7 cal/g, or 149 J/g
Moles of naphthalene = 0.300 mol

STEP 2: Identify answer and units.

Heat = ?? cal or J

STEP 3: Identify conversion factors. First convert moles of naphthalene to grams using the molar mass (128 g/mol) as a conversion factor. Then use the heat of fusion as a conversion factor to calculate the total heat necessary to melt the mass of naphthalene.

$$(0.300 \text{ mol naphthalene})\left(\frac{128.0 \text{ g}}{1 \text{ mol}}\right) = 38.4 \text{ g napthalene}$$

Heat of fusion = 35.7 cal/g or 149 J/g

STEP 4: Solve. Multiplying the mass of naphthalene by the heat of fusion then gives the answer.

$$(38.4 \text{ g naphthalene})\left(\frac{35.7 \text{ cal}}{1 \text{ g naphthalene}}\right) = 1370 \text{ cal} = 1.37 \text{ kcal, or}$$

$$(38.4 \text{ g naphthalene})\left(\frac{149 \text{ J}}{1 \text{ g naphthalene}}\right) = 5720 \text{ J} = 5.72 \text{ kJ}$$

BALLPARK CHECK The calculated result agrees with our estimate (1.4 kcal or 6.0 kJ)

PROBLEM 8.23
How much heat in kilocalories is required to melt and boil 1.50 mol of isopropyl alcohol (rubbing alcohol; molar mass = 60.0 g/mol)? The heat of fusion and heat of vaporization of isopropyl alcohol are given in Table 8.5.

PROBLEM 8.24
How much heat in kilojoules is released by the condensation of 2.5 mol of steam? The heat of vaporization is given in Table 8.5.

PROBLEM 8.25
The physical state of CO_2 depends on the temperature and pressure (see Chemistry in Action: CO_2 as an Environmentally Friendly Solvent on p. 245). In what state would you expect to find CO_2 at 50 atm and 25 °C?

CHEMISTRY IN ACTION

CO₂ as an Environmentally Friendly Solvent

When you think of CO_2 you most likely think of the gas that is absorbed by plants for photosynthesis or exhaled by animals during respiration. You have also probably seen CO_2 in the form of dry ice, that very cold solid that sublimes to a gas. But how can CO_2 be a solvent? After all, carbon dioxide is a gas, not a liquid, at room temperature. Furthermore, CO_2 at atmospheric pressure does not become liquid even when cooled. When the temperature drops to $-78\ °C$ at 1 atm pressure, CO_2 goes directly from gas to solid (dry ice) without first becoming liquid. Only when the pressure is raised does liquid CO_2 exist. At a room temperature of $22.4\ °C$, a pressure of 60 atm is needed to force gaseous CO_2 molecules close enough together so they condense to a liquid. Even as a liquid, though, CO_2 is not a particularly good solvent. Only when it enters an unusual and rarely seen state of matter called the *supercritical state* does CO_2 become a remarkable solvent.

To understand the supercritical state of matter, consider the two factors that determine the physical state of a substance: temperature and pressure. In the solid state, molecules are packed closely together and do not have enough kinetic energy to overcome the intermolecular forces. If we increase the temperature, however, we can increase the kinetic energy so that the molecules can move apart and produce a phase change to either a liquid or a gas. In the gas state, molecules are too far apart to interact, but increasing the pressure will force molecules closer together, and, eventually, intermolecular attractions between molecules will cause them to condense into a liquid or solid state. This dependence of the physical state on temperature and pressure is represented by a *phase diagram*, such as the one shown here for CO_2.

The supercritical state represents a situation that is intermediate between liquid and gas. There is *some* space between molecules, but not much. The molecules are too far apart to be truly a liquid, yet they are too close together to be truly a gas. Supercritical CO_2 exists above the *critical point*, when the pressure is above 72.8 atm and the temperature is above $31.2\ °C$. This pressure is high enough to force molecules close together and prevent them from expanding into the gas state. Above this temperature, however, the molecules have too much kinetic energy to condense into the liquid state.

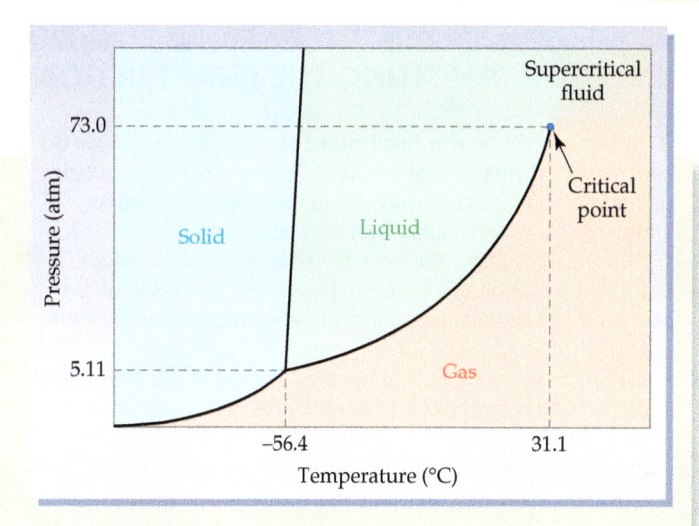

Because open spaces already exist between CO_2 molecules, it is energetically easy for dissolved molecules to slip in, and supercritical CO_2 is therefore an extraordinarily good solvent. Among its many applications, supercritical CO_2 is used in the beverage and food-processing industries to decaffeinate coffee beans and to obtain spice extracts from vanilla, pepper, cloves, nutmeg, and other seeds. In the cosmetics and perfume industry, fragrant oils are extracted from flowers using supercritical CO_2. Perhaps the most important future application is the use of carbon dioxide for dry-cleaning clothes, thereby replacing environmentally harmful chlorinated solvents.

The use of supercritical CO_2 as a solvent has many benefits, including the fact that it is nontoxic and nonflammable. Most important, though, is that the technology is environmentally friendly. Industrial processes using CO_2 are designed as closed systems so that the CO_2 is recaptured after use and continually recycled. No organic solvent vapors are released into the atmosphere and no toxic liquids seep into groundwater supplies, as can occur with current procedures using chlorinated organic solvents. The future looks bright for this new technology.

See Chemistry in Action Problems 8.104 and 8.105 at the end of the chapter.

SUMMARY: REVISITING THE CHAPTER GOALS

1. What are the major intermolecular forces, and how do they affect the states of matter? There are three major types of *intermolecular forces*, which act to hold molecules near one another in solids and liquids. *Dipole–dipole forces* are the electrical attractions that occur between polar molecules. *London dispersion forces* occur between all molecules as a result of temporary molecular polarities due to unsymmetrical electron distribution. These forces increase in strength with molecular weight and with the surface area of molecules. *Hydrogen bonding*, the strongest of the three intermolecular forces, occurs between a hydrogen atom bonded to O, N, or F and a nearby O, N, or F atom (see *Problems 34–37, 116*).

2. How do scientists explain the behavior of gases? According to the *kinetic-molecular theory of gases*, the physical behavior of gases can be explained by assuming that they consist of particles moving rapidly at random, separated from other particles by great distances, and colliding without loss of energy. Gas pressure is the result of molecular collisions with a surface (see *Problems 29, 30, 40, 41, 53, 59, 68, 102, 103, 106*).

3. How do gases respond to changes in temperature, pressure, and volume? *Boyle's law* says that the volume of a fixed amount of gas at constant temperature is inversely proportional to its pressure ($P_1V_1 = P_2V_2$). *Charles's law* says that the volume of a fixed amount of gas at constant pressure is directly proportional to its Kelvin temperature ($V_1/T_1 = V_2/T_2$). *Gay-Lussac's law* says that the pressure of a fixed amount of gas at constant volume is directly proportional to its Kelvin temperature ($P_1/T_1 = P_2/T_2$). Boyle's law, Charles's law, and Gay-Lussac's law together give the *combined gas law* ($P_1V_1/T_1 = P_2V_2/T_2$), which applies to changing conditions for a fixed quantity of gas. *Avogadro's law* says that equal volumes of gases at the same temperature and pressure contain the same number of moles ($V_1/n_1 = V_2/n_2$) (see *Problems 26, 27, 32, 38–75, 107, 111, 112, 115*).

4. What is the ideal gas law? The four gas laws together give the *ideal gas law*, $PV = nRT$, which relates the effects of temperature, pressure, volume, and molar amount. At 0 °C and 1 atm pressure, called *standard temperature and pressure* (STP), 1 mol of any gas (6.02×10^{23} molecules) occupies a volume of 22.4 L (see *Problems 76–85, 108–110, 113–115, 118, 119*).

5. What is partial pressure? The amount of pressure exerted by an individual gas in a mixture is called the *partial pressure* of the gas. According to *Dalton's law*, the total pressure exerted by the mixture is equal to the sum of the partial pressures of the individual gases (see *Problems 33, 86–89, 117*).

6. What are the various kinds of solids, and how do they differ? Solids are either crystalline or amorphous. *Crystalline solids* are those whose constituent particles have an ordered arrangement; *amorphous solids* lack internal order and do not have sharp melting points. There are several kinds of crystalline solids: *Ionic solids* are those such as sodium chloride, whose constituent particles are ions. *Molecular solids* are those such as ice, whose constituent particles are molecules held together by intermolecular forces. *Covalent network solids* are those such as diamond, whose atoms are linked together by covalent bonds into a giant three-dimensional array. *Metallic solids*, such as silver or iron, also consist of large arrays of atoms, but their crystals have metallic properties such as electrical conductivity (see *Problems 96–99*).

7. What factors affect a change of state? When a solid is heated, particles begin to move around freely at the *melting point*, and the substance becomes liquid. The amount of heat necessary to melt a given amount of solid at its melting point is its *heat of fusion*. As a liquid is heated, molecules escape from the surface of a liquid until an equilibrium is reached between liquid and gas, resulting in a *vapor pressure* of the liquid. At a liquid's *boiling point*, its vapor pressure equals atmospheric pressure, and the entire liquid is converted into gas. The amount of heat necessary to vaporize a given amount of liquid at its boiling point is called its *heat of vaporization* (see *Problems 27, 28, 31, 90–95, 98, 99, 104*).

KEY WORDS

Amorphous solid, *p. 241*
Avogadro's law, *p. 232*
Boiling point (bp), *p. 215*
Boyle's law, *p. 226*
Change of state, *p. 214*
Charles's law, *p. 229*
Combined gas law, *p. 231*
Crystalline solid, *p. 241*
Dalton's law, *p. 236*
Dipole–dipole force, *p. 216*
Gas constant (R), *p. 234*

Gas laws, *p. 225*
Gay-Lussac's law, *p. 230*
Heat of fusion, *p. 242*
Heat of vaporization, *p. 242*
Hydrogen bond, *p. 217*
Ideal gas, *p. 221*
Ideal gas law, *p. 233*
Intermolecular force, *p. 216*
Kinetic–molecular theory of gases, *p. 220*
London dispersion force, *p. 217*

Melting point (mp), *p. 215*
Normal boiling point, *p. 238*
Partial pressure, *p. 236*
Pressure (P), *p. 221*
Standard temperature and pressure (STP), *p. 233*
Standard molar volume, *p. 233*
Vapor, *p. 237*
Vapor pressure, *p. 237*

CONCEPT MAP: GASES, LIQUIDS, AND SOLIDS

Concept Map. The physical state of matter (solid, liquid, gas) depends on the strength of the intermolecular forces between molecules compared to the kinetic energy of the molecules. When the kinetic energy (i.e, temperature) is greater than the forces holding molecules in a given state, then a phase change occurs. Thus, the physical properties of matter (melting and boiling points, etc.) depend on the strength of the intermolecular forces between molecules, which depend on chemical structure and molecular shape. These relationships are reflected here in Figure 8.25.

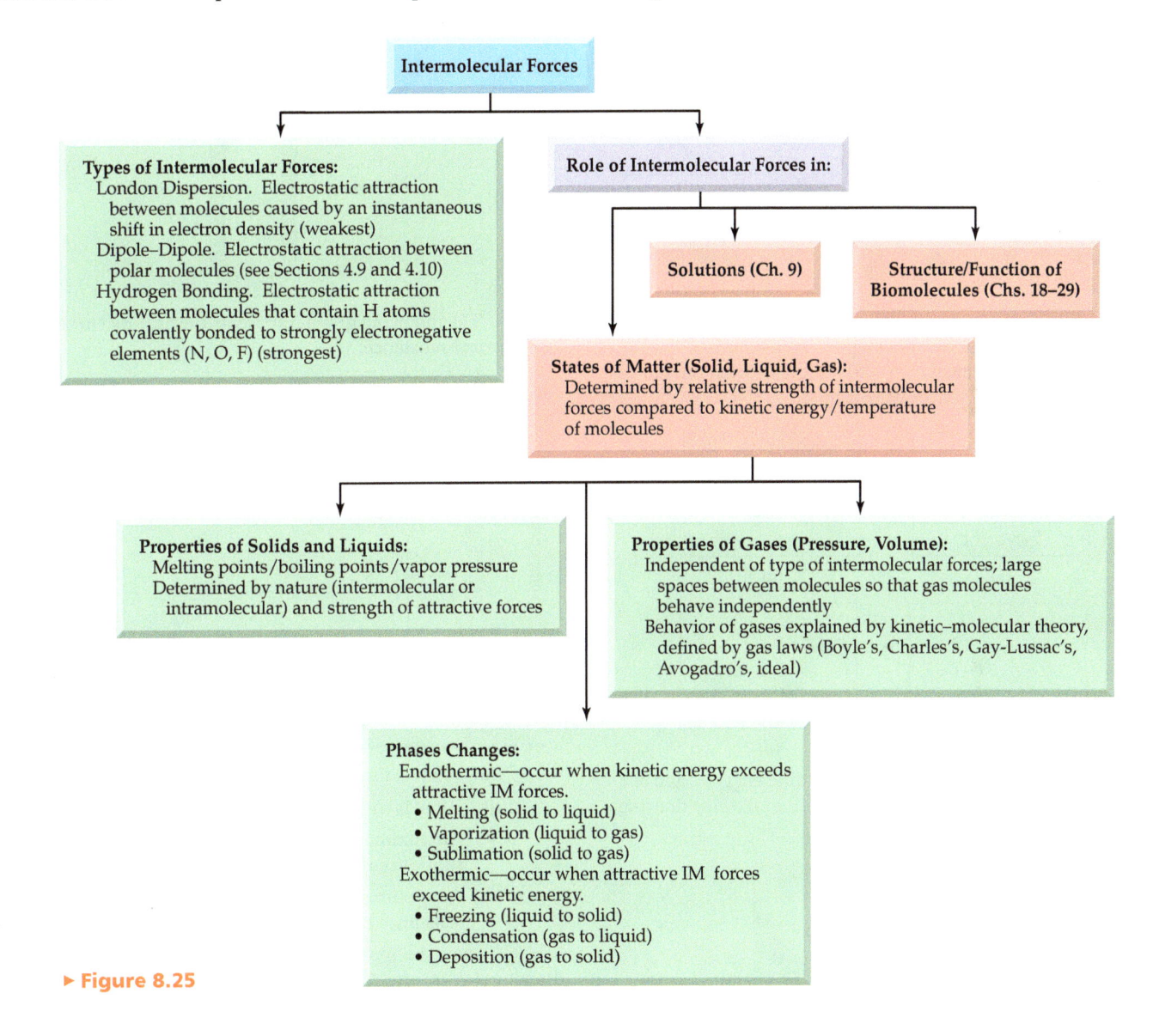

```
                        ┌─────────────────────┐
                        │ Intermolecular Forces│
                        └──────────┬──────────┘
                  ┌────────────────┴────────────────┐
                  ▼                                 ▼
   ┌──────────────────────────┐        ┌──────────────────────────┐
   │ Types of Intermolecular  │        │ Role of Intermolecular   │
   │ Forces:                  │        │ Forces in:               │
   └──────────────────────────┘        └──────────────────────────┘
```

Types of Intermolecular Forces:
- London Dispersion. Electrostatic attraction between molecules caused by an instantaneous shift in electron density (weakest)
- Dipole–Dipole. Electrostatic attraction between polar molecules (see Sections 4.9 and 4.10)
- Hydrogen Bonding. Electrostatic attraction between molecules that contain H atoms covalently bonded to strongly electronegative elements (N, O, F) (strongest)

Role of Intermolecular Forces in:
- Solutions (Ch. 9)
- Structure/Function of Biomolecules (Chs. 18–29)

States of Matter (Solid, Liquid, Gas): Determined by relative strength of intermolecular forces compared to kinetic energy/temperature of molecules

Properties of Solids and Liquids:
Melting points/boiling points/vapor pressure Determined by nature (intermolecular or intramolecular) and strength of attractive forces

Properties of Gases (Pressure, Volume):
Independent of type of intermolecular forces; large spaces between molecules so that gas molecules behave independently
Behavior of gases explained by kinetic–molecular theory, defined by gas laws (Boyle's, Charles's, Gay-Lussac's, Avogadro's, ideal)

Phases Changes:
Endothermic—occur when kinetic energy exceeds attractive IM forces.
- Melting (solid to liquid)
- Vaporization (liquid to gas)
- Sublimation (solid to gas)

Exothermic—occur when attractive IM forces exceed kinetic energy.
- Freezing (liquid to solid)
- Condensation (gas to liquid)
- Deposition (gas to solid)

▶ **Figure 8.25**

UNDERSTANDING KEY CONCEPTS

8.26 Assume that you have a sample of gas in a cylinder with a movable piston, as shown in the following drawing:

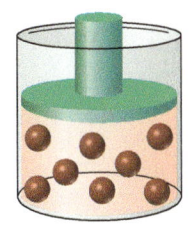

Redraw the apparatus to show what the sample will look like after the following changes:

(a) The temperature is increased from 300 K to 450 K at constant pressure.

(b) The pressure is increased from 1 atm to 2 atm at constant temperature.

(c) The temperature is decreased from 300 K to 200 K and the pressure is decreased from 3 atm to 2 atm.

CHAPTER 8 Gases, Liquids, and Solids

8.27 Assume that you have a sample of gas at 350 K in a sealed container, as represented in part (a). Which of the drawings (b)–(d) represents the gas after the temperature is lowered from 350 K to 150 K if the gas has a boiling point of 200 K? Which drawing represents the gas at 150 K if the gas has a boiling point of 100 K?

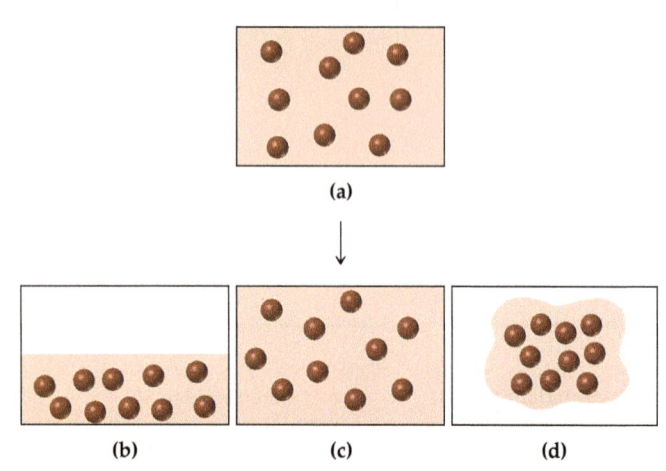

8.28 Assume that drawing (a) represents a sample of H_2O at 200 K. Which of the drawings (b)–(d) represents what the sample will look like when the temperature is raised to 300 K?

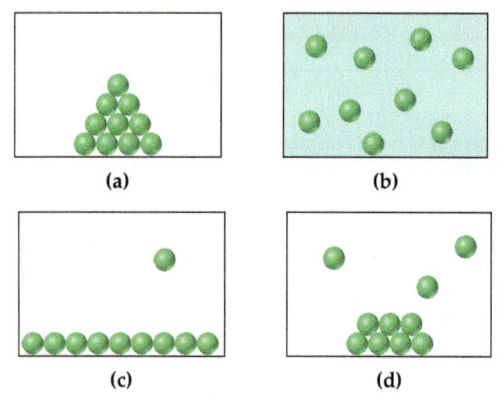

8.29 Three bulbs, two of which contain different gases and one of which is empty, are connected as shown in the following drawing:

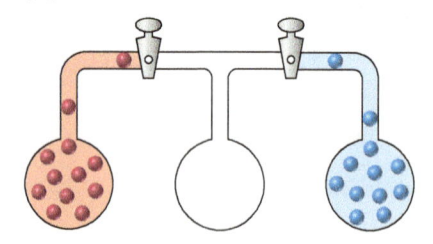

Redraw the apparatus to represent the gases after the stopcocks are opened and the system is allowed to come to equilibrium.

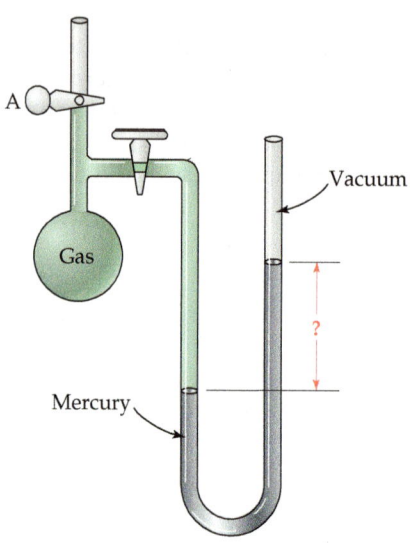

8.30 Redraw the following open-ended manometer to show what it would look like when stopcock A is opened.

8.31 The following graph represents the heating curve of a hypothetical substance:

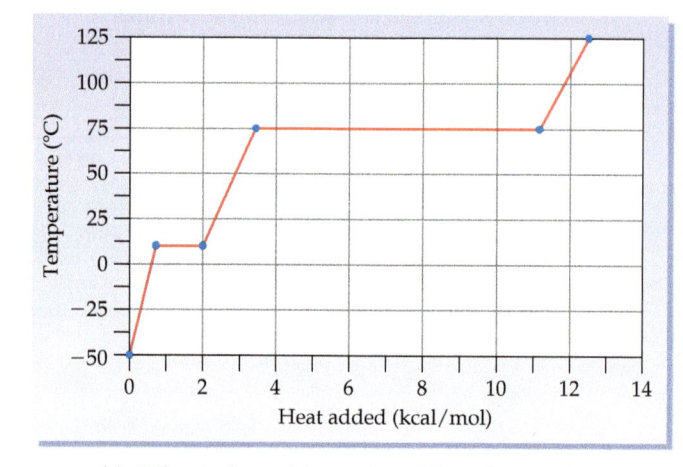

(a) What is the melting point of the substance?
(b) What is the boiling point of the substance?
(c) Approximately what is the heat of fusion for the substance in kcal/mol?
(d) Approximately what is the heat of vaporization for the substance in kcal/mol?

8.32 Show the approximate level of the movable piston in drawings (a)–(c) after the indicated changes have been made to the gas.

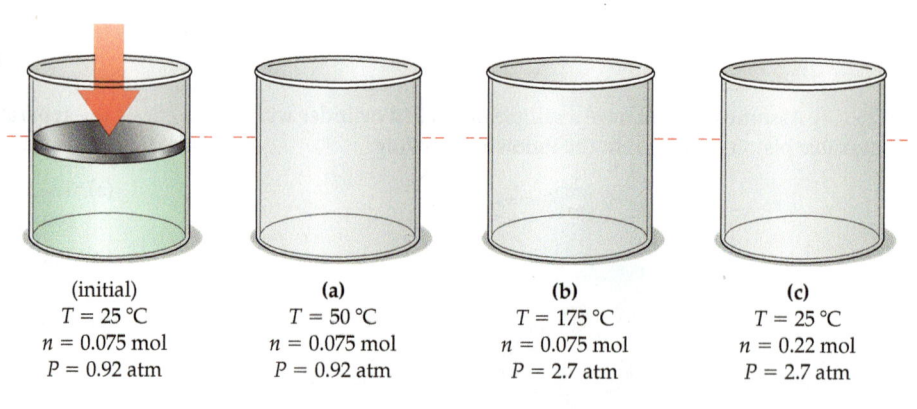

(initial)
T = 25 °C
n = 0.075 mol
P = 0.92 atm

(a)
T = 50 °C
n = 0.075 mol
P = 0.92 atm

(b)
T = 175 °C
n = 0.075 mol
P = 2.7 atm

(c)
T = 25 °C
n = 0.22 mol
P = 2.7 atm

8.33 The partial pressure of the blue gas in the container represented in the picture is 240 mmHg. What are the partial pressures of the yellow and red gases? What is the total pressure inside the container?

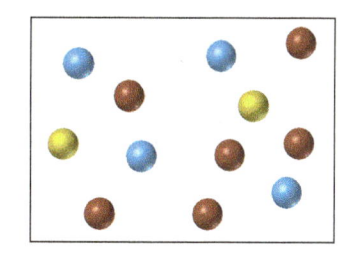

ADDITIONAL PROBLEMS

INTERMOLECULAR FORCES

8.34 What characteristic must a compound have to experience the following intermolecular forces?

(a) London dispersion forces   (b) Dipole–dipole forces
(c) Hydrogen bonding

8.35 Identify the predominant intermolecular force in each of the following substances.

(a) N_2 (b) HCN   (c) CCl_4
(d) NH_3 (e) CH_3Cl (f) CH_3COOH

8.36 Dimethyl ether (CH_3OCH_3) and ethanol (C_2H_5OH) have the same formula (C_2H_6O), but the boiling point of dimethyl ether is $-25\,°C$, while that of ethanol is $78\,°C$. Explain this difference in boiling points.

8.37 Iodine is a solid at room temperature (mp = $113.5\,°C$) while bromine is a liquid (mp = $-7\,°C$). Explain this difference in terms of intermolecular forces.

GASES AND PRESSURE

8.38 How is 1 atm of pressure defined?

8.39 List four common units for measuring pressure.

8.40 What are the four assumptions of the kinetic–molecular theory of gases?

8.41 How does the kinetic–molecular theory of gases explain gas pressure?

8.42 Convert the following values into mmHg:

(a) Standard pressure   (b) 25.3 psi   (c) 7.5 atm
(d) 28.0 in. Hg   (e) 41.8 Pa

8.43 Atmospheric pressure at the top of Mt. Whitney in California is 440 mmHg.

(a) How many atmospheres is this?
(b) How many pascals is this?

8.44 What is the pressure (in mmHg) inside a container of gas connected to a mercury-filled, open-ended manometer of the sort shown in Figure 8.10 when the level in the arm connected to the container is 17.6 cm lower than the level in the arm open to the atmosphere and the atmospheric pressure reading outside the apparatus is 754.3 mmHg? What is the pressure inside the container in atm?

8.45 What is the pressure (in atmospheres) inside a container of gas connected to a mercury-filled, open-ended manometer of the sort shown in Figure 8.10 when the level in the arm connected to the container is 28.3 cm higher than the level in the arm open to the atmosphere, and the atmospheric pressure reading outside the apparatus is 1.021 atm? What is the pressure in mmHg?

BOYLE'S LAW

8.46 What is Boyle's law, and what variables must be kept constant for the law to hold?

8.47 Which assumption(s) of the kinetic–molecular theory explain the behavior of gases described by Boyle's Law? Explain your answer.

8.48 The pressure of gas in a 600.0 mL cylinder is 65.0 mmHg. What is the new volume when the pressure is increased to 385 mmHg?

8.49 The volume of a balloon is 2.85 L at 1.00 atm. What pressure is required to compress the balloon to a volume of 1.70 L?

8.50 The use of chlorofluorocarbons (CFCs) as refrigerants and propellants in aerosol cans has been discontinued as a result of concerns about the ozone layer. If an aerosol can contained 350 mL of CFC gas at a pressure of 5.0 atm, what volume would this gas occupy at 1.0 atm?

8.51 A balloon occupies a volume of 1.25 L at sea level where the ambient pressure is 1 atm. What volume would the balloon occupy at an altitude of 35,000 ft, where the air pressure is only 220 mmHg?

CHARLES'S LAW

8.52 What is Charles's law, and what variables must be kept constant for the law to hold?

8.53 Which assumption(s) of the kinetic–molecular theory explain the behavior of gases described by Charles's Law? Explain your answer.

8.54 A hot-air balloon has a volume of 960 L at 291 K. To what temperature (in °C) must it be heated to raise its volume to 1200 L, assuming the pressure remains constant?

8.55 A hot-air balloon has a volume of 875 L. What is the original temperature of the balloon if its volume changes to 955 L when heated to 56 °C?

8.56 A gas sample has a volume of 185 mL at 38 °C. What is its volume at 97 °C?

8.57 A balloon has a volume of 43.0 L at 25 °C. What is its volume at 2.8 °C?

GAY-LUSSAC'S LAW

8.58 What is Gay-Lussac's law, and what variables must be kept constant for the law to hold?

8.59 Which assumption(s) of the kinetic–molecular theory explain the behavior of gases described by Gay-Lussac's Law? Explain your answer.

8.60 A glass laboratory flask is filled with gas at 25 °C and 0.95 atm pressure, sealed, and then heated to 117 °C. What is the pressure inside the flask?

8.61 An aerosol can has an internal pressure of 3.85 atm at 25 °C. What temperature is required to raise the pressure to 18.0 atm?

COMBINED GAS LAW

8.62 A gas has a volume of 2.84 L at 1.00 atm and 0 °C. At what temperature does it have a volume of 7.50 L at 520 mmHg?

8.63 A compressed-air tank carried by scuba divers has a volume of 6.80 L and a pressure of 120 atm at 20 °C. What is the volume of air in the tank at 0 °C and 1.00 atm pressure (STP)?

8.64 When H_2 gas was released by the reaction of HCl with Zn, the volume of H_2 collected was 75.4 mL at 23 °C and 748 mmHg. What is the volume of the H_2 at 0 °C and 1.00 atm pressure (STP)?

8.65 What is the effect on the volume of a gas if you simultaneously:

(a) Halve its pressure and double its Kelvin temperature?

(b) Double its pressure and double its Kelvin temperature?

8.66 What is the effect on the pressure of a gas if you simultaneously:

(a) Halve its volume and double its Kelvin temperature?

(b) Double its volume and halve its Kelvin temperature?

8.67 A small cylinder of helium gas used for filling balloons has a volume of 2.30 L and a pressure of 1850 atm at 25 °C. How many balloons can you fill if each one has a volume of 1.5 L and a pressure of 1.25 atm at 25 °C?

AVOGADRO'S LAW AND STANDARD MOLAR VOLUME

8.68 Explain Avogadro's law using the kinetic–molecular theory of gases.

8.69 What conditions are defined as standard temperature and pressure (STP)?

8.70 How many molecules are in 1.0 L of O_2 at STP or 1.0 L? How may grams of O_2?

8.71 How many moles of gas are in a volume of 48.6 L at STP?

8.72 What is the mass of CH_4 in a sample that occupies a volume of 16.5 L at STP?

8.73 Assume that you have 1.75 g of the deadly gas hydrogen cyanide, HCN. What is the volume of the gas at STP?

8.74 A typical room is 4.0 m long, 5.0 m wide, and 2.5 m high. What is the total mass of the oxygen in the room assuming that the gas in the room is at STP and that air contains 21% oxygen and 79% nitrogen?

8.75 What is the total volume and number of moles of nitrogen in the room described in Problem 8.74?

IDEAL GAS LAW

8.76 What is the ideal gas law?

8.77 How does the ideal gas law differ from the combined gas law?

8.78 Which sample contains more molecules: 2.0 L of Cl_2 at STP, or 3.0 L of CH_4 at 300 K and 1150 mmHg? Which sample weighs more?

8.79 Which sample contains more molecules: 2.0 L of CO_2 at 300 K and 500 mmHg, or 1.5 L of N_2 at 57 °C and 760 mmHg? Which sample weighs more?

8.80 If 2.3 mol of He has a volume of 0.15 L at 294 K, what is the pressure in atm? In psi?

8.81 If 3.5 mol of O_2 has a volume of 27.0 L at a pressure of 1.6 atm, what is its temperature in °C?

8.82 If 15.0 g of CO_2 gas has a volume of 0.30 L at 310 K, what is its pressure in mmHg?

8.83 If 20.0 g of N_2 gas has a volume of 4.00 L and a pressure of 6.0 atm, what is its temperature in degrees celsius?

8.84 If 18.0 g of O_2 gas has a temperature of 350 K and a pressure of 550 mmHg, what is its volume?

8.85 How many moles of a gas will occupy a volume of 0.55 L at a temperature of 347 K and a pressure of 2.5 atm?

DALTON'S LAW AND PARTIAL PRESSURE

8.86 What is meant by *partial pressure*?

8.87 What is Dalton's law?

8.88 If the partial pressure of oxygen in air at 1.0 atm is 160 mmHg, what is its partial pressure on the summit of Mt. Whitney, where atmospheric pressure is 440 mmHg? Assume that the percent oxygen is the same.

8.89 Scuba divers who suffer from decompression sickness are treated in hyperbaric chambers using heliox (21% oxygen, 79% helium), at pressures up to 120 psi. Calculate the partial pressure of O_2 (in mmHg) in a hyperbaric chamber under these conditions.

LIQUIDS

8.90 What is the vapor pressure of a liquid?

8.91 What is a liquid's heat of vaporization?

8.92 What is the effect of pressure on a liquid's boiling point?

8.93 Which of the following substances would you expect to have the higher vapor pressure: CH_3OH or CH_3Cl? Explain

8.94 The heat of vaporization of water is 9.72 kcal/mol.

(a) How much heat (in kilocalories) is required to vaporize 3.00 mol of H_2O?

(b) How much heat (in kilocalories) is released when 320 g of steam condenses?

8.95 Patients with a high body temperature are often given "alcohol baths." The heat of vaporization of isopropyl alcohol (rubbing alcohol) is 159 cal/g. How much heat is removed from the skin by the evaporation of 190 g (about 1/2 a cup) of isopropyl alcohol?

SOLIDS

8.96 What is the difference between an amorphous and a crystalline solid?

8.97 List three kinds of crystalline solids, and give an example of each.

8.98 The heat of fusion of acetic acid, the principal organic component of vinegar, is 45.9 cal/g. How much heat (in kilocalories) is required to melt 1.75 mol of solid acetic acid?

8.99 The heat of fusion of sodium metal is 630 cal/mol. How much heat (in kilocalories) is required to melt 262 g of sodium?

CHEMISTRY IN ACTION

8.100 What evidence is there that global warming is occurring? [*Greenhouse Gases and Global Warming, p. 224*]

8.101 What are the three most important greenhouse gases? [*Greenhouse Gases and Global Warming, p. 224*]

8.102 What is the difference between a systolic and a diastolic pressure reading? Is a blood pressure of 180/110 within the normal range? [*Blood Pressure, p. 228*]

8.103 Convert the blood pressure reading in Problem 8.102 to atm. [*Blood Pressure, p. 228*]

8.104 What is a supercritical fluid? [*CO_2 as an Environmentally Friendly Solvent, p. 245*]

8.105 What are the environmental advantages of using supercritical CO_2 in place of chlorinated organic solvents? [*CO_2 as an Environmentally Friendly Solvent, p. 245*]

GENERAL QUESTIONS AND PROBLEMS

8.106 Use the kinetic–molecular theory to explain why gas pressure increases if the temperature is raised and the volume is kept constant.

8.107 Hydrogen and oxygen react according to the equation $2 H_2(g) + O_2(g) \longrightarrow 2 H_2O(g)$. According to Avogadro's law, how many liters of hydrogen are required to react with 2.5 L of oxygen at STP?

8.108 If 3.0 L of hydrogen and 1.5 L of oxygen at STP react to yield water, how many moles of water are formed? What gas volume does the water have at a temperature of 100 °C and 1 atm pressure?

8.109 Approximately 240 mL/min of CO_2 is exhaled by an average adult at rest. Assuming a temperature of 37 °C and 1 atm pressure, how many moles of CO_2 is this?

8.110 How many grams of CO_2 are exhaled by an average resting adult in 24 hours? (See Problem 8.109.)

8.111 Imagine that you have two identical containers, one containing hydrogen at STP and the other containing oxygen at STP. How can you tell which is which without opening them?

8.112 When fully inflated, a hot-air balloon has a volume of 1.6×10^5 L at an average temperature of 375 K and 0.975 atm. Assuming that air has an average molar mass of 29 g/mol, what is the density of the air in the hot-air balloon? How does this compare with the density of air at STP?

8.113 A 10.0 g sample of an unknown gas occupies 14.7 L at a temperature of 25 °C and a pressure of 745 mmHg. How many moles of gas are in the sample? What is the molar mass of the gas?

8.114 One mole of any gas has a volume of 22.4 L at STP. What are the molecular weights of the following gases, and what are their densities in grams per liter at STP?

(a) CH_4 (b) CO_2 (c) O_2

8.115 Gas pressure outside the space shuttle is approximately 1×10^{-14} mm Hg at a temperature of approximately 1 K. If the gas is almost entirely hydrogen atoms (H, not H_2), what volume of space is occupied by 1 mol of atoms? What is the density of H gas in atoms per liter?

8.116 Ethylene glycol, $C_2H_6O_2$, has one OH bonded to each carbon.

(a) Draw the Lewis dot structure of ethylene glycol.

(b) Draw the Lewis dot structure of chloroethane, C_2H_5Cl.

(c) Chloroethane has a slightly higher molar mass than ethylene glycol, but a much lower boiling point (3 °C versus 198 °C). Explain.

8.117 A rule of thumb for scuba diving is that the external pressure increases by 1 atm for every 10 m of depth. A diver using a compressed air tank is planning to descend to a depth of 25 m.

(a) What is the external pressure at this depth? (Remember that the pressure at sea level is 1 atm.)

(b) Assuming that the tank contains 20% oxygen and 80% nitrogen, what is the partial pressure of each gas in the diver's lungs at this depth?

8.118 The *Rankine* temperature scale used in engineering is to the Fahrenheit scale as the Kelvin scale is to the Celsius scale. That is, 1 Rankine degree is the same size as 1 Fahrenheit degree, and 0 °R = absolute zero.

(a) What temperature corresponds to the freezing point of water on the Rankine scale?

(b) What is the value of the gas constant R on the Rankine scale in $(L \cdot atm)/(°R \cdot mol)$?

8.119 Isooctane, C_8H_{18}, is the component of gasoline from which the term *octane rating* derives.

(a) Write a balanced equation for the combustion of isooctane to yield CO_2 and H_2O.

(b) Assuming that gasoline is 100% isooctane and that the density of isooctane is 0.792 g/mL, what mass of CO_2 (in kilograms) is produced each year by the annual U.S. gasoline consumption of 4.6×10^{10} L?

(c) What is the volume (in liters) of this CO_2 at STP?

CHAPTER 9

Solutions

CONTENTS

- **9.1** Mixtures and Solutions
- **9.2** The Solution Process
- **9.3** Solid Hydrates
- **9.4** Solubility
- **9.5** The Effect of Temperature on Solubility
- **9.6** The Effect of Pressure on Solubility: Henry's Law
- **9.7** Units of Concentration
- **9.8** Dilution
- **9.9** Ions in Solution: Electrolytes
- **9.10** Electrolytes in Body Fluids: Equivalents and Milliequivalents
- **9.11** Properties of Solutions
- **9.12** Osmosis and Osmotic Pressure
- **9.13** Dialysis

◄ The giant sequoia relies on osmotic pressure—a colligative property of solutions—to transport water and nutrients from the roots to the treetops 300 ft up.

CHAPTER GOALS

1. **What are solutions, and what factors affect solubility?**
   THE GOAL: Be able to define the different kinds of mixtures and explain the influence on solubility of solvent and solute structure, temperature, and pressure. (◀◀ B., E.)

2. **How is the concentration of a solution expressed?**
   THE GOAL: Be able to define, use, and convert between the most common ways of expressing solution concentrations.

3. **How are dilutions carried out?**
   THE GOAL: Be able to calculate the concentration of a solution prepared by dilution and explain how to make a desired dilution.

4. **What is an electrolyte?**
   THE GOAL: Be able to recognize strong and weak electrolytes and nonelectrolytes, and express electrolyte concentrations. (◀◀ A.)

5. **How do solutions differ from pure solvents in their behavior?**
   THE GOAL: Be able to explain vapor-pressure lowering, boiling-point elevation, and freezing-point depression for solutions. (◀◀ F., G.)

6. **What is osmosis?**
   THE GOAL: Be able to describe osmosis and some of its applications.

CONCEPTS TO REVIEW ◀◀

A. Ions and Ionic Compounds
(Sections 3.1, 3.4)

B. Enthalpy Changes
(Section 7.2)

C. Chemical Equilibrium
(Section 7.7)

D. Le Châtelier's Principle
(Section 7.9)

E. Intermolecular Forces and Hydrogen Bonds
(Section 8.2)

F. Partial Pressure of Gases
(Section 8.11)

G. Vapor Pressure
(Section 8.12)

Up to this point, we have been concerned primarily with pure substances, both elements and compounds. In day-to-day life, however, most of the materials we come in contact with are mixtures. Air, for example, is a gaseous mixture of primarily oxygen and nitrogen; blood is a liquid mixture of many different components; and many rocks are solid mixtures of different minerals. In this chapter, we look closely at the characteristics and properties of mixtures, with particular attention to the uniform mixtures we call *solutions*.

9.1 Mixtures and Solutions

As we saw in Section 1.3, a *mixture* is an intimate combination of two or more substances, both of which retain their chemical identities. Mixtures can be classified as either *heterogeneous* or *homogeneous*, as indicated in Figure 9.1, depending on their appearance. **Heterogeneous mixtures** are those in which the mixing is not uniform and which therefore have regions of different composition. Rocky Road ice cream, for example, is a heterogeneous mixture, with something different in every spoonful. Granite and many other rocks are also heterogeneous, having a grainy character due to the heterogeneous mixing of different minerals. **Homogeneous mixtures** are those in which the mixing *is* uniform and that therefore have the same composition throughout. Seawater, a homogeneous mixture of soluble ionic compounds in water, is an example.

Homogeneous mixtures can be further classified as either *solutions* or *colloids*, according to the size of their particles. **Solutions**, the most important class of homogeneous mixtures, contain particles the size of a typical ion or small molecule—roughly 0.1–2 nm in diameter. **Colloids**, such as milk and fog, are also homogeneous in appearance but contain larger particles than solutions—in the range 2–500 nm diameter.

Liquid solutions, colloids, and heterogeneous mixtures can be distinguished in several ways. For example, liquid solutions are transparent (although they may be colored). Colloids may appear transparent if the particle size is small, but they have a murky or opaque appearance if the particle size is larger. Neither solutions nor small-particle colloids separate on standing, and the particles in both are too small to be removed by filtration. Heterogeneous mixtures and large-particle colloids, also known as "suspensions," are murky or opaque and their particles will slowly settle on prolonged standing. House paint is an example.

Heterogeneous mixture A nonuniform mixture that has regions of different composition.

Homogeneous mixture A uniform mixture that has the same composition throughout.

Solution A homogeneous mixture that contains particles the size of a typical ion or small molecule.

Colloid A homogeneous mixture that contains particles that range in diameter from 2 to 500 nm.

253

254 CHAPTER 9 Solutions

▶ **Figure 9.1**
Classification of mixtures.
The components in heterogeneous mixtures are not uniformly mixed, and the composition varies with location within the mixture. In homogeneous mixtures, the components are uniformly mixed at the molecular level.

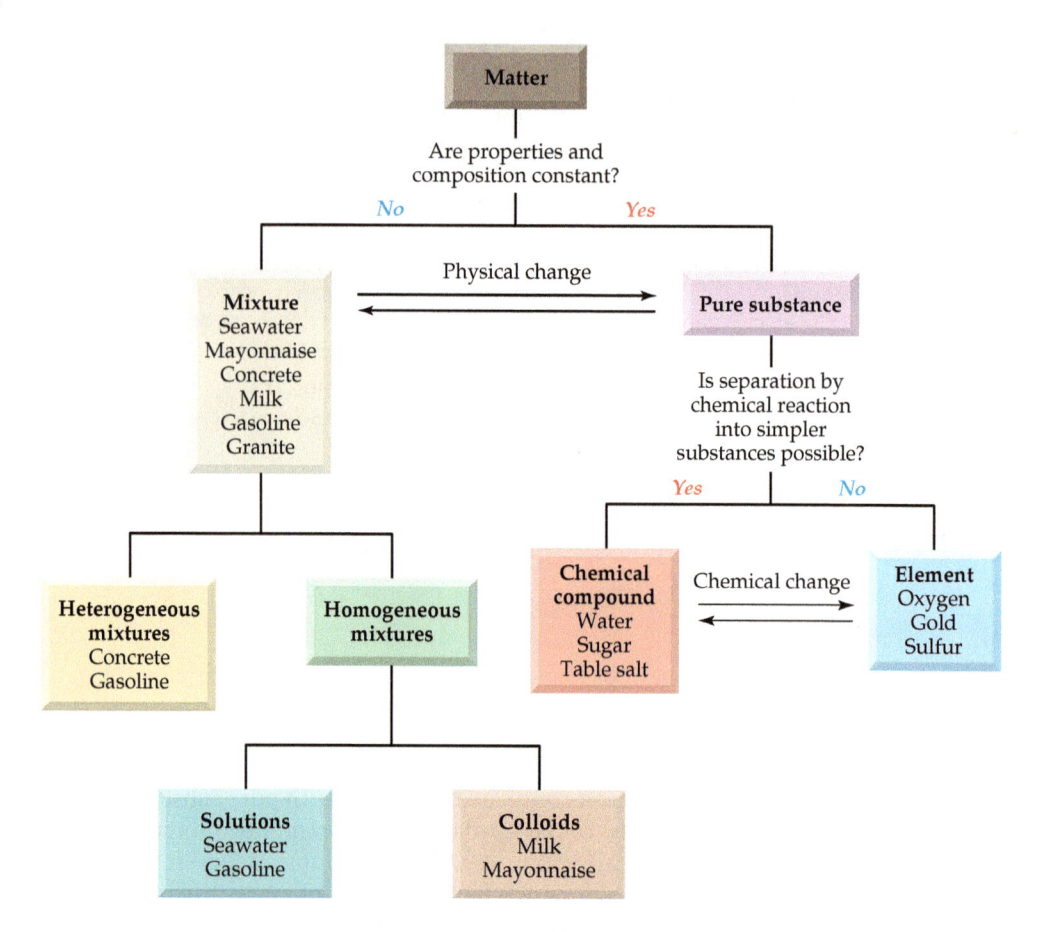

Table 9.1 gives some examples of solutions, colloids, and heterogeneous mixtures. It is interesting to note that blood has characteristics of all three. About 45% by volume of blood consists of suspended red and white cells, which settle slowly on standing; the remaining 55% is *plasma*, which contains ions in solution and colloidal protein molecules.

TABLE 9.1 Some Characteristics of Solutions, Colloids, and Heterogeneous Mixtures

Type of Mixture	Particle Size	Examples	Characteristics
Solution	<2.0 nm	Air, seawater, gasoline, wine	Transparent to light; does not separate on standing; nonfilterable
Colloid	2.0–500 nm	Butter, milk, fog, pearl	Often murky or opaque to light; does not separate on standing; nonfilterable
Heterogeneous	>500 nm	Blood, paint, aerosol sprays	Murky or opaque to light; separates on standing; filterable

Although we usually think of solids dissolved in liquids when we talk about solutions, solutions actually occur in all three phases of matter (Table 9.2). Metal alloys like 14-karat gold (58% gold with silver and copper) and brass (10–40% zinc with copper), for instance, are solutions of one solid with another. For solutions in which a gas or solid is dissolved in a liquid, the dissolved substance is called the **solute** and the liquid is called the **solvent**. In seawater, for example, the dissolved salts would be the solutes and water would be the solvent. When one liquid is dissolved in another, the minor component is usually considered the solute and the major component is the solvent.

Solute A substance that is dissolved in a solvent.

Solvent The substance in which another substance (the solute) is dissolved.

TABLE 9.2 **Some Different Types of Solutions**

Type of Solution	Example
Gas in gas	Air (O_2, N_2, Ar, and other gases)
Gas in liquid	Seltzer water (CO_2 in water)
Gas in solid	H_2 in palladium metal
Liquid in liquid	Gasoline (mixture of hydrocarbons)
Liquid in solid	Dental amalgam (mercury in silver)
Solid in liquid	Seawater (NaCl and other salts in water)
Solid in solid	Metal alloys such as 14-karat gold (Au, Ag, and Cu)

PROBLEM 9.1
Classify the following liquid mixtures as heterogeneous or homogeneous. Further classify each homogeneous mixture as a solution or colloid.
(a) Orange juice   (b) Apple juice
(c) Hand lotion   (d) Tea

9.2 The Solution Process

What determines whether a substance is soluble in a given liquid? Solubility depends primarily on the strength of the attractions between solute and solvent particles relative to the strengths of the attractions within the pure substances. Ethyl alcohol is soluble in water, for example, because hydrogen bonding (Section 8.2) is nearly as strong between water and ethyl alcohol molecules as it is between water molecules alone or ethyl alcohol molecules alone.

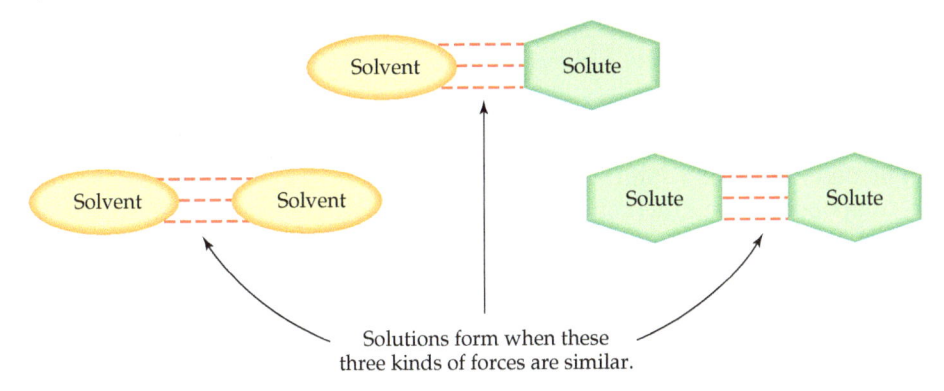

Solvent ---- Solute

Solvent ---- Solvent   Solute ---- Solute

Solutions form when these three kinds of forces are similar.

A good rule of thumb for predicting solubility is that "like dissolves like," meaning that substances with similar intermolecular forces form solutions with one another, whereas substances with different intermolecular forces do not (Section 8.2).

Polar solvents dissolve polar and ionic solutes; nonpolar solvents dissolve nonpolar solutes. Thus, a polar, hydrogen-bonding compound like water dissolves ethyl alcohol and sodium chloride, whereas a nonpolar organic compound like hexane (C_6H_{14}) dissolves other nonpolar organic compounds like fats and oils. Water and oil, however, do not dissolve one another, as summed up by the old saying, "Oil and water don't mix." The intermolecular forces between water molecules are so strong that after an oil–water mixture is shaken, the water layer re-forms, squeezing out the oil molecules.

Water solubility is not limited to ionic compounds and ethyl alcohol. Many polar organic substances, such as sugars, amino acids, and even some proteins, dissolve in water. In addition, small, moderately polar organic molecules such as chloroform ($CHCl_3$) are soluble in water to a limited extent. When mixed with water, a small amount of the organic compound dissolves, but the remainder forms a separate liquid

Solvation The clustering of solvent molecules around a dissolved solute molecule or ion.

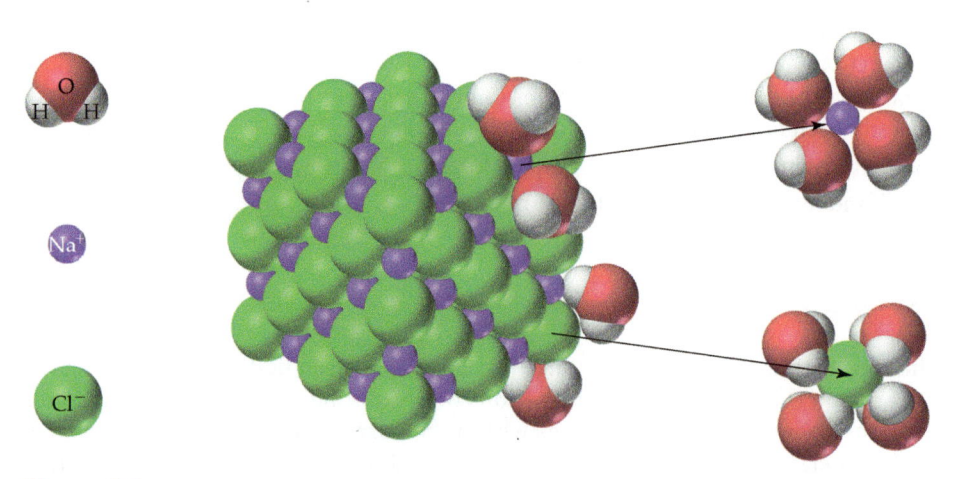

layer. As the number of carbon atoms in organic molecules increases, though, water solubility decreases.

The process of dissolving an ionic solid in a polar liquid can be visualized as shown in Figure 9.2 for sodium chloride. When NaCl crystals are put in water, ions at the crystal surface come into contact with polar water molecules. Positively charged Na^+ ions are attracted to the negatively polarized oxygen of water, and negatively charged Cl^- ions are attracted to the positively polarized hydrogens. The combined forces of attraction between an ion and several water molecules pull the ion away from the crystal, exposing a fresh surface, until ultimately the crystal dissolves. Once in solution, Na^+ and Cl^- ions are completely surrounded by solvent molecules, a phenomenon called **solvation** (or, specifically for water, *hydration*). The water molecules form a loose shell around the ions, stabilizing them by electrical attraction.

▲ **Figure 9.2**
Dissolution of an NaCl crystal in water.
Polar water molecules surround the individual Na^+ and Cl^- ions at an exposed edge or corner, pulling them from the crystal surface into solution and surrounding them. Note how the negatively polarized oxygens of water molecules cluster around Na^+ ions and the positively polarized hydrogens cluster around Cl^- ions.

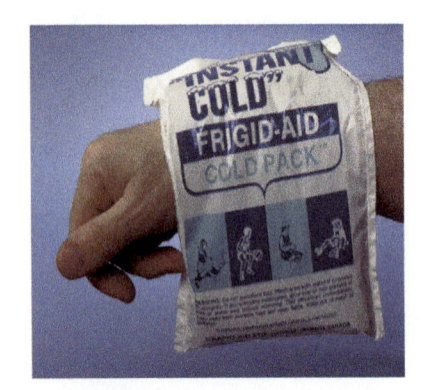

▲ Instant cold packs used to treat muscle strains and sprains often take advantage of the endothermic enthalpy of a solution of salts such as ammonium nitrate.

The dissolution of a solute in a solvent is a physical change, because the solution components retain their chemical identities. When sugar dissolves in water, for example, the individual sugar and water molecules still have the same chemical formulas as in the pure or undissolved state. Like all chemical and physical changes, the dissolution of a substance in a solvent has associated with it a heat change, or *enthalpy* change (Section 7.2). Some substances dissolve exothermically, releasing heat and warming the resultant solution, whereas other substances dissolve endothermically, absorbing heat and cooling the resultant solution. Calcium chloride, for example, *releases* 19.4 kcal/mol (81.2 kJ/mol) of heat energy when it dissolves in water, but ammonium nitrate (NH_4NO_3) *absorbs* 6.1 kcal/mol (25.5 kJ/mol) of heat energy. Athletes and others take advantage of both situations when they use instant hot packs or cold packs to treat injuries. Both hot and cold packs consist of a pouch of water and a dry chemical, such as $CaCl_2$ or $MgSO_4$ for hot packs, and NH_4NO_3 for cold packs. Squeezing the pack breaks the pouch and the solid dissolves, either raising or lowering the temperature.

Worked Example 9.1 Formation of Solutions

Which of the following pairs of substances would you expect to form solutions?
  (a) Carbon tetrachloride (CCl_4) and hexane (C_6H_{14}).
  (b) Octane (C_8H_{18}) and methyl alcohol (CH_3OH).

ANALYSIS Identify the kinds of intermolecular forces in each substance (Section 8.2). Substances with similar intermolecular forces tend to form solutions.

SOLUTION

(a) Hexane contains only C—H and C—C bonds, which are nonpolar. Carbon tetrachloride contains polar C—Cl bonds, but they are distributed symmetrically in the tetrahedral molecule so that it too is nonpolar. The major intermolecular force for both compounds is London dispersion forces, so they will form a solution.

(b) Octane contains only C—H and C—C bonds and so is nonpolar; the major intermolecular force is dispersion. Methyl alcohol contains polar C—O and O—H bonds; it is polar and forms hydrogen bonds. The intermolecular forces for the two substances are so dissimilar that they do not form a solution.

PROBLEM 9.2
Which of the following pairs of substances would you expect to form solutions?
(a) CCl_4 and water
(b) Benzene (C_6H_6) and $MgSO_4$
(c) Hexane (C_6H_{14}) and heptane (C_7H_{16})
(d) Ethyl alcohol (C_2H_5OH) and heptanol ($C_7H_{15}OH$)

9.3 Solid Hydrates

Some ionic compounds attract water strongly enough to hold on to water molecules even when crystalline, forming what are called *solid hydrates*. For example, the plaster of Paris used to make decorative objects and casts for broken limbs is calcium sulfate hemihydrate, $CaSO_4 \cdot \frac{1}{2}H_2O$. The dot between $CaSO_4$ and $\frac{1}{2}H_2O$ in the formula indicates that for every two $CaSO_4$ formula units in the crystal there is also one water molecule present.

$$CaSO_4 \cdot \tfrac{1}{2}H_2O \quad \text{A solid hydrate}$$

After being ground up and mixed with water to make plaster, $CaSO_4 \cdot \frac{1}{2}H_2O$ gradually changes into the crystalline dihydrate $CaSO_4 \cdot 2\,H_2O$, known as *gypsum*. During the change, the plaster hardens and expands in volume, causing it to fill a mold or shape itself closely around a broken limb. Table 9.3 lists some other ionic compounds that are handled primarily as hydrates.

TABLE 9.3 Some Common Solid Hydrates

Formula	Name	Uses
$AlCl_3 \cdot 6\,H_2O$	Aluminum chloride hexahydrate	Antiperspirant
$CaSO_4 \cdot 2\,H_2O$	Calcium sulfate dihydrate (gypsum)	Cements, wallboard molds
$CaSO_4 \cdot \frac{1}{2}H_2O$	Calcium sulfate hemihydrate (plaster of Paris)	Casts, molds
$CuSO_4 \cdot 5\,H_2O$	Copper(II) sulfate pentahydrate (blue vitriol)	Pesticide, germicide, topical fungicide
$MgSO_4 \cdot 7\,H_2O$	Magnesium sulfate heptahydrate (epsom salts)	Laxative, anticonvulsant
$Na_2B_4O_7 \cdot 10\,H_2O$	Sodium tetraborate decahydrate (borax)	Cleaning compounds, fireproofing agent
$Na_2S_2O_3 \cdot 5\,H_2O$	Sodium thiosulfate pentahydrate (hypo)	Photographic fixer

258 CHAPTER 9 Solutions

Hygroscopic Having the ability to pull water molecules from the surrounding atmosphere.

Still other ionic compounds attract water so strongly that they pull water vapor from humid air to become hydrated. Compounds that show this behavior, such as calcium chloride ($CaCl_2$), are called **hygroscopic** and are often used as drying agents. You might have noticed a small bag of a hygroscopic compound (probably silica gel, SiO_2) included in the packing material of a new MP3 player, camera, or other electronic device to keep humidity low during shipping.

> **PROBLEM 9.3**
> Write the formula of sodium sulfate decahydrate, known as Glauber's salt and used as a laxative.
>
> **PROBLEM 9.4**
> What mass of Glauber's salt must be used to provide 1.00 mol of sodium sulfate?

9.4 Solubility

We saw in Section 9.2 that ethyl alcohol is soluble in water because hydrogen bonding is nearly as strong between water and ethyl alcohol molecules as it is between water molecules alone or ethyl alcohol molecules alone. So similar are the forces in this particular case, in fact, that the two liquids are **miscible**, or mutually soluble in all proportions. Ethyl alcohol will continue to dissolve in water no matter how much is added.

Miscible Mutually soluble in all proportions.

Most substances, however, reach a solubility limit beyond which no more will dissolve in solution. Imagine, for instance that you are asked to prepare a saline solution (aqueous NaCl). You might measure out some water, add solid NaCl, and stir the mixture. Dissolution occurs rapidly at first but then slows down as more and more NaCl is added. Eventually the dissolution stops because an equilibrium is reached when the numbers of Na^+ and Cl^- ions leaving a crystal and going into solution are equal to the numbers of ions returning from solution to the crystal. At this point, the solution is said to be **saturated**. A maximum of 35.8 g of NaCl will dissolve in 100 mL of water at 20 °C. Any amount above this limit simply sinks to the bottom of the container and sits there.

Saturated solution A solution that contains the maximum amount of dissolved solute at equilibrium.

The equilibrium reached by a saturated solution is like the equilibrium reached by a reversible reaction (Section 7.7). Both are dynamic situations in which no *apparent* change occurs because the rates of forward and backward processes are equal. Solute particles leave the solid surface and reenter the solid from solution at the same rate.

$$\text{Solid solute} \underset{\text{Crystallize}}{\overset{\text{Dissolve}}{\rightleftharpoons}} \text{Solution}$$

Solubility The maximum amount of a substance that will dissolve in a given amount of solvent at a specified temperature.

The maximum amount of a substance that will dissolve in a given amount of a solvent at a given temperature, usually expressed in grams per 100 mL (g/100 mL), is called the substance's **solubility**. Solubility is a characteristic property of a specific solute–solvent combination, and different substances have greatly differing solubilities. Only 9.6 g of sodium hydrogen carbonate will dissolve in 100 mL of water at 20 °C, for instance, but 204 g of sucrose will dissolve under the same conditions.

9.5 The Effect of Temperature on Solubility

As anyone who has ever made tea or coffee knows, temperature often has a dramatic effect on solubility. The compounds in tea leaves or coffee beans, for instance, dissolve easily in hot water but not in cold water. The effect of temperature is different for every substance, however, and is usually unpredictable. As shown in Figure 9.3(a), the solubilities of most molecular and ionic solids increase with increasing temperature, but the solubilities of others (NaCl) are almost unchanged, and the solubilities of still others [$Ce_2(SO_4)_3$] decrease with increasing temperature.

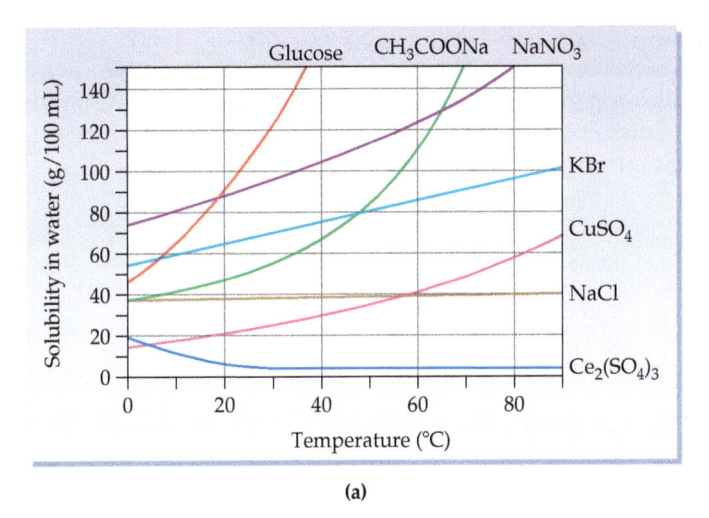

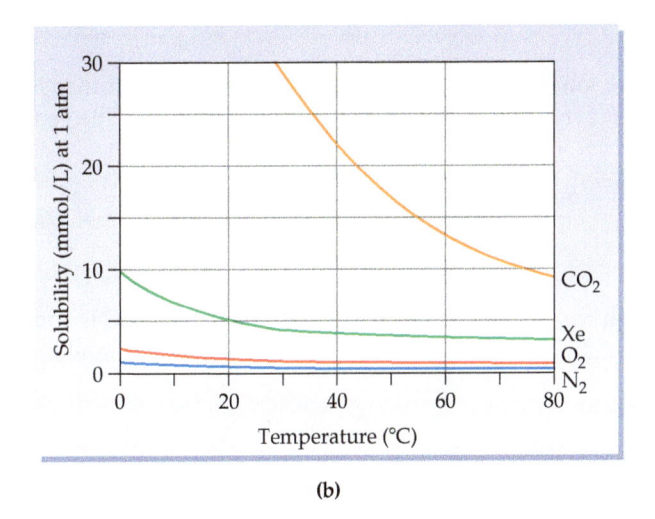

▲ Figure 9.3
Solubilities of some (a) solids and (b) gases, in water as a function of temperature.
Most solid substances become more soluble as temperature rises (although the exact relationship is usually complex), while the solubility of gases decreases.

Supersaturated solution A solution that contains more than the maximum amount of dissolved solute; a nonequilibrium situation.

Solids that are more soluble at high temperature than at low temperature can sometimes form what are called **supersaturated solutions**, which contain even more solute than a saturated solution. Suppose, for instance, that a large amount of a substance is dissolved at a high temperature. As the solution cools, the solubility decreases and the excess solute should precipitate to maintain equilibrium. But if the cooling is done very slowly, and if the container stands quietly, crystallization might not occur immediately and a supersaturated solution might result. Such a solution is unstable, however, and precipitation can occur dramatically when a tiny seed crystal is added or container disturbed to initiate crystallization (Figure 9.4).

Unlike solids, the influence of temperature on the solubility of gases *is* predictable: Addition of heat decreases the solubility of most gases, as seen in Figure 9.3(b) (helium is the only common exception). One result of this temperature-dependent decrease in gas solubility can sometimes be noted in a stream or lake near the outflow of warm water from an industrial operation. As water temperature increases, the concentration of dissolved oxygen in the water decreases, killing fish that cannot tolerate the lower oxygen levels.

▲ **Figure 9.4**
A supersaturated solution of sodium acetate in water.
When a tiny seed crystal is added, larger crystals rapidly grow and precipitate from the solution until equilibrium is reached.

Worked Example 9.2 Solubility of Gases: Effect of Temperature

From the following graph of solubility versus temperature for O_2, estimate the concentration of dissolved oxygen in water at 25 °C and at 35 °C. By what percentage does the concentration of O_2 change?

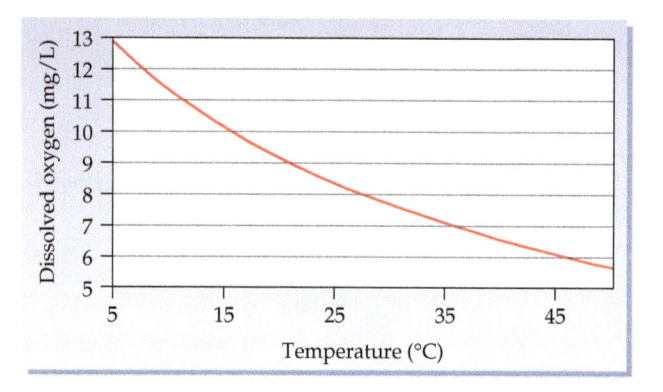

ANALYSIS The solubility of O_2 (on the *y*-axis) can be determined by finding the appropriate temperature (on the *x*-axis) and extrapolating. The percent change is calculated as

$$\frac{(\text{Solubility at 25 °C}) - (\text{Solubility at 35 °C})}{(\text{Solubility at 25 °C})} \times 100$$

SOLUTION
From the graph we estimate that the solubility of O_2 at 25 °C is approximately 8.3 mg/L and at 35 °C is 7.0 mg/L. The percent change in solubility is

$$\frac{8.3 - 7.0}{8.3} \times 100 = 16\%$$

PROBLEM 9.5
A solution is prepared by dissolving 12.5 g of KBr in 20 mL of water at 60 °C (see Figure 9.3). Is this solution saturated, unsaturated, or supersaturated? What will happen if the solution is cooled to 10 °C?

9.6 The Effect of Pressure on Solubility: Henry's Law

Pressure has virtually no effect on the solubility of a solid or liquid, but it has a strong effect on the solubility of a gas. According to **Henry's law**, the solubility (or concentration) of a gas in a liquid is directly proportional to the partial pressure of the gas over the liquid. If the partial pressure of the gas doubles, solubility doubles; if the gas pressure is halved, solubility is halved (Figure 9.5).

▶▶ Recall from Section 8.11 that each gas in a mixture exerts a partial pressure independent of other gases present (Dalton's law of partial pressures).

▶ **Figure 9.5**
Henry's law.
The solubility of a gas is directly proportional to its partial pressure. An increase in pressure causes more gas molecules to enter solution until equilibrium is restored between the dissolved and undissolved gas.

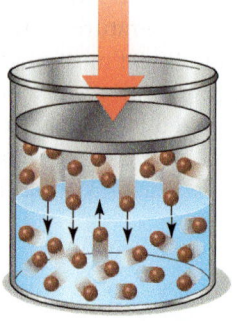

(a) Equilibrium   (b) Pressure increase   (c) Equilibrium restored

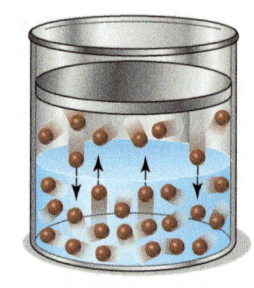

SECTION 9.6 The Effect of Pressure on Solubility: Henry's Law

Henry's law The solubility (or concentration) of a gas is directly proportional to the partial pressure of the gas if the temperature is constant. That is, concentration (C) divided by pressure (P) is constant when T is constant,

or $\dfrac{C}{P_{gas}} = k$ (At a constant temperature)

Henry's law can be explained using Le Châtelier's principle. In the case of a saturated solution of a gas in a liquid, an equilibrium exists whereby gas molecules enter and leave the solution at the same rate. When the system is stressed by increasing the pressure of the gas, more gas molecules go into solution to relieve that increase. Conversely, when the pressure of the gas is decreased, more gas molecules come out of solution to relieve the decrease.

▶▶ Le Châtelier's principle states that when a system at equilibrium is placed under stress, the equilibrium shifts to relieve that stress (Section 7.9).

$$\text{Gas + Solvent} \underset{}{\overset{[\text{Pressure increases} \longrightarrow]}{\rightleftharpoons}} \text{Solution}$$

As an example of Henry's law in action, think about the fizzing that occurs when you open a bottle of soft drink or champagne. The bottle is sealed under greater than 1 atm of CO_2 pressure, causing some of the CO_2 to dissolve. When the bottle is opened, however, CO_2 pressure drops and gas comes fizzing out of solution.

Writing Henry's law in the form $P_{gas} = C/k$ shows that partial pressure can be used to express the concentration of a gas in a solution, a practice especially common in health-related sciences. Table 9.4 gives some typical values and illustrates the convenience of having the same unit for concentration of a gas in both air and blood. Compare the oxygen partial pressures in saturated alveolar air (air in the lungs) and in arterial blood, for instance. The values are almost the same because the gases dissolved in blood come to equilibrium with the same gases in the lungs.

TABLE 9.4 Partial Pressures and Normal Gas Concentrations in Body Fluids

Sample	P_{N_2}	P_{O_2}	P_{CO_2}	P_{H_2O}
Inspired air (dry)	597	159	0.3	3.7
Alveolar air (saturated)	573	100	40	47
Expired air (saturated)	569	116	28	47
Arterial blood	573	95	40	
Venous blood	573	40	45	
Peripheral tissues	573	40	45	

Partial Pressure (mmHg)

If the partial pressure of a gas over a solution changes while the temperature is constant, the new solubility of the gas can be found easily. Because C/P is a constant value at constant temperature, Henry's law can be restated to show how one variable changes if the other changes:

$$\dfrac{C_1}{P_1} = \dfrac{C_2}{P_2} = k \quad \text{(Where } k \text{ is constant at a fixed temperature)}$$

Worked Example 9.3 gives an illustration of how to use this equation.

Worked Example 9.3 Solubility of Gases: Henry's Law

At a partial pressure of oxygen in the atmosphere of 159 mmHg, the solubility of oxygen in blood is 0.44 g/100 mL. What is the solubility of oxygen in blood at 11,000 ft, where the partial pressure of O_2 is 56 mmHg?

ANALYSIS According to Henry's law, the solubility of the gas divided by its pressure is constant:

$$\frac{C_1}{P_1} = \frac{C_2}{P_2}$$

Of the four variables in this equation, we know P_1, C_1, and P_2, and we need to find C_2.

BALLPARK ESTIMATE The pressure drops by a factor of about 3 (from 159 mmHg to 56 mmHg). Since the ratio of solubility to pressure is constant, the solubility must also drop by a factor of 3 (from 0.44 g/100 mL to about 0.15 g/100 mL).

SOLUTION

STEP 1: Identify known information. We have values for P_1, C_1, and P_2.

$P_1 = 159$ mmHg
$C_1 = 0.44$ g/100 mL
$P_2 = 56$ mmHg

STEP 2: Identify answer and units. We are looking for the solubility of O_2 (C_2) at a partial pressure P_2.

Solubility of O_2, $C_2 =$?? g/100 mL

STEP 3: Identify conversion factors or equations. In this case, we restate Henry's law to solve for C_2.

$$\frac{C_1}{P_1} = \frac{C_2}{P_2} \Rightarrow C_2 = \frac{C_1 P_2}{P_1}$$

STEP 4: Solve. Substitute the known values into the equation and calculate C_2.

$$C_2 = \frac{C_1 P_2}{P_1} = \frac{(0.44 \text{ g}/100 \text{ mL})(56 \text{ mmHg})}{159 \text{ mmHg}} = 0.15 \text{ g}/100 \text{ mL}$$

BALLPARK CHECK The calculated answer matches our estimate.

PROBLEM 9.6
At 20 °C and a partial pressure of 760 mmHg, the solubility of CO_2 in water is 0.169 g/100 mL at this temperature. What is the solubility of CO_2 at 2.5×10^4 mmHg?

PROBLEM 9.7
At a total atmospheric pressure of 1.00 atm, the partial pressure of CO_2 in air is approximately 4.0×10^{-4} atm. Using the data in Problem 9.6, what is the solubility of CO_2 in an open bottle of seltzer water at 20 °C?

PROBLEM 9.8
The atmospheric pressure at the top of Mt. Everest is only 265 mmHg. If the atmospheric composition is 21% oxygen, calculate the partial pressure of O_2 at this altitude and determine the percent saturation of hemoglobin under these conditions (see Chemistry in Action: Breathing and Oxygen Transport on p. 263).

9.7 Units of Concentration

Although we speak casually of a solution of, say, orange juice as either "dilute" or "concentrated," laboratory work usually requires an exact knowledge of a solution's concentration. As indicated in Table 9.5 on page 264, there are several common methods for expressing concentration. The units differ, but all the methods describe how much solute is present in a given quantity of solution.

CHEMISTRY IN ACTION

Breathing and Oxygen Transport

Like all other animals, humans need oxygen. When we breathe, the freshly inspired air travels through the bronchial passages and into the lungs. The oxygen then diffuses through the delicate walls of the approximately 150 million alveolar sacs of the lungs and into arterial blood, which transports it to all body tissues.

Only about 3% of the oxygen in blood is dissolved; the rest is chemically bound to *hemoglobin* molecules, large proteins with *heme* groups embedded in them. Each hemoglobin molecule contains four heme groups, and each heme group contains an iron atom that is able to bind 1 O_2 molecule. Thus, a single hemoglobin molecule can bind up to 4 molecules of oxygen. The entire system of oxygen transport and delivery in the body depends on the pickup and release of O_2 by hemoglobin (Hb) according to the following series of equilibria:

$$O_2(\text{lungs}) \rightleftharpoons O_2(\text{blood}) \quad (\text{Henry's law})$$
$$Hb + 4\,O_2(\text{blood}) \rightleftharpoons Hb(O_2)_4$$
$$Hb(O_2)_4 \rightleftharpoons Hb + 4\,O_2\,(\text{cell})$$

The delivery of oxygen depends on the concentration of O_2 in the various tissues, as measured by partial pressure (P_{O_2}, Table 9.4). The amount of oxygen carried by hemoglobin at any given value of P_{O_2} is usually expressed as a percent saturation and can be found from the curve shown in the accompanying figure. When $P_{O_2} = 100$ mmHg, the saturation in the lungs is 97.5%, meaning that each hemoglobin is carrying close to its maximum of 4 O_2 molecules. When $P_{O_2} = 26$ mmHg, however, the saturation drops to 50%.

So, how does the body ensure that enough oxygen is available to the various tissues? When large amounts of oxygen are needed—during a strenuous workout, for example—oxygen is released from hemoglobin to the hardworking, oxygen-starved muscle cells, where P_{O_2} is low. Increasing the supply of oxygen to the blood (by breathing harder and faster) shifts all the equilibria toward the right, according to Le Châtelier's principle (Section 7.9), to supply the additional O_2 needed by the muscles.

What about people living at high altitudes? In Leadville, CO, for example, where the altitude is 10,156 ft, the P_{O_2} in the lungs is only about 68 mmHg. Hemoglobin is only 90% saturated with O_2 at this pressure, meaning that less oxygen is available for delivery to the tissues. The body responds by producing erythropoietin (EPO), a hormone that stimulates the bone marrow to produce more red blood cells and hemoglobin molecules. The increase in Hb provides more capacity for O_2 transport and drives the Hb + O_2 equilibria to the right.

▲ At high altitudes, the partial pressure of oxygen in the air is too low to saturate hemoglobin sufficiently. Additional oxygen is therefore needed.

World-class athletes use the mechanisms of increased oxygen transport associated with higher levels of hemoglobin to enhance their performance. High-altitude training centers have sprung up, with living and training regimens designed to increase blood EPO levels. Unfortunately, some athletes have also tried to "cheat" by using injections of EPO and synthetic analogs, and "blood doping" to boost performance. This has led the governing bodies of many sports federations, including the Olympic Committee, to start testing for such abuse.

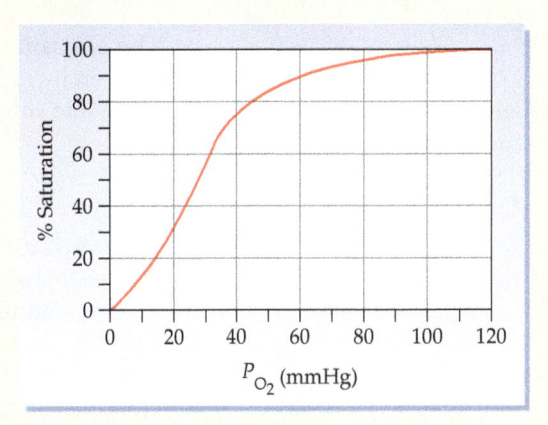

▲ An oxygen-carrying curve for hemoglobin. The percent saturation of the oxygen binding sites on hemoglobin depends on the partial pressure of oxygen P_{O_2}.

See Chemistry in Action Problem 9.90 at the end of the chapter.

264 CHAPTER 9 Solutions

TABLE **9.5** **Some Units for Expressing Concentration**

Concentration Measure	Solute Measure	Solution Measure
Percent		
Mass/mass percent, (m/m)%	Mass (g)	Mass (g)
Volume/volume percent, (v/v)%	Volume*	Volume*
Mass/volume percent, (m/v)%	Mass (g)	Volume (mL)
Parts per million, ppm	Parts*	10^6 parts*
Parts per billion, ppb	Parts*	10^9 parts*
Molarity, M	Moles	Volume (L)

*Any units can be used as long as they are the same for both solute and solution.

Let us look at each of the concentration measures listed in Table 9.5 individually, beginning with *percent concentrations*.

Percent Concentrations

Percent concentrations express the amount of solute in one hundred units of solution. The amount of solute and the amount of solution can be represented in units of mass or volume. For solid solutions, such as a metal alloy, concentrations are typically expressed as **mass/mass percent concentration, (m/m)%**:

mass/mass percent concentration, (m/m)% Concentration expressed as the number of grams of solute per 100 grams of solution.

$$(m/m)\% \text{ concentration} = \frac{\text{Mass of solute (g)}}{\text{Mass of solution (g)}} \times 100\%$$

For example, the mass percent of copper in a red-gold ring that contains 19.20 g of gold and 4.80 g of copper would be calculated as:

$$(m/m)\% \text{ Cu} = \frac{\text{mass of Cu (g)}}{\text{mass of Cu (g)} + \text{mass of Au (g)}} \times 100\%$$

$$= \frac{4.80 \text{ g}}{4.80 \text{ g} + 19.20 \text{ g}} \times 100\% = 20.0\%$$

The concentration of a solution made by dissolving one liquid in another is often given by expressing the volume of solute as a percentage of the volume of final solution—the **volume/volume percent concentration, (v/v)%**.

volume/volume percent concentration, (v/v)% Concentration expressed as the number of milliliters of solute dissolved in 100 mL of solution.

$$(v/v)\% \text{ concentration} = \frac{\text{Volume of solute (mL)}}{\text{Volume of solution (mL)}} \times 100\%$$

For example, if 10.0 mL of ethyl alcohol is dissolved in enough water to give 100.0 mL of solution, the ethyl alcohol concentration is (10.0 mL/100.0 mL) × 100% = 10.0% (v/v).

A third common method for expressing percent concentration is to give the number of grams (mass) as a percentage of the number of milliliters (volume) of the final solution—called the **mass/volume percent concentration, (m/v)%**. Mathematically, (m/v)% concentration is found by taking the number of grams of solute per milliliter of solution and multiplying by 100%:

mass/volume percent concentration, (m/v)% Concentration expressed as the number of grams of solute per 100 mL of solution.

$$(m/v)\% \text{ concentration} = \frac{\text{Mass of solute (g)}}{\text{Volume of solution (mL)}} \times 100\%$$

For example, if 15 g of glucose is dissolved in enough water to give 100 mL of solution, the glucose concentration is 15 g/100 mL or 15% (m/v):

$$\frac{15 \text{ g glucose}}{100 \text{ mL solution}} \times 100\% = 15\% \text{ (m/v)}$$

To prepare 100 mL of a specific mass/volume solution, the weighed solute is dissolved in just enough solvent to give a final volume of 100 mL, not in an initial volume of 100 mL solvent. (If the solute is dissolved in 100 mL of solvent, the final volume of the solution will likely be a bit larger than 100 mL, since the volume of the solute is included.) In practice, the appropriate amount of solute is weighed and placed in a *volumetric flask*, as shown in Figure 9.6. Enough solvent is then added to dissolve the solute, and further solvent is added until an accurately calibrated final volume is reached. The solution is then shaken until it is uniformly mixed. Worked Examples 9.4–9.7 illustrate how percent concentrations can be calculated for a solution, or how the percent concentration can be used as a conversion factor to determine the amount of solute in a given amount of solution.

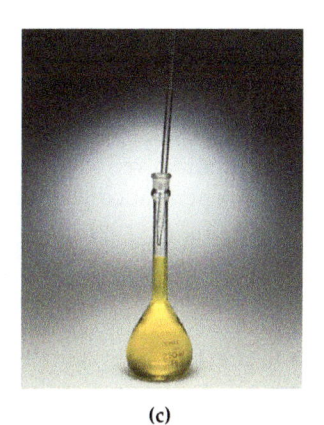

(a) (b) (c)

◀ **Figure 9.6**
Preparing a solution of known mass/volume percent concentration, (m/v)%.
(a) A measured number of grams of solute is placed in a volumetric flask. (b) Enough solvent is added to dissolve the solute by swirling. (c) Further solvent is carefully added until the calibration mark on the neck of the flask is reached, and the solution is shaken until uniform.

Worked Example 9.4 Mass Percent as Conversion Factor: Mass of Solution to Mass of Solute

The percentage of gold in jewelry is typically reported in carats, with 24 carats representing 100% gold. A sample of 18-carat gold would contain 18 grams of gold in 24 grams of metal, which would equal a (m/m)% of 75%. Calculate the mass of gold in a 5.05 g ring that is 18-carat gold.

ANALYSIS We are given a concentration and the total mass of the sample solution (the gold alloy in the ring), and we need to find the mass of gold by rearranging the equation for (m/m)% concentration.

BALLPARK ESTIMATE A 75% (m/m) solution contains 75 g for every 100 g of solution, so 10 g contains 7.5 g. The mass of the ring is a little more than 5 g (or half of 10 g) so the amount of gold in the ring will be slightly more than half of 7.5 g, or ~3.8 g gold.

SOLUTION

$$(5.05 \text{ g})\left(\frac{75 \text{ g Au}}{100 \text{ g solution}}\right) = 3.79 \text{ g Au}$$

BALLPARK CHECK The calculated answer is consistent with our estimate of 3.8 g gold.

Worked Example 9.5 Volume Percent as Conversion Factor: Volume of Solution to Volume of Solute

How many milliliters of methyl alcohol are needed to prepare 75 mL of a 5.0% (v/v) solution?

ANALYSIS We are given a solution volume (75 mL) and a concentration [5.0% (v/v), meaning 5.0 mL solute/100 mL solution]. The concentration acts as a conversion factor for finding the amount of methyl alcohol needed.

BALLPARK ESTIMATE A 5% (v/v) solution contains 5 mL of solute in 100 mL of solution, so the amount of solute in 75 mL of solution must be about three-fourths of 5 mL, which means between 3 and 4 mL.

SOLUTION

$$(75 \text{ mL solution})\left(\frac{5.0 \text{ mL methyl alcohol}}{100 \text{ mL solution}}\right) = 3.8 \text{ mL methyl alcohol}$$

BALLPARK CHECK The calculated answer is consistent with our estimate of between 3 and 4 mL.

Worked Example 9.6 Solution Concentration: Mass/Volume Percent

A solution of heparin sodium, an anticoagulant for blood, contains 1.8 g of heparin sodium dissolved to make a final volume of 15 mL of solution. What is the mass/volume percent concentration of this solution?

ANALYSIS Mass/volume percent concentration is defined as the mass of the solute in grams divided by the volume of solution in milliliters and multiplied by 100%.

BALLPARK ESTIMATE The mass of solute (1.8 g) is smaller than the volume of solvent (15 mL) by a little less than a factor of 10. The weight/volume percent should thus be a little greater than 10%.

SOLUTION

$$(m/v)\% \text{ concentration} = \frac{1.8 \text{ g heparin sodium}}{15 \text{ mL}} \times 100\% = 12\% \text{ (m/v)}$$

BALLPARK CHECK The calculated (m/v)% is reasonably close to our original estimate of 10%.

Worked Example 9.7 Mass/Volume Percent as Conversion Factor: Volume to Mass

How many grams of NaCl are needed to prepare 250 mL of a 1.5% (m/v) saline solution?

ANALYSIS We are given a concentration and a volume, and we need to find the mass of solute by rearranging the equation for (m/v)% concentration.

BALLPARK ESTIMATE The desired (m/v)% value, 1.5%, is between 1 and 2%. For a volume of 250 mL, we would need 2.5 g of solute for a 1% (m/v) solution and 5.0 g of solute for a 2% solution. Thus, for our 1.5% solution, we need a mass midway between 2.5 and 5.0 g, or about 3.8 g.

SOLUTION

$$\text{Since } (m/v)\% = \frac{\text{Mass of solute in g}}{\text{Volume of solution in mL}} \times 100\%$$

$$\text{then Mass of solute in g} = \frac{(\text{Volume of solution in mL})[(m/v)]\%}{100\%}$$

$$= \frac{(250)(1.5\%)}{100\%} = 3.75 \text{ g} = 3.8 \text{ g NaCl}$$

(2 significant figures)

BALLPARK CHECK The calculated answer matches our estimate.

SECTION 9.7 Units of Concentration 267

PROBLEM 9.9
A metal alloy contains 15.8% nickel (m/m)%. What mass of the metal alloy would contain 36.5 g of nickel?

PROBLEM 9.10
How would you use a 500.0 mL volumetric flask to prepare a 7.5% (v/v) solution of acetic acid in water?

PROBLEM 9.11
In clinical lab reports, some concentrations are given in mg/dL. Convert a Ca^{2+} concentration of 8.6 mg/dL to mass/volume percent.

PROBLEM 9.12
What amounts of solute or solvent are needed to prepare the following solutions?
(a) Mass of glucose needed to prepare 125.0 mL of 16% (m/v) glucose ($C_6H_{12}O_6$)
(b) Volume of water needed to prepare a 2.0% (m/v) KCl solution using 1.20 g KCl

Parts per Million (ppm) or Parts per Billion (ppb)

The concentration units mass/mass percent (m/m)%, volume/volume percent (v/v)%, and mass/volume percent (w/v)% can also be defined as *parts per hundred* (pph) since 1% means one item per 100 items. When concentrations are very small, as often occurs in dealing with trace amounts of pollutants or contaminants, it is more convenient to use **parts per million (ppm)** or **parts per billion (ppb)**. The "parts" can be in any unit of either mass or volume as long as the units of both solute and solvent are the same:

$$\text{ppm} = \frac{\text{Mass of solute (g)}}{\text{Mass of solution (g)}} \times 10^6 \quad \text{or} \quad \frac{\text{Volume of solute (mL)}}{\text{Volume of solution (mL)}} \times 10^6$$

$$\text{ppb} = \frac{\text{Mass of solute (g)}}{\text{Mass of solution (g)}} \times 10^9 \quad \text{or} \quad \frac{\text{Volume of solute (mL)}}{\text{Volume of solution (mL)}} \times 10^9$$

Parts per million (ppm) Number of parts per one million (10^6) parts.

Parts per billion (ppb) Number of parts per one billion (10^9) parts.

To take an example, the maximum allowable concentration in air of the organic solvent benzene (C_6H_6) is currently set by government regulation at 1 ppm. A concentration of 1 ppm means that if you take a million "parts" of air in any unit—say, mL—then 1 of those parts is benzene vapor and the other 999,999 parts are other gases:

$$1 \text{ ppm} = \frac{1 \text{ mL}}{1,000,000 \text{ mL}} \times 10^6$$

Because the density of water is approximately 1.0 g/mL at room temperature, 1.0 L (or 1000 mL) of an aqueous solution weighs 1000 g. Therefore, when dealing with very dilute concentrations of solutes dissolved in water, ppm is equivalent to mg solute/L solution, and ppb is equivalent to μg solute/L solution. To demonstrate that these units are equivalent, the conversion from ppm to mg/L is as follows:

$$1 \text{ ppm} = \left(\frac{1 \text{ g solute}}{10^6 \text{ g solution}}\right)\left(\frac{1 \text{ mg solute}}{10^{-3} \text{ g solute}}\right)\left(\frac{10^3 \text{ g solution}}{1 \text{ L solution}}\right) = \frac{1 \text{ mg solute}}{1 \text{ L solution}}$$

Worked Example 9.8 ppm as Conversion Factor: Mass of Solution to Mass of Solute

The maximum allowable concentration of chloroform, $CHCl_3$, in drinking water is 100 ppb. What is the maximum amount (in grams) of chloroform allowed in a glass containing 400 g (400 mL) of water?

ANALYSIS We are given a solution amount (400 g) and a concentration (100 ppb). This concentration of 100 ppb means

$$100 \text{ ppb} = \frac{\text{Mass of solute (g)}}{\text{Mass of solution (g)}} \times 10^9$$

This equation can be rearranged to find the mass of solute.

BALLPARK ESTIMATE A concentration of 100 ppb means there are 100×10^{-9} g (1×10^{-7} g) of solute in 1 g of solution. In 400 g of solution, we should have 400 times this amount, or $400 \times 10^{-7} = 4 \times 10^{-5}$ g.

SOLUTION

$$\text{Mass of solute (g)} = \frac{\text{Mass of solution (g)}}{10^9} \times 100 \text{ ppb}$$

$$= \frac{400 \text{ g}}{10^9} \times 100 \text{ ppb} = 4 \times 10^{-5} \text{ g (or 0.04 mg)}$$

BALLPARK CHECK The calculated answer matches our estimate.

PROBLEM 9.13
What is the concentration in ppm of sodium fluoride in tap water that has been fluoridated by the addition of 32 mg of NaF for every 20 kg of solution?

PROBLEM 9.14
The maximum amounts of lead and copper allowed in drinking water are 0.015 mg/kg for lead and 1.3 mg/kg for copper. Express these values in parts per million, and tell the maximum amount of each (in grams) allowed in 100 g of water.

Mole/Volume Concentration: Molarity

We saw in Chapter 6 that the various relationships between amounts of reactants and products in chemical reactions are calculated in *moles* (Sections 6.1–6.3). Thus, the most generally useful means of expressing concentration in the laboratory is **molarity (M)**, the number of moles of solute dissolved per liter of solution. For example, a solution made by dissolving 1.00 mol (58.5 g) of NaCl in enough water to give 1.00 L of solution has a concentration of 1.00 mol/L, or 1.00 M. The molarity of any solution is found by dividing the number of moles of solute by the number of liters of solution (solute + solvent):

Molarity (M) Concentration expressed as the number of moles of solute per liter of solution.

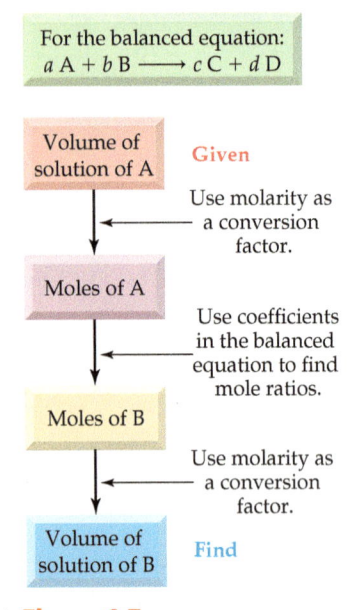

$$\text{Molarity (M)} = \frac{\text{Moles of solute}}{\text{Liters of solution}}$$

Note that a solution of a given molarity is prepared by dissolving the solute in enough solvent to give a *final* solution volume of 1.00 L, not by dissolving it in an *initial* volume of 1.00 L. If an initial volume of 1.00 L was used, the final solution volume might be a bit larger than 1.00 L because of the additional volume of the solute. In practice, solutions are prepared using a volumetric flask, as shown previously in Figure 9.6.

Molarity can be used as a conversion factor to relate the volume of a solution to the number of moles of solute it contains. If we know the molarity and volume of a solution, we can calculate the number of moles of solute. If we know the number of moles of solute and the molarity of the solution, we can find the solution's volume.

$$\text{Molarity} = \frac{\text{Moles of solute}}{\text{Volume of solution (L)}}$$

$$\text{Moles of solute} = \text{Molarity} \times \text{Volume of solution}$$

$$\text{Volume of solution} = \frac{\text{Moles of solute}}{\text{Molarity}}$$

The flow diagram in Figure 9.7 shows how molarity is used in calculating the quantities of reactants or products in a chemical reaction, and Worked Examples 9.10 and 9.11 show how the calculations are done. Note that Problem 9.17 employs *millimolar* (mM) concentrations, which are useful in healthcare fields for expressing low concentrations such as are often found in body fluids (1 mM = 0.001 M).

▲ **Figure 9.7**
Molarity and conversions.
A flow diagram summarizing the use of molarity for conversions between solution volume and moles to find quantities of reactants and products for chemical reactions in solution.

For the balanced equation:
$a\text{A} + b\text{B} \longrightarrow c\text{C} + d\text{D}$

Volume of solution of A — *Given*
↓ Use molarity as a conversion factor.
Moles of A
↓ Use coefficients in the balanced equation to find mole ratios.
Moles of B
↓ Use molarity as a conversion factor.
Volume of solution of B — *Find*

SECTION 9.7 Units of Concentration 269

Worked Example 9.9 Solution Concentration: Molarity

What is the molarity of a solution made by dissolving 2.355 g of sulfuric acid (H_2SO_4) in water and diluting to a final volume of 50.0 mL? The molar mass of H_2SO_4 is 98.1 g/mol.

ANALYSIS Molarity is defined as moles of solute per liter of solution: M = mol/L. Thus, we must first find the number of moles of sulfuric acid by doing a mass to mole conversion, and then divide the number of moles by the volume of the solution.

BALLPARK ESTIMATE The molar mass of sulfuric acid is about 100 g/mol, so 2.355 g is roughly 0.025 mol. The volume of the solution is 50.0 mL, or 0.05 L, so we have about 0.025 mol of acid in 0.05 L of solution, which is a concentration of about 0.5 M.

SOLUTION

STEP 1: Identify known information. We know the mass of sulfuric acid and the final volume of solution.

Mass of H_2SO_4 = 2.355 g
Volume of solution = 50.0 mL

STEP 2: Identify answer including units. We need to find the molarity (M) in units of moles per liter.

$$\text{Molarity} = \frac{\text{Moles } H_2SO_4}{\text{Liters of solution}}$$

STEP 3: Identify conversion factors and equations. We know both the amount of solute and the volume of solution, but first we must make two conversions: convert mass of H_2SO_4 to moles of H_2SO_4, using molar mass as a conversion factor, and convert volume from milliliters to liters.

$$(2.355 \text{ g } H_2SO_4)\left(\frac{1 \text{ mol } H_2SO_4}{98.1 \text{ g } H_2SO_4}\right) = 0.0240 \text{ mol } H_2SO_4$$

$$(50.0 \text{ mL})\left(\frac{1 \text{ L}}{1000 \text{ mL}}\right) = 0.0500 \text{ L}$$

STEP 4: Solve. Substitute the moles of solute and volume of solution into the molarity expression.

$$\text{Molarity} = \frac{0.0240 \text{ mol } H_2SO_4}{0.0500 \text{ L}} = 0.480 \text{ M}$$

BALLPARK CHECK The calculated answer is close to our estimate, which was 0.5 M.

Worked Example 9.10 Molarity as Conversion Factor: Molarity to Mass

A blood concentration of 0.065 M ethyl alcohol (EtOH) is sufficient to induce a coma. At this concentration, what is the total mass of alcohol (in grams) in an adult male whose total blood volume is 5.6 L? The molar mass of ethyl alcohol is 46.0 g/mol. (Refer to the flow diagram in Figure 9.7 to identify which conversions are needed.)

ANALYSIS We are given a molarity (0.065 M) and a volume (5.6 L), which allows us to calculate the number of moles of alcohol in the blood. A mole to mass conversion then gives the mass of alcohol.

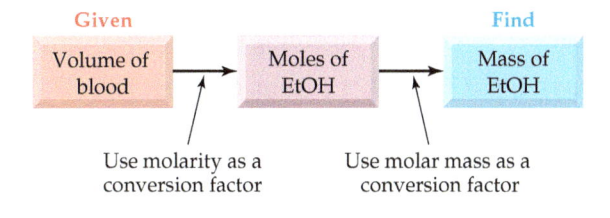

```
Given                              Find
Volume of  →  Moles of  →  Mass of
  blood        EtOH          EtOH
          ↑              ↑
   Use molarity as a   Use molar mass as a
   conversion factor   conversion factor
```

SOLUTION

$$(5.6 \text{ L blood})\left(\frac{0.065 \text{ mol EtOH}}{1 \text{ L blood}}\right) = 0.36 \text{ mol EtOH}$$

$$(0.36 \text{ mol EtOH})\left(\frac{46.0 \text{ g EtOH}}{1 \text{ mol EtOH}}\right) = 17 \text{ g EtOH}$$

270 CHAPTER 9 Solutions

Worked Example 9.11 Molarity as Conversion Factor: Molarity to Volume

In our stomachs, gastric juice that is about 0.1 M in HCl aids in digestion. How many milliliters of gastric juice will react completely with an antacid tablet that contains 500 mg of magnesium hydroxide? The molar mass of $Mg(OH)_2$ is 58.3 g/mol, and the balanced equation is

$$2\, HCl(aq) + Mg(OH)_2(aq) \longrightarrow MgCl_2(aq) + 2\, H_2O(l)$$

ANALYSIS We are given the molarity of HCl and need to find the volume. We first convert the mass of $Mg(OH)_2$ to moles and then use the coefficients in the balanced equation to find the moles of HCl that will react. Once we have the moles of HCl and the molarity in moles per liter, we can find the volume.

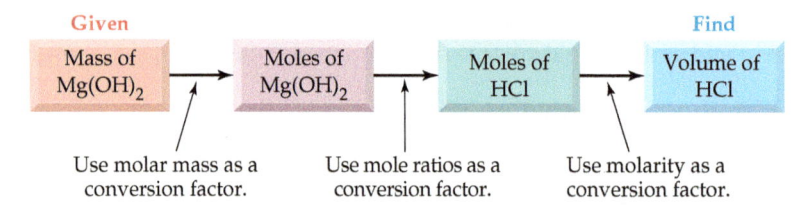

Given: Mass of $Mg(OH)_2$ → Moles of $Mg(OH)_2$ → Moles of HCl → Find: Volume of HCl

Use molar mass as a conversion factor. Use mole ratios as a conversion factor. Use molarity as a conversion factor.

SOLUTION

$$[500\ mg\ Mg(OH)_2]\left(\frac{1\ g}{1000\ mg}\right)\left[\frac{1\ mol\ Mg(OH)_2}{58.3\ g\ Mg(OH)_2}\right] = 0.008\ 58\ mol\ Mg(OH)_2$$

$$[0.008\ 58\ mol\ Mg(OH)_2]\left[\frac{2\ mol\ HCl}{1\ mol\ Mg(OH)_2}\right]\left(\frac{1\ L\ HCl}{0.1\ mol\ HCl}\right) = 0.2\ L\ (200\ mL)$$

PROBLEM 9.15
What is the molarity of a solution that contains 50.0 g of vitamin B_1 hydrochloride (molar mass = 337 g/mol) in 160 mL of solution?

PROBLEM 9.16
How many moles of solute are present in the following solutions?
(a) 175 mL of 0.35 M $NaNO_3$
(b) 480 mL of 1.4 M HNO_3

PROBLEM 9.17
The concentration of cholesterol ($C_{27}H_{46}O$) in blood is approximately 5.0 mM. How many grams of cholesterol are in 250 mL of blood?

PROBLEM 9.18
Calcium carbonate reacts with HCl according to the following equation:

$$2\, HCl(aq) + CaCO_3(aq) \longrightarrow CaCl_2(aq) + H_2O(l) + CO_2(g)$$

(a) How many moles of HCl are in 65 mL of 0.12 M HCl?
(b) What mass of calcium carbonate (in grams) is needed for complete reaction with the HCl in (a)?

9.8 Dilution

Many solutions, from orange juice to chemical reagents, are stored in high concentrations and then prepared for use by *dilution*—that is, by adding additional solvent to lower the concentration. For example, you might make up 1/2 gal of orange juice by adding water to a canned concentrate. In the same way, you might buy a medicine or chemical reagent as a concentrated solution and dilute it before use.

The key fact to remember about dilution is that the amount of *solute* remains constant; only the *volume* is changed by adding more solvent. If, for example, the initial and final concentrations are given in molarity, then we know that the number of moles of solute is the same both before and after dilution and can be determined by multiplying molarity times volume:

$$\text{Number of moles} = \text{Molarity (mol/L)} \times \text{Volume (L)}$$

$$M = \text{moles/volume}$$

Because the number of moles remains constant, we can set up the following equation, where M_c and V_c refer to the concentrated solution (before dilution), and M_d and V_d refer to the solution after dilution:

$$\text{Moles of solute} = M_c V_c = M_d V_d$$

This equation can be rewritten to solve for M_d, the concentration of the solution after dilution:

$$M_d = M_c \times \frac{V_c}{V_d} \quad \text{where} \quad \frac{V_c}{V_d} \text{ is a } dilution\ factor$$

The equation shows that the concentration after dilution (M_d) can be found by multiplying the initial concentration (M_c) by a **dilution factor**, which is simply the ratio of the initial and final solution volumes (V_c/V_d). If, for example, the solution volume *increases* by a factor of 5, from 10 mL to 50 mL, then the concentration must *decrease* to one-fifth of its initial value because the dilution factor is 10 mL/50 mL, or 1/5. Worked Example 9.12 shows how to use this relationship for calculating dilutions.

Dilution factor The ratio of the initial and final solution volumes (V_c/V_d).

The relationship between concentration and volume can also be used to find what volume of initial solution to start with to achieve a given dilution:

$$\text{Since} \quad M_c V_c = M_d V_d$$

$$\text{then} \quad V_c = V_d \times \frac{M_d}{M_c}$$

In this case, V_c is the initial volume that must be diluted to prepare a less concentrated solution with volume V_d. The initial volume is found by multiplying the final volume (V_d) by the ratio of the final and initial concentrations (M_d/M_c). For example, to decrease the concentration of a solution to 1/5 its initial value, the initial volume must be 1/5 the desired final volume. Worked Example 9.13 gives a sample calculation.

Although the preceding discussion, and the following Worked Examples, use concentration units of molarity, the dilution equation can be generalized to allow for the use of other concentration units. A more general equation would be $C_c V_c = C_d V_d$, where C refers to other concentration units, such as ppm, or m/v%.

Worked Example 9.12 Dilution of Solutions: Concentration

What is the final concentration if 75 mL of a 3.5 M glucose solution is diluted to a volume of 450 mL?

ANALYSIS The number of moles of solute is constant, so

$$M_c V_c = M_d V_d$$

Of the four variables in this equation, we know the initial concentration M_c (3.5 M), the initial volume V_c (75 mL), and the final volume V_d (450 mL), and we need to find the final concentration M_d.

BALLPARK ESTIMATE The volume increases by a factor of 6, from 75 mL to 450 mL, so the concentration must decrease by a factor of 6, from 3.5 M to about 0.6 M.

SOLUTION
Solving the above equation for M_d and substituting in the known values gives

$$M_d = \frac{M_c V_c}{V_d} = \frac{(3.5 \text{ M glucose})(75 \text{ mL})}{450 \text{ mL}} = 0.58 \text{ M glucose}$$

BALLPARK CHECK The calculated answer is close to our estimate of 0.6 M.

Worked Example **9.13** Dilution of Solutions: Volume

Aqueous NaOH can be purchased at a concentration of 1.0 M. How would you use this concentrated solution to prepare 750 mL of 0.32 M NaOH?

ANALYSIS The number of moles of solute is constant, so

$$M_c V_c = M_d V_d$$

Of the four variables in this equation, we know the initial concentration M_c (1.0 M), the final volume V_d (750 mL), and the final concentration M_d (0.32 M), and we need to find the initial volume V_c.

BALLPARK ESTIMATE We want the solution concentration to decrease by a factor of about 3, from 1.0 M to 0.32 M, which means we need to dilute the 1.0 M solution by a factor of 3. This means the final volume must be about three times greater than the initial volume. Because our final volume is to be 750 mL, we must start with an initial volume of about 250 mL.

SOLUTION
Solving the above equation for V_1 and substituting in the known values gives

$$V_c = \frac{V_d M_d}{M_c} = \frac{(750 \text{ mL})(0.32 \text{ M})}{1.0 \text{ M}} = 240 \text{ mL}$$

To prepare the desired solution, dilute 240 mL of 1.0 M NaOH with water to make a final volume of 750 mL.

BALLPARK CHECK The calculated answer (240 mL) is reasonably close to our estimate of 250 mL.

PROBLEM 9.19
Aqueous ammonia is commercially available at a concentration of 16.0 M. How much of the concentrated solution would you use to prepare 500.0 mL of a 1.25 M solution?

PROBLEM 9.20
The Environmental Protection Agency has set the limit for arsenic in drinking water at 0.010 ppm. To what volume would you need to dilute 1.5 L of water containing 5.0 ppm arsenic to reach the acceptable limit?

9.9 Ions in Solution: Electrolytes

Look at Figure 9.8, which shows a light bulb connected to a power source through a circuit that is interrupted by two metal strips dipped into a beaker of liquid. When the strips are dipped into pure water, the bulb remains dark, but when they are dipped into an aqueous NaCl solution, the circuit is closed and the bulb lights. This simple demonstration shows that ionic compounds in aqueous solution can conduct electricity.

▶▶▶ As we learned in Section 3.1, electricity can only flow through a medium containing charged particles that are free to move.

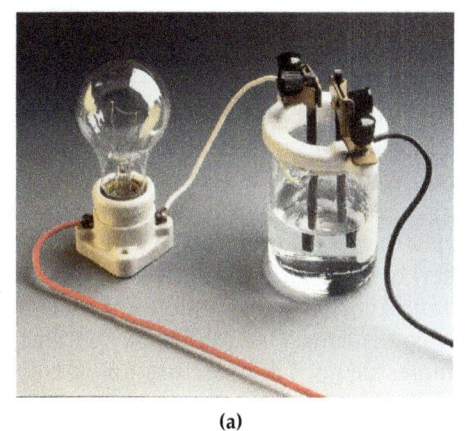

SECTION 9.10 Electrolytes in Body Fluids: Equivalents and Milliequivalents

◀ Figure 9.8
A simple demonstration shows that electricity can flow through a solution of ions.
(a) With pure water in the beaker, the circuit is incomplete, no electricity flows, and the bulb does not light. (b) With a concentrated NaCl solution in the beaker, the circuit is complete, electricity flows, and the light bulb glows.

(a)   (b)

Substances like NaCl that conduct an electric current when dissolved in water are called **electrolytes**. Conduction occurs because negatively charged Cl^- anions migrate through the solution toward the metal strip connected to the positive terminal of the power source, whereas positively charged Na^+ cations migrate toward the strip connected to the negative terminal. As you might expect, the ability of a solution to conduct electricity depends on the concentration of ions in solution. Distilled water contains virtually no ions and is nonconducting; ordinary tap water contains low concentrations of dissolved ions (mostly Na^+, K^+, Mg^{2+}, Ca^{2+}, and Cl^-) and is weakly conducting; and a concentrated solution of NaCl is strongly conducting.

Ionic substances like NaCl that ionize completely when dissolved in water are called **strong electrolytes**, and molecular substances like acetic acid (CH_3CO_2H) that are only partially ionized are **weak electrolytes**. Molecular substances like glucose that do not produce ions when dissolved in water are **nonelectrolytes**.

Electrolyte A substance that produces ions and therefore conducts electricity when dissolved in water.

Strong electrolyte; completely ionized
$$NaCl(s) \xrightarrow[\text{in water}]{\text{Dissolve}} Na^+(aq) + Cl^-(aq)$$

Weak electrolyte; partly ionized
$$CH_3CO_2H(l) \xrightleftharpoons[\text{in water}]{\text{Dissolve}} CH_3CO_2^-(aq) + H^+(aq)$$

Nonelectrolyte; not ionized
$$Glucose(s) \xrightleftharpoons[\text{in water}]{\text{Dissolve}} Glucose(aq)$$

Strong electrolyte A substance that ionizes completely when dissolved in water.

Weak electrolyte A substance that is only partly ionized in water.

Nonelectrolyte A substance that does not produce ions when dissolved in water.

9.10 Electrolytes in Body Fluids: Equivalents and Milliequivalents

What happens if NaCl and KBr are dissolved in the same solution? Because the cations (K^+ and Na^+) and anions (Cl^- and Br^-) are all mixed together and no reactions occur between them, an identical solution could just as well be made from KCl and NaBr. Thus, we can no longer speak of having a NaCl + KBr solution; we can only speak of having a solution with four different ions in it.

A similar situation exists for blood and other body fluids, which contain many different anions and cations. Since they are all mixed together, it is difficult to "assign" specific cations to specific anions or to talk about specific ionic compounds. Instead, we are interested only in individual ions and in the total numbers of positive and negative charges. To discuss such mixtures, we use a new term—*equivalents* of ions.

For ions, one **equivalent (Eq)** is equal to the number of ions that carry 1 mol of charge. Of more practical use is the unit **gram-equivalent (g-Eq)**, which is the amount of ion (in grams) that contains one mole of charge. It can be calculated simply as the molar mass of the ion divided by the absolute value of its charge.

Equivalent For ions, the amount equal to 1 mol of charge.

Gram-equivalent For ions, the molar mass of the ion divided by the ionic charge.

$$\text{One gram-equivalent of ion} = \frac{\text{Molar mass of ion (g)}}{\text{Charge on ion}}$$

274 CHAPTER 9 Solutions

If the ion has a charge of +1 or −1, 1 gram-equivalent of the ion is simply the molar mass of the ion in grams. Thus, 1 gram-equivalent of Na^+ is 23 g, and 1 gram-equivalent of Cl^- is 35.5 g. If the ion has a charge of +2 or −2, however, 1 gram-equivalent is equal to the ion's formula weight in grams divided by 2. Thus, 1 gram-equivalent of Mg^{2+} is $(24.3 \text{ g})/2 = 12.2$ g, and 1 gram-equivalent of CO_3^{2-} is $[12.0 \text{ g} + (3 \times 16.0 \text{ g})]/2 = 30.0$ g. The gram-equivalent is a useful conversion factor when converting from volume of solution to mass of ions, as seen in Worked Example 9.14.

The number of equivalents of a given ion per liter of solution can be found by multiplying the molarity of the ion (moles per liter) by the charge on the ion. Because ion concentrations in body fluids are often low, clinical chemists find it more convenient to talk about *milliequivalents* of ions rather than equivalents. One milliequivalent (mEq) of an ion is 1/1000 of an equivalent. For example, the normal concentration of Na^+ in blood is 0.14 Eq/L, or 140 mEq/L.

$$1 \text{ mEq} = 0.001 \text{ Eq} \qquad 1 \text{ Eq} = 1000 \text{ mEq}$$

Note that the gram-equivalent for an ion can now be expressed as grams per equivalent or as mg per mEq.

Average concentrations of the major electrolytes in blood plasma are given in Table 9.6. As you might expect, the total milliequivalents of positively and negatively charged electrolytes must be equal to maintain electrical neutrality. Adding the milliequivalents of positive and negative ions in Table 9.6, however, shows a higher concentration of positive ions than negative ions. The difference, called the *anion gap*, is made up by the presence of negatively charged proteins and the anions of organic acids.

TABLE 9.6 Concentrations of Major Electrolytes in Blood Plasma

Cation	Concentration (mEq/L)
Na^+	136–145
Ca^{2+}	4.5–6.0
K^+	3.6–5.0
Mg^{2+}	3

Anion	Concentration (mEq/L)
Cl^-	98–106
HCO_3^-	25–29
SO_4^{2-} and HPO_4^{2-}	2

Worked Example 9.14 Equivalents as Conversion Factors: Volume to Mass

The normal concentration of Ca^{2+} in blood is 5.0 mEq/L. How many milligrams of Ca^{2+} are in 1.00 L of blood?

ANALYSIS We are given a volume and a concentration in milliequivalents per liter, and we need to find an amount in milligrams. Thus, we need to calculate the gram-equivalent for Ca^{2+} and then use concentration as a conversion factor between volume and mass, as indicated in the following flow diagram:

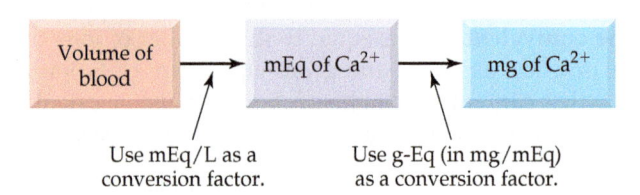

Volume of blood → mEq of Ca^{2+} → mg of Ca^{2+}

Use mEq/L as a conversion factor.

Use g-Eq (in mg/mEq) as a conversion factor.

BALLPARK ESTIMATE The molar mass of calcium is 40.08 g/mol, and the calcium ion carries a charge of 2+. Thus, 1 g-Eq of Ca^{2+} equals about 20 g/Eq or 20 mg/mEq. This means that the 5.0 mEq of Ca^{2+} ions in 1.00 L of blood corresponds to a mass of 5.0 mEq Ca^{2+} × 20 mg/mEq = 100 mg Ca^{2+}.

SOLUTION

$$(1.00 \text{ L blood})\left(\frac{5.0 \text{ mEq } Ca^{2+}}{1.0 \text{ L blood}}\right)\left(\frac{20.04 \text{ mg } Ca^{2+}}{1 \text{ mEq } Ca^{2+}}\right) = 100 \text{ mg } Ca^{2+}$$

BALLPARK CHECK The calculated answer (100 mg of Ca^{2+} in 1.00 L of blood) matches our estimate.

PROBLEM 9.21
How many grams are in 1 Eq of the following ions? How many grams in 1 mEq?
(a) K^+ (b) Br^- (c) Mg^{2+} (d) SO_4^{2-} (e) Al^{3+} (f) PO_4^{3-}

PROBLEM 9.22
Look at the data in Table 9.6, and calculate how many milligrams of Mg^{2+} are in 250 mL of blood.

PROBLEM 9.23
A typical sports drink for electrolyte replacement contains 20 mEq/L of Na^+ and 10 mEq/L of K^+ ions (see Chemistry in Action: Electrolytes, Fluid Replacement, and Sports Drinks on p. 276). Convert these concentrations to m/v%.

9.11 Properties of Solutions

The properties of solutions are similar in many respects to those of pure solvents, but there are also some interesting and important differences. One such difference is that solutions have higher boiling points than the pure solvents; another is that solutions have lower freezing points. Pure water boils at 100.0 °C and freezes at 0.0 °C, for example, but a 1.0 M solution of NaCl in water boils at 101.0 °C and freezes at −3.7 °C.

The elevation of boiling point and the lowering of freezing point for a solution as compared with a pure solvent are examples of **colligative properties**—properties that depend on the *concentration* of a dissolved solute but not on its chemical identity. Other colligative properties are a lower vapor pressure for a solution compared with the pure solvent and *osmosis*, the migration of solvent molecules through a semipermeable membrane.

Colligative property A property of a solution that depends only on the number of dissolved particles, not on their chemical identity.

Colligative Properties

- Vapor pressure is lower for a solution than for a pure solvent.
- Boiling point is higher for a solution than for a pure solvent.
- Freezing point is lower for a solution than for a pure solvent.
- Osmosis occurs when a solution is separated from a pure solvent by a semipermeable membrane.

Vapor-Pressure Lowering in Solutions

We said in Section 8.13 that the vapor pressure of a liquid depends on the equilibrium between molecules entering and leaving the liquid surface. Only those molecules at the surface of the liquid that are sufficiently energetic will evaporate. If, however, some of the liquid (solvent) molecules at the surface are replaced by other (solute) particles that do not evaporate, then the rate of evaporation of solvent molecules decreases and the

CHEMISTRY IN ACTION

Electrolytes, Fluid Replacement, and Sports Drinks

Electrolytes are essential in many physiological processes, and significant changes in electrolyte levels can be potentially life-threatening if not addressed quickly. Heavy and continuous diarrhea from conditions such as cholera can result in dehydration and very low sodium levels in the body (hyponatremia). Restoration of electrolytes can be accomplished by oral rehydration therapy (ORT). The introduction of ORT in developing countries decreased infant mortality from diarrhea, which had previously been the leading cause of death in children under 5 years of age. A typical ORT solution contains sodium (75 mEq/L), potassium (75b mEq/L), chloride (65 mEq/L), citrate (10 mEq/L), and glucose (75 mmol/L). Heavy sweating during strenuous exercise can also lead to dehydration and loss of electrolytes.

The composition of sweat is highly variable, but the typical concentration for the Na^+ ion is about 30–40 mEq/L, and that of K^+ ion is about 5–10 mEq/L. In addition, there are small amounts of other metal ions, such as Mg^{2+}, and there are sufficient Cl^- ions (35–50 mEq/L) to balance the positive charge of all these cations. If water and electrolytes are not replaced, dehydration, hyperthermia and heat stroke, dizziness, nausea, muscle cramps, impaired kidney function, and other difficulties ensue. As a rule of thumb, a sweat loss equal to 5% of body weight—about 3.5 L for a 150 lb person—is the maximum amount that can be safely allowed for a well-conditioned athlete.

Plain water works perfectly well to replace sweat lost during short bouts of activity up to a few hours in length, but a carbohydrate–electrolyte beverage, or "sports drink," is much superior for rehydrating during and after longer activity in which substantial amounts of electrolytes have been lost. Some of the better known sports drinks are little more than overpriced sugar–water solutions, but others are carefully formulated and highly effective for fluid replacement. Nutritional research has shown that a serious sports drink should meet the following criteria. There are several dry-powder mixes on the market to choose from.

- The drink should contain 6–8% of soluble complex carbohydrates (about 15 g per 8 oz serving) and only a small amount of simple sugar for taste. The complex carbohydrates, which usually go by the name "maltodextrin," provide a slow release of glucose into the bloodstream. Not only does the glucose provide a steady source of energy, it also enhances the absorption of water from the stomach.

▲ Drinking water to replace fluids is adequate for short periods of activity, but extended exercise requires replacement of fluid and electrolytes, such as those found in sports drinks.

- The drink should contain electrolytes to replenish those lost in sweat. Concentrations of approximately 20 mEq/L for Na^+ ions, 10 mEq/L for K^+ ion, and 4 mEq/L for Mg^{2+} ions are recommended. These amounts correspond to about 100 mg sodium, 100 mg potassium, and 25 mg magnesium per 8 oz serving.
- The drink should be noncarbonated because carbonation can cause gastrointestinal upset during exercise, and it should not contain caffeine, which acts as a diuretic.
- The drink should taste good so the athlete will want to drink it. Thirst is a poor indicator of fluid requirements, and most people will drink less than needed unless a beverage is flavored.

In addition to complex carbohydrates, electrolytes, and flavorings, some sports drinks also contain vitamin A (as beta-carotene), vitamin C (ascorbic acid), and selenium, which act as antioxidants to protect cells from damage. Some drinks also contain the amino acid glutamine, which appears to lessen lactic acid buildup in muscles and thus helps muscles bounce back more quickly after an intense workout.

See Chemistry in Actions Problems 9.91 and 9.92 at the end of the chapter.

vapor pressure of a solution is lower than that of the pure solvent (Figure 9.9). Note that the *identity* of the solute particles is irrelevant; only their concentration matters.

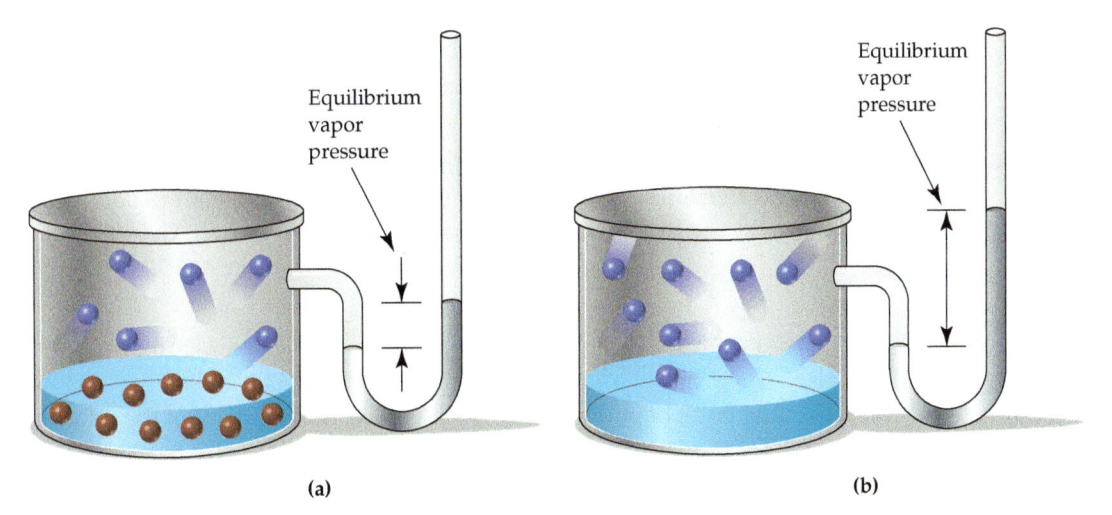

▲ Figure 9.9
Vapor-pressure lowering of solution.
(a) The vapor pressure of a solution is lower than (b) the vapor pressure of the pure solvent because fewer solvent molecules are able to escape from the surface of the solution.

Boiling Point Elevation of Solutions

One consequence of the vapor-pressure–lowering for a solution is that the boiling point of the solution is higher than that of the pure solvent. Recall from Section 8.13 that boiling occurs when the vapor pressure of a liquid reaches atmospheric pressure. But because the vapor pressure of a solution is lower than that of the pure solvent at a given temperature, the solution must be heated to a higher temperature for its vapor pressure to reach atmospheric pressure. Figure 9.10 shows a close-up plot of vapor pressure versus temperature for pure water and for a 1.0 M NaCl solution. The vapor pressure of pure water reaches atmospheric pressure (760 mmHg) at 100.0 °C, but the vapor pressure of the NaCl solution does not reach the same point until 101.0 °C.

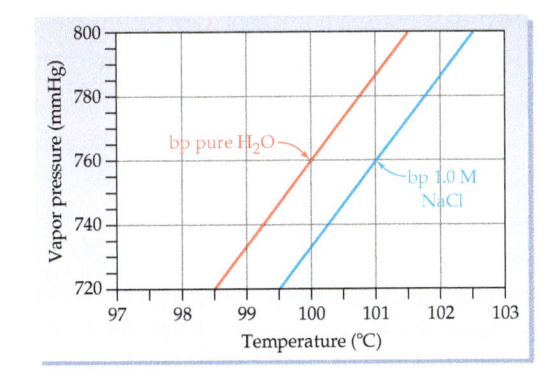

◀ Figure 9.10
Vapor pressure and temperature. A close-up plot of vapor pressure versus temperature for pure water (red curve) and for a 1.0 M NaCl solution (blue curve). Pure water boils at 100.0 °C, but the solution does not boil until 101.0 °C.

For each mole of solute particles added, regardless of chemical identity, the boiling point of 1 kg of water is raised by 0.51 °C, or

$$\Delta T_{boiling} = \left(0.51 \,°C \frac{kg \, water}{mol \, particles}\right)\left(\frac{mol \, particles}{kg \, water}\right)$$

The addition of 1 mol of a molecular substance like glucose to 1 kg of water therefore raises the boiling point from 100.0 °C to 100.51 °C. The addition of 1 mol of NaCl per kilogram of water, however, raises the boiling point by $2 \times 0.51 \,°C = 1.02 \,°C$ because the solution contains 2 mol of solute particles—Na^+ and Cl^- ions.

Worked Example 9.15 Properties of Solutions: Boiling Point Elevation

What is the boiling point of a solution of 0.75 mol of KBr in 1.0 kg of water?

ANALYSIS The boiling point increases 0.51 °C for each mole of solute per kilogram of water. Since KBr is a strong electrolyte, there are 2 moles of ions (K^+ and Br^-) for every 1 mole of KBr that dissolves.

BALLPARK ESTIMATE The boiling point will increase about 0.5 °C for every 1 mol of ions in 1 kg of water. Since 0.75 mol of KBr produce 1.5 mol of ions, the boiling point should increase by (1.5 mol ions) × (0.5 °C/mol ions) = 0.75 °C.

SOLUTION

$$\Delta T_{boiling} = \left(0.51\,°C\frac{\text{kg water}}{\text{mol ions}}\right)\left(\frac{2\,\text{mol ions}}{1\,\text{mol KBr}}\right)\left(\frac{0.75\,\text{mol KBr}}{1.0\,\text{kg water}}\right) = 0.77\,°C$$

The normal boiling point of pure water is 100 °C, so the boiling point of the solution increases to 100.77 °C.

BALLPARK CHECK The 0.77 °C increase is consistent with our estimate of 0.75 °C.

PROBLEM 9.24
A solution is prepared by dissolving 0.67 mol of $MgCl_2$ in 0.50 kg of water.
(a) How many moles of ions are present in solution?
(b) What is the change in the boiling point of the aqueous solution?

PROBLEM 9.25
When 1.0 mol of HF is dissolved in 1.0 kg of water, the boiling point of the resulting solution is 100.5 °C. Is HF a strong or weak electrolyte? Explain.

KEY CONCEPT PROBLEM 9.26

The following diagram shows plots of vapor pressure versus temperature for a solvent and a solution.
(a) Which curve represents the pure solvent and which the solution?
(b) What is the approximate boiling point elevation for the solution?
(c) What is the approximate concentration of the solution in mol/kg, if 1 mol of solute particles raises the boiling point of 1 kg of solvent by 3.63 °C?

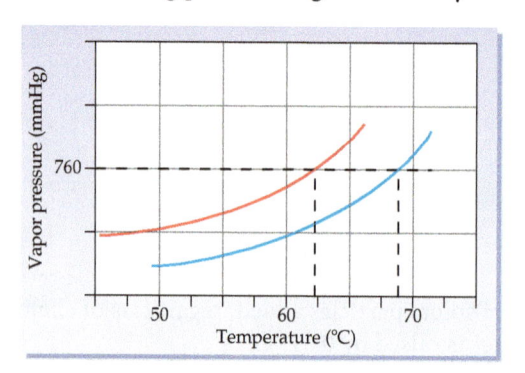

Freezing Point Depression of Solutions

Just as solutions have lower vapor pressure and consequently higher boiling points than pure solvents, they also have lower freezing points. Motorists in cold climates take advantage of this effect when they add "antifreeze" to the water in automobile cooling systems. Antifreeze is a nonvolatile solute, usually ethylene glycol ($HOCH_2CH_2OH$), that is added in sufficient concentration to lower the freezing point below the lowest expected outdoor temperature. In the same way, salt sprinkled on icy roads lowers the freezing point of ice below the road temperature and thus causes ice to melt.

Freezing point depression has much the same cause as vapor pressure lowering and boiling point elevation. Solute molecules are dispersed between solvent molecules throughout the solution, thereby making it more difficult for solvent molecules to come together and organize into ordered crystals.

For each mole of nonvolatile solute particles, the freezing point of 1 kg of water is lowered by 1.86 °C, or

$$\Delta T_{freezing} = \left(-1.86\,°C\,\frac{\text{kg water}}{\text{mol particles}}\right)\left(\frac{\text{mol particles}}{\text{kg water}}\right)$$

Thus, addition of 1 mol of antifreeze to 1 kg of water lowers the freezing point from 0.00 °C to −1.86 °C, and addition of 1 mol of NaCl (2 mol of particles) to 1 kg of water lowers the freezing point from 0.00 °C to −3.72 °C.

Worked Example 9.16 Properties of Solutions: Freezing Point Depression

The cells of a tomato contain mostly an aqueous solution of sugar and other substances. If a typical tomato freezes at −2.5 °C, what is the concentration of dissolved particles in the tomato cells (in moles of particles per kg of water)?

ANALYSIS The freezing point decreases by 1.86 °C for each mole of solute dissolved in 1 kg of water. We can use the decrease in freezing point (2.5 °C) to find the amount of solute per kg of water.

BALLPARK ESTIMATE The freezing point will decrease by about 1.9 °C for every 1 mol of solute particles in 1 kg of water. To lower the freezing point by 2.5 °C (about 30% more) will require about 30% more solute, or 1.3 mol.

SOLUTION

$$\Delta T_{freezing} = -2.5\,°C$$

$$= \left(-1.86\,°C\,\frac{\text{kg water}}{\text{mol solute particles}}\right)\left(\frac{??\text{ mol solute particles}}{1.0\text{ kg water}}\right)$$

We can rearrange this expression to

$$(-2.5\,°C)\left(\frac{1}{-1.86\,°C}\,\frac{\text{mol solute particles}}{\text{kg water}}\right) = 1.3\,\frac{\text{mol solute particles}}{\text{kg water}}$$

BALLPARK CHECK The calculated answer agrees with our estimate of 1.3 mol/kg.

PROBLEM 9.27
What is the freezing point of a solution of 1.0 mol of glucose in 1.0 kg of water?

PROBLEM 9.28
When 0.5 mol of a certain ionic substance is dissolved in 1.0 kg of water, the freezing point of the resulting solution is −2.8 °C. How many ions does the substance give when it dissolves?

9.12 Osmosis and Osmotic Pressure

Certain materials, including those that make up the membranes around living cells, are *semipermeable*. They allow water and other small molecules to pass through, but they block the passage of large solute molecules or ions. When a solution and a pure solvent, or two solutions of different concentration, are separated

280 CHAPTER 9 Solutions

Osmosis The passage of solvent through a semipermeable membrane separating two solutions of different concentration.

by a semipermeable membrane, solvent molecules pass through the membrane in a process called **osmosis**. Although the passage of solvent through the membrane takes place in both directions, passage from the pure solvent side to the solution side is favored and occurs more often. As a result, the amount of liquid on the pure solvent side decreases, the amount of liquid on the solution side increases, and the concentration of the solution decreases.

For the simplest explanation of osmosis, let us look at what happens on the molecular level. As shown in Figure 9.11, a solution inside a bulb is separated by a semipermeable membrane from pure solvent in the outer container. Solvent molecules in the outer container, because of their somewhat higher concentration, approach the membrane more frequently than do molecules in the bulb, thereby passing through more often and causing the liquid level in the attached tube to rise.

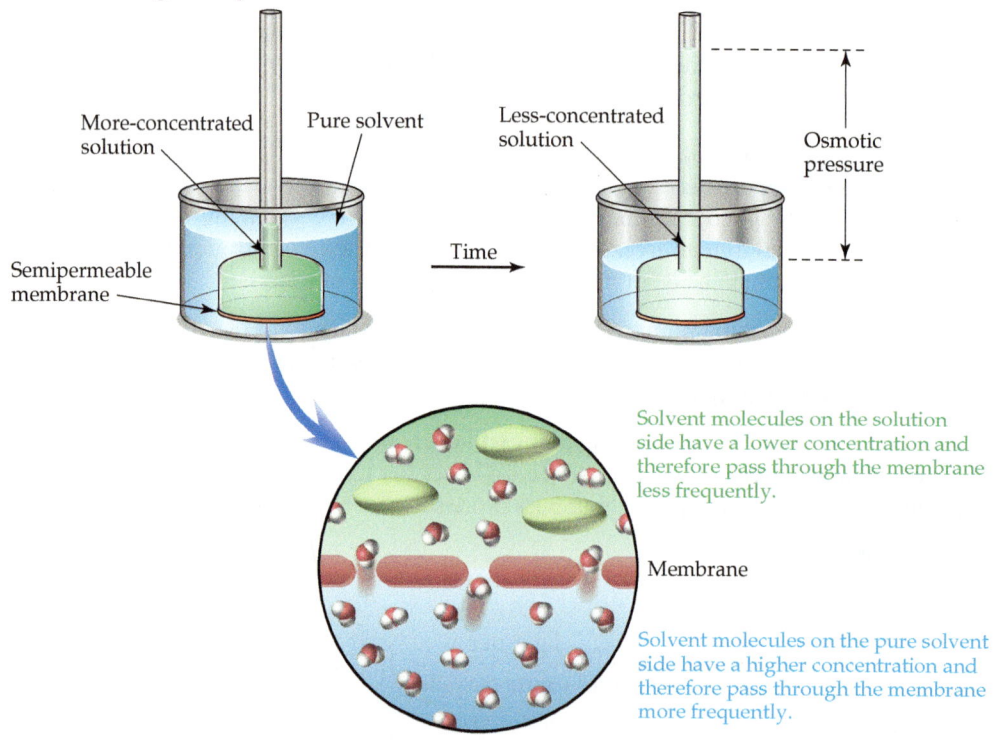

▶ **Figure 9.11**
The phenomenon of osmosis.
A solution inside the bulb is separated from pure solvent in the outer container by a semipermeable membrane. Solvent molecules in the outer container have a higher concentration than molecules in the bulb and therefore pass through the membrane more frequently. The liquid in the tube therefore rises until an equilibrium is reached. At equilibrium, the osmotic pressure exerted by the column of liquid in the tube is sufficient to prevent further net passage of solvent.

More-concentrated solution | Pure solvent | Less-concentrated solution | Osmotic pressure

Semipermeable membrane

Time →

Solvent molecules on the solution side have a lower concentration and therefore pass through the membrane less frequently.

Membrane

Solvent molecules on the pure solvent side have a higher concentration and therefore pass through the membrane more frequently.

As the liquid in the tube rises, its increased weight creates an increased pressure that pushes solvent back through the membrane until the rates of forward and reverse passage become equal and the liquid level stops rising. The amount of pressure necessary to achieve this equilibrium is called the **osmotic pressure** (π) of the solution and can be determined from the expression

Osmotic pressure The amount of external pressure that must be applied to a solution to prevent the net movement of solvent molecules across a semipermeable membrane.

$$\pi = \left(\frac{n}{V}\right)RT$$

where n is the number of moles of particles in the solution, V is the solution volume, R is the gas constant (Section 8.10), and T is the absolute temperature of the solution. Note the similarity between this equation for the osmotic pressure of a solution and the equation for the pressure of an ideal gas, $P = (n/V)RT$. In both cases, the pressure has units of atmospheres.

Osmotic pressures can be extremely high, even for relatively dilute solutions. The osmotic pressure of a 0.15 M NaCl solution at 25 °C, for example, is 7.3 atm, a value that supports a difference in water level of approximately 250 ft!

As with other colligative properties, the amount of osmotic pressure depends only on the concentration of solute particles, not on their identity. Thus, it is convenient to use a new unit, *osmolarity* (osmol), to describe the concentration of particles in solution. The **osmolarity** of a solution is equal to the number of moles of dissolved particles (ions or molecules) per liter of solution. A 0.2 M glucose solution, for instance, has

Osmolarity (osmol) The sum of the molarities of all dissolved particles in a solution.

an osmolarity of 0.2 osmol, but a 0.2 M solution of NaCl has an osmolarity of 0.4 osmol because it contains 0.2 mol of Na$^+$ ions and 0.2 mol of Cl$^-$ ions.

Osmosis is particularly important in living organisms because the membranes around cells are semipermeable. The fluids both inside and outside cells must therefore have the same osmolarity to prevent buildup of osmotic pressure and consequent rupture of the cell membrane.

In blood, the plasma surrounding red blood cells has an osmolarity of approximately 0.30 osmol and is said to be **isotonic** with (that is, has the same osmolarity as) the cell contents. If the cells are removed from plasma and placed in 0.15 M NaCl (called *physiological saline solution*), they are unharmed because the osmolarity of the saline solution (0.30 osmol) is the same as that of plasma. If, however, red blood cells are placed in pure water or in any solution with an osmolarity much lower than 0.30 osmol (a **hypotonic** solution), water passes through the membrane into the cell, causing the cell to swell up and burst, a process called *hemolysis*.

Finally, if red blood cells are placed in a solution having an osmolarity greater than the cell contents (a **hypertonic** solution), water passes out of the cells into the surrounding solution, causing the cells to shrivel, a process called *crenation*. Figure 9.12 shows red blood cells under all three conditions: isotonic, hypotonic, and hypertonic. Therefore, it is critical that any solution used intravenously be isotonic to prevent red blood cells from being destroyed.

Isotonic Having the same osmolarity.

Hypotonic Having an osmolarity *less than* the surrounding blood plasma or cells.

Hypertonic Having an osmolarity *greater than* the surrounding blood plasma or cells.

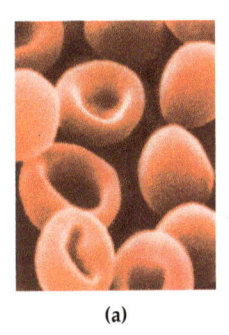

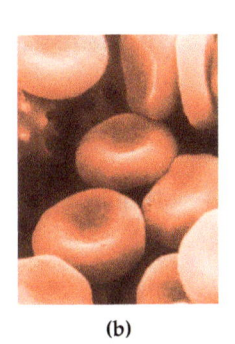

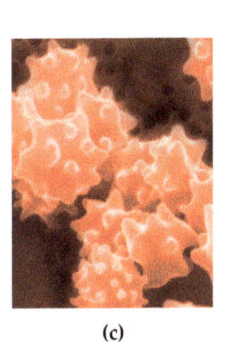

(a)   (b)   (c)

◄ **Figure 9.12**
Red blood cells.
In an isotonic solution the blood cells are normal in appearance (a), but the cells in a hypotonic solution (b) are swollen because of water gain, and those in a hypertonic solution (c) are shriveled because of water loss.

Worked Example 9.17 Properties of Solutions: Osmolarity

The solution of glucose commonly used intravenously has a concentration of 5.0% (m/v) glucose. What is the osmolarity of this solution? The molar mass of glucose is 180 g/mol.

ANALYSIS Since glucose is a molecular substance that does not give ions in solution, the osmolarity of the solution is the same as the molarity. Recall from Section 9.7 that a solution of 5.0% (m/v) glucose has a concentration of 5.0 g glucose per 100 mL of solution, which is equivalent to 50 g per liter of solution. Thus, finding the molar concentration of glucose requires a mass to mole conversion.

BALLPARK ESTIMATE One liter of solution contains 50 g of glucose (MW = 180 g/mol). Thus, 50 g of glucose is equal to a little more than 0.25 mol, so a solution concentration of 50 g/L is equal to about 0.25 osmol, or 0.25 M.

SOLUTION

STEP 1: Identify known information. We know the (m/v)% concentration of the glucose solution.

$$5.0\% \, (m/v) = \frac{5.0 \text{ g glucose}}{100 \text{ mL solution}} \times 100\%$$

STEP 2: Identify answer and units. We are looking for osmolarity, which in this case is equal to the molarity of the solution because glucose is a molecular substance and does not dissociate into ions.

Osmarity = Molarity = ?? mol/liter

282 CHAPTER 9 Solutions

STEP 3: Identify conversion factors. The (m/v)% concentration is defined as grams of solute per 100 mL of solution, and molarity is defined as moles of solute per liter of solution. We will need to convert from milliliters to liters and then use molar mass to convert grams of glucose to moles of glucose.

$$\frac{\text{g glucose}}{100 \text{ mL}} \times \frac{1000 \text{ mL}}{\text{L}} \longrightarrow \frac{\text{g glucose}}{\text{L}}$$

$$\frac{\text{g glucose}}{\text{L}} \times \frac{1 \text{ mol glucose}}{180 \text{ g glucose}} \longrightarrow \frac{\text{moles glucose}}{\text{L}}$$

STEP 4: Solve. Starting with the (m/v)% glucose concentration, we first find the number of grams of glucose in 1 L of solution and then convert to moles of glucose per liter.

$$\left(\frac{5.0 \text{ g glucose}}{100 \text{ mL solution}}\right)\left(\frac{1000 \text{ mL}}{1 \text{ L}}\right) = \frac{50 \text{ g glucose}}{\text{L solution}}$$

$$\left(\frac{50 \text{ g glucose}}{1 \text{ L}}\right)\left(\frac{1 \text{ mol}}{180 \text{ g}}\right) = 0.28 \text{ M glucose} = 0.28 \text{ osmol}$$

BALLPARK CHECK The calculated osmolarity is reasonably close to our estimate of 0.25 osmol.

Worked Example 9.18 Properties of Solutions: Osmolarity

What mass of NaCl is needed to make 1.50 L of a 0.300 osmol solution? The molar mass of NaCl is 58.44 g/mol.

ANALYSIS Since NaCl is an ionic substance that produces 2 mol of ions (Na^+, Cl^-) when it dissociates, the osmolarity of the solution is twice the molarity. From the volume and the osmolarity we can determine the moles of NaCl needed and then perform a mole to mass conversion.

SOLUTION

STEP 1: Identify known information. We know the volume and the osmolarity of the final NaCl solution.

$V = 1.50 \text{ L}$

$0.300 \text{ osmol} = \left(\frac{0.300 \text{ mol ions}}{\text{L}}\right)$

STEP 2: Identify answer and units. We are looking for the mass of NaCl.

Mass of NaCl = ?? g

STEP 3: Identify conversion factors. Starting with osmolarity in the form (moles NaCl/L), we can use volume to determine the number of moles of solute. We can then use molar mass for the mole to mass conversion.

$$\left(\frac{\text{moles NaCl}}{\text{L}}\right) \times (\text{L}) = \text{moles NaCl}$$

$$(\text{moles NaCl}) \times \left(\frac{\text{g NaCl}}{\text{mole NaCl}}\right) = \text{g NaCl}$$

STEP 4: Solve. Use the appropriate conversions, remembering that NaCl produces two ions per formula unit, to find the mass of NaCl.

$$\left(\frac{0.300 \text{ mol ions}}{\text{L}}\right)\left(\frac{1 \text{ mol NaCl}}{2 \text{ mol ions}}\right)(1.50 \text{ L}) = 0.225 \text{ mol NaCl}$$

$$(0.225 \text{ mol NaCl})\left(\frac{58.44 \text{ g NaCl}}{\text{mol NaCl}}\right) = 13.1 \text{ g NaCl}$$

> **PROBLEM 9.29**
> What is the osmolarity of the following solutions?
> (a) 0.35 M KBr
> (b) 0.15 M glucose + 0.05 M K_2SO_4
>
> **PROBLEM 9.30**
> A typical oral rehydration solution (ORS) for infants contains 90 mEq/L Na^+, 20 mEq/L K^+, 110 mEq/L Cl^-, and 2.0% (m/v) glucose (MW = 180 g/mol).
> (a) Calculate the concentration of each ORS component in units of molarity.
> (b) What is the osmolarity of the solution, and how does it compare with the osmolarity of blood plasma?

9.13 Dialysis

Dialysis is similar to osmosis, except that the pores in a dialysis membrane are larger than those in an osmotic membrane so that both solvent molecules and small solute particles can pass through, but large colloidal particles such as proteins cannot pass. (The exact dividing line between a "small" molecule and a "large" one is imprecise, and dialysis membranes with a variety of pore sizes are available.) Dialysis membranes include animal bladders, parchment, and cellophane.

Perhaps the most important medical use of dialysis is in artificial kidney machines, where *hemodialysis* is used to cleanse the blood of patients whose kidneys malfunction (Figure 9.13). Blood is diverted from the body and pumped through a long cellophane dialysis tube suspended in an isotonic solution formulated to contain many of the same components as blood plasma. These substances—glucose, NaCl, $NaHCO_3$, and KCl—have the same concentrations in the dialysis solution as they do in blood so that they have no net passage through the membrane.

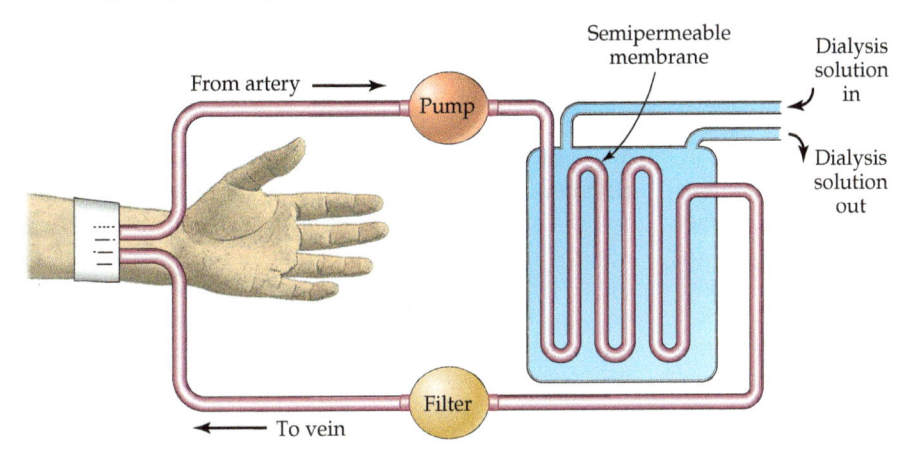

◀ **Figure 9.13**
Operation of a hemodialysis unit used for purifying blood. Blood is pumped from an artery through a coiled semipermeable membrane of cellophane. Small waste products pass through the membrane and are washed away by an isotonic dialysis solution.

Small waste materials such as urea pass through the dialysis membrane from the blood to the solution side where they are washed away, but cells, proteins, and other important blood components are prevented from passing through the membrane because of their larger size. In addition, the dialysis fluid concentration can be controlled so that imbalances in electrolytes are corrected. The wash solution is changed every 2 h, and a typical hemodialysis procedure lasts for 4–7 h.

As noted above, colloidal particles are too large to pass through a semipermeable membreane. Protein molecules, in particular, do not cross semipermeable membranes and thus play an essential role in determining the osmolarity of body fluids. The distribution of water and solutes across the capillary walls that separate blood plasma from the fluid

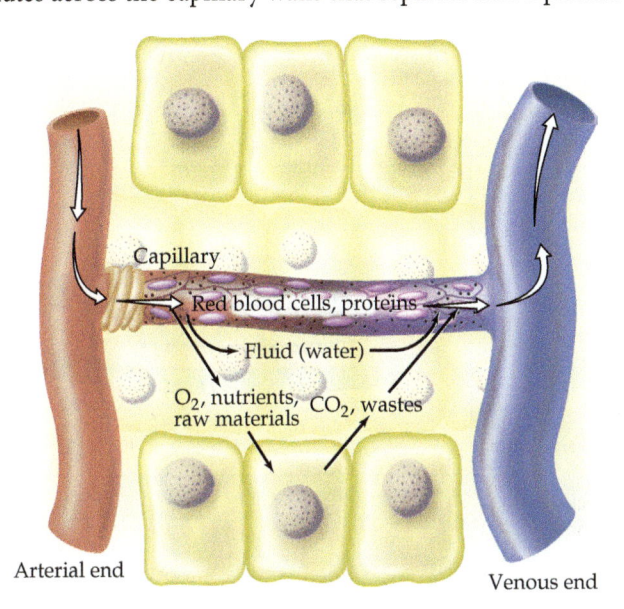

◀ The delivery of oxygen and nutrients to the cells and the removal of waste products are regulated by osmosis.

surrounding cells is controlled by the balance between blood pressure and osmotic pressure. The pressure of blood inside the capillary tends to push water out of the plasma (filtration), but the osmotic pressure of colloidal protein molecules tends to draw water into the plasma (reabsorption). The balance between the two processes varies with location in the body. At the arterial end of a capillary, where blood pumped from the heart has a higher pressure, filtration is favored, At the venous end, where blood pressure is lower, reabsorption is favored, causing waste products from metabolism to enter the bloodstream, to be removed by the kidneys.

CHEMISTRY IN ACTION

Timed-Release Medications

There is much more in most medications than medicine. Even something as simple as a generic aspirin tablet contains a binder to keep it from crumbling, a filler to bring it to the right size and help it disintegrate in the stomach, and a lubricant to keep it from sticking to the manufacturing equipment. Timed-release medications are more complex still.

The widespread use of timed-release medication dates from the introduction of Contac decongestant in 1961. The original idea was simple: tiny beads of medicine were encapsulated by coating them with varying thicknesses of a slow-dissolving polymer. Those beads with a thinner coat dissolve and release their medicine more rapidly; those with a thicker coat dissolve more slowly. Combining the right number of beads with the right thicknesses into a single capsule makes possible the gradual release of medication over a predictable time.

The technology of timed-release medications has become much more sophisticated in recent years, and the kinds of medications that can be delivered have become more numerous. Some medicines, for instance, either damage the stomach lining or are destroyed by the highly acidic environment in the stomach but can be delivered safely if given an *enteric coating*. The enteric coating is a polymeric material formulated so that it is stable in acid but reacts and is destroyed when it passes into the more basic environment of the intestines.

More recently, dermal patches have been developed to deliver drugs directly by diffusion through the skin. Patches are available to treat conditions from angina to motion sickness, as well as nicotine patches to help reduce cigarette cravings. One clever new device for timed release of medication through the skin uses the osmotic effect to force a drug from its reservoir. Useful only for drugs that do not dissolve in water, the device is divided into two compartments, one containing medication covered by a perforated membrane and the other containing a hygroscopic material (Section 9.3) covered by a semipermeable membrane. As moisture from the air diffuses through the membrane into the compartment with the hygroscopic material, the buildup of osmotic pressure squeezes the medication out of the other compartment through tiny holes.

▲ The small beads of medicine are coated with different thicknesses of a slow-dissolving polymer so that they dissolve and release medicine at different times.

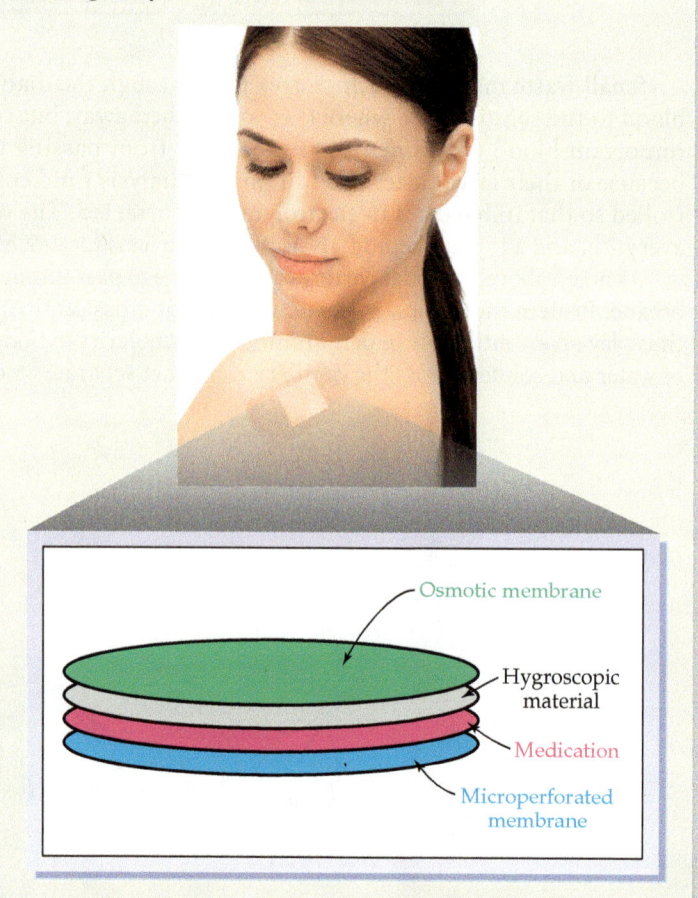

See Chemistry in Action Problem 9.93 at the end of the chapter.

SUMMARY: REVISITING THE CHAPTER GOALS

1. What are solutions, and what factors affect solubility? Mixtures are classified as either *heterogeneous*, if the mixing is nonuniform, or *homogeneous*, if the mixing is uniform. *Solutions* are homogeneous mixtures that contain particles the size of ions and molecules (<2.0 nm diameter), whereas larger particles (2.0–500 nm diameter) are present in *colloids*.

The maximum amount of one substance (the *solute*) that can be dissolved in another (the *solvent*) is called the substance's *solubility*. Substances tend to be mutually soluble when their intermolecular forces are similar. The solubility in water of a solid often increases with temperature, but the solubility of a gas always decreases with temperature. Pressure significantly affects gas solubilities, which are directly proportional to their partial pressure over the solution (*Henry's law*) (see Problems 36–43, 94, 105).

2. How is the concentration of a solution expressed? The concentration of a solution can be expressed in several ways, including molarity, weight/weight percent composition, weight/volume percent composition, and parts per million (or billion). Osmolarity is used to express the total concentration of dissolved particles (ions and molecules). Molarity, which expresses concentration as the number of moles of solute per liter of solution, is the most useful method when calculating quantities of reactants or products for reactions in aqueous solution (see Problems 44–65, 86, 88, 89, 91, 94–105, 107, 108).

3. How are dilutions carried out? A dilution is carried out by adding more solvent to an existing solution. Only the amount of solvent changes; the amount of solute remains the same. Thus, the molarity times the volume of the dilute solution is equal to the molarity times the volume of the concentrated solution: $M_c V_c = M_d V_d$ (see Problems 35, 66–71, 98).

4. What is an electrolyte? Substances that form ions when dissolved in water and whose water solutions therefore conduct an electric current are called *electrolytes*. Substances that ionize completely in water are *strong electrolytes*, those that ionize partially are *weak electrolytes*, and those that do not ionize are *nonelectrolytes*. Body fluids contain small amounts of many different electrolytes, whose concentrations are expressed as moles of ionic charge, or equivalents, per liter (see Problems 32, 33, 72–79, 97, 108).

5. How do solutions differ from pure solvents in their behavior? In comparing a solution to a pure solvent, the solution has a lower vapor pressure at a given temperature, a higher boiling point, and a lower melting point. Called *colligative properties*, these effects depend only on the number of dissolved particles, not on their chemical identity (see Problems 32, 33, 43, 80–83, 108).

6. What is osmosis? *Osmosis* occurs when solutions of different concentration are separated by a semipermeable membrane that allows solvent molecules to pass but blocks the passage of solute ions and molecules. Solvent flows from the more dilute side to the more concentrated side until sufficient *osmotic pressure* builds up and stops the flow. An effect similar to osmosis occurs when membranes of larger pore size are used. In *dialysis*, the membrane allows the passage of solvent and small dissolved molecules but prevents passage of proteins and larger particles (see Problems 31, 84, 85, 87).

KEY WORDS

Colligative property, *p. 275*

Colloid, *p. 253*

Dilution factor, *p. 271*

Electrolyte, *p. 273*

Equivalent (Eq), *p. 273*

Gram-equivalent (g-Eq), *p. 273*

Henry's law, *p. 261*

Heterogeneous mixture, *p. 253*

Homogeneous mixture, *p. 253*

Hygroscopic, *p. 258*

Hypertonic, *p. 281*

Hypotonic, *p. 281*

Isotonic, *p. 281*

Mass/mass percent concentration, (m/m)%, *p. 264*

mass/volume percent concentration, (m/v)%, *p. 264*

Miscible, *p. 258*

Molarity (M), *p. 268*

Nonelectrolyte, *p. 273*

Osmolarity (osmol), *p. 280*

Osmosis, *p. 280*

Osmotic pressure, *p. 280*

Parts per billion (ppb), *p. 267*

Parts per million (ppm), *p. 267*

Saturated solution, *p. 258*

Solubility, *p. 258*

Solute, *p. 254*

Solution, *p. 253*

Solvation, *p. 256*

Solvent, *p. 254*

Strong electrolyte, *p. 273*

Supersaturated solution, *p. 259*

Volume/volume percent concentration, (v/v)%, *p. 264*

Weak electrolyte, *p. 273*

CONCEPT MAP: SOLUTIONS

Formation of a solution depends on many factors, including the attractive forces between solute and solvent particles, temperature, and pressure (gases). The extent to which a solute dissolves in solution can be expressed either qualitatively or using quantitative concentration units. The most common concentration unit in chemical applications is molarity (moles of solute/L solution), which is also useful in quantitative relationships involving reactions that take place in solution. Colligative properties of solution, including boiling and freezing points, will vary with the amount of solute dissolved in solution. These relationships are illustrated in the concept map in Figure 9.14.

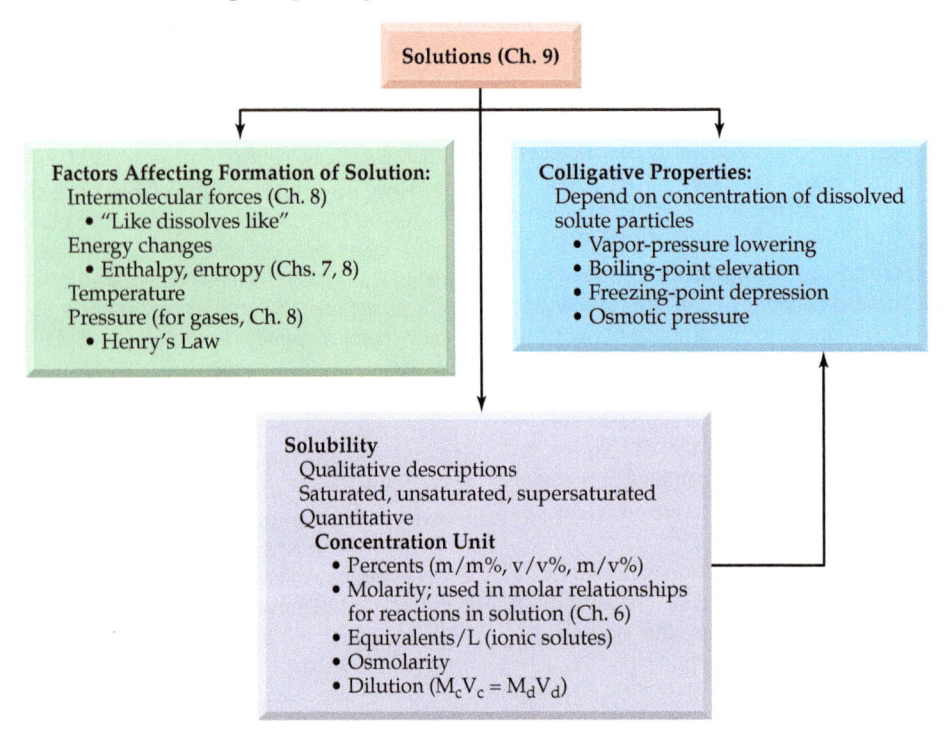

▲ Figure 9.14

UNDERSTANDING KEY CONCEPTS

9.31 Assume that two liquids are separated by a semipermeable membrane, with pure solvent on the right side, and a solution of a solute on the left side. Make a drawing that shows the situation after equilibrium is reached.

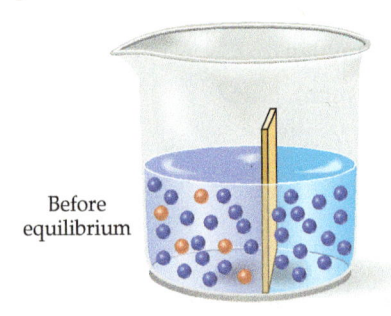

9.32 When 1 mol of HCl is added to 1 kg of water, the boiling point increases by 1.0 °C, but when 1 mol of acetic acid, CH_3CO_2H, is added to 1 kg of water, the boiling point increases by only 0.5 °C. Explain.

9.33 HF is a weak electrolyte and HBr is a strong electrolyte. Which of the curves in the figure represents the change in the boiling point of an aqueous solution when 1 mole of HF is added to 1 kg of water, and which represents the change when 1 mol of HBr is added?

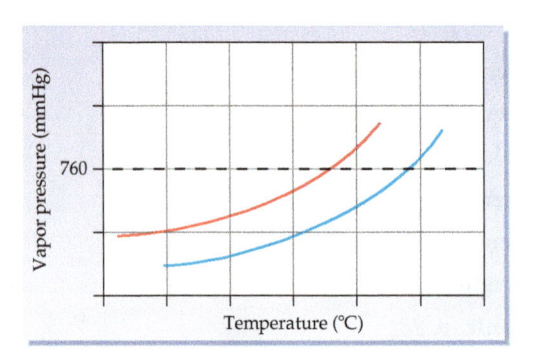

9.34 Assume that you have two full beakers, one containing pure water (blue) and the other containing an equal volume of a 10% (w/v) solution of glucose (green). Which of the drawings (a)–(c) best represents the two beakers after they have stood uncovered for several days and partial evaporation has occurred? Explain.

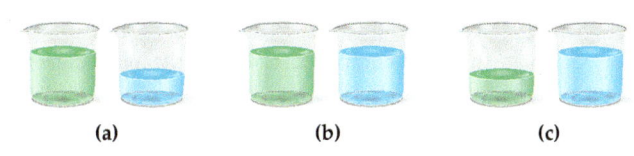

9.35 A beaker containing 150.0 mL of 0.1 M glucose is represented by (a). Which of the drawings (b)–(d) represents the solution that results when 50.0 mL is withdrawn from (a) and then diluted by a factor of 4?

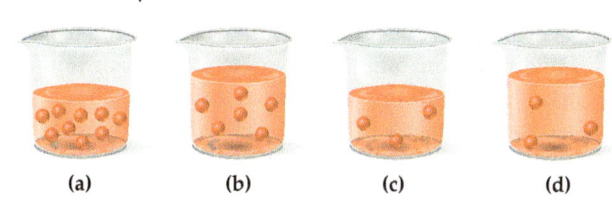

ADDITIONAL PROBLEMS

SOLUTIONS AND SOLUBILITY

9.36 What is the difference between a homogeneous mixture and a heterogeneous one?

9.37 How can you tell a solution from a colloid?

9.38 What characteristic of water allows it to dissolve ionic solids?

9.39 Why does water not dissolve motor oil?

9.40 Which of the following are solutions?

(a) Italian salad dressing   (b) Rubbing alcohol
(c) Algae in pond water   (d) Black coffee

9.41 Based on the predominant intermolecular forces, which of the following pairs of liquids are likely to be miscible?

(a) H_2SO_4 and H_2O (b) C_8H_{18} and C_6H_6
(c) CH_2Cl_2, and H_2O (d) CS_2 and CCl_4

9.42 The solubility of NH_3 gas in water at an NH_3 pressure of 760.0 mmHg is 51.8 g/100 mL. What is the solubility of NH_3 if its partial pressure is reduced to 225.0 mmHg?

9.43 The solubility of CO_2 gas in water is 0.15 g/100 mL at a CO_2 pressure of 760 mmHg. What is the solubility of CO_2 in a soft drink (which is mainly water) that was bottled under a CO_2 pressure of 4.5 atm?

CONCENTRATION AND DILUTION OF SOLUTIONS

9.44 Is a solution highly concentrated if it is saturated? Is a solution saturated if it is highly concentrated?

9.45 How is mass/volume percent concentration defined and for what types of solutions is it typically used?

9.46 How is molarity defined?

9.47 How is volume/volume percent concentration defined and for what types of solutions is it typically used?

9.48 How would you prepare 750.0 mL of a 6.0% (v/v) ethyl alcohol solution?

9.49 A dilute aqueous solution of boric acid, H_3BO_3 is often used as an eyewash. How would you prepare 500.0 mL of a 0.50% (m/v) boric acid solution?

9.50 Describe how you would prepare 250 mL of a 0.10 M NaCl solution.

9.51 Describe how you would prepare 1.50 L of a 7.50% (m/v) $Mg(NO_3)_2$ solution.

9.52 What is the mass/volume percent concentration of the following solutions?

(a) 0.078 mol KCl in 75 mL of solution
(b) 0.044 mol sucrose $(C_{12}H_{22}O_{11})$ in 380 mL of solution

9.53 The concentration of glucose in blood is approximately 90 mg/100 mL. What is the mass/volume percent concentration of glucose? What is the molarity of glucose?

9.54 How many moles of each substance are needed to prepare the following solutions?

(a) 50.0 mL of 8.0% (m/v) KCl (MW = 74.55 g/mol)
(b) 200.0 mL of 7.5% (m/v) acetic acid (MW = 60.05 g/mol)

9.55 Which of the following solutions is more concentrated?

(a) 0.50 M KCl or 5.0% (m/v) KCl
(b) 2.5% (m/v) $NaHSO_4$ or 0.025 M $NaHSO_4$

9.56 If you had only 23 g of KOH remaining in a bottle, how many milliliters of 10.0% (m/v) solution could you prepare? How many milliliters of 0.25 M solution?

9.57 Over-the-counter hydrogen peroxide (H_2O_2) solutions are 3% (m/v). What is this concentration in moles per liter?

9.58 The lethal dosage of potassium cyanide (KCN) in rats is 10 mg KCN per kilogram of body weight. What is this concentration in parts per million?

9.59 The maximum concentration set by the U.S. Environmental Protection Agency for lead in drinking water is 15 ppb. (*Hint*: 1 ppb = $1 \mu g/L$)

(a) What is this concentration in milligrams per liter?
(b) How many liters of water contaminated at this maximum level must you drink to consume 1.0 μg of lead?

9.60 What is the molarity of the following solutions?

(a) 12.5 g $NaHCO_3$ in 350.0 mL solution
(b) 45.0 g H_2SO_4 in 300.0 mL solution
(c) 30.0 g NaCl dissolved to make 500.0 mL solution

9.61 How many grams of solute are in the following solutions?

(a) 200 mL of 0.30 M acetic acid, CH_3CO_2H
(b) 1.50 L of 0.25 M NaOH
(c) 750 mL of 2.5 M nitric acid, HNO_3

9.62 How many milliliters of a 0.75 M HCl solution do you need to obtain 0.0040 mol of HCl?

9.63 Nalorphine, a relative of morphine, is used to combat withdrawal symptoms in heroin users. How many milliliters of a 0.40% (m/v) solution of nalorphine must be injected to obtain a dose of 1.5 mg?

9.64 A flask containing 450 mL of 0.50 M H_2SO_4 was accidentally knocked to the floor. How many grams of $NaHCO_3$ do you need to put on the spill to neutralize the acid according to the following equation?

$$H_2SO_4(aq) + 2\,NaHCO_3(aq) \longrightarrow$$
$$Na_2SO_4(aq) + 2\,H_2O(l) + 2\,CO_2(g)$$

9.65 Sodium thiosulfate ($Na_2S_2O_3$), the major component in photographic fixer solution, reacts with silver bromide to dissolve it according to the following reaction:

$$AgBr(s) + 2\,Na_2S_2O_3(aq) \longrightarrow$$
$$Na_3Ag(S_2O_3)_2(aq) + NaBr(aq)$$

  (a) How many moles of $Na_2S_2O_3$ would be required to react completely with 0.450 g of AgBr?
  (b) How many mL of 0.02 M $Na_2S_2O_3$ contain this number of moles?

9.66 What is the final volume of an orange juice prepared from 100.0 mL of orange juice concentrate if the final juice is to be 20.0% of the strength of the original?

9.67 What is the final volume of NaOH solution prepared from 100.0 mL of 0.500 M NaOH if you wanted the final concentration to be 0.150 M?

9.68 An aqueous solution that contains 285 ppm of potassium nitrate (KNO_3) is being used to feed plants in a garden. What volume of this solution is needed to prepare 2.0 L of a solution that is 75 ppm in KNO_3?

9.69 What is the concentration of a NaCl solution, in (m/v)%, prepared by diluting 65 mL of a saturated solution, which has a concentration of 37 (m/v)%, to 480 mL?

9.70 Concentrated (12.0 M) hydrochloric acid is sold for household and industrial purposes under the name "muriatic acid." How many milliliters of 0.500 M HCl solution can be made from 25.0 mL of 12.0 M HCl solution?

9.71 Dilute solutions of $NaHCO_3$ are sometimes used in treating acid burns. How many milliliters of 0.100 M $NaHCO_3$ solution are needed to prepare 750.0 mL of 0.0500 M $NaHCO_3$ solution?

ELECTROLYTES

9.72 What is an electrolyte?

9.73 Give an example of a strong electrolyte and a nonelectrolyte.

9.74 What does it mean when we say that the concentration of Ca^{2+} in blood is 3.0 mEq/L?

9.75 What is the total anion concentration (in mEq/L) of a solution that contains 5.0 mEq/L Na^+, 12.0 mEq/L Ca^{2+}, and 2.0 mEq/L Li^+?

9.76 Kaochlor, a 10% (m/v) KCl solution, is an oral electrolyte supplement administered for potassium deficiency. How many milliequivalents of K^+ are in a 30 mL dose?

9.77 Calculate the gram-equivalent for each of the following ions:
  (a) Ca^{2+} (b) K^+
  (c) SO_4^{2-} (d) PO_4^{3-}

9.78 Look up the concentration of Cl^- ion in blood in Table 9.6. How many milliliters of blood would be needed to obtain 1.0 g of Cl^- ions?

9.79 Normal blood contains 3 mEq/L of Mg^{2+}. How many milligrams of Mg^{2+} are present in 150.0 mL of blood?

PROPERTIES OF SOLUTIONS

9.80 Which lowers the freezing point of 2.0 kg of water more, 0.20 mol NaOH or 0.20 mol $Ba(OH)_2$? Both compounds are strong electrolytes. Explain.

9.81 Which solution has the higher boiling point, 0.500 M glucose or 0.300 M KCl? Explain.

9.82 Methanol, CH_3OH, is sometimes used as an antifreeze for the water in automobile windshield washer fluids. How many moles of methanol must be added to 5.00 kg of water to lower its freezing point to $-10.0\,°C$? (For each mole of solute, the freezing point of 1 kg of water is lowered 1.86 °C.)

9.83 Hard candy is prepared by dissolving pure sugar and flavoring in water and heating the solution to boiling. What is the boiling point of a solution produced by adding 650 g of cane sugar (molar mass 342.3 g/mol) to 1.5 kg of water? (For each mole of nonvolatile solute, the boiling point of 1 kg of water is raised 0.51 °C.)

OSMOSIS

9.84 Why do red blood cells swell up and burst when placed in pure water?

9.85 What does it mean when we say that a 0.15 M NaCl solution is isotonic with blood, whereas distilled water is hypotonic?

9.86 Which of the following solutions has the higher osmolarity?
  (a) 0.25 M KBr or 0.20 M Na_2SO_4
  (b) 0.30 M NaOH or 3.0% (m/v) NaOH

9.87 Which of the following solutions will give rise to a greater osmotic pressure at equilibrium: 5.00 g of NaCl in 350.0 mL water or 35.0 g of glucose in 400.0 mL water? For NaCl, MW = 58.5 amu; for glucose, MW = 180 amu.

9.88 A pickling solution for preserving food is prepared by dissolving 270 g of NaCl in 3.8 L of water. Calculate the osmolarity of the solution.

9.89 An isotonic solution must be approximately 0.30 osmol. How much KCl is needed to prepare 175 mL of an isotonic solution?

CHEMISTRY IN ACTION

9.90 How does the body increase oxygen availability at high altitude? [*Breathing and Oxygen Transport, p. 263*]

9.91 What are the major electrolytes in sweat, and what are their approximate concentrations in mEq/L? [*Electrolytes, Fluid Replacement, and Sports Drinks, p. 276*]

9.92 Why is a sports drink more effective than plain water for rehydration after extended exercise? [*Electrolytes, Fluid Replacement, and Sports Drinks, p. 276*]

9.93 How does an enteric coating on a medication work? [*Timed-Release Medications, p. 284*]

GENERAL QUESTIONS AND PROBLEMS

9.94 Hyperbaric chambers, which provide high pressures (up to 6 atm) of either air or pure oxygen, are used to treat a variety of conditions, ranging from decompression sickness in deep-sea divers to carbon monoxide poisoning.

(a) What is the partial pressure of O_2 (in millimeters of Hg) in a hyperbaric chamber pressurized to 5 atm with air that is 18% in O_2?

(b) What is the solubility of O_2 (in grams per 100 mL) in the blood at this partial pressure? The solubility of O_2 is 2.1 g/100 mL for $P_{O_2} = 1$ atm.

9.95 Express the solubility of O_2 in Problem 9.94(b) in units of molarity.

9.96 Uric acid, the principal constituent of some kidney stones, has the formula $C_5H_4N_4O_3$. In aqueous solution, the solubility of uric acid is only 0.067 g/L. Express this concentration in (m/v)%, in parts per million, and in molarity.

9.97 Emergency treatment of cardiac arrest victims sometimes involves injection of a calcium chloride solution directly into the heart muscle. How many grams of $CaCl_2$ are administered in an injection of 5.0 mL of a 5.0% (m/v) solution? How many milliequivalents of Ca^{2+}?

9.98 Nitric acid, HNO_3, is available commercially at a concentration of 16 M.

(a) What volume would you need to obtain 0.150 mol HNO_3?

(b) To what volume must you dilute this volume of HNO_3 from part (a) to prepare a 0.20 M solution?

9.99 One test for vitamin C (ascorbic acid, $C_6H_8O_6$) is based on the reaction of the vitamin with iodine:

$$C_6H_8O_6(aq) + I_2(aq) \longrightarrow C_6H_6O_6(aq) + 2\,HI(aq)$$

(a) A 25.0 mL sample of a fruit juice requires 13.0 mL of 0.0100 M I_2 solution for reaction. How many moles of ascorbic acid are in the sample?

(b) What is the molarity of ascorbic acid in the fruit juice?

(c) The Food and Drug Administration recommends that 60 mg of ascorbic acid be consumed per day. How many milliliters of the fruit juice in part (a) must a person drink to obtain the recommended dosage?

9.100 *Ringer's solution*, used in the treatment of burns and wounds, is prepared by dissolving 8.6 g of NaCl, 0.30 g of KCl, and 0.33 g of $CaCl_2$ in water and diluting to a volume of 1.00 L. What is the molarity of each component?

9.101 What is the osmolarity of Ringer's solution (see Problem 9.100)? Is it hypotonic, isotonic, or hypertonic with blood plasma (0.30 osmol)?

9.102 The typical dosage of statin drugs for the treatment of high cholesterol is 10 mg. Assuming a total blood volume of 5.0 L, calculate the (m/v)% concentration of drug in the blood in units of g/100 mL.

9.103 Assuming the density of blood in healthy individuals is approximately 1.05 g/mL, report the concentration of drug in Problem 9.102 in units of ppm.

9.104 In all 50 states, a person with a blood alcohol concentration of 0.080% (v/v) is considered legally drunk. What volume of total alcohol does this concentration represent, assuming a blood volume of 5.0 L?

9.105 Ammonia, NH_3, is very soluble in water (51.8 g/L at 20 °C and 760 mmHg).

(a) Show how NH_3 can hydrogen bond to water.

(b) What is the solubility of ammonia in water in moles per liter?

9.106 Cobalt(II) chloride, a blue solid, can absorb water from the air to form cobalt(II) chloride hexahydrate, a pink solid. The equilibrium is so sensitive to moisture in the air that $CoCl_2$ is used as a humidity indicator.

(a) Write a balanced equation for the equilibrium. Be sure to include water as a reactant to produce the hexahydrate.

(b) How many grams of water are released by the decomposition of 2.50 g of cobalt(II) chloride hexahydrate?

9.107 How many milliliters of 0.150 M $BaCl_2$ are needed to react completely with 35.0 mL of 0.200 M Na_2SO_4? How many grams of $BaSO_4$ will be formed?

9.108 Many compounds are only partially dissociated into ions in aqueous solution. Trichloroacetic acid (CCl_3CO_2H), for instance, is partially dissociated in water according to the equation

$$CCl_3CO_2H(aq) \rightleftharpoons H^+(aq) + CCl_3CO_2^-(aq)$$

For a solution prepared by dissolving 1.00 mol of trichloroacetic acid in 1.00 kg of water, 36.0% of the trichloroacetic acid dissociates to form H^+ and $CCl_3CO_2^-$ ions.

(a) What is the total concentration of dissolved ions and molecules in 1 kg of water?

(b) What is the freezing point of this solution? (The freezing point of 1 kg of water is lowered 1.86 °C for each mole of solute particles.)

CHAPTER 10

Acids and Bases

CONTENTS

- 10.1 Acids and Bases in Aqueous Solution
- 10.2 Some Common Acids and Bases
- 10.3 The Brønsted–Lowry Definition of Acids and Bases
- 10.4 Acid and Base Strength
- 10.5 Acid Dissociation Constants
- 10.6 Water as Both an Acid and a Base
- 10.7 Measuring Acidity in Aqueous Solution: pH
- 10.8 Working with pH
- 10.9 Laboratory Determination of Acidity
- 10.10 Buffer Solutions
- 10.11 Acid and Base Equivalents
- 10.12 Some Common Acid–Base Reactions
- 10.13 Titration
- 10.14 Acidity and Basicity of Salt Solutions

◀ Acids are found in many of the foods we eat, including tomatoes, peppers, and these citrus fruits.

CHAPTER GOALS

1. **What are acids and bases?**
   THE GOAL: Be able to recognize acids and bases and write equations for common acid–base reactions. (◀◀ A.)

2. **What effect does the strength of acids and bases have on their reactions?**
   THE GOAL: Be able to interpret acid strength using acid dissociation constants K_a and predict the favored direction of acid–base equilibria. (◀◀ B, C.)

3. **What is the ion-product constant for water?**
   THE GOAL: Be able to write the equation for this constant and use it to find the concentration of H_3O^+ or OH^-. (◀◀ C.)

4. **What is the pH scale for measuring acidity?**
   THE GOAL: Be able to explain the pH scale and find pH from the H_3O^+ concentration. (◀◀ D.)

5. **What is a buffer?**
   THE GOAL: Be able to explain how a buffer maintains pH and how the bicarbonate buffer functions in the body. (◀◀ C.)

6. **How is the acid or base concentration of a solution determined?**
   THE GOAL: Be able to explain how a titration procedure works and use the results of a titration to calculate acid or base concentration in a solution. (◀◀ A, E.)

CONCEPTS TO REVIEW

A. Acids, Bases, and Neutralization Reactions
(Sections 3.11 and 5.5)

B. Reversible Reactions and Chemical Equilibrium
(Section 7.7)

C. Equilibrium Equations and Equilibrium Constants
(Section 7.8)

D. Units of Concentration; Molarity
(Section 9.7)

E. Ion Equivalents
(Section 9.10)

Acids! The word evokes images of dangerous, corrosive liquids that eat away everything they touch. Although a few well-known substances such as sulfuric acid (H_2SO_4) do indeed fit this description, most acids are relatively harmless. In fact, many acids, such as ascorbic acid (vitamin C), are necessary for life. We have already touched on the subject of acids and bases on several occasions, but the time has come for a more detailed study.

10.1 Acids and Bases in Aqueous Solution

Let us take a moment to review what we said about acids and bases in Sections 3.11 and 5.10 before going on to a more systematic study:

- An acid is a substance that produces hydrogen ions, H^+, when dissolved in water.
- A base is a substance that produces hydroxide ions, OH^-, when dissolved in water.
- The neutralization reaction of an acid with a base yields water plus a *salt*, an ionic compound composed of the cation from the base and the anion from the acid.

The above definitions of acids and bases were proposed in 1887 by the Swedish chemist Svante Arrhenius and are useful for many purposes. The definitions are limited, however, because they refer only to reactions that take place in aqueous solutions. (We will see shortly how the definitions can be broadened.) Another issue is that the H^+ ion is so reactive it does not exist in water. Instead, H^+ reacts with H_2O to give the **hydronium ion**, H_3O^+, as mentioned in Section 3.11. When gaseous HCl dissolves in water, for instance, H_3O^+ and Cl^- are formed. As described in Section 4.9, electrostatic potential maps show that the hydrogen of HCl is positively polarized and electron-poor (blue), whereas the oxygen of water is negatively polarized and electron-rich (red):

Hydronium ion The H_3O^+ ion, formed when an acid reacts with water.

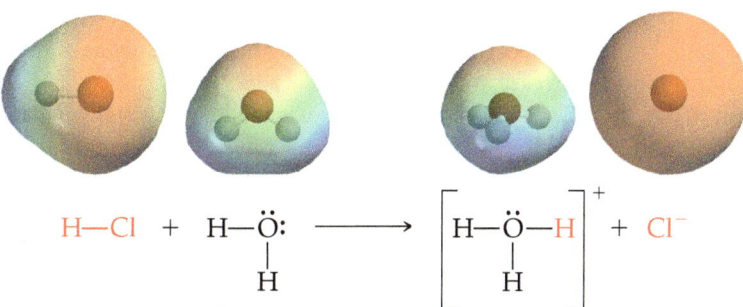

$$H-Cl \;+\; H-\ddot{O}-H \;\longrightarrow\; \left[H-\ddot{O}-H\right]^+ \;+\; Cl^-$$

Thus, the Arrhenius definition is updated to acknowledge that an acid yields H_3O^+ in water rather than H^+. In practice, however, the notations H_3O^+ and $H^+(aq)$ are often used interchangeably.

The Arrhenius definition of a base is correct as far as it goes, but it is important to realize that the OH^- ions "produced" by the base can come from either of two sources. Metal hydroxides, such as NaOH, KOH, and $Ba(OH)_2$, are ionic compounds that already contain OH^- ions and merely release those ions when they dissolve in water. Some molecular compounds such as ammonia, however, are not ionic and contain no OH^- ions in their structure. Nonetheless, they can act as bases to produce OH^- ions in reactions with water, as will be seen in Section 10.3.

10.2 Some Common Acids and Bases

Acids and bases are present in a variety of foods and consumer products. Acids generally have a sour taste, and nearly every sour food contains an acid: Lemons, oranges, and grapefruit contain citric acid, for instance, and sour milk contains lactic acid. Bases are not so obvious in foods, but most of us have them stored under the kitchen or bathroom sink. Bases are present in many household cleaning agents, from perfumed bar soap, to ammonia-based window cleaners, to the substance you put down the drain to dissolve hair, grease, and other materials that clog it.

Some of the most common acids and bases are listed below. It is a good idea at this point to learn their names and formulas, because we will refer to them often.

- **Sulfuric acid, H_2SO_4,** is probably the most important raw material in the chemical and pharmaceutical industries, and it is manufactured in greater quantity worldwide than any other industrial chemical. Over 45 million tons are prepared in the United States annually for use in many hundreds of industrial processes, including the preparation of phosphate fertilizers. Its most common consumer use is as the acid found in automobile batteries. As anyone who has splashed battery acid on his or her skin or clothing knows, sulfuric acid is highly corrosive and can cause painful burns.
- **Hydrochloric acid, HCl,** or *muriatic acid*, as it was historically known, has many industrial applications, including its use in metal cleaning and in the manufacture of high-fructose corn syrup. Aqueous HCl is also present as "stomach acid" in the digestive systems of most mammals.
- **Phosphoric acid, H_3PO_4,** is used in vast quantities in the manufacture of phosphate fertilizers. In addition, it is also used as an additive in foods and toothpastes. The tart taste of many soft drinks is due to the presence of phosphoric acid.
- **Nitric acid, HNO_3,** is a strong oxidizing agent that is used for many purposes, including the manufacture of ammonium nitrate fertilizer and military explosives. When spilled on the skin, it leaves a characteristic yellow coloration because of its reaction with skin proteins.
- **Acetic acid, CH_3CO_2H,** is the primary organic constituent of vinegar. It also occurs in all living cells and is used in many industrial processes such as the preparation of solvents, lacquers, and coatings.
- **Sodium hydroxide, NaOH,** also called *caustic soda* or *lye*, is the most commonly used of all bases. Industrially, it is used in the production of aluminum from its ore, in the production of glass, and in the manufacture of soap from animal fat. Concentrated solutions of NaOH can cause severe burns if allowed to sit on the skin for long. Drain cleaners often contain NaOH because it reacts with the fats and proteins found in grease and hair.
- **Calcium hydroxide, $Ca(OH)_2$,** or *slaked lime*, is made industrially by treating lime (CaO) with water. It has many applications, including its use in mortars and cements. An aqueous solution of $Ca(OH)_2$ is often called *limewater*.
- **Magnesium hydroxide, $Mg(OH)_2$,** or *milk of magnesia*, is an additive in foods, toothpaste, and many over-the-counter medications. Antacids such as Rolaids™, Mylanta™, and Maalox™, for instance, all contain magnesium hydroxide.

▲ Common household cleaners typically contain bases (NaOH, NH_3). Soap is manufactured by the reaction of vegetable oils and animal fats with the bases NaOH and KOH.

- **Ammonia, NH_3,** is used primarily as a fertilizer, but it also has many other industrial applications, including the manufacture of pharmaceuticals and explosives. A dilute solution of ammonia is frequently used around the house as a glass cleaner.

10.3 The Brønsted–Lowry Definition of Acids and Bases

The Arrhenius definition of acids and bases discussed in Section 10.1 applies only to processes that take place in an aqueous solution. A far more general definition was proposed in 1923 by the Danish chemist Johannes Brønsted and the English chemist Thomas Lowry. A **Brønsted–Lowry acid** is any substance that is able to give a hydrogen ion, H^+, to another molecule or ion. A hydrogen *atom* consists of a proton and an electron, so a hydrogen *ion*, H^+, is simply a proton. Thus, we often refer to acids as *proton donors*. The reaction need not occur in water, and a Brønsted–Lowry acid need not give appreciable concentrations of H_3O^+ ions in water.

Different acids can supply different numbers of H^+ ions, as we saw in Section 3.11. Acids with one proton to donate, such as HCl or HNO_3, are called *monoprotic acids*; H_2SO_4 is a *diprotic acid* because it has two protons to donate, and H_3PO_4 is a *triprotic acid* because it has three protons to donate. Notice that the acidic H atoms (that is, the H atoms that are donated as protons) are bonded to electronegative atoms, such as chlorine or oxygen.

Brønsted–Lowry acid A substance that can donate a hydrogen ion, H^+, to another molecule or ion.

Hydrochloric acid (monoprotic) 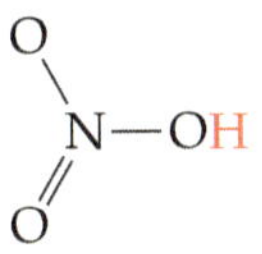   Nitric acid (monoprotic) 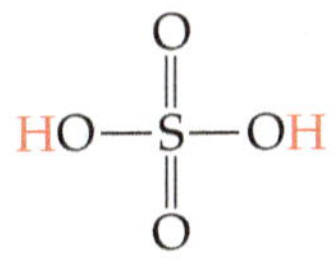   Sulfuric acid (diprotic) 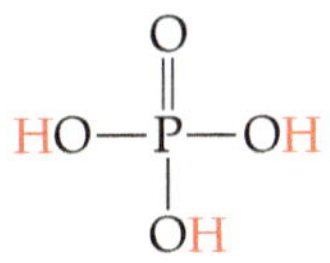   Phosphoric acid (triprotic)

Acetic acid (CH_3CO_2H), an example of an organic acid, actually has a total of 4 hydrogens, but only the one bonded to the electronegative oxygen is positively polarized and therefore acidic. The 3 hydrogens bonded to carbon are not acidic. Most organic acids are similar in that they contain many hydrogen atoms, but only the one in the $-CO_2H$ group (blue in the electrostatic potential map) is acidic:

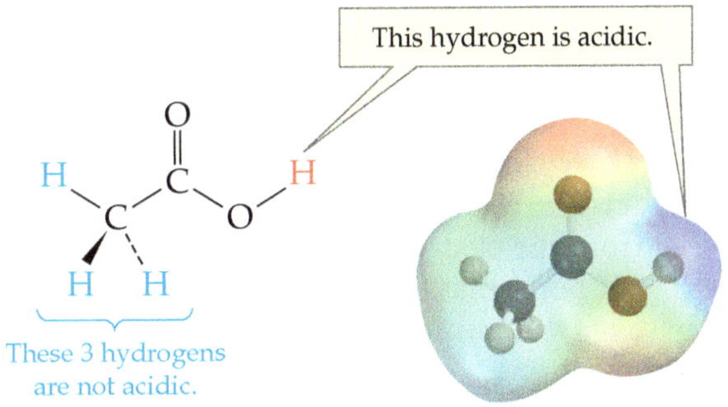

Acetic acid will react with water to produce H_3O^+ ions (Arrhenius acid definition) by donating a proton (Brønsted–Lowry acid definition) to water, as shown:

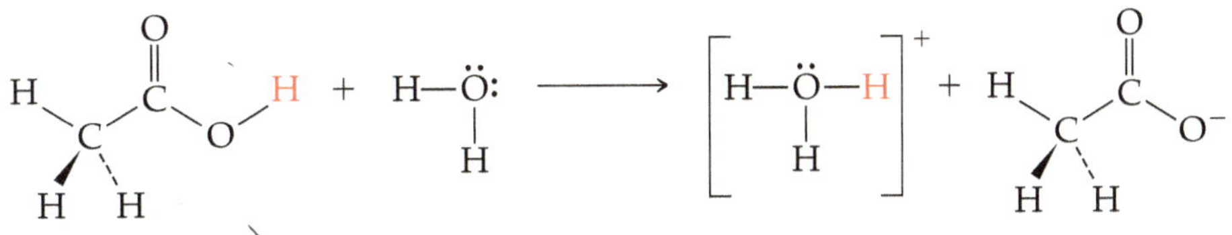

Whereas a Brønsted–Lowry acid is a substance that *donates* H^+ ions, a **Brønsted–Lowry base** is a substance that *accepts* H^+ ions from an acid. Ammonia will react

Brønsted–Lowry base A substance that can accept H^+ ions from an acid.

with water to produce OH⁻ ions (Arrhenius base definition) by accepting a proton (Brønsted–Lowry base definition), as shown:

$$H-\overset{H}{\underset{H}{\ddot{N}}}-H(g) + H_2O(l) \rightleftharpoons H-\overset{H}{\underset{H}{\overset{+}{N}}}-H(aq) + OH^-(aq)$$

This OH⁻ ion comes from H₂O.

As with the acids, reactions involving Brønsted–Lowry bases need not occur in water, and the Brønsted–Lowry base need not give appreciable concentrations of OH⁻ ions in water. Gaseous NH₃, for example, acts as a base to accept H⁺ from gaseous HCl and yield the ionic solid NH₄⁺ Cl⁻:

$$H\overset{\ddot{N}}{\underset{H}{\diagup}}_H \;+\; H-Cl \longrightarrow \left[H\overset{H}{\underset{H}{\overset{|}{\diagup N\diagdown}}}_H \right]^+ \;+\; Cl^-$$

Base   Acid

Putting the acid and base definitions together, *an acid–base reaction is one in which a proton is transferred.* The general reaction between proton-donor acids and proton-acceptor bases can be represented as

Electrons on base form bond with H⁺ from acid.

$$B: \;+\; H-A \;\rightleftharpoons\; B\overset{+}{-}H \;+\; A^-$$

$$B:^- \;+\; H-A \;\rightleftharpoons\; B-H \;+\; A^-$$

where the abbreviation HA represents a Brønsted–Lowry acid and B: or B:⁻ represents a Brønsted–Lowry base. Notice in these acid–base reactions that both electrons in the product B—H bond come from the base, as indicated by the curved arrow flowing from the electron pair of the base to the hydrogen atom of the acid. Thus, the B—H bond that forms is a coordinate covalent bond. In fact, a Brønsted–Lowry base *must* have such a lone pair of electrons; without them, it could not accept H⁺ from an acid.

▶▶ Recall from Section 4.4 that a coordinate covalent bond is one where both electrons are donated by the same atom.

A base can either be neutral (B:) or negatively charged (B:⁻). If the base is neutral, then the product has a positive charge (BH⁺) after H⁺ has been added. Ammonia is an example:

$$H-\overset{H}{\underset{H}{\overset{|}{N}}}: \;+\; H-A \;\rightleftharpoons\; H-\overset{H}{\underset{H}{\overset{+}{N}}}-H \;+\; :A^-$$

Adding an H⁺ creates positive charge.

Ammonia (a neutral base, B:)   Ammonium ion

If the base is negatively charged, then the product is neutral (BH). Hydroxide ion is an example:

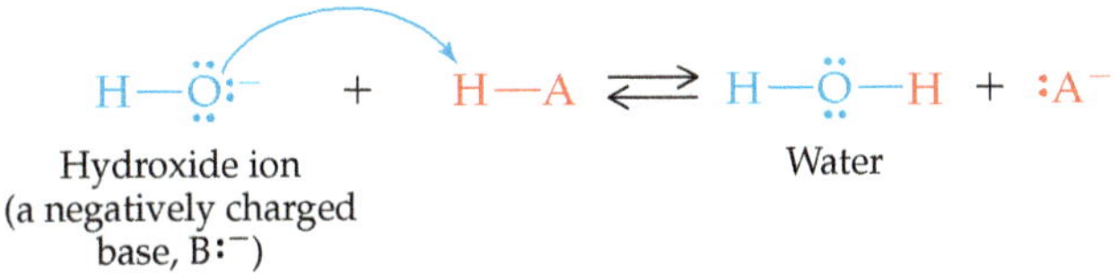

$$H-\ddot{\underset{..}{O}}:^- + H-A \rightleftharpoons H-\ddot{\underset{..}{O}}-H + :A^-$$

Hydroxide ion
(a negatively charged base, B:⁻)

Water

An important consequence of the Brønsted–Lowry definitions is that the *products* of an acid–base reaction can also behave as acids and bases. Many acid–base reactions are reversible, although in some cases the equilibrium constant for the reaction is quite large. For example, suppose we have as a forward reaction an acid HA donating a proton to a base B to produce A^-. This product A^- is a base because it can act as a proton acceptor in the reverse reaction. At the same time, the product BH^+ acts as an acid because it may donate a proton in the reverse reaction:

▶▶ When the equilibrium constant for a reaction is greater than 1, the forward reaction is favored. When the equilibrium constant is less than 1, the reverse reaction is favored (Section 7.8).

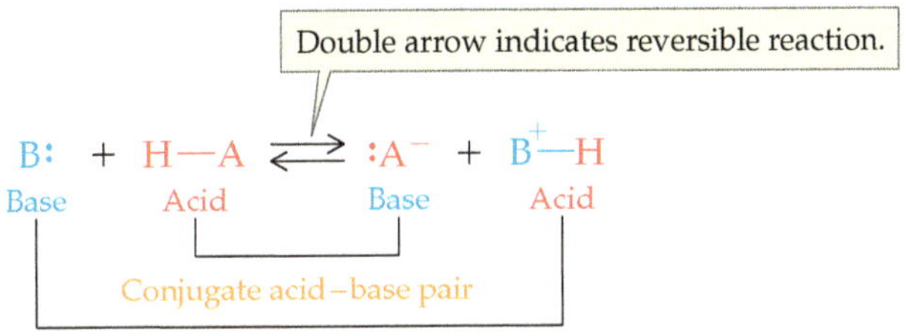

Double arrow indicates reversible reaction.

$$\underset{\text{Base}}{B:} + \underset{\text{Acid}}{H-A} \rightleftharpoons \underset{\text{Base}}{:A^-} + \underset{\text{Acid}}{B\overset{+}{-}H}$$

Conjugate acid–base pair

Pairs of chemical species such as B, BH^+ and HA, A^- are called **conjugate acid–base pairs**. They are species that are found on opposite sides of a chemical reaction whose formulas differ by only one H^+. Thus, the product anion A^- is the **conjugate base** of the reactant acid HA, and HA is the **conjugate acid** of the base A^-. Similarly, the reactant B is the conjugate base of the product acid BH^+, and BH^+ is the conjugate acid of the base B. The number of protons in a conjugate acid–base pair is always one greater than the number of protons in the base of the pair. To give some examples, acetic acid and acetate ion, the hydronium ion and water, and the ammonium ion and ammonia all make conjugate acid–base pairs:

Conjugate acid–base pair Two substances whose formulas differ by only a hydrogen ion, H^+.

Conjugate base The substance formed by loss of H^+ from an acid.

Conjugate acid The substance formed by addition of H^+ to a base.

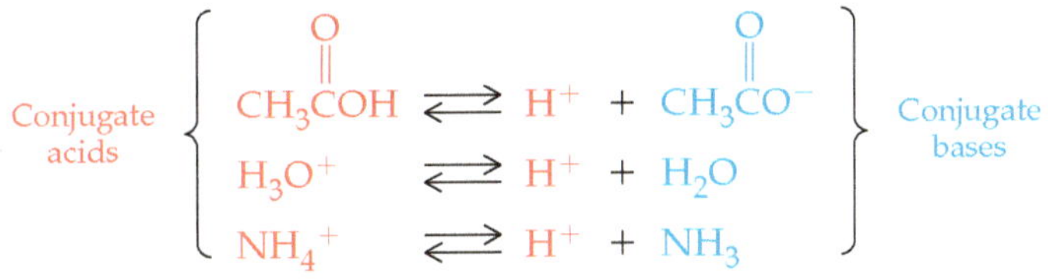

Conjugate acids $\begin{Bmatrix} CH_3COH \rightleftharpoons H^+ + CH_3CO^- \\ H_3O^+ \rightleftharpoons H^+ + H_2O \\ NH_4^+ \rightleftharpoons H^+ + NH_3 \end{Bmatrix}$ Conjugate bases

Worked Example 10.1 Acids and Bases: Identifying Brønsted–Lowry Acids and Bases

Identify each of the following as a Brønsted–Lowry acid or base:

(a) PO_4^{3-} (b) $HClO_4$ (c) CN^-

ANALYSIS A Brønsted–Lowry acid must have a hydrogen that it can donate as H^+, and a Brønsted–Lowry base must have an atom with a lone pair of electrons that can bond to H^+. Typically, a Brønsted–Lowry base is an anion derived by loss of H^+ from an acid.

SOLUTION

(a) The phosphate anion (PO_4^{3-}) has no proton to donate, so it must be a Brønsted–Lowry base. It is derived by loss of 3 H^+ ions from phosphoric acid, H_3PO_4.
(b) Perchloric acid ($HClO_4$) is a Brønsted–Lowry acid because it can donate an H^+ ion.
(c) The cyanide ion (CN^-) has no proton to donate, so it must be a Brønsted-Lowry base. It is derived by loss removal of an H^+ ion from hydrogen cyanide, HCN.

CHAPTER 10 Acids and Bases

Worked Example 10.2 Acids and Bases: Identifying Conjugate Acid–Base Pairs

Write formulas for
(a) The conjugate acid of the cyanide ion, CN^-
(b) The conjugate base of perchloric acid, $HClO_4$

ANALYSIS A conjugate acid is formed by adding H^+ to a base; a conjugate base is formed by removing H^+ from an acid.

SOLUTION
(a) HCN is the conjugate acid of CN^-
(b) ClO_4^- is the conjugate base of $HClO_4$.

PROBLEM 10.1
Which of the following would you expect to be Brønsted–Lowry acids?
(a) HCO_2H (b) H_2S (c) $SnCl_2$

PROBLEM 10.2
Which of the following would you expect to be Brønsted–Lowry bases?
(a) SO_3^{2-} (b) Ag^+ (c) F^-

PROBLEM 10.3
Write formulas for:
(a) The conjugate acid of HS^-
(b) The conjugate acid of PO_4^{3-}
(c) The conjugate base of H_2CO_3
(d) The conjugate base of NH_4^+

KEY CONCEPT PROBLEM 10.4

For the reaction shown here, identify the Brønsted–Lowry acids, bases, and conjugate acid–base pairs.

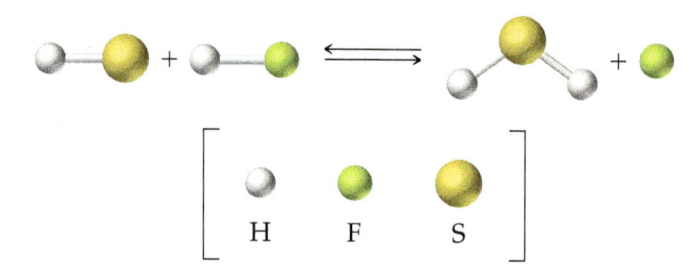

[H   F   S]

10.4 Acid and Base Strength

Some acids and bases, such as sulfuric acid (H_2SO_4), hydrochloric acid (HCl), or sodium hydroxide (NaOH), are highly corrosive. They react readily and, in contact with skin, can cause serious burns. Other acids and bases are not nearly as reactive. Acetic acid (CH_3COOH, the major component in vinegar) and phosphoric acid (H_3PO_4) are found in many food products. Why are some acids and bases relatively "safe," while others must be handled with extreme caution? The answer lies in how easily they produce the active ions for an acid (H^+) or a base (OH^-).

As indicated in Table 10.1, acids differ in their ability to give up a proton. The six acids at the top of the table are **strong acids**, meaning that they give up a proton easily and are essentially 100% **dissociated**, or split apart into ions, in water. Those remaining are **weak acids**, meaning that they give up a proton with difficulty and are substantially less than 100% dissociated in water. In a similar way, the conjugate bases at the

Strong acid An acid that gives up H^+ easily and is essentially 100% dissociated in water.

Dissociation The splitting apart of an acid in water to give H^+ and an anion.

Weak acid An acid that gives up H^+ with difficulty and is less than 100% dissociated in water.

TABLE 10.1 Relative Strengths of Acids and Conjugate Bases

		ACID			CONJUGATE BASE	
Increasing acid strength ↑	Strong acids: 100% dissociated	Perchloric acid Sulfuric acid Hydriodic acid Hydrobromic acid Hydrochloric acid Nitric acid	$HClO_4$ H_2SO_4 HI HBr HCl HNO_3	ClO_4^- HSO_4^- I^- Br^- Cl^- NO_3^-	Perchlorate ion Hydrogen sulfate ion Iodide ion Bromide ion Chloride ion Nitrate ion	Little or no reaction as bases
		Hydronium ion	H_3O^+	H_2O	Water	
	Weak acids	Hydrogen sulfate ion Phosphoric acid Nitrous acid Hydrofluoric acid Acetic acid	HSO_4^- H_3PO_4 HNO_2 HF CH_3COOH	SO_4^{2-} $H_2PO_4^-$ NO_2^- F^- CH_3COO^-	Sulfate ion Dihydrogen phosphate ion Nitrite ion Fluoride ion Acetate ion	Very weak bases
	Very weak acids	Carbonic acid Dihydrogen phosphate ion Ammonium ion Hydrocyanic acid Bicarbonate ion Hydrogen phosphate ion	H_2CO_3 $H_2PO_4^-$ NH_4^+ HCN HCO_3^- HPO_4^{2-}	HCO_3^- HPO_4^{2-} NH_3 CN^- CO_3^{2-} PO_4^{3-}	Bicarbonate ion Hydrogen phosphate ion Ammonia Cyanide ion Carbonate ion Phosphate ion	Weak bases
		Water	H_2O	OH^-	Hydroxide ion	Strong base ↓ Increasing base strength

top of the table are **weak bases** because they have little affinity for a proton, and the conjugate bases at the bottom of the table are **strong bases** because they grab and hold a proton tightly.

Note that diprotic acids, such as sulfuric acid H_2SO_4, undergo two stepwise dissociations in water. The first dissociation yields HSO_4^- and occurs to the extent of nearly 100%, so H_2SO_4 is a strong acid. The second dissociation yields SO_4^{2-} and takes place to a much lesser extent because separation of a positively charged H^+ from the negatively charged HSO_4^- anion is difficult. Thus, HSO_4^- is a weak acid:

$$H_2SO_4(l) + H_2O(l) \longrightarrow H_3O^+(aq) + HSO_4^-(aq)$$
$$HSO_4^-(aq) + H_2O(l) \rightleftharpoons H_3O^+(aq) + SO_4^{2-}(aq)$$

Perhaps the most striking feature of Table 10.1 is the inverse relationship between acid strength and base strength. **The stronger the acid, the weaker its conjugate base; the weaker the acid, the stronger its conjugate base.** HCl, for example, is a strong acid, so Cl^- is a very weak base. H_2O, however, is a very weak acid, so OH^- is a strong base.

Why is there an inverse relationship between acid strength and base strength? To answer this question, think about what it means for an acid or base to be strong or weak. A strong acid H—A is one that readily gives up a proton, meaning that its conjugate base A^- has little affinity for the proton. But this is exactly the definition of a weak base—a substance that has little affinity for a proton. As a result, the reverse

Weak base A base that has only a slight affinity for H^+ and holds it weakly.

Strong base A base that has a high affinity for H^+ and holds it tightly.

reaction occurs to a lesser extent, as indicated by the size of the forward and reverse arrows in the reaction:

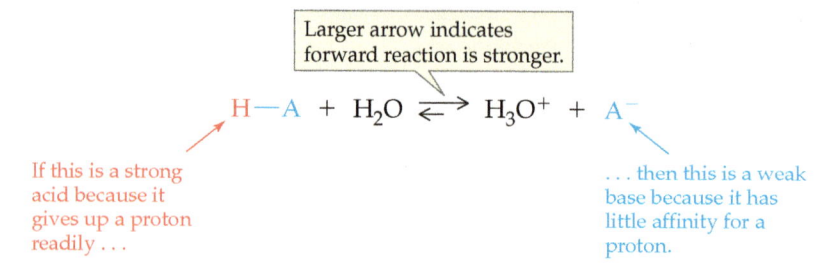

$$H-A + H_2O \rightleftharpoons H_3O^+ + A^-$$

Larger arrow indicates forward reaction is stronger.

If this is a strong acid because it gives up a proton readily...

...then this is a weak base because it has little affinity for a proton.

In the same way, a weak acid is one that gives up a proton with difficulty, meaning that its conjugate base has a high affinity for the proton. But this is just the definition of a strong base—a substance that has a high affinity for the proton. The reverse reaction now occurs more readily.

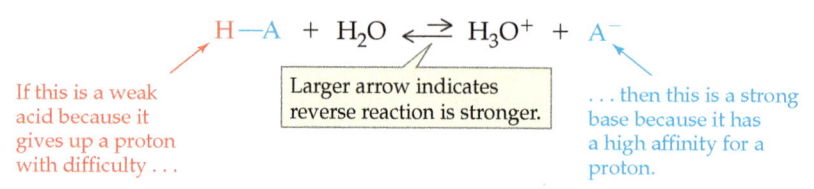

$$H-A + H_2O \rightleftharpoons H_3O^+ + A^-$$

Larger arrow indicates reverse reaction is stronger.

If this is a weak acid because it gives up a proton with difficulty...

...then this is a strong base because it has a high affinity for a proton.

Knowing the relative strengths of different acids as shown in Table 10.1 makes it possible to predict the direction of proton-transfer reactions. *An acid–base proton-transfer equilibrium always favors reaction of the stronger acid with the stronger base and formation of the weaker acid and base.* That is, the proton always leaves the stronger acid (whose weaker conjugate base cannot hold the proton) and always ends up in the weaker acid (whose stronger conjugate base holds the proton tightly). Put another way, in a contest for the proton, the stronger base always wins.

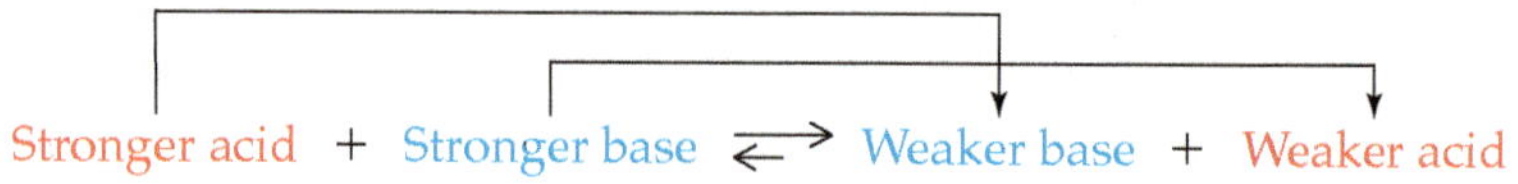

Stronger acid + Stronger base $\rightleftharpoons$ Weaker base + Weaker acid

To try out this rule, compare the reactions of acetic acid with water and with hydroxide ion. The idea is to write the equation, identify the acid on each side of the arrow, and then decide which acid is stronger and which is weaker. For example, the reaction of acetic acid with water to give acetate ion and hydronium ion is favored in the reverse direction, because acetic acid is a weaker acid than H_3O^+:

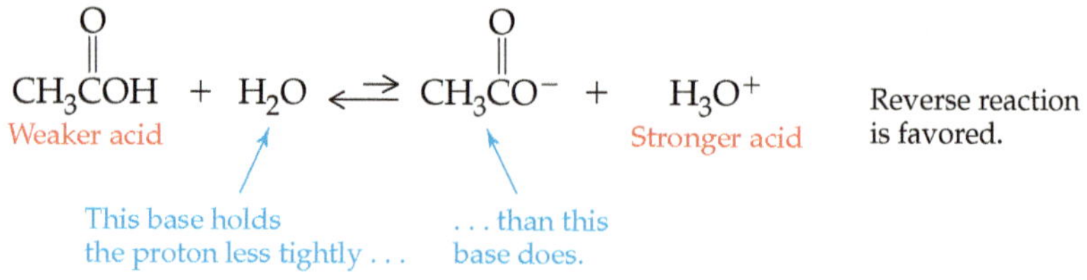

$$CH_3COH + H_2O \rightleftharpoons CH_3CO^- + H_3O^+$$

Weaker acid — Stronger acid — Reverse reaction is favored.

This base holds the proton less tightly... ...than this base does.

On the other hand, the reaction of acetic acid with hydroxide ion to give acetate ion and water is favored in the forward direction, because acetic acid is a stronger acid than H_2O:

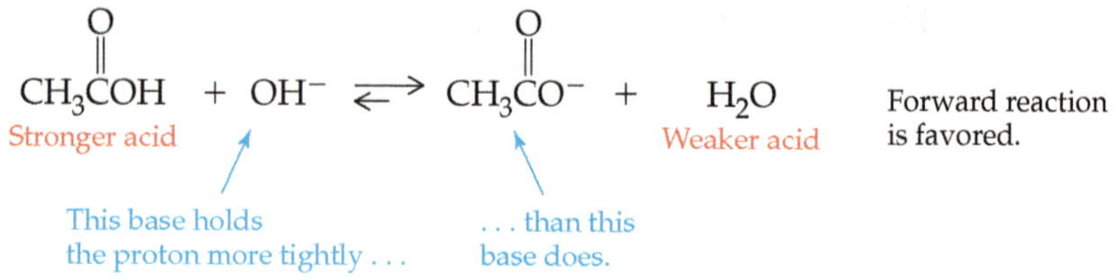

$$CH_3COH + OH^- \rightleftharpoons CH_3CO^- + H_2O$$

Stronger acid — Weaker acid — Forward reaction is favored.

This base holds the proton more tightly... ...than this base does.

CHEMISTRY IN ACTION

GERD—Too Much Acid or Not Enough?

Strong acids are very caustic substances that can dissolve even metals, and no one would think of ingesting them. However, the major component of the gastric juices secreted in the stomach is hydrochloric acid—a strong acid—and the acidic environment in the stomach is vital to good health and nutrition.

Stomach acid is essential for the digestion of proteins and for the absorption of certain micronutrients, such as calcium, magnesium, iron, and vitamin B$_{12}$. It also creates a sterile environment in the gut by killing yeast and bacteria that may be ingested. If these gastric juices leak up into the esophagus, the tube through which food and drink enter the stomach, they can cause the burning sensation in the chest or throat known as either heartburn or acid indigestion. Persistent irritation of the esophagus is known as gastro-esophageal reflux disease (GERD) and, if untreated, can lead to more serious health problems.

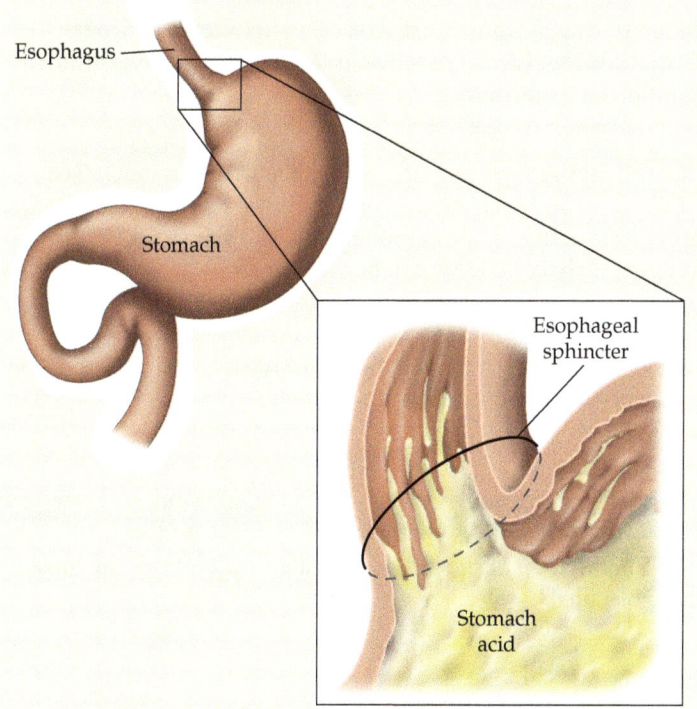

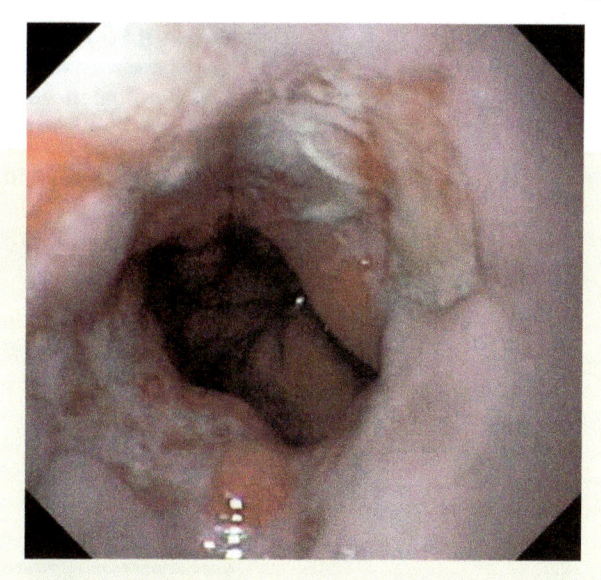

▲ If not treated, GERD can cause ulcers and scarring of esophageal tissue.

Hydrogen ions and chloride ions are secreted separately from the cytoplasm of parietal cells lining the stomach and then combine to form HCl that is usually close to 0.10 M. The HCl is then released into the stomach cavity, where the concentration is diluted to about 0.01–0.001 M. Unlike the esophagus, the stomach is coated by a thick mucus layer that protects the stomach wall from damage by this caustic solution.

Those who suffer from acid indigestion can obtain relief by using over-the-counter antacids, such as TUMS™ or Rolaids™ (see Section 10.12, p. 316). Chronic conditions such as GERD, however, are often treated with prescription medications. GERD can be treated by two classes of drugs. Proton-pump inhibitors (PPI), such as Prevacid™ and Prilosec™, prevent the production of the H$^+$ ions in the parietal cells, while H$_2$-receptor blockers (Tagamet™, Zantac™, and Pepcid™) prevent the release of stomach acid into the lumen. Both drugs effectively decrease the production of stomach acid to ease the symptoms of GERD.

Ironically, GERD can also be caused by not having enough stomach acid—a condition known as *hypochlorhydria*. The valve that controls the release of stomach contents to the small intestine is triggered by acidity. If this valve fails to open because the stomach is not acidic enough, the contents of the stomach can be churned back up into the esophagus.

See Chemistry in Action Problems 10.94 and 10.95 at the end of the chapter.

▲ The burning sensation and other symptoms associated with GERD are caused by the reflux of the acidic contents of the stomach into the esophagus.

Worked Example 10.3 Acid/Base Strength: Predicting Direction of H-transfer Reactions

Write a balanced equation for the proton-transfer reaction between phosphate ion (PO_4^{3-}) and water, and determine in which direction the equilibrium is favored.

ANALYSIS Look in Table 10.1 to see the relative acid and base strengths of the species involved in the reaction. The acid–base proton-transfer equilibrium will favor reaction of the stronger acid and formation of the weaker acid.

SOLUTION
Phosphate ion is the conjugate base of a weak acid (HPO_4^{2-}) and is therefore a relatively strong base. Table 10.1 shows that HPO_4^{2-} is a stronger acid than H_2O, and OH^- is a stronger base than PO_4^{3-}, so the reaction is favored in the reverse direction:

$$PO_4^{3-}(aq) + H_2O(l) \rightleftharpoons HPO_4^{2-}(aq) + OH^-(aq)$$
Weaker base Weaker acid Stronger acid Stronger base

PROBLEM 10.5
Use Table 10.1 to identify the stronger acid in the following pairs:
(a) H_2O or NH_4^+ (b) H_2SO_4 or CH_3CO_2H (c) HCN or H_2CO_3

PROBLEM 10.6
Use Table 10.1 to identify the stronger base in the following pairs:
(a) F^- or Br^- (b) OH^- or HCO_3^-

PROBLEM 10.7
Write a balanced equation for the proton-transfer reaction between a hydrogen phosphate ion and a hydroxide ion. Identify each acid–base pair, and determine in which direction the equilibrium is favored.

PROBLEM 10.8
Hydrochloric acid is the primary component of gastric juice in the stomach (see Chemistry in Action: GERD—Too Much Acid or Not Enough? on p. 299). The reaction between hydrochloric acid and the carbonate ion, the primary active ingredient in antacid tablets such as TUMS®, can be written as

$$HCl(aq) + CO_3^{2-}(aq) \rightleftharpoons HCO_3^-(aq) + Cl^-(aq)$$

Identify the conjugate acid–base pairs in the reaction, and rewrite the arrows in the reaction to indicate if the forward or reverse reaction is favored.

KEY CONCEPT PROBLEM 10.9

From this electrostatic potential map of the amino acid alanine, identify the most acidic hydrogens in the molecule:

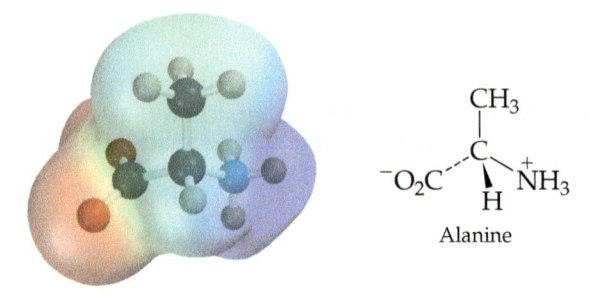

Alanine

10.5 Acid Dissociation Constants

The reaction of a weak acid with water, like any chemical equilibrium, can be described by an equilibrium equation (Section 7.8), where square brackets indicate the concentrations of the enclosed species in molarity (moles per liter).

For the reaction $HA(aq) + H_2O(l) \rightleftharpoons H_3O^+(aq) + A^-(aq)$

We have $K = \dfrac{[H_3O^+][A^-]}{[HA][H_2O]}$

Because water is a solvent as well as a participant for the reaction, its concentration is essentially constant and has no effect on the equilibrium. Therefore, we usually put the equilibrium constant K and the water concentration $[H_2O]$ together to make a new constant called the **acid dissociation constant**, (K_a). The acid dissociation constant is simply the hydronium ion concentration $[H_3O^+]$ times the conjugate base concentration $[A^-]$ divided by the undissociated acid concentration $[HA]$:

Acid dissociation constant $K_a = K[H_2O] = \dfrac{[H_3O^+][A^-]}{[HA]}$

Acid dissociation constant, (K_a) The equilibrium constant for the dissociation of an acid (HA), equal to $[H^+][A^-]/[HA]$.

For a strong acid, the H_3O^+ and A^- concentrations are much larger than the HA concentration, so K_a is very large. In fact, the K_a values for strong acids such as HCl are so large that it is difficult and not very useful to measure them. For a weak acid, however, the H_3O^+ and A^- concentrations are smaller than the HA concentration, so K_a is small. Table 10.2 gives K_a values for some common acids and illustrates several important points:

- Strong acids have K_a values much greater than 1 because dissociation is favored.
- Weak acids have K_a values much less than 1 because dissociation is not favored.
- Donation of each successive H^+ from a polyprotic acid is more difficult than the one before it, so K_a values become successively lower.
- Most organic acids, which contain the $-CO_2H$ group, have K_a values near 10^{-5}.

TABLE 10.2 Some Acid Dissociation Constants, K_a, at 25 °C

Acid	K_a	Acid	K_a
Hydrofluoric acid (HF)	3.5×10^{-4}	*Polyprotic acids*	
Hydrocyanic acid (HCN)	4.9×10^{-10}	Sulfuric acid	
Ammonium ion (NH_4^+)	5.6×10^{-10}	H_2SO_4	Large
		HSO_4^-	1.2×10^{-2}
Organic acids		Phosphoric acid	
Formic acid (HCOOH)	1.8×10^{-4}	H_3PO_4	7.5×10^{-3}
Acetic acid (CH_3COOH)	1.8×10^{-5}	$H_2PO_4^-$	6.2×10^{-8}
Propanoic acid (CH_3CH_2COOH)	1.3×10^{-5}	HPO_4^{2-}	2.2×10^{-13}
		Carbonic acid	
Ascorbic acid (vitamin C)	7.9×10^{-5}	H_2CO_3	4.3×10^{-7}
		HCO_3^-	5.6×10^{-11}

PROBLEM 10.10
Benzoic acid ($C_7H_5CO_2H$) has $K_a = 6.5 \times 10^{-5}$ and citric acid ($C_6H_8O_7$) has $K_a = 7.2 \times 10^{-4}$. Which is the stronger conjugate base, benzoate ($C_7H_5CO_2^-$) or citrate ($C_6H_7O_7^-$)?

10.6 Water as Both an Acid and a Base

Water is neither an acid nor a base in the Arrhenius sense because it does not contain appreciable concentrations of either H_3O^+ or OH^-. In the Brønsted–Lowry sense, however, water can act as *both* an acid and a base. When in contact with a base, water reacts as a Brønsted–Lowry acid and *donates* a proton to the base. In its reaction with ammonia, for example, water donates H^+ to ammonia to form the ammonium ion:

$$NH_3 + H_2O \longrightarrow NH_4^+ + OH^-$$

Ammonia (base)   Water (acid)   Ammonium ion (acid)   Hydroxide ion (base)

When in contact with an acid, water reacts as a Brønsted–Lowry base and *accepts* H^+ from the acid. This, of course, is exactly what happens when an acid such as HCl dissolves in water, as discussed in Section 10.1.

Water uses two electrons to form a bond to H^+.

$$H-\ddot{O}: + H-Cl \longrightarrow H-\overset{+}{\underset{H}{\ddot{O}}}-H + Cl^-$$
$$\phantom{H-\ddot{O}:}|$$
$$\phantom{H-\ddot{O}:}H$$

Water (A base)   (An acid)   Hydronium ion

Amphoteric A substance that can react as either an acid or a base.

Substances like water, which can react as either an acid or a base depending on the circumstances, are said to be **amphoteric** (am-pho-**tare**-ic). When water acts as an acid, it donates H^+ and becomes OH^-; when it acts as a base, it accepts H^+ and becomes H_3O^+. (*Note:* HCO_3^-, $H_2PO_4^-$ and HPO_4^{2-} are also amphoteric.)

Dissociation of Water

We have seen how water can act as an acid when a base is present and as a base when an acid is present. But what about when no other acids or bases are present? In this case, one water molecule acts as an acid while another water molecule acts as a base, reacting to form the hydronium and hydroxide ions:

$$H_2O(l) + H_2O(l) \rightleftharpoons H_3O^+(aq) + OH^-(aq)$$

Because each dissociation reaction yields 1 H_3O^+ ion and 1 OH^- ion, the concentrations of the 2 ions are identical. Also, the equilibrium arrows indicate that this reaction favors reactants, so that not many H_3O^+ and OH^- ions are present at equilibrium. At 25 °C, the concentration of each is 1.00×10^{-7} M. We can write the equilibrium constant expression for the dissociation of water as

$$K = \frac{[H_3O^+][OH^-]}{[H_2O][H_2O]}$$

where $[H_3O^+] = [OH^-] = 1.00 \times 10^{-7}$ M   (at 25 °C)

▶▶ Refer to discussion of equilibria involving pure liquids and solids in Section 7.8.

Ion-product constant for water (K_w) The product of the H_3O^+ and OH^- molar concentrations in water or any aqueous solution ($K_w = [H_3O^+][OH^-] = 1.00 \times 10^{-14}$).

As a pure substance the concentration of water is essentially constant. We can therefore put the water concentrations $[H_2O]$ together to make a new constant called the **ion-product constant for water** (K_w), which is simply the H_3O^+ concentration times the OH^- concentration. At 25 °C, $K_w = 1.00 \times 10^{-14}$.

Ion-product constant for water $K_w = K[H_2O][H_2O]$
$$= [H_3O^+][OH^-]$$
$$= 1.0 \times 10^{-14} \quad \text{(at 25 °C)}$$

The importance of the equation $K_w = [H_3O^+][OH^-]$ is that it applies to all aqueous solutions, not just to pure water. Since the product of $[H_3O^+]$ times $[OH^-]$ is always constant for any solution, we can determine the concentration of one species if we know the concentration of the other. If an acid is present in solution, for instance, so that $[H_3O^+]$ is large, then $[OH^-]$ must be small. If a base is present in solution so that $[OH^-]$ is large, then $[H_3O^+]$ must be small. For example, for a 0.10 M HCl solution, we know that $[H_3O^+] = 0.10$ M because HCl is 100% dissociated. Thus, we can calculate that $[OH^-] = 1.0 \times 10^{-13}$ M:

$$\text{Since } K_w \times [H_3O^+][OH^-] = 1.00 \times 10^{-14}$$

$$\text{we have } [OH^-] = \frac{K_w}{[H_3O^+]} = \frac{1.00 \times 10^{-14}}{0.10} = 1.0 \times 10^{-13} \text{ M}$$

Similarly, for a 0.10 M NaOH solution, we know that $[OH^-] = 0.10$ M, so $[H_3O^+] = 1.0 \times 10^{-13}$ M:

$$[H_3O^+] = \frac{K_w}{[OH^-]} = \frac{1.00 \times 10^{-14}}{0.10} = 1.0 \times 10^{-13} \text{ M}$$

Solutions are identified as acidic, neutral, or basic (*alkaline*) according to the value of their H_3O^+ and OH^- concentrations:

Acidic solution: $[H_3O^+] > 10^{-7}$ M and $[OH^-] < 10^{-7}$ M
Neutral solution: $[H_3O^+] = 10^{-7}$ M and $[OH^-] = 10^{-7}$ M
Basic solution: $[H_3O^+] < 10^{-7}$ M and $[OH^-] > 10^{-7}$ M

Worked Example 10.4 Water Dissociation Constant: Using K_w to Calculate $[OH^-]$

Milk has an H_3O^+ concentration of 4.5×10^{-7} M. What is the value of $[OH^-]$? Is milk acidic, neutral, or basic?

ANALYSIS The OH^- concentration can be found by dividing K_w by $[H_3O^+]$. An acidic solution has $[H_3O^+] > 10^{-7}$ M, a neutral solution has $[H_3O^+] = 10^{-7}$ M, and a basic solution has $[H_3O^+] < 10^{-7}$ M.

BALLPARK ESTIMATE Since the H_3O^+ concentration is slightly *greater* than 10^{-7} M, the OH^- concentration must be slightly *less* than 10^{-7} M, on the order of 10^{-8}.

SOLUTION

$$[OH^-] = \frac{K_w}{[H_3O^+]} = \frac{1.00 \times 10^{-14}}{4.5 \times 10^{-7}} = 2.2 \times 10^{-8} \text{ M}$$

Milk is slightly acidic because its H_3O^+ concentration is slightly larger than 1×10^{-7} M.

BALLPARK CHECK The OH^- concentration is of the same order of magnitude as our estimate.

PROBLEM 10.11
Identify the following solutions as either acidic or basic. What is the value of $[OH^-]$ in each?
(a) Household ammonia, $[H_3O^+] = 3.1 \times 10^{-12}$ M
(b) Vinegar, $[H_3O^+] = 4.0 \times 10^{-3}$ M

10.7 Measuring Acidity in Aqueous Solution: pH

In many fields, from medicine to chemistry to winemaking, it is necessary to know the exact concentration of H_3O^+ or OH^- in a solution. If, for example, the H_3O^+ concentration in blood varies only slightly from a value of 4.0×10^{-8} M, death can result.

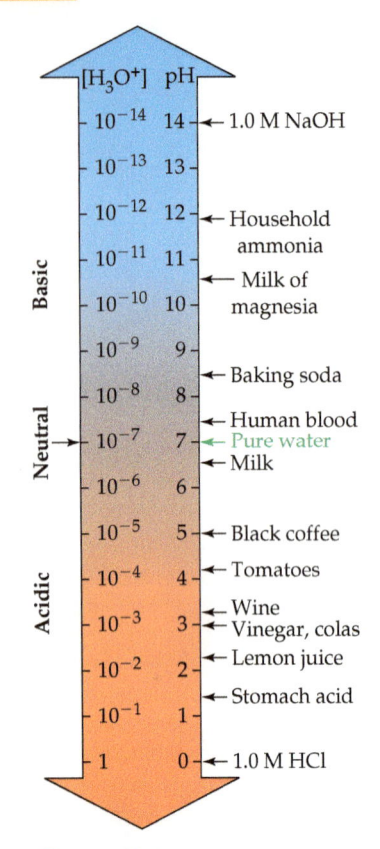

304 CHAPTER 10 Acids and Bases

Although correct, it is nevertheless awkward or, in some instances inconvenient, to refer to low concentrations of H_3O^+ using molarity. Fortunately, there is an easier way to express and compare H_3O^+ concentrations—the *pH scale*.

The pH of an aqueous solution is a number, usually between 0 and 14, that indicates the H_3O^+ concentration of the solution. A pH smaller than 7 corresponds to an acidic solution, a pH larger than 7 corresponds to a basic solution, and a pH of exactly 7 corresponds to a neutral solution. The pH scale and pH values of some common substances are shown in Figure 10.1.

Mathematically, a **p function** is defined as the negative common logarithm of some variable. The **pH** of a solution, therefore, is the negative common logarithm of the H_3O^+ concentration:

$$pH = -\log[H^+] \text{ (or } [H_3O^+])$$

If you have studied logarithms, you may remember that the common logarithm of a number is the power to which 10 must be raised to equal the number. The pH definition can therefore be restated as

$$[H_3O^+] = 10^{-pH}$$

For example, in neutral water at 25 °C, where $[H_3O^+] = 1 \times 10^{-7}$ M, the pH is 7; in a strong acid solution where $[H_3O^+] = 1 \times 10^{-1}$ M, the pH is 1; and in a strong base solution where $[H_3O^+] = 1 \times 10^{-14}$ M, the pH is 14:

Acidic solution: pH < 7, $[H_3O^+] > 1 \times 10^{-7}$ M
Neutral solution: pH = 7, $[H_3O^+] = 1 \times 10^{-7}$ M
Basic solution: pH > 7, $[H_3O^+] < 1 \times 10^{-7}$ M

Keep in mind that the pH scale covers an enormous range of acidities because it is a *logarithmic* scale, which involves powers of 10 (Figure 10.2). A change of only 1 pH unit means a 10-fold change in $[H_3O^+]$, a change of 2 pH units means a 100-fold change in $[H_3O^+]$, and a change of 12 pH units means a change of 10^{12} (a trillion) in $[H_3O^+]$.

To get a feel for the size of the quantities involved, think of a typical backyard swimming pool, which contains about 100,000 L of water. You would have to add only 0.10 mol of HCl (3.7 g) to lower the pH of the pool from 7.0 (neutral) to 6.0, but you would have to add 10,000 mol of HCl (370 kg!) to lower the pH of the pool from 7.0 to 1.0.

The logarithmic pH scale is a convenient way of reporting the relative acidity of solutions, but using logarithms can also be useful when calculating H_3O^+ and OH^- concentrations. Remember that the equilibrium between H_3O^+ and OH^- in aqueous solutions is expressed by K_w, where

$$K_w = [H_3O^+][OH^-] = 1 \times 10^{-14} \quad (\text{at } 25 \,°C)$$

If we convert this equation to its negative logarithmic form, we obtain

$$-\log(K_w) = -\log[H_3O^+] - \log[OH^-]$$
$$-\log(1 \times 10^{-14}) = -\log[H_3O^+] - \log[OH^-]$$
$$\text{or} \quad 14.00 = pH + pOH$$

The logarithmic form of the K_w equation can simplify the calculation of solution pH from OH^- concentration, as demonstrated in Worked Example 10.7.

▲ **Figure 10.1**
The pH scale and the pH values of some common substances.
A low pH corresponds to a strongly acidic solution, a high pH corresponds to a strongly basic solution, and a pH of 7 corresponds to a neutral solution.

p function The negative common logarithm of some variable, $pX = -\log(X)$.

pH A measure of the acid strength of a solution; the negative common logarithm of the H_3O^+ concentration.

Worked Example 10.5 Measuring Acidity: Calculating pH from $[H_3O^+]$

The H_3O^+ concentration in coffee is about 1×10^{-5} M. What pH is this?

ANALYSIS The pH is the negative common logarithm of the H_3O^+ concentration: $pH = -\log[H_3O^+]$.

SOLUTION
Since the common logarithm of 1×10^{-5} M is -5.0, the pH is 5.0.

Worked Example 10.6 Measuring Acidity: Calculating [H₃O⁺] from pH

Lemon juice has a pH of about 2. What $[H_3O^+]$ is this?

ANALYSIS In this case, we are looking for the $[H_3O^+]$, where $[H_3O^+] = 10^{-pH}$.

SOLUTION
Since pH = 2.0, $[H_3O^+] = 10^{-2} = 1 \times 10^{-2}$ M.

Worked Example 10.7 Measuring Acidity: Using K_w to Calculate [H₃O⁺] and pH

A cleaning solution is found to have $[OH^-] = 1 \times 10^{-3}$ M. What is the pH?

ANALYSIS To find pH, we must first find the value of $[H_3O^+]$ by using the equation $[H_3O^+] = K_w/[OH^-]$. Alternatively, we can calculate the pOH of the solution and then use the logarithmic form of the K_w equation: pH = 14.00 − pOH.

SOLUTION
Rearranging the K_w equation, we have

$$[H_3O^+] = \frac{K_w}{[OH^-]} = \frac{1.00 \times 10^{-14}}{1 \times 10^{-3}} = 1 \times 10^{-11} \text{ M}$$

$$pH = -\log(1 \times 10^{-11}) = 11.0$$

Using the logarithmic form of the K_w equation, we have

$$pH = 14.0 - pOH = 14.0 - (-\log[OH^-])$$
$$pH = 14.0 - (-\log(1 \times 10^{-3}))$$
$$pH = 14.0 - 3.0 = 11.0$$

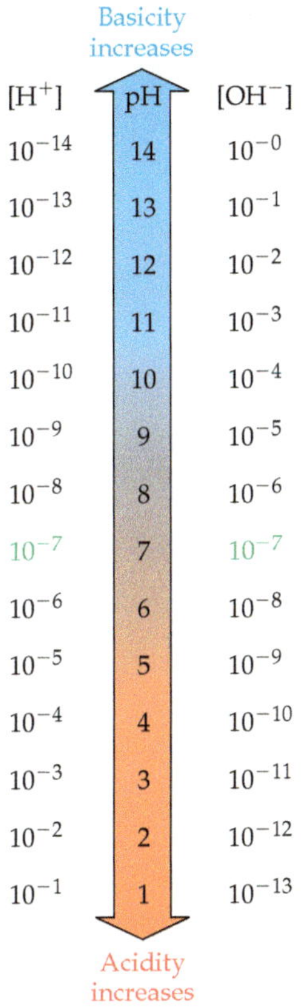

▲ **Figure 10.2**
The relationship of the pH scale to H^+ and OH^- concentrations.

Worked Example 10.8 Measuring Acidity: Calculating pH of Strong Acid Solutions

What is the pH of a 0.01 M solution of HCl?

ANALYSIS To find pH, we must first find the value of $[H_3O^+]$.

SOLUTION
Since HCl is a strong acid (Table 10.1), it is 100% dissociated, and the H_3O^+ concentration is the same as the HCl concentration: $[H_3O^+] = 0.01$ M, or 1×10^{-2} M, and pH = 2.0.

PROBLEM 10.12
Calculate the pH of the solutions in Problem 10.11.

PROBLEM 10.13
Give the hydronium ion and hydroxide ion concentrations of solutions with the following values of pH. Which of the solutions is most acidic? Which is most basic?
(a) pH 13.0   (b) pH 3.0   (c) pH 8.0

PROBLEM 10.14
Which solution would have the higher pH: 0.010 M HNO_2 or 0.010 M HNO_3? Explain.

10.8 Working with pH

Converting between pH and H_3O^+ concentration is easy when the pH is a whole number, but how do you find the H_3O^+ concentration of blood, which has a pH of 7.4, or the pH of a solution with $[H_3O^+] = 4.6 \times 10^{-3}$ M? Sometimes it is sufficient to make an estimate. The pH of blood (7.4) is between 7 and 8, so the H_3O^+ concentration of blood must be between 1×10^{-7} and 1×10^{-8} M. To be exact about finding pH values, though, requires a calculator.

Converting from pH to $[H_3O^+]$ requires finding the *antilogarithm* of the negative pH, which is done on many calculators with an "INV" key and a "log" key. Converting from $[H_3O^+]$ to pH requires finding the logarithm, which is commonly done with a "log" key and an "expo" or "EE" key for entering exponents of 10. Consult your calculator instructions if you are not sure how to use these keys. Remember that the sign of the number given by the calculator must be changed from minus to plus to get the pH.

The H_3O^+ concentration in blood with pH = 7.4 is

$$[H_3O^+] = \text{antilog}(-7.4) = 4 \times 10^{-8} \text{ M}$$

The pH of a solution with $[H_3O^+] = 4.6 \times 10^{-3}$ M is

$$\text{pH} = -\log(4.6 \times 10^{-3}) = -(-2.34) = 2.34$$

A note about significant figures: an antilogarithm contains the same number of significant figures as the original number has to the right of the decimal point. A logarithm contains the same number of digits to the right of the decimal point as the number of significant figures in the original number.

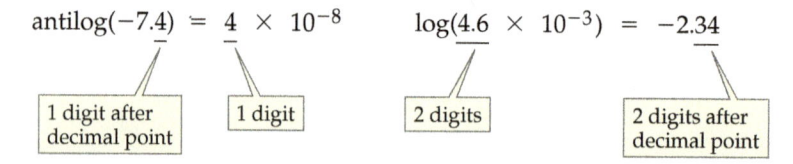

$$\text{antilog}(-7.4) = 4 \times 10^{-8} \qquad \log(4.6 \times 10^{-3}) = -2.34$$

- 1 digit after decimal point
- 1 digit
- 2 digits
- 2 digits after decimal point

Worked Example 10.9 Working with pH: Converting a pH to $[H_3O^+]$

Soft drinks usually have a pH of approximately 3.1. What is the $[H_3O^+]$ concentration in a soft drink?

ANALYSIS To convert from a pH value to an $[H_3O^+]$ concentration requires using the equation $[H_3O^+] = 10^{-\text{pH}}$, which requires finding an antilogarithm on a calculator.

BALLPARK ESTIMATE Because the pH is between 3.0 and 4.0, the $[H_3O^+]$ must be between 1×10^{-3} and 1×10^{-4}. A pH of 3.1 is very close to 3.0, so the $[H_3O^+]$ must be just slightly below 1×10^{-3} M.

SOLUTION
Entering the negative pH on a calculator (-3.1) and pressing the "INV" and "log" keys gives the answer 7.943×10^{-4}, which must be rounded off to 8×10^{-4} because the pH has only one digit to the right of the decimal point.

BALLPARK CHECK The calculated $[H_3O^+]$ of 8×10^{-4} M is between 1×10^{-3} M and 1×10^{-4} M and, as we estimated, just slightly below 1×10^{-3} M. (Remember, 8×10^{-4} is 0.8×10^{-3}.)

Worked Example 10.10 Working with pH: Calculating pH for Strong Acid Solutions

What is the pH of a 0.0045 M solution of $HClO_4$?

ANALYSIS Finding pH requires first finding $[H_3O^+]$ and then using the equation $\text{pH} = -\log[H_3O^+]$. Since $HClO_4$ is a strong acid (see Table 10.1), it is 100% dissociated, and so the H_3O^+ concentration is the same as the $HClO_4$ concentration.

BALLPARK ESTIMATE Because $[H_3O^+] = 4.5 \times 10^{-3}$ M is close to midway between 1×10^{-2} M and 1×10^{-3} M, the pH must be close to the midway point between 2.0 and 3.0. (Unfortunately, because the logarithm scale is not linear, trying to estimate the midway point is not a simple process.)

SOLUTION
$[H_3O^+] = 0.0045$ M $= 4.5 \times 10^{-3}$ M. Taking the negative logarithm gives pH $= 2.35$.

BALLPARK CHECK The calculated pH is consistent with our estimate.

Worked Example 10.11 Working with pH: Calculating pH for Strong Base Solutions

What is the pH of a 0.0032 M solution of NaOH?

ANALYSIS Since NaOH is a strong base, the OH⁻ concentration is the same as the NaOH concentration. Starting with the OH⁻ concentration, finding pH requires either using the K_w equation to find $[H_3O^+]$ or calculating pOH and then using the logarithmic form of the K_w equation.

BALLPARK ESTIMATE Because $[OH^-] = 3.2 \times 10^{-3}$ M is close to midway between 1×10^{-2} M and 1×10^{-3} M, the pOH must be close to the midway point between 2.0 and 3.0. Subtracting the pOH from 14 would therefore yield a pH between 11 and 12.

SOLUTION

$$[OH^-] = 0.0032 \text{ M} = 3.2 \times 10^{-3} \text{ M}$$

$$[H_3O^+] = \frac{K_w}{(3.2 \times 10^{-3})} = 3.1 \times 10^{-12} \text{ M}$$

Taking the negative logarithm gives pH $= -\log(3.1 \times 10^{-12}) = 11.51$. Alternatively, we can calculate pOH and subtract from 14.00 using the logarithmic form of the K_w equation. For $[OH^-] = 0.0032$ M,

$$\text{pOH} = -\log(3.2 \times 10^{-3}) = 2.49$$
$$\text{pH} = 14.00 - 2.49 = 11.51$$

Since the given OH⁻ concentration included two significant figures, the final pH includes two significant figures beyond the decimal point.

BALLPARK CHECK The calculated pH is consistent with our estimate.

PROBLEM 10.15
Identify the following solutions as acidic or basic, estimate $[H_3O^+]$ and $[OH^-]$ values for each, and rank them in order of increasing acidity:
(a) Saliva, pH $= 6.5$
(b) Pancreatic juice, pH $= 7.9$
(c) Orange juice, pH $= 3.7$
(d) Wine, pH $= 3.5$

PROBLEM 10.16
Calculate the pH of the following solutions and report it to the correct number of significant figures:
(a) Seawater with $[H_3O^+] = 5.3 \times 10^{-9}$ M
(b) A urine sample with $[H_3O^+] = 8.9 \times 10^{-6}$ M

PROBLEM 10.17
What is the pH of a 0.0025 M solution of HCl?

10.9 Laboratory Determination of Acidity

The pH of water is an important indicator of water quality in applications ranging from swimming pool and spa maintenance to municipal water treatment. There are several ways to measure the pH of a solution. The simplest but least accurate method is to use an **acid–base indicator**, a dye that changes color depending on the pH of the solution. For example, the well-known dye *litmus* is red below pH 4.8 but blue above pH 7.8 and the indicator *phenolphthalein* (fee-nol-**thay**-lean) is colorless below pH 8.2 but red above pH 10. To make pH determination particularly easy, test kits are available that contain a mixture of indicators known as *universal indicator* to give approximate pH measurements in the range 2–10 (Figure 10.3a). Also available are rolls of "pH paper," which make it possible to determine pH simply by putting a drop of solution on the paper and comparing the color that appears to the color on a calibration chart (Figure 10.3b).

Acid–base indicator A dye that changes color depending on the pH of a solution.

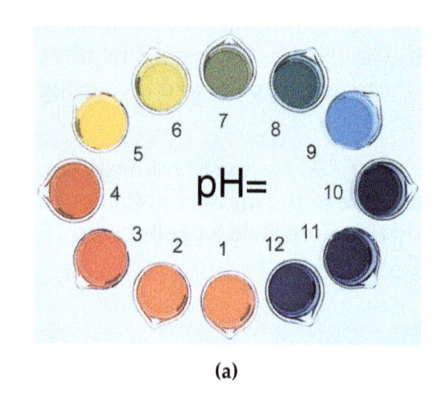

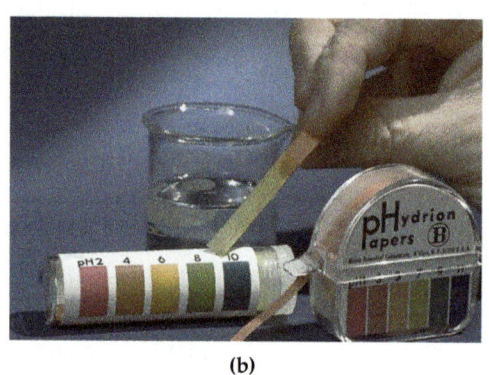

▶ **Figure 10.3**
Finding pH.
(a) The color of universal indicator in solutions of known pH from 1 to 12. (b) Testing pH with a paper strip. Comparing the color of the strip with the code on the package gives the approximate pH.

A much more accurate way to determine pH uses an electronic pH meter like the one shown in Figure 10.4. Electrodes are dipped into the solution, and the pH is read from the meter.

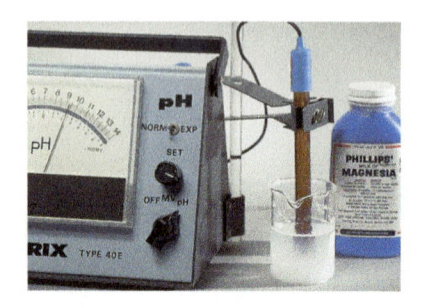

10.10 Buffer Solutions

Much of the body's chemistry depends on maintaining the pH of blood and other fluids within narrow limits. This is accomplished through the use of **buffers**—combinations of substances that act together to prevent a drastic change in pH.

Most buffers are mixtures of a weak acid and a roughly equal concentration of its conjugate base—for example, a solution that contains 0.10 M acetic acid and 0.10 M acetate ion. If a small amount of OH^- is added to a buffer solution, the pH increases, but not by much because the acid component of the buffer neutralizes the added OH^-. If a small amount of H_3O^+ is added to a buffer solution, the pH decreases, but again not by much because the base component of the buffer neutralizes the added H_3O^+.

To see why buffer solutions work, look at the equation for the acid dissociation constant of an acid HA.

▲ **Figure 10.4**
Using a pH meter to obtain an accurate reading of pH.
Is milk of magnesia acidic or basic?

Buffer A combination of substances that act together to prevent a drastic change in pH; usually a weak acid and its conjugate base.

$$\text{For the reaction:} \quad HA(aq) + H_2O(l) \rightleftharpoons A^-(aq) + H_3O^+(aq)$$

$$\text{we have} \quad K_a = \frac{[H_3O^+][A^-]}{[HA]}$$

Rearranging this equation shows that the value of $[H_3O^+]$, and thus the pH, depends on the ratio of the undissociated acid concentration to the conjugate base concentration, $[HA]/[A^-]$:

$$[H_3O^+] = K_a \frac{[HA]}{[A^-]}$$

In the case of the acetic acid–acetate ion buffer, for instance, we have

$$CH_3CO_2H(aq) + H_2O(l) \rightleftharpoons H_3O^+(aq) + CH_3CO_2^-(aq)$$
$$(0.10 \text{ M}) \qquad\qquad\qquad\qquad\qquad (0.10 \text{ M})$$

$$\text{and} \quad [H_3O^+] = K_a \frac{[CH_3CO_2H]}{[CH_3CO_2^-]}$$

Initially, the pH of the 0.10 M acetic acid–0.10 M acetate ion buffer solution is 4.74. When acid is added, most will be removed by reaction with $CH_3CO_2^-$. The equilibrium reaction shifts to the left, and as a result the concentration of CH_3CO_2H increases and the concentration of $CH_3CO_2^-$ decreases. As long as the changes in $[CH_3CO_2H]$ and $[CH_3CO_2^-]$ are relatively small, however, the ratio of $[CH_3CO_2H]$ to $[CH_3CO_2^-]$ changes only slightly, and there is little change in the pH.

When base is added to the buffer, most will be removed by reaction with CH_3CO_2H. The equilibrium shifts to the right, and so the concentration of CH_3CO_2H decreases and the concentration of $CH_3CO_2^-$ increases. Here too, though, as long as the concentration changes are relatively small, there is little change in the pH.

The ability of a buffer solution to resist changes in pH when acid or base is added is illustrated in Figure 10.5. Addition of 0.010 mol of H_3O^+ to 1.0 L of pure water changes the pH from 7 to 2, and addition of 0.010 mol of OH^- changes the pH from 7 to 12. A similar addition of acid to 1.0 L of a 0.10 M acetic acid–0.10 M acetate ion buffer, however, changes the pH from only 4.74 to 4.68, and addition of base changes the pH from only 4.74 to 4.85.

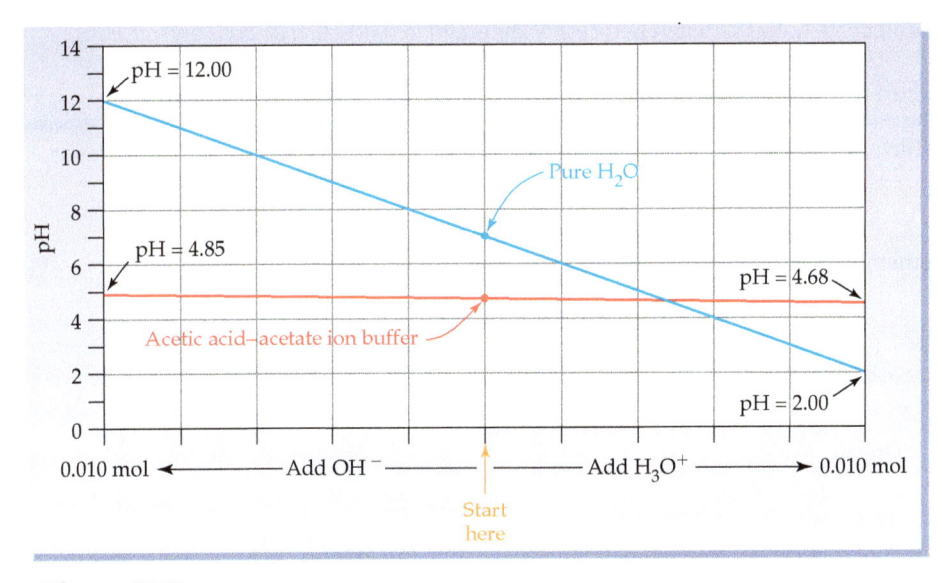

▲ **Figure 10.5**
A comparison of the change in pH.
When 0.010 mol of acid or 0.010 mol of base are added to 1.0 L of pure water and to 1.0 L of a 0.10 M acetic acid–0.10 M acetate ion buffer, the pH of the water varies between 12 and 2, while the pH of the buffer varies only between 4.85 and 4.68.

As we did with K_w, we can convert the rearranged K_a equation to its logarithmic form to obtain

$$pH = pK_a - \log\left(\frac{[HA]}{[A^-]}\right)$$

$$\text{or} \quad pH = pK_a + \log\left(\frac{[A^-]}{[HA]}\right)$$

This expression is known as the **Henderson–Hasselbalch equation** and is very useful in buffer applications, particularly in biology and biochemistry. Examination of the Henderson–Hasselbalch equation provides useful insights into how to prepare a buffer and into the factors that affect the pH of a buffer solution.

Henderson–Hasselbalch equation
The logarithmic form of the K_a equation for a weak acid, used in applications involving buffer solutions.

The effective pH range of a buffer will depend on the pK_a of the acid HA and on the relative concentrations of HA and conjugate base A^-. In general, the most effective buffers meet the following conditions:

- The pK_a for the weak acid should be close to the desired pH of the buffer solution.
- The ratio of [HA] to [A^-] should be close to 1, so that neither additional acid nor additional base changes the pH of the solution dramatically.
- The molar amounts of HA and A^- in the buffer should be approximately 10 times greater than the molar amounts of either acid or base you expect to add so that the ratio [A^-]/[HA] does not undergo a large change.

The pH of body fluids is maintained by three major buffer systems. Two of these buffers, the carbonic acid–bicarbonate ($H_2CO_3 - HCO_3^-$) system and the dihydrogen phosphate–hydrogen phosphate ($H_2PO_4^- - HPO_4^{2-}$) system, depend on weak acid–conjugate base interactions exactly like those of the acetate buffer system described previously:

$$H_2CO_3(aq) + H_2O(l) \rightleftharpoons HCO_3^-(aq) + H_3O^+(aq) \quad pK_a = 6.37$$
$$H_2PO_4^-(aq) + H_2O(l) \rightleftharpoons HPO_4^{2-}(aq) + H_3O^+(aq) \quad pK_a = 7.21$$

The third buffer system depends on the ability of proteins to act as either proton acceptors or proton donors at different pH values.

LOOKING AHEAD In Chapter 29, we will see how the regulation of blood pH by the bicarbonate buffer system is particularly important in preventing *acidosis* and *alkalosis*.

Worked Example 10.12 Buffers: Selecting a Weak Acid for a Buffer Solution

Which of the organic acids in Table 10.2 would be the most appropriate for preparing a pH 4.15 buffer solution?

ANALYSIS The pH of the buffer solution depends on the pK_a of the weak acid. Remember that $pK_a = -\log(K_a)$.

SOLUTION
The K_a and pK_a values for the four organic acids in Table 10.2 are tabulated below. The ascorbic acid ($pK_a = 4.10$) will produce a buffer solution closest to the desired pH of 4.15.

Organic Acid	K_a	pK_a
Formic acid (HCOOH)	1.8×10^{-4}	3.74
Acetic acid (CH_3COOH)	1.8×10^{-5}	4.74
Propanoic acid (CH_3CH_2COOH)	1.3×10^{-5}	4.89
Ascorbic acid (vitamin C)	7.9×10^{-5}	4.10

Worked Example 10.13 Buffers: Calculating the pH of a Buffer Solution

What is the pH of a buffer solution that contains 0.100 M HF and 0.120 M NaF? The K_a of HF is 3.5×10^{-4}, and so $pK_a = 3.46$.

ANALYSIS The Henderson–Hasselbalch equation can be used to calculate the pH of a buffer solution: $pH = pK_a + \log\left(\dfrac{[F^-]}{[HF]}\right)$.

BALLPARK ESTIMATE If the concentrations of F^- and HF were equal, the log term in our equation would be zero, and the pH of the solution would be equal to the pK_a for HF, which means pH = 3.46. However, since the concentration of the conjugate base ([F^-] = 0.120 M) is slightly higher than the concentration of the conjugate acid ([HF] = 0.100 M), then the pH of the buffer solution will be slightly higher (more basic) than the pK_a.

SOLUTION

$$pH = pK_a + \log\left(\frac{[F^-]}{[HF]}\right)$$

$$pH = 3.46 + \log\left(\frac{0.120}{0.100}\right) = 3.46 + 0.08 = 3.54$$

BALLPARK CHECK The calculated pH of 3.54 is consistent with the prediction that the final pH will be slightly higher than the pK_a of 3.46.

Worked Example 10.14 Buffers: Measuring the Effect of Added Base on pH

What is the pH of 1.00 L of the 0.100 M hydrofluoric acid–0.120 M fluoride ion buffer system described in Worked Example 10.13 after 0.020 mol of NaOH is added?

ANALYSIS Initially, the 0.100 M HF–0.120 M NaF buffer has pH = 3.54, as calculated in Worked Example 10.13. The added base will react with the acid as indicated in the neutralization reaction,

$$HF(aq) + OH^-(aq) \longrightarrow H_2O(l) + F^-(aq)$$

which means [HF] decreases and [F⁻] increases. With the pK_a and the concentrations of HF and F⁻ known, pH can be calculated using the Henderson–Hasselbalch equation.

BALLPARK ESTIMATE After the neutralization reaction, there is more conjugate base (F⁻) and less conjugate acid (HF), and so we expect the pH to increase slightly from the initial value of 3.54.

SOLUTION
When 0.020 mol of NaOH is added to 1.00 L of the buffer, the HF concentration *decreases* from 0.100 M to 0.080 M as a result of an acid–base reaction. At the same time, the F⁻ concentration *increases* from 0.120 M to 0.140 M because additional F⁻ is produced by the neutralization. Using these new values gives

$$pH = 3.46 + \log\left(\frac{0.140}{0.080}\right) = 3.46 + 0.24 = 3.70$$

The addition of 0.020 mol of base causes the pH of the buffer to rise only from 3.54 to 3.70.

BALLPARK CHECK The final pH, 3.70, is slightly more basic than the initial pH of 3.54, consistent with our prediction.

PROBLEM 10.18
What is the pH of 1.00 L of the 0.100 M hydrofluoric acid–0.120 M fluoride ion buffer system described in Worked Example 10.13 after 0.020 mol of HNO_3 is added?

PROBLEM 10.19
The ammonia/ammonium buffer system is sometimes used to optimize polymerase chain reactions (PCR) used in DNA studies. The equilibrium for this buffer can be written as

$$NH_4^+(aq) + H_2O(l) \rightleftharpoons H_3O^+(aq) + NH_3(aq)$$

Calculate the pH of a buffer that contains 0.050 M ammonium chloride and 0.080 M ammonia. The K_a of ammonium is 5.6×10^{-10}.

PROBLEM 10.20
What is the ratio of bicarbonate ion to carbonic acid $([HCO_3^-]/[H_2CO_3])$ in blood serum that has a pH of 7.40? (see Chemistry in Action: Buffers in the Body: Acidosis and Alkalosis on p. 312).

CHEMISTRY IN ACTION

Buffers in the Body: Acidosis and Alkalosis

A group of teenagers at a rock concert experience a collective fainting spell. A person taking high doses of aspirin for chronic pain appears disoriented and is having trouble breathing. A person with type 1 diabetes complains of tiredness and stomach pains. An athlete who recently completed a highly strenuous workout suffers from muscle cramps and nausea. A patient on an HIV drug regimen experiences increasing weakness and numbness in the hands and feet. What do all these individuals have in common? They are all suffering from abnormal fluctuations in blood pH, resulting in conditions known as *acidosis* (pH < 7.35) or *alkalosis* (pH > 7.45).

Each of the fluids in our bodies has a pH range suited to its function, as shown in the accompanying table. The stability of cell membranes, the shapes of huge protein molecules that must be folded in certain ways to function, and the activities of enzymes are all dependent on appropriate H_3O^+ concentrations. Blood plasma and the interstitial fluid surrounding cells, which together compose one-third of body fluids, have a slightly basic pH with a normal range of 7.35–7.45. The highly complex series of reactions and equilibria that take place throughout the body are very sensitive to pH—variations of even a few tenths of a pH unit can produce severe physiological symptoms.

pH of Body Fluids

Fluid	pH
Blood plasma	7.4
Interstitial fluid	7.4
Cytosol	7.0
Saliva	5.8–7.1
Gastric juice	1.6–1.8
Pancreatic juice	7.5–8.8
Intestinal juice	6.3–8.0
Urine	4.6–8.0
Sweat	4.0–6.8

Maintaining the pH of blood serum in its optimal range is accomplished by the carbonic acid–bicarbonate buffer system (Section 10.10), which depends on the relative amounts of CO_2 and bicarbonate dissolved in the blood. Because carbonic acid is unstable and therefore in equilibrium with CO_2 and water, there is an extra step in the bicarbonate buffer mechanism:

$$CO_2(aq) + H_2O(l) \rightleftharpoons H_2CO_3(aq) \rightleftharpoons HCO_3^-(aq) + H_3O^+(aq)$$

As a result, the bicarbonate buffer system is intimately related to the elimination of CO_2, which is continuously produced in cells and transported to the lungs to be exhaled. Anything that significantly shifts the balance between dissolved CO_2 and HCO_3^- can upset these equilibria and raise or lower the pH. How does this happen, and how does the body compensate?

▲ Hyperventilation, the rapid breathing due to excitement or stress, removes CO_2 and increases blood pH resulting in respiratory alkalosis.

The relationships between the bicarbonate buffer system, the lungs, and the kidneys are shown in the figure on the next page. Under normal circumstances, the reactions shown in the figure are in equilibrium. Addition of excess acid (red arrows) causes formation of H_2CO_3 and results in lowering of H_3O^+ concentration. Removal of acid (blue arrows) causes formation of more H_3O^+ by dissociation of H_2CO_3. The maintenance of pH by this mechanism is supported by a reserve of bicarbonate ions in body fluids. Such a buffer can accommodate large additions of H_3O^+ before there is a significant change in the pH.

Additional backup to the bicarbonate buffer system is provided by the kidneys. Each day a quantity of acid equal to that produced in the body is excreted in the urine. In the process, the kidney returns HCO_3^- to the extracellular fluids, where it becomes part of the bicarbonate reserve.

Respiratory acidosis can be caused by a decrease in respiration, which leads to a buildup of excess CO_2 in the blood and a corresponding decrease in pH. This could be caused by a blocked air passage due to inhaled food—removal of the blockage restores normal breathing and a return to the optimal pH. *Metabolic acidosis* results from an excess of other acids in the blood that reduce the bicarbonate concentration. High doses of aspirin (acetylsalicylic acid, Section 17.5), for example, increase the hydronium ion concentration and decrease the pH. Strenuous exercise generates excess lactate in the muscles, which is released into the bloodstream (Section 23.11). The liver converts lactate into glucose, which is the body's major source of energy; this process consumes bicarbonate ions, which decreases the pH. Some HIV drug therapies can damage cellular mitochondria (Section 21.3), resulting in a buildup of lactic acid in the cells and bloodstream. In the case of a person with diabetes, lack of insulin causes the body to start burning fat, which generates ketones and keto acids (Chapter 16), organic compounds that lower the blood pH.

The body attempts to correct acidosis by increasing the rate and depth of respiration—breathing faster "blows off" CO_2, shifting the CO_2–bicarbonate equilibrium to the left and raising the pH. The net effect is rapid reversal of the acidosis.

SECTION 10.11 Acid and Base Equivalents 313

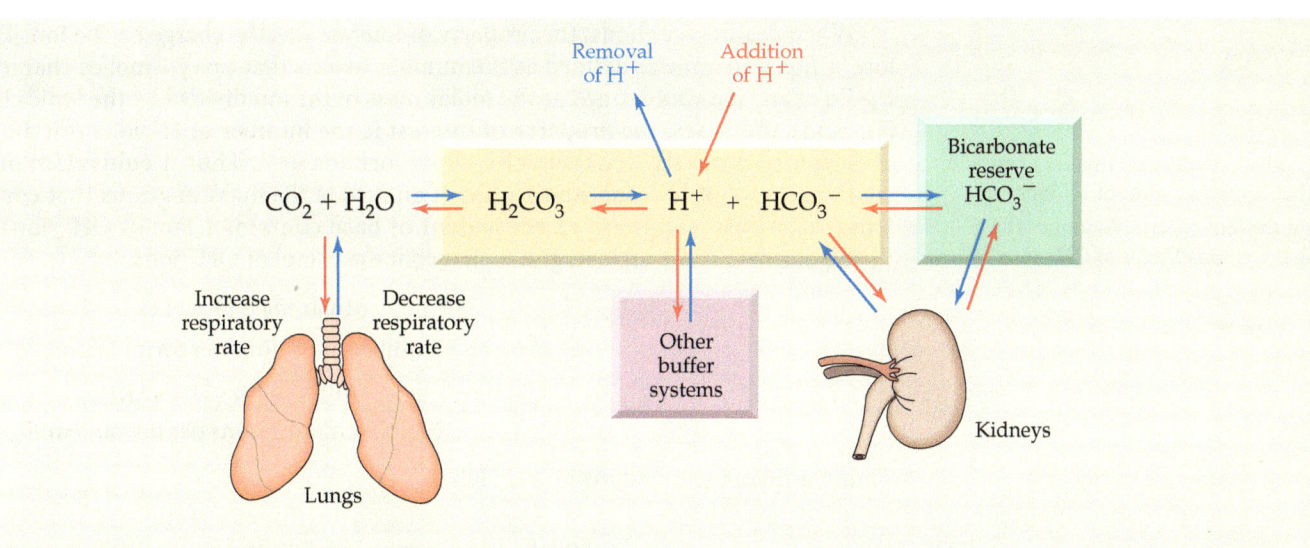

Although this may be sufficient for cases of respiratory acidosis, it provides only temporary relief for metabolic acidosis. A long-term solution depends on removal of excess acid by the kidneys, which can take several hours.

What about our teenage fans? In their excitement they have hyperventilated—their increased breathing rate has removed too much CO_2 from their blood and they are suffering from *respiratory alkalosis*. The body responds by "fainting" to decrease respiration and restore the CO_2 levels in the blood. When they regain consciousness, they will be ready to rock once again.

See Chemistry in Action Problems 10.96 and 10.97 at the end of the chapter.

KEY CONCEPT PROBLEM 10.21

A buffer solution is prepared using CN^- (from NaCN salt) and HCN in the amounts indicated. The K_a for HCN is 4.9×10^{-10}. Calculate the pH of the buffer solution.

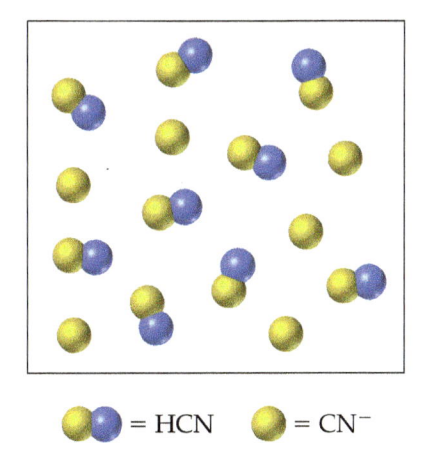

10.11 Acid and Base Equivalents

We said in Section 9.10 that it is sometimes useful to think in terms of ion *equivalents* (Eq) and *gram-equivalents* (g-Eq) when we are primarily interested in an ion itself rather than the compound that produced the ion. For similar reasons, it can also be useful to consider acid or base equivalents and gram-equivalents.

Equivalent of acid Amount of an acid that contains 1 mole of H⁺ ions.

Equivalent of base Amount of base that contains 1 mole of OH⁻ ions.

When dealing with ions, the property of interest was the charge on the ion. Therefore, 1 Eq of an ion was defined as the number of ions that carry 1 mol of charge, and 1 g-Eq of any ion was defined as the molar mass of the ion divided by the ionic charge. For acids and bases, the property of interest is the number of H⁺ ions (for an acid) or the number of OH⁻ ions (for a base) per formula unit. Thus, 1 **equivalent of acid** contains 1 mol of H⁺ ions, and 1 g-Eq of an acid is the mass in grams that contains 1 mol of H⁺ ions. Similarly, 1 **equivalent of base** contains 1 mol of OH⁻ ions, and 1 g-Eq of a base is the mass in grams that contains 1 mol of OH⁻ ions:

$$\text{One gram-equivalent of acid} = \frac{\text{Molar mass of acid (g)}}{\text{Number of H}^+ \text{ ions per formula unit}}$$

$$\text{One gram-equivalent of base} = \frac{\text{Molar mass of base (g)}}{\text{Number of OH}^- \text{ ions per formula unit}}$$

Thus 1 g-Eq of the monoprotic acid HCl is

$$1 \text{ g-Eq HCl} = \frac{36.5 \text{ g}}{1 \text{ H}^+ \text{ per HCl}} = 36.5 \text{ g}$$

which is equal to molar mass of the acid, but one gram-equivalent of the diprotic acid H_2SO_4 is

$$1 \text{ g-Eq H}_2\text{SO}_4 = \frac{98.0 \text{ g}}{2 \text{ H}^+ \text{ per H}_2\text{SO}_4} = 49.0 \text{ g}$$

which is the molar mass divided by 2, because 1 mol of H_2SO_4 contains 2 mol of H⁺.

$$\text{One equivalent of H}_2\text{SO}_4 = \frac{\text{Molar mass of H}_2\text{SO}_4}{2} = \frac{98.0 \text{ g}}{2} = 49.0 \text{ g}$$

Divide by 2 because H_2SO_4 is diprotic.

Using acid–base equivalents has two practical advantages: First, they are convenient when only the acidity or basicity of a solution is of interest rather than the identity of the acid or base. Second, they show quantities that are chemically equivalent in their properties; 36.5 g of HCl and 49.0 g of H_2SO_4 are chemically equivalent quantities because each reacts with 1 Eq of base. *One equivalent of any acid neutralizes one equivalent of any base.*

Because acid–base equivalents are so useful, clinical chemists sometimes express acid and base concentrations in *normality* rather than molarity. The **normality (N)** of an acid or base solution is defined as the number of equivalents (or milliequivalents) of acid or base per liter of solution. For example, a solution made by dissolving 1.0 g-Eq (49.0 g) of H_2SO_4 in water to give 1.0 L of solution has a concentration of 1.0 Eq/L, which is 1.0 N. Similarly, a solution that contains 0.010 Eq/L of acid is 0.010 N and has an acid concentration of 10 mEq/L:

Normality (N) A measure of acid (or base) concentration expressed as the number of acid (or base) equivalents per liter of solution.

$$\text{Normality (N)} = \frac{\text{Equivalents of acid or base}}{\text{Liters of solution}}$$

The values of molarity (M) and normality (N) are the same for monoprotic acids, such as HCl, but are not the same for diprotic or triprotic acids. A solution made by diluting 1.0 g-Eq (49.0 g = 0.50 mol) of the diprotic acid H_2SO_4 to a volume of 1.0 L has a *normality* of 1.0 N but a *molarity* of 0.50 M. For any acid or base, normality is always equal to molarity times the number of H⁺ or OH⁻ ions produced per formula unit:

Normality of acid = (Molarity of acid) × (Number of H⁺ ions produced per formula unit)
Normality of base = (Molarity of base) × (Number of OH⁻ ions produced per formula unit)

SECTION 10.11 Acid and Base Equivalents 315

Worked Example 10.15 Equivalents: Mass to Equivalent Conversion for Diprotic Acid

How many equivalents are in 3.1 g of the diprotic acid H_2S? The molar mass of H_2S is 34.0 g.

ANALYSIS The number of acid or base equivalents is calculated by doing a gram to mole conversion using molar mass as the conversion factor and then multiplying by the number of H^+ ions produced.

BALLPARK ESTIMATE The 3.1 g is a little less than 0.10 mol of H_2S. Since it is a diprotic acid, (two H^+ per mole), this represents a little less than 0.2 Eq of H_2S.

SOLUTION

$$(3.1 \text{ g } H_2S)\left(\frac{1 \text{ mol } H_2S}{34.0 \text{ g } H_2S}\right)\left(\frac{2 \text{ Eq } H_2S}{1 \text{ mol } H_2S}\right) = 0.18 \text{ Eq } H_2S$$

BALLPARK CHECK The calculated value of 0.18 is consistent with our prediction of a little less than 0.2 Eq of H_2S.

Worked Example 10.16 Equivalents: Calculating Equivalent Concentrations

What is the normality of a solution made by diluting 6.5 g of H_2SO_4 to a volume of 200 mL? What is the concentration of this solution in milliequivalents per liter? The molar mass of H_2SO_4 is 98.0 g.

ANALYSIS Calculate how many equivalents of H_2SO_4 are in 6.5 g by using the molar mass of the acid as a conversion factor and then determine the normality of the acid.

SOLUTION

STEP 1: Identify known information. We know the molar mass of H_2SO_4, the mass of H_2SO_4 to be dissolved, and the final volume of solution.

MW of H_2SO_4 = 98.0 g/mol
Mass of H_2SO_4 = 6.5 g
Volume of solution = 200 mL

STEP 2: Identify answer including units. We need to calculate the normality of the final solution.

Normality = ?? (equiv./L)

STEP 3: Identify conversion factors. We will need to convert the mass of H_2SO_4 to moles, and then to equivalents of H_2SO_4. We will then need to convert volume from mL to L.

$$(6.5 \text{ g } H_2SO_4)\left(\frac{1 \text{ mol } H_2SO_4}{98.0 \text{ g } H_2SO_4}\right)\left(\frac{2 \text{ Eq } H_2SO_4}{1 \text{ mol } H_2SO_4}\right)$$
$$= 0.132 \text{ Eq } H_2SO_4 \text{ (don't round yet!)}$$

$$(200 \text{ mL})\left(\frac{1 \text{ L}}{1000 \text{ mL}}\right) = 0.200 \text{ L}$$

STEP 4: Solve. Dividing the number of equivalents by the volume yields the Normality.

$$\frac{0.132 \text{ Eq } H_2SO_4}{0.200 \text{ L}} = 0.66 \text{ N}$$

The concentration of the sulfuric acid solution is 0.66 N, or 660 mEq/L.

PROBLEM 10.22

How many equivalents are in the following?
(a) 5.0 g HNO_3
(b) 12.5 g $Ca(OH)_2$
(c) 4.5 g H_3PO_4

PROBLEM 10.23

What are the normalities of the solutions if each sample in Problem 10.22 is dissolved in water and diluted to a volume of 300.0 mL?

10.12 Some Common Acid–Base Reactions

Among the most common of the many kinds of Brønsted–Lowry acid–base reactions are those of an acid with hydroxide ion, an acid with bicarbonate or carbonate ion, and an acid with ammonia or a related nitrogen-containing compound. Let us look briefly at each of the three types.

Reaction of Acids with Hydroxide Ion

One equivalent of an acid reacts with 1 Eq of a metal hydroxide to yield water and a salt in a neutralization reaction:

$$HCl(aq) + KOH(aq) \longrightarrow H_2O(l) + KCl(aq)$$
(An acid) (A base)   (Water)   (A salt)

Such reactions are usually written with a single arrow because their equilibria lie far to the right and they have very large equilibrium constants ($K = 5 \times 10^{15}$; Section 7.8). The net ionic equation (Section 5.8) for all such reactions makes clear why acid–base equivalents are useful and why the properties of the acid and base disappear in neutralization reactions: The equivalent ions for the acid (H^+) and the base (OH^-) are used up in the formation of water.

$$H^+(aq) + OH^-(aq) \longrightarrow H_2O(l)$$

> **PROBLEM 10.24**
> Maalox, an over-the-counter antacid, contains aluminum hydroxide, $Al(OH)_3$, and magnesium hydroxide, $Mg(OH)_2$. Write balanced equations for the reaction of both with stomach acid (HCl).

Reaction of Acids with Bicarbonate and Carbonate Ion

Bicarbonate ion reacts with acid by accepting H^+ to yield carbonic acid, H_2CO_3. Similarly, carbonate ion accepts 2 protons in its reaction with acid. Carbonic acid is unstable, however, rapidly decomposing to carbon dioxide gas and water:

$$H^+(aq) + HCO_3^-(aq) \longrightarrow [H_2CO_3(aq)] \longrightarrow H_2O(l) + CO_2(g)$$
$$2\,H^+(aq) + CO_3^{2-}(aq) \longrightarrow [H_2CO_3(aq)] \longrightarrow H_2O(l) + CO_2(g)$$

Most metal carbonates are insoluble in water—marble, for example, is almost pure calcium carbonate, $CaCO_3$—but they nevertheless react easily with aqueous acid. In fact, geologists often test for carbonate-bearing rocks by putting a few drops of aqueous HCl on the rock and watching to see if bubbles of CO_2 form (Figure 10.6). This reaction is also responsible for the damage to marble and limestone artwork caused by acid rain (See Chemistry in Action: Acid Rain on p. 320). The most common application involving carbonates and acid, however, is the use of antacids that contain carbonates, such as TUMS™ or Rolaids™, to neutralize excess stomach acid.

▲ **Figure 10.6**
Marble.
Marble, which is primarily $CaCO_3$, releases bubbles of CO_2 when treated with hydrochloric acid.

> **PROBLEM 10.25**
> Write a balanced equation for each of the following reactions:
> (a) $HCO_3^-(aq) + H_2SO_4(aq) \longrightarrow ?$
> (b) $CO_3^{2-}(aq) + HNO_3(aq) \longrightarrow ?$

Reaction of Acids with Ammonia

Acids react with ammonia to yield ammonium salts, such as ammonium chloride, NH₄Cl, most of which are water-soluble:

$$NH_3(aq) + HCl(aq) \rightarrow NH_4Cl(aq)$$

Living organisms contain a group of compounds called *amines*, which contain nitrogen atoms bonded to carbon. Amines react with acids just as ammonia does, yielding water-soluble salts. Methylamine, for example, an organic compound found in rotting fish, reacts with HCl:

```
      H  H                         H  H
      |  |                         |  |   +
  H—C—N:   +   H—Cl    →      H—C—N—H      Cl⁻
      |  |                         |  |
      H  H                         H  H
   Methylamine                 Methylammonium chloride
```

LOOKING AHEAD ▶▶ In Chapter 15, we will see that amines occur in all living organisms, both plant and animal, as well as in many pharmaceutical agents. Amines called amino acids form the building blocks from which proteins are made, as we will see in Chapter 18.

PROBLEM 10.26
What products would you expect from the reaction of ammonia and sulfuric acid in aqueous solution?

$$2\,NH_3(aq) + H_2SO_4(aq) \longrightarrow ?$$

PROBLEM 10.27
Show how ethylamine ($C_2H_5NH_2$) reacts with hydrochloric acid to form an ethylammonium salt.

10.13 Titration

Determining the pH of a solution gives the solution's H_3O^+ concentration but not necessarily its total acid concentration. That is because the two are not the same thing. The H_3O^+ concentration gives only the amount of acid that has dissociated into ions, whereas total acid concentration gives the sum of dissociated plus undissociated acid. In a 0.10 M solution of acetic acid, for instance, the total acid concentration is 0.10 M, yet the H_3O^+ concentration is only 0.0013 M (pH = 2.89) because acetic acid is a weak acid that is only about 1% dissociated.

The total acid or base concentration of a solution can be found by carrying out a **titration** procedure, as shown in Figure 10.7. Let us assume, for instance, that we want to find the acid concentration of an HCl solution. (Likewise, we might need to find the base concentration of an NaOH solution.) We begin by measuring out a known volume of the HCl solution and adding an acid–base indicator. Next, we fill a calibrated glass tube called a *buret* with an NaOH solution of known concentration, and we slowly add the NaOH to the HCl until neutralization is complete (the *end point*), identified by a color change in the indicator.

Reading from the buret gives the volume of the NaOH solution that has reacted with the known volume of HCl. Knowing both the concentration and volume of the NaOH solution then allows us to calculate the molar amount of NaOH, and the coefficients in the balanced equation allow us to find the molar amount of HCl that has been neutralized. Dividing the molar amount of HCl by the volume of the HCl solution

Titration A procedure for determining the total acid or base concentration of a solution.

318 CHAPTER 10 Acids and Bases

▶ **Figure 10.7**
Titration of an acid solution of unknown concentration with a base solution of known concentration.
(a) A measured volume of the acid solution is placed in the flask along with an indicator. (b) The base of known concentration is then added from a buret until the color change of the indicator shows that neutralization is complete (the *end point*).

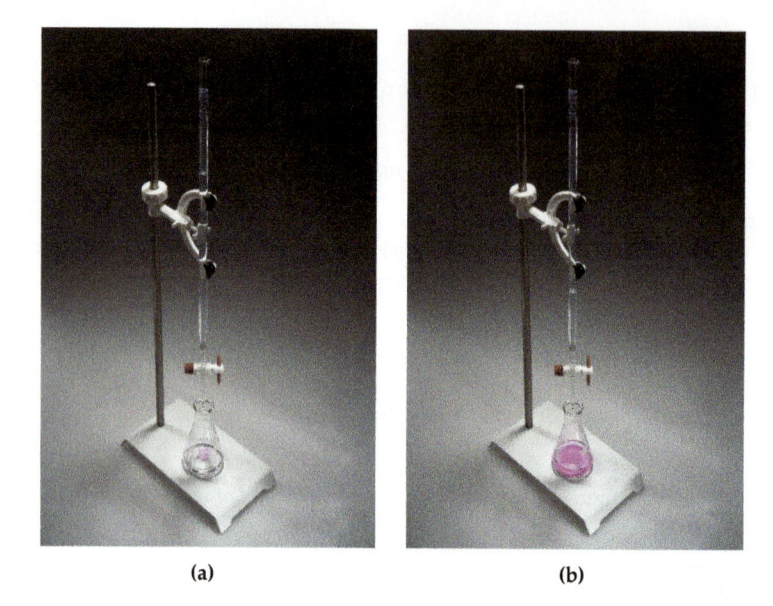

(a)   (b)

gives the concentration. The calculation thus involves mole–volume conversions just like those done in Section 9.7. Figure 10.8 shows a flow diagram of the strategy, and Worked Example 10.17 shows how to calculate total acid concentration.

When the titration involves a neutralization reaction in which one mole of acid reacts with one mole of base, such as that shown in Figure 10.8, then the moles of acid and base needed for complete reaction can be represented as

$$M_{acid} \times V_{acid} = M_{base} \times V_{base}$$

When the coefficients for the acid and base in the balanced neutralization reaction are not the same, such as in the reaction of a diprotic acid (H_2SO_4) with a monoprotic base (NaOH), then we can use equivalents of acid and base instead of moles, and Normality instead of Molarity:

$$(Eq)_{acid} = (Eq)_{base}$$
$$N_{acid} \times V_{acid} = N_{base} \times V_{base}.$$

We can convert between Normality and Molarity as described in Section 10.11.

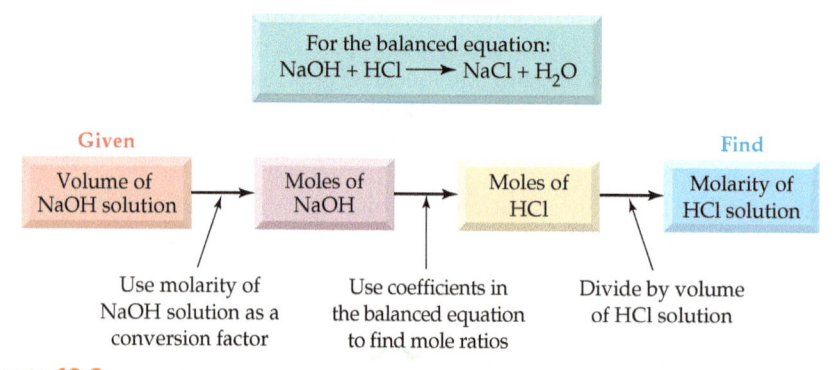

For the balanced equation:
NaOH + HCl ⟶ NaCl + H₂O

Given: Volume of NaOH solution → Moles of NaOH → Moles of HCl → **Find**: Molarity of HCl solution

Use molarity of NaOH solution as a conversion factor | Use coefficients in the balanced equation to find mole ratios | Divide by volume of HCl solution

▲ **Figure 10.8**
A flow diagram for an acid–base titration.
This diagram summarizes the calculations needed to determine the concentration of an HCl solution by titration with an NaOH solution of known concentration. The steps are similar to those shown in Figure 9.7.

SECTION 10.13 Titration 319

Worked Example 10.17 Titrations: Calculating Total Acid Concentration

When a 5.00 mL sample of household vinegar (dilute aqueous acetic acid) is titrated, 44.5 mL of 0.100 M NaOH solution is required to reach the end point. What is the acid concentration of the vinegar in moles per liter, equivalents per liter, and milli-equivalents per liter? The neutralization reaction is

$$CH_3CO_2H(aq) + NaOH(aq) \longrightarrow CH_3CO_2^-Na^+(aq) + H_2O(l)$$

ANALYSIS To find the molarity of the vinegar, we need to know the number of moles of acetic acid dissolved in the 5.00 mL sample. Following a flow diagram similar to Figure 10.8, we use the volume and molarity of NaOH to find the number of moles. From the chemical equation, we use the mole ratio to find the number of moles of acid, and then divide by the volume of the acid solution. Because acetic acid is a monoprotic acid, the normality of the solution is numerically the same as its molarity.

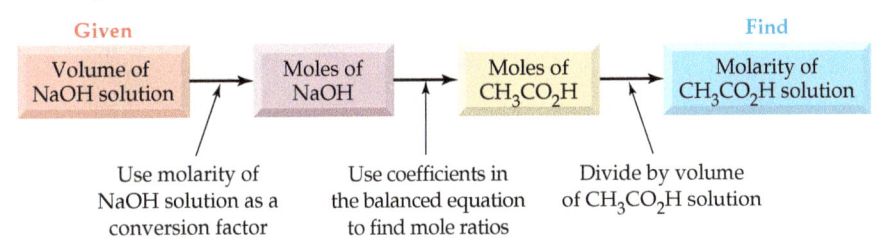

Given: Volume of NaOH solution → Moles of NaOH → Moles of CH_3CO_2H → **Find**: Molarity of CH_3CO_2H solution

- Use molarity of NaOH solution as a conversion factor
- Use coefficients in the balanced equation to find mole ratios
- Divide by volume of CH_3CO_2H solution

BALLPARK ESTIMATE The 5.00 mL of vinegar required nearly nine times as much NaOH solution (44.5 mL) for complete reaction. Since the neutralization stoichiometry is 1:1, the molarity of the acetic acid in the vinegar must be nine times greater than the molarity of NaOH, or approximately 0.90 M.

SOLUTION
Substitute the known information and appropriate conversion factors into the flow diagram, and solve for the molarity of the acetic acid:

$$(44.5 \text{ mL NaOH}) \left(\frac{0.100 \text{ mol NaOH}}{1000 \text{ mL}} \right) \left(\frac{1 \text{ mol } CH_3CO_2H}{1 \text{ mol NaOH}} \right)$$

$$\times \left(\frac{1}{0.005\ 00 \text{ L}} \right) = 0.890 \text{ M } CH_3CO_2H$$

$$= 0.890 \text{ N } CH_3CO_2H$$

Expressed in milliequivalents, this concentration is

$$\frac{0.890 \text{ Eq}}{\text{L}} \times \frac{1000 \text{ mEq}}{1 \text{ Eq}} = 890 \text{ mEq/L}$$

BALLPARK CHECK The calculated result (0.890 M) is very close to our estimate of 0.90 M.

PROBLEM 10.28
A titration is carried out to determine the concentration of the acid in an old bottle of aqueous HCl whose label has become unreadable. What is the HCl concentration if 58.4 mL of 0.250 M NaOH is required to titrate a 20.0 mL sample of the acid?

CHEMISTRY IN ACTION

Acid Rain

As the water that evaporates from oceans and lakes condenses into raindrops, it dissolves small quantities of gases from the atmosphere. Under normal conditions, rain is slightly acidic, with a pH close to 5.6, because of atmospheric CO_2 that dissolves to form carbonic acid:

$$CO_2(aq) + H_2O(l) \rightleftharpoons H_2CO_3(aq) \rightleftharpoons HCO_3^-(aq) + H_3O^+(aq)$$

In recent decades, however, the acidity of rainwater in many industrialized areas of the world has increased by a factor of over 100, to a pH between 3 and 3.5.

The primary cause of this so-called *acid rain* is industrial and automotive pollution. Each year, large power plants and smelters pour millions of tons of sulfur dioxide (SO_2) gas into the atmosphere, where some is oxidized by air to produce sulfur trioxide (SO_3). Sulfur oxides then dissolve in rain to form dilute sulfurous acid (H_2SO_3) and sulfuric acid (H_2SO_4):

$$SO_2(g) + H_2O(l) \longrightarrow H_2SO_3(aq)$$
$$SO_3(g) + H_2O(l) \longrightarrow H_2SO_4(aq)$$

Nitrogen oxides produced by the high-temperature reaction of N_2 with O_2 in coal-burning plants and in automobile engines further contribute to the problem. Nitrogen dioxide (NO_2) dissolves in water to form dilute nitric acid (HNO_3) and nitric oxide (NO):

$$3\,NO_2(g) + H_2O(l) \longrightarrow 2\,HNO_3(aq) + NO(g)$$

Oxides of both sulfur and nitrogen have always been present in the atmosphere, produced by such natural sources as volcanoes and lightning bolts, but their amounts have increased dramatically over the last century because of industrialization. The result is a notable decrease in the pH of rainwater in more densely populated regions, including Europe and the eastern United States.

Many processes in nature require such a fine pH balance that they are dramatically upset by the shift that has occurred in the pH of rain. Some watersheds contain soils that have high "buffering capacity" and so are able to neutralize acidic compounds in acid rain. Other areas, such as the northeastern United States and eastern Canada, where soil-buffering capacity is poor, have experienced negative ecological effects. Acid rain releases aluminum salts from soil, and the ions then wash into streams. The low pH and increased aluminum levels are so toxic to fish and other organisms that many lakes and streams in these areas are devoid of aquatic life. Massive tree die-offs have occurred throughout central and eastern Europe as acid rain has lowered the pH of the soil and has leached nutrients from leaves.

Fortunately, acidic emissions in the United States have been greatly reduced in recent years as a result of the Clean Air Act

▲ This limestone statue adorning the Rheims Cathedral in France has been severely eroded by acid rain.

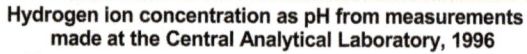

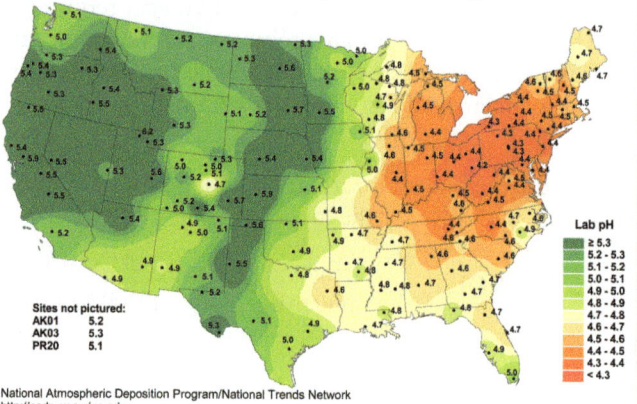

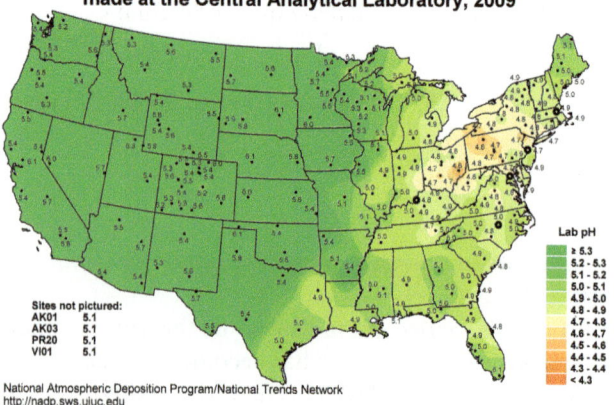

▲ These maps compare the average pH of precipitation in the United States in 1996 and in 2009. During this period, total acid deposition in much of the eastern United States decreased substantially.

Amendments of 1990. Industrial emissions of SO_2 and nitrogen oxides decreased by over 40% from 1990 to 2007, resulting in a decrease in acid rain depositions, particularly in the eastern United States and Canada (see accompanying figure). While significant reductions have been realized, most environmental scientists agree that additional reductions in these pollutant emissions are necessary to ensure the recovery of affected lakes and streams.

See Chemistry in Action Problems 10.98 and 10.99 at the end of the chapter.

PROBLEM 10.29
How many milliliters of 0.150 M NaOH are required to neutralize 50.0 mL of 0.200 M H_2SO_4? The balanced neutralization reaction is:

$$H_2SO_4(aq) + 2\,NaOH(aq) \longrightarrow Na_2SO_4(aq) + 2\,H_2O(l).$$

PROBLEM 10.30
A 21.5 mL sample of a KOH solution of unknown concentration requires 16.1 mL of 0.150 M H_2SO_4 solution to reach the end point in a titration.
(a) How many moles of H_2SO_4 were necessary to reach the end point? How many equivalents?
(b) What is the molarity of the KOH solution?

PROBLEM 10.31
Titration of a 50.00 ml sample of acid rain required 9.30 mL of 0.0012 M NaOH to reach the end point. What was the total $[H_3O^+]$ in the rain sample? What was the pH? (see Chemistry in Action: Acid Rain on p. 320).

10.14 Acidity and Basicity of Salt Solutions

It is tempting to think of all salt solutions as neutral; after all, they come from the neutralization reaction between an acid and a base. In fact, salt solutions can be neutral, acidic, or basic, depending on the ions present, because some ions react with water to produce H_3O^+ and some ions react with water to produce OH^-. To predict the acidity of a salt solution, it is convenient to classify salts according to the acid and base from which they are formed in a neutralization reaction. The classification and some examples are given in Table 10.3.

TABLE 10.3 Acidity and Basicity of Salt Solutions

Anion Derived from Acid That Is:	Cation Derived from Base That Is:	Solution	Example
Strong	Weak	Acidic	NH_4Cl, NH_4NO_3
Weak	Strong	Basic	$NaHCO_3, KCH_3CO_2$
Strong	Strong	Neutral	$NaCl, KBr, Ca(NO_3)_2$
Weak	Weak	More information needed	

The general rule for predicting the acidity or basicity of a salt solution is that the stronger partner from which the salt is formed dominates. That is, a salt formed from a strong acid and a weak base yields an acidic solution because the strong acid dominates; a salt formed from a weak acid and a strong base yields a basic solution because the base dominates; and a salt formed from a strong acid and a strong base yields a neutral solution because neither acid nor base dominates. Here are some examples.

Salt of Strong Acid + Weak Base ⟶ Acidic Solution

A salt such as NH₄Cl, which can be formed by reaction of a strong acid (HCl) with a weak base (NH₃), yields an acidic solution. The Cl⁻ ion does not react with water, but the NH₄⁺ ion is a weak acid that gives H₃O⁺ ions:

$$NH_4^+(aq) + H_2O(l) \rightleftharpoons NH_3(aq) + H_3O^+(aq)$$

Salt of Weak Acid + Strong Base ⟶ Basic Solution

A salt such as sodium bicarbonate, which can be formed by reaction of a weak acid (H₂CO₃) with a strong base (NaOH), yields a basic solution. The Na⁺ ion does not react with water, but the HCO₃⁻ ion is a weak base that gives OH⁻ ions:

$$HCO_3^-(aq) + H_2O(l) \rightleftharpoons H_2CO_3(aq) + OH^-(aq)$$

Salt of Strong Acid + Strong Base ⟶ Neutral Solution

A salt such as NaCl, which can be formed by reaction of a strong acid (HCl) with a strong base (NaOH), yields a neutral solution. Neither the Cl⁻ ion nor the Na⁺ ion reacts with water.

Salt of Weak Acid + Weak Base

Both cation and anion in this type of salt react with water, so we cannot predict whether the resulting solution will be acidic or basic without quantitative information. The ion that reacts to the greater extent with water will govern the pH—it may be either the cation or the anion.

Worked Example 10.18 Acidity and Basicity of Salt Solutions

> Predict whether the following salts produce an acidic, basic, or neutral solution:
> (a) BaCl₂   (b) NaCN   (c) NH₄NO₃
>
> **ANALYSIS** Look in Table 10.1 to see the classification of acids and bases as strong or weak.
>
> **SOLUTION**
> (a) BaCl₂ gives a neutral solution because it is formed from a strong acid (HCl) and a strong base [Ba(OH)₂].
> (b) NaCN gives a basic solution because it is formed from a weak acid (HCN) and a strong base (NaOH).
> (c) NH₄NO₃ gives an acidic solution because it is formed from a strong acid (HNO₃) and a weak base (NH₃).

PROBLEM 10.32
Predict whether the following salts produce an acidic, basic, or neutral solution:
(a) K₂SO₄   (b) Na₂HPO₄   (c) MgF₂   (d) NH₄Br

SUMMARY: REVISITING THE CHAPTER GOALS

1. What are acids and bases? According to the *Brønsted–Lowry definition*, an acid is a substance that donates a hydrogen ion (a proton, H⁺) and a base is a substance that accepts a hydrogen ion. Thus, the generalized reaction of an acid with a base involves the reversible transfer of a proton:

$$B: + H{-}A \rightleftharpoons A:^- + H{-}B^+$$

In aqueous solution, water acts as a base and accepts a proton from an acid to yield a *hydronium ion*, H₃O⁺. Reaction of an acid with a metal hydroxide, such as KOH, yields water and a salt; reaction with bicarbonate ion (HCO₃⁻) or carbonate ion (CO₃²⁻) yields water, a salt, and CO₂ gas; and reaction with ammonia yields an ammonium salt (see Problems 33, 37, 38, 42, 43, 60, 94, 100, 102).

2. What effect does the strength of acids and bases have on their reactions? Different acids and bases differ in their ability to give up or accept a proton. A *strong acid* gives up a proton easily and is 100% *dissociated* in aqueous solution; a *weak acid* gives up a proton with difficulty, is only slightly dissociated in water, and establishes an equilibrium between dissociated and undissociated forms. Similarly, a *strong base* accepts and holds a proton readily, whereas a *weak base* has a low affinity for a proton and establishes an equilibrium in aqueous solution. The two substances that are related by the gain or loss of a proton are called a *conjugate acid–base pair*. The exact strength of an acid is defined by an *acid dissociation constant*, K_a:

For the reaction $HA + H_2O \rightleftharpoons H_3O^+ + A^-$

we have $K_a = \dfrac{[H_3O^+][A^-]}{[HA]}$

A proton-transfer reaction always takes place in the direction that favors formation of the weaker acid (see Problems 34–36, 38–41, 44–55, 58–65, 99, 104, 108).

3. What is the ion-product constant for water? Water is *amphoteric*; that is, it can act as either an acid or a base. Water also dissociates slightly into H_3O^+ ions and OH^- ions; the product of whose concentrations in any aqueous solution is the *ion-product constant for water*, $K_w = [H_3O^+][OH^-] = 1.00 \times 10^{-14}$ at 25 °C (see Problems 56, 69–71, 101).

4. What is the pH scale for measuring acidity? The acidity or basicity of an aqueous solution is given by its *pH*, defined as the negative logarithm of the hydronium ion concentration, $[H_3O^+]$. A pH below 7 means an acidic solution; a pH equal to 7 means a neutral solution; and a pH above 7 means a basic solution (see Problems 57, 61–71, 76, 78, 94, 96–101, 104, 110).

5. What is a buffer? The pH of a solution can be controlled through the use of a *buffer* that acts to remove either added H_3O^+ ions or added OH^- ions. Most buffer solutions consist of roughly equal amounts of a weak acid and its conjugate base. The bicarbonate buffer present in blood and the hydrogen phosphate buffer present in cells are particularly important examples (see Problems 72–79, 105, 107).

6. How is the acid or base concentration of a solution determined? Acid (or base) concentrations are determined in the laboratory by *titration* of a solution of unknown concentration with a base (or acid) solution of known strength until an indicator signals that neutralization is complete (see Problems 80–93, 103, 106, 109, 110).

CONCEPT MAP: ACIDS AND BASES

Acids and Bases

Definitions
Arrhenius (Chs. 3, 5)
  Acids produce H_3O^+ ions in water
  Bases produce OH^- ions in water
Brønsted–Lowry
  Acids are proton donors
  Bases are proton acceptors

Concentrations of Acids/Bases
pH scale (pH = –log($[H_3O^+]$))
Molarity (mol/L) vs. Normality (Eq/L) (Ch. 9)
Determined by titrations
Neturalization reaction (Chs. 3, 5)
$N_{acid} \times V_{acid} = N_{base} \times V_{base}$

Strength of Acids/Bases
Strong acids/bases = 100% ionized (Ch. 3)
Weak acids/bases
  Reversible reactions (Ch. 7)
    Extent of ionization indicated by K_a (Ch. 7)
  Conjugate acid/base pairs
  Acid/base behavior of salts (basic anions, acidic cations)
Buffers
  Contains weak acid + conjugate base
  (or weak base + conjugate acid)
  Resists changes in pH when other acides or bases are added
Biochemical applications (Chs. 18, 19, 29)

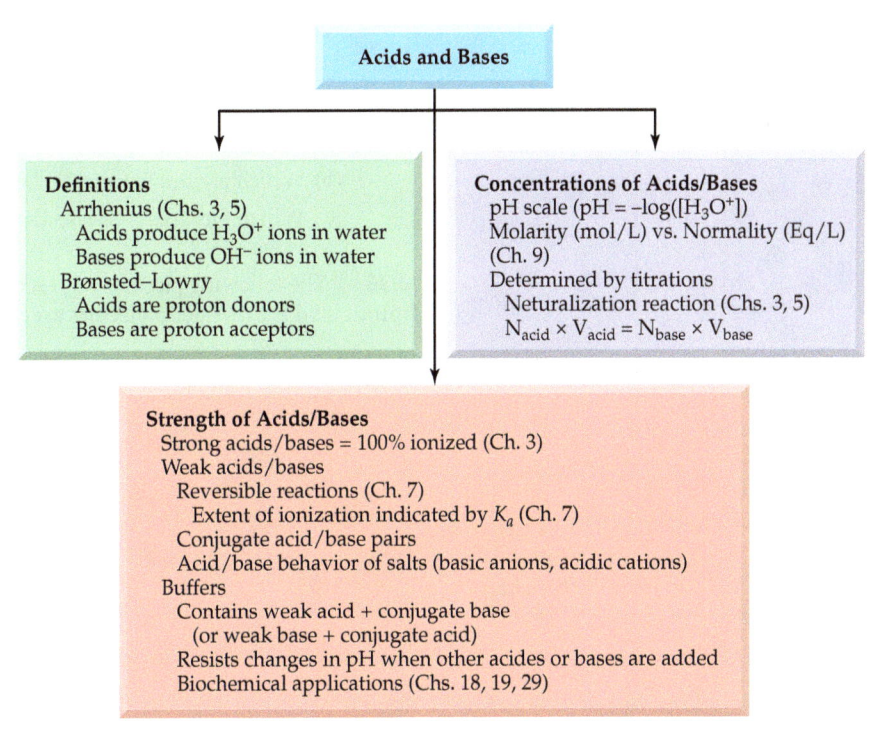

▲ Figure 10.9

Acids and bases play important roles in many chemical and biochemical processes, and many common substances are classified as acids or bases. Acid and base behavior is related to the ability to exchange protons, or to form H_3O^+ or OH^- ions, respectively, in water. Strong acids and bases ionize completely in aqueous solution, while weak acids/bases ionize only partially and establish an equilibrium with their conjugates. The relationship between these concepts and some of their practical and/or quantitative applications are illustrated in Figure 10.9.

324 CHAPTER 10 Acids and Bases

KEY WORDS

Acid dissociation constant (K_a), p. 301
Acid–base indicator, p. 308
Amphoteric, p. 302
Brønsted–Lowry acid, p. 293
Brønsted–Lowry base, p. 293
Buffer, p. 308
Conjugate acid, p. 295
Conjugate acid–base pair, p. 295
Conjugate base, p. 295

Dissociation, p. 296
Equivalent of acid, p. 314
Equivalent of base, p. 314
Gram-equivalent of acid, p. 314
Gram-equivalent of base, p. 314
Henderson–Hasselbalch equation, p. 309
Hydronium ion, p. 291
Ion-product constant for water (K_w), p. 302

Normality (N), p. 314
p function, p. 304
pH, p. 304
Strong acid, p. 296
Strong base, p. 297
Titration, p. 317
Weak acid, p. 296
Weak base, p. 297

UNDERSTANDING KEY CONCEPTS

10.33 An aqueous solution of OH⁻, represented as a blue sphere, is allowed to mix with a solution of an acid H_nA, represented as a red sphere. Three possible outcomes are depicted by boxes (1)–(3), where the green spheres represent A^{n-}, the anion of the acid:

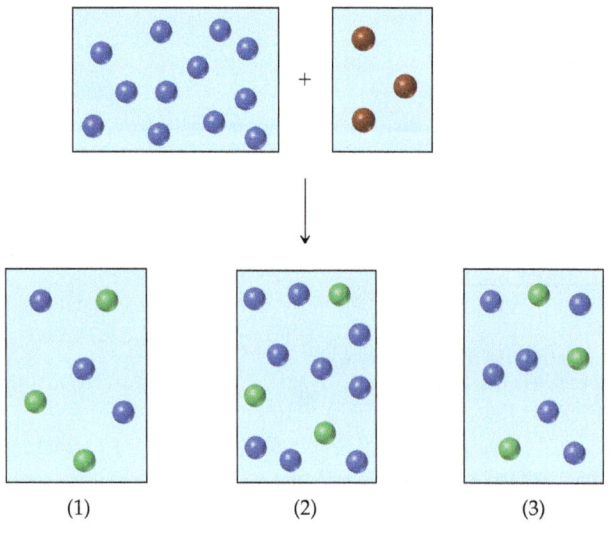

Which outcome corresponds to the following reactions?

(a) $HF + OH^- \longrightarrow H_2O + F^-$
(b) $H_2SO_3 + 2\ OH^- \longrightarrow 2\ H_2O + SO_3^{2-}$
(c) $H_3PO_4 + 3\ OH^- \longrightarrow 3\ H_2O + PO_4^{3-}$

10.34 Electrostatic potential maps of acetic acid (CH_3CO_2H) and ethyl alcohol (CH_3CH_2OH) are shown. Identify the most acidic hydrogen in each, and tell which of the two is likely to be the stronger acid.

CH₃CO₂H   CH₃CH₂OH

10.35 The following pictures represent aqueous acid solutions. Water molecules are not shown.

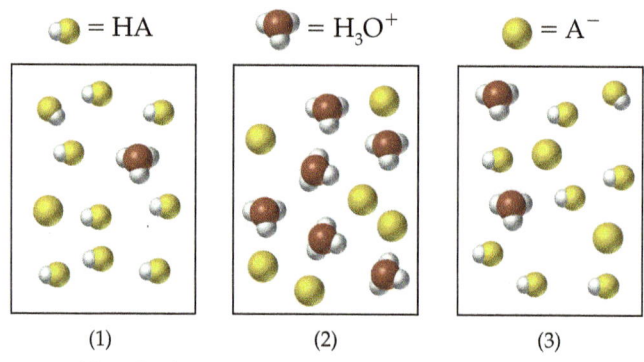

○ = HA   ● = H₃O⁺   ○ = A⁻

(a) Which picture represents the weakest acid?
(b) Which picture represents the strongest acid?
(c) Which picture represents the acid with the smallest value of K_a?

10.36 The following pictures represent aqueous solutions of a diprotic acid H_2A. Water molecules are not shown.

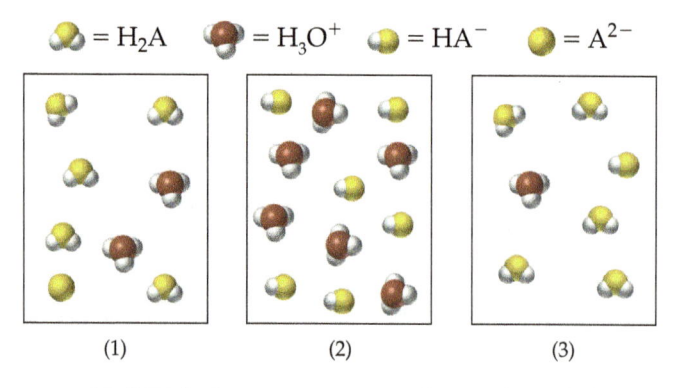

○ = H₂A   ● = H₃O⁺   ○ = HA⁻   ○ = A²⁻

(a) Which picture represents a solution of a weak diprotic acid?
(b) Which picture represents an impossible situation?

10.37 Assume that the red spheres in the buret represent H_3O^+ ions, the blue spheres in the flask represent OH^- ions, and you are carrying out a titration of the base with the acid. If the volumes in the buret and the flask are identical and the concentration of the acid in the buret is 1.00 M, what is the concentration of the base in the flask?

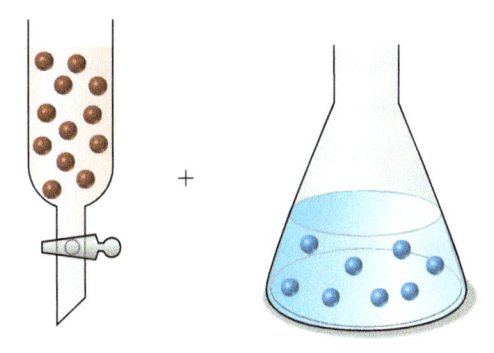

ADDITIONAL PROBLEMS

ACIDS AND BASES

10.38 What happens when a strong acid such as HBr is dissolved in water?

10.39 What happens when a weak acid such as CH_3CO_2H is dissolved in water?

10.40 What happens when a strong base such as KOH is dissolved in water?

10.41 What happens when a weak base such as NH_3 is dissolved in water?

10.42 What is the difference between a monoprotic acid and a diprotic acid? Give an example of each.

10.43 What is the difference between H^+ and H_3O^+?

10.44 Which of the following are strong acids? Look at Table 10.1 if necessary.

(a) $HClO_4$ (b) H_2CO_3 (c) H_3PO_4
(d) NH_4^+ (e) HI (f) $H_2PO_4^-$

10.45 Which of the following are weak bases? Look at Table 10.1 if necessary.

(a) NH_3 (b) $Ca(OH)_2$ (c) HPO_4^{2-}
(d) LiOH (e) CN^- (f) NH_2^-

BRØNSTED–LOWRY ACIDS AND BASES

10.46 Identify the following substances as a Brønsted–Lowry base, a Brønsted–Lowry acid, or neither:

(a) HCN (b) $CH_3CO_2^-$ (c) $AlCl_3$
(d) H_2CO_3 (e) Mg^{2+} (f) $CH_3NH_3^+$

10.47 Label the Brønsted–Lowry acids and bases in the following equations, and tell which substances are conjugate acid–base pairs.

(a) $CO_3^{2-}(aq) + HCl(aq) \longrightarrow$
$HCO_3^-(aq) + Cl^-(aq)$
(b) $H_3PO_4(aq) + NH_3(aq) \longrightarrow$
$H_2PO_4^-(aq) + NH_4^+(aq)$
(c) $NH_4^+(aq) + CN^-(aq) \rightleftharpoons NH_3(aq) + HCN(aq)$
(d) $HBr(aq) + OH^-(aq) \longrightarrow H_2O(l) + Br^-(aq)$
(e) $H_2PO_4^-(aq) + N_2H_4(aq) \rightleftharpoons$
$HPO_4^{2-}(aq) + N_2H_5^+(aq)$

10.48 Write the formulas of the conjugate acids of the following Brønsted–Lowry bases:

(a) $ClCH_2CO_2^-$ (b) C_5H_5N
(c) SeO_4^{2-} (d) $(CH_3)_3N$

10.49 Write the formulas of the conjugate bases of the following Brønsted–Lowry acids:

(a) HCN (b) $(CH_3)_2NH_2^+$
(c) H_3PO_4 (d) $HSeO_3^-$

10.50 The hydrogen-containing anions of many polyprotic acids are amphoteric. Write equations for HCO_3^- and $H_2PO_4^-$ acting as bases with the strong acid HCl and as acids with the strong base NaOH.

10.51 Write balanced equations for proton-transfer reactions between the listed pairs. Indicate the conjugate pairs, and determine the favored direction for each equilibrium.

(a) HCl and PO_4^{3-} (b) HCN and SO_4^{2-}
(c) $HClO_4$ and NO_2^- (d) CH_3O^- and HF

10.52 Sodium bicarbonate ($NaHCO_3$), also known as baking soda, is a common home remedy for acid indigestion and is also used to neutralize acid spills in the laboratory. Write a balanced chemical equation for the reaction of sodium bicarbonate with

(a) Gastric juice (HCl) (b) Sulfuric acid (H_2SO_4)

10.53 Refer to Section 10.12 to write balanced equations for the following acid–base reactions:

(a) $LiOH + HNO_3 \longrightarrow$ (b) $BaCO_3 + HI \longrightarrow$
(c) $H_3PO_4 + KOH \longrightarrow$ (d) $Ca(HCO_3)_2 + HCl \longrightarrow$
(e) $Ba(OH)_2 + H_2SO_4 \longrightarrow$

ACID AND BASE STRENGTH: K_a AND pH

10.54 How is K_a defined? Write the equation for K_a for the generalized acid HA.

10.55 Rearrange the equation you wrote in Problem 10.54 to solve for $[H_3O^+]$ in terms of K_a.

10.56 How is K_w defined, and what is its numerical value at 25 °C?

10.57 How is pH defined?

10.58 A solution of 0.10 M HCl has a pH = 1.00, whereas a solution of 0.10 M CH_3COOH has a pH = 2.88. Explain.

10.59 Calculate $[H_3O^+]$ for the 0.10 M CH_3COOH solution in Problem 10.58. What percent of the weak acid is dissociated?

10.60 Write the expressions for the acid dissociation constants for the three successive dissociations of phosphoric acid, H_3PO_4, in water.

10.61 Based on the K_a values in Table 10.1, rank the following solutions in order of increasing pH: 0.10 M HCOOH, 0.10 M HF, 0.10 M H_2CO_3, 0.10 M HSO_4^-, 0.10 M NH_4^+.

10.62 The electrode of a pH meter is placed in a sample of urine, and a reading of 7.9 is obtained. Is the sample acidic, basic, or neutral? What is the concentration of H_3O^+ in the urine sample?

10.63 A 0.10 M solution of the deadly poison hydrogen cyanide, HCN, has a pH of 5.2. Calculate the $[H_3O^+]$ of the solution. Is HCN a strong or a weak acid?

10.64 Human sweat can have a pH ranging from 4.0–6.8. Calculate the range of $[H_3O^+]$ in normal human sweat. How many orders of magnitude does this range represent?

10.65 Saliva has a pH range of 5.8–7.1. Approximately what is the H_3O^+ concentration range of saliva?

10.66 What is the approximate pH of a 0.02 M solution of a strong monoprotic acid? Of a 0.02 M solution of a strong base, such as KOH?

10.67 Calculate the pOH of each solution in Problems 10.62–10.65.

10.68 Without using a calculator, match the H_3O^+ concentrations of the following solutions, (a)–(d), to the corresponding pH, i–iv:

(a) Fresh egg white: $[H_3O^+] = 2.5 \times 10^{-8}$ M
(b) Apple cider: $[H_3O^+] = 5.0 \times 10^{-4}$ M
(c) Household ammonia: $[H_3O^+] = 2.3 \times 10^{-12}$ M
(d) Vinegar (acetic acid): $[H_3O^+] = 4.0 \times 10^{-3}$ M

i. pH = 3.30   ii. pH = 2.40   iii. pH = 11.64   iv. pH = 7.60

10.69 What are the OH^- concentration and pOH for each solution in Problem 10.68? Rank the solutions according to increasing acidity.

10.70 What are the H_3O^+ and OH^- concentrations of solutions that have the following pH values?

(a) pH 4   (b) pH 11   (c) pH 0
(d) pH 1.38   (e) pH 7.96

10.71 About 12% of the acid in a 0.10 M solution of a weak acid dissociates to form ions. What are the H_3O^+ and OH^- concentrations? What is the pH of the solution?

BUFFERS

10.72 What are the two components of a buffer system? How does a buffer work to hold pH nearly constant?

10.73 Which system would you expect to be a better buffer: HNO_3 + $Na^+NO_3^-$, or CH_3CO_2H + $CH_3CO_2^-Na^+$? Explain.

10.74 The pH of a buffer solution containing 0.10 M acetic acid and 0.10 M sodium acetate is 4.74.

(a) Write the Henderson–Hasselbalch equation for this buffer.
(b) Write the equations for reaction of this buffer with a small amount of HNO_3 and with a small amount of NaOH.

10.75 Which of the following buffer systems would you use if you wanted to prepare a solution having a pH of approximately 9.5?

(a) 0.08 M $H_2PO_4^-$ / 0.12 M HPO_4^{2-}
(b) 0.08 M NH_4^+ / 0.12 M NH_3

10.76 What is the pH of a buffer system that contains 0.200 M hydrocyanic acid (HCN) and 0.150 M sodium cyanide (NaCN)? The pK_a of hydrocyanic acid is 9.31.

10.77 Consider 1.00 L of the buffer system described in Problem 10.76.

(a) What are the [HCN] and [CN^-] after 0.020 mol of HCl is added? What is the pH?
(b) What are the [HCN] and [CN^-] after 0.020 mol of NaOH is added? What is the pH?

10.78 What is the pH of a buffer system that contains 0.15 M NH_4^+ and 0.10 M NH_3? The pK_a of NH_4^+ is 9.25.

10.79 How many moles of NaOH must be added to 1.00 L of the solution described in Problem 10.78 to increase the pH to 9.25? (Hint: What is the $[NH_3]/[NH_4^+]$ when the pH = pK_a?)

CONCENTRATIONS OF ACID AND BASE SOLUTIONS

10.80 What does it mean when we talk about acid *equivalents* and base *equivalents*?

10.81 How does normality compare to molarity for monoprotic and polyprotic acids??

10.82 Calculate the gram-equivalent for each of the following acids and bases.

(a) HNO_3 (b) H_3PO_4 (c) KOH   (d) $Mg(OH)_2$

10.83 What mass of each of the acids and bases in Problem 10.82 is needed to prepare 500 mL of 0.15 N solution?

10.84 How many milliliters of 0.0050 N KOH are required to neutralize 25 mL of 0.0050 N H_2SO_4? To neutralize 25 mL of 0.0050 M H_2SO_4?

10.85 How many equivalents are in 75.0 mL of 0.12 M H_2SO_4 solution? In 75.0 mL of a 0.12 M H_3PO_4 solution?

10.86 How many equivalents of an acid or base are in the following?

(a) 0.25 mol $Mg(OH)_2$
(b) 2.5 g $Mg(OH)_2$
(c) 15 g CH_3CO_2H

10.87 What mass of citric acid (triprotic, $C_6H_5O_7H_3$) contains 152 mEq of citric acid?

10.88 What are the molarity and the normality of a solution made by dissolving 5.0 g of $Ca(OH)_2$ in enough water to make 500.0 mL of solution?

10.89 What are the molarity and the normality of a solution made by dissolving 25 g of citric acid (triprotic, $C_6H_5O_7H_3$) in enough water to make 800 mL of solution?

10.90 Titration of a 12.0 mL solution of HCl requires 22.4 mL of 0.12 M NaOH. What is the molarity of the HCl solution?

10.91 How many equivalents are in 15.0 mL of 0.12 M Ba(OH)$_2$ solution? What volume of 0.085 M HNO$_3$ is required to reach the end point when titrating 15.0 mL of this solution?

10.92 Titration of a 10.0 mL solution of KOH requires 15.0 mL of 0.0250 M H$_2$SO$_4$ solution. What is the molarity of the KOH solution?

10.93 If 35.0 mL of a 0.100 N acid solution is needed to reach the end point in titration of 21.5 mL of a base solution, what is the normality of the base solution?

CHEMISTRY IN ACTION

10.94 The concentration of HCl when released to the stomach cavity is diluted to between 0.01 and 0.001 M [*GERD—Too Much Acid or Not Enough?* p. 299]

(a) What is the pH range in the stomach cavity?

(b) Write a balanced equation for the neutralization of stomach acid by NaHCO$_3$.

(c) How many grams of NaHCO$_3$ are required to neutralize 15.0 mL of a solution having a pH of 1.8?

10.95 What are the functions of the acidic gastric juices in the stomach? [*GERD—Too Much Acid or Not Enough?* p. 299]

10.96 Metabolic acidosis is often treated by administering bicarbonate intravenously. Explain how this treatment can increase blood serum pH. [*Buffers in the Body: Acidosis and Alkalosis,* p. 312]

10.97 Which body fluid is most acidic? Which is most basic? [*Buffers in the Body: Acidosis and Alkalosis,* p. 312]

10.98 Rain typically has a pH of about 5.6. What is the H$_3$O$^+$ concentration in rain? [*Acid Rain,* p. 320]

10.99 Acid rain with a pH as low as 1.5 has been recorded in West Virginia. [*Acid Rain,* p. 320]

(a) What is the H$_3$O$^+$ concentration in this acid rain?

(b) How many grams of HNO$_3$ must be dissolved to make 25 L of solution that has a pH of 1.5?

GENERAL QUESTIONS AND PROBLEMS

10.100 A solution is prepared by bubbling 15.0 L of HCl(g) at 25 °C and 1 atm into 250.0 mL of water.

(a) Assuming all the HCl dissolves in the water, how many moles of HCl are in solution?

(b) What is the pH of the solution?

10.101 The dissociation of water into H$_3$O$^+$ and OH$^-$ ions depends on temperature. At 0 °C the [H$_3$O$^+$] = 3.38 × 10^{-8} M, at 25 °C the [H$_3$O$^+$] = 1.00 × 10^{-7} M, and at 50 °C the [H$_3$O$^+$] = 2.34 × 10^{-7} M.

(a) Calculate the pH of water at 0 °C and 50 °C.

(b) What is the value of K_w at 0 °C and 50 °C?

(c) Is the dissociation of water endothermic or exothermic?

10.102 Alka-Seltzer™, a drugstore antacid, contains a mixture of NaHCO$_3$, aspirin, and citric acid, C$_6$H$_5$O$_7$H$_3$. Why does Alka-Seltzer™ foam and bubble when dissolved in water? Which ingredient is the antacid?

10.103 How many milliliters of 0.50 M NaOH solution are required to titrate 40.0 mL of a 0.10 M H$_2$SO$_4$ solution to an end point?

10.104 Which solution contains more acid, 50 mL of a 0.20 N HCl solution or 50 mL of a 0.20 N acetic acid solution? Which has a higher hydronium ion concentration? Which has a lower pH?

10.105 One of the buffer systems used to control the pH of blood involves the equilibrium between H$_2$PO$_4^-$ and HPO$_4^{2-}$. The pK_a for H$_2$PO$_4^-$ is 7.21.

(a) Write the Henderson–Hasselbalch equation for this buffer system.

(b) What HPO$_4^{2-}$ to H$_2$PO$_4^-$ ratio is needed to maintain the optimum blood pH of 7.40?

10.106 A 0.15 N solution of HCl is used to titrate 30.0 mL of a Ca(OH)$_2$ solution of unknown concentration. If 140.0 mL of HCl is required, what is the normality of the Ca(OH)$_2$ solution? What is the molarity?

10.107 Which of the following combinations produces an effective buffer solution? Assuming equal concentrations of each acid and its conjugate base, calculate the pH of each buffer solution.

(a) NaF and HF   (b) HClO$_4$ and NaClO$_4$

(c) NH$_4$Cl and NH$_3$ (d) KBr and HBr

10.108 One method of analyzing ammonium salts is to treat them with NaOH and then heat the solution to remove the NH$_3$ gas formed.

$$NH_4^+(aq) + OH^-(aq) \longrightarrow NH_3(g) + H_2O(l)$$

(a) Label the Brønsted–Lowry acid–base pairs.

(b) If 2.86 L of NH$_3$ at 60 °C and 755 mmHg is produced by the reaction of NH$_4$Cl, how many grams of NH$_4$Cl were in the original sample?

10.109 One method of reducing acid rain is "scrubbing" the combustion products before they are emitted from power plant smoke stacks. The process involves addition of an aqueous suspension of lime (CaO) to the combustion chamber and stack, where the lime reacts with SO$_2$ to give calcium sulfite (CaSO$_3$):

$$CaO(aq) + SO_2(g) \longrightarrow CaSO_3(aq)$$

(a) How much lime (in g) is needed to remove 1 mol of SO$_2$?

(b) How much lime (in kg) is needed to remove 1 kg of SO$_2$?

10.110 Sodium oxide, Na$_2$O, reacts with water to give NaOH.

(a) Write a balanced equation for the reaction.

(b) What is the pH of the solution prepared by allowing 1.55 g of Na$_2$O to react with 500.0 mL of water? Assume that there is no volume change.

(c) How many milliliters of 0.0100 M HCl are needed to neutralize the NaOH solution prepared in (b)?

CHAPTER 11

Nuclear Chemistry

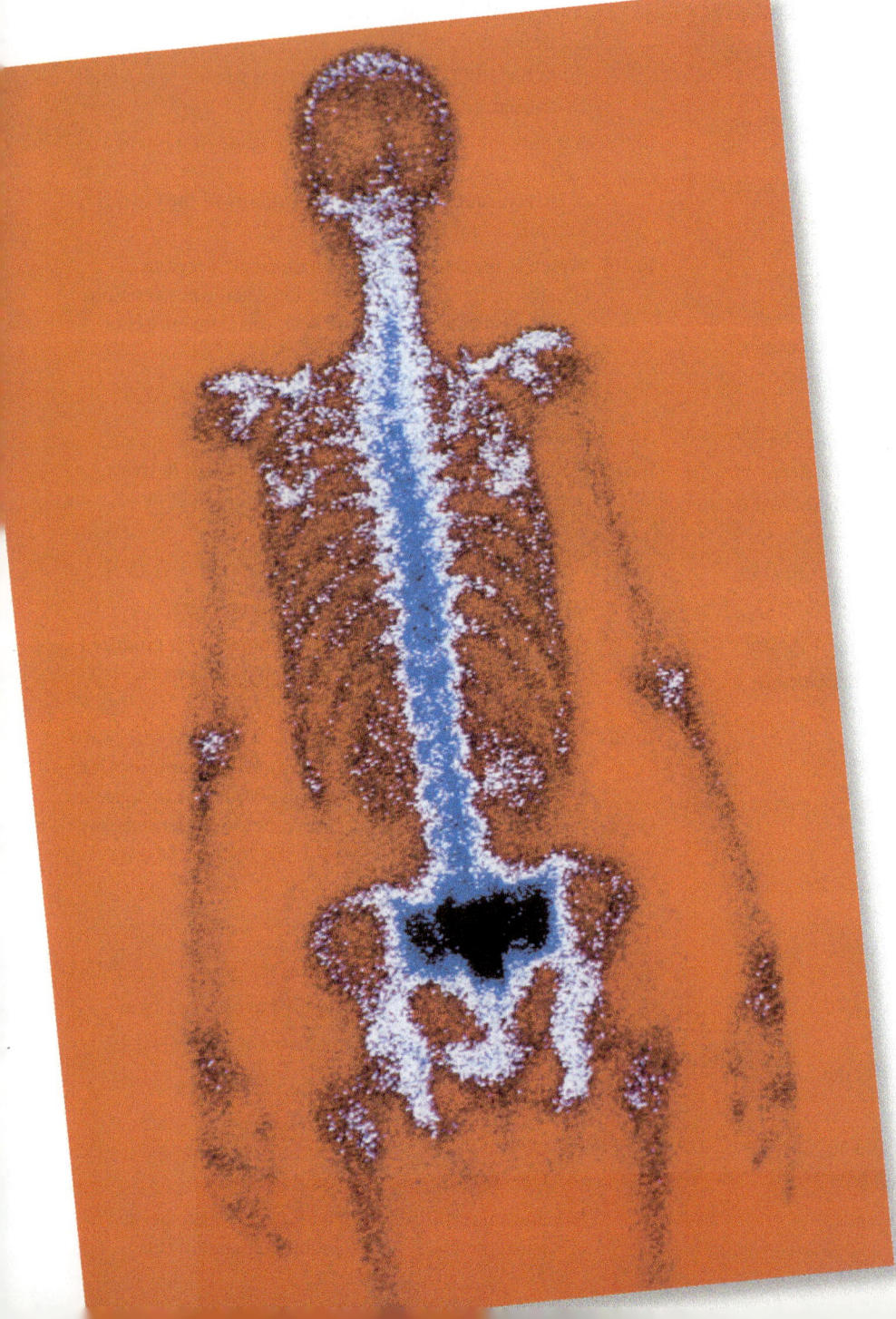

CONTENTS

- **11.1** Nuclear Reactions
- **11.2** The Discovery and Nature of Radioactivity
- **11.3** Stable and Unstable Isotopes
- **11.4** Nuclear Decay
- **11.5** Radioactive Half-Life
- **11.6** Radioactive Decay Series
- **11.7** Ionizing Radiation
- **11.8** Detecting Radiation
- **11.9** Measuring Radiation
- **11.10** Artificial Transmutation
- **11.11** Nuclear Fission and Nuclear Fusion

◀ Many medical diagnostic techniques, including this total body bone scan, take advantage of the properties of radioactive isotopes.

CHAPTER GOALS

1. **What is a nuclear reaction, and how are equations for nuclear reactions balanced?**
   THE GOAL: Be able to write and balance equations for nuclear reactions. (◂◂ A, B, C.)

2. **What are the different kinds of radioactivity?**
   THE GOAL: Be able to list the characteristics of three common kinds of radiation—α, β, and γ (alpha, beta, and gamma).

3. **How are the rates of nuclear reactions expressed?**
   THE GOAL: Be able to explain half-life and calculate the quantity of a radioisotope remaining after a given number of half-lives.

4. **What is ionizing radiation?**
   THE GOAL: Be able to describe the properties of the different types of ionizing radiation and their potential for harm to living tissue.

5. **How is radioactivity measured?**
   THE GOAL: Be able to describe the common units for measuring radiation.

6. **What is transmutation?**
   THE GOAL: Be able to explain nuclear bombardment and balance equations for nuclear bombardment reactions. (◂◂ A, B, and C.)

7. **What are nuclear fission and nuclear fusion?**
   THE GOAL: Be able to explain nuclear fission and nuclear fusion.

CONCEPTS TO REVIEW

A. Atomic Theory
(Section 2.1)

B. Elements and Atomic Number
(Section 2.2)

C. Isotopes
(Section 2.3)

In all of the reactions we have discussed thus far, only the *bonds* between atoms have changed; the chemical identities of atoms themselves have remained unchanged. Anyone who reads the paper or watches television knows, however, that atoms *can* change, often resulting in the conversion of one element into another. Atomic weapons, nuclear energy, and radioactive radon gas in our homes are all topics of societal importance, and all involve *nuclear chemistry*—the study of the properties and reactions of atomic nuclei.

11.1 Nuclear Reactions

Recall from Section 2.2 that an atom is characterized by its *atomic number, Z,* and its *mass number, A.* The atomic number, written below and to the left of the element symbol, gives the number of protons in the nucleus and identifies the element. The mass number, written above and to the left of the element symbol, gives the total number of **nucleons**, a general term for both protons (p) and neutrons (n). The most common isotope of carbon, for example, has 12 nucleons: 6 protons and 6 neutrons: $^{12}_{6}C$.

Mass number $\rightarrow$ $^{12}_{6}C$ $\leftarrow$ Atomic number
Carbon-12

6 protons
6 neutrons
12 nucleons

Nucleon A general term for both protons and neutrons.

Atoms with identical atomic numbers but different mass numbers are called *isotopes*, and the nucleus of a specific isotope is called a **nuclide**. Thirteen isotopes of carbon are known—two occur commonly (^{12}C and ^{13}C) and one (^{14}C) is produced in small amounts in the upper atmosphere by the action of neutrons from cosmic rays on ^{14}N. The remaining 10 carbon isotopes have been produced artificially. Only the two commonly occurring isotopes are stable indefinitely; the others undergo spontaneous **nuclear reactions**, which change their nuclei. Carbon-14, for example, is an unstable isotope that slowly decomposes and is converted to nitrogen-14 plus an electron, a process we can write as

$$^{14}_{6}C \longrightarrow \,^{14}_{7}N + \,^{0}_{-1}e$$

The electron is often written as $^{0}_{-1}e$, where the superscript 0 indicates that the mass of an electron is essentially zero when compared with that of a proton or neutron, and the subscript -1 indicates that the charge is -1. (The subscript in this instance is not

▸▸ The different isotopes of an atom each have the same number of protons and only differ in their number of neutrons (Section 2.3).

Nuclide The nucleus of a specific isotope of an element.

Nuclear reaction A reaction that changes an atomic nucleus, usually causing the change of one element into another.

a true atomic number; in Section 11.4 the purpose of representing the electron this way will become clear.)

Nuclear reactions, such as the spontaneous decay of ^{14}C, are distinguished from chemical reactions in several ways:

- A *nuclear* reaction involves a change in an atom's nucleus, usually producing a different element. A *chemical* reaction, by contrast, involves only a change in distribution of the outer-shell electrons around the atom and never changes the nucleus itself or produces a different element.
- Different isotopes of an element have essentially the same behavior in chemical reactions but often have completely different behavior in nuclear reactions.
- The rate of a nuclear reaction is unaffected by a change in temperature or pressure or by the addition of a catalyst.
- The nuclear reaction of an atom is essentially the same whether it is in a chemical compound or in an uncombined, elemental form.
- The energy change accompanying a nuclear reaction can be up to several million times greater than that accompanying a chemical reaction. The nuclear transformation of 1.0 g of uranium-235 releases 3.4×10^8 kcal (1.4×10^9 kJ), for example, whereas the chemical combustion of 1.0 g of methane releases only 12 kcal (50 kJ).

11.2 The Discovery and Nature of Radioactivity

The discovery of *radioactivity* dates to the year 1896 when the French physicist Henri Becquerel made a remarkable observation. While investigating the nature of phosphorescence—the luminous glow of some minerals and other substances that remains when the light is suddenly turned off—Becquerel happened to place a sample of a uranium-containing mineral on top of a photographic plate that had been wrapped in black paper and put in a drawer to protect it from sunlight. On developing the plate, Becquerel was surprised to find a silhouette of the mineral. He concluded that the mineral was producing some kind of unknown radiation, which passed through the paper and exposed the photographic plate.

Marie Sklodowska Curie and her husband, Pierre, took up the challenge and began a series of investigations into this new phenomenon, which they termed **radioactivity**. They found that the source of the radioactivity was the element uranium (U) and that two previously unknown elements, which they named polonium (Po) and radium (Ra), were also radioactive. For these achievements, Becquerel and the Curies shared the 1903 Nobel Prize in physics.

Further work on radioactivity by the English scientist Ernest Rutherford established that there were at least two types of radiation, which he named *alpha* (α) and *beta* (β) after the first two letters of the Greek alphabet. Shortly thereafter, a third type of radiation was found and named for the third Greek letter, *gamma* (γ).

Subsequent studies showed that when the three kinds of radiation are passed between two plates with opposite electrical charges, each is affected differently. Alpha radiation bends toward the negative plate and must therefore have a positive charge. Beta radiation, by contrast, bends toward the positive plate and must have a negative charge, whereas gamma radiation does not bend toward either plate and has no charge (Figure 11.1).

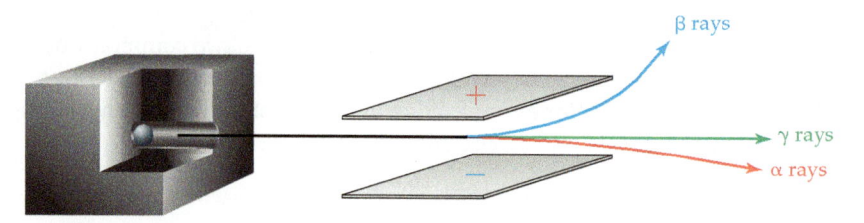

Radioactivity The spontaneous emission of radiation from a nucleus.

▲ **Figure 11.1**
The effect of an electric field on α, β, and γ, radiation.
The radioactive source in the shielded box emits radiation, which passes between the two electrically charged plates. Alpha radiation is deflected toward the negative plate, β radiation is deflected toward the positive plate, and γ radiation is not deflected.

Another difference among the three kinds of radiation soon became apparent when it was discovered that alpha and beta radiations are composed of small particles with a measurable mass, whereas **gamma (γ) radiation** consists of high-energy electromagnetic waves and has no mass. Rutherford was able to show that a **beta (β) particle** is an electron (e^-) and that an **alpha (α) particle** is actually a helium nucleus, He^{2+}. (Recall that a helium *atom* consists of two protons, two neutrons, and two electrons. When the two electrons are removed, the remaining helium nucleus, or α particle, has only the two protons and two neutrons.)

Yet a third difference among the three kinds of radiation is their penetrating power. Because of their relatively large mass, α particles move slowly (up to about 1/10 the speed of light) and can be stopped by a few sheets of paper or by the top layer of skin. Beta particles, because they are much lighter, move at up to 9/10 the speed of light and have about 100 times the penetrating power of α particles. A block of wood or heavy protective clothing is necessary to stop β radiation, which can otherwise penetrate the skin and cause burns and other damage. Gamma rays move at the speed of light (3.00×10^8 m/s) and have about 1000 times the penetrating power of α particles. A lead block several inches thick is needed to stop γ radiation, which can otherwise penetrate and damage the body's internal organs.

The characteristics of the three kinds of radiation are summarized in Table 11.1. Note that an α particle, even though it is an ion with a +2 charge, is usually written using the symbol ^4_2He without the charge. A β particle is usually written $^0_{-1}e$, as noted previously.

Gamma (γ) radiation Radioactivity consisting of high-energy light waves.

▶▶ See Chemistry in Action: Atoms and Light on p. 66 in Chapter 2 for a discussion of gamma rays and the rest of the electromagnetic spectrum.

Beta (β) particle An electron (e^-), emitted as radiation.

Alpha (α) particle A helium nucleus (He^{2+}), emitted as α radiation.

TABLE 11.1 Characteristics of α, β, and γ Radiation

Type of Radiation	Symbol	Charge	Composition	Mass (AMU)	Velocity	Relative Penetrating Power
Alpha	α, ^4_2He	+2	Helium nucleus	4	Up to 10% speed of light	Low (1)
Beta	β, $^0_{-1}e$	−1	Electron	1/1823	Up to 90% speed of light	Medium (100)
Gamma	γ, $^0_0γ$	0	High-energy radiation	0	Speed of light (3.00×10^8 m/s)	High (1000)

11.3 Stable and Unstable Isotopes

Every element in the periodic table has at least one radioactive isotope, or **radioisotope**, and more than 3300 radioisotopes are known. Their radioactivity is the result of having unstable nuclei, although the exact causes of this instability are not fully understood. Radiation is emitted when an unstable radioactive nucleus, or **radionuclide**, spontaneously changes into a more stable one.

For elements in the first few rows of the periodic table, stability is associated with a roughly equal number of neutrons and protons (Figure 11.2). Hydrogen, for example, has stable ^1_1H (protium) and ^2_1H (deuterium) isotopes, but its ^3_1H isotope (tritium) is radioactive. As elements get heavier, the number of neutrons relative to protons in stable nuclei increases. Lead-208 ($^{208}_{82}Pb$), for example, the most abundant stable isotope of lead, has 126 neutrons and 82 protons in its nuclei. Nevertheless, of the 35 known isotopes of lead, only 3 are stable whereas 32 are radioactive. In fact, there are only 264 stable isotopes among all the elements. All isotopes of elements with atomic numbers higher than that of bismuth (83) are radioactive.

Most of the more than 3300 known radioisotopes have been made in high-energy particle accelerators by reactions that will be described in Section 11.10. Such isotopes are called *artificial radioisotopes* because they are not found in nature. All isotopes of the transuranium elements (those heavier than uranium) are artificial. The much smaller number of radioactive isotopes found in Earth's crust, such as $^{238}_{92}U$, are called *natural radioisotopes*.

Radioisotope A radioactive isotope.

Radionuclide The nucleus of a radioactive isotope.

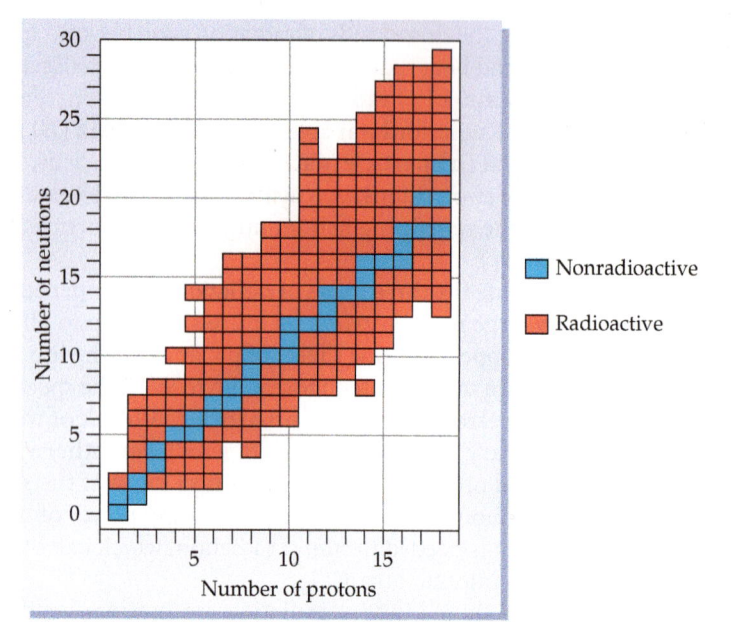

▲ **Figure 11.2**
A plot of the numbers of neutrons and protons for known isotopes of the first 18 elements. Stable (nonradioactive) isotopes of these elements have equal or nearly equal numbers of neutrons and protons.

Aside from their radioactivity, different radioisotopes of the same element have the same chemical properties as stable isotopes, which accounts for their great usefulness as *tracers*. A chemical compound tagged with a radioactive atom undergoes exactly the same reactions as its nonradioactive counterpart. The difference is that the tagged compound can be located with a radiation detector and its location determined, as discussed in Chemistry in Action: Body Imaging on page 348.

11.4 Nuclear Decay

Think for a minute about the consequences of α and β radiation. If radioactivity involves the spontaneous emission of a small particle from an unstable atomic nucleus, then the nucleus itself must undergo a change. With that understanding of radioactivity came the startling discovery that atoms of one element can change into atoms of another element, something that had previously been thought impossible. The spontaneous emission of a particle from an unstable nucleus is called **nuclear decay**, or *radioactive decay*, and the resulting change of one element into another is called **transmutation**.

Nuclear decay: Radioactive element ⟶ New element + Emitted particle

We now look at what happens to a nucleus when nuclear decay occurs.

Nuclear decay The spontaneous emission of a particle from an unstable nucleus.

Transmutation The change of one element into another.

Alpha Emission

When an atom of uranium-238 ($^{238}_{92}U$) emits an α particle, the nucleus loses 2 protons and 2 neutrons. Because the number of protons in the nucleus has now changed from 92 to 90, the *identity* of the atom has changed from uranium to thorium. Furthermore, since the total number of nucleons has decreased by 4, uranium-238 has become thorium-234 ($^{234}_{90}Th$) (Figure 11.3).

Note that the equation for a nuclear reaction is not balanced in the usual chemical sense because the kinds of atoms are not the same on both sides of the arrow. Instead, we say that a nuclear equation is balanced when the number of nucleons on both sides of the equation is the same and when the sums of the charges on the nuclei plus any ejected subatomic particles (protons or electrons) are the same on both sides of the

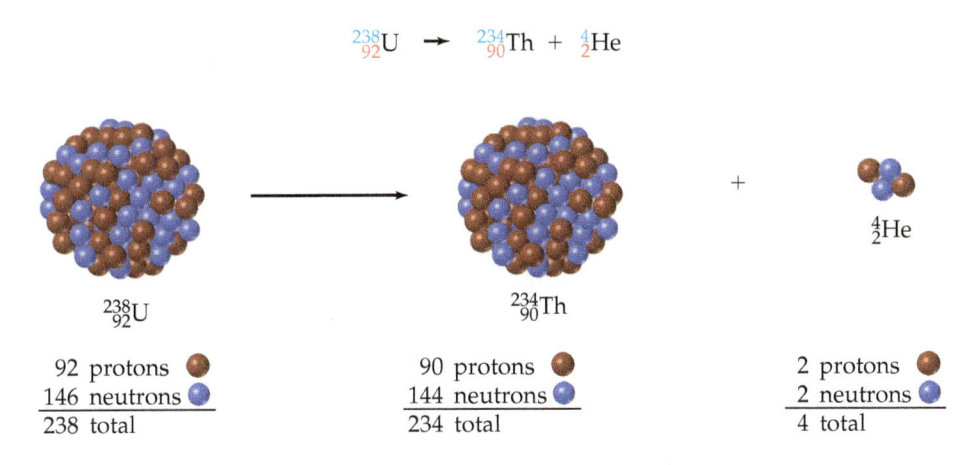

SECTION 11.4 Nuclear Decay   333

$$^{238}_{92}U \longrightarrow ^{234}_{90}Th + ^{4}_{2}He$$

◀ **Figure 11.3**
Alpha emission.
Emission of an α particle from an atom of uranium-238 produces an atom of thorium-234.

$^{238}_{92}U$
92 protons
146 neutrons
238 total

$^{234}_{90}Th$
90 protons
144 neutrons
234 total

$^{4}_{2}He$
2 protons
2 neutrons
4 total

equation. In the decay of $^{238}_{92}U$ to give $^{4}_{2}He$ and $^{234}_{90}Th$, for example, there are 238 nucleons and 92 nuclear charges on both sides of the nuclear equation.

92 protons
146 neutrons
238 nucleons
$\longrightarrow ^{238}_{92}U \longrightarrow ^{4}_{2}He + ^{234}_{90}Th$

90 protons
144 neutrons
234 nucleons

2 protons
2 neutrons
4 nucleons

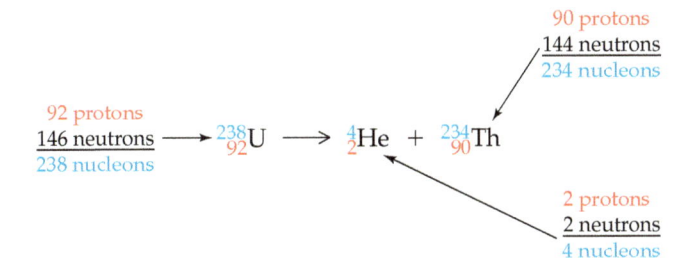

Worked Example 11.1 Balancing Nuclear Reactions: Alpha Emission

Polonium-208 is one of the α emitters studied by Marie Curie. Write the equation for the α decay of polonium-208, and identify the element formed.

ANALYSIS Look up the atomic number of polonium (84) in the periodic table, and write the known part of the nuclear equation, using the standard symbol for polonium-208:

$$^{208}_{84}Po \longrightarrow ^{4}_{2}He + \text{?}$$

Then, calculate the mass number and atomic number of the product element, and write the final equation.

SOLUTION
The mass number of the product is 208 − 4 = 204, and the atomic number is 84 − 2 = 82. A look at the periodic table identifies the element with atomic number 82 as lead (Pb).

$$^{208}_{84}Po \longrightarrow ^{4}_{2}He + ^{204}_{82}Pb$$

Check your answer by making sure that the mass numbers and atomic numbers on the two sides of the equation are balanced:

Mass numbers: 208 = 4 + 204   Atomic numbers: 84 = 2 + 82

PROBLEM 11.1
High levels of radioactive radon-222 ($^{222}_{86}Rn$) have been found in many homes built on radium-containing rock, leading to the possibility of health hazards. What product results from α emission by radon-222?

PROBLEM 11.2
What isotope of radium (Ra) is converted into radon-222 by α emission?

Beta Emission

Whereas α emission leads to the loss of two protons and two neutrons from the nucleus, β emission involves the *decomposition* of a neutron to yield an electron and a proton. This process can be represented as

$$^{1}_{0}n \longrightarrow \, ^{1}_{1}p + \, ^{0}_{-1}e$$

where the electron ($^{0}_{-1}e$) is ejected as a β particle, and the proton is retained by the nucleus. Note that the electrons emitted during β radiation come from the *nucleus* and not from the occupied orbitals surrounding the nucleus. The decomposition of carbon-14 to form nitrogen-14 in Section 11.1 is an example of beta decay.

The net result of β emission is that the atomic number of the atom *increases* by 1 because there is a new proton. The mass number of the atom remains the same, however, because a neutron has changed into a proton, leaving the total number of nucleons unchanged. For example, iodine-131 ($^{131}_{53}I$), a radioisotope used in detecting thyroid problems, undergoes nuclear decay by β emission to yield xenon-131 ($^{131}_{54}Xe$):

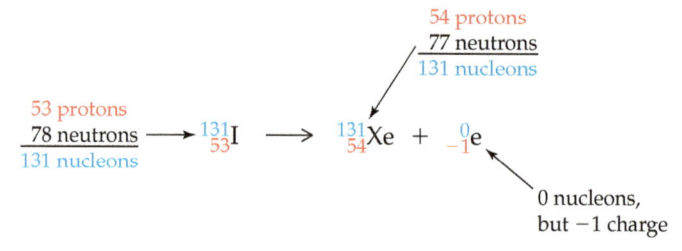

$$^{131}_{53}I \longrightarrow \, ^{131}_{54}Xe + \, ^{0}_{-1}e$$

(53 protons, 78 neutrons, 131 nucleons → 54 protons, 77 neutrons, 131 nucleons + 0 nucleons, but −1 charge)

Note that the superscripts (mass numbers) are balanced in this equation because a β particle has a mass near zero, and the subscripts are balanced because a β particle has a charge of −1.

Worked Example 11.2 Balancing Nuclear Reactions: Beta Emission

Write a balanced nuclear equation for the β decay of chromium-55.

ANALYSIS Write the known part of the nuclear equation:

$$^{55}_{24}Cr \longrightarrow \, ^{0}_{-1}e + \, ?$$

Then calculate the mass number and atomic number of the product element, and write the final equation.

SOLUTION
The mass number of the product stays at 55, and the atomic number increases by 1, 24 + 1 = 25, so the product is manganese-55.

$$^{55}_{24}Cr \longrightarrow \, ^{0}_{-1}e + \, ^{55}_{25}Mn$$

Check your answer by making sure that the mass numbers and atomic numbers on the two sides of the equation are balanced:

Mass numbers: 55 = 0 + 55   Atomic numbers: 24 = −1 + 25

PROBLEM 11.3
Strontium-89 is a short-lived β emitter often used in the treatment of bone tumors. Write a nuclear equation for the decay of strontium-89.

PROBLEM 11.4
Write nuclear equations for the formation of each of the following nuclides by β emission.

(a) $^{3}_{2}He$ (b) $^{210}_{83}Bi$ (c) $^{20}_{10}Ne$

Gamma Emission

Emission of γ rays, unlike the emission of α and β particles, causes no change in mass or atomic number because γ rays are simply high-energy electromagnetic waves. Although γ emission can occur alone, it usually accompanies α or β emission as a mechanism for the new nucleus that results from a transmutation to release some extra energy.

Since γ emission affects neither mass number nor atomic number, it is often omitted from nuclear equations. Nevertheless, γ rays are of great importance. Their penetrating power makes them by far the most dangerous kind of external radiation for humans and also makes them useful in numerous medical applications. Cobalt-60, for example, is used in cancer therapy as a source of penetrating γ rays that kill cancerous tissue.

$$^{60}_{27}\text{Co} \longrightarrow {}^{60}_{28}\text{Ni} + {}^{0}_{-1}\text{e} + {}^{0}_{0}\gamma$$

Positron Emission

In addition to α, β, and γ radiation, there is another common type of radioactive decay process called *positron emission*, which involves the conversion of a proton in the nucleus into a neutron plus an ejected **positron**, ${}^{0}_{1}\text{e}$ or β^{+}. A positron, which can be thought of as a "positive electron," has the same mass as an electron but a positive charge. This process can be represented as

Positron A "positive electron," which has the same mass as an electron but a positive charge.

$$^{1}_{1}\text{p} \longrightarrow {}^{1}_{0}\text{n} + {}^{0}_{1}\text{e}$$

The result of positron emission is a decrease in the atomic number of the product nucleus because a proton has changed into a neutron, but no change in the mass number. Potassium-40, for example, undergoes positron emission to yield argon-40, a nuclear reaction important in geology for dating rocks. Note once again that the sum of the two subscripts on the right of the nuclear equation ($18 + 1 = 19$) is equal to the subscript in the ${}^{40}_{19}\text{K}$ nucleus on the left.

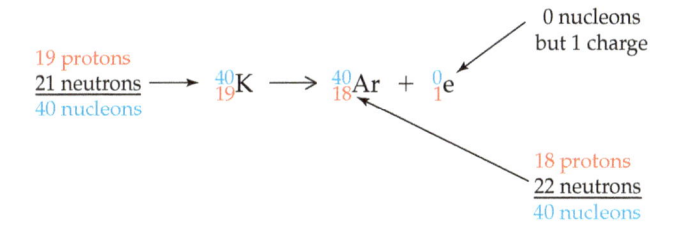

19 protons
21 neutrons $\longrightarrow {}^{40}_{19}\text{K} \longrightarrow {}^{40}_{18}\text{Ar} + {}^{0}_{1}\text{e}$
40 nucleons

0 nucleons but 1 charge

18 protons
22 neutrons
40 nucleons

Electron Capture

Electron capture, symbolized E.C., is a process in which the nucleus captures an inner-shell electron from the surrounding electron cloud, thereby converting a proton into a neutron, and energy is released in the form of gamma rays. The mass number of the product nucleus is unchanged, but the atomic number decreases by 1, just as in positron emission. The conversion of mercury-197 into gold-197 is an example:

Electron capture (E.C.) A process in which the nucleus captures an inner-shell electron from the surrounding electron cloud, thereby converting a proton into a neutron.

80 protons
117 neutrons   Inner-shell
197 nucleons   electron

79 protons
118 neutrons
197 nucleons

$$^{197}_{80}\text{Hg} + {}^{0}_{-1}\text{e} \longrightarrow {}^{197}_{79}\text{Au}$$

Do not plan on using this reaction to get rich, however. Mercury-197 is not one of the naturally occurring isotopes of Hg and is typically produced by transmutation reactions as discussed in Section 11.10.

In Figure 11.2 we see that most of the stable isotopes of the lighter elements have nearly the same number of neutrons and protons. With this fact in mind, we can often predict the most likely decay mode: unstable isotopes that have more protons than neutrons are more likely to undergo β decay to convert a proton to a neutron, while unstable isotopes having more neutrons than protons are more likely to undergo either positron emission or electron capture to convert a neutron to a proton. Also, the very heavy isotopes (Z > 83) will most likely undergo α-decay to lose both neutrons and protons to decrease the atomic number. Characteristics of the five kinds of radioactive decay processes are summarized in Table 11.2.

TABLE 11.2 A Summary of Radioactive Decay Processes

Process	Symbol	Change in Atomic Number	Change in Mass Number	Change in Number of Neutrons
α emission	$^{4}_{2}He$ or α	−2	−4	−2
β emission	$^{0}_{-1}e$ or β^{-}*	+1	0	−1
γ emission	$^{0}_{0}\gamma$ or γ	0	0	0
Positron emission	$^{0}_{1}e$ or β^{+}*	−1	0	+1
Electron capture	E.C.	−1	0	+1

*Superscripts are used to indicate the charge associated with the two forms of beta decay; β^{-}, or a beta particle, carries a −1 charge, while β^{+}, or a positron, carries a +1 charge.

Worked Example 11.3 Balancing Nuclear Reactions: Electron Capture, Positron Emission

Write balanced nuclear equations for the following processes:

(a) Electron capture by polonium-204: $^{204}_{84}Po + ^{0}_{-1}e \longrightarrow ?$
(b) Positron emission from xenon-118: $^{118}_{54}Xe \longrightarrow ^{0}_{1}e + ?$

ANALYSIS The key to writing nuclear equations is to make sure that the number of nucleons is the same on both sides of the equation and that the number of charges is the same.

SOLUTION

(a) In electron capture, the mass number is unchanged and the atomic number decreases by 1, giving bismuth-204: $^{204}_{84}Po + ^{0}_{-1}e \longrightarrow ^{204}_{83}Bi$.

Check your answer by making sure that the number of nucleons and the number of charges are the same on both sides of the equation:

   Mass number: 204 + 0 = 204   Atomic number: 84 + (−1) = 83

(b) In positron emission, the mass number is unchanged and the atomic number decreases by 1, giving iodine-118: $^{118}_{54}Xe \longrightarrow ^{0}_{1}e + ^{118}_{53}I$.

CHECK! Mass number: 118 = 0 + 118   Atomic number: 54 = 1 + 53

PROBLEM 11.5
Write nuclear equations for positron emission from the following radioisotopes:
(a) $^{38}_{20}Ca$ (b) $^{118}_{54}Xe$ (c) $^{79}_{37}Rb$

PROBLEM 11.6
Write nuclear equations for the formation of the following radioisotopes by electron capture:
(a) $^{62}_{29}Cu$ (b) $^{110}_{49}In$ (c) $^{81}_{35}Br$

KEY CONCEPT PROBLEM 11.7

The red arrow in this graph indicates the changes that occur in the nucleus of an atom during a nuclear reaction. Identify the isotopes involved as product and reactant, and name the type of decay process.

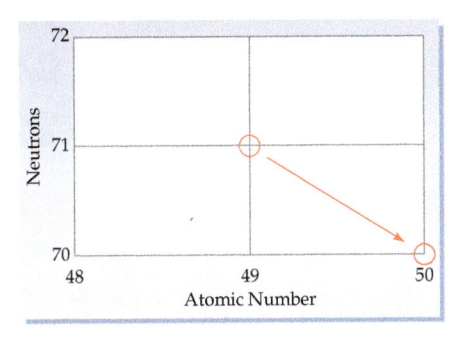

11.5 Radioactive Half-Life

The rate of radioactive decay varies greatly from one radioisotope to another. Some radioisotopes, such as uranium-238, decay at a barely perceptible rate over billions of years, but others, such as carbon-17, decay within thousandths of a second.

Rates of nuclear decay are measured in units of **half-life ($t_{1/2}$)**, defined as the amount of time required for one-half of a radioactive sample to decay. For example, the half-life of iodine-131 is 8.021 days. If today you have 1.000 g of $^{131}_{53}$I, then 8.021 days from now you will have only 50% of that amount (0.500 g) because one-half of the sample will have decayed into $^{131}_{54}$Xe. After 8.021 more days (16.063 days total), you will have only 25% (0.250 g) of your original $^{131}_{53}$I sample; after another 8.021 days (24.084 days total), you will have only 12.5% (0.125 g); and so on. Each passage of a half-life causes the decay of one-half of whatever sample remains. The half-life of any particular isotope is the same no matter what the size of the sample, the temperature, or any other external conditions. There is no known way to slow down, speed up, or otherwise change the characteristics of radioactive decay.

Half-life ($t_{1/2}$) The amount of time required for one-half of a radioactive sample to decay.

$$1.000 \text{ g } ^{131}_{53}\text{I} \xrightarrow{8 \text{ days}} 0.500 \text{ g } ^{131}_{53}\text{I} \xrightarrow{8 \text{ days}} 0.250 \text{ g } ^{131}_{53}\text{I} \xrightarrow{8 \text{ days}} 0.125 \text{ g } ^{131}_{53}\text{I} \longrightarrow$$

	One half-life	Two half-lives (16 days total)	Three half-lives (24 days total)
100%	50% remaining	25% remaining	12.5% remaining

The fraction of radioisotope remaining after the passage of each half-life is represented by the curve in Figure 11.4 and can be calculated as

$$\text{fraction remaining} = (0.5)^n$$

where n is the number of half-lives that have elapsed.

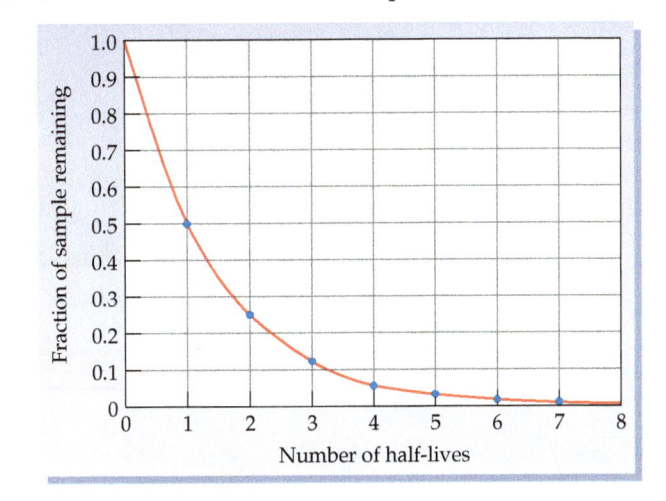

◀ **Figure 11.4**
The decay of a radioactive nucleus over time.
All nuclear decays follow this curve, whether the half-lives are measured in years, days, minutes, or seconds. That is, the fraction of sample remaining after one half-life is 0.50, the fraction remaining after two half-lives is 0.25, the fraction remaining after three half-lives is 0.125, and so on.

CHEMISTRY IN ACTION

Medical Uses of Radioactivity

The origins of nuclear medicine date from 1901, when the French physician Henri Danlos first used radium in the treatment of a tubercular skin lesion. Since that time, the use of radioactivity has become a crucial part of modern medical care, both diagnostic and therapeutic. Current nuclear techniques can be grouped into three classes: (1) *in vivo* procedures, (2) radiation therapy, and (3) imaging procedures. The first two are described here, and the third one is described on page 348 in the Chemistry in Action: Body Imaging.

In Vivo Procedures

In vivo studies—those that take place inside the body—are carried out to assess the functioning of a particular organ or body system. A *radiopharmaceutical* agent is administered, and its path in the body—whether absorbed, excreted, diluted, or concentrated—is determined by analysis of blood or urine samples.

Among the many *in vivo* procedures utilizing radioactive agents is a simple method for the determination of whole-blood volume, a common indicator used in the diagnosis of congestive heart failure, hypertension, and renal failure. A known quantity of red blood cells labeled with radioactive chromium-51 is injected into the patient and allowed to circulate to be distributed evenly throughout the body. After a suitable interval, a blood sample is taken and blood volume is calculated by comparing the concentration of labeled cells in the blood with the quantity of labeled cells injected. This and similar procedures are known as *isotope dilution* and are described by

$$R_{sample} = R_{tracer}\left(\frac{W_{sample}}{W_{system} + W_{tracer}}\right)$$

where R_{sample} is the counting rate (a measure of radioactivity) of the analyzed sample, R_{tracer} is the counting rate of the tracer added to the system, and W refers to either the mass or volume of the analyzed sample, added tracer, or total system as indicated.

Therapeutic Procedures

Therapeutic procedures—those in which radiation is purposely used as a weapon to kill diseased tissue—involve either external or internal sources of radiation. External radiation therapy for the treatment of cancer is often carried out with γ rays emanating from a cobalt-60 source. The highly radioactive source is shielded by a thick lead container and has a small opening directed toward the site of the tumor. By focusing the radiation beam on the tumor, the tumor receives the full exposure while exposure of surrounding parts of the body is minimized. Nevertheless, enough healthy tissue is affected so that most patients treated in this manner suffer the effects of radiation sickness.

Internal radiation therapy is a much more selective technique than external therapy. In the treatment of thyroid disease, for example, a radioactive substance such as iodine-131 is administered. This powerful β emitter is incorporated into the iodine-containing hormone thyroxine, which concentrates in the thyroid gland. Because β particles penetrate no farther than several millimeters, the localized ^{131}I produces a high radiation dose that destroys only the surrounding diseased tissue. To treat some tumors, such as those in the female reproductive system, a radioactive source is placed physically close to the tumor for a specific amount of time.

Boron neutron-capture therapy (BNCT) is a relatively new technique in which boron-containing drugs are administered to a patient and concentrate in the tumor site. The tumor is then irradiated with a neutron beam from a nuclear reactor. The boron absorbs a neutron and undergoes transmutation to produce an alpha particle and a lithium nucleus. These highly energetic particles have very low penetrating power and can kill nearby tumor tissue while sparing the healthy surrounding tissue. Because one disadvantage of BNCT is the need for access to a nuclear reactor, this treatment is available only in limited locations.

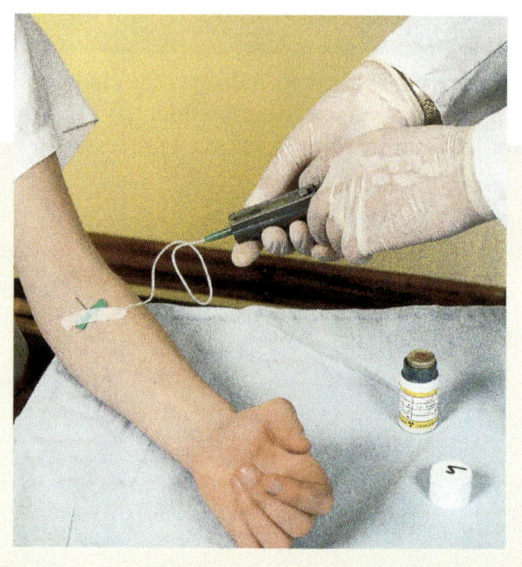

▲ A person's blood volume can be found by injecting a small amount of radioactive chromium-51 and measuring the dilution factor.

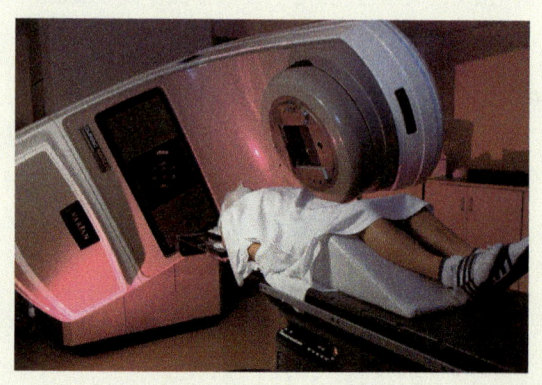

▲ A cancer patient receiving radiotherapy treatments, a therapeutic application of radioactive isotopes.

See Chemistry in Action Problems 11.70 and 11.71 at the end of the chapter.

One of the better known half-life applications is radiocarbon dating to determine the age of archaeological artifacts. The method is based on the slow and constant production of radioactive carbon-14 atoms in the upper atmosphere by bombardment of nitrogen atoms with neutrons from cosmic rays. Carbon-14 atoms combine with oxygen to yield $^{14}CO_2$, which slowly mixes with ordinary $^{12}CO_2$ and is then taken up by plants during photosynthesis. When these plants are eaten by animals, carbon-14 enters the food chain and is distributed evenly throughout all living organisms.

As long as a plant or animal is living, a dynamic equilibrium is established in which the organism excretes or exhales the same amount of ^{14}C that it takes in. As a result, the ratio of ^{14}C to ^{12}C in the living organism is the same as that in the atmosphere—about 1 part in 10^{12}. When the plant or animal dies, however, it no longer takes in more ^{14}C. Thus, the $^{14}C/^{12}C$ ratio in the organism slowly decreases as ^{14}C undergoes radioactive decay. At 5730 years (one ^{14}C half-life) after the death of the organism, the $^{14}C/^{12}C$ ratio has decreased by a factor of 2; at 11,460 years after death, the $^{14}C/^{12}C$ ratio has decreased by a factor of 4; and so on. By measuring the amount of ^{14}C remaining in the traces of any once-living organism, archaeologists can determine how long ago the organism died. The accuracy of the technique lessens as a sample gets older, but artifacts with an age of 1000–20,000 years can be dated with reasonable accuracy.

The half-lives of some useful radioisotopes are given in Table 11.3. As you might expect, radioisotopes that are used internally for medical applications have fairly short half-lives so that they decay rapidly and do not remain in the body for prolonged periods.

TABLE 11.3 Half-Lives of Some Useful Radioisotopes

Radioisotope	Symbol	Radiation	Half-Life	Use
Tritium	$^{3}_{1}H$	β	12.33 years	Biochemical tracer
Carbon-14	$^{14}_{6}C$	β	5730 years	Archaeological dating
Sodium-24	$^{24}_{11}Na$	β	14.959 hours	Examining circulation
Phosphorus-32	$^{32}_{15}P$	β	14.262 days	Leukemia therapy
Potassium-40	$^{40}_{19}K$	β, β^+	1.277×10^9 years	Geological dating
Cobalt-60	$^{60}_{27}Co$	β, γ	5.271 years	Cancer therapy
Arsenic-74	$^{74}_{33}As$	β^+	17.77 days	Locating brain tumors
Technetium-99m*	$^{99m}_{43}Tc$	γ	6.01 hours	Brain scans
Iodine-131	$^{131}_{53}I$	β	8.021 days	Thyroid therapy
Uranium-235	$^{235}_{92}U$	α, γ	7.038×10^8 years	Nuclear reactors

*The *m* in technetium-99m stands for *metastable*, meaning that the nucleus undergoes γ emission but does not change its mass number or atomic number.

Worked Example 11.4 Nuclear Reactions: Half-Life

Phosphorus-32, a radioisotope used in leukemia therapy, has a half-life of about 14 days. Approximately what percentage of a sample remains after 8 weeks?

ANALYSIS Determine how many half-lives have elapsed. For an integral number of half-lives, we can multiply the starting amount (100%) by 1/2 for each half-life that has elapsed.

SOLUTION
Since one half-life of $^{32}_{15}P$ is 14 days (2 weeks), 8 weeks represents four half-lives. The fraction that remains after 8 weeks is thus

$$\text{Final Percentage} = 100\% \times (0.5)^4 = 100\% \times \overbrace{\left(\frac{1}{2} \times \frac{1}{2} \times \frac{1}{2} \times \frac{1}{2}\right)}^{\text{Four half-lives}}$$
$$= 100\% \times \frac{1}{16} = 6.25\%$$

> **Worked Example 11.5** Nuclear Reactions: Half-Life
>
> As noted in Table 11.3, iodine-131 has a half-life of about 8 days. Approximately what fraction of a sample remains after 20 days?
>
> **ANALYSIS** Determine how many half-lives have elapsed. For a non-integral number (i.e., fraction) of half-lives, use the equation below to determine the fraction of radioisotope remaining.
>
> $$\text{fraction remaining} = (0.5)^n$$
>
> **BALLPARK ESTIMATE** Since the half-life of iodine-131 is 8 days, an elapsed time of 20 days is 2.5 half-lives. The fraction remaining should be between 0.25 (fraction remaining after two half-lives) and 0.125 (fraction remaining after three half-lives). Since the relationship between the number of half-lives and fraction remaining is not linear (see Figure 11.4), the fraction remaining will not be exactly halfway between these values but instead will be slightly closer to the lower fraction, say 0.17.
>
> **SOLUTION**
>
> $$\text{fraction remaining} = (0.5)^n = (0.5)^{2.5} = 0.177$$
>
> **BALLPARK CHECK** The fraction remaining is close to our estimate of 0.17.

PROBLEM 11.8

The half-life of carbon-14, an isotope used in archaeological dating, is 5730 years. What percentage of $^{14}_{6}C$ remains in a sample estimated to be 17,000 years old?

PROBLEM 11.9

A 1.00 mL sample of red blood cells containing chromium-51 as a tracer was injected into a patient. After several hours a 5.00 mL sample of blood was drawn and its activity compared to the activity of the injected tracer sample. If the collected sample activity was 0.10% of the original tracer, calculate the total blood volume of the patient (see Chemistry in Action: Medical Uses of Radioactivity, p. 338).

KEY CONCEPT PROBLEM 11.10

What is the half-life of the radionuclide that shows the following decay curve?

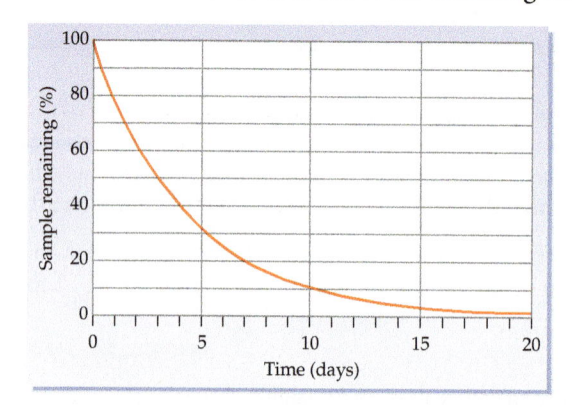

11.6 Radioactive Decay Series

When a radioactive isotope decays, nuclear change occurs and a different element is formed. Often, this newly formed nucleus is stable, but sometimes the product nucleus is itself radioactive and undergoes further decay. In fact, some radioactive nuclei undergo an extended **decay series** of nuclear disintegrations before they ultimately reach a nonradioactive product. This is particularly true for the isotopes of heavier elements. Uranium-238, for example, undergoes a series of 14 sequential nuclear reactions, ultimately stopping at lead-206 (Figure 11.5).

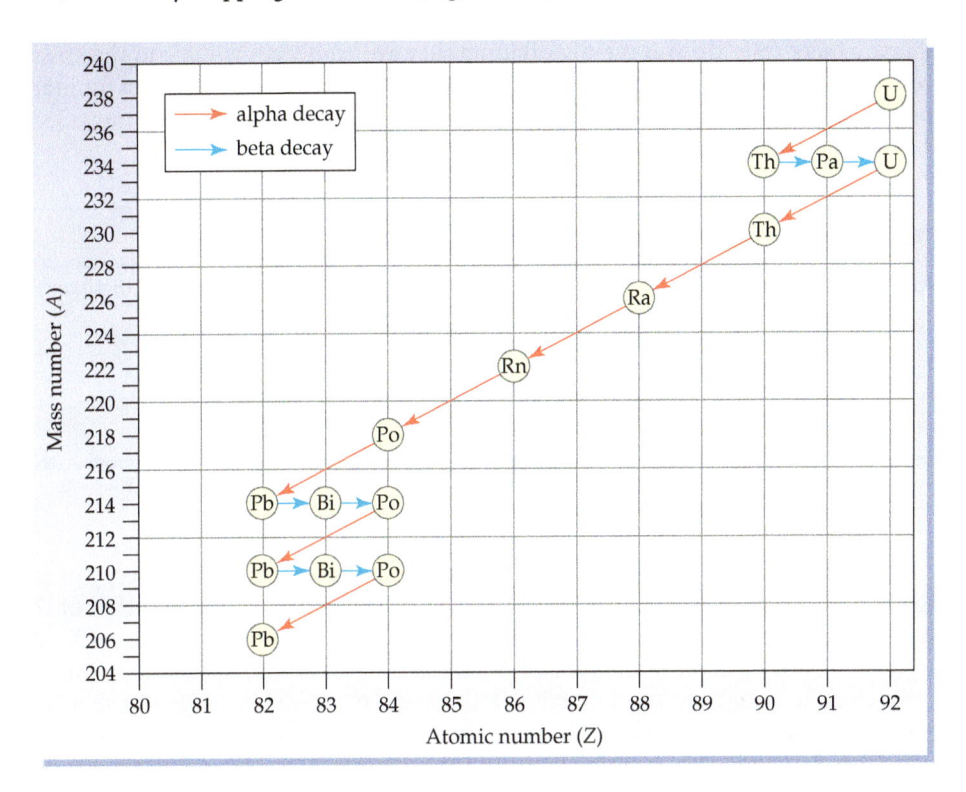

Decay series A sequential series of nuclear disintegrations leading from a heavy radioisotope to a nonradioactive product.

◀ **Figure 11.5**
The decay series from $^{238}_{92}$U to $^{206}_{82}$Pb. Each isotope except for the last is radioactive and undergoes nuclear decay. The long slanted arrows represent α emissions, and the short horizontal arrows represent β emissions.

One of the intermediate radionuclides in the uranium-238 decay series is radium-226. Radium-226 has a half-life of 1600 years and undergoes α decay to produce radon-222, a gas. Rocks, soil, and building materials that contain radium are sources of radon-222, which can seep through cracks in basements and get into the air inside homes and other buildings. Radon itself is a gas that passes in and out of the lungs without being incorporated into body tissue. If, however, a radon-222 atom should happen to undergo alpha decay while in the lungs, the result is the solid decay product polonium-218. Further decay of the ^{218}Po emits α particles, which can damage lung tissue.

11.7 Ionizing Radiation

High-energy radiation of all kinds is often grouped together under the name **ionizing radiation**. This includes not only α particles, β particles, and γ rays but also *X rays* and *cosmic rays*. **X rays** are like γ rays; they have no mass and consist of high-energy electromagnetic radiation. The only difference between them is that the energy of X rays is somewhat less than that of γ rays (see Chemistry in Action: Atoms and Light in Chapter 2). **Cosmic rays** are not rays at all but are a mixture of high-energy particles that shower Earth from outer space. They consist primarily of protons, along with some α and β particles.

The interaction of any kind of ionizing radiation with a molecule knocks out an orbital electron, converting the atom or molecule into an extremely reactive ion:

$$\text{Molecule} \xrightarrow{\text{ionizing radiation}} \text{Ion} + e^-$$

Ionizing radiation A general name for high-energy radiation of all kinds.

X rays Electromagnetic radiation with an energy somewhat less than that of γ rays.

Cosmic rays A mixture of high-energy particles—primarily of protons and various atomic nuclei—that shower Earth from outer space.

This reactive ion can react with other molecules nearby, creating still other fragments that can in turn cause further reactions. In this manner, a large dose of ionizing radiation can destroy the delicate balance of chemical reactions in living cells, ultimately causing the death of an organism.

A small dose of ionizing radiation may not cause visible symptoms but can nevertheless be dangerous if it strikes a cell nucleus and damages the genetic machinery inside. The resultant changes might lead to a genetic mutation, to cancer, or to cell death. The nuclei of rapidly dividing cells, such as those in bone marrow, the lymph system, the lining of the intestinal tract, or an embryo, are the most readily damaged. Because cancer cells are also rapidly dividing they are highly susceptible to the effects of ionizing radiation, which is why radiation therapy is an effective treatment for many types of cancer (see Chemistry in Action: Medical Uses of Radioactivity on p. 338). Some properties of ionizing radiation are summarized in Table 11.4.

TABLE 11.4 Some Properties of Ionizing Radiation

Type of Radiation	Energy Range*	Penetrating Distance in Water**
α	3–9 MeV	0.02–0.04 mm
β	0–3 MeV	0–4 mm
X	100 eV–10 keV	0.01–1 cm
γ	10 keV–10 MeV	1–20 cm

* The energies of subatomic particles are often measured in electron volts (eV): 1 eV = 6.703×10^{-19} cal, or 2.805×10^{-18} J.
** Distance at which one-half of the radiation is stopped.

The effects of ionizing radiation on the human body vary with the energy of the radiation, its distance from the body, the length of exposure, and the location of the source outside or inside the body. When coming from outside the body, γ rays and X rays are potentially more harmful than α and β particles because they pass through clothing and skin and into the body's cells. Alpha particles are stopped by clothing and skin, and β particles are stopped by wood or several layers of clothing. These types of radiation are much more dangerous when emitted within the body, however, because all their radiation energy is given up to the immediately surrounding tissue. Alpha emitters are especially hazardous internally and are almost never used in medical applications.

Health professionals who work with X rays or other kinds of ionizing radiation protect themselves by surrounding the source with a thick layer of lead or other dense material. Protection from radiation is also afforded by controlling the distance between the worker and the radiation source because radiation intensity (I) decreases with the square of the distance from the source. The intensities of radiation at two different distances, 1 and 2, are given by the equation

$$\frac{I_1}{I_2} = \frac{d_2^2}{d_1^2}$$

For example, suppose a source delivers 16 units of radiation at a distance of 1.0 m. Doubling the distance to 2.0 m decreases the radiation intensity to one-fourth:

$$\frac{16 \text{ units}}{I_2} = \frac{(2 \text{ m})^2}{(1 \text{ m})^2}$$

$$I_2 = 16 \text{ units} \times \frac{1 \text{ m}^2}{4 \text{ m}^2} = 4 \text{ units}$$

Worked Example 11.6 Ionizing Radiation: Intensity versus Distance from the Source

If a radiation source gives 75 units of radiation at a distance of 2.4 m, at what distance does the source give 25 units of radiation?

ANALYSIS Radiation intensity (I) decreases with the square of the distance (d) from the source according to the equation

$$\frac{I_1}{I_2} = \frac{d_2^2}{d_1^2}$$

We know three of the four variables in this equation (I_1, I_2, and d_1), and we need to find d_2.

BALLPARK ESTIMATE In order to decrease the radiation intensity from 75 units to 25 units (a factor of 3), the distance must *increase* by a factor of $\sqrt{3} = 1.7$. Thus, the distance should increase from 2.4 m to about 4 m.

SOLUTION

STEP 1: Identify known information. We know three of the four variables.

$I_1 = 75$ units
$I_2 = 25$ units
$d_1 = 2.4$ m

STEP 2: Identify answer and units.

$d_2 = ???$ m

STEP 3: Identify equation. Rearrange the equation relating intensity and distance to solve for d_2.

$$\frac{I_1}{I_2} = \frac{d_2^2}{d_1^2}$$

$$d_2^2 = \frac{I_1 d_1^2}{I_2} \quad \Rightarrow \quad d_2 = \sqrt{\frac{I_1 d_1^2}{I_2}}$$

STEP 4: Solve. Substitute in known values so that unwanted units cancel.

$$d_2 = \sqrt{\frac{(75 \text{ units})(2.4 \text{ m})^2}{(25 \text{ units})}} = 4.2 \text{ m}$$

BALLPARK CHECK The calculated result is consistent with our estimate of about 4 m.

PROBLEM 11.11
A β-emitting radiation source gives 250 units of radiation at a distance of 4.0 m. At what distance does the radiation drop to one-tenth its original value?

11.8 Detecting Radiation

Small amounts of naturally occurring radiation have always been present, but people have been aware of it only within the past 100 years. The problem is that radiation is invisible. We cannot see, hear, smell, touch, or taste radiation, no matter how high the dose. We can, however, detect radiation by taking advantage of its ionizing properties.

The simplest device for detecting exposure to radiation is the photographic film badge worn by people who routinely work with radioactive materials. The film is protected from exposure to light, but any other radiation striking the badge causes the film to fog (remember Becquerel's discovery). At regular intervals, the film is developed and compared with a standard to indicate the radiation exposure.

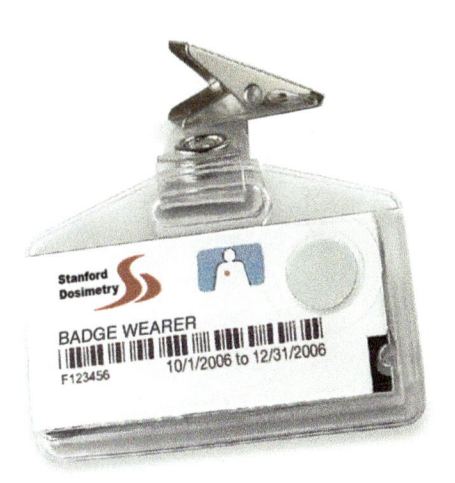

The most versatile method for measuring radiation in the laboratory is the *scintillation counter*, a device in which a substance called a *phosphor* emits a flash of light when struck by radiation. The number of flashes are counted electronically and converted into an electrical signal.

▲ This photographic film badge is a common device for monitoring radiation exposure.

Perhaps the best-known method for detecting and measuring radiation is the *Geiger counter*, an argon-filled tube containing two electrodes (Figure 11.6). The inner walls of the tube are coated with an electrically conducting material and given a negative charge, and a wire in the center of the tube is given a positive charge. As radiation enters the tube through a thin window, it strikes and ionizes argon atoms, which briefly conduct a tiny electric current between the walls and the center electrode. The passage of the current is detected, amplified, and used to produce a clicking sound or to register on a meter. The more radiation that enters the tube, the more frequent the clicks. Geiger counters are useful for seeking out a radiation source in a large area and for gauging the intensity of emitted radiation.

▶ Figure 11.6

A Geiger counter for measuring radiation.
As radiation enters the tube through a thin window, it ionizes argon atoms and produces electrons that conduct a tiny electric current between the walls and the center electrode. The current flow then registers on the meter.

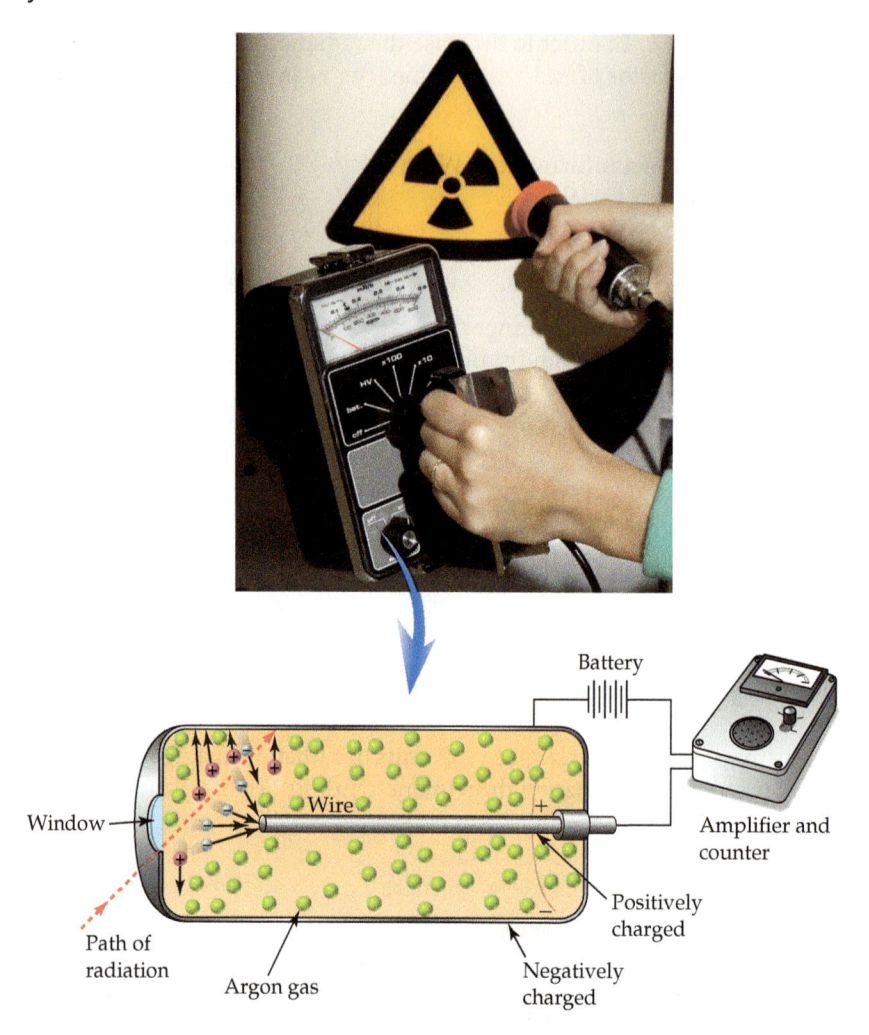

11.9 Measuring Radiation

Radiation intensity is expressed in different ways, depending on what characteristic of the radiation is measured (Table 11.5). Some units measure the number of nuclear decay events, while others measure exposure to radiation or the biological consequences of radiation.

TABLE 11.5 Common Units for Measuring Radiation

Unit	Quantity Measured	Description
Curie (Ci)	Decay events	Amount of radiation equal to 3.7×10^{10} disintegrations per second
Roentgen (R)	Ionizing intensity	Amount of radiation producing 2.1×10^9 charges per cubic centimeter of dry air
Rad	Energy absorbed per gram of tissue	1 rad = 1 R
Rem	Tissue damage	Amount of radiation producing the same damage as 1 R of X rays
Sievert (Sv)	Tissue damage	1 Sv = 100 rem

CHEMISTRY IN ACTION

Irradiated Food

The idea of irradiating food to kill harmful bacteria is not new; it goes back almost as far as the earliest studies on radiation. Not until the 1940s did serious work get under way, however, when U.S. Army scientists found that irradiation increased the shelf-life of ground beef. Nevertheless, widespread civilian use of the technique has been a long time in coming, spurred on in recent years by outbreaks of food poisoning that resulted in several deaths.

The principle of food irradiation is simple: exposure of contaminated food to ionizing radiation—usually γ rays produced by cobalt-60 or cesium-137—destroys the genetic material of any bacteria or other organisms present, thereby killing them. Irradiation will not, however, kill viruses or prions (see Chemistry in Action: Prions: Proteins That Cause Disease in Chapter 18), the cause of "mad-cow" disease. The amount of radiation depends on the desired effect. For example, to delay ripening of fruit may require a dose of 0.25 – 0.75 kGy, while sterilization of packaged meat requires a much higher dose of 25 – 70 kGy. The food itself undergoes little if any change when irradiated and does not itself become radioactive. The only real argument against food irradiation, in fact, is that it is *too* effective. Knowing that irradiation will kill nearly all harmful organisms, a food processor might be tempted to cut back on normal sanitary practices!

Food irradiation has been implemented to a much greater extent in Europe than in the United States. The largest marketers of irradiated food are Belgium, France, and the Netherlands, which irradiate between 10,000 and 20,000 tons of food per year. Currently, over 40 countries permit food irradiation and over 500,000 metric tons of food are treated annually worldwide.

▲ **Irradiating food kills bacteria and extends shelf life. Most irradiated food products are labeled with the Radura symbol (in green) to inform the public that the food product was exposed to radiation.**

One of the major concerns in the United States is the possible generation of *radiolytic products*, compounds formed in food by exposure to ionizing radiation. The U.S. Food and Drug Administration, after studying the matter extensively, has declared that food irradiation is safe and that it does not appreciably alter the vitamin or other nutritional content of food. Spices, fruits, pork, and vegetables were approved for irradiation in 1986, followed by poultry in 1990 and red meat, particularly ground beef, in 1997. In 2000, approval was extended to whole eggs and sprouting seeds. Should the food industry adopt irradiation of meat as its standard practice, occurrences of *E. coli* and *salmonella* contaminations, resulting in either massive product recalls or serious health concerns for consumers will become a thing of the past.

See Chemistry in Action Problems 11.72 and 11.73 at the end of the chapter.

- **Curie** The *curie* (Ci), the *millicurie* (mCi), and the *microcurie* (μCi) measure the number of radioactive disintegrations occurring each second in a sample. One curie is the decay rate of 1 g of radium, equal to 3.7×10^{10} disintegrations per second; 1 mCi = 0.001 Ci = 3.7×10^7 disintegrations per second; and 1 μCi = 0.000 001 Ci = 3.7×10^4 disintegrations per second.

    The dosage of a radioactive substance administered orally or intravenously is usually given in millicuries. To calculate the size of a dose, it is necessary to determine the decay rate of the isotope solution per milliliter. Because the emitter concentration is constantly decreasing as it decays, the activity must be measured immediately before administration. Suppose, for example, that a solution containing iodine-131 for a thyroid-function study is found to have a decay rate of 0.020 mCi/mL and the dose administered is to be 0.050 mCi. The amount of the solution administered must be

$$\frac{0.05 \text{ mCi}}{\text{Dose}} \times \frac{1 \text{ mL } ^{131}\text{I solution}}{0.020 \text{ mCi}} = 2.5 \text{ mL } ^{131}\text{I solution/dose}$$

- **Roentgen** The *roentgen* (R) is a unit for measuring the ionizing intensity of γ or X radiation. In other words, the roentgen measures the capacity of the radiation for affecting matter. One roentgen is the amount of radiation that produces 2.1×10^9 units of charge in 1 cm^3 of dry air at atmospheric pressure. Each collision of ionizing radiation with an atom produces one ion, or one unit of charge.

- **Rad** The *rad* (radiation absorbed dose) is a unit for measuring the energy absorbed per gram of material exposed to a radiation source and is defined as the absorption of 1×10^{-5} J of energy per gram. The energy absorbed varies with the type of material irradiated and the type of radiation. For most purposes, though, the roentgen and the rad are so close that they can be considered identical when used for X rays and γ rays: 1 R = 1 rad.
- **Rem** The *rem* (roentgen equivalent for man) measures the amount of tissue damage caused by radiation. One rem is the amount of radiation that produces the same effect as 1 R of X rays. Rems are the preferred units for medical purposes because they measure equivalent doses of different kinds of radiation. The rem is calculated as

$$\text{Rems} = \text{rads} \times \text{RBE}$$

where RBE is a *relative biological effectiveness* factor, which takes into account the differences in energy and of the different types of radiation. Although the actual biological effects of radiation depend greatly on both the source and the energy of the radiation, the RBE of X rays, γ rays, and β particles are essentially equivalent (RBE = 1), while the accepted RBE for α particles is 20. For example, 1 rad of α radiation causes 20 times more tissue damage than 1 rad of γ rays, but 1 rem of α radiation and 1 rem of γ rays cause the same amount of damage. Thus, the rem takes both ionizing intensity and biological effect into account, whereas the rad deals only with intensity.
- **SI Units** In the SI system, the *becquerel* (Bq) is defined as one disintegration per second. The SI unit for energy absorbed is the *gray* (Gy; 1 Gy = 100 rad). For radiation dose, the SI unit is the *sievert* (Sv), which is equal to 100 rem.

The biological consequences of different radiation doses are given in Table 11.6. Although the effects seem frightening, the average radiation dose received annually by most people is only about 0.27 rem. About 80% of this *background radiation* comes from natural sources (rocks and cosmic rays); the remaining 20% comes from consumer products and from medical procedures such as X rays. The amount due to emissions from nuclear power plants and to fallout from testing of nuclear weapons in the 1950s is barely detectable.

TABLE 11.6 Biological Effects of Short-Term Radiation on Humans

Dose (rem)	Biological Effects
0–25	No detectable effects
25–100	Temporary decrease in white blood cell count
100–200	Nausea, vomiting, longer-term decrease in white blood cells
200–300	Vomiting, diarrhea, loss of appetite, listlessness
300–600	Vomiting, diarrhea, hemorrhaging, eventual death in some cases
Above 600	Eventual death in nearly all cases

PROBLEM 11.12
Radiation released during the 1986 Chernobyl nuclear power plant disaster is expected to increase the background radiation level worldwide by about 5 mrem. By how much will this increase the annual dose of the average person? Express your answer as a percentage.

PROBLEM 11.13
A solution of selenium-75, a radioisotope used in the diagnosis of pancreatic disease, is found just prior to administration to have an activity of 44 μCi/mL. If 3.98 mL were delivered intravenously to the patient, what dose of Se-75 (in μCi) did the patient receive?

PROBLEM 11.14
A typical food irradiation application for the inhibition of sprout formation in potatoes applies a dose of 0.20 kGy. What is this dose in units of rad if the radiation is predominantly γ rays? If it is predominantly α particles? (See Chemistry in Action: Irradiated Food on p. 345.)

11.10 Artificial Transmutation

Very few of the approximately 3300 known radioisotopes occur naturally. Most are made from stable isotopes by **artificial transmutation**, the change of one atom into another brought about by nuclear bombardment reactions.

When an atom is bombarded with a high-energy particle, such as a proton, a neutron, an α particle, or even the nucleus of another element, an unstable nucleus is created in the collision. A nuclear change then occurs, and a different element is produced. For example, transmutation of ^{14}N to ^{14}C occurs in the upper atmosphere when neutrons produced by cosmic rays collide with atmospheric nitrogen. In the collision, a neutron dislodges a proton (^{1}H) from the nitrogen nucleus as the neutron and nucleus fuse together:

$$^{14}_{7}\text{N} + ^{1}_{0}\text{n} \longrightarrow ^{14}_{6}\text{C} + ^{1}_{1}\text{H}$$

Artificial transmutation can lead to the synthesis of entirely new elements never before seen on Earth. In fact, all the *transuranium elements*—those elements with atomic numbers greater than 92—have been produced by bombardment reactions. For example, plutonium-241 (^{241}Pu) can be made by bombardment of uranium-238 with α particles:

$$^{238}_{92}\text{U} + ^{4}_{2}\text{He} \longrightarrow ^{241}_{94}\text{Pu} + ^{1}_{0}\text{n}$$

Plutonium-241 is itself radioactive, with a half-life of 14.35 years, decaying by β emission to yield americium-241, which in turn decays by α emission with a half-life of 432.2 years. (If the name *americium* sounds vaguely familiar, it is because this radioisotope is used in smoke detectors.)

$$^{241}_{94}\text{Pu} \longrightarrow ^{241}_{95}\text{Am} + ^{0}_{-1}\text{e}$$

Note that all the equations just given for artificial transmutations are balanced. The sum of the mass numbers and the sum of the charges are the same on both sides of each equation.

Artificial transmutation The change of one atom into another brought about by a nuclear bombardment reaction.

▲ Smoke detectors contain a small amount of americium-241. The α particles emitted by this radioisotope ionize the air within the detector, causing it to conduct a tiny electric current. When smoke enters the chamber, conductivity drops and an alarm is triggered.

Worked Example 11.7 Balancing Nuclear Reactions: Transmutation

Californium-246 is formed by bombardment of uranium-238 atoms. If 4 neutrons are also formed, what particle is used for the bombardment?

ANALYSIS First write an incomplete nuclear equation incorporating the known information:

$$^{238}_{92}\text{U} + ? \longrightarrow ^{246}_{98}\text{Cf} + 4^{1}_{0}\text{n}$$

Then find the numbers of nucleons and charges necessary to balance the equation. In this instance, there are 238 nucleons on the left and $246 + 4 = 250$ nucleons on the right, so the bombarding particle must have $250 - 238 = 12$ nucleons. Furthermore, there are 92 nuclear charges on the left and 98 on the right, so the bombarding particle must have $98 - 92 = 6$ protons.

SOLUTION
The missing particle is $^{12}_{6}$C.

$$^{238}_{92}\text{U} + ^{12}_{6}\text{C} \longrightarrow ^{246}_{98}\text{Cf} + 4^{1}_{0}\text{n}$$

CHEMISTRY IN ACTION

Body Imaging

We are all familiar with the appearance of a standard X-ray image, produced when X rays pass through the body and the intensity of the radiation that exits is recorded on film. X-ray imaging is, however, only one of a host of noninvasive imaging techniques that are now in common use.

Among the most widely used imaging techniques are those that give diagnostic information about the health of various parts of the body by analyzing the distribution pattern of a radioactively tagged substance in the body. A radiopharmaceutical agent that is known to concentrate in a specific organ or other body part is injected into the body, and its distribution pattern is monitored by an external radiation detector such as a γ ray camera. Depending on the medical condition, a diseased part might concentrate more of the radiopharmaceutical than normal and thus show up on the film as a radioactive hot spot against a cold background. Alternatively, the diseased part might concentrate less of the radiopharmaceutical than normal and thus show up as a cold spot on a hot background.

Among the radioisotopes most widely used for diagnostic imaging is technetium-99m, whose short half-life of only 6 hours minimizes the patient's exposure to radioactivity. Enhanced body images, such as the brain scan shown in the accompanying photograph, are an important tool in the diagnosis of cancer and many other medical conditions.

Several other techniques now used in medical diagnosis are made possible by *tomography*, a technique in which computer processing allows production of images through "slices" of the body. In X-ray tomography, commonly known as *CAT* or *CT* scanning (computerized tomography), the X-ray source and an array of detectors move rapidly in a circle around a patient's body, collecting up to 90,000 readings. CT scans can detect structural abnormalities such as tumors without the use of radioactive materials.

Combining tomography with radioisotope imaging gives cross-sectional views of regions that concentrate a radioactive substance.

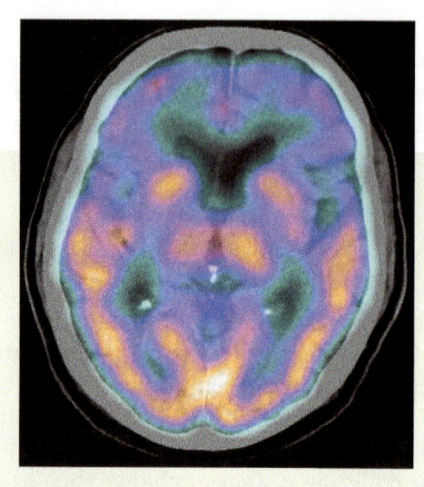

▲ This enhanced image of the brain of a 72-year old male was obtained using PET following injection of 20 mCi of Tc-99m. The results can be used to distinguish between dementia and depression.

One such technique, *positron emission tomography* (PET), utilizes radioisotopes that emit positrons and ultimately yield γ rays. Oxygen-15, nitrogen-13, carbon-11, and fluorine-18 are commonly used for PET because they can be readily incorporated into many physiologically active compounds. An ^{18}F-labeled glucose derivative, for instance, is useful for imaging brain regions that respond to various stimuli. The disadvantage of PET scans is that the necessary radioisotopes are so short-lived that they must be produced on-site immediately before use. The cost of PET is therefore high, because a hospital must install and maintain the necessary nuclear facility.

Magnetic resonance imaging (MRI) is a medical imaging technique that uses powerful magnetic and radio-frequency fields to interact with specific nuclei in the body (usually the nuclei of hydrogen atoms) to generate images in which the contrast between soft tissues is much better than that seen with CT. The original name for this technique was *nuclear* magnetic resonance imaging, but the *nuclear* was eliminated because in the public mind this word conjured up negative images of ionizing radiation. Ironically, MRI does not involve any nuclear radiation at all.

See Chemistry in Action Problems 11.74 and 11.75 at the end of the chapter.

PROBLEM 11.15
What isotope results from α decay of the americium-241 in smoke detectors?

PROBLEM 11.16
The element berkelium, first prepared at the University of California at Berkeley in 1949, is made by α bombardment of $^{241}_{95}$Am. Two neutrons are also produced during the reaction. What isotope of berkelium results from this transmutation? Write a balanced nuclear equation.

PROBLEM 11.17
Write a balanced nuclear equation for the reaction of argon-40 with a proton:

$$^{40}_{18}\text{Ar} + ^{1}_{1}\text{H} \longrightarrow ? + ^{1}_{0}\text{n}$$

> **PROBLEM 11.18**
> Technetium-99*m* (Tc-99*m*) is used extensively in diagnostic applications, including positron emission tomography (PET) scans (see Chemistry in Action: Body Imaging on p. 348). The half-life of Tc-99*m* is 6 hours. How long will it take for the Tc-99*m* activity to decrease to 0.1% of its original activity?

11.11 Nuclear Fission and Nuclear Fusion

In the preceding section, we saw that particle bombardment of various elements causes artificial transmutation and results in the formation of new, usually heavier elements. Under very special conditions with a very few isotopes, however, different kinds of nuclear events occur. Certain very heavy nuclei can split apart, and certain very light nuclei can fuse together. The two resultant processes—**nuclear fission** for the fragmenting of heavy nuclei and **nuclear fusion** for the joining together of light nuclei—have changed the world since their discovery in the late 1930s and early 1940s.

Nuclear fission The fragmenting of heavy nuclei.

Nuclear fusion The joining together of light nuclei.

The huge amounts of energy that accompany these nuclear processes are the result of mass-to-energy conversions and are predicted by Einstein's equation

$$E = mc^2$$

where E = energy, m = mass change associated with the nuclear reaction, and c = the speed of light (3.0×10^8 m/s). Based on this relationship, a mass change as small as 1 µg results in a release of 2.15×10^4 kcal (9.00×10^4 kJ) of energy!

Nuclear Fission

Uranium-235 is the only naturally occurring isotope that undergoes nuclear fission. When this isotope is bombarded by a stream of relatively slow-moving neutrons, its nucleus splits to give isotopes of other elements. The split can take place in more than 400 ways, and more than 800 different fission products have been identified. One of the more frequently occurring pathways generates barium-142 and krypton-91, along with 2 additional neutrons plus the 1 neutron that initiated the fission:

$$^{1}_{0}n + ^{235}_{92}U \longrightarrow ^{142}_{56}Ba + ^{91}_{36}Kr + 3\ ^{1}_{0}n$$

As indicated by the balanced nuclear equation above, *one* neutron is used to initiate fission of a ^{235}U nucleus, but *three* neutrons are released. Thus, a nuclear **chain reaction** can be started: 1 neutron initiates one fission that releases 3 neutrons. Those 3 neutrons initiate three new fissions that release 9 neutrons. The 9 neutrons initiate nine fissions that release 27 neutrons, and so on at an ever-faster pace (Figure 11.7). It is worth noting that the neutrons produced by fission reactions are highly energetic. They possess penetrating power greater than α and β particles, but less than γ rays. In a nuclear fission reactor, the neutrons must first be slowed down to allow them to react. If the sample size is small, many of the neutrons escape before initiating additional fission events, and the chain reaction stops. If a sufficient amount of ^{235}U is present, however—an amount called the **critical mass**—then the chain reaction becomes self-sustaining. Under high-pressure conditions that confine the ^{235}U to a small volume, the chain reaction occurs so rapidly that a nuclear explosion results. For ^{235}U, the critical mass is about 56 kg, although the amount can be reduced to approximately 15 kg by placing a coating of ^{238}U around the ^{235}U to reflect back some of the escaping neutrons.

Chain reaction A reaction that, once started, is self-sustaining.

Critical mass The minimum amount of radioactive material needed to sustain a nuclear chain reaction.

An enormous quantity of heat is released during nuclear fission—the fission of just 1.0 g of uranium-235 produces 3.4×10^8 kcal (1.4×10^9 kJ) for instance. This heat can be used to convert water to steam, which can be harnessed to turn huge generators and produce electric power. Although the United States, France, and Japan are responsible for nearly 50% of all nuclear power generated worldwide, only about 19% of the

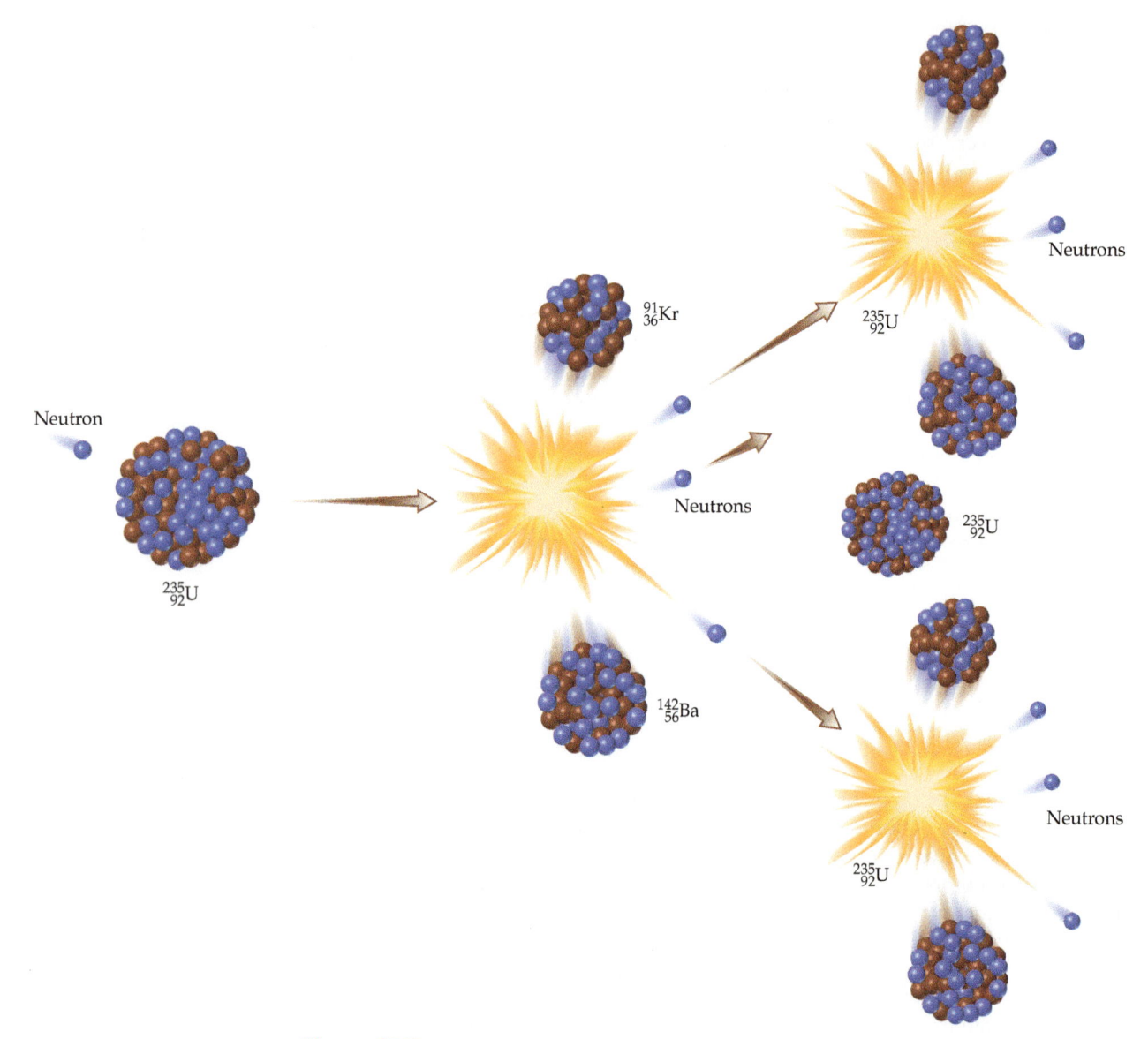

▲ **Figure 11.7**
A chain reaction.
Each fission event produces additional neutrons that induce more fissions. The rate of the process increases at each stage. Such chain reactions usually lead to the formation of many different fission products in addition to the two indicated.

electricity consumed in the United States is nuclear-generated. In France, nearly 80% of electricity is generated by nuclear power plants.

Two major objections that have caused much public debate about nuclear power plants are safety and waste disposal. Although a nuclear explosion is not possible under the conditions that typically exist in a power plant, there is a serious potential radiation hazard should an accident rupture the containment vessel holding the nuclear fuel and release radioactive substances to the environment. There have been several such instances in the last 35 years, most notably Three Mile Island in Pennsylvania (1979), Chernobyl in the Ukraine (1986), and the more recent Fukushima reactor damaged by the tsunami in Japan (2011). Perhaps even more important is the problem posed by disposal of radioactive wastes from nuclear plants. Many of these wastes have such long half-lives that hundreds or even thousands of years must elapse before they will be safe for humans to approach. How to dispose of such hazardous materials safely is an unsolved problem.

PROBLEM 11.19

What other isotope besides tellurium-137 is produced by nuclear fission of uranium-235?

$$^{235}_{92}U + ^{1}_{0}n \longrightarrow ^{137}_{52}Te + 2\,^{1}_{0}n + ?$$

Nuclear Fusion

Just as heavy nuclei such as ^{235}U release energy when they undergo *fission*, very light nuclei such as the isotopes of hydrogen release enormous amounts of energy when they undergo *fusion*. In fact, it is just such a fusion reaction of hydrogen nuclei to produce helium that powers our sun and other stars. Among the processes thought to occur in the sun are those in the following sequence leading to helium-4:

$$^{1}_{1}H + ^{2}_{1}H \longrightarrow ^{3}_{2}He$$
$$^{3}_{2}He + ^{3}_{2}He \longrightarrow ^{4}_{2}He + 2\,^{1}_{1}H$$
$$^{3}_{2}He + ^{1}_{1}H \longrightarrow ^{4}_{2}He + ^{0}_{1}e$$

Under the conditions found in stars, where the temperature is on the order of 2×10^7 K and pressures approach 10^5 atmospheres, nuclei are stripped of all their electrons and have enough kinetic energy that nuclear fusion readily occurs. The energy of our sun, and all the stars, comes from thermonuclear fusion reactions in their core that fuse hydrogen and other light elements, transmuting them into heavier elements. On Earth, however, the necessary conditions for nuclear fusion are not easily created. For more than 50 years scientists have been trying to create the necessary conditions for fusion in laboratory reactors, including the Tokamak Fusion Test Reactor (TFTR) at Princeton, New Jersey, and the Joint European Torus (JET) at Culham, England. Recent advances in reactor design have raised hopes that a commercial fusion reactor will be realized within the next 20 years.

If the dream becomes reality, controlled nuclear fusion can provide the ultimate cheap, clean power source. The fuel is deuterium (^2H), available in the oceans in limitless amounts, and there are few radioactive by-products.

PROBLEM 11.20

One of the possible reactions for nuclear fusion involves the collision of 2 deuterium nuclei. Complete the reaction by identifying the missing particle:

$$^{2}_{1}H + ^{2}_{1}H \longrightarrow ^{1}_{0}n + ?$$

SUMMARY: REVISITING THE CHAPTER GOALS

1. What is a nuclear reaction, and how are equations for nuclear reactions balanced? A *nuclear reaction* is one that changes an atomic nucleus, causing the change of one element into another. Loss of an α particle leads to a new atom whose atomic number is 2 less than that of the starting atom. Loss of a β particle leads to an atom whose atomic number is 1 greater than that of the starting atom:

$$\alpha \text{ emission: } ^{238}_{92}U \longrightarrow ^{234}_{90}Th + ^{4}_{2}He$$
$$\beta \text{ emission: } ^{131}_{53}I \longrightarrow ^{131}_{54}Xe + ^{0}_{-1}e$$

A nuclear reaction is balanced when the sum of the *nucleons* (protons and neutrons) is the same on both sides of the reaction arrow and when the sum of the charges on the nuclei plus any ejected subatomic particles is the same (*see* Problems 22, 24, 26, 38, 40, 41, 44–53, 81, 82, 84, 85, 90–95).

2. What are the different kinds of radioactivity? *Radioactivity* is the spontaneous emission of radiation from the nucleus of an unstable atom. The three major kinds of radiation are called *alpha* (α), *beta* (β), and *gamma* (γ). Alpha radiation consists of helium nuclei, small particles containing 2 protons and

2 neutrons ($^4_2$He); β radiation consists of electrons ($^{\ 0}_{-1}$e); and γ radiation consists of high-energy light waves. Every element in the periodic table has at least one radioactive isotope, or *radioisotope* (see Problems 22, 25, 27, 29, 30–32, 40, 41, 44–47, 49, 81, 82, 93).

3. **How are the rates of nuclear reactions expressed?** The rate of a nuclear reaction is expressed in units of *half-life* ($t_{1/2}$), where one half-life is the amount of time necessary for one half of the radioactive sample to decay (see Problems 21, 23, 28, 29, 54–59, 77, 83, 85).

4. **What is ionizing radiation?** High-energy radiation of all types—α particles, β particles, γ rays, and X rays—is called *ionizing radiation*. When any of these kinds of radiation strikes an atom, it dislodges an orbital electron and gives a reactive ion that can be lethal to living cells. Gamma rays and X rays are the most penetrating and most harmful types of external radiation; α and β particles are the most dangerous types of internal radiation because of their high energy and the resulting damage to surrounding tissue (see Problems 33–37, 63, 65, 72, 76, 84, 86, 87).

5. **How is radioactivity measured?** Radiation intensity is expressed in different ways according to the property being measured. The *curie* (Ci) measures the number of radioactive disintegrations per second in a sample; the *roentgen* (R) measures the ionizing ability of radiation. The *rad* measures the amount of radiation energy absorbed per gram of tissue; and the *rem* measures the amount of tissue damage caused by radiation. Radiation effects become noticeable with a human exposure of 25 rem and become lethal at an exposure above 600 rem (see Problems 60–69, 79, 80).

6. **What is transmutation?** *Transmutation* is the change of one element into another brought about by a nuclear reaction. Most known radioisotopes do not occur naturally but are made by bombardment of an atom with a high-energy particle. In the ensuing collision between particle and atom, a nuclear change occurs and a new element is produced by *artificial transmutation* (see Problems 38, 39, 48, 50, 51, 53, 90, 94, 95).

7. **What are nuclear fission and nuclear fusion?** With a very few isotopes, including $^{235}_{92}$U, the nucleus is split apart by neutron bombardment to give smaller fragments. A large amount of energy is released during this *nuclear fission*, leading to use of the reaction for generating electric power. *Nuclear fusion* results when small nuclei such as those of tritium ($^3_1$H) and deuterium ($^2_1$H) combine to give a heavier nucleus (see Problems 42, 43, 48, 88, 91, 92).

KEY WORDS

Alpha (α) particle, *p. 331*
Artificial transmutation, *p. 347*
Beta (β) particle, *p. 331*
Chain reaction, *p. 349*
Cosmic rays, *p. 341*
Critical mass, *p. 349*
Decay series, *p. 341*
Electron capture (E.C.), *p. 335*

Gamma (γ) radiation, *p. 331*
Half-life ($t_{1/2}$) *p. 337*
Ionizing radiation, *p. 341*
Nuclear decay, *p. 332*
Nuclear fission, *p. 349*
Nuclear fusion, *p. 349*
Nuclear reaction, *p. 329*
Nucleon, *p. 329*

Nuclide, *p. 329*
Positron, *p. 335*
Radioactivity, *p. 330*
Radioisotope, *p. 331*
Radionuclide, *p. 331*
Transmutation, *p. 332*
X rays, *p. 341*

UNDERSTANDING KEY CONCEPTS

11.21 Magnesium-28 decays by β emission to give aluminum-28. If yellow spheres represent $^{28}_{12}$Mg atoms and blue spheres represent $^{28}_{13}$Al atoms, how many half-lives have passed in the following sample?

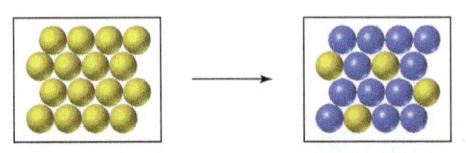

11.22 Write a balanced nuclear equation to represent the decay reaction described in Problem 11.21.

11.23 Refer to Figure 11.4 and then make a drawing similar to those in Problem 11.21 representing the decay of a sample of $^{28}_{12}$Mg after approximately four half-lives have passed.

11.24 Write the symbol of the isotope represented by the following drawing. Blue spheres represent neutrons and red spheres represent protons.

11.25 Shown in the following graph is a portion of the decay series for plutonium-241 ($^{241}_{94}$Pu). The series has two kinds of arrows: shorter arrows pointing right and longer arrows pointing left. Which arrow corresponds to an α emission, and which to a β emission? Explain.

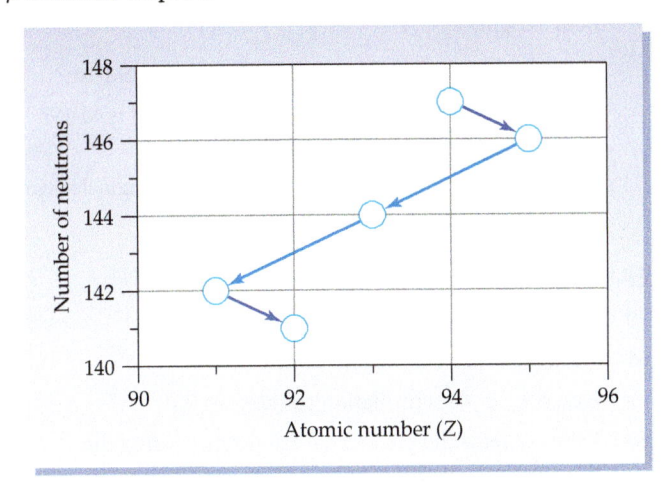

11.26 Identify and write the symbol for each of the five nuclides in the decay series shown in Problem 11.25.

11.27 Identify the isotopes involved, and tell the type of decay process occurring in the following nuclear reaction:

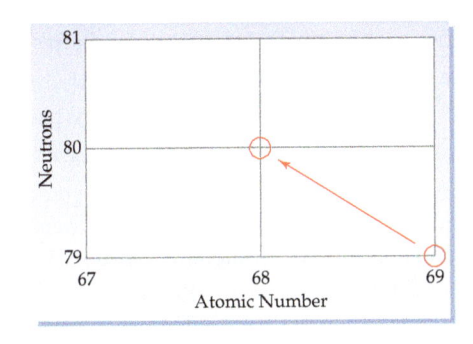

11.28 What is the half-life of the radionuclide that shows the following decay curve?

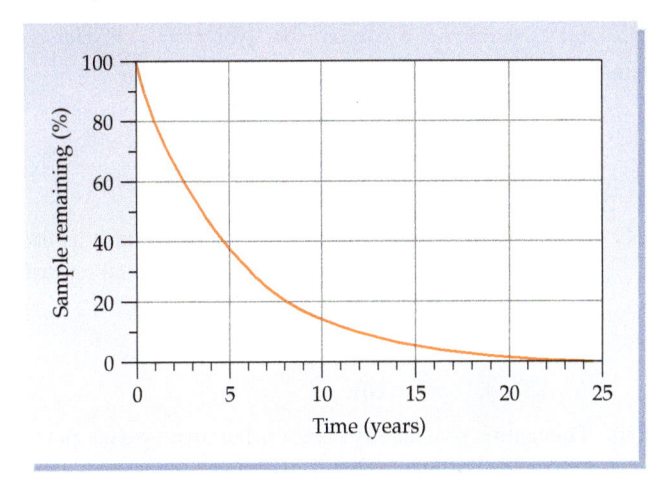

11.29 What is wrong with the following decay curve? Explain.

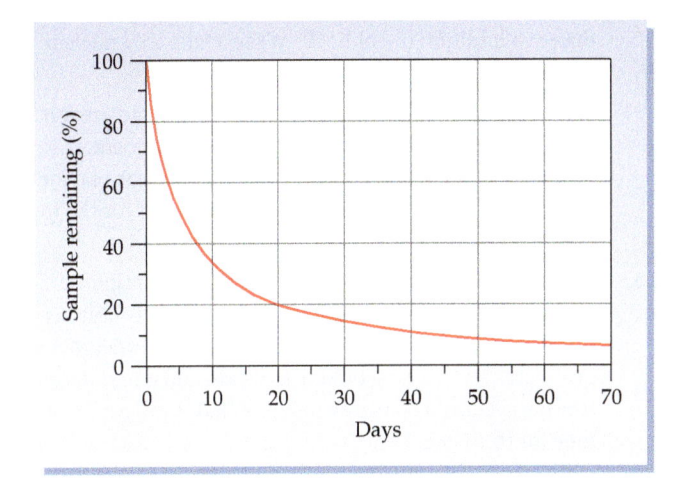

ADDITIONAL PROBLEMS

RADIOACTIVITY

11.30 What does it mean to say that a substance is radioactive?

11.31 Describe how α radiation, β radiation, γ radiation, positron emission, and electron capture differ.

11.32 List three of the five ways in which a nuclear reaction differs from a chemical reaction.

11.33 What happens when ionizing radiation strikes an atom in a chemical compound?

11.34 How does ionizing radiation lead to cell damage?

11.35 What are the main sources of background radiation?

11.36 How can a nucleus emit an electron during β decay when there are no electrons present in the nucleus to begin with?

11.37 What is the difference between an α particle and a helium atom?

NUCLEAR DECAY AND TRANSMUTATION

11.38 What does it mean to say that a nuclear equation is balanced?

11.39 What are transuranium elements, and how are they made?

11.40 What happens to the mass number and atomic number of an atom that emits an α particle? A β particle?

11.41 What happens to the mass number and atomic number of an atom that emits a γ ray? A positron?

11.42 How does nuclear fission differ from normal radioactive decay?

11.43 What characteristic of uranium-235 fission causes a chain reaction?

11.44 What products result from radioactive decay of the following β emitters?

(a) $^{35}_{16}S$ (b) $^{24}_{10}Ne$ (c) $^{90}_{38}Sr$

11.45 What radioactive nuclides will produce the following products following α decay?

(a) $^{186}_{76}Os$ (b) $^{204}_{85}At$ (c) $^{241}_{94}Pu$

11.46 Identify the starting radioisotopes needed to balance each of these nuclear reactions:

(a) $? + ^{4}_{2}He \longrightarrow ^{113}_{49}In$ (b) $? + ^{4}_{2}He \longrightarrow ^{13}_{7}N + ^{1}_{0}n$

11.47 Identify the radioisotope product needed to balance each of these nuclear reactions:

(a) $^{26}_{11}\text{Na} \longrightarrow \ ? + ^{0}_{-1}\text{e}$ (b) $^{212}_{83}\text{Bi} \longrightarrow \ ? + ^{4}_{2}\text{He}$

11.48 Balance the following equations for the nuclear fission of $^{235}_{92}\text{U}$:

(a) $^{235}_{92}\text{U} + ^{1}_{0}\text{n} \longrightarrow ^{160}_{62}\text{Sm} + ^{72}_{30}\text{Zn} + ?\ ^{1}_{0}\text{n}$

(b) $^{235}_{92}\text{U} + ^{1}_{0}\text{n} \longrightarrow ^{87}_{35}\text{Br} + ? + 3\ ^{1}_{0}\text{n}$

11.49 Complete the following nuclear equations and identify each as α decay, β decay, positron emission, or electron capture:

(a) $^{126}_{50}\text{Sn} \longrightarrow \ ? + ^{126}_{51}\text{Sb}$

(b) $^{210}_{88}\text{Ra} \longrightarrow \ ? + ^{206}_{86}\text{Rn}$

(c) $^{76}_{36}\text{Kr} + ? \longrightarrow ^{76}_{35}\text{Br}$

11.50 For centuries, alchemists dreamed of turning base metals into gold. The dream finally became reality when it was shown that mercury-198 can be converted into gold-198 when bombarded by neutrons. What small particle is produced in addition to gold-198? Write a balanced nuclear equation for the reaction.

11.51 Cobalt-60 (half-life = 5.3 years) is used to irradiate food, to treat cancer, and to disinfect surgical equipment. It is produced by irradiation of cobalt-59 in a nuclear reactor. It decays to nickel-60. Write nuclear equations for the formation and decay reactions of cobalt-60.

11.52 Bismuth-212 attaches readily to monoclonal antibodies and is used in the treatment of various cancers. This bismuth-212 is formed after the parent isotope undergoes a decay series consisting of four α decays and one β decay. (the decays could be in any order). What is the parent isotope for this decay series?

11.53 Meitnerium-266 ($^{266}_{109}\text{Mt}$) was prepared in 1982 by bombardment of bismuth-209 atoms with iron-58. What other product must also have been formed? Write a balanced nuclear equation for the transformation.

HALF-LIFE

11.54 What does it mean when we say that strontium-90, a waste product of nuclear power plants, has a half-life of 28.8 years?

11.55 How many half lives must pass for the mass of a radioactive sample to decrease to 35% of the original mass? To 10%?

11.56 Selenium-75, a β emitter with a half-life of 120 days, is used medically for pancreas scans.

(a) Approximately how long would it take for a 0.050 g sample of selenium-75 to decrease to 0.010 g?

(b) Approximately how much selenium-75 would remain from a 0.050 g sample that has been stored for one year? (*Hint*: How many half-lives are in one year?)

11.57 Approximately how long would it take a sample of selenium-75 to lose 75% of its radioactivity? To lose 99%? (See Problem 11.56.)

11.58 The half-life of mercury-197 is 64.1 hours. If a patient undergoing a kidney scan is given 5.0 ng of mercury-197, how much will remain after 7 days? After 30 days?

11.59 Gold-198, a β emitter used to treat leukemia, has a half-life of 2.695 days. The standard dosage is about 1.0 mCi/kg body weight.

(a) What is the product of the β emission of gold-198?

(b) How long does it take a 30.0 mCi sample of gold-198 to decay so that only 3.75 mCi remains?

(c) How many millicuries are required in a single dosage administered to a 70.0 kg adult?

MEASURING RADIOACTIVITY

11.60 Describe how a Geiger counter works.

11.61 Describe how a film badge works.

11.62 Describe how a scintillation counter works.

11.63 Why are rems the preferred units for measuring the health effects of radiation?

11.64 Approximately what amount (in rems) of short-term exposure to radiation produces noticeable effects in humans?

11.65 Match each unit in the left column with the property being measured in the right column:

1. curie   (a) Ionizing intensity of radiation
2. rem   (b) Amount of tissue damage
3. rad   (c) Number of disintegrations per second
4. roentgen (d) Amount of radiation per gram of tissue

11.66 Technetium-99*m* is used for radioisotope-guided surgical biopsies of certain bone cancers. A patient must receive an injection of 28 mCi of technetium-99*m* 6–12 hours before surgery. If the activity of the solution is 15 mCi, what volume should be injected?

11.67 Sodium-24 is used to study the circulatory system and to treat chronic leukemia. It is administered in the form of saline (NaCl) solution, with a therapeutic dosage of 180 μCi/kg body weight.

(a) What dosage (in mCi) would be administered to a 68 kg adult patient?

(b) How many milliliters of a 6.5 mCi/mL solution are needed to treat a 68 kg adult?

11.68 A selenium-75 source is producing 300 rem at a distance of 2.0 m?

(a) What is its intensity at 16 m?

(b) What is its intensity at 25 m?

11.69 If a radiation source has an intensity of 650 rem at 1.0 m, what distance is needed to decrease the intensity of exposure to below 25 rem, the level at which no effects are detectable?

CHEMISTRY IN ACTION

11.70 What are the three main classes of techniques used in nuclear medicine? Give an example of each. [*Medical Uses of Radioactivity, p. 338*]

11.71 A 2 mL solution containing 1.25 μCi/mL is injected into the bloodstream of a patient. After dilution, a 1.00 mL sample is withdrawn and found to have an activity of

2.6 × 10⁻⁴ μCi. Calculate total blood volume. [*Medical Uses of Radioactivity, p. 338*]

11.72 What is the purpose of food irradiation, and how does it work? [*Irradiated Food, p. 345*]

11.73 What kind of radiation is used to treat food? [*Irradiated Food, p. 345*]

11.74 What are the advantages of CT and PET relative to conventional X rays? [*Body Imaging, p. 348*]

11.75 What advantages does MRI have over CT and PET imaging? [*Body Imaging, p. 348*]

GENERAL QUESTIONS AND PROBLEMS

11.76 Film badge dosimeters typically include filters to target specific types of radiation. A film badge is constructed that includes a region containing a tin foil filter, a region containing a plastic film filter, and a region with no filter. Which region monitors exposure to α-radiation? Which monitors exposure to β-radiation? Which monitors γ-radiation? Explain.

11.77 Some dried beans with a $^{14}C/^{12}C$ ratio one-eighth of the current value are found in an old cave. How old are the beans?

11.78 Harmful chemical spills can often be cleaned up by treatment with another chemical. For example, a spill of H_2SO_4 might be neutralized by addition of $NaHCO_3$. Why is it that the harmful radioactive wastes from nuclear power plants cannot be cleaned up as easily?

11.79 Why is a scintillation counter or Geiger counter more useful for determining the existence and source of a new radiation leak than a film badge?

11.80 A Geiger counter records an activity of 28 counts per minute (cpm) when located at a distance of 10 m. What will be the activity (in cpm) at a distance of 5 m?

11.81 Most of the stable isotopes for elements lighter than Ca-40 have equal numbers of protons and neutrons in the nucleus. What would be the most probable decay mode for an isotope that had more protons than neutrons? More neutrons than protons?

11.82 Technetium-99*m*, used for brain scans and to monitor heart function, is formed by decay of molybdenum-99.

(a) By what type of decay does ^{99}Mo produce ^{99m}Tc?

(b) Molybdenum-99 is formed by neutron bombardment of a natural isotope. If one neutron is absorbed and there are no other by-products of this process, from what isotope is ^{99}Mo formed?

11.83 The half-life of technetium-99*m* (Problem 11.82) is 6.01 hours. If a sample with an initial activity of 15 μCi is injected into a patient, what is the activity in 24 hours, assuming that none of the sample is excreted?

11.84 Plutonium-238 is an α emitter used to power batteries for heart pacemakers.

(a) Write the balanced nuclear equation for this emission.

(b) Why is a pacemaker battery enclosed in a metal case before being inserted into the chest cavity?

11.85 Sodium-24, a beta-emitter used in diagnosing circulation problems, has a half-life of 15 hours.

(a) Write the balanced nuclear equation for this emission.

(b) What fraction of sodium-24 remains after 50 hours?

11.86 High levels of radioactive fallout after the 1986 accident at the Chernobyl nuclear power plant in what is now Ukraine resulted in numerous miscarriages in humans and many instances of farm animals born with severe defects. Why are embryos and fetuses particularly susceptible to the effects of radiation?

11.87 One way to demonstrate the dose factor of ionizing radiation (penetrating distance × ionizing energy) is to think of radiation as cookies. Imagine that you have four cookies—an α cookie, a β cookie, a γ cookie, and a neutron cookie. Which one would you eat, which would you hold in your hand, which would you put in your pocket, and which would you throw away?

11.88 What are the main advantages of nuclear fission relative to nuclear fusion as an energy source? What are the drawbacks?

11.89 Although turning lead into gold in a nuclear reactor is technologically feasible (Problem 11.50), it is not economical. It is far easier to convert gold into lead. The process involves a series of neutron bombardments, and can be summarized as

$$^{197}_{79}Au + ?\, ^{1}_{0}n \longrightarrow\, ^{204}_{82}Pb + ?\, ^{0}_{-1}e$$

How many neutrons and β particles are involved?

11.90 Balance the following transmutation reactions:

(a) $^{253}_{99}Es + ? \longrightarrow\, ^{256}_{101}Md + \, ^{1}_{0}n$

(b) $^{250}_{98}Cf + \, ^{11}_{5}B \longrightarrow ? + 4\, ^{1}_{0}n$

11.91 The most abundant isotope of uranium, ^{238}U, does not undergo fission. In a *breeder reactor*, however, a ^{238}U atom captures a neutron and emits 2 beta particles to make a fissionable isotope of plutonium, which can then be used as fuel in a nuclear reactor. Write the balanced nuclear equation.

11.92 Boron is used in *control rods* for nuclear reactors because it can absorb neutrons to keep a chain reaction from becoming supercritical, and decays by emitting alpha particles. Balance the equation:

$$^{10}_{5}B + \, ^{1}_{0}n \longrightarrow ? + \, ^{4}_{2}He$$

11.93 Thorium-232 decays by a 10-step series, ultimately yielding lead-208. How many α particles and how many β particles are emitted?

11.94 Californium-246 is formed by bombardment of uranium-238 atoms. If four neutrons are formed as by-products, what particle is used for the bombardment?

11.95 The most recently discovered element 117 (Ununseptium, Uus) was synthesized by nuclear transmutation reactions in which berkelium-249 was bombarded with calcium-48. Two isotopes of Uus were identified:

$$^{48}_{20}Ca + \, ^{249}_{97}Bk \longrightarrow\, ^{294}_{117}Uus + ?\, ^{1}_{0}n$$

$$^{48}_{20}Ca + \, ^{249}_{97}Bk \longrightarrow\, ^{293}_{117}Uus + ?\, ^{1}_{0}n$$

How many neutrons are produced in each reaction?

CHAPTER 12

Introduction to Organic Chemistry: Alkanes

CONTENTS

- 12.1 The Nature of Organic Molecules
- 12.2 Families of Organic Molecules: Functional Groups
- 12.3 The Structure of Organic Molecules: Alkanes and Their Isomers
- 12.4 Drawing Organic Structures
- 12.5 The Shapes of Organic Molecules
- 12.6 Naming Alkanes
- 12.7 Properties of Alkanes
- 12.8 Reactions of Alkanes
- 12.9 Cycloalkanes
- 12.10 Drawing and Naming Cycloalkanes

◄ The gasoline, kerosene, and other products of this petroleum refinery are primarily mixtures of simple organic compounds called alkanes.

CHAPTER GOALS

1. **What are the basic properties of organic compounds?**
   THE GOAL: Be able to identify organic compounds and the types of bonds contained in them. (◀◀ A, B, D, E.)

2. **What are functional groups, and how are they used to classify organic molecules?**
   THE GOAL: Be able to classify organic molecules into families by functional group. (◀◀ C, B, E.)

3. **What are isomers?**
   THE GOAL: Be able to recognize and draw constitutional isomers. (◀◀ C, D.)

4. **How are organic molecules drawn?**
   THE GOAL: Be able to convert between structural formulas and condensed or line structures. (◀◀ C, D.)

5. **What are alkanes and cycloalkanes, and how are they named?**
   THE GOAL: Be able to name an alkane or cycloalkane from its structure, or write the structure, given the name. (◀◀ C.)

6. **What are the general properties and chemical reactions of alkanes?**
   THE GOAL: Be able to describe the physical properties of alkanes and the products formed in the combustion and halogenation reactions of alkanes. (◀◀ E.)

CONCEPTS TO REVIEW

A. Covalent Bonds
(Sections 4.1 and 4.2)

B. Multiple Covalent Bonds
(Section 4.3)

C. Drawing Lewis Structures
(Section 4.7)

D. VSEPR and Molecular Shapes
(Section 4.8)

E. Polar Covalent Bonds
(Section 4.9)

The study of chemistry progressed in the 1700s as scientists isolated substances from the world around them and examined their properties. Researchers began to notice differences between the properties of compounds obtained from living sources and those obtained from minerals. As a result, the term *organic chemistry* was introduced to describe the study of compounds derived from living organisms, while *inorganic chemistry* was used to refer to the study of compounds from minerals.

It was long believed that organic compounds could only be obtained from a living source; this concept, known as *vitalism*, hindered the study of these types of molecules because vitalist chemists believed that organic materials could not be synthesized from inorganic components. It wasn't until 1828 that Friedrich Wöhler first prepared an organic compound, urea, from an inorganic salt, ammonium cyanate, disproving the theory of vitalism and truly pioneering the field of organic chemistry.

Today, we know that there are no fundamental differences between organic and inorganic compounds: The same scientific principles are applicable to both. The only common characteristic of compounds from living sources is that they contain the element carbon as their primary component. Thus, organic chemistry is now defined as the study of carbon-based compounds.

Why is carbon special? The answer derives from its position in the periodic table. As a group 4A nonmetal, carbon atoms have the unique ability to form four strong covalent bonds. Also, unlike atoms of other elements, carbon atoms can readily form strong bonds with other carbon atoms to produce long chains and rings. As a result, only carbon is able to form such a diverse and immense array of compounds, from methane with 1 carbon atom to DNA with billions of carbons.

12.1 The Nature of Organic Molecules

Let us begin a study of **organic chemistry**—the chemistry of carbon compounds—by reviewing what we have seen in earlier chapters about the structures of organic molecules:

- **Carbon is tetravalent; it always forms four bonds** (Section 4.2). In methane, for example, carbon is connected to 4 hydrogen atoms:

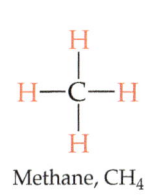

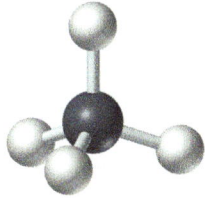

Methane, CH_4

Organic chemistry The study of carbon compounds.

- **Organic molecules have covalent bonds** (Section 4.2). In ethane, for example, the bonds result from the sharing of 2 electrons, either between 2 C atoms or a C and an H atom:

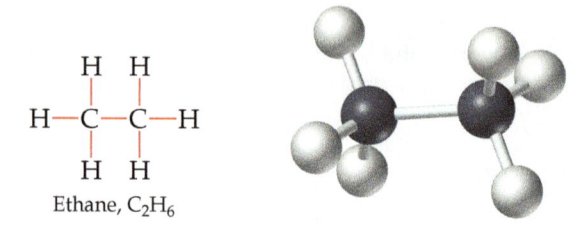

  Ethane, C_2H_6

- **When carbon bonds to a more electronegative element, polar covalent bonds result** (Section 4.9). In chloromethane, for example, the electronegative chlorine atom attracts electrons more strongly than carbon, resulting in polarization of the C—Cl bond so that carbon and hydrogens have a partial positive charge, $\delta+$, and chlorine has a partial negative charge, $\delta-$. It is useful to think of polar covalent bonds in this manner, as it will later help to explain their reactivity. In electrostatic potential maps (Section 4.9), the chlorine atom is therefore in the red region of the map and the carbon atom in the blue region:

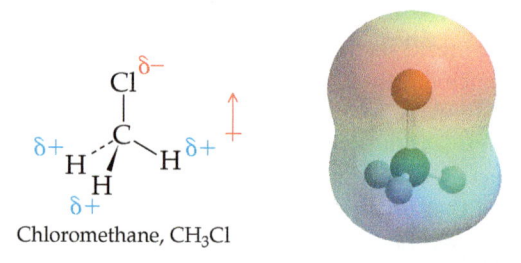

  Chloromethane, CH_3Cl

- **Carbon forms multiple covalent bonds by sharing more than 2 electrons with a neighboring atom** (Section 4.3). In ethylene, for example, the 2 carbon atoms share 4 electrons in a double bond; in acetylene (also called ethyne), the 2 carbons share 6 electrons in a triple bond:

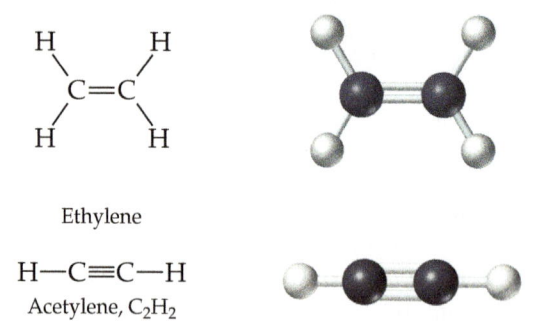

  Ethylene

  H—C≡C—H
  Acetylene, C_2H_2

- **Organic molecules have specific three-dimensional shapes** (Section 4.8). When carbon is bonded to 4 atoms, as in methane, CH_4, the bonds are oriented toward the four corners of a regular tetrahedron with carbon in the center. Such three-dimensionality is commonly shown using normal lines for bonds in the plane of the page, dashed lines for bonds receding behind the page, and wedged lines for bonds coming out of the page:

- **Organic molecules often contain nitrogen and oxygen in addition to carbon and hydrogen** (Section 4.7). Nitrogen can form single, double, and triple bonds to carbon, while oxygen can form single and double bonds. Hydrogen can only form single bonds to carbon:

$$\text{C---N} \quad \text{C---O} \quad \text{C---H}$$
$$\text{C=N} \quad \text{C=O}$$
$$\text{C}\equiv\text{N}$$

Covalent bonding makes organic compounds quite different from the inorganic compounds we have been concentrating on up to this point. For example, inorganic compounds such as NaCl have high melting points and high boiling points because they consist of large numbers of oppositely charged ions held together by strong electrical attractions. By contrast, organic compounds consist of atoms joined by covalent bonds, forming individual molecules. Because the organic molecules are attracted to one another only by weak non-ionic intermolecular forces, organic compounds generally have lower melting and boiling points than inorganic salts. As a result, many simple organic compounds are liquids or low melting solids at room temperature, and a few are gases.

Other important differences between organic and inorganic compounds include solubility and electrical conductivity. Whereas many inorganic compounds dissolve in water to yield solutions of ions that conduct electricity, most organic compounds are insoluble in water, and almost all of those that are soluble do not conduct electricity. Only small polar organic molecules, such as glucose and ethyl alcohol, or large molecules with many polar groups, such as some proteins, interact with water molecules through both dipole–dipole interactions and/or hydrogen bonding and thus dissolve in water. This lack of water solubility for organic compounds has important practical consequences, varying from the difficulty in removing greasy dirt and cleaning up environmental oil spills to drug delivery.

▶▶ Other unique properties of ionic compounds are discussed in Section 3.4.

▶▶ Recall from Section 8.11 the various intermolecular forces: dipole–dipole forces, London dispersion forces, and hydrogen bonds.

▶▶ Section 9.9 explores how anions and cations in solution conduct electric current.

▶▶ Recall from Section 9.2 that a compound is only soluble when the intermolecular forces between solvent and solute are comparable in strength to the intermolecular forces of the pure solvent or solute.

◀ Oil spills can be a serious environmental problem because oil is insoluble in water.

LOOKING AHEAD ▶▶ The interior of a living cell is largely a water solution that contains many hundreds of different compounds. In later chapters, we will see how cells use membranes composed of water-insoluble organic molecules to enclose their watery interiors and to regulate the flow of substances across the cell boundary.

12.2 Families of Organic Molecules: Functional Groups

More than 18 *million* organic compounds are described in the scientific literature. Each of these 18 million compounds has unique chemical and physical properties, and many of them also have unique biological properties (both desired and undesired). How can we ever understand them all?

360 CHAPTER 12 Introduction to Organic Chemistry: Alkanes

Chemists have learned through experience that organic compounds can be classified into families according to their structural features, and that the chemical behavior of family members is often predictable based on their specific grouping of atoms. Instead of 18 million compounds with seemingly random chemical reactivity, there are just a few general families of organic compounds whose chemistry falls into simple patterns.

The structural features that allow us to classify organic compounds into distinct chemical families are called **functional groups**. A functional group is an atom or group of atoms that has a characteristic physical and chemical behavior. Each functional group is always part of a larger molecule, and a molecule may have more than one class of functional group present, as we shall soon see. An important property of functional groups is that a given functional group *tends to undergo the same reactions in every molecule that contains it*. For example, the carbon–carbon double bond is a common functional group. Ethylene (C_2H_4), the simplest compound with a double bond, undergoes many chemical reactions similar to those of oleic acid ($C_{18}H_{34}O_2$), a much larger and more complex compound that also contains a double bond. Both, for example, react with hydrogen gas in the same manner, as shown in Figure 12.1. These identical reactions with hydrogen are typical: *The chemistry of an organic molecule is primarily determined by the functional groups it contains, not by its size or complexity.*

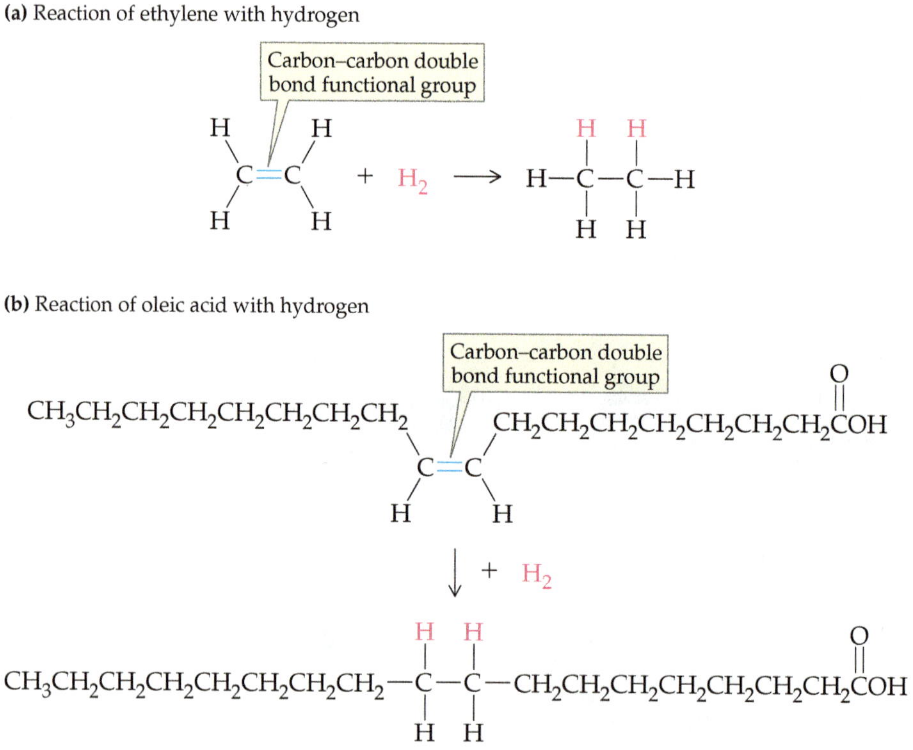

> **Functional group** An atom or group of atoms within a molecule that has a characteristic physical and chemical behavior.

> ▶ **Figure 12.1**
> **The reactions of (a) ethylene and (b) oleic acid with hydrogen.** The carbon–carbon double-bond functional group adds 2 hydrogen atoms in both cases, regardless of the complexity of the rest of the molecule.

(a) Reaction of ethylene with hydrogen

(b) Reaction of oleic acid with hydrogen

Table 12.1 lists some of the most important families of organic molecules and their distinctive functional groups. Compounds that contain a C=C double bond, for instance, are in the *alkene* family; compounds that have an —OH group bound to a tetravalent carbon are in the *alcohol* family; and so on. To aid in identifying the organic functional groups you will encounter, we have included an Organic Functional Group Flow Scheme (Figure 12.5) at the end of the chapter; it should be used in conjunction with Table 12.1. You will find it helpful as you proceed through the remainder of this text.

Much of the chemistry discussed in this and the next five chapters is the chemistry of the families listed in Table 12.1, so it is best to learn the names and become familiar with their structures now. Note that they fall into four groups:

> **Hydrocarbon** An organic compound that contains only carbon and hydrogen.

- The first four families in Table 12.1 are **hydrocarbons**, organic compounds that contain only carbon and hydrogen. *Alkanes* have only single bonds and contain

SECTION 12.2 Families of Organic Molecules: Functional Groups 361

TABLE 12.1 Some Important Families of Organic Molecules

FAMILY NAME	FUNCTIONAL GROUP STRUCTURE*	SIMPLE EXAMPLE	LINE STRUCTURE	NAME ENDING
Alkane	Contains only C—H and C—C single bonds	$CH_3CH_2CH_3$ Propane		-ane
Alkene	\C=C/	$H_2C=CH_2$ Ethylene		-ene
Alkyne	—C≡C—	H—C≡C—H Acetylene (Ethyne)	H≡H	-yne
Aromatic	(benzene ring)	Benzene		None
Alkyl halide	—C—X (X=F, Cl, Br, I)	CH_3CH_2Cl Ethyl chloride	Cl	None
Alcohol	—C—O—H	CH_3CH_2OH Ethyl alcohol (Ethanol)	OH	-ol
Ether	—C—O—C—	CH_3CH_2—O—CH_2CH_3 Diethyl ether		None
Amine	—C—N	$CH_3CH_2NH_2$ Ethylamine	NH_2	-amine
Aldehyde	—C—C(=O)—H	CH_3—C(=O)—H Acetaldehyde (Ethanal)		-al
Ketone	—C—C(=O)—C—	CH_3—C(=O)—CH_3 Acetone		-one
Carboxylic acid	—C—C(=O)—OH	CH_3—C(=O)—OH Acetic acid	OH	-ic acid
Anhydride	—C—C(=O)—O—C(=O)—C—	CH_3—C(=O)—O—C(=O)—CH_3 Acetic anhydride		None
Ester	—C—C(=O)—O—C—	CH_3—C(=O)—O—CH_3 Methyl acetate	OCH_3	-ate
Amide	—C—C(=O)—NH_2, —C—C(=O)—N—H, —C—C(=O)—N—	CH_3—C(=O)—NH_2 Acetamide	NH_2	-amide
Thiol	—C—SH	CH_3CH_2SH Ethyl thiol	SH	None
Disulfide	C—S—S—C	CH_3SSCH_3 Dimethyl disulfide	S—S	None
Sulfide	C—S—C	$CH_3CH_2SCH_3$ Ethyl methyl sulfide	S	None

*The bonds whose connections are not specified are assumed to be attached to carbon or hydrogen atoms in the rest of the molecule.

no functional groups. As we will see, the absence of functional groups makes alkanes relatively unreactive. *Alkenes* contain a carbon–carbon double-bond functional group; *alkynes* contain a carbon–carbon triple-bond functional group; and *aromatic* compounds contain a six-membered ring of carbon atoms with three alternating double bonds.

- The next four families in Table 12.1 have functional groups that contain only single bonds and have a carbon atom bonded to an electronegative atom. *Alkyl halides* have a carbon–halogen bond; *alcohols* have a carbon–oxygen bond; *ethers* have two carbons bonded to the same oxygen; and *amines* have a carbon–nitrogen bond.
- The next six families in Table 12.1 have functional groups that contain a carbon–oxygen double bond: *aldehydes, ketones, carboxylic acids, anhydrides, esters,* and *amides*.
- The remaining three families in Table 12.1 have functional groups that contain sulfur: *thioalcohols* (known simply as *thiols*), *sulfides*, and *disulfides*. These three families play an important role in protein function (Chapter 18).

Worked Example 12.1 Molecular Structures: Identifying Functional Groups

To which family of organic compounds do the following compounds belong? Explain.

(a) $H-\underset{\underset{H}{|}}{\overset{\overset{H}{|}}{C}}-\underset{}{\overset{H}{C}}=\underset{}{\overset{H}{C}}-\underset{\underset{H}{|}}{\overset{\overset{H}{|}}{C}}-H$

(b) $H-\underset{\underset{H}{|}}{\overset{\overset{H}{|}}{C}}-\underset{\underset{H}{|}}{\overset{\overset{H}{|}}{C}}-\underset{\underset{O-H}{|}}{\overset{\overset{H}{|}}{C}}-\underset{\underset{H}{|}}{\overset{\overset{H}{|}}{C}}-H$

(c) (benzene ring with CH_2-CH_3 substituent)

(d) $H-\underset{\underset{H}{|}}{\overset{\overset{H}{|}}{C}}-\underset{\underset{H}{|}}{\overset{\overset{H}{|}}{C}}-\underset{\underset{O}{\|}}{C}-\underset{\underset{H}{|}}{\overset{\overset{H}{|}}{C}}-\underset{\underset{H}{|}}{\overset{\overset{H}{|}}{C}}-H$

(e) $CH_3-\underset{\underset{}{|}}{\overset{\overset{NH_2}{|}}{CH}}-\underset{\underset{O}{\|}}{C}-NH_2$

(f) (phenyl)$-CH_2-\underset{\underset{S-S-CH_3}{|}}{CH}-CH_3$

ANALYSIS Use the Organic Functional Group Flow Scheme (Figure 12.5, see end of chapter) and Table 12.1 to identify each functional group, and name the corresponding family to which the compound belongs. Begin by determining what elements are present and whether multiple bonds are present.

SOLUTION
(a) This compound contains only carbon and hydrogen atoms, so it is a *hydrocarbon*. There is only one carbon–carbon double bond, so it is an *alkene*.
(b) This compound contains an oxygen and has only single bonds. The presence of the O—H group bonded to tetravalent carbon identifies this compound as an *alcohol*.
(c) This compound also contains only carbon and hydrogen atoms, which identifies it as a *hydrocarbon*. It has three double bonds in a ring. The six-membered carbon ring with alternating double bonds also identifies this compound as an *aromatic* hydrocarbon compound.
(d) This molecule contains an oxygen that is double bonded to a carbon (a *carbonyl group*, discussed in Chapter 16), and there is no singly bound oxygen or nitrogen also connected to the carbon. The carbon–oxygen double bond is connected to two other carbons (as opposed to a hydrogen); that identifies this compound as a *ketone*.
(e) Many of the organic molecules we will come across have more than one functional group present in the same molecule; in these cases we will classify the molecule as belonging to multiple functional group families. This molecule contains oxygen and nitrogen in addition to carbon and hydrogen, so it is not a hydrocarbon. The presence of the carbonyl group further classifies this molecule, but here we run into a problem: one —NH₂ is attached to the

carbonyl but the other —NH₂ is not. This leads us to conclude that there are two functional groups present: an *amide* and an *amine*:

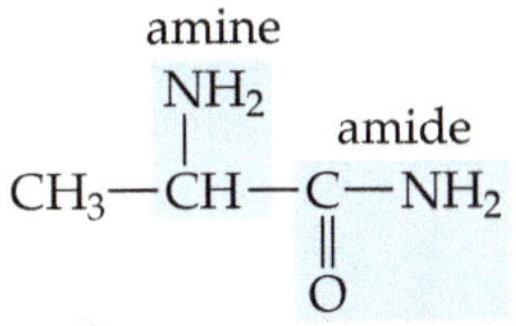

$$\underset{\text{amine}}{\text{NH}_2} \\ CH_3-CH-\underset{\underset{O}{\|}}{C}-NH_2 \;\; \text{amide}$$

(f) This molecule also contains two functional groups: a ring containing alternating carbon–carbon single and double bonds as well as an S—S group. From our flow scheme, we trace the double bond to indicate we have an aromatic hydrocarbon, while the sulfurs indicate the presence of a disulfide:

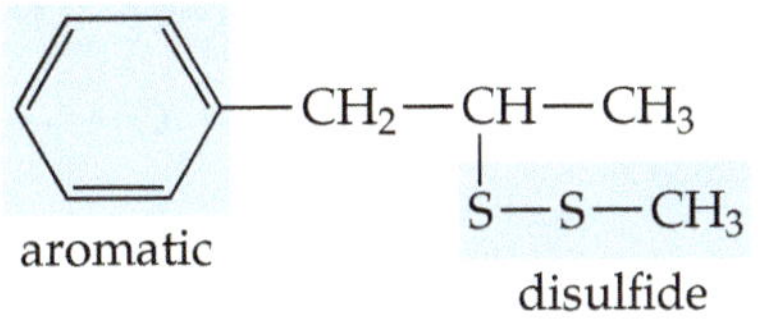

$$\underset{\text{aromatic}}{\bigcirc}-CH_2-\underset{\underset{\text{disulfide}}{S-S-CH_3}}{CH}-CH_3$$

Worked Example 12.2 Molecular Structures: Drawing Functional Groups

Given the family of organic compounds to which the compound belongs, propose structures for compounds having the following chemical formulas.

(a) An amine having the formula C_2H_7N
(b) An alkyne having the formula C_3H_4
(c) An ether having the formula $C_4H_{10}O$

ANALYSIS Identify the functional group for each compound from Table 12.1. Once the atoms in this functional group are eliminated from the chemical formula, the remaining structure can be determined. (Remember that each carbon atom forms four bonds, nitrogen forms three bonds, oxygen forms two bonds, and hydrogen forms only one bond.)

SOLUTION

(a) Amines have a C—NH₂ group. Eliminating these atoms from the formula leaves 1 C atom and 5 H atoms. Since only the carbons are capable of forming more than one bond, the 2 C atoms must be bonded together. The remaining H atoms are then bonded to the carbons until each C has 4 bonds.

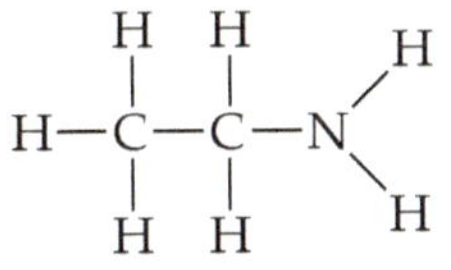

$$H-\underset{H}{\overset{H}{|}}C-\underset{H}{\overset{H}{|}}C-N\underset{H}{\overset{H}{\diagup}}$$

(b) The alkynes contain a C≡C bond. This leaves 1 C atom and 4 H atoms. Attach this C to one of the carbons in the triple bond, and then distribute the H atoms until each carbon has a full complement of four bonds.

$$H-\underset{H}{\overset{H}{|}}C-C\equiv C-H$$

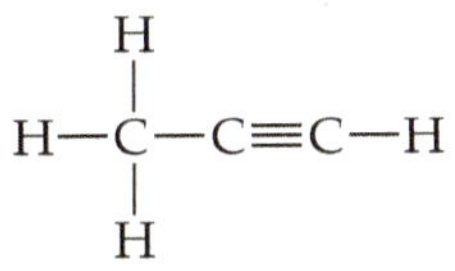

(c) The ethers contain a C—O—C group. Eliminating these atoms leaves 2 C atoms and 10 H atoms. The C atoms can be distributed on either end of the ether group, and the H atoms are then distributed until each carbon atom has a full complement of four bonds.

$$H-\underset{H}{\overset{H}{|}}C-\underset{H}{\overset{H}{|}}C-O-\underset{H}{\overset{H}{|}}C-\underset{H}{\overset{H}{|}}C-H \quad \text{or} \quad H-\underset{H}{\overset{H}{|}}C-\underset{H}{\overset{H}{|}}C-\underset{H}{\overset{H}{|}}C-O-\underset{H}{\overset{H}{|}}C-H$$

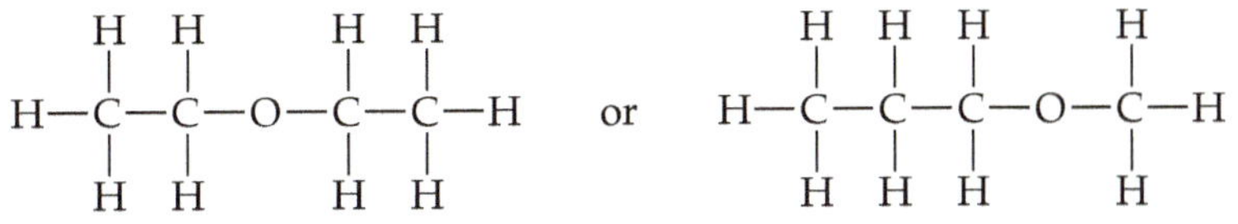

PROBLEM 12.1
Many organic compounds contain more than one functional group. Locate and identify the functional groups in (a) lactic acid, from sour milk; (b) methyl methacrylate, used in making Lucite and Plexiglas; and (c) phenylalanine, an amino acid found in proteins.

(a) 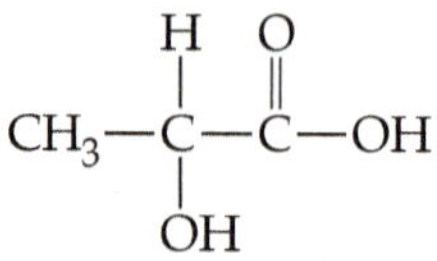 CH₃—CH(OH)—C(=O)—OH

(b) CH₂=C(CH₃)—C(=O)—O—CH₃

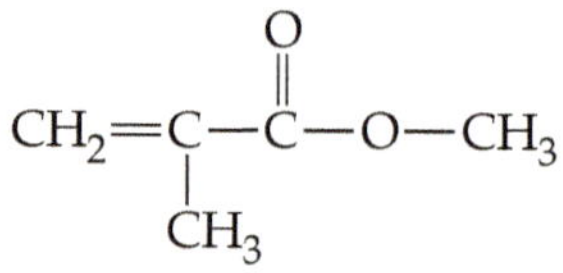

(c) phenyl—CH₂—CH(NH₂)—C(=O)—OH

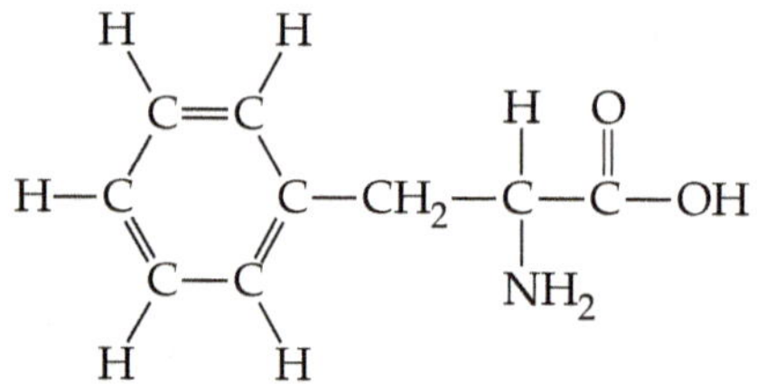

PROBLEM 12.2
Propose structures for molecules that fit the following descriptions:
(a) C_3H_6O containing an aldehyde functional group
(b) C_3H_6O containing a ketone functional group
(c) $C_3H_6O_2$ containing a carboxylic acid functional group

12.3 The Structure of Organic Molecules: Alkanes and Their Isomers

Alkane A hydrocarbon that has only single bonds.

Hydrocarbons that contain only single bonds belong to the family of organic molecules called **alkanes**. Imagine how 1 carbon and 4 hydrogens can combine, and you will realize there is only one possibility: methane, CH_4. Now, imagine how 2 carbons and 6 hydrogens can combine—only ethane, CH_3CH_3, is possible. Likewise, with the combination of 3 carbons with 8 hydrogens—only propane, $CH_3CH_2CH_3$, is possible. The general rule for *all* hydrocarbons except methane is that each carbon *must* be bonded to at least one other carbon. The carbon atoms bond together to form the "backbone" of the compound, with the hydrogens on the periphery. The general formula for alkanes is C_nH_{2n+2}, where *n* is the number of carbons in the compound.

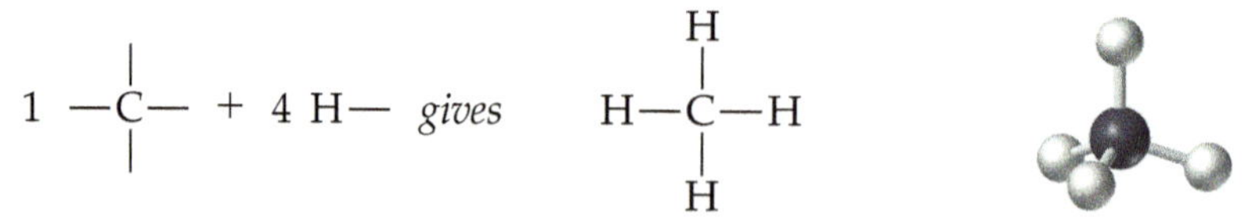

1 —C— + 4 H— gives H—C—H (Methane)

2 —C— + 6 H— gives H—C—C—H (Ethane)

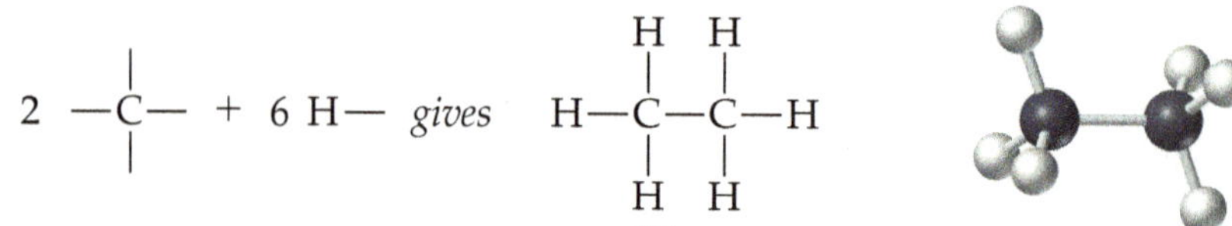

3 —C— + 8 H— gives H—C—C—C—H (Propane)

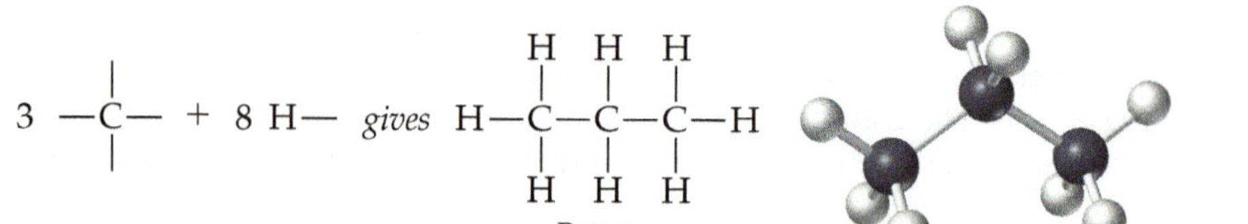

As larger numbers of carbons and hydrogens combine, the ability to form *isomers* arises. Compounds that have the same molecular formula but different structural for-

SECTION 12.3 The Structure of Organic Molecules: Alkanes and Their Isomers 365

mulas are called **isomers** of one another. For example, there are two ways in which molecules that have the formula C_4H_{10} can be formed. The 4 carbons can either be joined in a contiguous row or have a branched arrangement:

Isomers Compounds with the same molecular formula but different structures.

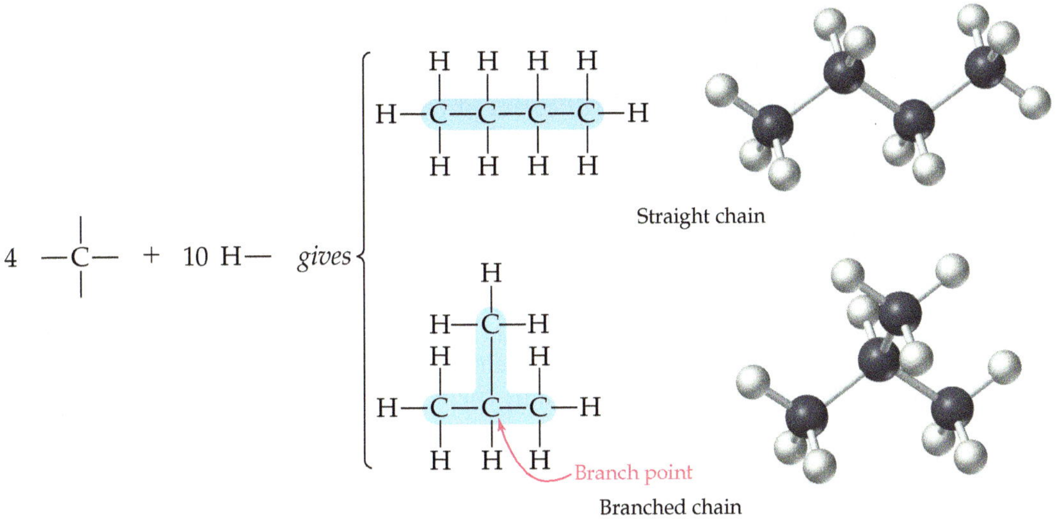

$$4 \ -\overset{|}{\underset{|}{C}}- \ + \ 10 \ H- \quad gives$$

H H H H
| | | |
H—C—C—C—C—H
| | | |
H H H H

Straight chain

```
          H
          |
      H—C—H
  H   H   H
  |   |   |
H—C—C—C—H
  |   |   |
  H   H   H
          └─ Branch point
```

Branched chain

The same is seen with the molecules that have the formula C_5H_{12}, for which three isomers are possible:

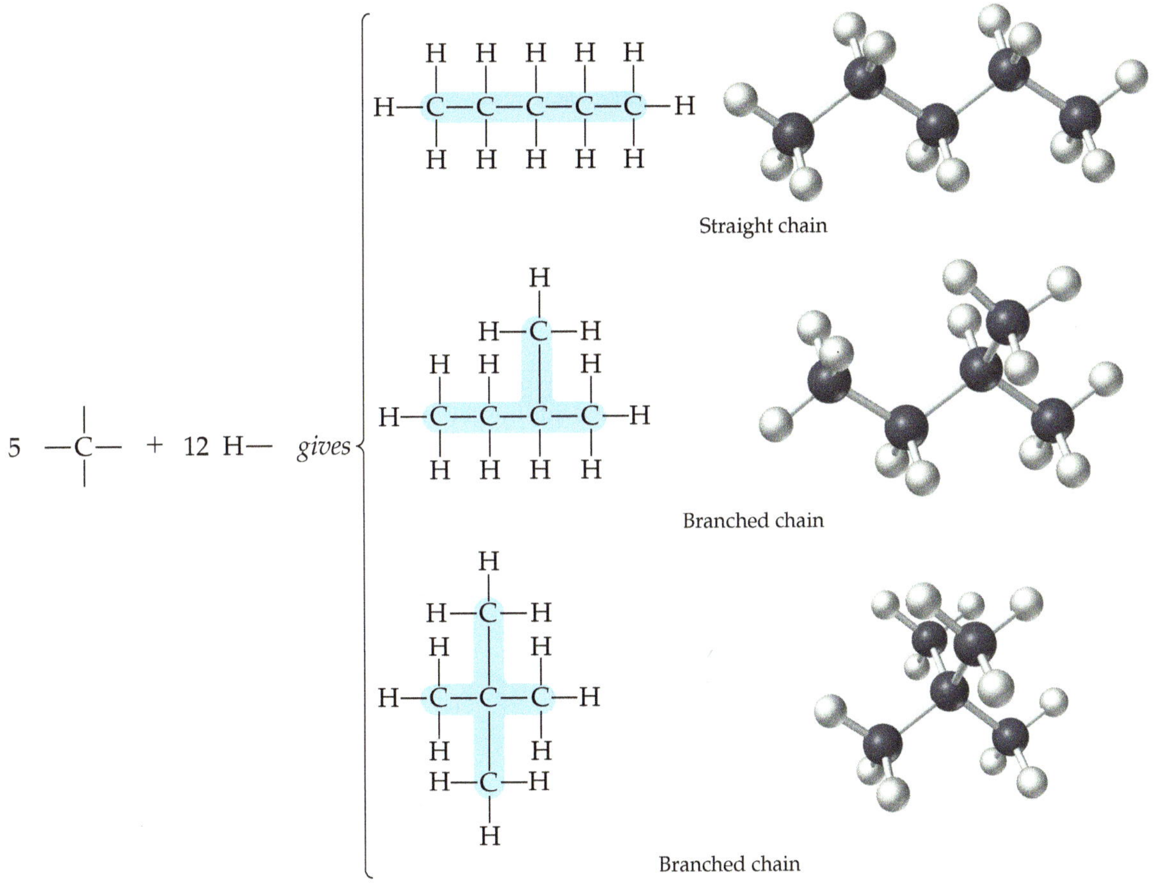

$$5 \ -\overset{|}{\underset{|}{C}}- \ + \ 12 \ H- \quad gives$$

H H H H H
| | | | |
H—C—C—C—C—C—H
| | | | |
H H H H H

Straight chain

```
              H
              |
          H—C—H
  H   H   |   H
  |   |   |   |
H—C—C—C—C—H
  |   |   |   |
  H   H   H   H
```

Branched chain

```
              H
              |
          H—C—H
  H       |   H
  |       |   |
H—C———C———C—H
  |       |   |
  H   H—C—H   H
              |
              H
```

Branched chain

Compounds with all their carbons connected in a continuous chain are called **straight-chain alkanes**; those with a branching connection of carbons are called **branched-chain alkanes**. Note that in a straight-chain alkane, you can draw a line through all the carbon atoms without lifting your pencil from the paper. In a branched-chain alkane, however, you must either lift your pencil from the paper or retrace your steps to draw a line through all the carbons.

Straight-chain alkane An alkane that has all its carbons connected in a row.

Branched-chain alkane An alkane that has a branching connection of carbons.

366 CHAPTER 12 Introduction to Organic Chemistry: Alkanes

Constitutional isomers Compounds with the same molecular formula but different connections among their atoms.

The two isomers of C_4H_{10} and the three isomers of C_5H_{12} shown above are **constitutional isomers**—compounds with the same molecular formula but with different connections among their constituent atoms. Needless to say, the number of possible alkane isomers grows rapidly as the number of carbon atoms increases.

Constitutional isomers of a given molecular formula are chemically distinct from one another. They have different structures, physical properties (such as melting and boiling points), and potentially different physiological properties. When the molecular formula contains atoms other than carbon and hydrogen, the constitutional isomers obtained can also be **functional group isomers**: isomers that differ in both molecular connection and family classification. In these cases the differences between isomers can be dramatic. For example, ethyl alcohol and dimethyl ether both have the formula C_2H_6O, but ethyl alcohol is a liquid with a boiling point of 78.5 °C and dimethyl ether is a gas with a boiling point of +23 °C. While ethyl alcohol is a depressant of the central nervous system, dimethyl ether is a nontoxic compound with anesthetic properties at high concentrations. Clearly, molecular formulas by themselves are not very useful in organic chemistry; a knowledge of structure is also necessary.

Functional group isomer Isomers having the same chemical formula but belonging to different chemical families due to differences in bonding; ethyl alcohol and dimethyl ether are examples of functional group isomers.

Ethyl alcohol
C_2H_6O

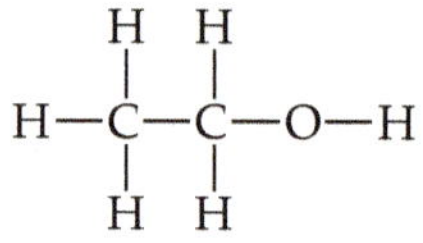

Dimethyl ether
C_2H_6O

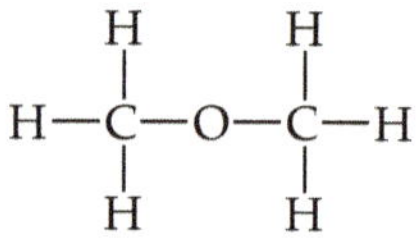

Worked Example 12.3 Molecular Structures: Drawing Isomers

Draw all isomers that have the formula C_6H_{14}.

ANALYSIS Knowing that all the carbons must be bonded together to form the molecule, find all possible arrangements of the 6 carbon atoms. Begin with the isomer that has all 6 carbons in a straight chain, then draw the isomer that has 5 carbons in a straight chain, using the remaining carbon to form a branch, then repeat for the isomer having 4 carbons in a straight chain and 2 carbons in branches. Once each carbon backbone is drawn, arrange the hydrogens around the carbons to complete the structure. (Remember that each carbon can only have *four* bonds total.)

SOLUTION
The straight-chain isomer contains all 6 carbons bonded to form a chain with no branches. The branched isomers are drawn by starting with either a 5-carbon chain or a 4-carbon chain, and adding the extra carbons as branches in the middle of the chain. Hydrogens are added until each carbon has a full complement of four bonds.

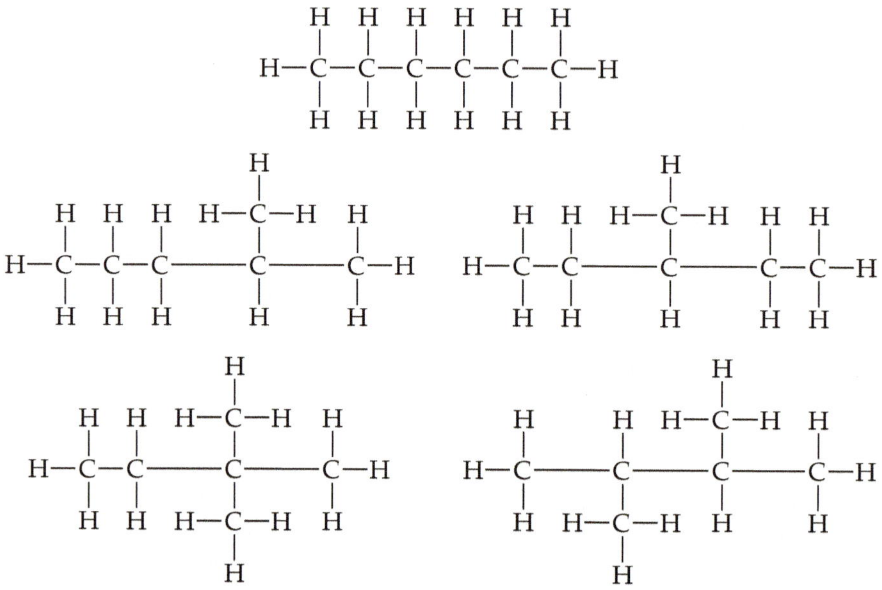

PROBLEM 12.3
Draw the straight-chain isomer with the formula (a) C_7H_{16}; (b) C_9H_{20}.

PROBLEM 12.4
Draw the two branched-chain isomers with the formula C_7H_{16}, where the longest chain in the molecule is 6 carbons long.

12.4 Drawing Organic Structures

Drawing structural formulas that show every atom and every bond in a molecule is both time-consuming and awkward, even for relatively small molecules. Much easier is the use of **condensed structures**, which are simpler but still show the essential information about which functional groups are present and how atoms are connected. In condensed structures, C—C and C—H single bonds are not necessarily shown; rather, they are "understood." If a carbon atom has 3 hydrogens bonded to it, we write CH_3; if the carbon has 2 hydrogens bonded to it, we write CH_2; and so on. For example, the 4-carbon, straight-chain alkane called butane and its branched-chain isomer (2-methylpropane) can be written as the following condensed structures:

Condensed structure A shorthand way of drawing structures in which C—C and C—H bonds are understood rather than shown.

▶▶ Condensed structures were explored in Section 4.7.

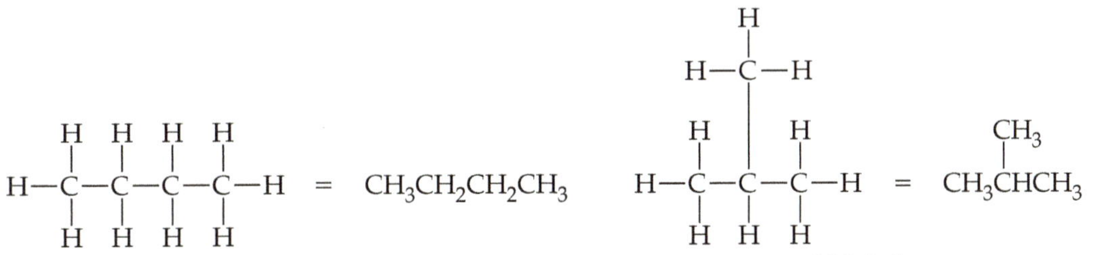

Butane: Structural formula = $CH_3CH_2CH_2CH_3$ Condensed formula

2-Methylpropane: Structural formula = CH_3CHCH_3 with CH_3 branch — Condensed formula

Note in these condensed structures for butane and 2-methylpropane that the horizontal bonds between carbons are not usually shown—the CH_3 and CH_2 units are simply placed next to one another—but that the vertical bond in 2-methylpropane *is* shown for clarity.

Occasionally, as a further simplification, not all the CH_2 groups (called **methylenes**) are shown. Instead, CH_2 is shown once in parentheses, with a subscript indicating the number of methylene units strung together. For example, the 6-carbon straight-chain alkane (hexane) can be written as

Methylene Another name for a CH_2 unit.

$$CH_3CH_2CH_2CH_2CH_2CH_3 = CH_3(CH_2)_4CH_3$$

Worked Example 12.4 Molecular Structures: Writing Condensed Structures

Write condensed structures for the isomers from Worked Example 12.3.

ANALYSIS Eliminate all horizontal bonds, substituting reduced formula components (CH_3, CH_2, and so on) for each carbon in the compound. Show vertical bonds to branching carbons for clarity.

SOLUTION

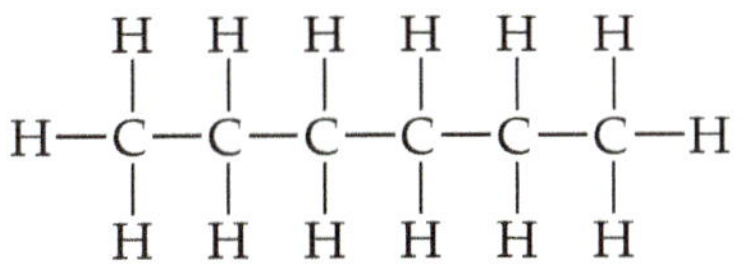

H—C—C—C—C—C—C—H (with H's) → $CH_3CH_2CH_2CH_2CH_2CH_3$ or $CH_3(CH_2)_4CH_3$

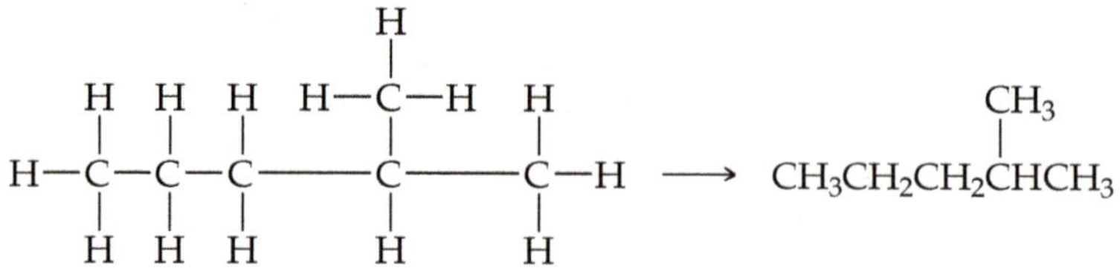

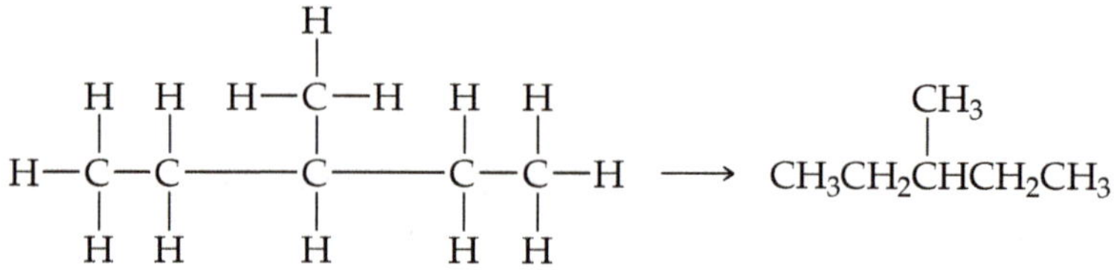

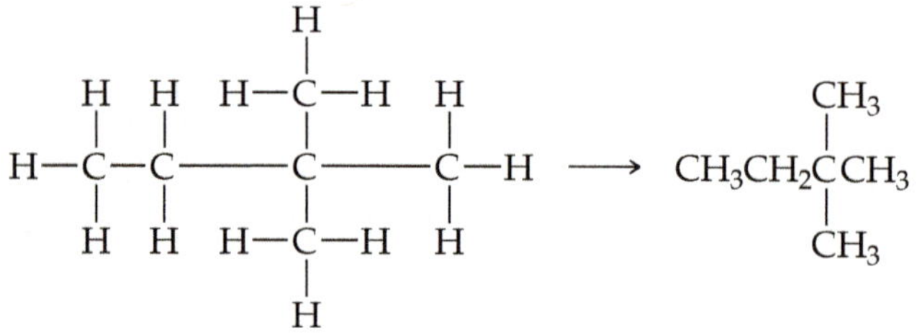

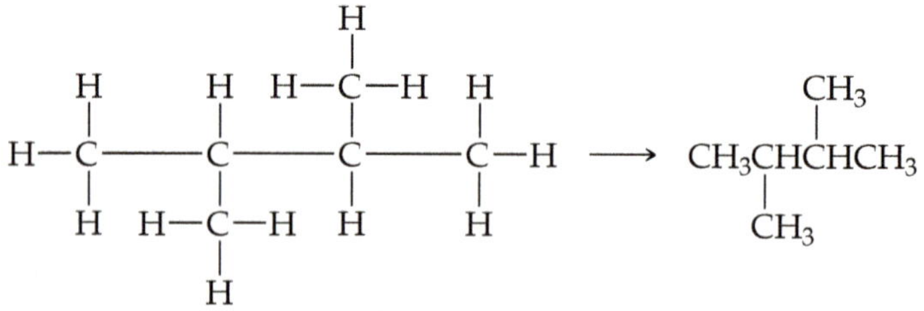

368 CHAPTER 12 Introduction to Organic Chemistry: Alkanes

$$\text{CH}_3\text{CH}_2\text{CH}_2\text{CHCH}_3 \text{ with } \text{CH}_3 \text{ branch}$$

$$\text{CH}_3\text{CH}_2\text{CHCH}_2\text{CH}_3 \text{ with } \text{CH}_3 \text{ branch}$$

$$\text{CH}_3\text{CH}_2\text{CCH}_3 \text{ with two } \text{CH}_3 \text{ branches}$$

$$\text{CH}_3\text{CHCHCH}_3 \text{ with two } \text{CH}_3 \text{ branches}$$

PROBLEM 12.5
Draw the following three isomers of C_5H_{12} as condensed structures:

(a) 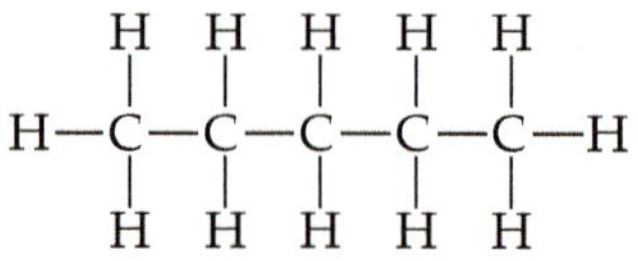 Pentane

(b) 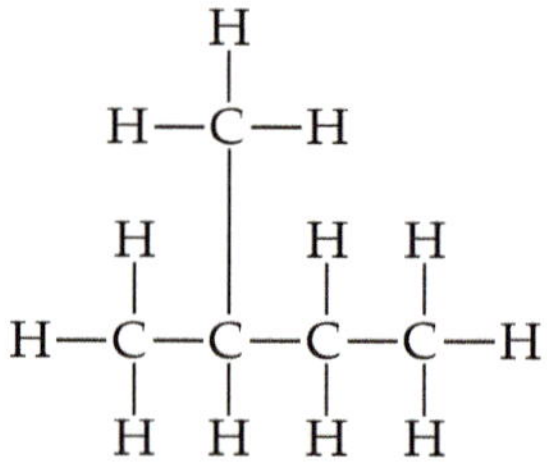 2-Methylbutane

(c) 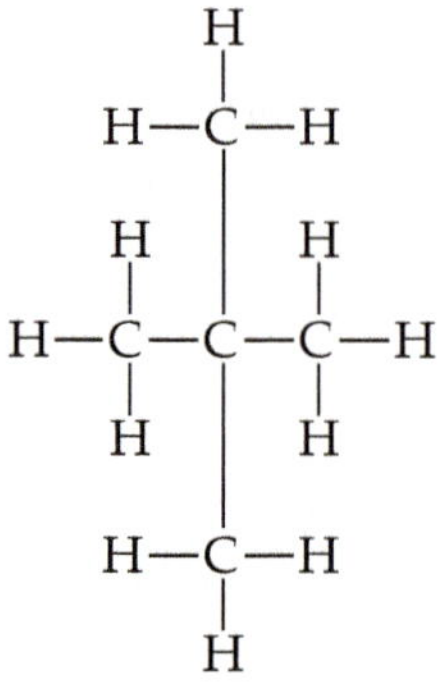 2,2-Dimethylpropane

Another way of representing organic molecules is to use **line (or line-angle) structures**, which are structures in which the symbols C and H do not appear. Instead, a chain of carbon atoms and their associated hydrogens are represented by a zigzag arrangement of short lines, with any branches off the main chain represented by additional lines. The line structure for 2-methylbutane, for instance, is

$$\text{same as} \quad \begin{array}{c} CH_3 \\ | \\ CH_3CHCH_2CH_3 \end{array}$$

Line structure Also known as line-angle structure; a shorthand way of drawing structures in which carbon and hydrogen atoms are not explicitly shown. Instead, a carbon atom is understood to be wherever a line begins or ends and at every intersection of two lines, and hydrogens are understood to be wherever they are needed to have each carbon form four bonds.

Line structures are a simple and quick way to represent organic molecules without the clutter arising from showing all carbons and hydrogens present. Chemists, biologists, pharmacists, doctors, and nurses all use line structures to conveniently convey to one another very complex organic structures. Another advantage is that a line structure gives a more realistic depiction of the angles seen in a carbon chain.

Drawing a molecule in this way is simple, provided one follows these guidelines:

1. Each carbon–carbon bond is represented by a line.
2. Anywhere a line ends or begins, as well as any vertex where two lines meet, represents a carbon atom.
3. Any atom other than another carbon or a hydrogen attached to a carbon must be shown.
4. Since a neutral carbon atom forms four bonds, all bonds not shown for any carbon are understood to be the number of carbon–hydrogen bonds needed to have the carbon form four bonds.

Converting line structures to structural formulas or to condensed structures is simply a matter of correctly interpreting each line ending and each intersection in a line structure. For example, the common pain reliever ibuprofen has the condensed and line structures

Finally, it is important to note that chemists and biochemists often use a mixture of structural formulas, condensed structures, and line structures to represent the molecules they study. As you progress through this textbook, you will see many complicated molecules represented in this way, so it is a good idea to get used to thinking interchangeably in all three formats.

Worked Example 12.5 Molecular Structures: Converting Condensed Structures to Line Structures

Convert the following condensed structures to line structures:

(a) $CH_3CH_2CHCHCH_2CH_3$ with CH_3 above and CH_3 below

(b) $CH_3CHCH-CCH_2CH_3$ with OH, Cl above and CH_3, CH_3 below

ANALYSIS Find the longest continuous chain of carbon atoms in the condensed structure. Begin the line structure by drawing a zigzag line in which the number of vertices plus line ends equals the number of carbon atoms in the chain. Show branches coming off the main chain by drawing vertical lines at the vertices as needed. Show all atoms that are not carbons or are not hydrogens attached to carbons.

SOLUTION

(a) Begin by drawing a zigzag line in which the total number of ends + vertices equals the number of carbons in the longest chain (here 6, with the carbons numbered for clarity):

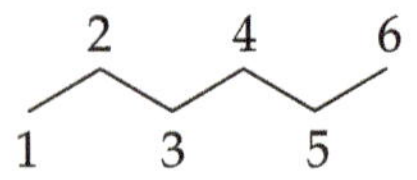

Looking at the condensed structure, you see CH_3 groups on carbons 3 and 4; these two methyl groups are represented by lines coming off those carbons in the line structure:

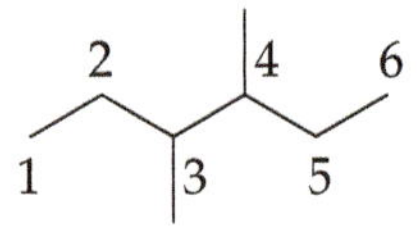

This is the complete line structure. Notice that the hydrogens are not shown, but understood. For example, carbon 4 has three bonds shown: one to carbon 3, one to carbon 5, and one to the branch methyl group; the fourth bond this carbon must have is understood to be to a hydrogen.

(b) Proceed as in (a), drawing a zigzag line for the longest chain of carbon atoms, which again contains 6 carbons. Next draw a line coming off each carbon bonded to a CH_3 group (carbons 3 and 4). Both the OH and the Cl groups must be shown to give the final structure:

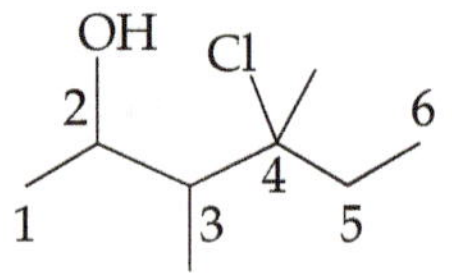

Note from this line structure that it does not matter in such a two-dimensional drawing what direction you show for a group that branches off the main chain, as long as it is attached to the correct carbon. This is true for condensed structures as well. Quite often, the direction that a group is shown coming off a main chain of carbon atoms is chosen simply for aesthetic reasons.

Worked Example 12.6 Molecular Structures: Converting Line Structures to Condensed Structures

Convert the following line structures to condensed structures:

(a)

(b)

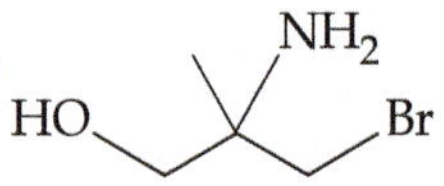

ANALYSIS Convert all vertices and line ends to carbons. Write in any noncarbon atoms and any hydrogens bonded to a noncarbon atom. Add hydrogens as needed so that each carbon has four groups attached. Remove lines connecting carbons except for branches.

SECTION 12.4 Drawing Organic Structures 371

SOLUTION

(a) Anywhere a line ends and anywhere two lines meet, write a C:

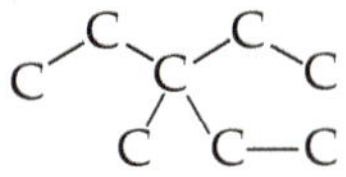

Because there are no atoms other than carbons and hydrogens in this molecule, the next step is to add hydrogens as needed to have four bonds for each carbon:

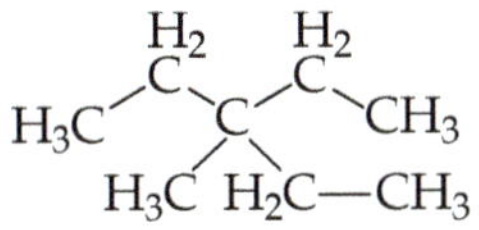

Finally, eliminate all lines except for branches to get the condensed structure:

$$\begin{array}{c} CH_3 \\ | \\ CH_3CH_2-C-CH_2CH_3 \\ | \\ CH_2CH_3 \end{array}$$

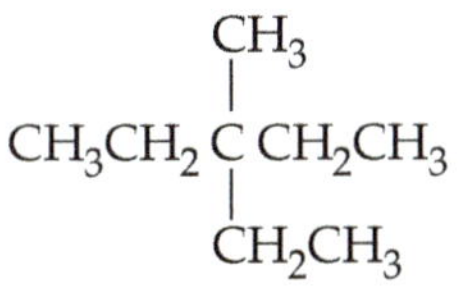

(b) Begin the condensed structure with a drawing showing a carbon at each line end and at each intersection of two lines:

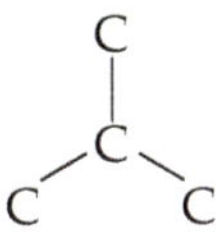

Next, write in all the noncarbon atoms and the hydrogen bonded to the oxygen. Then, add hydrogens so that each carbon forms four bonds:

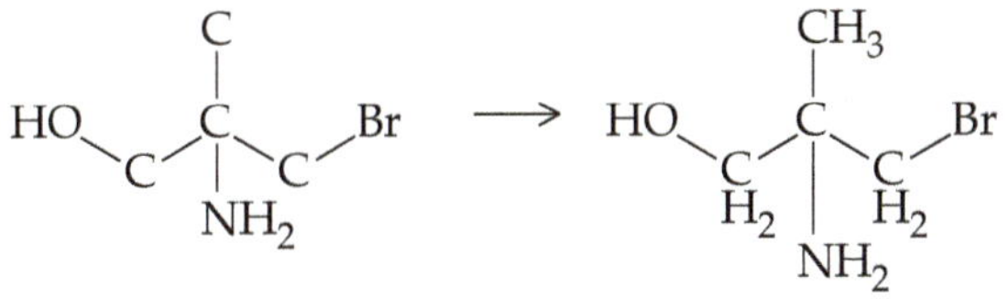

Eliminate all lines except for branches for the completed condensed structure:

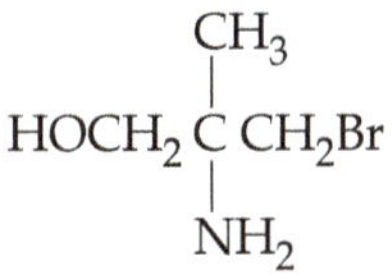

$$\begin{array}{c} CH_3 \\ | \\ HOCH_2-C-CH_2Br \\ | \\ NH_2 \end{array}$$

PROBLEM 12.6

Convert the following condensed structures to line structures:

(a)
$$\begin{array}{c} CH_2CH_3 \\ | \\ CH_3CH_2C-CH_2CH_2CH_3 \\ | \\ CH_2OH \end{array}$$

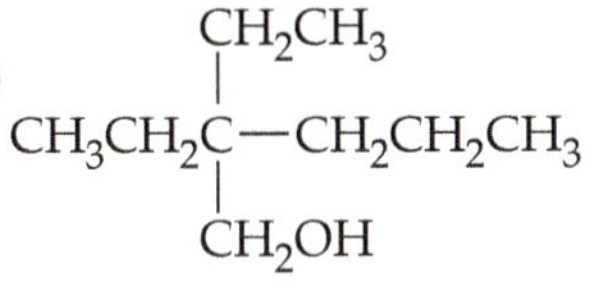

(b)
$$\begin{array}{c} CH_2CH_3 \quad CH_3 \\ | \quad\quad\quad | \\ CH_3CH\ CH\ CH_2CHCH_3 \\ | \\ CH_3CHCH_2CH_3 \end{array}$$

(c)
$$\begin{array}{c} Br \quad\quad\quad Cl \\ | \quad\quad\quad | \\ CH_3C-CHCH_2CH_2CHCH_2CH_3 \\ | \quad | \\ CH_3\ CHCH_2CH_3 \end{array}$$

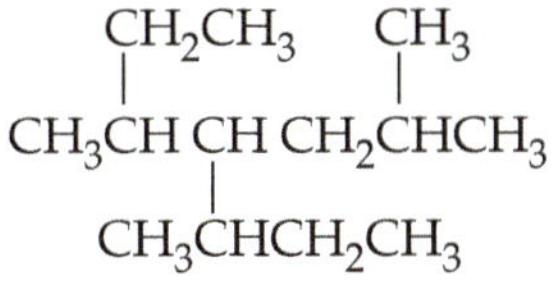

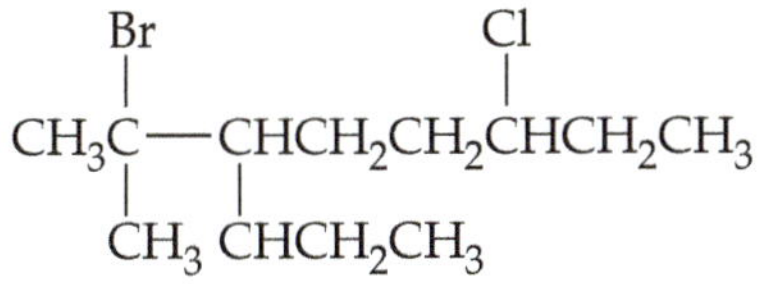

> **PROBLEM 12.7**
> Convert the following line structures to condensed structures: 
>
> (a) [line structure with Cl substituent]
>
> (b) [line structure]
>
> **PROBLEM 12.8**
> Draw both condensed and line structures for the chemicals listed in Problem 12.2.

12.5 The Shapes of Organic Molecules

Every carbon atom in an alkane has its four bonds pointing toward the four corners of a tetrahedron, but chemists do not usually worry about three-dimensional shapes when writing condensed structures. Condensed structures do not imply any particular three-dimensional shape; they only indicate the connections between atoms without specifying geometry. Line structures do try to give some limited feeling for the shape of a molecule, but even here, the ability to show three-dimensional shape is limited unless dashed and wedged lines are used for the bonds (Section 4.8).

Butane, for example, has no one single shape because *rotation* takes place around carbon–carbon single bonds. The two parts of a molecule joined by a carbon–carbon single bond in a noncyclic structure are free to spin around the bond, giving rise to an infinite number of possible three-dimensional geometries, or **conformations**. The various conformations of a molecule such as butane are called **conformers** of one another. Conformers differ from one another as a result of rotation around carbon–carbon single bonds. Although the conformers of a given molecule have different three-dimensional shapes and different energies, the conformers cannot be separated from one another. A given butane molecule might be in its fully extended conformation at one instant but in a twisted conformation an instant later (Figure 12.2). An actual sample of butane contains a great many molecules that are constantly changing conformation. At any given instant, however, most of the molecules have the least crowded, lowest-energy extended conformation shown in Figure 12.2a. The same is true for all other alkanes: At any given instant, most molecules are in the least crowded conformation.

Conformation The specific three-dimensional arrangement of atoms in a molecule achieved specifically through rotations around carbon–carbon single bonds.

Conformer Molecular structures having identical connections between atoms and directly interconvertible through C—C bond rotations; that is, they represent identical compounds.

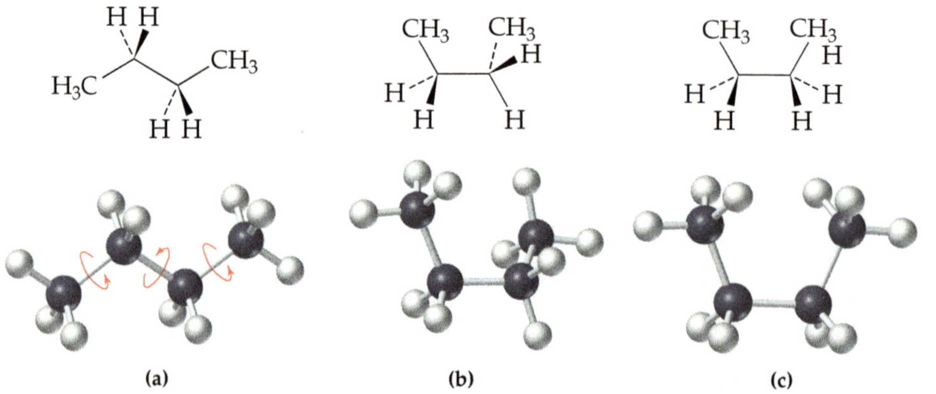

▶ **Figure 12.2**
Some conformations of butane (there are many others as well). The least crowded, extended conformation in (a) is the lowest-energy one, while the eclipsed conformation shown in (c) is the highest-energy one.

As long as any two structures have identical connections between atoms, and are interconvertible either by "flipping" the molecule or through C—C bond rotations, they are conformers of each other and represent the same compound, no matter how the structures are drawn. It is important to remember that no bonds are broken and reformed when interconverting conformers. Sometimes, you have to mentally rotate structures to see whether they are conformers or actually different molecules. To see in the longest continuous chain; and the suffix identifies what family the molecule belongs to:

Prefix—Parent—Suffix

Where are substituents located? — How many carbons? — What family does the molecule belong to?

Straight-chain alkanes are named by counting the number of carbon atoms and adding the family suffix *-ane*. With the exception of the first four compounds—*meth*ane, *eth*ane, *prop*ane, and *but*ane—whose parent names have historical origins, the alkanes are named from Greek numbers according to the number of carbons present (Table 12.2). Thus, *pent*ane is the 5-carbon alkane, *hex*ane is the 6-carbon alkane, and so on. Straight-chain alkanes have no substituents, so prefixes are not needed. The first ten alkane names are so common that they should be memorized.

TABLE 12.2 Names of Straight-Chain Alkanes

Number of Carbons	Structure	Name
1	CH_4	Methane
2	CH_3CH_3	Ethane
3	$CH_3CH_2CH_3$	Propane
4	$CH_3CH_2CH_2CH_3$	Butane
5	$CH_3CH_2CH_2CH_2CH_3$	Pentane
6	$CH_3CH_2CH_2CH_2CH_2CH_3$	Hexane
7	$CH_3CH_2CH_2CH_2CH_2CH_2CH_3$	Heptane
8	$CH_3CH_2CH_2CH_2CH_2CH_2CH_2CH_3$	Octane
9	$CH_3CH_2CH_2CH_2CH_2CH_2CH_2CH_2CH_3$	Nonane
10	$CH_3CH_2CH_2CH_2CH_2CH_2CH_2CH_2CH_2CH_3$	Decane

Substituents, such as —CH_3 and —CH_2CH_3, that branch off the main chain are called **alkyl groups**. An alkyl group can be thought of as the part of an alkane that remains when 1 hydrogen atom is removed to create an available bonding site. For example, removal of a hydrogen from methane gives the **methyl group**, —CH_3, and removal of a hydrogen from ethane gives the **ethyl group**, —CH_2CH_3. Notice that these alkyl groups are named simply by replacing the *-ane* ending of the parent alkane with an *-yl* ending:

Alkyl group The part of an alkane that remains when a hydrogen atom is removed.

Methyl group The —CH_3 alkyl group.

Ethyl group The —CH_2CH_3 alkyl group.

Alkyl groups are derived from a parent alkane.

Methane →(Remove one H)→ —CH_3 Methyl group

Ethane →(Remove one H)→ —CH_2CH_3 Ethyl group

Both methane and ethane have only one "kind" of hydrogen. It does not matter which of the 4 methane hydrogens is removed, so there is only one possible methyl group. Similarly, it does not matter which of the 6 equivalent ethane hydrogens is removed, so only one ethyl group is possible.

376 CHAPTER 12 Introduction to Organic Chemistry: Alkanes

Propyl group The straight-chain alkyl group —CH₂CH₂CH₃.

Isopropyl group The branched-chain alkyl group —CH(CH₃)₂.

The situation is more complex for larger alkanes, which contain more than one kind of hydrogen. Propane, for example, has two different kinds of hydrogens. Removal of any one of the 6 hydrogens attached to an end carbon yields a straight-chain alkyl group called **propyl**, whereas removal of either one of the 2 hydrogens attached to the central carbon yields a branched-chain alkyl group called **isopropyl**:

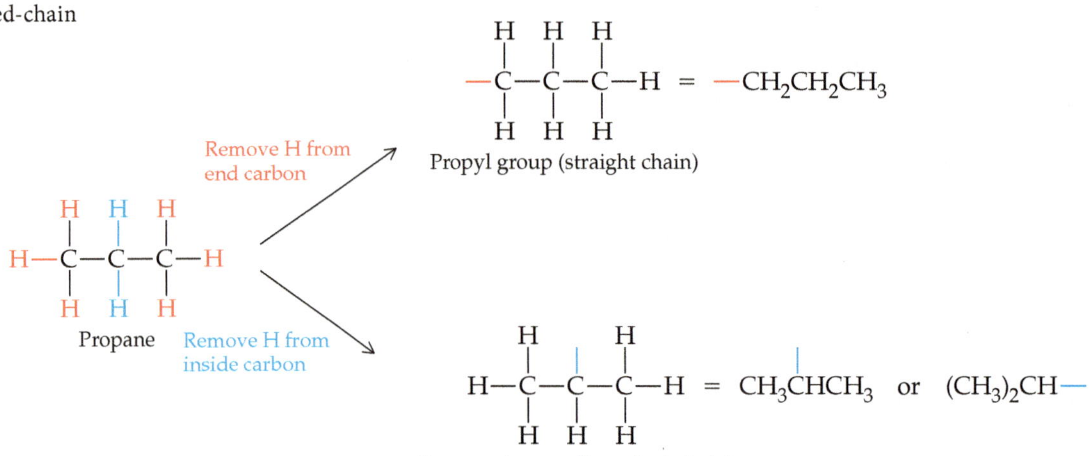

$$-\overset{H}{\underset{H}{C}}-\overset{H}{\underset{H}{C}}-\overset{H}{\underset{H}{C}}-H \;=\; -CH_2CH_2CH_3$$
Propyl group (straight chain)

Propane — Remove H from end carbon / Remove H from inside carbon

$$H-\overset{H}{\underset{H}{C}}-\overset{\;}{\underset{\;}{C}}-\overset{H}{\underset{H}{C}}-H \;=\; CH_3CHCH_3 \;\; \text{or} \;\; (CH_3)_2CH-$$
Isopropyl group (branched chain)

It is important to realize that alkyl groups are not compounds but rather are simply partial structures that help us name compounds. The names of some common alkyl groups are listed in Figure 12.3; you will want to commit them to memory.

▶ **Figure 12.3**
The most common alkyl groups found in organic molecules are shown here; the red bond shows the attachment the group has to the rest of the molecule.*

Some Common Alkyl Groups*

| CH₃— | CH₃CH₂— | CH₃CH₂CH₂— | CH₃CH— (with CH₃ above) |
| Methyl | Ethyl | n-Propyl | Isopropyl |

| CH₃CH₂CH₂CH₂— | CH₃CHCH₂CH₃ | CH₃CHCH₂— (with CH₃ above) | CH₃CCH₃ (with CH₃ above and CH₃ below) |
| n-Butyl | sec-Butyl | Isobutyl | tert-Butyl |

*The red bond shows the connection to the rest of the molecule.

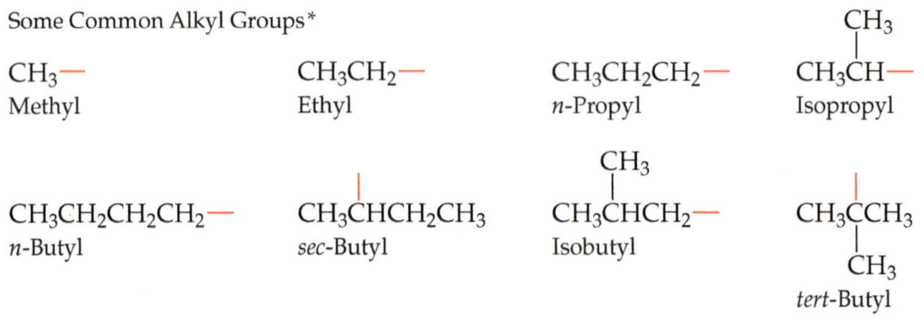

Notice that four butyl (4-carbon) groups are listed in Figure 12.3: butyl, sec-butyl, isobutyl, and tert-butyl. The prefix sec- stands for secondary, and the prefix tert- stands for tertiary, referring to the number of other carbon atoms attached to the branch point. There are four possible substitution patterns for carbons attached to four atoms and these are designated primary, secondary, tertiary, and quaternary. It is important to note that these designations strictly apply to carbons having only single bonds. A **primary (1°) carbon atom** has 1 other carbon attached to it (typically indicated as an —R group in the molecular structure), a **secondary (2°) carbon atom** has 2 other carbons attached, a **tertiary (3°) carbon atom** has 3 other carbons attached, and a **quaternary (4°) carbon atom** has 4 other carbons attached:

Primary (1°) carbon atom A carbon atom with 1 other carbon attached to it.

Secondary (2°) carbon atom A carbon atom with 2 other carbons attached to it.

Tertiary (3°) carbon atom A carbon atom with 3 other carbons attached to it.

Quaternary (4°) carbon atom A carbon atom with 4 other carbons attached to it.

$$R-\underset{H}{\overset{H}{C}}-H \qquad R-\underset{H}{\overset{R}{C}}-H \qquad R-\underset{R}{\overset{R}{C}}-H \qquad R-\underset{R}{\overset{R}{C}}-R$$

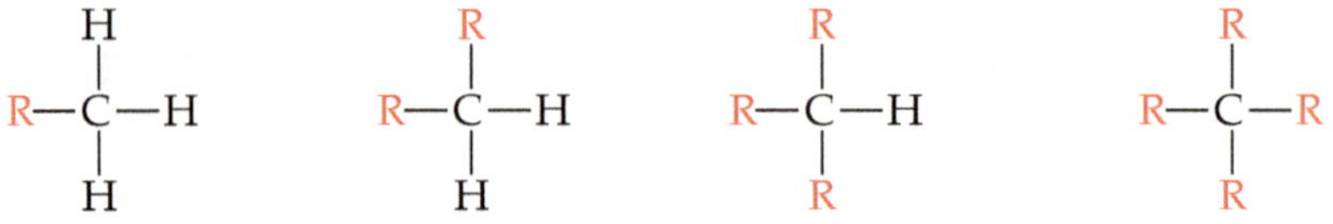

Primary carbon (1°) has one other carbon attached.

Secondary carbon (2°) has two other carbons attached.

Tertiary carbon (3°) has three other carbons attached.

Quaternary carbon (4°) has four other carbons attached.

*The symbol **R** is used here and in later chapters as a general abbreviation for any organic substituent.* You should think of it as representing the **R**est of the molecule, which we are not bothering to specify. The R is used to allow you to focus on a particular structural feature of a molecule without the "clutter" of the other atoms in the molecule detracting from it. The R might represent a methyl, ethyl, or propyl group, or any of a vast number of other possibilities. For example, the generalized formula R—OH for an alcohol might refer to an alcohol as simple as CH_3OH or CH_3CH_2OH or one as complicated as cholesterol, shown here:

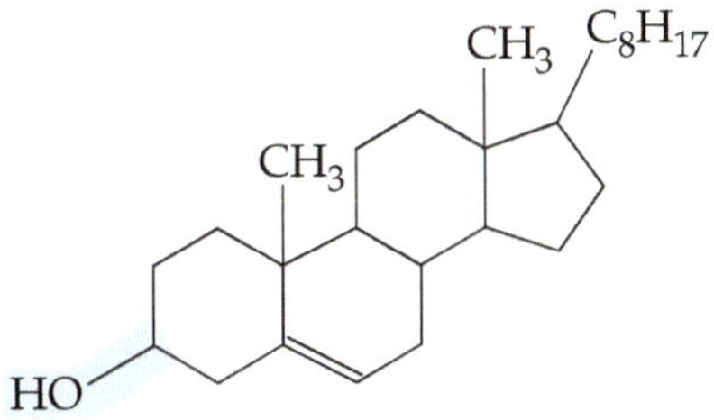

Branched-chain alkanes can be named by following four steps:

STEP 1: Name the main chain. Find the longest continuous chain of carbons, and name the chain according to the number of carbon atoms it contains. The longest chain may not be immediately obvious because it is not always written on one line; you may have to "turn corners" to find it.

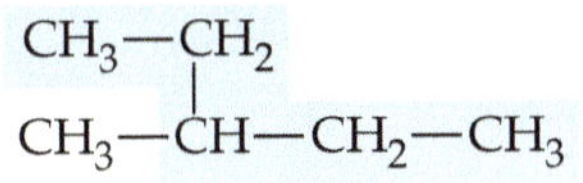

$CH_3—CH_2$
$|$
$CH_3—CH—CH_2—CH_3$

Name as a substituted pentane, not as a substituted butane, because the *longest* chain has five carbons.

STEP 2: Number the carbon atoms in the main chain, beginning at the end nearer the first branch point:

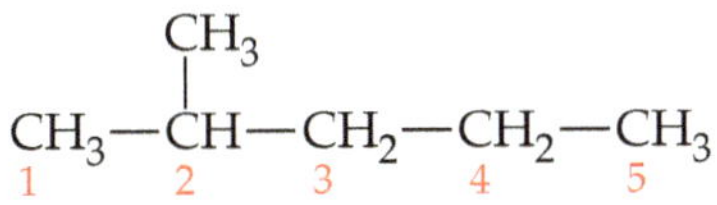

CH_3
$|$
$CH_3—CH—CH_2—CH_2—CH_3$
1   2   3   4   5

The first (and only) branch occurs at C2 if we start numbering from the left, but would occur at C4 if we started from the right by mistake.

STEP 3: Identify the branching substituents, and number each according to its point of attachment to the main chain:

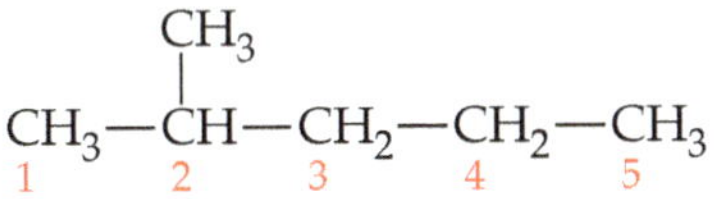

CH_3
$|$
$CH_3—CH—CH_2—CH_2—CH_3$
1   2   3   4   5

The main chain is a pentane. There is one —CH_3 substituent group connected to C2 of the chain.

If there are two substituents on the same carbon, assign the same number to both. There must always be as many numbers in the name as there are substituents.

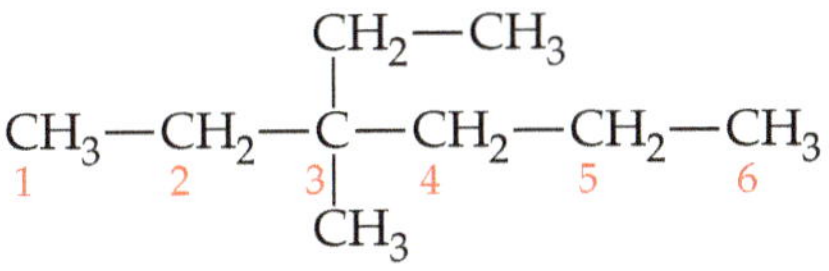

$CH_2—CH_3$
$|$
$CH_3—CH_2—C—CH_2—CH_2—CH_3$
1   2   3| 4   5   6
CH_3

The main chain is a hexane. There are two substituents, a —CH_3 and a —CH_2CH_3, both connected to C3 of the chain.

STEP 4: Write the name as a single word, using hyphens to separate the numbers from the different prefixes and commas to separate numbers, if necessary. If two or more different substituent groups are present, cite them in alphabetical order. If two or more

378 CHAPTER 12 Introduction to Organic Chemistry: Alkanes

identical substituents are present, use one of the prefixes *di-*, *tri-*, *tetra-*, and so forth, but do not use these prefixes for alphabetizing purposes.

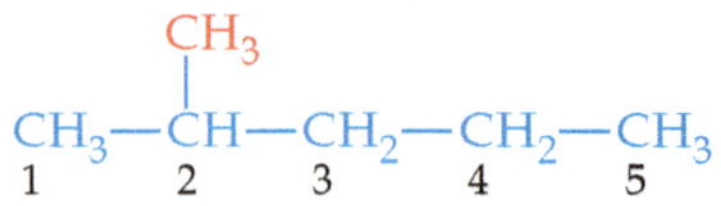

$$CH_3-\underset{\underset{1}{|}}{\overset{CH_3}{\underset{2}{C}H}}-\underset{3}{CH_2}-\underset{4}{CH_2}-\underset{5}{CH_3}$$

2-Methylpentane (a 5-carbon main chain with a 2-methyl substituent)

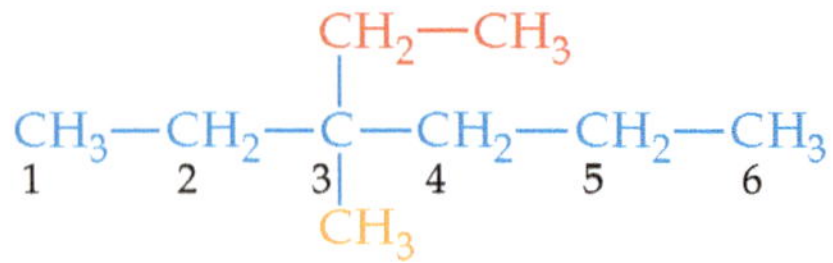

$$CH_3-\underset{1}{CH_2}-\underset{\underset{3}{\overset{|}{C}}}{\overset{CH_2-CH_3}{\underset{|}{C}}}-\underset{4}{CH_2}-\underset{5}{CH_2}-\underset{6}{CH_3}$$
$$CH_3$$

3-Ethyl-3-methylhexane (a 6-carbon main chain with 3-ethyl and 3-methyl substituents cited alphabetically)

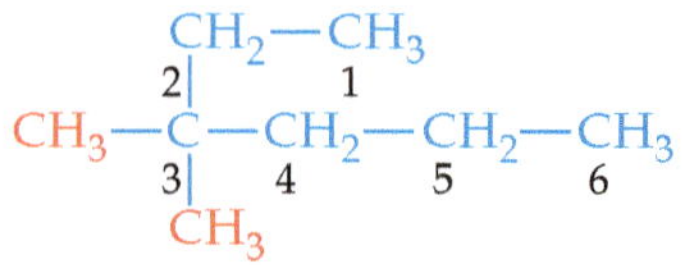

$$\overset{CH_2-CH_3}{\underset{2\ \ \ \ 1}{|}}$$
$$CH_3-\underset{\underset{3}{\overset{|}{C}}}{C}-\underset{4}{CH_2}-\underset{5}{CH_2}-\underset{6}{CH_3}$$
$$CH_3$$

3,3-Dimethylhexane (a 6-carbon main chain with two 3-methyl substitutents)

Worked Example 12.9 Naming Organic Compounds: Alkanes

What is the IUPAC name of the following alkanes?

(a) 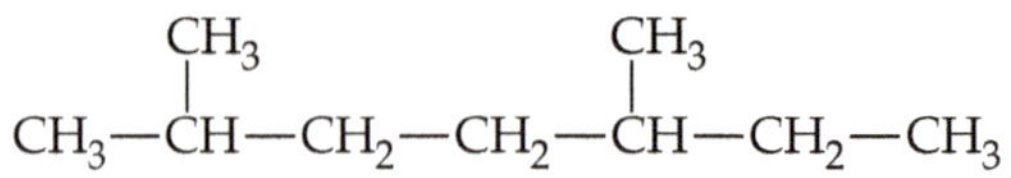 $CH_3-\overset{\overset{CH_3}{|}}{CH}-CH_2-CH_2-\overset{\overset{CH_3}{|}}{CH}-CH_2-CH_3$

(b)

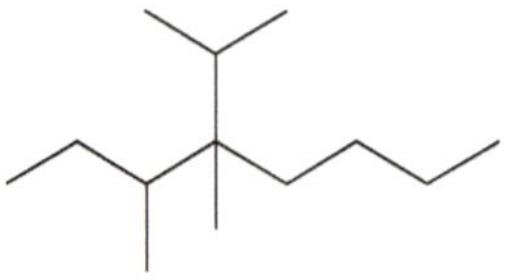

ANALYSIS Follow the four steps outlined in the text.

SOLUTION

(a) **STEP 1:** The longest continuous chain of carbon atoms is seven, so the main chain is a *hept*ane.

STEP 2: Number the main chain beginning at the end nearer the first branch:

$$CH_3-\underset{1}{}\underset{2}{\overset{\overset{CH_3}{|}}{CH}}-\underset{3}{CH_2}-\underset{4}{CH_2}-\underset{5}{\overset{\overset{CH_3}{|}}{CH}}-\underset{6}{CH_2}-\underset{7}{CH_3}$$

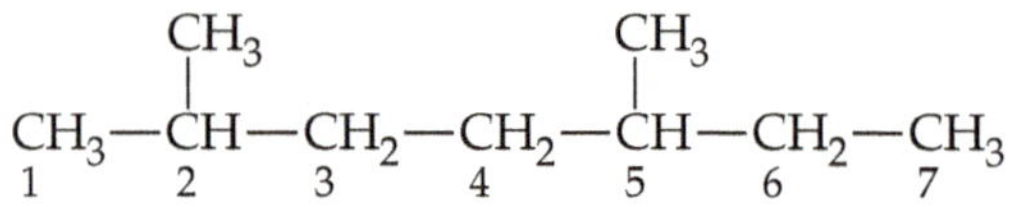

STEP 3: Identify and number the substituents (a 2-methyl and a 5-methyl in this case):

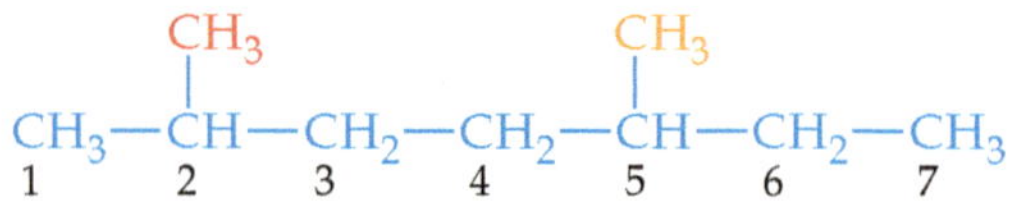

$$CH_3-\underset{1}{}\underset{2}{\overset{\overset{CH_3}{|}}{CH}}-\underset{3}{CH_2}-\underset{4}{CH_2}-\underset{5}{\overset{\overset{CH_3}{|}}{CH}}-\underset{6}{CH_2}-\underset{7}{CH_3}$$

Substituents: 2-methyl and 5-methyl

STEP 4: Write the name as one word, using the prefix *di-* because there are two methyl groups. Separate the two numbers by a comma, and use a hyphen between the numbers and the word.

Name: 2, 5-Dimethylheptane

(b) **STEP 1:** The longest continuous chain of carbon atoms is eight, so the main chain is an *oct*ane.

STEP 2: Number the main chain beginning at the end nearer the first branch:

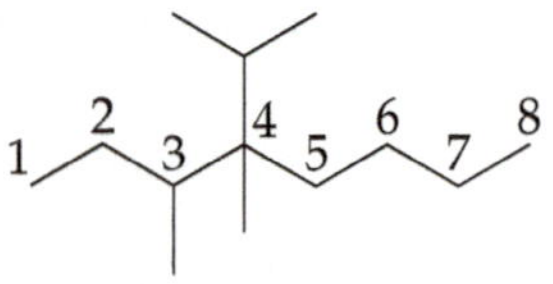

SECTION 12.6 Naming Alkanes 379

STEP 3: Identify and number the substituents:

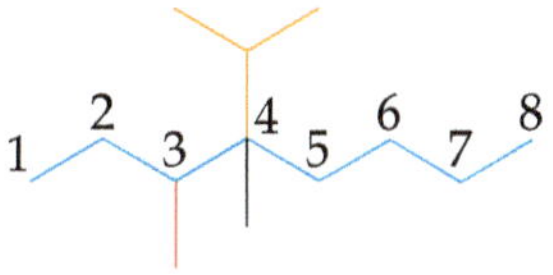

3-methyl, 4-methyl, 4-isopropyl

STEP 4: Write the name as one word, again using the prefix *di*- because there are two methyl groups.

Name: 3, 4-Dimethyl-4-isopropyloctane

Worked Example 12.10 Molecular Structure: Identifying 1°, 2°, 3°, and 4° Carbons

Identify each carbon atom in the following molecule as primary, secondary, tertiary, or quaternary.

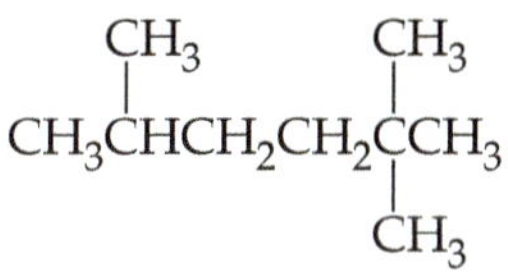

$$CH_3CHCH_2CH_2CCH_3$$
with CH_3 substituents

ANALYSIS Look at each carbon atom in the molecule, count the number of other carbon atoms attached, and make the assignment accordingly: primary (1 carbon attached); secondary (2 carbons attached); tertiary (3 carbons attached); quaternary (4 carbons attached).

SOLUTION

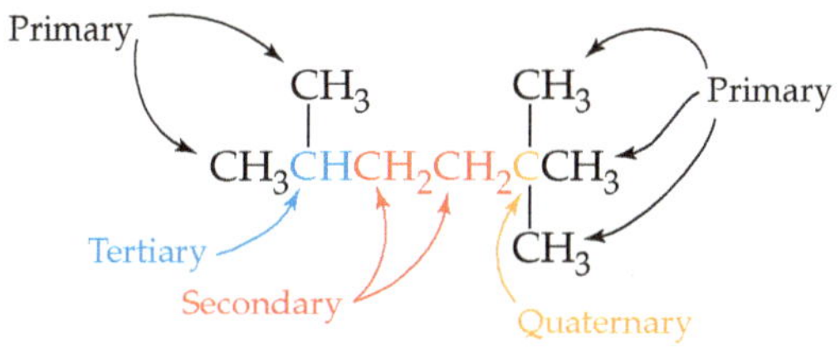

Primary, Tertiary, Secondary, Quaternary, Primary labeled on the structure.

Worked Example 12.11 Molecular Structures: Drawing Condensed Structures from Names

Draw condensed and line structures corresponding to the following IUPAC names:

(a) 2,3-Dimethylpentane
(b) 3-Ethylheptane

ANALYSIS Starting with the parent chain, add the named alkyl substituent groups to the appropriately numbered carbon atom(s).

SOLUTION

(a) The parent chain has 5 carbons (*pentane*), with two methyl groups ($-CH_3$) attached to the second and third carbon in the chain:

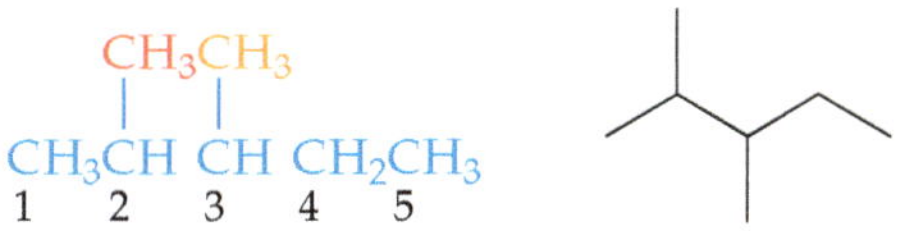

$$CH_3CH\ CH\ CH_2CH_3$$
$$\ \ 1\ \ \ 2\ \ \ 3\ \ \ 4\ \ \ 5$$
with CH_3 groups on C2 and C3

(b) The parent chain has 7 carbons (*hept*ane), with one ethyl group (—CH₂CH₃) attached to the third carbon in the chain:

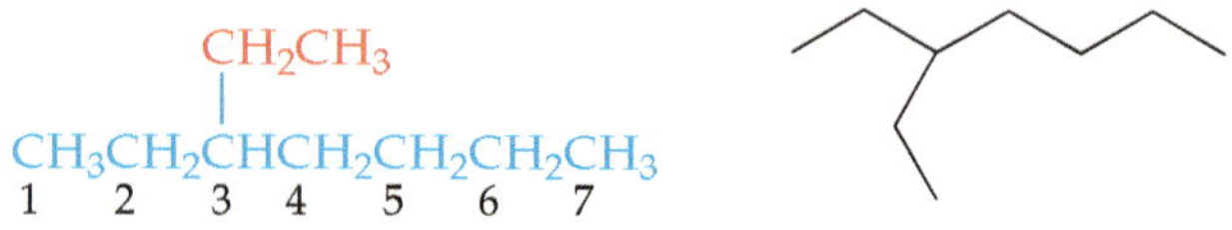

$$\text{CH}_3\text{CH}_2\overset{\underset{|}{\text{CH}_2\text{CH}_3}}{\text{CH}}\text{CH}_2\text{CH}_2\text{CH}_2\text{CH}_3$$
$$\;1\quad 2\quad 3\quad 4\quad 5\quad 6\quad 7$$

PROBLEM 12.11
Identify each carbon in the molecule shown in Worked Example 12.9b as primary, secondary, tertiary, or quaternary.

PROBLEM 12.12
What are the IUPAC names of the following alkanes?

(a) 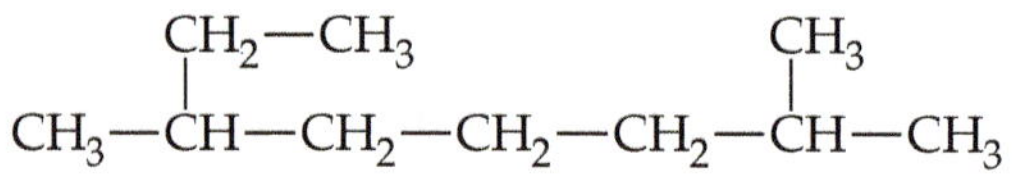 $\text{CH}_3-\overset{\underset{|}{\text{CH}_2-\text{CH}_3}}{\text{CH}}-\text{CH}_2-\text{CH}_2-\text{CH}_2-\overset{\underset{|}{\text{CH}_3}}{\text{CH}}-\text{CH}_3$

(b) 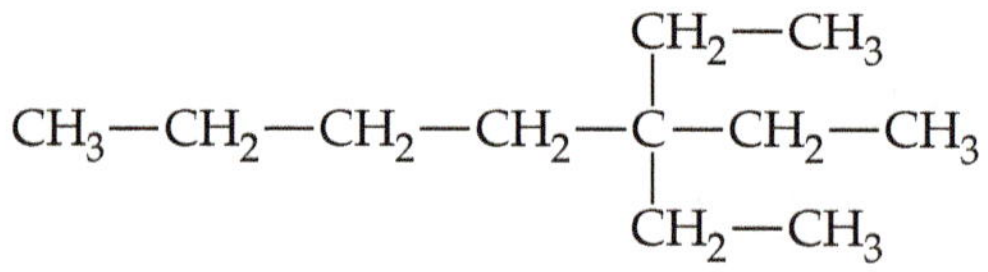 $\text{CH}_3-\text{CH}_2-\text{CH}_2-\text{CH}_2-\overset{\overset{\text{CH}_2-\text{CH}_3}{|}}{\underset{\underset{\text{CH}_2-\text{CH}_3}{|}}{\text{C}}}-\text{CH}_2-\text{CH}_3$

PROBLEM 12.13
Draw both condensed and line structures corresponding to the following IUPAC names and label each carbon as primary, secondary, tertiary, or quaternary:

(a) 3-Methylhexane   (b) 3,4-Dimethyloctane   (c) 2,2,4-Trimethylpentane

PROBLEM 12.14
Draw and name alkanes that meet the following descriptions:
(a) A 5-carbon alkane with a tertiary carbon atom
(b) A 7-carbon alkane that has both a tertiary and a quaternary carbon atom

---

🔑 KEY CONCEPT PROBLEM 12.15
What are the IUPAC names of the following alkanes?

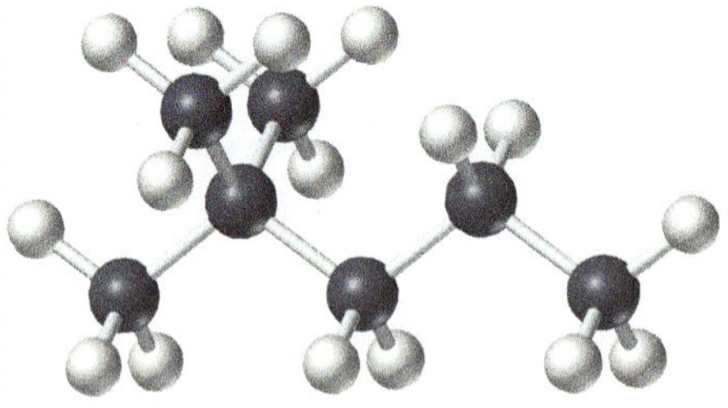

(a) 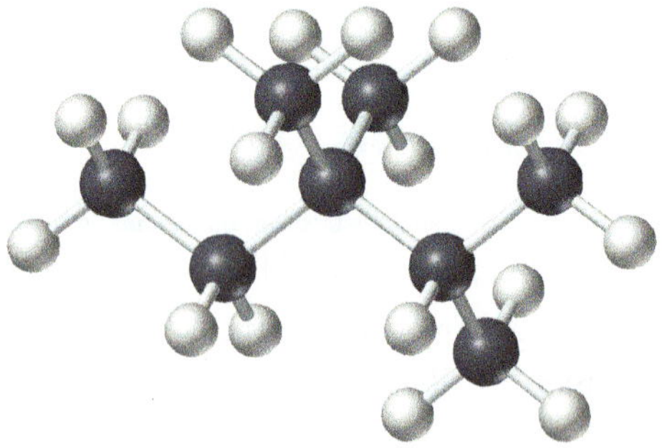   (b)

12.7 Properties of Alkanes

▶▶ Review the effects of London dispersion forces on molecules in Section 8.11.

Alkanes contain only nonpolar C—C and C—H bonds, so the only intermolecular forces influencing them are weak London dispersion forces. The effect of these forces is shown in the regularity with which the melting and boiling points of straight-chain alkanes increase with molecular size (Figure 12.4). The first four alkanes—methane, ethane, propane, and butane—are gases at room temperature and pressure. Alkanes

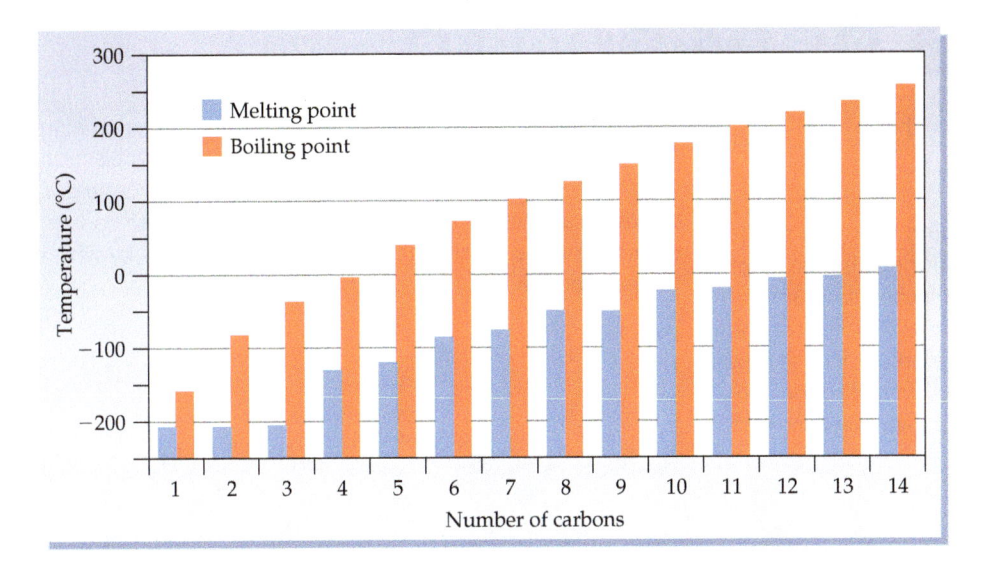

◀ **Figure 12.4**
The boiling and melting points for the C_1—C_{14} straight-chain alkanes increase with molecular size.

with 5–15 carbon atoms are liquids; those with 16 or more carbon atoms are generally low-melting, waxy solids.

In keeping with their low polarity, alkanes are insoluble in water but soluble in nonpolar organic solvents, including other alkanes. Because alkanes are generally less dense than water, they float on its surface. Low-molecular-weight alkanes are volatile and must be handled with care because their vapors are flammable. Mixtures of alkane vapors and air can explode when ignited by a single spark.

The physiological effects of alkanes are limited. Methane, ethane, and propane gases are nontoxic, but the danger of inhaling them lies in potential suffocation due to lack of oxygen. Breathing the vapor of larger alkanes in large concentrations can induce loss of consciousness. There is also a danger in breathing droplets of liquid alkanes because they dissolve nonpolar substances in lung tissue and cause pneumonia-like symptoms.

Mineral oil, petroleum jelly, and paraffin wax are mixtures of higher alkanes. All are harmless to body tissue and are used in numerous food and medical applications. Mineral oil passes through the body unchanged and is sometimes used as a laxative. Petroleum jelly (sold as Vaseline) softens, lubricates, and protects the skin. Paraffin wax is used in candle making, on surfboards, and in home canning. See Chemistry in Action box on page 385 for more surprising uses of alkanes.

▶▶ Recall from Section 9.2 the rule of thumb when predicting solubility: "like dissolves like."

Properties of Alkanes:

- Odorless or mild odor; colorless; tasteless; nontoxic
- Nonpolar; insoluble in water but soluble in nonpolar organic solvents; less dense than water
- Flammable; otherwise not very reactive

12.8 Reactions of Alkanes

Alkanes do not react with acids, bases, or most other common laboratory *reagents* (a substance that causes a reaction to occur). Their only major reactions are with oxygen (combustion) and with halogens (halogenation). Both of these reaction types have complicated mechanisms and occur through the intermediacy of free radicals (Section 13.8).

Combustion

Most of you probably get to school everyday using some sort of transportation that uses gasoline, which is a mixture of alkanes. To power a vehicle, that mixture of alkanes must be converted into energy. The reaction of an alkane with oxygen is called **combustion**, an oxidation reaction that commonly takes place in a controlled manner

Combustion A chemical reaction that produces a flame, usually because of burning with oxygen.

382 CHAPTER 12 Introduction to Organic Chemistry: Alkanes

MASTERING REACTIONS

Organic Chemistry and the Curved Arrow Formalism

Starting with this chapter and continuing on through the remainder of this text, you will be exploring the world of organic chemistry and its close relative, biochemistry. Both of these areas of chemistry are much more "visual" than those you have been studying; organic chemists, for example, look at how and why reactions occur by examining the flow of electrons. For example, consider the following reaction of 2-iodopropane with sodium cyanide:

$$(CH_3)_2CHI + NaCN \longrightarrow (CH_3)_2CHCN + NaI$$

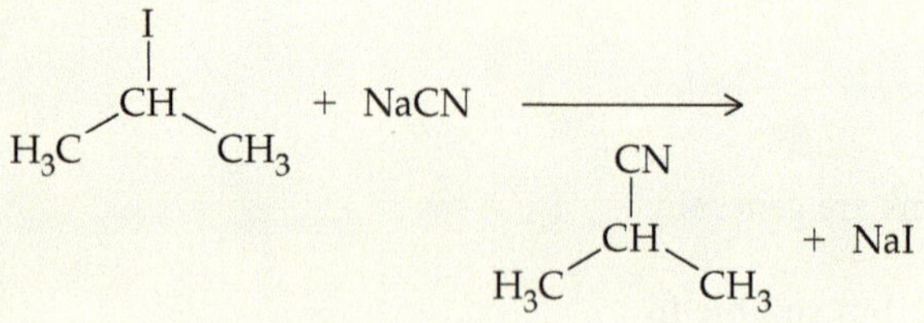

This seemingly simple process (known as a *substitution reaction*, discussed in Chapter 13) is not adequately described by the equation. To help to understand what may really be going on, organic chemists use what is loosely described as "electron pushing" and have adopted what is known as *curved arrow formalism* to represent it. The movement of electrons is depicted using curved arrows, where the number of electrons corresponds to the head of the arrow. Single-headed arrows represent movement of one electron, while a double-headed arrow indicates the movement of two:

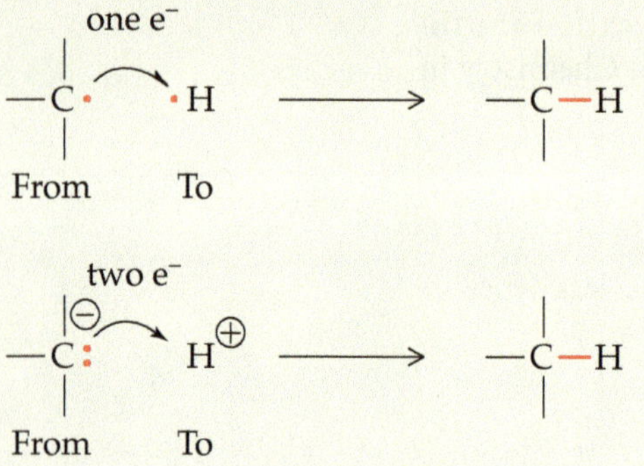

one e⁻: —C• •H ⟶ —C—H (From / To)

two e⁻: —C:⁻ H⁺ ⟶ —C—H (From / To)

The convention is to show the movement *from* an area of high electron density (the start of the arrow) *to* one of lower electron density (the head of the arrow). Using curved arrow formalism, we can examine the reaction of 2-iodopropane with sodium cyanide in more detail. There are two distinct paths by which this reaction can occur:

Path 1

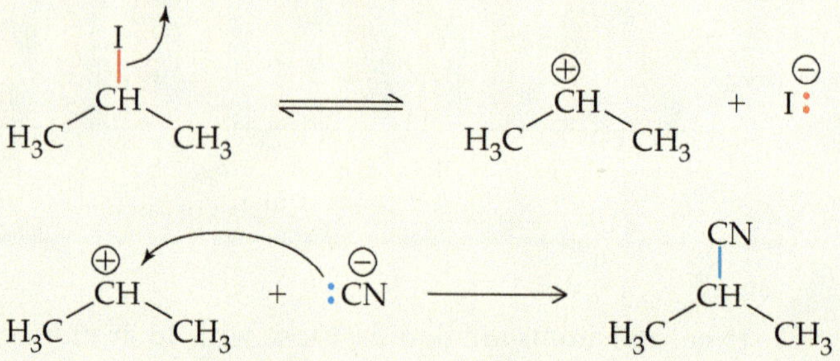

(CH_3)_2CHI ⇌ (CH_3)_2CH⁺ + I⁻

(CH_3)_2CH⁺ + :CN⁻ ⟶ (CH_3)_2CHCN

Path 2

(CH_3)_2CHI + :CN⁻ ⟶ (CH_3)_2CHCN + I⁻

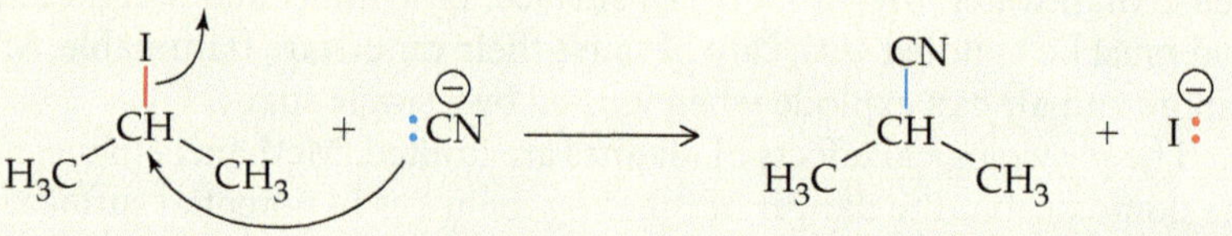

Notice that while both pathways lead ultimately to the same product, the curved arrow formalism shows us that they have significantly different ways of occurring. Although it is not important right now to understand which of the two paths is actually operative (it turns out to be a function of solvent), concentrations, catalysts, temperature, and other conditions) it is important that you get used to thinking of reactions as an "electron flow" of sorts. Throughout the next six chapters, you will see more of these "Mastering Reactions" boxes; they are intended to give you a little more insight into the otherwise seemingly random reactions that organic molecules undergo.

See Mastering Reactions Problems 12.76 and 12.77 at the end of the chapter.

▶▶ Combustion reactions are exothermic, as we learned in Section 7.3.

in an engine or furnace. Carbon dioxide and water are the products of complete combustion of any hydrocarbon, and a large amount of heat is released (ΔH is a negative number.) Some examples were given in Table 7.1.

$$CH_4(g) + 2\,O_2(g) \longrightarrow CO_2(g) + 2\,H_2O(g) \quad \Delta H = -213 \text{ kcal/mol } (-891 \text{ kJ/mol})$$

When hydrocarbon combustion is incomplete because of faulty engine or furnace performance, carbon monoxide and carbon-containing soot are among the products. Carbon monoxide is a highly toxic and dangerous substance, especially so because it has no odor and can easily go undetected (See the Chemistry in Action feature "CO and NO: Pollutants or Miracle Molecules?" in Chapter 4). Breathing air that contains as little as 2% CO for only one hour can cause respiratory and nervous system damage or death. The supply of oxygen to the brain is cut off by carbon monoxide because it binds strongly to blood hemoglobin at the site where oxygen is normally bound. By contrast with CO, CO_2 is nontoxic and causes no harm, except by suffocation when present in high concentration.

> **PROBLEM 12.16**
> Write a balanced equation for the complete combustion of ethane with oxygen.

The second notable reaction of alkanes is *halogenation*, the replacement of an alkane hydrogen by a chlorine or bromine in a process initiated by heat or light. Halogenation is important because it is used to prepare both a number of molecules that are key industrial solvents (such as dichloromethane, chloroform, and carbon tetrachloride) as well as others (such as bromoethane) that are used for the preparation of other larger organic molecules. In a halogenation reaction, only one H at a time is replaced; however, if allowed to react for a long enough time, all H's will be replaced with halogens. Complete chlorination of methane, for example, yields carbon tetrachloride:

$$CH_4 + 4\,Cl_2 \xrightarrow{\text{Heat or light}} CCl_4 + 4\,HCl$$

Although the above equation for the reaction of methane with chlorine is balanced, it does not fully represent what actually happens. In fact, this reaction, like many organic reactions, yields a mixture of products:

$$CH_4 + Cl_2 \longrightarrow CH_3Cl + HCl \xrightarrow{Cl_2} CH_2Cl_2 + HCl \xrightarrow{Cl_2} CHCl_3 + HCl \xrightarrow{Cl_2} CCl_4 + HCl$$

CH_3Cl, chloromethane
CH_2Cl_2, dichloromethane
$CHCl_3$, chloroform
CCl_4, carbon tetrachloride

When we write the equation for an organic reaction, our attention is usually focused on converting a particular reactant into a desired product; any minor by-products and inorganic compounds (such as the HCl formed in the chlorination of methane) are often of little interest and are ignored. Thus, it is not always necessary to balance the equation for an organic reaction as long as the reactant, the major product, and any necessary reagents and conditions are shown. A chemist who plans to convert methane into bromomethane might therefore, write the equation as

$$CH_4 \xrightarrow[\text{Light, heat}]{Br_2} CH_3Br$$

Like many equations for organic reactions, this equation is not balanced.

In using this convention, it is customary to put reactants and reagents above the arrow and conditions, solvents, and catalysts below the arrow.

> **PROBLEM 12.17**
> Write the structures of all six possible products with either 1 or 2 chlorine atoms that form in the reaction of propane with Cl_2.

12.9 Cycloalkanes

The organic compounds described thus far have all been open-chain, or *acyclic*, alkanes. **Cycloalkanes**, which contain rings of carbon atoms, are also well known and are widespread throughout nature. To form a closed ring requires an additional

Cycloalkane An alkane that contains a ring of carbon atoms.

C—C bond and the loss of 2 H atoms; the general formula for cycloalkanes, therefore, is C_nH_{2n}, which, as we will find in the next chapter, is the same as that for alkenes. Compounds of all ring sizes from 3 through 30 and beyond have been prepared in the laboratory. The two simplest cycloalkanes—cyclopropane and cyclobutane—contain 3 and 4 carbon atoms, respectively:

$$\begin{array}{c} CH_2 \\ / \ \backslash \\ H_2C-CH_2 \end{array}$$
Cyclopropane
(mp −128 °C, bp −33 °C)

$$\begin{array}{c} H_2C-CH_2 \\ |\quad\quad | \\ H_2C-CH_2 \end{array}$$
Cyclobutane
(mp −50 °C, bp −12 °C)

Note that if we flatten the rings in cyclopropane and cyclobutane, the C—C—C bond angles are 60° and 90°, respectively—values that are considerably compressed from the normal tetrahedral value of 109.5°. As a result, these compounds are less stable and more reactive than other cycloalkanes. The five-membered (cyclopentane) ring has nearly ideal bond angles, and so does the six-membered (cyclohexane) ring. Both cyclopentane and cyclohexane accomplish this nearly ideal state by adopting a puckered, nonplanar shape called a *chair conformation*, further discussion of which, while important, is beyond the scope of this textbook. Both cyclopentane and cyclohexane rings are therefore stable, and many naturally occurring and biochemically active molecules, such as the steroids (Chapter 28), contain such rings.

Cyclic and acyclic alkanes are similar in many of their properties. Cyclopropane and cyclobutane are gases at room temperature, whereas larger cycloalkanes are liquids or solids. Like alkanes, cycloalkanes are nonpolar, insoluble in water, and flammable. Because of their cyclic structures, however, cycloalkane molecules are more rigid and less flexible than their open-chain counterparts. Rotation is not possible around the carbon–carbon bonds in cycloalkanes without breaking open the ring.

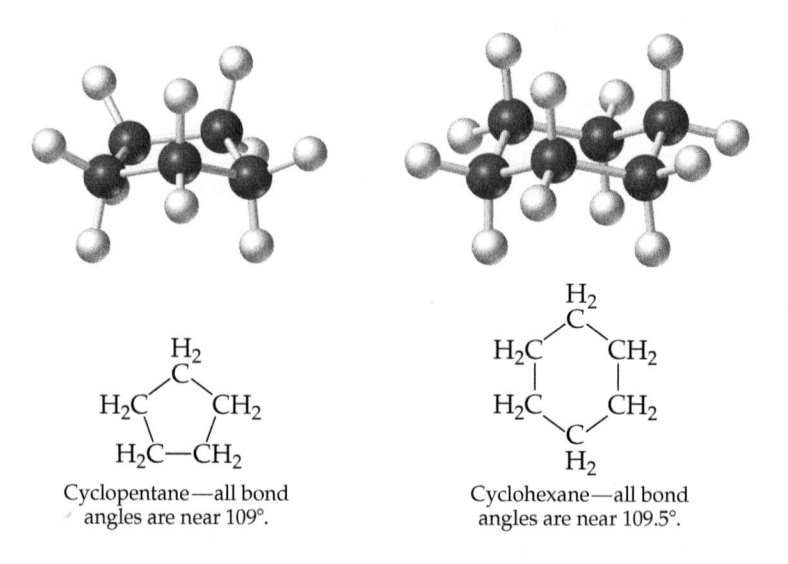

$$\begin{array}{c} H_2 \\ C \\ H_2C \quad CH_2 \\ | \quad\quad\quad | \\ H_2C-CH_2 \end{array}$$
Cyclopentane—all bond angles are near 109°.

$$\begin{array}{c} H_2 \\ C \\ H_2C \quad\quad CH_2 \\ | \quad\quad\quad\quad | \\ H_2C \quad\quad CH_2 \\ C \\ H_2 \end{array}$$
Cyclohexane—all bond angles are near 109.5°.

CHEMISTRY IN ACTION

Surprising Uses of Petroleum

Whenever the word "petroleum" is mentioned, the first thing that comes to most people's mind is gasoline and oil. Petroleum, arising from the decay of ancient plants and animals, is found deep below the earth's crust; it is, simply put, a mixture of hydrocarbons of varying sizes. Petroleum's worth as both a portable, energy-dense fuel and as the starting point of many industrial chemicals makes it one of the world's most important commodities. Current estimates put 90% of vehicular fuel needs worldwide as being met by oil. In addition, 40% of total energy consumption in the United States is petroleum-based, but petroleum is responsible for only 2% of electricity generation. In an effort to create a "greener" environment and more sustainable energy, a great fervor has developed to find alternative energy sources. Wind, solar, biodiesel, and geothermal are but a few of the emerging technologies that have been pushed into the forefront; but a question arises: Can we completely eliminate the need for petroleum from our lives, even if we could find an alternative energy for transportation purposes?

Petrochemicals are chemical products derived specifically from petroleum and generally refer to those products that are not used for fuels. When crude oil is refined and cracked (the process during which complex organic molecules found in oil are converted into simpler molecules by the breaking of carbon–carbon bonds), a number of fractions having different boiling ranges are obtained. The lower boiling fractions contain the simplest alkanes and alkenes, many of which are important chemical feedstocks. A feedstock is a material that is used as the starting point for the preparation of more complex ones. The primary petrochemicals obtained can be broken down into three categories:

1. Alkenes (or olefins; Chapter 13): primarily ethylene, propylene, and butadiene. Ethylene and propylene are important sources of industrial chemicals and plastics products.
2. Aromatics (Chapter 13): most important among these are benzene, toluene, and the xylenes. These raw materials are used for making a variety of compounds, from dyes and synthetic detergents, to plastics and synthetic fibers, to pharmaceutical starting materials.
3. Synthesis gas: a mixture of carbon monoxide and hydrogen used to make methanol (which is used as both a solvent and feedstock for other products).

What specific types of products are made from these petrochemicals? Let's look at a few:

Lubricants such as light machine oils, motor oils, and greases are products used to keep almost all mechanical devices running smoothly and to prevent them from seizing up

▲ Petroleum jelly, originally an unwanted by-product of drilling, has found many uses in today's average household.

under high-use conditions. Look at the mixer or the blender in your kitchen: Lubricants are responsible for keeping it running efficiently and quietly. Wax is another raw petroleum product. These paraffin waxes are used to make candles and polishes as well as food packaging such as milk cartons. The shine you see on the fruit in your local supermarket is also a result of the use of wax.

Most of the rubber soles found on today's shoes are derived from butadiene. Natural rubber becomes sticky when hot and stiff when cold, but man-made rubber stays much more flexible. Car tires are also made from synthetic rubber, which makes them much safer to drive on. Today, the demand for synthetic rubber is four-times greater than for natural rubber.

Other alkenes obtained from the refining of petroleum, such as ethylene and propylene, are used to make plastics that are used in turn to make the clothes you wear, the carpet you walk on, the CDs and DVDs that entertain you, and hundreds of other products we take for granted.

One very interesting petroleum-derived material was once considered a nuisance by-product of oil drilling. "Black rod wax" is a paraffin-like substance that forms on oil-drilling rigs, causing the drills to malfunction. Workers had to scrape the thick, viscous material off to keep the drills running. However, they found that when applied to cuts and burns it would cause these injuries to heal faster. A young chemist named Robert Chesebrough obtained some of this material to see if it had commercial potential. After purification he obtained a light-colored gel that he named vaseline, or petroleum jelly. Chesebrough demonstrated the product by burning his skin with acid or an open flame, then spreading the ointment on his injuries and showing how his past injuries had healed by his miracle product.

As you can see, petroleum has many uses that are key in our everyday lives. Although lessening its use as a fuel for transportation can help to conserve what reserves we have, its complete elimination from our lives is, at this point in time, nearly impossible.

See Chemistry in Action Problems 12.74 and 12.75 at the end of the chapter.

12.10 Drawing and Naming Cycloalkanes

Even condensed structures become awkward when we work with large molecules that contain rings. Thus, line structures are used almost exclusively in drawing cycloalkanes, with *polygons* used for the cyclic parts of the molecules. A triangle represents cyclopropane, a square represents cyclobutane, a pentagon represents cyclopentane, and so on.

Cyclopropane   Cyclobutane   Cyclopentane   Cyclohexane   Cycloheptane

Methylcyclohexane, for example, looks like this in a line structure:

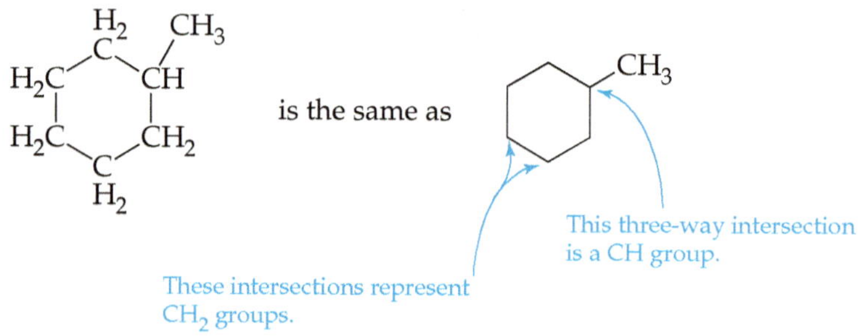

is the same as

This three-way intersection is a CH group.

These intersections represent CH_2 groups.

Cycloalkanes are named by a straightforward extension of the rules for naming open-chain alkanes. In most cases, only two steps are needed:

STEP 1: Use the cycloalkane name as the parent. That is, compounds are named as alkyl-substituted cycloalkanes rather than as cycloalkyl-substituted alkanes. If there is only one substituent on the ring, it is not even necessary to assign a number because all ring positions are identical.

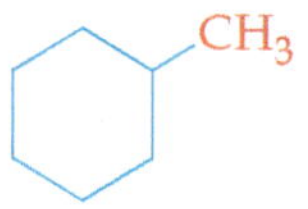

Parent compound: Cyclohexane
Name: Methylcyclohexane
(not cyclohexylmethane)

STEP 2: Identify and number the substituents. Start numbering at the group that has alphabetical priority, and proceed around the ring in the direction that gives the second substituent the lower possible number.

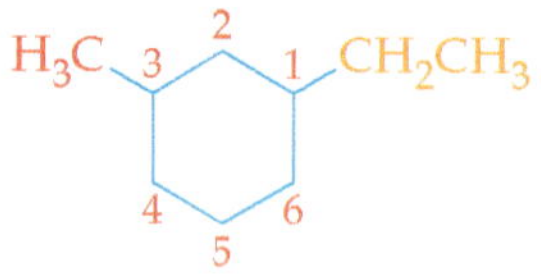

1-ethyl-3-methylcyclohexane
(not 1-ethyl-5-methylcyclohexane or
1-methyl-3-ethylcyclohexane or
1-methyl-5-ethylcyclohexane)

Worked Example 12.12 Naming Organic Compounds: Cycloalkanes

What is the IUPAC name of the following cycloalkane?

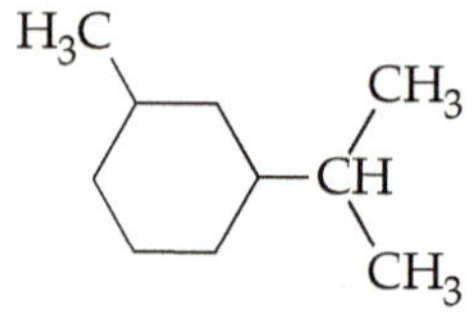

ANALYSIS First identify the parent cycloalkane, then add the positions and identity of any substituents.

SECTION 12.10 Drawing and Naming Cycloalkanes

SOLUTION

STEP 1: The parent cycloalkane contains 6 carbons (*hex*ane); hence, *cyclohexane*.

STEP 2: There are two substituents; a *methyl* (—CH₃) and an *isopropyl* (CH₃CHCH₃). Alphabetically, the isopropyl group is given priority (number 1); the methyl group is then found on the third carbon in the ring.

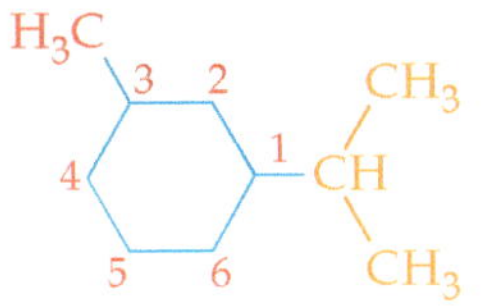

1-isopropyl-3-methylcyclohexane

Worked Example 12.13 Molecular Structures: Drawing Line Structures for Cycloalkanes

Draw a line structure for 1,4-dimethylcyclohexane.

ANALYSIS This structure consists of a 6-carbon ring with two methyl groups attached at positions 1 and 4. Draw a hexagon to represent a cyclohexane ring, and attach a —CH₃ group at an arbitrary position that becomes the first carbon in the chain, designated as C1. Then count around the ring to the fourth carbon (C4), and attach another —CH₃ group.

SOLUTION
Note that the second methyl group is written here as H₃C— because it is attached on the left side of the ring.

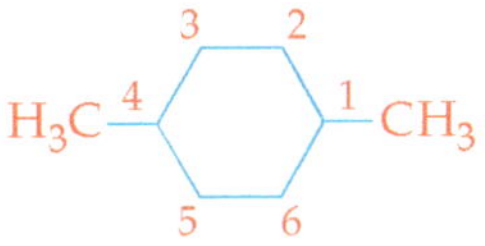

1,4-dimethylcyclohexane

PROBLEM 12.18
What are the IUPAC names of the following cycloalkanes?

(a) 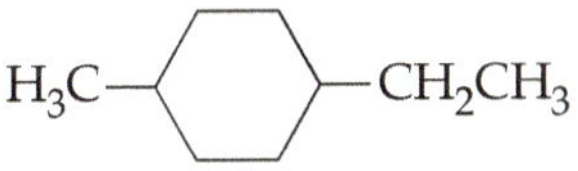 H₃C—⟨hexagon⟩—CH₂CH₃

(b) 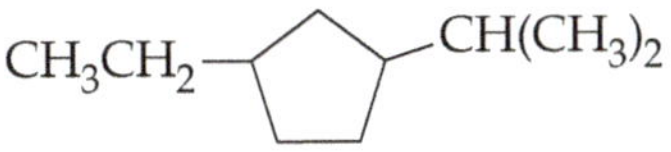 CH₃CH₂—⟨pentagon⟩—CH(CH₃)₂

PROBLEM 12.19
Draw line structures that represent the following IUPAC names:
(a) 1,1-Diethylcyclohexane
(b) 1,3,5-Trimethylcycloheptane

PROBLEM 12.20
In the box Chemistry in Action: Surprising Uses of Petroleum, three alkenes were mentioned as being important materials obtained from the refining of petroleum: ethylene, propylene, and butadiene.
(a) What consumer products are manufactured with ethylene and propylene?
(b) Butadiene is used in the manufacture of synthetic rubber. Why is this more desired than natural rubber?

KEY CONCEPT PROBLEM 12.21

What is the IUPAC name of the following cycloalkane?

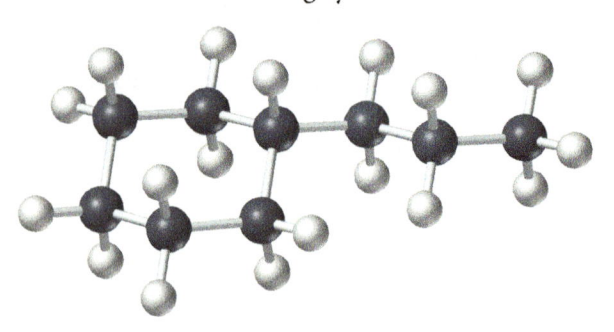

SUMMARY: REVISITING THE CHAPTER GOALS

1. What are the basic properties of organic compounds? Compounds made up primarily of carbon atoms are classified as organic. Many organic compounds contain carbon atoms that are joined in long chains by a combination of single (C—C), double (C=C), or triple (C≡C) bonds. In this chapter, we focused primarily on *alkanes*, hydrocarbon compounds that contain only single bonds between all C atoms (see Problems 29, 31, 32).

2. What are functional groups, and how are they used to classify organic molecules? Organic compounds can be classified into various families according to the functional groups they contain. A *functional group* is a part of a larger molecule and is composed of a group of atoms that has characteristic structure and chemical reactivity. A given functional group undergoes nearly the same chemical reactions in every molecule where it occurs (see Problems 25, 34–37, 66, 73).

3. What are isomers? *Isomers* are compounds that have the same formula but different structures. Isomers that differ in their connections among atoms are called *constitutional isomers*. When atoms other than carbon and hydrogen are present, the ability to have *functional group isomers* arises; these are molecules that, due to the differences in their connections, have not only different structures but also belong to different families of organic molecules (see Problems 28, 38–51).

4. How are organic molecules drawn? Organic compounds can be represented by *structural formulas* in which all atoms and bonds are shown, by *condensed structures* in which not all bonds are drawn, or by *line structures* in which the carbon skeleton is represented by lines and the locations of C and H atoms are understood (see Problems 22–24, 44, 45, 48, 49–51).

5. What are alkanes and cycloalkanes, and how are they named? Compounds that contain only carbon and hydrogen are called *hydrocarbons*, and hydrocarbons that have only single bonds are called *alkanes*. A *straight-chain alkane* has all its carbons connected in a row, a *branched-chain alkane* has a branching connection of atoms somewhere along its chain, and a *cycloalkane* has a ring of carbon atoms. Alkanes have the general formula C_nH_{2n+2}, whereas cycloalkanes have the formula C_nH_{2n}. Straight-chain alkanes are named by adding the family ending *-ane* to a parent; this tells how many carbon atoms are present. Branched-chain alkanes are named by using the longest continuous chain of carbon atoms for the parent and then identifying the *alkyl groups* present as branches off the main chain. The positions of the substituent groups on the main chain are identified by numbering the carbons in the chain so that the substituents have the lowest number. Cycloalkanes are named by adding *cyclo-* as a prefix to the name of the alkane (see Problems 25, 27, 52, 61, 67).

6. What are the general properties and chemical reactions of alkanes? Alkanes are generally soluble only in nonpolar organic solvents, have weak intermolecular forces, and are nontoxic. Their principal chemical reactions are *combustion*, a reaction with oxygen that gives carbon dioxide and water, and *halogenation*, a reaction in which hydrogen atoms are replaced by chlorine or bromine (see Problems 62–65, 70, 76, 77).

Key Words

CONCEPT MAP: INTRODUCTION TO ORGANIC CHEMISTRY FAMILIES

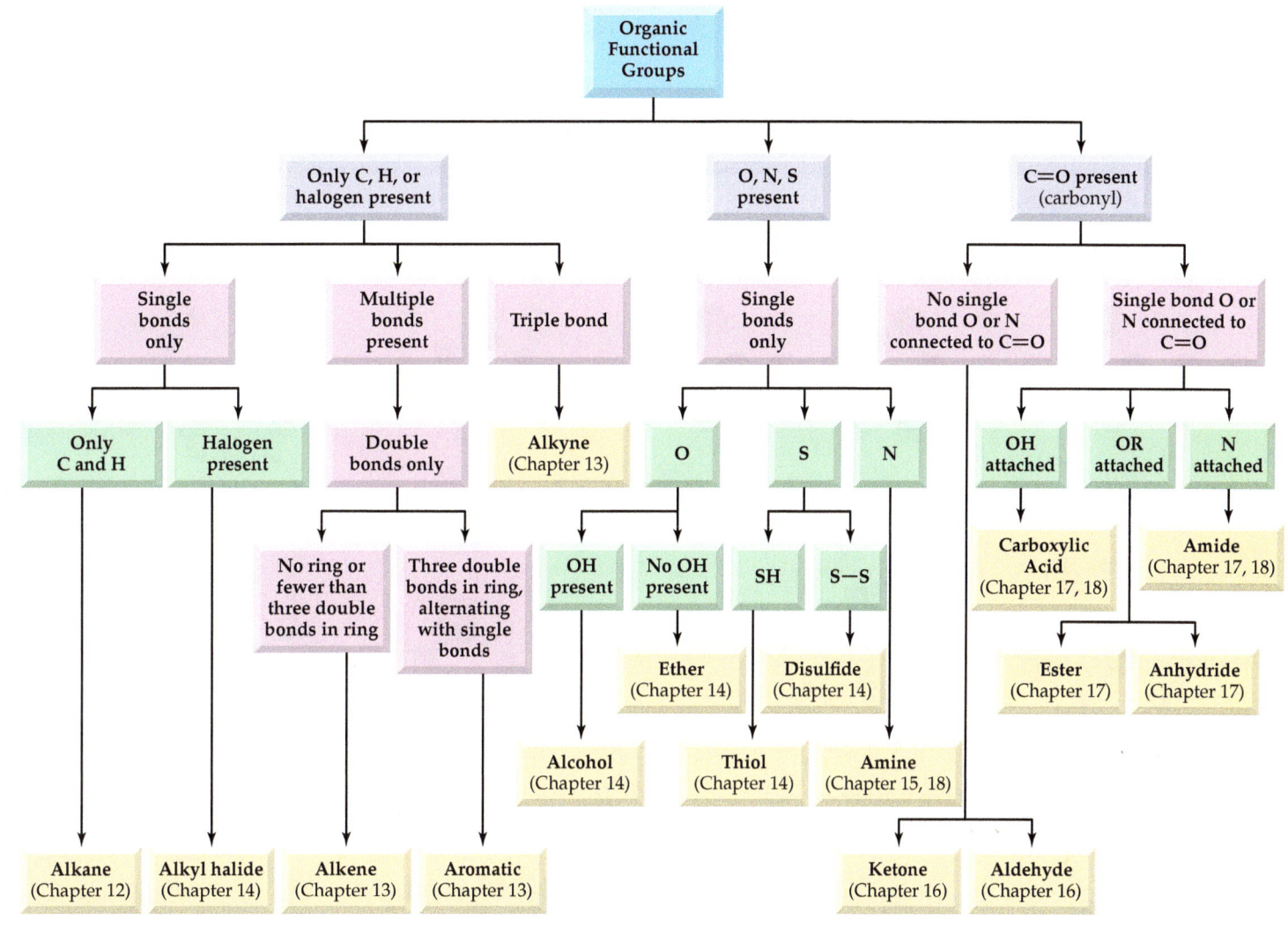

```
                          Organic
                         Functional
                           Groups
        ┌──────────────────┼──────────────────┐
  Only C, H, or        O, N, S           C=O present
  halogen present      present           (carbonyl)
   ┌────┼────┐            │              ┌──────┴──────┐
Single Multiple Triple  Single      No single      Single bond O or
bonds  bonds    bond    bonds       bond O or N    N connected to
only   present          only        connected to   C=O
                                    C=O
 ┌──┴──┐  ┌──┴──┐    │         ┌──┬──┐         ┌────┼────┐
Only  Halogen Double Alkyne    O  S  N         OH   OR   N
C and present bonds (Chapter                   attached attached attached
  H          only     13)
           ┌──┴──┐              ┌─┴─┐  ┌─┴─┐     │      │      │
        No ring  Three double  OH  No SH  S-S  Carboxylic    Amide
        or fewer bonds in ring,present OH                Acid     (Chapter 17, 18)
        than     alternating      present        (Chapter 17, 18)
        three    with single           │     Ether Disulfide  │
        double   bonds                 │   (Chapter (Chapter  ├──────┐
        bonds in                       │     14)      14)   Ester  Anhydride
        ring                           │                  (Chapter (Chapter
                                  Alcohol  Thiol  Amine    17)      17)
                                  (Chapter (Chapter (Chapter
                                    14)      14)    15, 18)
                                                     │
                                                 ┌───┴───┐
 Alkane  Alkyl halide  Alkene  Aromatic        Ketone  Aldehyde
(Chapter (Chapter     (Chapter (Chapter       (Chapter (Chapter
  12)      14)         13)      13)             16)      16)
```

▲ **Figure 12.5**

Functional Group Flow Scheme. Learning to classify organic molecules by the families to which they belong is a crucial skill you need to develop, since the chemistry that both organic and biological molecules undergo is directly related to their functional groups. The flow scheme in Figure 12.5 will aid you in this classification. First introduced in Section 12.2, it will be a key reference as you proceed through the rest of the chapters in this book. As we discuss each family in later chapters, sections of it will be reproduced and expanded to help also tie in the chemistry that those functional groups undergo.

KEY WORDS

Alkane, *p. 364*
Alkyl group, *p. 375*
Branched-chain alkane, *p. 365*
Combustion, *p.381*
Condensed structure, *p. 367*
Conformation, *p. 372*
Conformer, *p. 372*
Constitutional isomers, *p. 366*
Cycloalkane, *p. 383*

Ethyl group, *p. 375*
Functional group, *p. 360*
Functional group isomer, *p. 366*
Hydrocarbon, *p. 360*
Isomers, *p. 365*
Isopropyl group, *p. 376*
Line structure, *p. 369*
Methyl group, *p. 375*
Methylene group, *p. 367*

Organic chemistry, *p. 357*
Primary (1°) carbon atom, *p. 376*
Propyl group, *p. 376*
Quaternary (4°) carbon atom, *p. 376*
Secondary (2°) carbon atom, *p. 376*
Straight-chain alkane, *p. 365*
Substituent, *p. 374*
Tertiary (3°) carbon atom, *p. 376*

390 CHAPTER 12 Introduction to Organic Chemistry: Alkanes

SUMMARY OF REACTIONS

1. **Combustion of an alkane with oxygen (Section 12.8):**

   $CH_4 + 2 O_2 \longrightarrow CO_2 + 2 H_2O$

2. **Halogenation of an alkane to yield an alkyl halide (Section 12.8):**

   $CH_4 + Cl_2 \longrightarrow CH_3Cl + HCl$

UNDERSTANDING KEY CONCEPTS

12.22 How many hydrogen atoms are needed to complete the hydrocarbon formulas for the following carbon backbones?

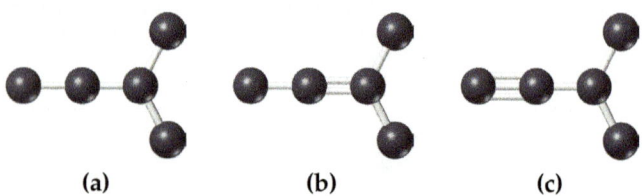

(a)   (b)   (c)

12.23 Convert the following models into condensed structures (black = C; white = H; red = O):

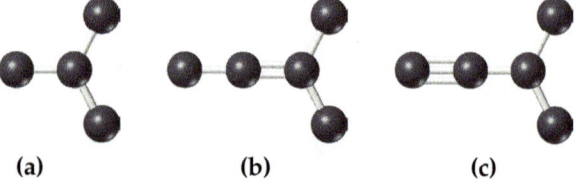

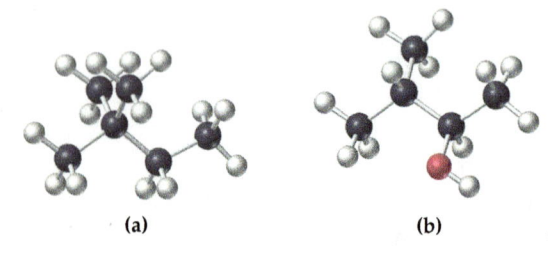

(a)   (b)

12.24 Convert the following models into line drawings (black = C; white = H; red = O; blue = N):

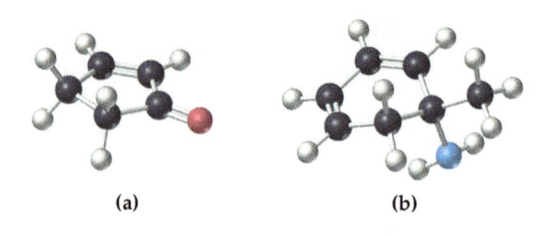

(a)   (b)

12.25 Convert the following models into line drawings and identify the functional groups in each:

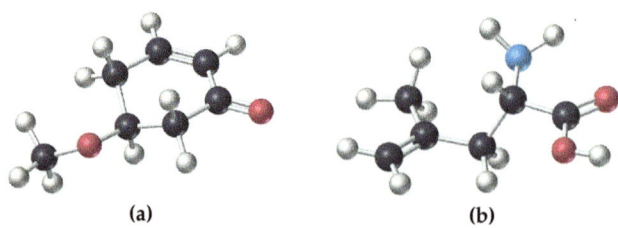

(a)   (b)

12.26 Give systematic names for the following alkanes:

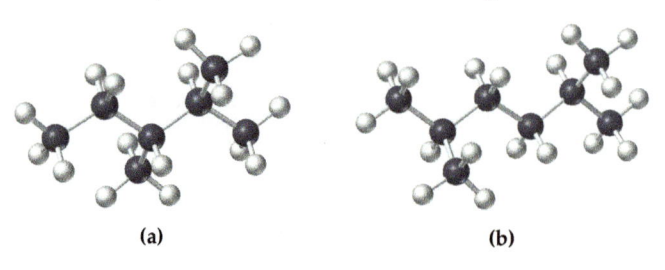

(a)   (b)

12.27 Give systematic names for the following cycloalkanes:

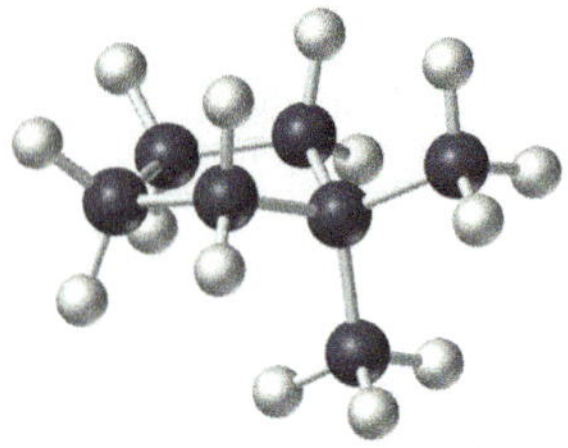

(a) 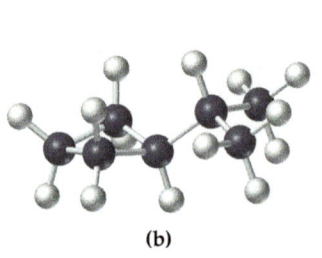   (b)

12.28 The following two compounds are isomers, even though both can be named 1,3-dimethylcyclopentane. What is the difference between them?

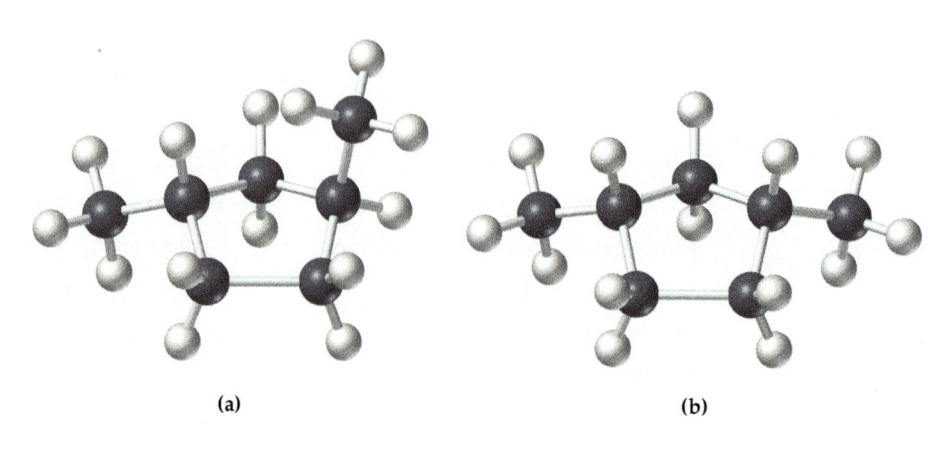

(a)   (b)

ADDITIONAL PROBLEMS

ORGANIC MOLECULES AND FUNCTIONAL GROUPS

12.29 What characteristics of carbon make possible the existence of so many different organic compounds?

12.30 What are functional groups, and why are they important?

12.31 Why are most organic compounds nonconducting and insoluble in water?

12.32 What is meant by the term *polar covalent bond*? Give an example of such a bond.

12.33 For each of the following, give an example of a member compound containing 5 carbons total:

(a) Alcohol (b) Amide (c) Carboxylic acid (d) Ester

12.34 Identify the circled functional groups in the following molecules:

(a)

(b)

12.35 Identify the functional groups in the following molecules:

(a) Vitamin A

(b) Ramipril (a new generation antihypertensive)

12.36 Propose structures for molecules that fit the following descriptions:

(a) An aldehyde with the formula $C_5H_{10}O$
(b) An ester with the formula $C_6H_{12}O_2$
(c) A compound with the formula C_3H_7NOS that is both an amide and a thiol

12.37 Propose structures for molecules that fit the following descriptions:

(a) An amide with the formula C_4H_9NO
(b) An aldehyde that has a ring of carbons, $C_6H_{10}O$
(c) An aromatic compound that is also an ether, $C_8H_{10}O$

ALKANES AND ISOMERS

12.38 What requirement must be met for two compounds to be isomers?

12.39 If one compound has the formula C_5H_{10} and another has the formula C_4H_{10}, are the two compounds isomers? Explain.

12.40 What is the difference between a secondary carbon and a tertiary carbon? What about the difference between a primary carbon and a quaternary carbon?

12.41 Why is it not possible for a compound to have a *quintary* carbon (five R groups attached to C)?

12.42 Give examples of compounds that meet the following descriptions:

(a) An alkane with 2 tertiary carbons
(b) A cycloalkane with only secondary carbons

12.43 Give examples of compounds that meet the following descriptions:

(a) A branched-chain alkane with only primary and quaternary carbons
(b) A cycloalkane with three substituents

12.44 (a) There are two isomers with the formula C_4H_{10}. Draw both the condensed and line structure for each isomer.

(b) Using the structures you drew in (a) as a starting point, draw both the condensed and line structures for the four isomeric alcohols having the chemical formula $C_4H_{10}O$.

(c) Using the structures you drew in (a) as a starting point, draw both the condensed and line structures for the three isomeric ethers having the chemical formula $C_4H_{10}O$.

12.45 Write condensed structures for the following molecular formulas. (You may have to use rings and/or multiple bonds in some instances.)

(a) C_2H_7N (b) C_4H_8 (Write the line structure as well.)
(c) C_2H_4O (d) CH_2O_2 (Write the line structure as well.)

12.46 How many isomers can you write that fit the following descriptions?

(a) Alcohols with formula $C_5H_{12}O$
(b) Amines with formula C_3H_9N
(c) Ketones with formula $C_5H_{10}O$

12.47 How many isomers can you write that fit the following descriptions?

(a) Aldehydes with formula $C_5H_{10}O$
(b) Esters with formula $C_4H_8O_2$
(c) Carboxylic acids with formula $C_4H_8O_2$

392 CHAPTER 12 Introduction to Organic Chemistry: Alkanes

12.48 Which of the following pairs of structures are identical, which are isomers, and which are unrelated?

(a) 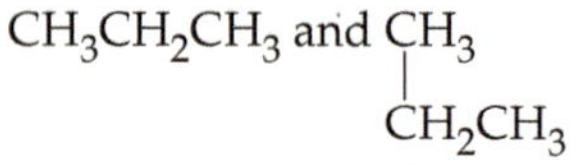 CH₃CH₂CH₃ and CH₃
                        |
                        CH₂CH₃

(b) 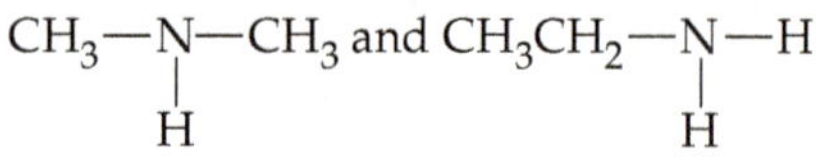 CH₃—N—CH₃ and CH₃CH₂—N—H
          | |
          H   H

(c) 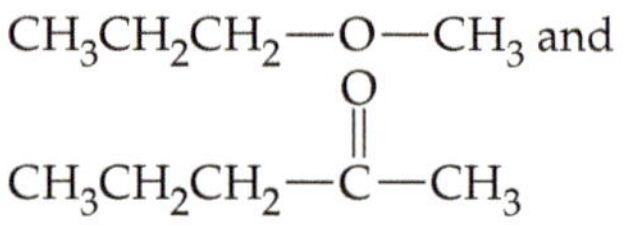 CH₃CH₂CH₂—O—CH₃ and
                 O
                 ‖
    CH₃CH₂CH₂—C—CH₃

(d) 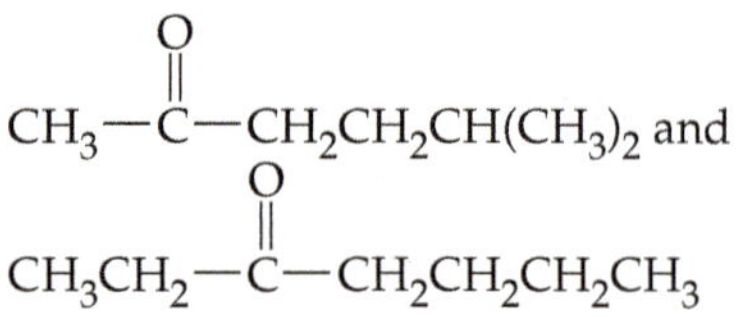 CH₃—C(=O)—CH₂CH₂CH(CH₃)₂ and
    CH₃CH₂—C(=O)—CH₂CH₂CH₂CH₃

(e) 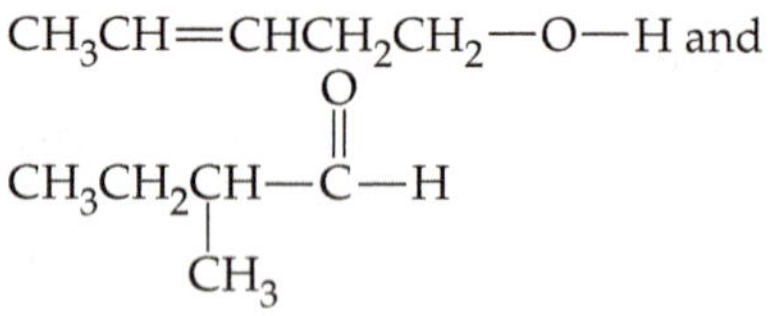 CH₃CH=CHCH₂CH₂—O—H and
                      O
                      ‖
    CH₃CH₂CH—C—H
           |
           CH₃

12.49 Which structure(s) in each group represent the same compound, and which represent isomers?

(a) 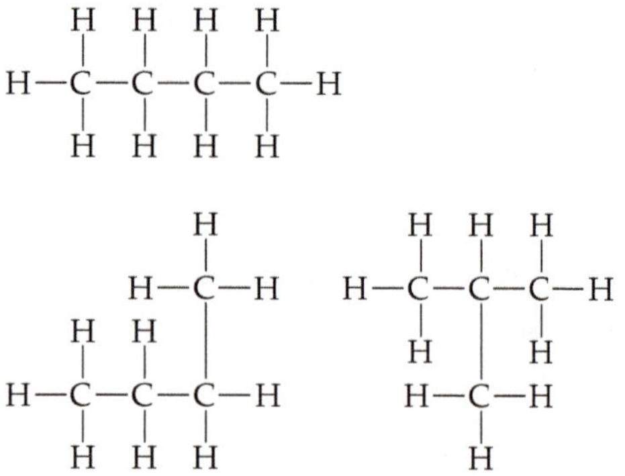 H—C—C—C—C—H (with H's on each carbon)

(b) 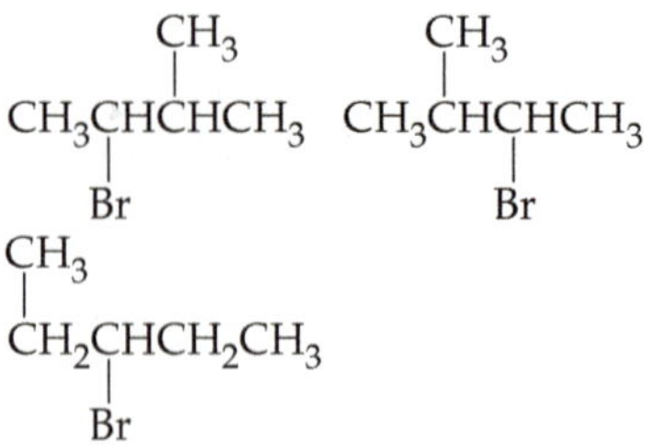 / (c) 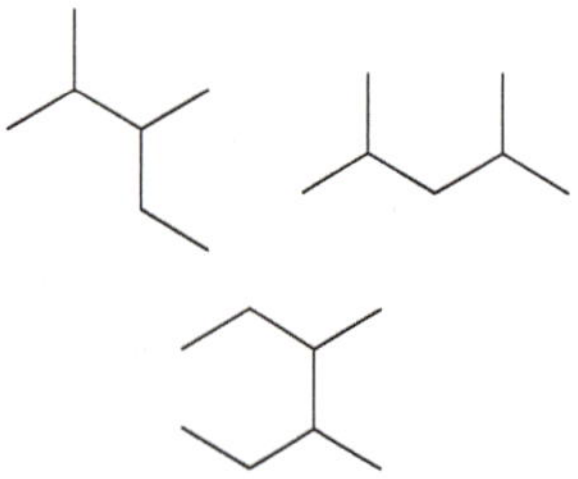 [structures shown]

CH₃CHCHCH₃   CH₃CHCHCH₃
  | |
  Br   Br
  CH₃
  |
  CH₂CHCH₂CH₃
       |
       Br

12.50 What is wrong with the following structures?

(a) CH₃=CHCH₂CH₂OH

(b) 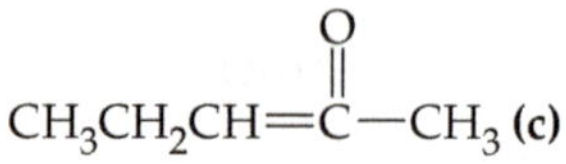 CH₃CH₂CH=C(=O)—CH₃ (c) CH₂CH₂CH₂C≡CCH₃
                                    |
                                    CH₃

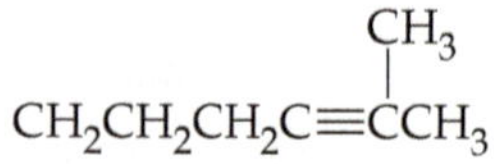

12.51 There are two things wrong with the following structure. What are they?

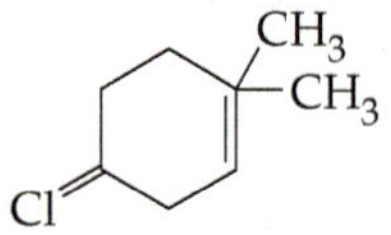

[cyclohexene with CH₃, CH₃ and Cl substituents]

ALKANE NOMENCLATURE

12.52 What are the IUPAC names of the following alkanes?

(a) 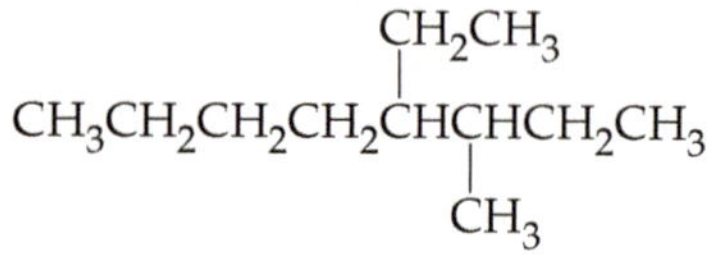 CH₃CH₂CH₂CH₂CHCHCH₂CH₃
                  | |
                  CH₂CH₃
                  |
                  CH₃

(b) 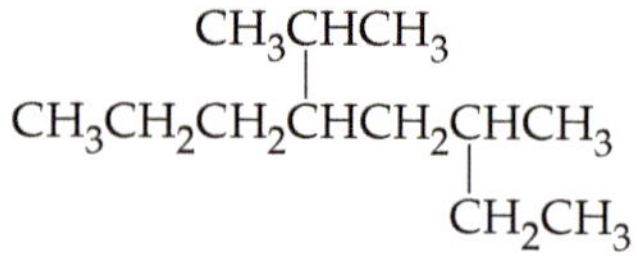 CH₃CH₂CH₂CHCH₂CHCH₃   (c) 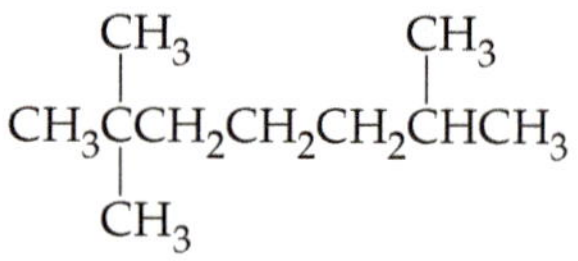 CH₃CCH₂CH₂CH₂CHCH₃
              | | | |
              CH₃CHCH₃   CH₃   CH₃
              | |
              CH₂CH₃   CH₃

(d) 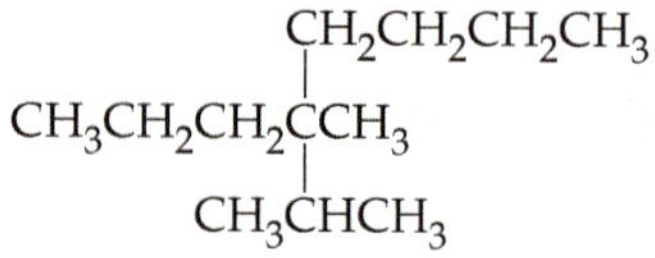 CH₃CH₂CH₂CCH₃   (e) 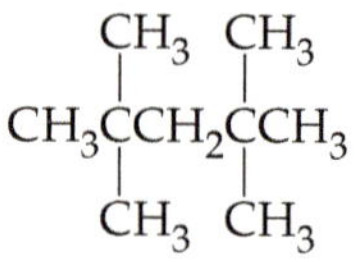 CH₃CCH₂CCH₃
              | | |
              CH₂CH₂CH₂CH₃   CH₃   CH₃
              | | |
              CH₃CHCH₃   CH₃   CH₃

(f) 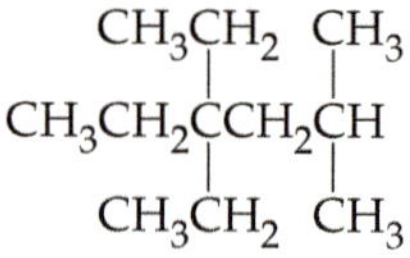 CH₃CH₂CCH₂CH   (g) 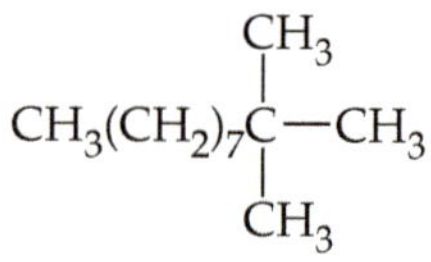 CH₃(CH₂)₇C—CH₃
         | | |
         CH₃CH₂ CH₃   CH₃
         | |
         CH₃CH₂ CH₃

12.53 Give IUPAC names for the five isomers with the formula C₆H₁₄.

12.54 Write condensed structures for the following compounds:

(a) 4-*tert*-Butyl-3,3,5-trimethylheptane
(b) 2,4-Dimethylpentane
(c) 4,4-Diethyl-3-methyloctane
(d) 3-Isopropyl-2,3,6,7-tetramethylnonane
(e) 3-Isobutyl-1-isopropyl-5-methylcycloheptane
(f) 1,1,3-Trimethylcyclopentane

12.55 Draw line structures for the following cycloalkanes:

(a) 1,1-Dimethylcyclopropane
(b) 1,2,3,4-Tetramethylcyclopentane
(c) Ethylhexane   (d) Cycloheptane
(e) 1-Methyl-3-propylcyclohexane
(f) 1-*sec*-Butyl-4-isopropylcyclooctane

12.56 Name the following cycloalkanes:

(a) 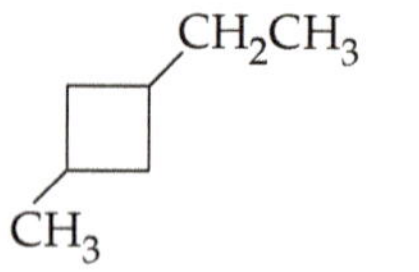 [cyclobutane with CH₂CH₃ and CH₃]
(b) 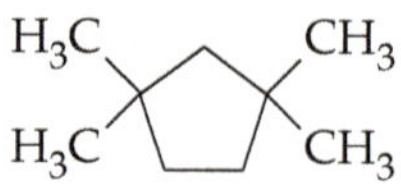 [cyclopentane with H₃C, H₃C, CH₃, CH₃]
(c) 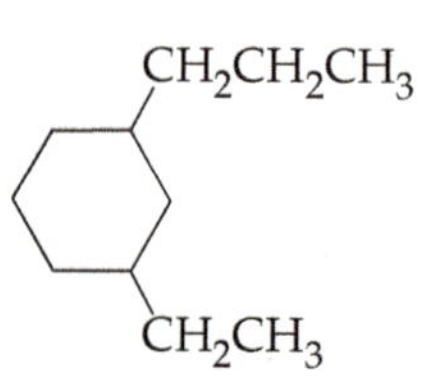 [cyclohexane with CH₂CH₂CH₃ and CH₂CH₃]
(d) 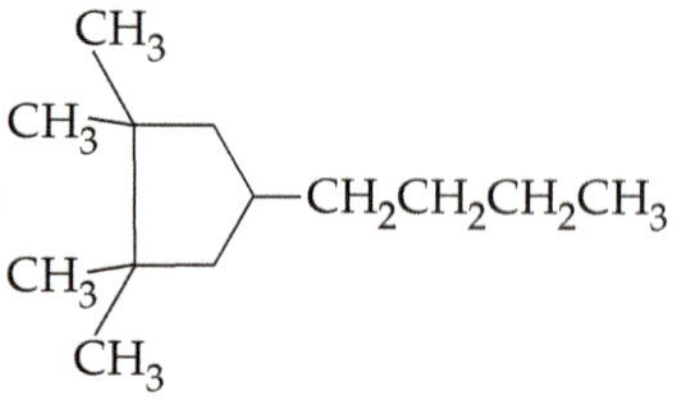 [cyclopentane with CH₃, CH₃, CH₃, CH₃ and CH₂CH₂CH₂CH₃]

12.57 Name the following cycloalkanes:

(a) 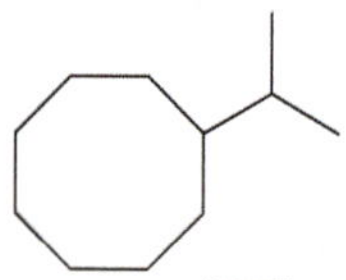 [cyclooctane with isopropyl]
(b) 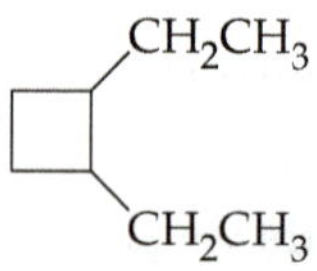 [cyclobutane with two CH₂CH₃ groups]
(c) 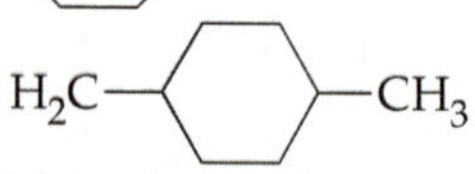 H₂C—[cyclohexane]—CH₃

12.58 The following names are incorrect. Tell what is wrong with each, and provide the correct names.

(a) CH₃CC(CH₃)(CH₃)CH₂CH₂CH₃ with CH₃ branches
2,2-Methylpentane

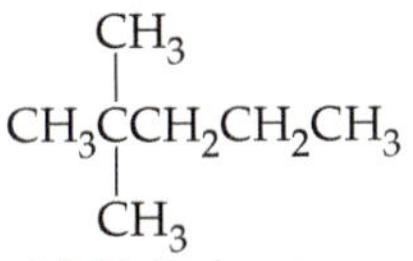

(b) (CH₃)₂CH—CH₂—CH(CH₃)₂
1,1-Diisopropylmethane

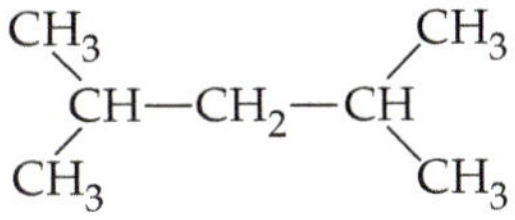

(c) CH₃CH(CH₃)CH₂—(cyclobutyl)
1-Cyclobutyl-2-methylpropane

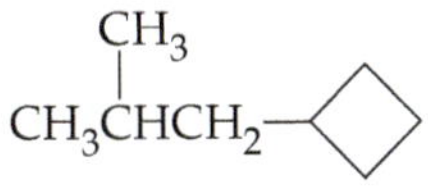

12.59 The following names are incorrect. Write the structural formula that agrees with the apparent name, and then write the correct name of the compound.

(a) 2-Ethylbutane
(b) 2-Isopropyl-2-methylpentane
(c) 5-Ethyl-1,1-methylcyclopentane
(d) 3-Ethyl-3,5,5-trimethylhexane
(e) 1,2-Dimethyl-4-ethylcyclohexane
(f) 2,4-Diethylpentane
(g) 5,5,6,6-Methyl-7,7-ethyldecane

12.60 Draw structures and give IUPAC names for the nine isomers of C_7H_{16}.

12.61 Draw the structural formulas and name all cyclic isomers with the formula C_5H_{10}.

REACTIONS OF ALKANES

12.62 Propane, commonly known as LP gas, burns in air to yield CO_2 and H_2O. Write a balanced equation for the reaction.

12.63 Write a balanced equation for the combustion of isooctane, C_8H_{16}, a component of gasoline.

12.64 Write the formulas of the three singly chlorinated isomers formed when 2,2-dimethylbutane reacts with Cl_2 in the presence of light.

12.65 Write the formulas of the seven doubly brominated isomers formed when 2,2-dimethylbutane reacts with Br_2 in the presence of light.

GENERAL QUESTIONS AND PROBLEMS

12.66 Identify the indicated functional groups in the following molecules:

(a) Testosterone, a male sex hormone 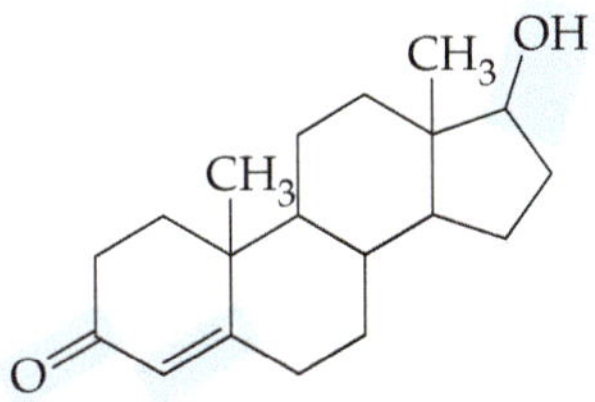
(b) Thienamycin, an antibiotic

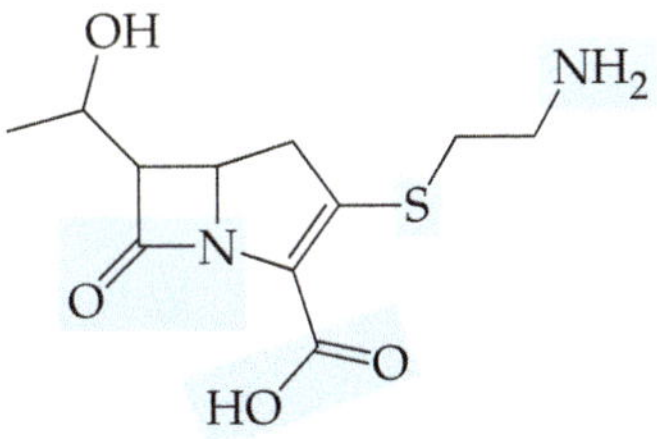

12.67 The line structure for aspartame is shown below:

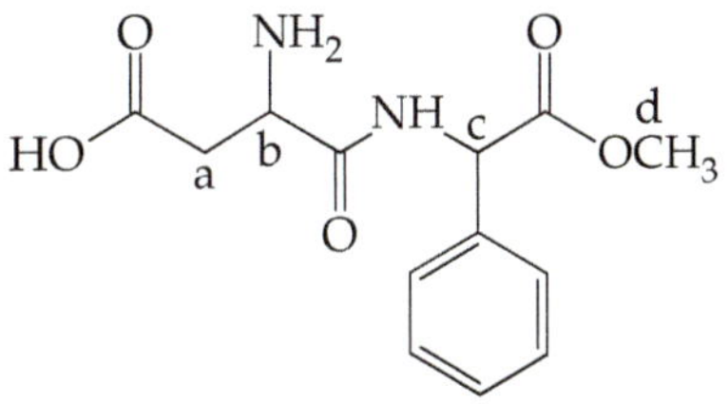

Identify carbons a–d as primary, secondary, tertiary, or quaternary.

12.68 Consider the compound shown in Problem 12.66a; how many tertiary carbons does it have?

12.69 If someone reported the preparation of a compound with the formula C_3H_9 most chemists would be skeptical. Why?

12.70 Most lipsticks are about 70% castor oil and wax. Why is lipstick more easily removed with petroleum jelly than with water?

12.71 When pentane is exposed to Br_2 in the presence of light, a halogenation reaction occurs. Write the formulas of:

(a) All possible products containing only one bromine.
(b) All possible products containing two bromines that are *not* on the same carbon.

12.72 Which do you think has a higher boiling point, pentane or neopentane (2,2-dimethylpropane)? Why?

12.73 Propose structures for the following:

(a) An aldehyde, C_4H_8O
(b) An iodo-substituted alkene, C_5H_9I
(c) A cycloalkane, C_7H_{14}
(d) A diene (dialkene), C_5H_8

CHEMISTRY IN ACTION

12.74 What is a chemical feedstock? [*Surprising Uses of Petroleum, p. 385*]

12.75 Why is the demand for synthetic rubber greater than that of natural rubber? [*Surprising Uses of Petroleum, p. 385*]

MASTERING REACTIONS

12.76 When ethyl alcohol is treated with acid, the initially formed intermediate is known as an oxonium ion:

$$CH_3-CH_2-\ddot{O}H + H^+ \rightleftharpoons CH_3-CH_2-\overset{H}{\underset{+}{\ddot{O}H}}$$

Using the curved arrow formalism, show how this process most likely occurs.

12.77 Consider the following two-step process:

$$CH_3-\ddot{S}H + {:}\ddot{O}H^- \rightleftharpoons CH_3-\ddot{S}{:}^- + H\ddot{O}H$$

$$CH_3-\ddot{S}{:}^- + CH_3-\ddot{I}{:} \longrightarrow CH_3-\ddot{S}-CH_3 + {:}\ddot{I}{:}^-$$

Using the curved arrow formalism, show how each step of this process is most likely to occur.

APPENDIX A
Scientific Notation

What Is Scientific Notation?

The numbers that you encounter in chemistry are often either very large or very small. For example, there are about 33,000,000,000,000,000,000,000 H$_2$O molecules in 1.0 mL of water, and the distance between the H and O atoms in an H$_2$O molecule is 0.000 000 000 095 7 m. These quantities are more conveniently written in *scientific notation* as 3.3×10^{22} molecules and 9.57×10^{-11} m, respectively. In scientific notation (also known as *exponential notation*), a quantity is represented as a number between 1 and 10 multiplied by a power of 10. In this kind of expression, the small raised number to the right of the 10 is the exponent.

Number	Exponential Form	Exponent
1,000,000	1×10^6	6
100,000	1×10^5	5
10,000	1×10^4	4
1,000	1×10^3	3
100	1×10^2	2
10	1×10^1	1
1		
0.1	1×10^{-1}	-1
0.01	1×10^{-2}	-2
0.001	1×10^{-3}	-3
0.000 1	1×10^{-4}	-4
0.000 01	1×10^{-5}	-5
0.000 001	1×10^{-6}	-6
0.000 000 1	1×10^{-7}	-7

Numbers greater than 1 have *positive* exponents, which tell how many times a number must be *multiplied* by 10 to obtain the correct value. For example, the expression 5.2×10^3 means that 5.2 must be multiplied by 10 three times:

$$5.2 \times 10^3 = 5.2 \times 10 \times 10 \times 10 = 5.2 \times 1000 = 5200$$

Note that doing this means moving the decimal point three places to the right:

$$5200.$$
$$123$$

The value of a positive exponent indicates *how many places to the right the decimal point must be moved* to give the correct number in ordinary decimal notation.

Numbers less than 1 have *negative exponents,* which tell how many times a number must be *divided* by 10 (or multiplied by one-tenth) to obtain the correct value. Thus, the expression 3.7×10^{-2} means that 3.7 must be divided by 10 two times:

$$3.7 \times 10^{-2} = \frac{3.7}{10 \times 10} = \frac{3.7}{100} = 0.037$$

Note that doing this means moving the decimal point two places to the left:

$$0.037$$
$$21$$

A-1

APPENDIX A SCIENTIFIC NOTATION

The value of a negative exponent indicates *how may places to the left the decimal point must be moved* to give the correct number in ordinary decimal notation.

Representing Numbers in Scientific Notation

How do you convert a number from ordinary notation to scientific notation? If the number is greater than or equal to 10, shift the decimal point to the *left* by n places until you obtain a number between 1 and 10. Then, multiply the result by 10^n. For example, the number 8137.6 is written in scientific notation as 8.1376×10^3:

$$8137.6 = 8.1376 \times 10^3$$

Shift decimal point to the left by 3 places to get a number between 1 and 10

Number of places decimal point was shifted to the left

When you shift the decimal point to the left by three places, you are in effect dividing the number by $10 \times 10 \times 10 = 1000 = 10^3$. Therefore, you must multiply the result by 10^3 so that the value of the number is unchanged.

To convert a number less than 1 to scientific notation, shift the decimal point to the *right* by n places until you obtain a number between 1 and 10. Then, multiply the result by 10^{-n}. For example, the number 0.012 is written in scientific notation as 1.2×10^{-2}:

$$0.012 = 1.2 \times 10^{-2}$$

Shift decimal point to the right by 2 places to get a number between 1 and 10

Number of places decimal point was shifted to the right

When you shift the decimal point to the right by two places, you are in effect multiplying the number by $10 \times 10 = 100 = 10^2$. Therefore, you must multiply the result by 10^{-2} so that the value of the number is unchanged. ($10^2 \times 10^{-2} = 10^0 = 1$.)

The following table gives some additional examples. To convert from scientific notation to ordinary notation, simply reverse the preceding process. Thus, to write the number 5.84×10^4 in ordinary notation, drop the factor of 10^4 and move the decimal point 4 places to the *right* ($5.84 \times 10^4 = 58{,}400$). To write the number 3.5×10^{-1} in ordinary notation, drop the factor of 10^{-1} and move the decimal point 1 place to the *left* ($3.5 \times 10^{-1} = 0.35$). Note that you don't need scientific notation for numbers between 1 and 10 because $10^0 = 1$.

Number	Scientific Notation
58,400	5.84×10^4
0.35	3.5×10^{-1}
7.296	$7.296 \times 10^0 = 7.296 \times 1$

Mathematical Operations with Scientific Notation

Addition and Subtraction in Scientific Notation

To add or subtract two numbers expressed in scientific notation, both numbers must have the same exponent. Thus, to add 7.16×10^3 and 1.32×10^2, first write the latter number as 0.132×10^3 and then add:

$$\begin{array}{r} 7.16 \times 10^3 \\ +0.132 \times 10^3 \\ \hline 7.29 \times 10^3 \end{array}$$

The answer has three significant figures. (Significant figures are discussed in Section 2.4.) Alternatively, you can write the first number as 71.6×10^2 and then add:

$$\begin{array}{r} 7.16 \times 10^2 \\ + \ 1.32 \times 10^2 \\ \hline 72.9 \ \times 10^2 = 7.29 \times 10^3 \end{array}$$

Subtraction of these two numbers is carried out in the same manner.

$$\begin{array}{r} 7.16 \ \times 10^3 \\ -0.132 \times 10^3 \\ \hline 7.03 \ \ \times 10^3 \end{array} \quad \text{or} \quad \begin{array}{r} 7.16 \times 10^2 \\ -1.32 \times 10^2 \\ \hline 70.3 \times 10^2 = 7.03 \times 10^3 \end{array}$$

Multiplication in Scientific Notation

To multiply two numbers expressed in scientific notation, multiply the factors in front of the powers of 10 and then add the exponents. For example,

$$(2.5 \times 10^4)(4.7 \times 10^7) = (2.5)(4.7) \times 10^{4+7} = 10 \times 10^{11} = 1.2 \times 10^{12}$$
$$(3.46 \times 10^5)(2.2 \times 10^{-2}) = (3.46)(2.2) \times 10^{5+(-2)} = 7.6 \times 10^3$$

Both answers have two significant figures.

Division in Scientific Notation

To divide two numbers expressed in scientific notation, divide the factors in front of the powers of 10 and then subtract the exponent in the denominator from the exponent in the numerator. For example,

$$\frac{3 \times 10^6}{7.2 \times 10^2} = \frac{3}{7.2} \times 10^{6-2} = 0.4 \times 10^4 = 4 \times 10^3 \ (1 \text{ significant figure})$$

$$\frac{7.50 \times 10^{-5}}{2.5 \times 10^{-7}} = \frac{7.50}{2.5} \times 10^{-5-(-7)} = 3.0 \times 10^2 \ (2 \text{ significant figures})$$

Scientific Notation and Electronic Calculators

With a scientific calculator you can carry out calculations in scientific notation. You should consult the instruction manual for your particular calculator to learn how to enter and manipulate numbers expressed in an exponential format. On most calculators, you enter the number $A \times 10^n$ by (i) entering the number A, (ii) pressing a key labeled EXP or EE, and (iii) entering the exponent n. If the exponent is negative, you press a key labeled $+/-$ before entering the value of n. (Note that you do not enter the number 10.) The calculator displays the number $A \times 10^n$ with the number A on the left followed by some space and then the exponent n. For example,

$$4.625 \times 10^2 \quad \text{is displayed as} \quad 4.625 \ 02$$

To add, subtract, multiply, or divide exponential numbers, use the same sequence of keystrokes as you would in working with ordinary numbers. When you add or subtract on a calculator, the numbers need not have the same exponent; the calculator automatically takes account of the different exponents. Remember, though, that the calculator often gives more digits in the answer than the allowed number of significant figures. It's sometimes helpful to outline the calculation on paper, as in the preceding examples, to keep track of the number of significant figures.

PROBLEM A.1
Perform the following calculations, expressing the results in scientific notation with the correct number of significant figures. (You don't need a calculator for these.)
(a) $(1.50 \times 10^4) + (5.04 \times 10^3)$
(b) $(2.5 \times 10^{-2}) - (5.0 \times 10^{-3})$

(c) $(6.3 \times 10^{15}) \times (10.1 \times 10^{3})$
(d) $(2.5 \times 10^{-3}) \times (3.2 \times 10^{-4})$
(e) $(8.4 \times 10^{4}) \div (3.0 \times 10^{6})$
(f) $(5.530 \times 10^{-2}) \div (2.5 \times 10^{-5})$

ANSWERS

(a) 2.00×10^{4} (b) 2.0×10^{-2} (c) 6.4×10^{19}
(d) 8.0×10^{-7} (e) 2.8×10^{-2} (f) 2.2×10^{3}

PROBLEM A.2

Perform the following calculations, expressing the results in scientific notation with the correct number of significant figures. (Use a calculator for these.)

(a) $(9.72 \times 10^{-1}) + (3.4823 \times 10^{2})$
(b) $(3.772 \times 10^{3}) - (2.891 \times 10^{4})$
(c) $(1.956 \times 10^{3}) \div (6.02 \times 10^{23})$
(d) $3.2811 \times (9.45 \times 10^{21})$
(e) $(1.0015 \times 10^{3}) \div (5.202 \times 10^{-9})$
(f) $(6.56 \times 10^{-6}) \times (9.238 \times 10^{-4})$

ANSWERS

(a) 3.4920×10^{2} (b) -2.514×10^{4} (c) 3.25×10^{-21}
(d) 3.10×10^{22} (e) 1.925×10^{11} (f) 6.06×10^{-9}

APPENDIX B
Conversion Factors

Length SI Unit: Meter (m)

    1 meter = 0.001 kilometer (km)

             = 100 centimeters (cm)

             = 1.0936 yards (yd)

    1 centimeter = 10 millimeters (mm)

                 = 0.3937 inch (in.)

    1 nanometer = 1×10^{-9} meter

    1 Angstrom (Å) = 1×10^{-10} meter

    1 inch = 2.54 centimeters

    1 mile = 1.6094 kilometers

Volume SI Unit: Cubic meter (m^3)

    1 cubic meter = 1000 liters (L)

    1 liter = 1000 cubic centimeters (cm^3)

          = 1000 milliliters (mL)

          = 1.056710 quarts (qt)

    1 cubic inch = 16.4 cubic centimeters

Temperature SI Unit: Kelvin (K)

    0 K = −273.15 °C

         = −459.67 °F

    °F = (9/5)°C + 32°; °F = (1.8 × °C) + 32°

    °C = (5/9)(°F − 32°); °C = $\dfrac{(°F - 32°)}{1.8}$

    K = °C + 273.15°

Mass SI Unit: Kilogram (kg)

    1 kilogram = 1000 grams (g)

                = 2.205 pounds (lb)

    1 gram = 1000 milligrams (mg)

           = 0.03527 ounce (oz)

    1 pound = 453.6 grams

    1 atomic mass unit = 1.66054×10^{-24} gram

Pressure SI Unit: Pascal (Pa)

    1 pascal = 9.869×10^{-6} atmosphere

    1 atmosphere = 101,325 pascals

                 = 760 mmHg (Torr)

                 = 14.70 lb/in^2

Energy SI Unit: Joule (J)

    1 joule = 0.23901 calorie (cal)

    1 calorie = 4.184 joules

    1 Calorie (nutritional unit) = 1000 calories

                                    = 1 kcal

Glossary

1,4 Link A glycosidic link between the hemiacetal hydroxyl group at C1 of one sugar and the hydroxyl group at C4 of another sugar.

Acetal A compound that has two ether-like —OR groups bonded to the same carbon atom.

Acetyl coenzyme A (acetyl-CoA) Acetyl-substituted coenzyme A—the common intermediate that carries acetyl groups into the citric acid cycle.

Acetyl group A $CH_3C=O$ group.

Achiral The opposite of chiral; having no right- or left-handedness and no nonsuper-imposable mirror images.

Acid A substance that provides H^+ ions in water.

Acid dissociation constant (K_a) The equilibrium constant for the dissociation of an acid (HA), equal to $[H^+][A^-]/[HA]$.

Acidosis The abnormal condition associated with a blood plasma pH below 7.35; may be respiratory or metabolic.

Acid-base indicator A dye that changes color depending on the pH of a solution.

Activation (of an enzyme) Any process that initiates or increases the action of an enzyme.

Activation energy (E_{act}) The amount of energy necessary for reactants to surmount the energy barrier to reaction; affects reaction rate.

Active site A pocket in an enzyme with the specific shape and chemical makeup necessary to bind a substrate.

Active transport Movement of substances across a cell membrane with the assistance of energy (for example, from ATP).

Actual Yield The amount of product actually formed in a reaction.

Acyl group An $RC=O$ group.

Addition reaction A general reaction type in which a substance X—Y adds to the multiple bond of an unsaturated reactant to yield a saturated product that has only single bonds.

Addition reaction, aldehydes and ketones Addition of an alcohol or other compound to the carbon-oxygen double bond to give a carbon-oxygen single bond.

Adenosine triphosphate (ATP) The principal energy-carrying molecule; removal of a phosphoryl group to give ADP releases free energy.

Aerobic In the presence of oxygen.

Agonist A substance that interacts with a receptor to cause or prolong the receptor's normal biochemical response.

Alcohol A compound that has an —OH group bonded to a saturated, alkane-like carbon atom, R—OH.

Alcoholic fermentation The anaerobic breakdown of glucose to ethanol plus carbon dioxide by the action of yeast enzymes.

Aldehyde A compound that has a carbonyl group bonded to one carbon and one hydrogen, RCHO.

Aldose A monosaccharide that contains an aldehyde carbonyl group.

Alkali metal An element in group 1A of the periodic table.

Alkaline earth metal An element in group 2A of the periodic table.

Alkaloid A naturally occurring nitrogen-containing compound isolated from a plant; usually basic, bitter, and poisonous.

Alkalosis The abnormal condition associated with a blood plasma pH above 7.45; may be respiratory or metabolic.

Alkane A hydrocarbon that has only single bonds.

Alkene A hydrocarbon that contains a carbon-carbon double bond.

Alkoxide ion The anion resulting from deprotonation of an alcohol, RO^-.

Alkoxy group An —OR group.

Alkyl group The part of an alkane that remains when a hydrogen atom is removed.

Alkyl halide A compound that has an alkyl group bonded to a halogen atom, R—X.

Alkyne A hydrocarbon that contains a carbon-carbon triple bond.

Allosteric control An interaction in which the binding of a regulator at one site on a protein affects the protein's ability to bind another molecule at a different site.

Allosteric enzyme An enzyme whose activity is controlled by the binding of an activator or inhibitor at a location other than the active site.

Alpha (α) particle A helium nucleus (He^{2+}), emitted as α-radiation.

Alpha- (α-) amino acid An amino acid in which the amino group is bonded to the carbon atom next to the —COOH group.

Alpha- (α-) helix Secondary protein structure in which a protein chain forms a right-handed coil stabilized by hydrogen bonds between peptide groups along its backbone.

Amide A compound that has a carbonyl group bonded to a carbon atom and a nitrogen atom group, $RCONR_2'$, where the R' groups may be alkyl groups or hydrogen atoms.

Amine A compound that has one or more organic groups bonded to nitrogen; primary, RNH_2; secondary, R_2NH; or tertiary, R_3N.

Amino acid A molecule that contains both an amino group and a carboxylic acid functional group.

Amino acid pool The entire collection of free amino acids in the body.

Amino group The —NH_2 functional group.

Amino-terminal (N-terminal) amino acid The amino acid with the free —NH_3^+ group at the end of a protein.

Ammonium ion A positive ion formed by addition of hydrogen to ammonia or an amine (may be primary, secondary, or tertiary).

Ammonium salt An ionic compound composed of an ammonium cation and an anion; an amine salt.

Amorphous solid A solid whose particles do not have an orderly arrangement.

Amphoteric Describing a substance that can react as either an acid or a base.

Anabolism Metabolic reactions that build larger biological molecules from smaller pieces.

Anaerobic In the absence of oxygen.

Anion A negatively charged ion.

Anomeric carbon atom The hemiacetal C atom in a cyclic sugar; the C atom bonded to an —OH group and an O in the ring.

Anomers Cyclic sugars that differ only in positions of substituents at the hemiacetal carbon (the anomeric carbon); the α form has the —OH on the opposite side from the —CH_2OH; the β form has the —OH on the same side as the —CH_2OH.

Antagonist A substance that blocks or inhibits the normal biochemical response of a receptor.

Antibody (immunoglobulin) Glycoprotein molecule that identifies antigens.

Anticodon A sequence of three ribonucleotides on tRNA that recognizes the complementary sequence (the codon) on mRNA.

Antigen A substance foreign to the body that triggers the immune response.

Antioxidant A substance that prevents oxidation by reacting with an oxidizing agent.

Aromatic The class of compounds containing benzene-like rings.

Artificial transmutation The change of one atom into another brought about by a nuclear bombardment reaction.

Atom The smallest and simplest particle of an element.

Atomic mass unit (amu) A convenient unit for describing the mass of an atom; 1 amu = 1/12 the mass of a carbon-12 atom.

Atomic number (Z) The number of protons in an atom.

Atomic theory A set of assumptions proposed by English scientist John Dalton to explain the chemical behavior of matter.

Atomic weight The weighted average mass of an element's atoms.

ATP synthase The enzyme complex in the inner mitochondrial membrane at which hydrogen ions cross the membrane and ATP is synthesized from ADP.

Autoimmune disease Disorder in which the immune system identifies normal body components as antigens and produces antibodies to them.

Avogadro's law Equal volumes of gases at the same temperature and pressure contain equal numbers of molecules (V/n = constant, or $V_1/n_1 = V_2/n_2$).

Avogadro's number (N_A) The number of units in 1 mole of anything; 6.02×10^{23}.

Balanced equation Describing a chemical equation in which the numbers and kinds of atoms are the same on both sides of the reaction arrow.

Base A substance that provides OH^- ions in water.

Base pairing The pairing of bases connected by hydrogen bonding (G-C and A-T), as in the DNA double helix.

Beta- (β-) Oxidation pathway A repetitive series of biochemical reactions that degrades fatty acids to acetyl-SCoA by removing carbon atoms two at a time.

Beta (β) particle An electron (e⁻), emitted as β radiation.

Beta- (β-) Sheet Secondary protein structure in which adjacent protein chains either in the same molecule or in different molecules are held in place by hydrogen bonds along the backbones.

Bile Fluid secreted by the liver and released into the small intestine from the gallbladder during digestion; contains bile acids, bicarbonate ion, and other electrolytes.

Bile acids Steroid acids derived from cholesterol that are secreted in bile.

Binary compound A compound formed by combination of two different elements.

Blood clot A network of fibrin fibers and trapped blood cells that forms at the site of blood loss.

Blood plasma Liquid portion of the blood: an extracellular fluid.

Blood serum Fluid portion of blood remaining after clotting has occurred.

Boiling point (bp) The temperature at which liquid and gas are in equilibrium.

Bond angle The angle formed by three adjacent atoms in a molecule.

Bond dissociation energy The amount of energy that must be supplied to break a bond and separate the atoms in an isolated gaseous molecule.

Bond length The optimum distance between nuclei in a covalent bond.

Boyle's law The pressure of a gas at constant temperature is inversely proportional to its volume (PV = constant, or $P_1 V_1 = P_2 V_2$).

Branched-chain alkane An alkane that has a branching connection of carbons.

Brønsted-Lowry acid A substance that can donate a hydrogen ion, H⁺, to another molecule or ion.

Brønsted-Lowry base A substance that can accept H⁺ from an acid.

Buffer A combination of substances that act together to prevent a drastic change in pH; usually a weak acid and its conjugate base.

Carbohydrate A member of a large class of naturally occurring polyhydroxy ketones and aldehydes.

Carbonyl compound Any compound that contains a carbonyl group C=O.

Carbonyl group A functional group that has a carbon atom joined to an oxygen atom by a double bond, C=O.

Carbonyl-group substitution reaction A reaction in which a new group replaces (substitutes for) a group attached to a carbonyl-group carbon in an acyl group.

Carboxyl group The —COOH functional group.

Carboxyl-terminal (C-terminal) amino acid The amino acid with the free —COO⁻ group at the end of a protein.

Carboxylate anion The anion that results from ionization of a carboxylic acid, RCOO⁻.

Carboxylic acid A compound that has a carbonyl group bonded to a carbon atom and an —OH group, RCOOH.

Carboxylic acid salt An ionic compound containing a carboxylic anion and a cation.

Catabolism Metabolic reaction pathways that break down food molecules and release biochemical energy.

Catalyst A substance that speeds up the rate of a chemical reaction but is itself unchanged.

Cation A positively charged ion.

Centromeres The central regions of chromosomes.

Chain reaction A reaction that, once started, is self-sustaining.

Change of state The conversion of a substance from one state to another—for example, from a liquid to a gas.

Charles's law The volume of a gas at constant pressure is directly proportional to its Kelvin temperature (V/T = constant, or $V_1/T_1 = V_2/T_2$).

Chemical change A change in the chemical makeup of a substance.

Chemical compound A pure substance that can be broken down into simpler substances by chemical reactions.

Chemical equation An expression in which symbols and formulas are used to represent a chemical reaction.

Chemical equilibrium A state in which the rates of forward and reverse reactions are the same.

Chemical formula A notation for a chemical compound using element symbols and subscripts to show how many atoms of each element are present.

Chemical reaction A process in which the identity and composition of one or more substances are changed.

Chemistry The study of the nature, properties, and transformations of matter.

Chiral carbon atom (chirality center) A carbon atom bonded to four different groups.

Chiral Having right- or left-handedness; able to have two different mirror-image forms.

Chromosome A complex of proteins and DNA; visible during cell division.

Cis-trans isomers Alkenes that have the same connections between atoms but differ in their three-dimensional structures because of the way that groups are attached to different sides of the double bond. The cis isomer has hydrogen atoms on the same side of the double bond; the trans isomer has them on opposite sides.

Citric acid cycle The series of biochemical reactions that breaks down acetyl groups to produce energy carried by reduced coenzymes and carbon dioxide.

Clones Identical copies of organisms, cells, or DNA segments from a single ancestor.

Codon A sequence of three ribonucleotides in the messenger RNA chain that codes for a specific amino acid; also the three nucleotide sequence (a stop codon) that stops translation.

Coefficient A number placed in front of a formula to balance a chemical equation.

Coenzyme An organic molecule that acts as an enzyme cofactor.

Cofactor A nonprotein part of an enzyme that is essential to the enzyme's catalytic activity; a metal ion or a coenzyme.

Colligative property A property of a solution that depends only on the number of dissolved particles, not on their chemical identity.

Colloid A homogeneous mixture that contains particles that range in diameter from 2 to 500 nm.

Combined gas law The product of the pressure and volume of a gas is proportional to its temperature (PV/T = constant, or $P_1 V_1 / T_1 = P_2 V_2 / T_2$).

Combustion A chemical reaction that produces a flame, usually because of burning with oxygen.

Competititve (enzyme) inhibition Enzyme regulation in which an inhibitor competes with a substrate for binding to the enzyme active site.

Concentration A measure of the amount of a given substance in a mixture.

Concentration gradient A difference in concentration within the same system.

Condensed structure A shorthand way of drawing structures in which C—C and C—H bonds are understood rather than shown.

Conformation The specific three-dimensional arrangement of atoms in a molecule at a given instant.

Conformers Molecular structures having identical connections between atoms.

Conjugate acid The substance formed by addition of H⁺ to a base.

Conjugate acid-base pair Two substances whose formulas differ by only a hydrogen ion, H⁺.

Conjugate base The substance formed by loss of H⁺ from an acid.

Conjugated protein A protein that incorporates one or more non-amino acid units in its structure.

Constitutional isomers Compounds with the same molecular formula but different connections among their atoms.

Conversion factor An expression of the relationship between two units.

Coordinate covalent bond The covalent bond that forms when both electrons are donated by the same atom.

Cosmic rays A mixture of high-energy particles—primarily of protons and various atomic nuclei—that shower the earth from outer space.

Covalent bond A bond formed by sharing electrons between atoms.

Critical mass The minimum amount of radioactive material needed to sustain a nuclear chain reaction.

Crystalline solid A solid whose atoms, molecules, or ions are rigidly held in an ordered arrangement.

Cycloalkane An alkane that contains a ring of carbon atoms.

Cycloalkene A cyclic hydrocarbon that contains a double bond.

Cytoplasm The region between the cell membrane and the nuclear membrane in a eukaryotic cell.

Cytosol The fluid part of the cytoplasm surrounding the organelles within a cell.

d-Block element A transition metal element that results from the filling of d orbitals.

D-Sugar Monosaccharide with the —OH group on the chiral carbon atom farthest from the carbonyl group pointing to the right in a Fischer projection.

Dalton's law The total pressure exerted by a mixture of gases is equal to the sum of the partial pressures exerted by each individual gas.

Decay series A sequential series of nuclear disintegrations leading from a heavy radioisotope to a nonradioactive product.

Degree of unsaturation The number of carbon-carbon double bonds in a molecule.

Dehydration The loss of water from an alcohol to yield an alkene.

Denaturation The loss of secondary, tertiary or quaternary protein structure due to disruption of noncovalent interactions and/or disulfide bonds that leaves peptide bond and primary structure intact.

Density The physical property that relates the mass of an object to its volume; mass per unit volume.

Deoxyribonucleotide A nucleotide containing 2-deoxy-D-ribose.

Diabetes mellitus A chronic condition due to either insufficient insulin or failure of insulin to activate crossing of cell membranes by glucose.

Diastereomers Stereoisomers that are not mirror images of each other.

Digestion A general term for the breakdown of food into small molecules.

Dilution factor The ratio of the initial and final solution volumes (V_1/V_2).

Dipole-dipole force The attractive force between positive and negative ends of polar molecules.

Disaccharide A carbohydrate composed of two monosaccharides.

Dissociation The splitting apart of an acid in water to give H^+ and an anion.

Disulfide A compound that contains a sulfur-sulfur bond, RS-SR.

Disulfide bond (in protein) An S-S bond formed between two cysteine side chains; can join two peptide chains together or cause a loop in a peptide chain.

DNA (deoxyribonucleic acid) The nucleic acid that stores genetic information; a polymer of deoxyribonucleotides.

Double bond A covalent bond formed by sharing two electron pairs.

Double helix Two strands coiled around each other in a screwlike fashion; in most organisms the two polynucleotides of DNA form a double helix.

Drug Any substance that alters body function when it is introduced from an external source.

Eicosanoid A lipid derived from a 20-carbon unsaturated carboxylic acid.

Electrolyte A substance that produces ions and therefore conducts electricity when dissolved in water.

Electron A negatively charged subatomic particle.

Electron affinity The energy released on adding an electron to a single atom in the gaseous state.

Electron capture A process in which the nucleus captures an inner-shell electron from the surrounding electron cloud, thereby converting a proton into a neutron.

Electron configuration The specific arrangement of electrons in an atom's shells and subshells.

Electron shell A grouping of electrons in an atom according to energy.

Electron subshell A grouping of electrons in a shell according to the shape of the region of space they occupy.

Electron-dot symbol An atomic symbol with dots placed around it to indicate the number of valence electrons.

Electron-transport chain The series of biochemical reactions that passes electrons from reduced coenzymes to oxygen and is coupled to ATP formation.

Electronegativity The ability of an atom to attract electrons in a covalent bond.

Element A fundamental substance that can't be broken down chemically into any simpler substance.

Elimination reaction A general reaction type in which a saturated reactant yields an unsaturated product by losing groups from two adjacent carbon atoms.

Enantiomers, optical isomers The two mirror-image forms of a chiral molecule.

Endergonic A nonspontaneous reaction or process that absorbs free energy and has a positive ΔG.

Endocrine system A system of specialized cells, tissues, and ductless glands that excretes hormones and shares with the nervous system the responsibility for maintaining constant internal body conditions and responding to changes in the environment.

Endothermic A process or reaction that absorbs heat and has a positive ΔH.

Energy The capacity to do work or supply heat.

Enthalpy A measure of the amount of energy associated with substances involved in a reaction.

Enthalpy change (ΔH) An alternative name for heat of reaction.

Entropy (S) The amount of disorder in a system.

Entropy change (ΔS) A measure of the increase in disorder ($\Delta S = +$) or decrease in disorder ($\Delta S = -$) as a chemical reaction or physical change occurs.

Enzyme A protein or other molecule that acts as a catalyst for a biological reaction.

Equilibrium constant (K) Value of the equilibrium constant expression for a given reaction.

Equivalent For ions, the amount equal to 1 mol of charge.

Equivalent of acid Amount of an acid that contains 1 mole of H^+ ions.

Equivalent of base Amount of base that contains 1 mole of OH^- ions.

Erythrocytes Red blood cells; transporters of blood gases.

Essential amino acid An amino acid that cannot be synthesized by the body and thus must be obtained in the diet.

Ester A compound that has a carbonyl group bonded to a carbon atom and an —OR' group, RCOOR'.

Esterification The reaction between an alcohol and a carboxylic acid to yield an ester plus water.

Ether A compound that has an oxygen atom bonded to two organic groups, R—O—R.

Ethyl group The —CH_2CH_3 alkyl group.

Exergonic A spontaneous reaction or process that releases free energy and has a negative ΔG.

Exon A nucleotide sequence in DNA that is part of a gene and codes for part of a protein.

Exothermic A process or reaction that releases heat and has a negative ΔH.

Extracellular fluid Fluid outside cells.

f-Block element An inner transition metal element that results from the filling of f orbitals.

Facilitated diffusion Passive transport across a cell membrane with the assistance of a protein that changes shape.

Factor-label method A problem-solving procedure in which equations are set up so that unwanted units cancel and only the desired units remain.

Fat A mixture of triacylglycerols that is solid because it contains a high proportion of saturated fatty acids.

Fatty acid A long-chain carboxylic acid; those in animal fats and vegetable oils often have 12–22 carbon atoms.

Feedback control Regulation of an enzyme's activity by the product of a reaction later in a pathway.

Fermentation The production of energy under anaerobic conditions.

Fibrin Insoluble protein that forms the fiber framework of a blood clot.

Fibrous protein A tough, insoluble protein whose protein chains form fibers or sheets.

Filtration (kidney) Filtration of blood plasma through a glomerulus and into a kidney nephron.

Fischer projection Structure that represents chiral carbon atoms as the intersections of two lines, with the horizontal lines representing bonds pointing out of the page and the vertical lines representing bonds pointing behind the page. For sugars, the aldehyde or ketone is at the top.

Formula unit The formula that identifies the smallest neutral unit of a compound.

Formula weight The sum of the atomic weights of the atoms in one formula unit of any compound.

Free radical An atom or molecule with an unpaired electron.

Free-energy change (ΔG) The criterion for spontaneous change (negative ΔG; $\Delta G = \Delta H - T\Delta S$).

Functional group An atom or group of atoms within a molecule that has a characteristic structure and chemical behavior.

Functional group isomer Isomers having the same chemical formula but belonging to different chemical families due to differences in bonding; ethyl alcohol and dimethyl ether are examples of functional group isomers.

Gamma (γ) radiation Radioactivity consisting of high-energy light waves.

Gas A substance that has neither a definite volume nor a definite shape.

Gas constant (R) The constant R in the ideal gas law, $PV = nRT$.

Gas laws A series of laws that predict the influence of pressure (P), volume (V), and temperature (T) on any gas or mixture of gases.

Gay-Lussac's law For a fixed amount of gas at a constant voume, pressure is directly proportional to the Kelvin temperature ($P/T =$ constant, or $P_1/T_1 = P_2/T_2$).

Gene Segment of DNA that directs the synthesis of a single polypeptide.

Genetic (enzyme) control Regulation of enzyme activity by control of the synthesis of enzymes.

Genetic code The sequence of nucleotides, coded in triplets (codons) in mRNA, that determines the sequence of amino acids in protein synthesis.

Genome All of the genetic material in the chromosomes of an organism; its size is given as the number of base pairs.

Genomics The study of whole sets of genes and their functions.

Globular protein A water-soluble protein whose chain is folded in a compact shape with hydrophilic groups on the outside.

Glomerular filtrate Fluid that enters the nephron from the glomerulus; filtered blood plasma.

Gluconeogenesis The biochemical pathway for the synthesis of glucose from non-carbohydrates, such as lactate, amino acids, or glycerol.

Glycerophospholipid (phosphoglyceride) A lipid in which glycerol is linked by ester bonds to two fatty acids and one phosphate, which is in turn linked by another ester bond to an amino alcohol (or other alcohol).

Glycogenesis The biochemical pathway for synthesis of glycogen.

Glycogenolysis The biochemical pathway for breakdown of glycogen to free glucose.

Glycol A dialcohol, or diol having the two —OH groups on adjacent carbons.

Glycolipid A lipid with a fatty acid bonded to the C2—NH$_2$ and a sugar bonded to the C1—OH group of sphingosine.

Glycolysis The biochemical pathway that breaks down a molecule of glucose into two molecules of pyruvate plus energy.

Glycoprotein A protein that contains a short carbohydrate chain.

Glycoside A cyclic acetal formed by reaction of a monosaccharide with an alcohol, accompanied by loss of H$_2$O.

Glycosidic bond Bond between the anomeric carbon atom of a monosaccharide and an —OR group.

Gram-equivalent For ions, the molar mass of the ion divided by the ionic charge.

Group One of the 18 vertical columns of elements in the periodic table.

Guanosine diphosphate (GDP) An energy-carrying molecule that can gain or lose a phosphoryl group to transfer energy.

Guanosine triphosphate (GTP) An energy-carrying molecule similar to ATP; removal of a phosphoryl group to give GDP releases free energy.

Half-life ($t_{1/2}$) The amount of time required for one-half of a radioactive sample to decay.

Halogen An element in group 7A of the periodic table.

Halogenation (alkene) The addition of Cl$_2$ or Br$_2$ to a multiple bond to give a 1,2-dihalide product.

Halogenation (aromatic) The substitution of a halogen group (—X) for a hydrogen on an aromatic ring.

Heat The kinetic energy transferred from a hotter object to a colder object when the two are in contact.

Heat of fusion The quantity of heat required to completely melt a substance once it has reached its melting point.

Heat of reaction (ΔH) The amount of heat absorbed or released in a reaction.

Heat of vaporization The quantity of heat needed to completely vaporize a liquid once it has reached its boiling point.

Hemiacetal A compound with both an alcohol-like —OH group and an ether-like —OR group bonded to the same carbon atom.

Hemostasis The stopping of bleeding.

Henderson-Hasselbalch equation The logarithmic form of the K_a equation for a weak acid, used in applications involving buffer solutions.

Henry's law The solubility of a gas in a liquid is directly proportional to its partial pressure over the liquid at constant temperature.

Heterocycle A ring that contains nitrogen or some other atom in addition to carbon.

Heterogeneous mixture A nonuniform mixture that has regions of different composition.

Heterogeneous nuclear RNA The initially synthesized mRNA strand containing both introns and exons.

Homogeneous mixture A uniform mixture that has the same composition throughout.

Hormone A chemical messenger secreted by cells of the endocrine system and transported through the bloodstream to target cells with appropriate receptors where it elicits a response.

Hydration The addition of water to a multiple bond to give an alcohol product.

Hydrocarbon An organic compound that contains only carbon and hydrogen.

Hydrogen bond The attraction between a hydrogen atom bonded to an electronegative O, N, or F atom and another nearby electronegative O, N, or F atom.

Hydrogenation The addition of H$_2$ to a multiple bond to give a saturated product.

Hydrohalogenation The addition of HCl or HBr to a multiple bond to give an alkyl halide product.

Hydrolysis A reaction in which a bond or bonds are broken and the H— and —OH of water add to the atoms of the broken bond or bonds.

Hydronium ion The H$_3$O$^+$ ion, formed when an acid reacts with water.

Hydrophilic Water-loving; a hydrophilic substance dissolves in water.

Hydrophobic Water-fearing; a hydrophobic substance does not dissolve in water.

Hygroscopic Having the ability to pull water molecules from the surrounding atmosphere.

Hyperglycemia Higher-than-normal blood glucose concentration.

Hypertonic Having an osmolarity greater than the surrounding blood plasma or cells.

Hypoglycemia Lower-than-normal blood glucose concentration.

Hypotonic Having an osmolarity less than the surrounding blood plasma or cells.

Ideal gas A gas that obeys all the assumptions of the kinetic-molecular theory.

Ideal gas law A general expression relating pressure, volume, temperature, and amount for an ideal gas: $PV = nRT$.

Immune response Defense mechanism of the immune system dependent on the recognition of specific antigens, including viruses, bacteria, toxic substances, and infected cells; either cell-mediated or antibody-mediated.

Induced-fit model A model of enzyme action in which the enzyme has a flexible active site that changes shape to best fit the substrate and catalyze the reaction.

Inflammation Result of the inflammatory response: includes swelling, redness, warmth, and pain.

Inflammatory response A nonspecific defense mechanism triggered by antigens or tissue damage.

Inhibition (of an enzyme) Any process that slows or stops the action of an enzyme.

Inner transition metal element An element in one of the 14 groups shown separately at the bottom of the periodic table.

Intermolecular force A force that acts between molecules and holds molecules close to one another in liquids and solids.

Interstitial fluid Fluid surrounding cells: an extracellular fluid.

Intracellular fluid Fluid inside cells.

Intron A portion of DNA between coding regions of a gene (exons); is transcribed and then removed from final messenger RNA.

Ion An electrically charged atom or group of atoms.

Ion-product constant for water (K_w) The product of the H$_3$O$^+$ and OH$^-$ molar concentrations in water or any aqueous solution ($K_w = [\text{H}_3\text{O}^+][\text{OH}^-] = 1.00 \times 10^{-14}$).

Ionic bond The electrical attractions between ions of opposite charge in a crystal.

Ionic compound A compound that contains ionic bonds.

Ionic equation An equation in which ions are explicitly shown.

Ionic solid A crystalline solid held together by ionic bonds.

Ionization energy The energy required to remove one electron from a single atom in the gaseous state.

Ionizing radiation A general name for high-energy radiation of all kinds.

Irreversible (enzyme) inhibition Enzyme deactivation in which an inhibitor forms covalent bonds to the active site, permanently blocking it.

Isoelectric point (pI) The pH at which a sample of an amino acid has equal number of + and − charges.

Isomers Compounds with the same molecular formula but different structures.

Isopropyl group The branched-chain alkyl group —CH(CH$_3$)$_2$.

Isotonic Having the same osmolarity.

Isotopes Atoms with identical atomic numbers but different mass numbers.

Ketoacidosis Lowered blood pH due to accumulation of ketone bodies.

Ketogenesis The synthesis of ketone bodies from acetyl-SCoA.

Ketone A compound that has a carbonyl group bonded to two carbons in organic groups that can be the same or different, R$_2$C=O, RCOR'.

Ketone bodies Compounds produced in the liver that can be used as fuel by muscle and brain tissue; 3-hydroxybutyrate, ace-toacetate, and acetone.

Ketose A monosaccharide that contains a ketone carbonyl group.

Kinetic energy The energy of an object in motion.

Kinetic-molecular theory (KMT) of gases A group of assumptions that explain the behavior of gases.

L-Sugar Monosaccharide with the —OH group on the chiral carbon atom farthest from the carbonyl group pointing to the left in a Fischer projection.

Law of conservation of energy Energy can be neither created nor destroyed in any physical or chemical change.

Law of conservation of mass Matter can be neither created nor destroyed in any physical or chemical change.

Le Châtelier's principle When a stress is applied to a system in equilibrium, the equilibrium shifts to relieve the stress.

Lewis base A compound containing an unshared pair of electrons.

Lewis structure A molecular representation that shows both the connections among atoms and the locations of lone-pair valence electrons.

Limiting reagent The reactant that runs out first in a chemical reaction.

Line structure A shorthand way of drawing structures in which atoms aren't shown; instead, a carbon atom is understood to be at every intersection of lines, and hydrogens are filled in mentally.

Lipid A naturally occurring molecule from a plant or animal that is soluble in nonpolar organic solvents.

Lipid bilayer The basic structural unit of cell membranes; composed of two parallel sheets of membrane lipid molecules arranged tail to tail.

Lipogenesis The biochemical pathway for synthesis of fatty acids from acetyl-CoA.

Lipoprotein A lipid-protein complex that transports lipids.

Liposome A spherical structure in which a lipid bilayer surrounds a water droplet.

Liquid A substance that has a definite volume but that changes shape to fit its container.

London dispersion force The short-lived attractive force due to the constant motion of electrons within molecules.

Lone pair A pair of electrons that is not used for bonding.

Main group element An element in one of the two groups on the left or the six groups on the right of the periodic table.

Markovnikov's rule In the addition of HX to an alkene, the H becomes attached to the carbon that already has the most H's, and the X becomes attached to the carbon that has fewer H's.

Mass A measure of the amount of matter in an object.

Mass/mass percent concentration [(m/m)%] Concentration expressed as the number of grams of solute per 100 grams of solution.

Mass number (A) The total number of protons and neutrons in an atom.

Mass/volume percent concentration [(m/v)%] Concentration expressed as the number of grams of solute per 100 mL of solution.

Matter The physical material that makes up the universe; anything that has mass and occupies space.

Melting point (mp) The temperature at which solid and liquid are in equilibrium.

Messenger RNA (mRNA) The RNA that carries the code transcribed from DNA and directs protein synthesis.

Metal A malleable element with a lustrous appearance that is a good conductor of heat and electricity.

Metalloid An element whose properties are intermediate between those of a metal and a nonmetal.

Methyl group The —CH_3 alkyl group.

Methylene Another name for a —CH_2 unit.

Micelle A spherical cluster formed by the aggregation of soap or detergent molecules so that their hydrophobic ends are in the center and their hydrophilic ends are on the surface.

Miscible Mutually soluble in all proportions.

Mitochondrial matrix The space surrounded by the inner membrane of a mitochondrion.

Mitochondrion (plural, mitochondria) An egg-shaped organelle where small molecules are broken down to provide the energy for an organism.

Mixture A blend of two or more substances, each of which retains its chemical identity.

Mobilization (of triacylglycerols) Hydrolysis of triacylglycerols in adipose tissue and release of fatty acids into the bloodstream.

Molar mass The mass in grams of one mole of a substance, numerically equal to the molecular weight.

Molarity (M) Concentration expressed as the number of moles of solute per liter of solution.

Mole The amount of a substance corresponding to 6.02×10^{23} units.

Molecular compound A compound that consists of molecules rather than ions.

Molecular formula A formula that shows the numbers and kinds of atoms in one molecule of a compound.

Molecular weight The sum of the atomic weights of the atoms in a molecule.

Molecule A group of atoms held together by covalent bonds.

Monomer A small molecule that is used to prepare a polymer.

Monosaccharide (simple sugar) A carbohydrate with 3–7 carbon atoms.

Mutagen A substance that causes mutations.

Mutarotation Change in rotation of plane-polarized light resulting from the equilibrium between cyclic anomers and the open-chain form of a sugar.

Mutation An error in base sequence that is carried along in DNA replication.

Native protein A protein with the shape (secondary, tertiary, and quaternary structure) in which it exists naturally in living organisms.

Net ionic equation An equation that does not include spectator ions.

Neurotransmitter A chemical messenger that travels between a neuron and a neighboring neuron or other target cell to transmit a nerve impulse.

Neutralization reaction The reaction of an acid with a base.

Neutron An electrically neutral subatomic particle.

Nitration The substitution of a nitro group (—NO_2) for a hydrogen on an aromatic ring.

Noble gas An element in group 8A of the periodic table.

Noncovalent forces Forces of attraction other than covalent bonds that can act between molecules or within molecules.

Nonelectrolyte A substance that does not produce ions when dissolved in water.

Nonessential amino acid One of 11 amino acids that are synthesized in the body and are therefore not necessary in the diet.

Nonmetal An element that is a poor conductor of heat and electricity.

Normal boiling point The boiling point at a pressure of exactly 1 atmosphere.

Normality (N) A measure of acid (or base) concentration expressed as the number of acid (or base) equivalents per liter of solution.

Nuclear decay The spontaneous emission of a particle from an unstable nucleus.

Nuclear fission The fragmenting of heavy nuclei.

Nuclear fusion The joining together of light nuclei.

Nuclear reaction A reaction that changes an atomic nucleus, usually causing the change of one element into another.

Nucleic acid A polymer of nucleotides.

Nucleon A general term for both protons and neutrons.

Nucleoside A 5-carbon sugar bonded to a cyclic amine base; like a nucleotide but missing the phosphate group.

Nucleotide A 5-carbon sugar bonded to a cyclic amine base and one phosphate group (a nucleoside monophosphate); monomer for nucleic acids.

Nucleus The dense, central core of an atom that contains protons and neutrons.

Nuclide The nucleus of a specific isotope of an element.

Octet rule Main-group elements tend to undergo reactions that leave them with 8 valence electrons.

Oil A mixture of triacylglycerols that is liquid because it contains a high proportion of unsaturated fatty acids.

Orbital A region of space within an atom where an electron in a given subshell can be found.

Organic chemistry The study of carbon compounds.

Osmolarity (osmol) The sum of the molarities of all dissolved particles in a solution.

Osmosis The passage of solvent through a semipermeable membrane separating two solutions of different concentration.

Osmotic pressure The amount of external pressure applied to the more concentrated solution to halt the passage of solvent molecules across a semipermeable membrane.

Oxidation The loss of one or more electrons by an atom.

Oxidation number A number that indicates whether an atom is neutral, electron-rich, or electron-poor.

Oxidation-Reduction, or Redox, reaction A reaction in which electrons are transferred from one atom to another.

Oxidative deamination Conversion of an amino acid —NH_2 group to an α-keto group, with removal of NH_4^+.

Oxidative phosphorylation The synthesis of ATP from ADP using energy released in the electron-transport chain.

Oxidizing agent A reactant that causes an oxidation by taking electrons from or increasing the oxidation number of another reactant.

p-Block element A main group element that results from the filling of p orbitals.

p function The negative common logarithm of some variable, $pX = -(\log X)$.

Partial pressure The pressure exerted by a gas in a mixture.

Parts per billion (ppb) Number of parts of solute (in mass or volume) per one billion parts of solution.

Parts per million (ppm) Number of parts of solute (in mass or volume) per one million parts of solution.

Passive transport Movement of a substance across a cell membrane without the use of energy, from a region of higher concentration to a region of lower concentration.

Pentose phosphate pathway The biochemical pathway that produces ribose (a pentose), NADPH, and other sugar phosphates from glucose; an alternative to glycolysis.

Peptide bond An amide bond that links two amino acids together.

Percent yield The percent of the theoretical yield actually obtained from a chemical reaction.

Period One of the seven horizontal rows of elements in the periodic table.

Periodic table A table of the elements in order of increasing atomic number and grouped according to their chemical similarities.

pH A measure of the acid strength of a solution; the negative common logarithm of the H_3O^+ concentration.

Phenol A compound that has an —OH group bonded directly to an aromatic, benzene-like ring, Ar—OH.

Phenyl The C_6H_5— group.

Phosphate ester A compound formed by reaction of an alcohol with phosphoric acid; may be a monoester, $ROPO_3H_2$; a diester, $(RO)_2PO_3H$; or a triester, $(RO)_3PO$; also may be a di- or triphosphate.

Phospholipid A lipid that has an ester link between phosphoric acid and an alcohol (glycerol or sphingosine).

Phosphoryl group The —PO_3^{2-} group in organic phosphates.

Phosphorylation Transfer of a phosphoryl group, —PO_3^{2-}, between organic molecules.

Physical change A change that does not affect the chemical makeup of a substance or object.

Physical quantity A physical property that can be measured.

Polar covalent bond A bond in which the electrons are attracted more strongly by one atom than by the other.

Polyatomic ion An ion that is composed of more than one atom.

Polymer A large molecule formed by the repetitive bonding together of many smaller molecules.

Polymorphism A variation in DNA sequence within a population.

Polysaccharide (complex carbohydrate) A carbohydrate that is a polymer of monosaccharides.

Polyunsaturated fatty acid A long-chain fatty acid that has two or more carbon-carbon double bonds.

Positron A "positive electron," which has the same mass as an electron but a positive charge.

Potential energy Energy that is stored because of position, composition, or shape.

Precipitate An insoluble solid that forms in solution during a chemical reaction.

Pressure The force per unit area pushing against a surface.

Primary carbon atom A carbon atom with one other carbon attached to it.

Primary protein structure The sequence in which amino acids are linked by peptide bonds in a protein.

Product A substance that is formed in a chemical reaction and is written on the right side of the reaction arrow in a chemical equation.

Property A characteristic useful for identifying a substance or object.

propyl group The straight-chain alkyl group —$CH_2CH_2CH_3$.

Protein A large biological molecule made of many amino acids linked together through amide (peptide) bonds.

Proton A positively charged subatomic particle.

Pure substance A substance that has uniform chemical composition throughout.

Quaternary ammonium ion A positive ion with four organic groups bonded to the nitrogen atom.

Quaternary ammonium salt An ionic compound composed of a quaternary ammonium ion and an anion.

Quaternary carbon atom A carbon atom with four other carbons attached to it.

Quaternary protein structure The way in which two or more protein chains aggregate to form large, ordered structures.

Radioactivity The spontaneous emission of radiation from a nucleus.

Radioisotope A radioactive isotope.

Radionuclide The nucleus of a radioactive isotope.

Reabsorption (kidney) Movement of solutes out of filtrate in a kidney tubule.

Reactant A substance that undergoes change in a chemical reaction and is written on the left side of the reaction arrow in a chemical equation.

Reaction mechanism A description of the individual steps by which old bonds are broken and new bonds are formed in a reaction.

Reaction rate A measure of how rapidly a reaction occurs.

Rearrangement reaction A general reaction type in which a molecule undergoes bond reorganization to yield an isomer.

Receptor A molecule or portion of a molecule with which a hormone, neurotransmitter, or other biochemically active molecule interacts to initiate a response in a target cell.

Recombinant DNA DNA that contains segments from two different species.

Reducing agent A reactant that causes a reduction by giving up electrons or increasing the oxidation number of another reactant.

Reducing sugar A carbohydrate that reacts in basic solution with a mild oxidizing agent.

Reduction The gain of one or more electrons by an atom.

Reductive deamination Conversion of an α-keto acid to an amino acid by reaction with NH_4^+.

Regular tetrahedron A geometric figure with four identical triangular faces.

Replication The process by which copies of DNA are made when a cell divides.

Residue (amino acid) An amino acid unit in a polypeptide.

Resonance The phenomenon where the true structure of a molecule is an average among two or more conventional structures.

Reversible reaction A reaction that can go in either the forward direction or the reverse direction, from products to reactants or reactants to products.

Ribonucleotide A nucleotide containing D-ribose.

Ribosomal RNA (rRNA) The RNA that is complexed with proteins in ribosomes.

Ribosome The structure in the cell where protein synthesis occurs; composed of protein and rRNA.

Ribozyme RNA that acts as an enzyme.

RNA (ribonucleic acids) The nucleic acids (messenger, transfer, and ribosomal) responsible for putting the genetic information to use in protein synthesis; polymers of ribonucleotides.

Rounding off A procedure used for deleting nonsignificant figures.

s-Block element A main group element that results from the filling of an s orbital.

Salt An ionic compound formed from reaction of an acid with a base.

Saponification The reaction of an ester with aqueous hydroxide ion to yield an alcohol and the metal salt of a carboxylic acid.

Saturated A molecule whose carbon atoms bond to the maximum number of hydrogen atoms.

Saturated fatty acid A long-chain carboxylic acid containing only carbon-carbon single bonds.

Saturated solution A solution that contains the maximum amount of dissolved solute at equilibrium.

Scientific Method Systematic process of observation, hypothesis, and experimentation to expand and refine a body of knowledge.

Scientific notation A number expressed as the product of a number between 1 and 10, times the number 10 raised to a power.

Second messenger Chemical messenger released inside a cell when a hydrophilic hormone or neurotransmitter interacts with a receptor on the cell surface.

Secondary carbon atom A carbon atom with two other carbons attached to it.

Secondary protein structure Regular and repeating structural patterns (for example, α-helix, β-sheet) created by hydrogen bonding between backbone atoms in neighboring segments of protein chains.

Secretion (kidney) Movement of solutes into filtrate in a kidney tubule.

SI units Units of measurement defined by the International System of Units.

Side chain (amino acid) The group bonded to the carbon next to the carboxyl group in an amino acid; different in different amino acids.

Significant figures The number of meaningful digits used to express a value.

Simple diffusion Passive transport by the random motion of diffusion through the cell membrane.

Simple protein A protein composed of only amino acid residues.

Single bond A covalent bond formed by sharing one electron pair.

Single-nucleotide polymorphism Common single-base-pair variation in DNA.

Soap The mixture of salts of fatty acids formed on saponification of animal fat.

Solid A substance that has a definite shape and volume.

Solubility The maximum amount of a substance that will dissolve in a given amount of solvent at a specified temperature.

Solute A substance dissolved in a liquid.

Solution A homogeneous mixture that contains particles the size of a typical ion or small molecule.

Solvation The clustering of solvent molecules around a dissolved solute molecule or ion.

Solvent The liquid in which another substance is dissolved.

Specific gravity The density of a substance divided by the density of water at the same temperature.

Specific heat The amount of heat that will raise the temperature of 1 g of a substance by 1 °C.

Specificity (enzyme) The limitation of the activity of an enzyme to a specific substrate, specific reaction, or specific type of reaction.

Spectator ion An ion that appears unchanged on both sides of a reaction arrow.

Sphingolipid A lipid derived from the amino alcohol sphingosine.

Spontaneous process A process or reaction that, once started, proceeds on its own without any external influence.

Standard molar volume The volume of one mole of a gas at standard temperature and pressure (22.4 L).

Standard temperature and pressure (STP) Standard conditions for a gas, defined as 0 °C (273 K) and 1 atm (760 mmHg) pressure.

State of matter The physical state of a substance as a solid, a liquid, or a gas.

Stereoisomers Isomers that have the same molecular and structural formulas, but different spatial arrangements of their atoms.

Sterol A lipid whose structure is based on the following tetracyclic (four-ring) carbon skeleton.

Straight-chain alkane An alkane that has all its carbons connected in a row.

Strong acid An acid that gives up H^+ easily and is essentially 100% dissociated in water.

Strong base A base that has a high affinity for H^+ and holds it tightly.

Strong electrolyte A substance that ionizes completely when dissolved in water.

Structural formula A molecular representation that shows the connections among atoms by using lines to represent covalent bonds.

Subatomic particles Three kinds of fundamental particles from which atoms are made: protons, neutrons, and electrons.

Substituent An atom or group of atoms attached to a parent compound.

Substitution reaction A general reaction type in which an atom or group of atoms in a molecule is replaced by another atom or group of atoms.

Substrate A reactant in an enzyme catalyzed reaction.

Sulfonation The substitution of a sulfonic acid group ($-SO_3H$) for a hydrogen on an aromatic ring.

Supersaturated solution A solution that contains more than the maximum amount of dissolved solute; a nonequilibrium situation.

Synapse The place where the tip of a neuron and its target cell lie adjacent to each other.

Telomeres The ends of chromosomes; in humans, contain long series of repeating groups of nucleotides.

Temperature The measure of how hot or cold an object is.

Tertiary carbon atom A carbon atom with three other carbons attached to it.

Tertiary protein structure The way in which an entire protein chain is coiled and folded into its specific three-dimensional shape.

Theoretical yield The amount of product formed assuming complete reaction of the limiting reagent.

Thiol A compound that contains an $-SH$ group, $R-SH$.

Titration A procedure for determining the total acid or base concentration of a solution.

Transamination The interchange of the amino group of an amino acid and the keto group of an α-keto acid.

Transcription The process by which the information in DNA is read and used to synthesize RNA.

Transfer RNA (tRNA) The RNA that transports amino acids into position for protein synthesis.

Transition metal element An element in one of the 10 smaller groups near the middle of the periodic table.

Translation The process by which RNA directs protein synthesis.

Transmutation The change of one element into another.

Triacylglycerol (triglyceride) A triester of glycerol with three fatty acids.

Triple bond A covalent bond formed by sharing three electron pairs.

Turnover number The maximum number of substrate molecules acted upon by one molecule of enzyme per unit time.

Uncompetitive (enzyme) inhibition Enzyme regulation in which an inhibitor binds to an enzyme elsewhere than at the active site, thereby changing the shape of the enzyme's active site and reducing its efficiency.

Unit A defined quantity used as a standard of measurement.

Unsaturated A molecule that contains a carbon–carbon multiple bond, to which more hydrogen atoms can be added.

Unsaturated fatty acid A long-chain carboxylic acid containing one or more carbon–carbon double bonds.

Urea cycle The cyclic biochemical pathway that produces urea for excretion.

Valence electron An electron in the outermost, or valence, shell of an atom.

Valence shell The outermost electron shell of an atom.

Valence-shell electron-pair repulsion (VSEPR) model A method for predicting molecular shape by noting how many electron charge clouds surround atoms and assuming that the clouds orient as far away from one another as possible.

Vapor The gas molecules in equilibrium with a liquid.

Vapor pressure The partial pressure of gas molecules in equilibrium with a liquid.

Vitamin An organic molecule, essential in trace amounts that must be obtained in the diet because it is not synthesized in the body.

Volume/volume percent concentration [(v/v)%] Concentration expressed as the number of milliliters of solute dissolved in 100 mL of solution.

Wax A mixture of esters of long-chain carboxylic acids with long-chain alcohols.

Weak acid An acid that gives up H^+ with difficulty and is less than 100% dissociated in water.

Weak base A base that has only a slight affinity for H^+ and holds it weakly.

Weak electrolyte A substance that is only partly ionized in water.

Weight A measure of the gravitational force that the earth or other large body exerts on an object.

Whole blood Blood plasma plus blood cells.

X rays Electromagnetic radiation with an energy somewhat less than that of γ rays.

Zwitterion A neutral dipolar ion that has one + charge and one − charge.

Zymogen A compound that becomes an active enzyme after undergoing a chemical change.

Answers to Selected Problems

Short answers are given for in-chapter problems, *Understanding Key Concepts* problems, and even-numbered end-of-chapter problems. Explanations and full answers for all problems are provided in the accompanying *Study Guide and Solutions Manual*.

Chapter 1

1.1 physical: (a), (d); chemical: (b), (c) **1.2** solid **1.3** mixture (heterogeneous): (a), (d); pure (element): (b), (c) **1.4** physical: (a), (c); chemical: (b), (d) **1.5** chemical change **1.6** (a) 2 (b) 1 (c) 6 (d) 5 (e) 4 (f) 3 **1.7** (a) 1 nitrogen atom, 3 hydrogen atoms (b) 1 sodium atom, 1 hydrogen atom, 1 carbon atom, 3 oxygen atoms (c) 8 carbon atoms, 18 hydrogen atoms (d) 6 carbon atoms, 8 hydrogen atoms, 6 oxygen atoms **1.8** Metalloids are at the boundary between metals and nonmetals. **1.9** metal; physical properties—solubility of compounds, tendency of elemental form to vaporize; chemical; reacts to form soluble compounds **1.10** (a) 0.01 m (b) 0.1 g (c) 1000 m (d) 0.000 001 s (e) 0.000 000 001 g **1.11** (a) 3 (b) 4 (c) 5 (d) exact **1.12** 32.3 °C; three significant figures **1.13** (a) 5.8×10^{-2} g (b) 4.6792×10^4 m (c) 6.072×10^{-3} cm (d) 3.453×10^2 kg **1.14** (a) 48,850 mg (b) 0.000 008 3 m (c) 0.0400 m **1.15** (a) 6.3000×10^5 (b) 1.30×10^3 (c) 7.942×10^{11} **1.16** (a) 2.30 g (b) 188.38 mL (c) 0.009 L (d) 1.000 kg **1.17** (a) 50.9 mL (b) 0.078 g (c) 11.9 m (d) 51 mg (e) 103 **1.18** (a) 454 g (b) 2.5 L (c) 105 qt **1.19** 795 mL **1.20** 7.36 m/s **1.21** (a) 10.6 mg/kg (b) 36 mg/kg **1.22** 331.0 K **1.23** 39 °C; 102 °F **1.24** 7,700 cal **1.25** 0.21 cal/g · °C **1.26** float; density = 0.637 g/cm³ **1.27** 8.392 mL **1.28** more dense **1.29** gases: helium (He), neon (Ne), argon (Ar), krypton (Kr), xenon (Xe), radon (Rn) coinage metals: copper (Cu), silver (Ag), gold (Au) **1.30** red: vanadium, metal; green: boron, metalloid; blue: bromine, nonmetal **1.31** Americium, a metal **1.32** (a) 0.978 (b) three (c) less dense **1.33** The smaller cylinder is more precise because the gradations are smaller. **1.34** 3 1/8 in.; 8.0 cm **1.35** start: 0.11 mL stop: 0.25 mL volume: 0.14 mL **1.36** higher in chloroform **1.38** physical: (a), (d); chemical (b), (c) **1.40** A gas has no definite shape or volume; a liquid has no definite shape but has a definite volume; a solid has a definite volume and a definite shape. **1.42** gas **1.44** mixture: (a), (b), (d), (f); pure: (c), (e) **1.46** (a) reactant: hydrogen peroxide; products: water, oxygen (b) compounds: hydrogen peroxide, water; element: oxygen **1.48** Metals: lustrous, malleable, conductors of heat and electricity; nonmetals: gases or brittle solids, nonconductors; metalloids: properties intermediate between metals and nonmetals. **1.50** (a) Gd (b) Ge (c) Tc (d) As (e) Cd **1.52** (a) Br (b) Mn (c) C (d) K **1.54** Carbon, hydrogen, nitrogen, oxygen; ten atoms **1.56** $C_{13}H_{18}O_2$ **1.58** A physical quantity consists of a number and a unit. **1.60** (a) cubic centimeter (b) decimeter (c) millimeter (d) nanoliter (e) milligram (f) cubic meter **1.62** 10^9 pg, 3.5×10^4 pg **1.64** (a) 9.457×10^3 (b) 7×10^{-5} (c) 2.000×10^{10} (d) 1.2345×10^{-2} (e) 6.5238×10^2 **1.66** (a) 6 (b) 3 (c) 3 (d) 4 (e) 1 to 5 (f) 2 or 3 **1.68** (a) 7,926 mi, 7,900 mi, 7,926.38 mi (b) 7.926381×10^3 mi **1.70** (a) 12.1 g (b) 96.19 cm (c) 263 mL (d) 20.9 mg **1.72** (a) 0.3614 cg (b) 0.0120 ML (c) 0.0144 mm (d) 60.3 ng (e) 1.745 dL (f) 1.5×10^3 cm **1.74** (a) 97.8 kg (b) 0.133 mL (c) 0.46 ng (d) 2.99 Mm **1.76** (a) 62.1 mi/hr (b) 91.1 ft/s **1.78** (a) 6×10^{-4} cm (b) 2×10^3 cells/cm; 4×10^3 cells/in. **1.80** 10 g **1.82** 6×10^{10} cells **1.84** 537 cal = 0.537 kcal **1.86** 0.092 cal/g · °C **1.88** Hg: 76 °C; Fe: 40.7 °C **1.90** 0.179 g/cm³ **1.92** 11.4 g/cm³ **1.94** 159 mL **1.96** 9 carbons, 8 hydrogens, 4 oxygens; 21 atoms; solid **1.98** −2 °C; 271 K **1.100** (a) BMI = 29 (b) BMI = 23.7 (c) BMI = 24.4; individual (a) **1.102** nonmetal, solid, nonconducting, not malleable **1.104** 3.12 in; 7.92 cm Discrepancies are due to rounding errors and changes in significant figures. **1.106** (a) 3.5×10^5 cal (1.46×10^6 J); (b) 9.86 °C **1.108** 3.9×10^{-2} g/dL iron, 8.3×10^{-3} g/dL calcium, 2.24×10^{-1} g/dL cholesterol **1.110** 7.8×10^6 mL/day **1.112** 0.13 g **1.114** 4.4 g; 0.0097 lb **1.116** 2200 mL **1.118** 2.2 tablespoons **1.120** iron **1.122** float

Chapter 2

2.1 1.39×10^{-8} g **2.2** 6.02×10^{23} atoms in all cases **2.3** When the mass in grams is numerically equal to the mass in amu, there are 6.02×10^{23} atoms. **2.4** 1.1×10^{-15} (or 1.1×10^{-13} %) **2.5** (a) Re (b) Sr (c) Te **2.6** 27 protons, 27 electrons, 33 neutrons **2.7** The answers agree. **2.8** $^{79}_{35}$Br, $^{81}_{35}$Br **2.9** $^{35}_{17}$Cl $^{37}_{17}$Cl **2.10** group 3A, period 3 **2.11** silver, calcium **2.12** nitrogen (2), phosphorus (3), arsenic (4), antimony (5), bismuth (6) **2.13** Metals: titanium, scandium; nonmetals: selenium, argon, astatine; metalloids: tellurium **2.14** (a) nonmetal, main group, noble gas (b) metal, main group (c) nonmetal, main group (d) metal, transition element **2.15** thirteen He-4 nuclei; four additional neutrons **2.16** (a) Na-23, Group 1A, third period, metal; (b) O-18, Group 6A, sixth period, nonmetal **2.17** 12, magnesium **2.18** sulfur; main group (6A); nonmetal; last electron found in a 3p orbital. **2.19** (a) $1s^2 2s^2 2p^2$ (b) $1s^2 2s^2 2p^6 3s^2 3p^3$ (c) $1s^2 2s^2 2p^6 3s^2 3p^5$ (d) $1s^2 2s^2 2p^6 3s^2 3p^6 4s^1$ **2.20** $4p^3$, all are unpaired **2.21** gallium **2.22** (a) $1s^2 2s^2 2p^5$; [He] $2s^2 2p^5$ (b) $1s^2 2s^2 2p^6 3s^2 3p^1$; [Ne] $3s^2 3p^1$ (c) $1s^2 2s^2 2p^6 3s^2 3p^6 4s^2 3d^{10} 4p^3$; [Ar] $4s^2 3d^{10} 4p^3$ **2.23** group 2A **2.24** group 7A, $1s^2 2s^2 2p^6 3s^2 3p^5$ **2.25** group 6A, $ns^2 np^4$ **2.26** ·Ẍ· **2.27** :R̈n: ·P̈b: :Ẍe: ·Ra· **2.28** red = 700 − 780 nm; blue = 400 − 480 nm; blue = higher energy **2.29**

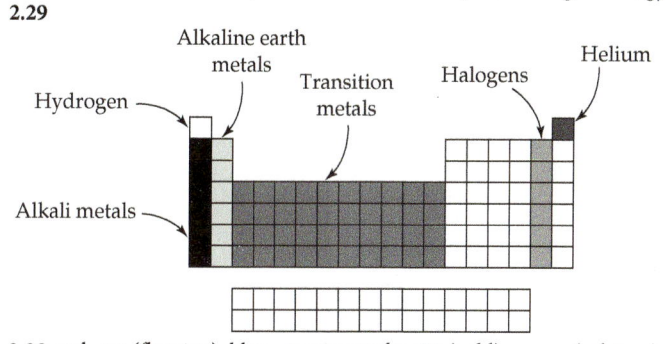

[periodic table diagram with labels: Alkali metals, Alkaline earth metals, Hydrogen, Transition metals, Halogens, Helium]

2.30 red: gas (fluorine); blue: atomic number 79 (gold); green: (calcium); beryllium, magnesium, strontium, barium, and radium are similar.
2.31

[periodic table diagram with labels: Two p electrons in 3rd shell, $ns^2 np^5$, Completely filled valence shell]

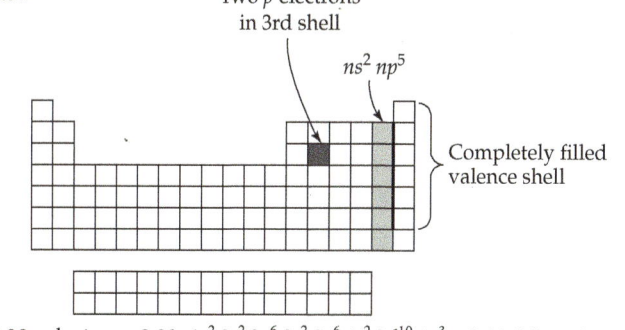

2.32 selenium **2.33** $1s^2 2s^2 2p^6 3s^2 3p^6 4s^2 3d^{10} 4p^3$ **2.34** Matter is composed of atoms. Atoms of different elements differ. Compounds consist of different atoms combined in specific proportions. Atoms do not change in chemical reactions. **2.36** (a) 3.4702×10^{-22} g (b) 2.1801×10^{-22} g (c) 6.6465×10^{-24} g **2.38** 14.01 g **2.40** 6.022×10^{23} atoms **2.42** protons (+ charge, 1 amu); neutrons (no charge, 1 amu); electrons (− charge, 0.0005 amu). **2.44** 18, 20, 22 **2.46** (a) and (c) **2.48** (a) $^{14}_{6}$C (b) $^{39}_{19}$K (c) $^{20}_{10}$Ne **2.50** $^{12}_{6}$C—six neutrons $^{13}_{6}$C—seven neutrons $^{14}_{6}$C—eight neutrons **2.52** 63.55 amu **2.54** Eight electrons are needed to fill the 3s and 3p subshells. **2.56** Am, metal **2.58** (a, b) transition metals (c) 3d **2.60** (a) Rb: (i), (v), (vii) (b) W: (i), (iv) (c) Ge: (iii), (v) (d) Kr: (ii), (v), (vi) **2.62** selenium **2.64** sodium, potassium, rubidium, cesium, francium **2.66** 2 **2.68** 2, 8, 18 **2.70** 3, 4, 5 **2.72** 10, neon

A-13

2.74 (a) two paired, two unpaired (b) four paired, one unpaired (c) two unpaired **2.76** 2, 1, 2, 1, 3, 3 **2.78** 2 **2.80** beryllium, 2s; arsenic, 4p **2.82** (a) 8 (b) 4 (c) 2 (d) 1 (e) 3 (f) 7 **2.84** A scanning tunneling microscope has much higher resolution. **2.86** H, He **2.88** (a) ultraviolet (b) gamma waves (c) X rays **2.90** He, Ne, Ar, Kr, Xe, Rn **2.92** Tellurium atoms have more neutrons than iodine atoms. **2.94** 1 (2 e), 2 (8 e), 3 (18 e), 4 (32 e), 5 (18 e), 6 (4 e) **2.96** 79.90 amu **2.98** Sr, metal, group 2A, period 5, 38 protons **2.100** 2, 8, 18, 18, 4; metal **2.102** (a) The 4s subshell fills before 3d (b) The 2s subshell fills before 2p. (c) Silicon has 14 electrons: $1s^2 2s^2 2p^6 3s^2 3p^2$ (d) The 3s electrons have opposite spins. **2.104** Electrons will fill or half-fill a d subshell instead of filling an s subshell of a higher shell. **2.106** 7p

Chapter 3

3.1 Mg^{2+} is a cation **3.2** S^{-2} is an anion **3.3** O^{2-} is an anion **3.4** less than Kr, but higher than most other elements **3.5** (a) B (b) Ca (c) Sc **3.6** (a) H (b) S (c) Cr **3.7** Common ionic substances: high-melting crystalline solids, good conductors of electricity when molten or in solution. Ionic liquids: low-melting, low volatility, high viscosity, low to moderate conductivity. **3.8** Potassium ($1s^2 2s^2 2p^6 3s^2 3p^6 4s^1$) can gain the argon configuration by losing 1 electron. **3.9** Aluminum must lose 3 electrons to form Al^{3+} **3.10** X: + ·Ÿ· ⟶ X^{2+} + :Ÿ:$^{2-}$ **3.11** cation **3.12** (a) Se + 2 e⁻ → Se^{2-} (b) Ba → Ba^{2+} + 2e⁻ (c) Br + e⁻ → Br⁻ **3.13** 3.4 gal **3.14** (a) copper(II) ion (b) fluoride ion (c) magnesium ion (d) sulfide ion **3.15** (a) Ag^+ (b) Fe^{2+} (c) Cu^+ (d) Te^{2-} **3.16** Na^+, sodium ion; K^+, potassium ion; Ca^{2+}, calcium ion; Cl^-, chloride ion **3.17** (a) nitrate ion (b) cyanide ion (c) hydroxide ion (d) hydrogen phosphate ion **3.18** Group 1 A: Na^+, K^+; Group 2A: Ca^{2+}, Mg^{2+}; transition metals: Fe^{2+}; halogens: Cl^- **3.19** (a) AgI (b) Ag_2O (c) Ag_3PO_4 **3.20** (a) Na_2SO_4 (b) $FeSO_4$ (c) $Cr_2(SO_4)_3$ **3.21** $(NH_4)_2CO_3$ **3.22** $Al_2(SO_4)_3$, $Al(CH_3CO_2)_3$ **3.23** blue: K_2S; red: $BaBr_2$; green: Al_2O_3 **3.24** Ca_3N_2 **3.25** silver(I) sulfide **3.26** (a) tin(IV) oxide (b) calcium cyanide (c) sodium carbonate (d) copper(I) sulfate (e) barium hydroxide (f) iron(II) nitrate **3.27** (a) Li_3PO_4 (b) $CuCO_3$ (c) $Al_2(SO_3)_3$ (d) CuF (e) $Fe_2(SO_4)_3$ (f) NH_4Cl **3.28** Cr_2O_3 chromium (III) oxide **3.29** Acids: (a), (d); bases (b), (c) **3.30** (a) HCl (b) H_2SO_4
3.31

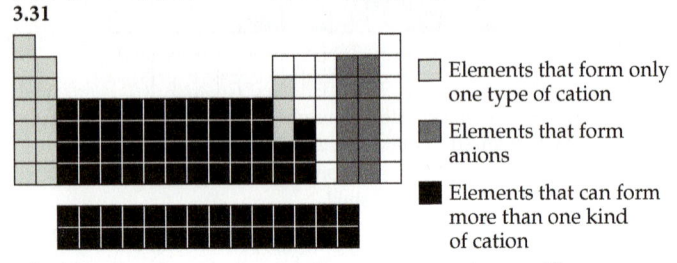

☐ Elements that form only one type of cation
■ Elements that form anions
▨ Elements that can form more than one kind of cation

All of the other elements form neither anions nor cations readily.

3.32

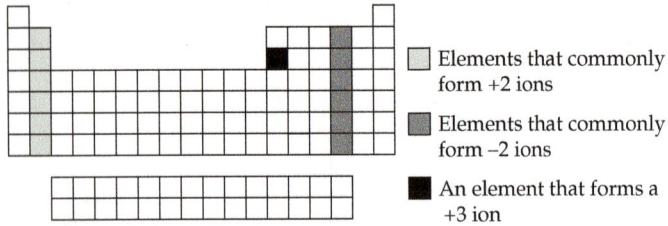

☐ Elements that commonly form +2 ions
▨ Elements that commonly form −2 ions
■ An element that forms a +3 ion

3.33 (a) O^{2-} (b) Na^+ (c) Ca^{2+} (d) Fe^{2+} **3.34** (a) sodium atom (larger) (b) Na^+ ion (smaller) **3.35** (a) chlorine atom (smaller) (b) Cl^- anion (larger) **3.36** iron (II) chloride or ferrous chloride, $FeCl_2$; iron (III) chloride or ferric chloride, $FeCl_3$; iron (II) oxide or ferrous oxide, FeO; iron (III) oxide or ferric oxide, Fe_2O_3; lead (II) chloride, $PbCl_2$; lead (IV) chloride, $PbCl_4$; lead (II) oxide, PbO; lead (IV) oxide, PbO_2 **3.37** (a) ZnS (b) $PbBr_2$ (c) CrF_3 (d) Al_2O_3 **3.38** (a) Ca → Ca^{2+} + 2 e⁻ (b) Au → Au^+ + e⁻ (c) F + e⁻ → F⁻ (d) Cr → Cr^{3+} + 3e⁻ **3.40** true: (d); false: (a), (b), (c) **3.42** Main group atoms undergo reactions that leave them with a noble gas electron configuration. **3.44** Se^{2-} **3.46** (a) Sr (b) Br **3.48** (a) $1s^2 2s^2 2p^6 3s^2 3p^6 4s^2 3d^{10} 4p^6$ (b) $1s^2 2s^2 2p^6 3s^2 3p^6 4s^2 3d^{10} 4p^6$

(c) $1s^2 2s^2 2p^6 3s^2 3p^6$ (d) $1s^2 2s^2 2p^6 3s^2 3p^6 4s^2 3d^{10} 4p^6 5s^2 4d^{10} 5p^6$ (e) $1s^2 2s^2 2p^6$ **3.50** (a) O (b) Li (c) Zn (d) N **3.52** none **3.54** Cr^{2+}: $1s^2 2s^2 2p^6 3s^2 3p^6 3d^4$; Cr^{3+}: $1s^2 2s^2 2p^6 3s^2 3p^6 3d^3$ **3.56** greater **3.58** (a) sulfide ion (b) tin(II) ion (c) strontium ion (d) magnesium ion (e) gold(I) ion **3.60** (a) Se^{2-} (b) O^{2-} (c) Ag^+ **3.62** (a) OH^- (b) HSO_4^- (c) $CH_3CO_2^-$ (d) MnO_4^- (e) OCl^- (f) NO_3^- (g) CO_3^{2-} (h) $Cr_2O_7^{2-}$ **3.64** (a) $Al_2(SO_4)_3$ (b) Ag_2SO_4 (c) $ZnSO_4$ (d) $BaSO_4$ **3.66** (a) $NaHCO_3$ (b) KNO_3 (c) $CaCO_3$ (d) NH_4NO_3

3.68

	S^{2-}	Cl^-	PO_4^{3-}	CO_3^{2-}
copper(II)	CuS	$CuCl_2$	$Cu_3(PO_4)_2$	$CuCO_3$
Ca^{2+}	CaS	$CaCl_2$	$Ca_3(PO_4)_2$	$CaCO_3$
NH_4^+	$(NH_4)_2S$	NH_4Cl	$(NH_4)_3PO_4$	$(NH_4)_2CO_3$
ferric ion	Fe_2S_3	$FeCl_3$	$FePO_4$	$Fe_2(CO_3)_3$

3.70 copper(II) sulfide, copper(II) chloride, copper(II) phosphate, copper(II) carbonate; calcium sulfide, calcium chloride, calcium phosphate, calcium carbonate; ammonium sulfide, ammonium chloride, ammonium phosphate, ammonium carbonate; iron(III) sulfide, iron(III) chloride, iron(III) phosphate, iron(III) carbonate **3.72** (a) magnesium carbonate (b) calcium acetate (c) silver(I) cyanide (d) sodium dichromate **3.74** $Ca_3(PO_4)_2$ **3.76** An acid gives H^+ ions in water; a base gives OH^- ions. **3.78** (a) $H_2CO_3 \rightarrow 2H^+ + CO_3^{2-}$ (b) HCN → H^+ + CN^- (c) $Mg(OH)_2 \rightarrow Mg^{2+}$ + 2 OH^- (d) KOH → K^+ + OH^- **3.80** The bulky cations prevent close packing of particles and prevent crystallization. **3.82** 2300 mg; 4 tsp **3.84** Sodium protects against fluid loss and is necessary for muscle contraction and transmission of nerve impulses. **3.86** 10 Ca^{2+}, 6 PO_4^{3-}, 2 OH^- **3.88** H^- has the helium configuration, $1s^2$ **3.90** (a) CrO_3 (b) VCl_5 (c) MnO_2 (d) MoS_2 **3.92** (a) −1 (b) 3 gluconate ions per iron(III) **3.94** (a) $Co(CN)_2$ (b) UO_3 (c) $SnSO_4$ (d) MnO_2 (e) K_3PO_4 (f) Ca_3P_2 (g) $LiHSO_4$ (h) $Al(OH)_3$ **3.96** (a) metal (b) nonmetal (c) X_2Y_3 (d) X: group 3A; Y: group 6A

Chapter 4

4.1 :Ï:Ï: ; xenon **4.2** (a) P 3, H 1 (b) Se 2, H 1 (c) H 1, Cl 1 (d) Si 4, F 1 **4.3** $PbCl_2$ = ionic; $PbCl_4$ = covalent **4.4** (a) CH_2Cl_2 (b) BH_3 (c) NI_3 (d) $SiCl_4$

4.5

4.6

4.7 $AlCl_3$ is a covalent compound, and Al_2O_3 is ionic

4.8

4.9 (a) (b) (c)

4.10
(a) (b) (c) (d)

4.11

ANSWERS TO SELECTED PROBLEMS   A-15

4.12 (a) $C_6H_{10}O_2$ (b) [structure]

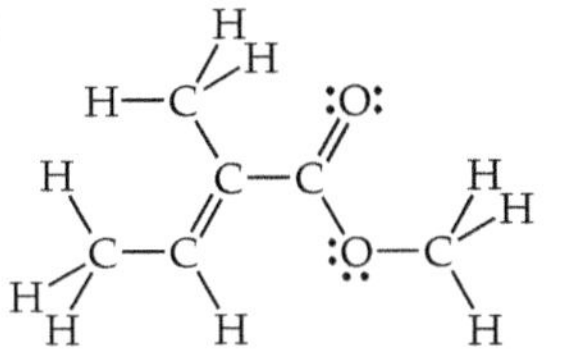

4.13 :C≡O: :N̈=O:
CO is reactive because it can form coordinate covalent bonds; NO is reactive because it has an unpaired electron.

4.14 [BF$_4$]$^-$ tetrahedral

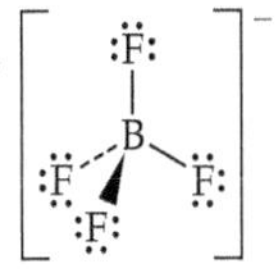

4.15 chloroform, $CHCl_3$—tetrahedral; dichloroethylene—planar
4.16 a = tetrahedral; b = trigonal planar   **4.17** Both are bent.
4.18 [structure with labels a-e]
(a) bent
(b) tetrahedral
(c) tetrahedral
(d) trigonal planar
(e) pyramidal

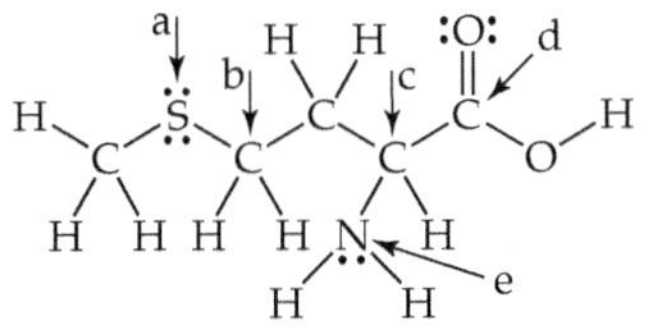

4.19 H = P < S < N < O
4.20 (a) polar covalent   (b) ionic   (c) covalent   (d) polar covalent
   $\overset{\delta^+}{I}—\overset{\delta^-}{Cl}$ $\overset{\delta^+}{P}—\overset{\delta^-}{Br}$

4.21 [structure of H$_2$C=O with dipole]

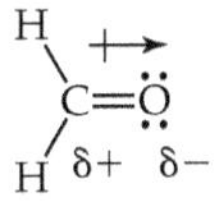

4.22 The carbons are tetrahedral; the oxygen is bent, the molecule is polar.
[structure of dimethyl ether]

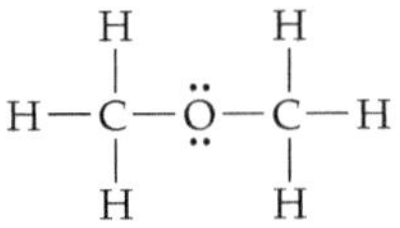

4.23 [structure with Li, C, H showing dipoles]

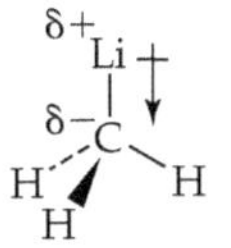

4.24 (a) disulfur dichloride   (b) iodine monochloride   (c) iodine trichloride
4.25 (a) SeF_4 (b) P_2O_5 (c) BrF_3
4.26 [structure of Geraniol]

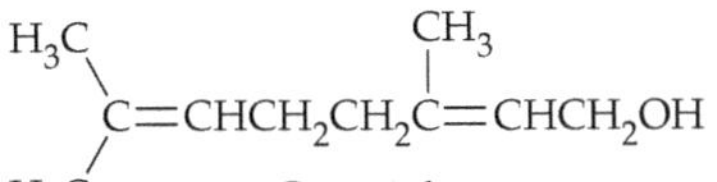

4.27 (a) tetrahedral   (b) pyramidal   (c) trigonal planar
4.28 (c) is square planar   **4.29** (a) $C_8H_9NO_2$
(b) [structure]
(c) All carbons are trigonal planar except the —CH$_3$ carbon. Nitrogen is pyramidal.

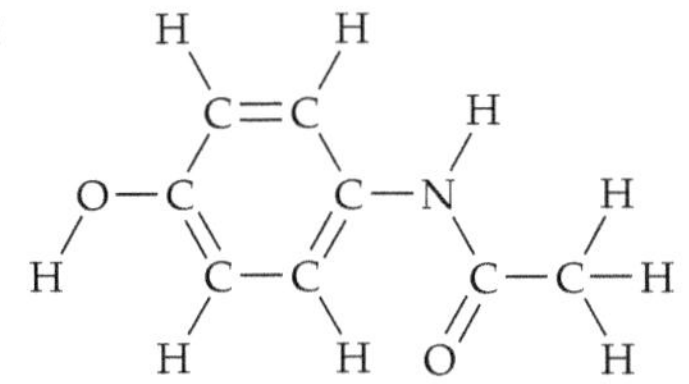

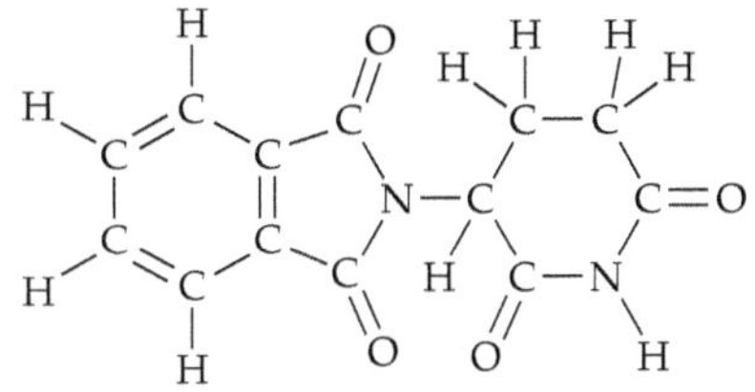

4.30 [structure]

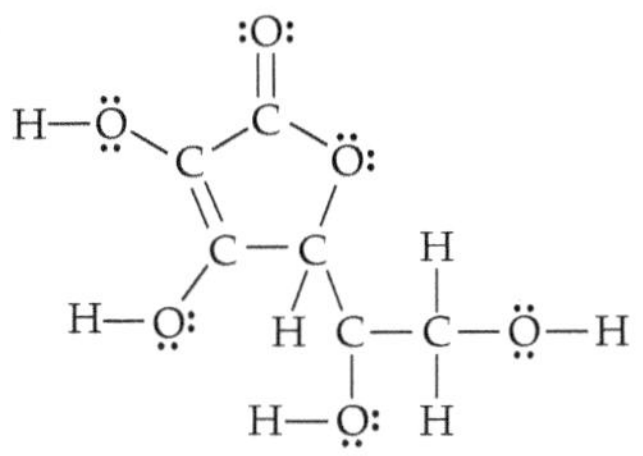

4.31 $C_{13}H_{10}N_2O_4$ [structure]

4.32 [structure with :Ö: ← electron-rich]

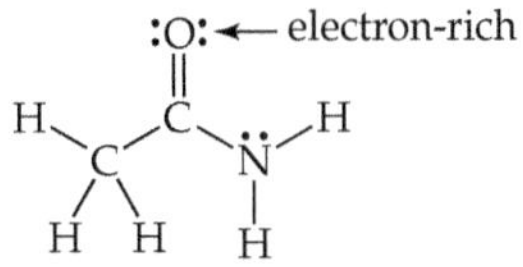

4.34 In a coordinate covalent bond, both electrons in the bond come from the same atom.   **4.36** covalent bonds: (a) (b); ionic bonds: (c) (d) (e)   **4.38** two covalent bonds   **4.40** (b), (c)   **4.42** $SnCl_4$
4.44 the N—O bond   **4.46** (a) A molecular formula shows the numbers and kinds of atoms; a structural formula shows how the atoms are bonded to one another. (b) A structural formula shows the bonds between atoms; a condensed structure shows atoms but not bonds. (c) A lone pair of valence electrons is not shared in a bond; a shared pair of electrons is shared between two atoms.   **4.48** (a) 10; triple bond (b) 18; double bond between N,O (c) 24; double bond between C,O (d) 20   **4.50** too many hydrogens
4.52 (a) H—Ö—N̈=Ö   (b) H—C(H)(H)—C≡N: (c) H—F̈:

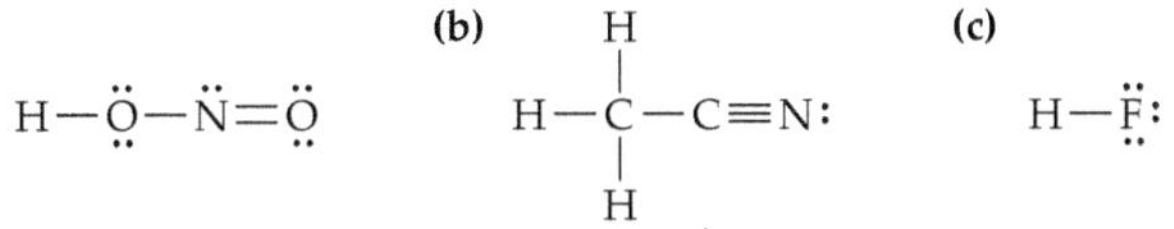

4.54 (a) $CH_3CH_2CH_3$ (b) $H_2C=CHCH_3$ (c) CH_3CH_2Cl
4.56 CH_3COOH
4.58
(a) H—Ö—N̈=Ö
(b) Ö=Ö—Ö:
(c) [structure]

4.60 [structure] Dimethyl ether

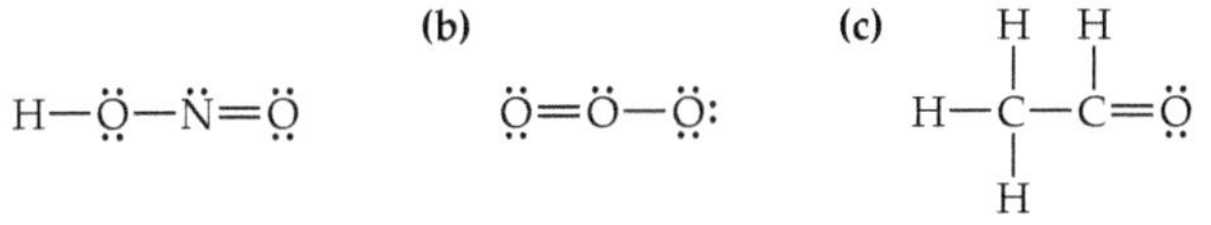

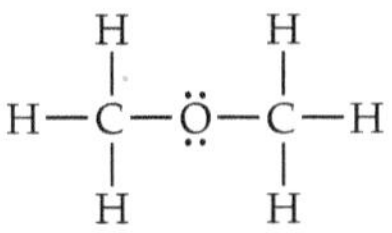

4.62 [structure of tetrachloroethylene] Tetrachloroethylene contains a double bond

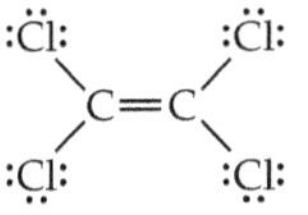

4.64 H—N̈—Ö—H
         |
         H

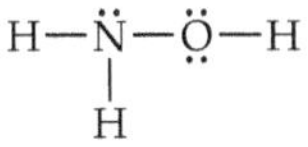

4.66 (a) [H—C(=O)—Ö:]$^-$ (b) [SO$_3$]$^{2-}$ (c) [:S̈—C≡N:]$^-$
(d) [PO$_4$]$^{3-}$ (e) [:Ö—Cl̈—Ö:]$^-$

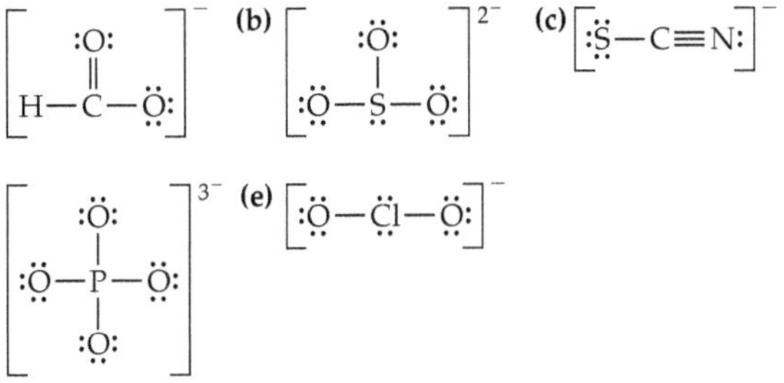

4.68 tetrahedral; pyramidal; bent   **4.70** (a), (b) tetrahedral (c), (d) trigonal planar (e) pyramidal   **4.72** All are trigonal planar, except for the —CH$_3$ carbon, which is tetrahedral.   **4.74** It should have low electronegativity, like other alkali metals.   **4.76** Cl > C > Cu > Ca > Cs
4.78 (a) $\overset{\delta^-}{O}—\overset{\delta^+}{Cl}$ (b) (c) (d) nonpolar   (e) $\overset{\delta^+}{C}—\overset{\delta^-}{O}$
4.80 $PH_3 < HCl < H_2O < CF_4$

A-16 ANSWERS TO SELECTED PROBLEMS

4.82 (a) H—Cl polar (b) PH₃ polar (c) OH₂ polar (d) nonpolar

4.84 S—H bonds are nonpolar. **4.86** (a) selenium dioxide (b) xenon tetroxide (c) dinitrogen pentasulfide (d) triphosphorus tetraselenide **4.88** (a) $SiCl_4$ (b) NaH (c) SbF_5 (d) OsO_4 **4.90** It relaxes arterial walls. **4.92** Carbohydrates, DNA, and proteins are all polymers that occur in nature. **4.94** $CH_3(CH_3)C=CHCH_2CH_2CH(CH_3)CH_2CH_2OH$

4.96 (a) [two structural formulas of H—C—C—C—H with OH and =O groups shown]

(b) The C=O carbons are trigonal planar; the other carbons are tetrahedral. (c) The C=O bonds are polar. **4.98** (a) C forms 4 bonds (b) N forms 3 bonds (c) S forms 2 bonds (d) could be correct **4.100** (b) tetrahedral (c) contains a coordinate covalent bond (d) has 19 p and 18 e⁻

[PH₄⁺ structure]

4.102 (a) calcium chloride (b) tellurium dichloride (c) boron trifluoride (d) magnesium sulfate (e) potassium oxide (f) iron(III) fluoride (g) phosphorus trifluoride

4.104 [Cl₂C(Cl)—C(H)—O—H structure]

4.106 [H—O—C(=O)—C(H)(H)—O—H structure]

4.108 (a) [Cl—C(=O)—O—C(H)₃ structure] (b) [H—C(H)(H)—C≡C—H structure]

Chapter 5

5.1 (a) Solid cobalt(II) chloride plus gaseous hydrogen fluoride yields solid cobalt(II) fluoride plus gaseous hydrogen chloride. (b) Aqueous lead(II) nitrate plus aqueous potassium iodide yields solid lead(II) iodide plus aqueous potassium nitrate. **5.2** balanced:(a), (c) **5.3** $3 O_2 \rightarrow 2 O_3$ **5.4** (a) $Ca(OH)_2 + 2 HCl \rightarrow CaCl_2 + 2 H_2O$ (b) $4 Al + 3 O_2 \rightarrow 2 Al_2O_3$ (c) $2 CH_3CH_3 + 7 O_2 \rightarrow 4 CO_2 + 6 H_2O$ (d) $2 AgNO_3 + MgCl_2 \rightarrow 2 AgCl + Mg(NO_3)_2$ **5.5** $2 A + B_2 \rightarrow A_2B_2$ **5.6** (a) precipitation (b) redox (c) acid–base neutralization **5.7** $6 CO_2 + 6 H_2O \rightarrow C_6H_{12}O_6 + 6 O_2$; redox reaction **5.8** Soluble: (b), (d); insoluble: (a), (c), (e) **5.9** (a) $NiCl_2(aq) + (NH_4)_2S(aq) \rightarrow NiS(s) + 2 NH_4Cl(aq)$; precipitation (b) $2 AgNO_3(aq) + CaBr_2(aq) \rightarrow Ca(NO_3)_2(aq) + 2 AgBr(s)$ **5.10** $CaCl_2(aq) + Na_2C_2O_4(aq) \rightarrow CaC_2O_4(s) + 2 NaCl(aq)$ (b) $2 AgNO_3(aq) + CaBr_2(aq) \rightarrow 2 AgBr(s) + CaNO_3(aq)$; precipitation **5.11** (a) $2 CsOH(aq) + H_2SO_4(aq) \rightarrow Cs_2SO_4(aq) + 2 H_2O(l)$ (b) $Ca(OH)_2(aq) + 2 CH_3CO_2H(aq) \rightarrow Ca(CH_3CO_2)_2(aq) + 2 H_2O(l)$ (c) $NaHCO_3(aq) + HBr(aq) \rightarrow NaBr(aq) + CO_2(g) + H_2O(l)$ **5.12** (a) oxidized reactant (reducing agent): Fe; reduced reactant (oxidizing agent): Cu^{2+} (b) oxidized reactant (reducing agent): Mg; reduced reactant (oxidizing agent): Cl_2; (c) oxidized reactant (reducing agent): Al; reduced reactant (oxidizing agent): Cr_2O_3; **5.13** $2 K(s) + Br_2(l) \rightarrow 2 KBr(s)$; oxidizing agent: Br_2; reducing agent: K **5.14** Li is oxidized and I_2 is reduced. **5.15** (a) V(III) (b) Sn(IV) (c) Cr(VI) (d) Cu(II) (e) Ni(II) **5.16** (a) not redox (b) Na oxidized from 0 to +1; H reduced from +1 to 0 (c) C oxidized from 0 to +4; O reduced from 0 to −2 (d) not redox (e) S oxidized from +4 to +6; Mn reduced from +7 to +2 **5.17** (b) oxidizing agent: H_2; reducing agent: Na (c) oxidizing agent: O_2; reducing agent: C (e) oxidizing agent: MnO_4^-; reducing agent: SO_2 **5.18** (a) $Zn(s) + Pb^{2+}(aq) \rightarrow Zn^{2+}(aq) + Pb(s)$

(b) $OH^-(aq) + H^+(aq) \rightarrow H_2O(l)$
(c) $2 Fe^{3+}(aq) + Sn^{2+}(aq) \rightarrow 2 Fe^{2+}(aq) + Sn^{4+}(aq)$
5.19 (a) redox (b) neutralization (c) redox **5.20** (d) **5.21** (c)
5.22 reactants: (d); products: (c) **5.23** (a) box 1 (b) box 2 (c) box 3
5.24 $2 Ag^+ + CO_3^{2-}$; $2 Ag^+ + CrO_4^{2-}$ **5.26** In a balanced equation, the numbers and kinds of atoms are the same on both sides of the reaction arrow. **5.28** (a) $SO_2(g) + H_2O(g) \rightarrow H_2SO_3(aq)$
(b) $2 K(s) + Br_2(l) \rightarrow 2 KBr(s)$
(c) $C_3H_8(g) + 5 O_2(g) \rightarrow 3 CO_2(g) + 4 H_2O(l)$
5.30 (a) $2 C_2H_6(g) + 7 O_2(g) \rightarrow 4 CO_2(g) + 6 H_2O(g)$ (b) balanced
(c) $2 Mg(s) + O_2(g) \rightarrow 2 MgO(s)$
(d) $2 K(s) + 2 H_2O(l) \rightarrow 2 KOH(aq) + H_2(g)$
5.32 (a) $Hg(NO_3)_2(aq) + 2 LiI(aq) \rightarrow 2 LiNO_3(aq) + HgI_2(s)$
(b) $I_2(s) + 5 Cl_2(g) \rightarrow 2 ICl_5(s)$ (c) $4 Al(s) + 3 O_2(g) \rightarrow 3 Al_2O_3(s)$
(d) $CuSO_4(aq) + 2 AgNO_3(aq) \rightarrow Ag_2SO_4(s) + Cu(NO_3)_2(aq)$
(e) $2 Mn(NO_3)_3(aq) + 3 Na_2S(aq) \rightarrow Mn_2S_3(s) + 6 NaNO_3(aq)$
5.34 (a) $2 C_4H_{10}(g) + 13 O_2(g) \rightarrow 8 CO_2(g) + 10 H_2O(l)$
(b) $C_2H_6O(g) + 3 O_2(g) \rightarrow 2 CO_2(g) + 3 H_2O(l)$
(c) $2 C_8H_{18}(g) + 25 O_2(g) \rightarrow 16 CO_2(g) + 18 H_2O(l)$
5.36 $4 HF + SiO_2 \rightarrow SiF_4 + 2 H_2O$
5.38 (a) redox (b) neutralization (c) precipitation (d) neutralization
5.40 (a) $Ba^{2+}(aq) + SO_4^{2-}(aq) \rightarrow BaSO_4(s)$
(b) $Zn(s) + 2 H^+(aq) \rightarrow Zn^{2+}(aq) + H_2(g)$
5.42 precipitation: (a) (d) (e); redox: (b) (c) **5.44** $Ba(NO_3)_2$
5.46 (a) $2 NaBr(aq) + Hg_2(NO_3)_2(aq) \rightarrow Hg_2Br_2(s) + 2 NaNO_3(aq)$
(d) $(NH_4)_2CO_3(aq) + CaCl_2(aq) \rightarrow CaCO_3(s) + 2 NH_4Cl(aq)$
(e) $2 KOH(aq) + MnBr_2(aq) \rightarrow Mn(OH)_2(s) + 2 KBr(aq)$
(f) $3 Na_2S(aq) + 2 Al(NO_3)_3(aq) \rightarrow Al_2S_3(s) + 6 NaNO_3(aq)$
5.48 (a) $2 Au^{3+}(aq) + 3 Sn(s) \rightarrow 3 Sn^{2+}(aq) + 2 Au(s)$
(b) $2 I^-(aq) + Br_2(l) \rightarrow 2 Br^-(aq) + I_2(s)$
(c) $2 Ag^+(aq) + Fe(s) \rightarrow Fe^{2+}(aq) + 2 Ag(s)$
5.50 (a) $Sr(OH)_2(aq) + FeSO_4(aq) \rightarrow SrSO_4(s) + Fe(OH)_2(s)$
(b) $S^{2-}(aq) + Zn^{2+}(aq) \rightarrow ZnS(s)$
5.52 Most easily oxidized: metals on left side; most easily reduced: groups 6A and 7A **5.54** oxidation number increases: (b),(c); oxidation number decreases (a), (d) **5.56** (a) Co: +3 (b) Fe: +2 (c) U: +6 (d) Cu: +2
(e) Ti: +4 (f) Sn: +2 **5.58** (a) oxidized: S; reduced: O (b) oxidized: Na; reduced: Cl (c) oxidized: Zn; reduced: Cu (d) oxidized: Cl; reduced: F
5.60 (a) $N_2O_4(l) + 2 N_2H_4(l) \rightarrow 3 N_2(g) + 4 H_2O(g)$
(b) $CaH_2(s) + 2 H_2O(l) \rightarrow Ca(OH)_2(aq) + 2 H_2(g)$
(c) $2 Al(s) + 6 H_2O(l) \rightarrow 2 Al(OH)_3(s) + 3 H_2(g)$
5.62 oxidizing agents: N_2O_4, H_2O; reducing agents: N_2H_4, CaH_2, Al
5.64 redox equation: $2 C_5H_4N_4 + 3 O_2 \rightarrow 2 C_5H_4N_4O_3$
5.66 Zn is the reducing agent, and Mn^{2+} is the oxidizing agent
5.68 $Li_2O(s) + H_2O(g) \rightarrow 2 LiOH(s)$; not a redox reaction
5.70 (a) $Al(OH)_3(aq) + 3 HNO_3(aq) \rightarrow Al(NO_3)_3(aq) + 3 H_2O(l)$ neutralization (b) $3 AgNO_3(aq) + FeCl_3(aq) \rightarrow 3 AgCl(s) + Fe(NO_3)_3(aq)$ precipitation (c) $(NH_4)_2Cr_2O_7(s) \rightarrow Cr_2O_3(s) + 4 H_2O(g) + N_2(g)$ redox (d) $Mn_2(CO_3)_3(s) \rightarrow Mn_2O_3(s) + 3 CO_2(g)$ redox
5.72 (a) $2 SO_2(g) + O_2(g) \rightarrow 2 SO_3(g)$
(b) $SO_3(g) + H_2O(g) \rightarrow H_2SO_4(l)$ (c) SO_2: +4; SO_3, H_2SO_4: +6
5.74 (a) +6 (b) −2 in ethanol; +4 in CO_2
(c) oxidizing agent: $Cr_2O_7^{2-}$; reducing agent: C_2H_5OH
5.76 $Fe^{3+}(aq) + 3 NaOH(aq) \rightarrow Fe(OH)_3(s) + 3 Na^+(aq)$
$Fe^{3+}(aq) + 3 OH^-(aq) \rightarrow Fe(OH)_3(s)$
5.78 $2 Bi^{3+}(aq) + 3 S^{2-}(aq) \rightarrow Bi_2S_3(s)$
5.80 $CO_2(g) + 2 NH_3(g) \rightarrow NH_2CONH_2(s) + H_2O(l)$
5.82 (a) reactants: I = −1, Mn = −4; products: I = 0, Mn = +2
(b) reducing agent = NaI; oxidizing agent = MnO_2

Chapter 6

6.1 (a) 206.0 amu (b) 232.0 amu **6.2** 1.71×10^{21} molecules **6.3** 0.15 g **6.4** 111.0 amu **6.5** 0.217 mol; 4.6 g **6.6** 5.00 g weighs more **6.7** If all of these were true, the estimate of Avogadro's Number would be larger.
6.8 (a) $Ni + 2 HCl \rightarrow NiCl_2 + H_2$; 4.90 mol (b) 6.00 mol
6.9 $6 CO_2 + 6 H_2O \rightarrow C_6H_{12}O_6 + 6 O_2$; 90.0 mol CO_2
6.10 (a) 39.6 mol (b) 13.8 g **6.11** 6.31 g WO_3; 0.165 g H_2
6.12 44.7 g; 57.0% **6.13** 47.3 g **6.14** 1.4×10^{-4} mol; 3.2×10^{-4} mol

6.15 A_2 **6.16** $C_5H_{11}NO_2S$; MW = 149.1 amu
6.17 (a) $A_2 + 3B \rightarrow 2AB_3$ **(b)** 2 mol AB_3; 0.67 mol AB_3
6.18 10 AB ($2B_2$ left over) **6.19** Blue is the limiting reagent, yield: 73%
6.20 22 g, 31 g **6.22** molecular weight = sum of the weights of individual atoms in a molecule; formula weight = sum of weights of individual atoms in a formula unit; molar mass = mass in grams of 6.022×10^{23} molecules or formula units of any substance **6.24** 5.25 mol ions **6.26** 10.6 g uranium **6.28 (a)** 1 mol **(b)** 1 mol **(c)** 2 mol **6.30** 6.44×10^{-4} mol
6.32 284.5 g **6.34 (a)** 0.0132 mol **(b)** 0.0536 mol **(c)** 0.0608 mol
(d) 0.0129 mol **6.36** 0.27 g; 9.0×10^{20} molecules aspirin
6.38 1.4×10^{-3} mol; 0.18 g **6.40 (a)** $C_4H_8O_2(l) + 2H_2(g) \rightarrow 2C_2H_6O(l)$ **(b)** 3.0 mol **(c)** 138 g **(d)** 12.5 g **(e)** 0.55 g
6.42 (a) $N_2(g) + 3H_2(g) \rightarrow 2NH_3(g)$ **(b)** 0.471 mol **(c)** 16.1 g
6.44 (a) $Fe_2O_3(s) + 3CO(g) \rightarrow 2Fe(s) + 3CO_2(g)$ **(b)** 1.59 g **(c)** 141 g
6.46 158 kg **6.48** 6×10^9 mol H_2SO_4; 6×10^8 kg H_2SO_4
6.50 17 mol SO_2 **6.52 (a)** CO is limiting **(b)** 11.4 g **(c)** 83.8%
6.54 (a) $CH_4(g) + 2Cl_2(g) \rightarrow CH_2Cl_2(l) + 2HCl(g)$ **(b)** 444 g
(c) 202 g **6.56 (a)** HNO_3 **(b)** 53.6 g (0.436 mol) **6.58** measuring the area of the oil, density of oil, molar mass of oil, mass of oil
6.60 $FeSO_4$; 151.9 g/mol; 91.8 mg Fe **6.62** 6×10^{13} molecules
6.64 (a) $C_{12}H_{22}O_{11}(s) \rightarrow 12C(s) + 11H_2O(l)$ **(b)** 25.3 g C
(c) 8.94 g H_2O **6.66 (a)** 6.40 g **(b)** 104 g
6.68 (a) $4NH_3(g) + 5O_2(g) \rightarrow 4NO(g) + 6H_2O(g)$ **(b)** 30.0 g NO
6.70 (a) $BaCl_2(aq) + Na_2SO_4(aq) \rightarrow BaSO_4(s) + 2NaCl(aq)$ **(b)** 45.0 g
6.72 (a) 45 g **(b)** 78%
6.74 (a) $P_4(s) + 10Cl_2(g) \rightarrow 4PCl_5(s)$ **(b)** 102 g PCl_5
6.76 $6NH_4ClO_4(s) + 10Al(s) \rightarrow 4Al_2O_3(s) + 2AlCl_3(s) + 12H_2O(g) + 3N_2(g)$ **(b)** 310 mol gases

Chapter 7
7.1 (a) $\Delta H = +652$ kcal/mol (2720 kJ/mol) **(b)** endothermic
7.2 (a) endothermic **(b)** 200 kcal; 836 kJ **(c)** 74.2 kcal; 310 kJ
7.3 91 kcal; 380 kJ **7.4** 303 kcal/mol; 6.4 kcal/g **7.5 (a)** increase
(b) decrease **(c)** decrease **7.6 (a)** 31.3 kcal/mol; 131 kJ/mol; not spontaneous **(b)** spontaneous at higher temperatures **7.7 (a)** +0.06 kcal/mol (+0.25 kJ/mol); nonspontaneous **(b)** 0.00 kcal/mol; equilibrium
(c) −0.05 kcal/mol (−0.21 kJ/mol); spontaneous **7.8 (a)** positive
(b) spontaneous at all temperatures
7.9

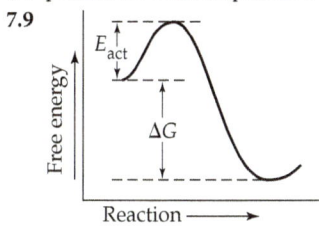

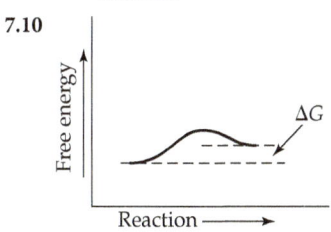

7.10

7.11 (a) see blue curve, Fig. 7.4 **(b)** increase the temperature, add a catalyst, increase the concentration of reagents **7.12** 1260 g
7.13 (a) $K = \dfrac{[NO_2]^2}{[N_2O_4]}$ **(b)** $K = \dfrac{[H_2O]^2}{[H_2S]^2[O_2]}$ **(c)** $K = \dfrac{[Br_2][F_2]^5}{[BrF_5]^2}$
7.14 (a) products strongly favored **(b)** reactants strongly favored
(c) products somewhat favored
7.15 $K = 29.0$
7.16 (a) $K = \dfrac{[AB]^2}{[A_2][B_2]}$; $K = \dfrac{[AB]^2}{[A_2][B]^2}$ **(b)** $K = 0.11$; $K = 0.89$
7.17 reaction favored by high pressure and low temperature
7.18 (a) favors reactants **(b)** favors product **(c)** favors product
7.19 $Cu_2O(s) + C(s) \rightarrow 2Cu(s) + CO(g)$; $\Delta G = -3.8$ kJ (−1.0 kcal)

7.20 ΔH is positive; ΔS is positive; ΔG is negative
7.21 ΔH is negative; ΔS is negative; ΔG is negative
7.22 (a) $2A_2 + B_2 \rightarrow 2A_2B$
(b) ΔH is negative; ΔS is negative; ΔG is negative
7.23 (a) blue curve represents faster reaction **(b)** red curve is spontaneous
7.24
(a)   **(b)**

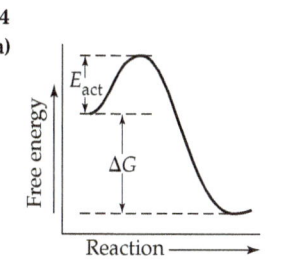

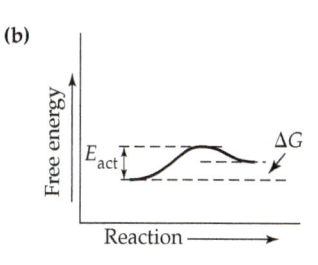

7.25 (a) positive **(b)** nonspontaneous at low temperature; spontaneous at high temperature **7.26** lower enthalpy for reactants
7.28 (a) $Br_2(l) + 7.4$ kcal/mol $\rightarrow Br_2(g)$ **(b)** 43 kcal **(c)** 15.9 kJ
7.30 (a) $2C_2H_2(g) + 5O_2(g) \rightarrow 4CO_2(g) + 2H_2O(g)$
(b) −579 kcal/mol (−2420 kJ/mol) **(c)** 11.2 kcal/g (47 kJ/g); one of the highest energy values **7.32 (a)** $C_6H_{12}O_6 + 6O_2 \rightarrow 6CO_2 + 6H_2O$
(b) -1.0×10^3 kcal/mol (-4.2×10^3 kJ/mol) **(c)** +57 kcal (+240 kJ)
7.34 increased disorder: **(a)**; decreased disorder: **(b)**, **(c)**
7.36 release or absorption of heat, and increase or decrease in entropy
7.38 ΔH is usually larger than $T\Delta S$
7.40 (a) endothermic **(b)** increases **(c)** $T\Delta S$ is larger than ΔH
7.42 (a) $H_2(g) + Br_2(l) \rightarrow 2HBr(g)$ **(b)** increases **(c)** yes, because ΔH is negative and ΔS is positive **(d)** $\Delta G = -25.6$ kcal/mol (−107 kJ/mol)
7.44 the amount of energy needed for reactants to surmount the barrier to reaction
7.46
(a)   **(b)**

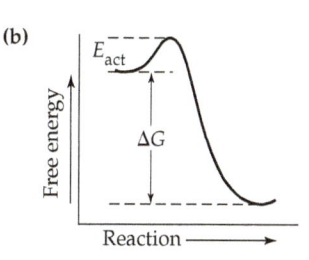

7.48 A catalyst lowers the activation energy. **7.50 (a)** yes **(b)** reaction rate is slow **7.52** At equilibrium, the rates of forward and reverse reactions are equal. Amounts of reactants and products need not be equal.
7.54 (a) $K = \dfrac{[CO_2]^2}{[CO]^2[O_2]}$ **(b)** $K = \dfrac{[H_2][MgCl_2]}{[HCl]^2}$
(c) $K = \dfrac{[H_3O^+][F^-]}{[HF]}$ **(d)** $K = \dfrac{[SO_2]}{[O_2]}$
7.56 $K = 7.19 \times 10^{-3}$; reactants are favored **7.58 (a)** 0.0869 mol/L
(b) 0.0232 mol/L **7.60** more reactant **7.62 (a)** endothermic
(b) reactants are favored **(c)** (1) favors ozone; (2) favors ozone; (3) favors O_2; (4) no effect; (5) favors ozone **7.64 (a)** decrease **(b)** no effect **(c)** increase
7.66 increase **7.68 (a)** increase **(b)** decrease **(c)** no effect **(d)** decrease
7.70 fat **7.72** thyroid, hypothalamus
7.74 (a) ADP + phosphoenolpyruvate $\rightarrow$ pyruvate + ATP
(b) −31.4 kJ/mol **7.76 (a)** $C_2H_5OH(l) + 3O_2(g) \rightarrow 2CO_2(g) + 3H_2O(g)$
(b) negative **(c)** 35.5 kcal **(d)** 5.63 g **(e)** 5.60 kcal/mL (23.4 kJ/mL)
7.78 (a) $Fe_3O_4(s) + 4H_2(g) \rightarrow 3Fe(s) + 4H_2O(g)$
$\Delta H = +36$ kcal/mol **(b)** 12 kcal (50 kJ) **(c)** 3.6 g H_2 **(d)** reactants
7.80 (a) :O: **(b)** −20 kcal/mol (−84 kJ/mol)
     ‖
     C
   ╱   ╲
 H_2N NH_2

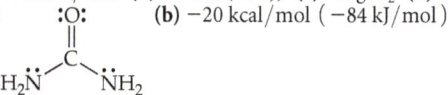

7.82 (a) $4NH_3(g) + 5O_2(g) \rightarrow 4NO(g) + 6H_2O(g) + $ heat
(b) $K = \dfrac{[NO]^4[H_2O]^6}{[NH_3]^2[O_2]^5}$ **(c)** (1) favors reactants (2) favors reactants
(3) favors reactants (4) favors products
7.84 (a) exergonic **(b)** $\Delta G = -10$ kcal/mol (−42 kJ/mol)
7.86 −1.91 kcal (−7.99 kJ/mol); exothermic

A-18 ANSWERS TO SELECTED PROBLEMS

Chapter 8

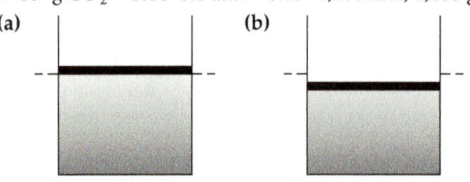

8.1 (a) disfavored by ΔH; favored by ΔS (b) $+0.02$ kcal/mol ($+0.09$ kJ/mol) (c) $\Delta H = -9.72$ kcal/mol (-40.6 kJ/mol); $\Delta S = -2.61$ cal/(mol·K) [-109 J/(mol·K)] **8.2** (a) decrease (b) increase **8.3** (a), (c) **8.4** (a) London forces (b) hydrogen bonds, dipole–dipole forces, London forces (c) dipole–dipole forces, London forces **8.5** 220 mmHg; 4.25 psi; 2.93×10^4 Pa **8.6** Atmospheric CO_2 levels would remain constant. **8.7** 1000 mmHg **8.8** 450 L **8.9** 1.3 L, 18 L **8.10** 2.16 psi/1.45 psi **8.11** 637 °C; -91 °C **8.12** 33 psi **8.13** 352 L **8.14** balloon (a) **8.15** 4.46×10^3 mol; 7.14×10^4 g CH_4; 1.96×10^5 g CO_2 **8.16** 5.0 atm **8.17** 1,100 mol; 4,400 g
8.18 (a) (b)

8.19 9.3 atm He; 0.19 atm O_2 **8.20** 75.4% N_2, 13.2% O_2, 5.3% CO_2, 6.2% H_2O **8.21** 35.0 mmHg **8.22** $P_{He} = 500$ mmHg; $P_{Xe} = 250$ mmHg **8.23** 1.93 kcal, 14.3 kcal **8.24** 102 kJ **8.25** gas
8.26 (a) (b) (c)

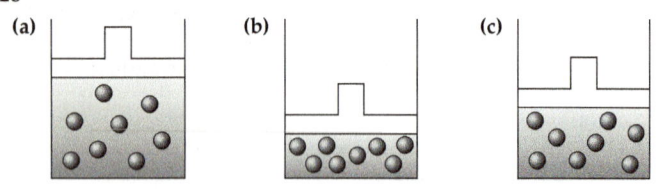

(a) volume increases by 50% (b) volume decreases by 50% (c) volume unchanged **8.27** (b); (c) **8.28** (c)

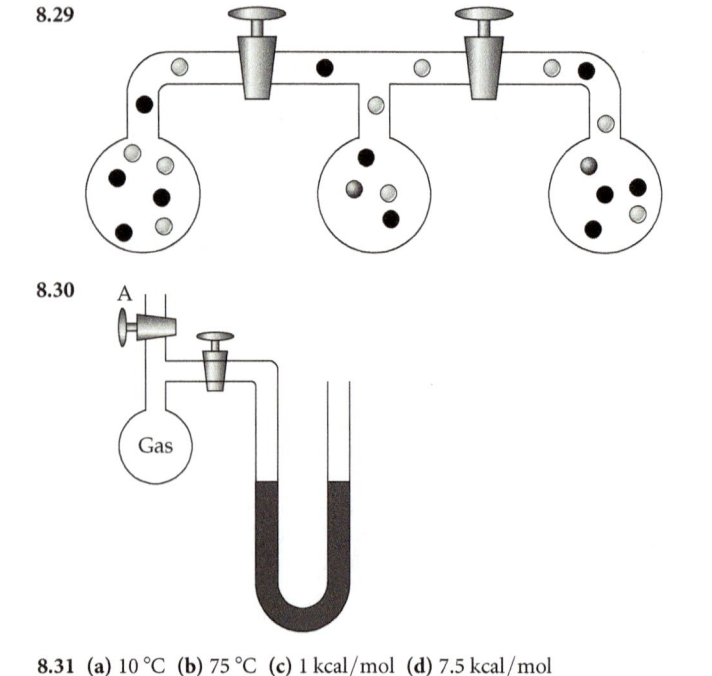

8.29

8.30

8.31 (a) 10 °C (b) 75 °C (c) 1 kcal/mol (d) 7.5 kcal/mol
8.32 (a) (b) (c)

8.33 red = 360 mmHg; yellow = 120 mmHg; total pressure = 720 mmHg **8.34** (a) all molecules (b) molecules with polar covalent bonds (c) molecules with —OH or —NH bonds **8.36** Ethanol forms hydrogen bonds. **8.38** One atmosphere is equal to exactly 760 mmHg. **8.40** (1) A gas consists of tiny particles moving at random with no forces between them. (2) The amount of space occupied by the gas particles is small. (3) The average kinetic energy of the gas particles is proportional to the Kelvin temperature. (4) Collisions between particles are elastic. **8.42** (a) 760 mmHg (b) 1310 mmHg (c) 5.7×10^3 mmHg (d) 711 mmHg (e) 0.314 mmHg **8.44** 930 mmHg; 1.22 atm **8.46** V varies inversely with P when n and T are constant. **8.48** 101 mL **8.50** 1.75 L **8.52** V varies directly with T when n and P are constant. **8.54** 364 K = 91 °C **8.56** 220 mL **8.58** P varies directly with T when n and V are constant. **8.60** 1.2 atm **8.62** 493 K = 220 °C **8.64** 68.4 mL **8.66** (a) P increases by factor of 4 (b) P decreases by factor of 4 **8.68** Because gas particles are so far apart and have no interactions, their chemical identity is unimportant. **8.70** 2.7×10^{22} molecules/L; 1.4 g **8.72** 11.8 g **8.74** 15 kg **8.76** $PV = nRT$ **8.78** Cl_2 has fewer molecules but weighs more. **8.80** 370 atm; 5400 psi **8.82** 2.2×10^4 mm Hg **8.84** 22.3 L **8.86** the pressure contribution of one component in a mixture of gases **8.88** 93 mmHg **8.90** the partial pressure of the vapor above the liquid **8.92** Increased pressure raises a liquid's boiling point; decreased pressure lowers it. **8.94** (a) 29.2 kcal (b) 173 kcal **8.96** Atoms in a crystalline solid have a regular, orderly arrangement. **8.98** 4.82 kcal **8.100** increase in atmospheric [CO_2]; increase in global temperatures **8.102** Systolic pressure is the maximum pressure just after contraction; diastolic pressure is the minimum pressure at the end of the heart cycle. **8.104** A supercritical fluid is intermediate in properties between liquid and gas. **8.106** As temperature increases, molecular collisions become more violent. **8.108** 0.13 mol; 4.0 L **8.110** 590 g/day **8.112** 0.92 g/L; less dense than air at STP **8.114** (a) 0.714 g/L (b) 1.96 g/L (c) 1.43 g/L

8.116 (a) (b)

(c) Ethylene glycol forms hydrogen bonds. **8.118** (a) 492 °R (b) $R = 0.0455$ (L·atm)/(mol·°R)

Chapter 9

9.1 (a) heterogeneous mixture (b) homogeneous solution (c) homogeneous colloid (d) homogeneous solution **9.2** (c), (d) **9.3** $Na_2SO_4 \cdot 10H_2O$ **9.4** 322 g **9.5** unsaturated; cooling makes the solution supersaturated **9.6** 5.6 g/100 mL **9.7** 6.8×10^{-5} g/100 mL **9.8** 56 mm Hg; ~90% saturated **9.9** 231 g **9.10** Place 38 mL acetic acid in flask and dilute to 500.0 mL. **9.11** 0.0086% (m/v) **9.12** (a) 20 g (b) 60 mL H_2O **9.13** 1.6 ppm **9.14** Pb: 0.015 ppm, 0.0015 mg; Cu: 1.3 ppm, 0.13 mg **9.15** 0.927 M **9.16** (a) 0.061 mol (b) 0.67 mol **9.17** 0.48 g **9.18** (a) 0.0078 mol (b) 0.39 g **9.19** 39.1 mL **9.20** 750 L **9.21** (a) 39.1 g; 39.1 mg (b) 79.9 g; 79.9 mg (c) 12.2 g; 12.2 mg (d) 48.0 g; 48.0 mg (e) 9.0 g; 9.0 mg (f) 31.7 g; 31.7 mg **9.22** 9.0 mg **9.23** $Na^+ = 0.046$ m/v %; $K^+ = 0.039$ m/v % **9.24** (a) 2.0 mol ions; (b) 2.0 °C **9.25** weak electrolyte **9.26** (a) red curve is pure solvent; green curve is solution (b) solvent bp = 62 °C; solution bp = 69 °C (c) 2 M **9.27** -1.9 °C **9.28** 3 ions/mol **9.29** (a) 0.70 osmol (b) 0.30 osmol **9.30** (a) 0.090 M Na^+; 0.020 M K^+; 0.110 M Cl^-; 0.11 M glucose (b) 0.33 osmol
9.31

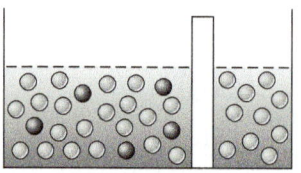

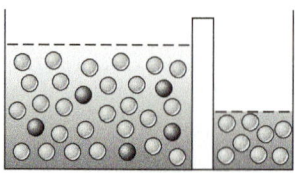

Before equilibrium   At equilibrium

9.32 HCl completely dissociates into ions; acetic acid dissociates only slightly. **9.33** Upper curve: HF; lower curve: HBr **9.34** (a) **9.35** (d) **9.36** homogeneous: mixing is uniform; heterogeneous: mixing is nonuniform **9.38** polarity **9.40** (b), (d) **9.42** 15.3 g/100 mL **9.44** Concentrated solutions can be saturated or not; saturated solutions can be concentrated or not. **9.46** Molarity is the number of moles of solute per liter of solution.

9.48 Dissolve 45.0 mL of ethyl alcohol in water and dilute to 750.0 mL
9.50 Dissolve 1.5 g NaCl in water to a final volume of 250 mL.
9.52 (a) 7.7% (m/v) **(b)** 3.9% (m/v) **9.54 (a)** 0.054 mol **(b)** 0.25 mol
9.56 230 mL, 1600 mL **9.58** 10 ppm **9.60 (a)** 0.425 M **(b)** 1.53 M
(c) 1.03 M **9.62** 5.3 mL **9.64** 38 g **9.66** 500 mL **9.68** 0.53 L
9.70 600 mL **9.72** a substance that conducts electricity when dissolved in water **9.74** Ca^{2+} concentration is 0.0015 M **9.76** 40 mEq **9.78** 0.28 L
9.80 $Ba(OH)_2$ **9.82** 26.9 mol **9.84** The inside of the cell has higher osmolarity than water, so water passes in and increases pressure. **9.86 (a)** 0.20 M Na_2SO_4 **(b)** 3% (m/v) NaOH **9.88** 2.4 osmol **9.90** The body manufactures more hemoglobin. **9.92** Sports drinks contain electrolytes, carbohydrates, and vitamins. **9.94 (a)** 680 mmHg **(b)** 1.9 g/100² mL
9.96 (a) 0.0067% (m/v) **(b)** 67 ppm **(c)** 0.000 40 M **9.98 (a)** 9.4 mL
(b) 0.75 L **9.100** NaCl: 0.147 M; KCl: 0.0040 M; $CaCl_2$: 0.0030 M
9.102 0.00020% (m/v) **9.104** 4.0 mL
9.106 (a) $CoCl_2(s) + 6 H_2O(l) \rightarrow CoCl_2 \cdot 6H_2O(s)$
(b) 1.13 g **9.110 (a)** 1.36 mol particles **(b)** 2.53 °C

Chapter 10

10.1 (a), (b) 10.2 (a), (c) 10.3 (a) H_2S **(b)** HPO_4^{2-} **(c)** HCO_3^- **(d)** NH_3
10.4 acids: HF, H_2S; bases: HS^-, F^-; conjugate acid–base pairs: H_2S and HS^-, HF and F^- **10.5 (a)** NH_4^+ **(b)** H_2SO_4 **(c)** H_2CO_3 **10.6 (a)** F^- **(b)** OH^-
10.7 $HPO_4^{2-} + OH^- \rightleftharpoons PO_4^{3-} + H_2O$; favored in forward direction
10.8 $HCl(aq) + CO_3^{2-}(aq) \rightleftharpoons HCO_3^-(aq) + Cl^-(aq)$ conjugate acid–base pairs: HCO_3^- and CO_3^{2-}, HCl and Cl^- **10.9** The —NH_3^+ hydrogens are most acidic **10.10** benzoate **10.11 (a)** basic, $[OH^-] = 3.2 \times 10^{-3}$ M
(b) acidic, $[OH^-] = 2.5 \times 10^{-12}$ M **10.12 (a)** 11.51 **(b)** 2.40
10.13 (a) $[H_3O^+] = 1 \times 10^{-13}$ M; $[OH^-] = 0.1$ M **(b)** $[H_3O^+] = 1 \times 10^{-3}$ M; $[OH^-] = 1 \times 10^{-11}$ M **(c)** $[H_3O^+] = 1 \times 10^{-8}$ M; $[OH^-] = 1 \times 10^{-6}$ M **(b)** is most acidic; **(a)** is most basic **10.14** 0.010 M HNO_2; weaker acid **10.15 (a)** acidic; $[H_3O^+] = 3 \times 10^{-7}$ M; $[OH^-] = 3 \times 10^{-8}$ M **(b)** most basic; $[H_3O^+] = 1 \times 10^{-8}$ M; $[OH^-] = 1 \times 10^{-6}$ M
(c) acidic; $[H_3O^+] = 2 \times 10^{-4}$ M; $[OH^-] = 5 \times 10^{-11}$ M
(d) most acidic; $[H_3O^+] = 3 \times 10^{-4}$ M; $[OH^-] = 3 \times 10^{-11}$ M
10.6 (a) 8.28 **(b)** 5.05 **10.17** 2.60 **10.18** 3.38 **10.19** 9.45
10.20 bicarbonate/carbonic acid = 10/1 **10.21** 9.13 **10.22 (a)** 0.079 Eq
(b) 0.338 Eq **(c)** 0.14 Eq **10.23 (a)** 0.26 N **(b)** 1.13 N **(c)** 0.47 N
10.24 $Al(OH)_3 + 3 HCl \rightarrow AlCl_3 + 3 H_2O$; $Mg(OH)_2 + 2 HCl \rightarrow MgCl_2 + 2 H_2O$ **10.25 (a)** $2 HCO_3^-(aq) + H_2SO_4(aq) \rightarrow 2 H_2O(l) + 2 CO_2(g) + SO_4^{2-}$ **(b)** $CO_3^{2-}(aq) + 2 HNO_3(aq) \rightarrow H_2O(l) + CO_2(g) + 2 NO_3^-(aq)$
10.26 $H_2SO_4(aq) + 2 NH_3(aq) \rightarrow (NH_4)_2SO_4(aq)$ **10.27** $CH_3CH_2NH_2 + HCl \rightarrow CH_3CH_2NH_3^+ Cl^-$ **10.28** 0.730 m **10.29** 133 mL
10.30 (a) 2.41×10^{-3} M; 4.83×10^{-3} Eq **(b)** 0.225 M
10.31 2.23×10^{-4} M; pH = 3.65 **10.32 (a)** neutral **(b)** basic **(c)** basic
(d) acidic **10.33 (a)** box 2 **(b)** box 3 **(c)** box 1 **10.34** The O—H hydrogen in each is most acidic; acetic acid **10.35 (a)** box 1 **(b)** box 2 **(c)** box 1
10.36 (a) box 3 **(b)** box 1 **10.37** 0.67 M **10.38** HBr dissociates into ions
10.40 KOH dissociates into ions **10.42** A monoprotic acid can donate one proton; a diprotic acid can donate two. **10.44 (a), (e) 10.46 (a)** acid
(b) base **(c)** neither **(d)** acid **(e)** neither **(f)** acid **10.48 (a)** CH_2ClCO_2H
(b) $C_5H_5NH^+$ **(c)** $HSeO_4^-$ **(d)** $(CH_3)_3NH^+$ **10.50 (a)** $HCO_3^- + HCl \rightarrow H_2O + CO_2 + Cl^-$; $HCO_3^- + NaOH \rightarrow H_2O + Na^+ + CO_3^{2-}$
(b) $H_2PO_4^- + HCl \rightarrow H_3PO_4 + Cl^-$; $H_2PO_4^- + NaOH \rightarrow H_2O + Na^+ + HPO_4^{2-}$ **10.52 (a)** $HCl + NaHCO_3 \rightarrow H_2O + CO_2 + NaCl$ **(b)** $H_2SO_4 + 2 NaHCO_3 \rightarrow 2 H_2O + 2 CO_2 + Na_2SO_4$
10.54 $K_a = \dfrac{[H_3O^+][A^-]}{[HA]}$ **10.56** $K_w = [H_3O^+][OH^-] = 1.0 \times 10^{-14}$
10.58 CH_3CO_2H is a weak acid and is only partially dissociated.
10.60
$K_a = \dfrac{[H_2PO_4^-][H_3O^+]}{[H_3PO_4]}$ $K_a = \dfrac{[HPO_4^-][H_3O^+]}{[H_2PO_4^-]}$ $K_a = \dfrac{[PO_4^{3-}][H_3O^+]}{[HPO_4^-]}$

10.62 basic; 1×10^{-8} **10.64** $[H_3O^+] = 1 \times 10^{-4}$ M — 1.6×10^{-7}; three orders of magnitude **10.66** pH = 1.7; pH = 12.3 **10.68 (a)** 7.60
(b) 3.30 **(c)** 11.64 **(d)** 2.40 **10.70 (a)** 1×10^{-4} M; 1×10^{-10} M;
(b) 1×10^{-11} M; 1×10^{-3} M **(c)** 1 M; 1×10^{-14} M **(d)** 4.2×10^{-2} M; 2.4×10^{-13} M **(e)** 1.1×10^{-8} M; 9.1×10^{-7} M **10.72** A buffer contains a weak acid and its anion. The acid neutralizes any added base, and the anion neutralizes any added acid.

10.74 (a) $pH = pK_a + \log \dfrac{[CH_3CO_2^-]}{[CH_3CO_2H]} = 4.74 + \log \dfrac{[0.100]}{[0.100]} = 4.74$
(b) $CH_3CO_2^- Na^+ + H_3O^+ \rightarrow CH_3CO_2H + Na^+$; $CH_3CO_2H + OH^- \rightarrow CH_3CO_2^- + H_2O$ **10.76** 9.19 **10.78** 9.07 **10.80** An equivalent is the formula weight in grams divided by the number of H_3O^+ or OH^- ions produced. **10.82** 63.0 g; 32.7 g; 56.1 g; 29.3 g **10.84** 25 mL; 50 mL
10.86 (a) 0.50 Eq **(b)** 0.084 Eq **(c)** 0.25 Eq **10.88** 0.13 M; 0.26 N
10.90 0.23 M **10.92** 0.075 M **10.94 (a)** pH = 2 to 3 **(b)** $NaHCO_3 + HCl \rightarrow CO_2 + H_2O + NaCl +$ **(c)** 20 mg **10.96** Intravenous bicarbonate neutralizes the hydrogen ions in the blood and restores pH.
10.98 3×10^{-6} M **10.100 (a)** 0.613 mol **(b)** 2.45 M **(c)** pH = 0.39
10.102 Citric acid reacts with sodium bicarbonate to release CO_2.
10.104 Both have the same amount of acid; HCl has higher $[H_3O^+]$ and lower pH. **10.106** 0.70 N; 0.35 M **10.108 (a)** NH_4^+, acid; OH^-, base; NH_3, conjugate base; H_2O, conjugate acid **(b)** 5.56 g
10.110 (a) $Na_2O(aq) + H_2O(l) \rightarrow 2 NaOH(aq)$ **(b)** 13.0 **(c)** 5.00 L

Chapter 11

11.1 $^{218}_{84}Po$ **11.2** $^{226}_{88}Ra$ **11.3** $^{89}_{38}Sr \rightarrow {}^{0}_{-1}e + {}^{89}_{39}Y$
11.4 (a) $^{3}_{1}H \rightarrow {}^{0}_{-1}e + {}^{3}_{2}He$ **(b)** $^{210}_{82}Pb \rightarrow {}^{0}_{-1}e + {}^{210}_{83}Bi$ **(c)** $^{20}_{9}F \rightarrow {}^{0}_{-1}e + {}^{20}_{10}Ne$
11.5 (a) $^{38}_{20}Ca \rightarrow {}^{0}_{1}e + {}^{38}_{19}K$ **(b)** $^{118}_{54}Xe \rightarrow {}^{0}_{1}e + {}^{118}_{53}I$ **(c)** $^{79}_{37}Rb \rightarrow {}^{0}_{1}e + {}^{79}_{36}Kr$
11.6 (a) $^{62}_{30}Zn + {}^{0}_{-1}e \rightarrow {}^{62}_{29}Cu$ **(b)** $^{110}_{50}Sn + {}^{0}_{-1}e \rightarrow {}^{110}_{49}In$ **(c)** $^{86}_{36}Kr + {}^{0}_{-1}e \rightarrow {}^{81}_{35}Br$
11.7 $^{120}_{49}In + {}^{0}_{-1}e \rightarrow {}^{120}_{50}Sn$ **11.8** 12% **11.9** 5.0 L **11.10** 3 days
11.11 13m **11.12** 2% **11.13** 175 μCi **11.14** 2×10^4 rem; 4×10^5 rem
11.15 $^{237}_{93}Np$ **11.16** $^{241}_{95}Am + {}^{4}_{2}He \rightarrow 2 {}^{1}_{0}n + {}^{243}_{97}Bk$ **11.17** $^{40}_{18}Ar + {}^{1}_{1}H \rightarrow {}^{1}_{0}n + {}^{40}_{19}K$ **11.18** 60 hours (10 half lives) **11.19** $^{235}_{92}U + {}^{1}_{0}n \rightarrow 2 {}^{1}_{0}n + {}^{137}_{52}Te + {}^{97}_{40}Zr$ **11.20** $^{3}_{2}He$ **11.21** 2 half-lives **11.22** $^{28}_{12}Mg \rightarrow {}^{0}_{-1}e + {}^{23}_{13}Al$
11.23

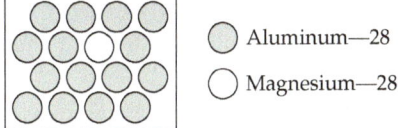

○ Aluminum—28

○ Magnesium—28

11.24 $^{14}_{6}C$ **11.25** The shorter arrow represent β emission; longer arrows represent α emission. **11.26** $^{241}_{94}Pu \rightarrow {}^{241}_{95}Am \rightarrow {}^{237}_{93}Np \rightarrow {}^{233}_{91}Pa \rightarrow {}^{233}_{92}U$
11.27 $^{148}_{69}Tm \rightarrow {}^{0}_{1}e \rightarrow {}^{148}_{68}Er$ or $^{148}_{69}Tm + {}^{0}_{-1}e \rightarrow {}^{148}_{68}Er$ **11.28** 3.5 years
11.29 The curve doesn't represent nuclear decay. **11.30** It emits radiation by decay of an unstable nucleus. **11.32** A nuclear reaction changes the identity of the atoms, is unaffected by temperature or catalysts, and often releases a large amount of energy. A chemical reaction does not change the identity of the atoms, is affected by temperature and catalysts, and involves relatively small energy changes. **11.34** by breaking bonds in DNA **11.36** A neutron decays to a proton and an electron. **11.38** The number of nucleons and the number of charges is the same on both sides. **11.40** α emission: Z decreases by 2 and A decreases by 4; β emission: Z increases by 1 and A is unchanged **11.42** In fission, a nucleus fragments to smaller pieces. **11.44 (a)** $^{35}_{17}Cl$ **(b)** $^{24}_{11}Na$ **(c)** $^{90}_{39}Y$ **11.46 (a)** $^{109}_{47}Ag$ **(b)** $^{10}_{5}B$
11.48 $4 {}^{1}_{0}n$ **(b)** $^{146}_{57}La$ **11.50** $^{198}_{80}Hg + {}^{1}_{0}n \rightarrow {}^{198}_{79}Au + {}^{1}_{1}H$; a proton
11.52 $^{228}_{90}Th$ **11.54** Half of a sample decays in that time. **11.56 (a)** 2.3 half-lives **(b)** 0.0063 g **11.58** 1 ng; 2×10^{-3} ng **11.60** The inside walls of a Geiger counter tube are negatively charged, and a wire in the center is positively charged. Radiation ionizes argon gas inside the tube, which creates a conducting path for current between the wall and the wire. **11.62** In a scintillation counter, a phosphor emits a flash of light when struck by radiation, and the flashes are counted. **11.64** more than 25 rems **11.66** 1.9 mL
11.68 (a) 4.7 rem **(b)** 1.9 rem **11.70** in vivo procedures, therapeutic procedures, boron neutron capture **11.72** Irradiation kills harmful microorganisms by destroying their DNA. **11.74** They yield more data, including three-dimensional images. **11.76** no filter — α radiation; plastic — β radiation; foil — γ radiation **11.78** Nuclear decay is an intrinsic property of a nucleus and is not affected by external conditions. **11.80** 112 cpm
11.82 (a) β emission **(b)** Mo-98 **11.84 (a)** $^{238}_{94}Pu \rightarrow {}^{4}_{2}He + {}^{234}_{92}U$
(b) for radiation shielding **11.86** Their cells divide rapidly. **11.88** advantages: few harmful byproducts, fuel is inexpensive; disadvantage: needs a high temperature **11.90 (a)** $^{253}_{99}Es + {}^{4}_{2}He \rightarrow {}^{256}_{101}Md + {}^{1}_{0}n$ **(b)** $^{250}_{98}Cf + {}^{11}_{5}B \rightarrow {}^{257}_{103}Lr + 4 {}^{1}_{0}n$ **11.92** $^{10}_{5}B + {}^{1}_{0}n \rightarrow {}^{7}_{3}Li + {}^{4}_{2}He$
11.94 $^{238}_{92}U + 3 {}^{4}_{2}He \rightarrow {}^{246}_{98}Cf + 4 {}^{1}_{0}n$

Chapter 12

12.1 (a) alcohol, carboxylic acid (b) double bond, ester (c) aromatic ring, amine, carboxylic acid

12.2 (a) CH_3CH_2CHO (b) CH_3COCH_3 (c) $CH_3CH_2CO_2H$ **12.3** (a) $CH_3CH_2CH_2CH_2CH_2CH_3$
(b) $CH_3CH_2CH_2CH_2CH_2CH_2CH_2CH_3$

12.4
$$CH_3CH_2CH_2CH_2\overset{\overset{CH_3}{|}}{C}HCH_3 \qquad CH_3CH_2CH_2\overset{\overset{CH_3}{|}}{C}HCH_2CH_3$$

12.5 (a) $CH_3CH_2CH_2CH_2CH_3$ Pentane
(b) $CH_3\overset{\overset{CH_3}{|}}{C}HCH_2CH_3$ 2-Methylbutane
(c) $CH_3\overset{\overset{CH_3}{|}}{\underset{\underset{CH_3}{|}}{C}}CH_3$ 2,2-Dimethylpropane

12.6 (a), (b), (c) [structures with OH, Br, Cl substituents]

12.7 (a)
$$CH_3CH_2\overset{\overset{H_3C}{|}}{\underset{\underset{\underset{CH_3}{|}}{CH_2CHCH_3}}{C}}-\overset{\overset{Cl}{|}}{C}HCH_2CH_3$$

(b)
$$CH_3\overset{\overset{H_3C\;\;CH_3}{\setminus\;\;/}}{C}-CH-\overset{\overset{H_3C\;\;CH_3}{\setminus\;\;/}}{\underset{\underset{CH_2CH_3}{|}}{C}}CH_3$$

12.8 (a) CH_3CH_2O [propanal structure]
(b) CH_3COCH_3 [acetone structure]
(c) $CH_3CH_2CO_2H$ [propanoic acid structure]

12.9 Structures (a) and (c) are identical, and are isomers of (b).

12.10
$CH_3CH_2CH_2CH_2CH_2\overset{\overset{CH_3}{|}}{C}HCH_3$ $CH_3CH_2CH_2CH_2\overset{\overset{CH_3}{|}}{C}HCH_2CH_3$ $CH_3CH_2CH_2\overset{\overset{CH_3}{|}}{C}HCH_2CH_2CH_3$

$CH_3CH_2CH_2CH_2\overset{\overset{CH_3}{|}}{\underset{\underset{CH_3}{|}}{C}}CH_3$ $CH_3CH_2CH_2\overset{\overset{CH_3}{|}}{\underset{\underset{CH_3}{|}}{C}}HCHCH_3$ $CH_3CH_2CH_2\overset{\overset{CH_3}{|}}{\underset{\underset{CH_3}{|}}{C}}CH_2CH_3$

$CH_3CH_2\overset{\overset{CH_3}{|}}{C}HCH_2\overset{\overset{CH_3}{|}}{C}HCH_3$ $CH_3\overset{\overset{CH_3}{|}}{C}HCH_2CH_2\overset{\overset{CH_3}{|}}{C}HCH_3$ $CH_3CH_2\overset{\overset{CH_3}{|}}{C}H\overset{\overset{CH_3}{|}}{\underset{\underset{CH_3}{|}}{C}}HCH_2CH_3$

12.11 [structure with labels p, s, t, q]

12.12 (a) 2,6-dimethyloctane (b) 3,3-diethylheptane

12.13
(a) $\underset{p\;\;s\;\;s\;\;t\;\;s\;\;p}{CH_3CH_2CH_2\overset{\overset{p\;CH_3}{|}}{C}HCH_2CH_3}$

(b) $\underset{p\;\;s\;\;s\;\;s\;\;\;t\;\;s\;\;p}{CH_3CH_2CH_2CH_2\overset{\overset{p\;CH_3}{\underset{\underset{p\;CH_3}{|}}{|\;\;t}}}{C}HCHCH_2CH_3}$

(c) $\underset{p\;\;t\;\;s\;\;\;\;p}{CH_3\overset{\overset{p\;CH_3}{|}}{C}HCH_2\overset{\overset{p\;CH_3}{|}}{\underset{\underset{q\;P\;CH_3}{|}}{C}}CH_3}$

ANSWERS TO SELECTED PROBLEMS A-21

12.14 (a) CH₃CH₂CH(CH₃)CH₃ 2-Methylbutane (b) (CH₃)₂CHC(CH₃)₃ 2,3,3-Trimethylbutane

12.15 (a) 2,2-dimethylpentane (b) 2,3,3-trimethylpentane
12.16 2 C₂H₆ + 7 O₂ → 4 CO₂ + 6 H₂O
12.17 CH₃CH₂CH₂Cl + CH₃CHClCH₃ + CH₃CCl(CH₃)CH₃ + CH₃CH₂CHCl₂ + CH₃CHClCH₂Cl + CH₂ClCH₂CH₂Cl

12.18 (a) 1-ethyl-4-methylcyclohexane (b) 1-ethyl-3-isopropylcyclopentane
12.19 (a) cyclohexane with two ethyl groups on same carbon (b) cycloheptane with methyl groups
12.20 (a) plastics (b) more resistant to heat and cold **12.21** propylcyclohexane **12.22** (a) 12 hydrogens (b) 10 hydrogens (c) 8 hydrogens
12.23 (a) CH₃CC(CH₃)CH₂CH₃ with CH₃ (b) CH₃CH(OH)CH(CH₃)CH₃
12.24 (a) cyclopentenone (b) cyclohexadiene with CH₃ and NH₂

12.25 (a) double bond, ketone, ether (b) double bond, amine, carboxylic acid
12.26 (a) 2,3-dimethylpentane (b) 2,5-dimethylhexane **12.27** (a) 1,1-dimethylcyclopentane (b) isopropylcyclobutane **12.28** The methyl groups are on the same side of the ring in one structure, and on opposite sides in the other. **12.30** groups of atoms that have a characteristic reactivity; chemistry of compounds is determined by their functional groups **12.32** A polar covalent bond is a covalent bond in which electrons are shared unequally.
12.34 (a) (i) amine; (ii) amide; (iii) ester; (iv) aldehyde (b) (v) ketone; (vi) aromatic ring; (vii) alcohol; (viii) carboxylic acid
12.36 (a) CH₃CH₂CH₂CH₂CHO Aldehyde
(b) CH₃CH₂CH₂C(O)—OCH₂CH₃ Ester
(c) HS—CH₂CH₂C(O)—NH₂ Amide, thiol

12.38 They must have the same molecular formula but different structures.
12.40 A primary carbon is bonded to one other carbon; a secondary carbon is bonded to two other carbons; a tertiary carbon is bonded to three other carbons; and a quaternary carbon is bonded to four other carbons.
12.42 (a) 2,3-dimethylbutane (b) cyclopentane

12.44
(a) CH₃CH₂CH₂CH₃   CH₃CH(CH₃)CH₃
(b) CH₃CH₂CH₂CH₂OH   CH₃CH₂CH(OH)CH₃   CH₃CH(CH₃)CH₂OH   CH₃C(CH₃)₂OH
(c) CH₃CH₂OCH₂CH₃   CH₃CH₂CH₂OCH₃   CH₃CH(CH₃)OCH₃

12.46
(a) CH₃CH₂CH₂CH₂OH   CH₃CH₂CH(OH)CH₃   CH₃CH(CH₃)CH₂OH
    CH₃C(OH)(CH₃)CH₃
(b) CH₃CH₂CH₂NH₂   CH₃CH(NH₂)CH₃   CH₃CH₂NHCH₃   (CH₃)₂NCH₃
(c) CH₃CH₂CH₂COCH₃   CH₃CH₂COCH₂CH₃   CH₃CH(CH₃)COCH₃

12.48 identical: (a); isomers: (b), (d), (e); unrelated: (c) **12.50** All have a carbon with five bonds. **12.52** (a) 4-ethyl-3-methyloctane (b) 5-isopropyl-3-methyloctane (c) 2,2,6-trimethylheptane (d) 4-isopropyl-4-methyloctane (e) 2,2,4,4-tetramethylpentane (f) 4,4-diethyl-2-methylhexane (g) 2,2-dimethyldecane
12.54
(a) CH₃CH₂C(C(CH₃)₃)(CH₃)—CH(CH₃)CH₂CH₃
(b) CH₃CH(CH₃)CH₂CH(CH₃)CH₃
(c) CH₃CH₂CH(CH₃)C(CH₂CH₃)(CH₂CH₃)CH₂CH₂CH₃
(d) CH₃CH(CH₃)CH(CH₃)CH₂CH₂CH(CH₃)CH(CH₃)CH₂CH₃
(e) cycloheptane with CH₃CH(CH₃)— and —CH₂CH(CH₃)CH₃ and —CH₃ substituents
(f) cyclopentane with (CH₃)₂ and CH₃ substituents

12.56 (a) 1-ethyl-3-methylcyclobutane (b) 1,1,3,3-tetramethylcyclopentane (c) 1-ethyl-3-propylcyclohexane (d) 4-butyl-1,1,2,2-tetramethylcyclopentane
12.58 (a) 2,2-dimethylpentane (b) 2,4-dimethylpentane (c) isobutylcyclobutane **12.60** heptane, 2-methylhexane, 3-methylhexane, 2,2-dimethylpentane, 2,3-dimethylpentane, 2,4-dimethylpentane, 3,3-dimethylpentane, 3-ethylpentane, 2,2,3-trimethylbutane

12.62 $C_3H_8 + 5\,O_2 \rightarrow 3\,CO_2 + 4\,H_2O$

12.64

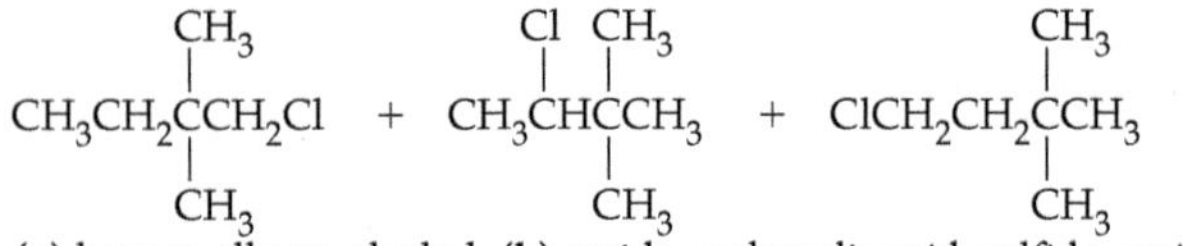

CH₃CH₂C(CH₃)₂CH₂Cl + CH₃CHClCH(CH₃)CH₃ + ClCH₂CH₂C(CH₃)₂CH₃

12.66 (a) ketone, alkene, alcohol (b) amide, carboxylic acid, sulfide, amine

12.68 four tertiary carbons   **12.70** Non-polar solvents dissolve non-polar substances.   **12.72** pentane; greater London forces   **12.74** A chemical feedstock is a simple organic chemical used as the starting material in many organic reactions.

12.76

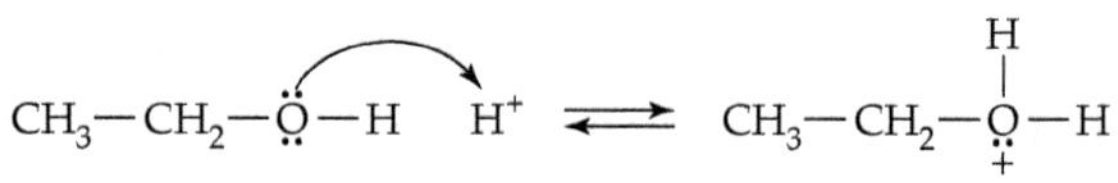

$CH_3-CH_2-\ddot{O}-H \quad H^+ \rightleftharpoons CH_3-CH_2-\overset{H}{\underset{+}{O}}-H$

Credits

Text and Art Credits

Chapter 8: 219, Adapted from NASA, Goddard Institute for Space Studies, Surface Temperature Analysis (GIS Temp), http://data.giss.nasa.gov/gistemp.

Photo Credits

Chapter 1: 2, imagebroker.net/SuperStock; **5**, Richard Megna/Fundamental Photographs; **8**, PhilSigin/iStockphoto; **12(a) top**, Norov Dmitriy/iStockphoto; **12(b) top**, Ben Mills; **12(c) top**, Shutterstock; **12(a) middle**, Andraž Cerar/Shutterstock; **12(b) middle**, Leeuwtje/iStockphoto; **12(c) middle**, Ben Mills; **12(a) bottom**, Russell Lappa/Photo Researchers, Inc.; **12(b) bottom**, Texas Instruments Inc.; **14(a)**, Richard Megna/Fundamental Photographs; **14(b)**, Richard Megna/Fundamental Photographs; **14(c)**, Richard Megna/Fundamental Photographs; **15**, eROMAZe/iStockphoto; **16**, Centers for Disease Control; **17**, Richard Megna/Fundamental Photographs; **19**, Pearson Education/McCracken Photographers; **21**, artkamalov/Shutterstock; **22**, Pearson Education/Eric Schrader; **24**, Pearson Education/Eric Schrader; **25**, tdbp/Alamy; **28**, Pearson Education/Michal Heron; **30**, Richard Megna/Fundamental Photographs; **31**, Stockbyte/Photolibrary; **33**, Claire VD/Shutterstock; **35**, Ivica Drusany/iStockphoto; **36**, BD Adams; **38**, Pearson Education/Eric Schrader.

Chapter 2: 44, Katie Dickinson/Shutterstock; **46**, AP Photo/Donna Carson; **48**, IBM Research, Almaden Research Center; **52**, Richard Megna/Fundamental Photographs; **54**, Richard Megna/Fundamental Photographs; **56**, NASA, ESA and H.E. Bond (STScI); **57**, Michael Neary Photography/iStockphoto; **66**, James Benet/iStockphoto.

Chapter 3: 72, Juan Jose Rodriguez Velandia/Shutterstock; **74**, Pearson Education/Eric Schrader; **76(a)**, Richard Megna/Fundamental Photographs; **76(b)**, Richard Megna/Fundamental Photographs; **77 top**, NASA; **77 bottom**, Richard Megna/Fundamental Photographs; **78 both**, Dimitris S. Argyropoulos; **83**, Daniella Zalcman/Shutterstock; **93**, Steve Gschmeissner/SPL/Photo Researchers, Inc.

Chapter 4: 98, David R. Frazier Photolibrary, Inc./Alamy; **100**, Martin Barraud/Alamy; **113**, Upsidedowndog/iStockphoto; **118**, Claudia Veja/Alamy; **125**, Yakobchuk Vasyl/Shutterstock.

Chapter 5: 132, Harald Sund/Brand X Pictures/Getty Images; **137**, David R. Frazier/Photo Researchers, Inc.; **138**, Richard Megna/Fundamental Photographs; **140**, Dr. P. Marazzi/Photo Researchers, Inc.; **143 left**, Richard Megna/Fundamental Photographs; **143 right**, Richard Megna/Fundamental Photographs; **147 top**, Luca DiCecco/Alamy; **147 bottom**, David Young-Wolff/PhotoEdit.

Chapter 6: 158, AFP/Getty Images/Newscom; **160**, Richard Megna/Fundamental Photographs; **164 left**, Library of Congress; **164 right**, Science Photo Library/Photo Researchers, Inc.; **172**, Jane Norton/Shutterstock.

Chapter 7: 178, Ron Lewis/Icon SMI/Newscom; **181**, Richard Megna/Fundamental Photographs; **185**, discpicture/Shutterstock; **186**, GeoStock/Getty Images; **187 left**, Aaron Amat/Shutterstock; **187 right**, Samuel Perry/Shutterstock; **194**, AC/General Motors/Peter Arnold Images/Photolibrary; **195**, Reuters/Vladimir Davydov; **196**, Photolibrary/Indexopen; **204**, Myles Dumas/iStockphoto.

Chapter 8: 212, Tony Waltham/AGE Fotostock; **221**, NASA; **227**, Stephen Sweet/iStockphoto; **228**, Laura Stone/Shutterstock; **238**, Richard Megna/Fundamental Photographs; **239 top**, Alexei Zaycev/iStockphoto; **239 middle**, Harry Taylor/Dorling Kindersley; **239 bottom**, AGE Fotostock; **240 left**, Jens Mayer/Shutterstock; **240 right**, Jonny Kristoffersson/iStockphoto.

Chapter 9: 252, Phil Schermeister/NGS Images; **256**, Pearson Education/Tom Bochsler; **259**, Richard Megna/Fundamental Photographs; **263**, AP Photo/Gurinder Osan; **265(a)**, Richard Megna/Fundamental Photographs; **265(b)**, Richard Megna/Fundamental Photographs; **265(c)**, Richard Megna/Fundamental Photographs; **273(a)**, Richard Megna/Fundamental Photographs; **273(b)**, Richard Megna/Fundamental Photographs; **276**, Jason Getz/Atlanta Journal-Constitution/MCT/Newscom; **281(a)**, Sam Singer; **281(b)**, Sam Singer; **281(c)**, Sam Singer; **284 left**, Martin Dohrn/SPL/Photo Researchers, Inc.; **284 right**, Lev Dolgachov/Shutterstock.

Chapter 10: 290, Liga Lauzuma/iStockphoto; **292**, Pearson Education/Eric Schrader; **299**, Gastrolab/Photo Researchers, Inc.; **308(a)**, Richard Megna/Fundamental Photographs; **308(b)**, Pearson Education/Tom Bochsler; **308 left**, Pearson Education/Tom Bochsler; **312**, Robert Caplin/Newscom; **316**, Pearson Education/Eric Schrader; **318(a)**, Richard Megna/Fundamental Photographs; **318(b)**, Richard Megna/Fundamental Photographs; **320 top**, RMAX/iStockphoto; **320 bottom**, National Atmospheric Deposition Program.

Chapter 11: 328, P. Berndt/Custom Medical Stock Photo/Newscom; **338 top**, Simon Fraser/RVI/SPL/Photo Researchers, Inc.; **338 bottom**, Media Minds/Alamy; **343**, Stanford Dosimetry, LLC; **344**, Mark Kostich/iStockphoto; **345**, Tony Freeman/PhotoEdit; **347**, Stephen Uber/iStockphoto; **348**, Custom Medical Stock Photo.

Chapter 12: 356, Kevin Burke/Corbis; **359**, Xinhua News Agency/Newscom; **385**, Chloe Johnson/Alamy.

Index

Note: Page numbers in **boldface** type indicate definitions of terms.

1,4 link, **676**
3TC (lamivudine), 795

Absolute alcohol, 435
ACE (angiotensin-converting enzyme) inhibitors, 607
Acetaldehyde
  in acetal formation, 501
  in ethyl alcohol metabolism, 446
  in hemiacetal formation, 500
  properties, 490, 493
  structure, 109, 488, 493
  uses, 125, 493
Acetal hydrolysis, 504–506, 507
Acetals, **501**–506, 674
Acetaminophen, 865
Acetanilide, 865
Acetic acid
  acidity, 293, 448
  carbonyl derivatives, 522
  naming of, 525
  properties, 520, 525
  structure, 525
  uses, 292
Acetoacetate, 765, 766, 767
Acetone
  in acetal formation, 501
  in hemiacetal formation, 500
  as ketone body, 765, 766
  odor on breath, 708, 767
  as product of *Clostridium* fermentation, 703
  properties, 216, 493–494, 499
  structure, 488, 493
  uses, 493–494
Acetyl acyl carrier protein (acetyl-ACP), 769
Acetylcholine
  breakdown of, 606
  drug interactions with, 858–859
  in irreversible inhibition, 606
  mechanism of, 857–858
  in neurotoxin mechanisms, 478
  structure, 478
  synthesis, 857
Acetylcholinesterase, 606
Acetyl-CoA, **630**
  breakdown in citric acid cycle, 641
  conversion of glucose into, 694
  conversion of pyruvate into, 701, 702, 704–705
  conversion to ketone bodies, 707, 759, 765
  in energy generation, 631, 758–759, 764
  in lipid metabolism, 759
  in lipogenesis, 768–769
  production of, 630, 631
Acetyl-coenzyme A. *See* Acetyl-CoA

Acetylene, 105, 398, 399
Acetyl groups, **518**, 640–641. *See also* Acetyl-CoA
Acetylsalicylic acid (ASA), 8
Achirality, **557**–559
Acid anhydrides, 516, 523
Acid–base balance, 884–886, 888–890
Acid–base indicators, **308**
Acid–base reactions
  carboxylic acids in, 526–527
  common reactions, 316–317
  definition of, 294–295
  neutralization reactions, 138, **141**–142, 318
Acid–base titration, 317–319
Acid dissociation constants (K_a), **301**, 526
Acid equivalents, 313–315
Acidity
  of aqueous solutions, 302–303
  of denaturation agents, 578
  determination of, 303–305, 308
  of salt solutions, 321–322
Acidosis, 312–313, **886**, 890
Acid rain, 320–321
Acids, **92**, 291–322. *See also* Acid–base reactions; Acidity; pH; Strong acids; Weak acids
  acid equivalents, 313–315
  anions derived from, 92
  in aqueous solution, 291–292
  basic concepts, 91–92
  buffer solutions, 308–311
  common, 292
  concentrations of, 317–319
  conjugate, 295
  corrosiveness of, 296
  definitions, 291–292, 293
  diprotic, 293, 315
  examples, 92
  in neutralization reactions, 141–142
  normality of, 314
  as proton donors, 293
  strength of, 296–300
  titration, 317–319
  unsaturated, 517–518, 519
  water as, 302
ACP (acyl carrier protein), 769
Acquired Immunodeficiency Syndrome (AIDS), 607, 795. *See also* HIV-1
Actinides, 53
Activation, enzyme, **602**
Activation energy (E_{act}), **191**–192, 193, 594
Active sites, **588**, 596, 603
Active transport, **743**, 744, 826
Actual yield, **169**–171
Acute toxicity, 499
Acyclic alkanes, 383, 384
Acyl carrier protein (ACP), 769
Acyl-CoA, 758, 762
Acyl-CoA acetyltransferase, 763

Acyl-CoA dehydrogenase, 762–763
Acylglycerols, 754
Acyl groups, **516**, 753
Acyl transferase, 761
Acyl transfer reactions, 516
Addition reactions, **403**
  alcohols, 500
  aldehydes, 500–506
  alkenes, 407–408, 409, 413, 414–415
  alkynes, 407–408
  carbonyls, 506–507
  ketones, 500–507
  Markovnikov's rule, 410–412, 415
Adenine
  base pairing of, 783–785
  in nucleotide chains, 782
  as purine derivative, 777
  structure, 421, 470
Adenosine, 637, 777, 778
Adenosine diphosphate (ADP), 632–634, 696, 697, 699. *See also* ATP–ADP conversion
Adenosine monophosphate (AMP), 697
Adenosine triphosphate (ATP), **629**. *See also* ATP–ADP conversion
Adenylate cyclase, 849
ADH (antidiuretic hormone), 845, 875–876, 890
Adipic acid, 538
Adipocytes, 759, 760, 761
Adipose tissue, 727, 728, 729, 755, 756
ADP (adenosine diphosphate), 632–634, 696, 697, 699. *See also* ATP–ADP conversion
ADP–ATP conversion. *See* ATP–ADP conversion
Adrenaline. *See* Epinephrine
Aerobic conditions, **701**, 702
Agglutination, 672
Agitation, mechanical, as denaturation agent, 578
Agonists, **858**–859
AIDS (Acquired Immunodeficiency Syndrome), 607, 795. *See also* HIV-1
α-Keratins, 571
Alanine, 557–558
Alanine aminotransferase (ALT), 829
Alanine transaminase (ALT), 601
Albumins, 755, 756, 757
Alcohol dehydrogenase, 446, 591
Alcohol family, 360
Alcoholic beverages, 195
Alcoholic fermentation, 435, 703–**704**
Alcoholism, 767, 862
Alcohols, **433**
  acidity, 448–450
  common, 434–436
  drawing, 451
  functional groups, 362
  hydrophilic/hydrophobic parts, 439

monosaccharide reactions with, 674–675
  naming, 436–438
  as organic molecule family, 361, 362
  phosphate esters of, 675–676
  primary, 437, 441, 496
  properties, 433–434, 439–440
  reactions of, 440–445, 528–529
  secondary, 437, 441, 444, 496
  tertiary, 437, 441
Aldehydes, **486**
  aromas, 485
  carbonyl groups in, 485
  common, 492–493
  functional groups, 362
  in monosaccharides, 658
  naming, 488
  as organic molecule family, 361, 362
  as products of alcohol oxidation, 444
  properties, 484, 490–492
  reactions of, 494–498, 500–506, 673–674
  structure, 486–487
  testing for, 495–496
  toxicity, 492
Aldohexose, 663
Aldolase, 696, 698
Aldoses, **658**, 674
Aldosterone, 739, 890
Aldotetrose, 660
Aleve®, 865
Alkali metals, **54**
  cation formation by, 75, 81
  compounds with halogens, 73–74
  electron-dot symbol, 79
  properties, 54
  as reducing agents, 144
  valence-shell electron configurations, 63, 79
Alkaline batteries, 147
Alkaline earth metals, **55**, 63, 81, 144
Alkaline phosphatase (ALP), 601
Alkaloids, **476**–477
Alkalosis, 312–313, **886**
Alkanes, 357–387
  acyclic, 383, 384
  cycloalkanes, **383**–384, 386–387
  isomers, 364–366
  naming, 374–380
  as organic molecule family, 360–362
  physiological effects of, 381
  properties, 380–381
  reactions of, 381–383
  saturated, 395–396
  structure, 364–366
  thermal "cracking," 396
Alkene families, 360
Alkene polymers, 415–418

I-1

INDEX

Alkenes, **396**
  addition reaction process, 414–415
  functional groups in, 362, 395
  halogenation of, 409
  hydration of, 413
  hydrogenation of, 407–408
  hydrohalogenation of, 410–412
  identifying reactions of, 405
  naming, 396–398
  as organic molecule family, 361, 362
  as petrochemical type, 385
  as plant hormones, 855
  polymerization of, 415–418
  as products of alcohol dehydration, 440
  properties, 402–403
  reactions of, 407–418
  structure, 398, 399–402
  unsaturated, 396
  uses, 385
Alkoxide ions, **448**–450
Alkoxy group, **450**
Alkyl bromides, 410
Alkyl chlorides, 410
Alkyl groups, **375**–376
Alkyl halides, 361, 362, **454**
Alkynes, **396**
  functional groups, 362, 395
  naming, 396–398
  as organic molecule family, 361, 362
  properties, 402–403
  reactions of, 407–408
  unsaturated, 396
Allergic responses, 880–881
Allopurinol, 140, 833
Allosteric control, **603**–604, 609, 698
Allosteric enzymes, **603**–604
Alloys, metal, 241, 242
ALP (alkaline phosphatase), 601
Alpha-amino acids, **551**
α-Amino acids, **551**
α-Amylase, 684, 693
Alpha decay, 341
Alpha emission, 332–333, 342
α-Helix structure, 217–218, **569**–570
Alpha-hydroxy acids, 528
α-Ketoglutarate, 641–642, 829
Alpha particles, **331**, 342
Alpha radiation, 330–331
Alpha rays, 330
ALT (alanine aminotransferase), 829
ALT (alanine transaminase), 601
Altitude, 263
Altman, Sidney, 790
Aluminum, 30, 81
Alzheimer's disease, 879
Amanita muscaria mushrooms, 499
Amides, **515**
  of fatty acids, 722–723
  formation of, 529, 530–531
  functional groups, 362
  hydrolysis of, 534–535, 536–537
  naming, 522
  as organic molecule family, 361, 362
  overview, 521–522

  properties, 521–523
  structure, 515–516, 521
  vs. amines, 522
Amine groups, 551
Amine hormones, 846, 847, 850–851
Amines, **461**, 461–478
  basicity, 467, 471–472, 522
  classifying, 461, 463–464
  drawing, 463–464
  functional groups, 362
  heterocyclic, 463, 464
  naming, 462
  as organic molecule family, 361, 362
  in plants, 476–477
  properties, 467–469
  reactions, 317, 471–472
  reactions with carboxylic acids, 528–529
  structure, 461–464, 467–468
  toxicity, 468
  vs. amides, 522
Amine salts, 474–475
Amino acid catabolism, 827, 828–830, 834–835
Amino acid metabolism, 826–828
Amino acid pool, **826**
Amino acids, **551**. *See also* Side chains, amino acid
  acid–base properties, 555–556
  active transport of, 826
  arrangement, 560–563
  in body fluids, 872
  carboxyl groups in, 551
  conditional, 836
  conversion of proteins to, 825–826
  conversion to glucose, 706, 713, 714
  derivatives, 846, 847, 850–851
  dietary supplements, 826
  essential, 564, **835**–838
  genetic codes for, 793
  glucogenic, 835
  handedness, 557–559, 662
  ketogenic, 835
  nonessential, **835**–838
  polymorphisms, 812–813
  in proteins, 552–554
  as source of pyruvate, 701
  storage, 767
  structure, 463
Aminobenzene, 422
Amino groups, **463**
Amino nitrogen, 828–830
Amino-terminal amino acids, **561**
Amitriptyline, 861
Ammonia
  as Brønsted-Lowry base, 293–294
  elimination of, 830–831
  reaction with acids, 317
  structure, 101, 116, 218
  toxicity of, 830
  uses, 293
Ammonium groups, 473
Ammonium ion, 106–107, **471**
Ammonium ions, quaternary, **462**
Ammonium salts, **474**
Ammonium salts, quaternary, **475**
Amorphous solids, **241**, 242

AMP (adenosine monophosphate), 697
Amphetamines, 473, 862
Amphoteric substances, **302**
Amylopectin, 684, 685
Amylose, 684
Anabolic steroids, 853
Anabolism, **630**, 758
Anaerobic conditions, **701**, 702, 703–704, 712
Anaerobic threshold, 712
Analgesics, 477
Anaphylactic shock, 850
Androgens, 739, 852
Androsterone, 739, 852
Anemia, 172
Anesthetics, 453, 454
Angiotensin, 561, 607
Angiotensin-converting enzyme (ACE) inhibitors, 607
Anhydrides, 361, 362
Aniline, 422
Animal fats
  composition, 726
  in diet, 728
  as lipids, 725
  properties, 727
  in soap, 731
Animal starch. *See* Glycogen
Anion gap, 274
Anionic detergents, 731
Anions, **74**
  in body fluids, 872
  formation by periodic table groups, 75, 81
  in human body, 86
  naming, 84
  octet rule for, 80
Anomeric carbon atoms, **666**
Anomers, **666**
Antacids, 292, 299
Antagonists, **858**–859
Antibiotics, 685–686
Antibodies, 672, **880**, 882
Antibody-mediated immune response, 880, 881–882
Anticoagulants, 681, 876
Anticodons, 796–**797**
Antidiuretic hormone (ADH), 564, 845, 846, 875–876, 890
Antigen–antibody complexes, 882
Antigenic shifts, 800
Antigens, **879**–880
Antihistamines, 461, 474, 860–861
Anti-inflammatories, 739, 746
Antilogarithms, 306
Antimycin A, 646
Antioxidants, **614**
  beta-carotene, 614, 648
  phenols, 449
  selenium, 614
  vitamin A, 648
  vitamin C, 610, 614, 648
  vitamin E, 449, 613, 614, 648
Antiseptics, 447
Ants, 514
Appetite regulation, 760
Arachidonic acid, 745, 746

Arginine, 836
Argon, 55, 66
Aristotle, 45
Aromas. *see* Odors
Aromatic acids, 517–518
Aromatic compounds, **418**
  amines, 462, 463, 471
  biological effects, 420
  functional groups, 362, 395
  naming, 421–423
  as organic molecule family, 361, 362
  properties, 420
  reactions of, 424–425
  reactivity, 419
  structure, 418–420, 423
Aromatics, 385
Arrhenius, Svante, 291–292
Arterial plaque, 757
Artificial radioisotopes, 331
Artificial transmutation, **347**
ASA (acetylsalicylic acid), 8
Ascorbic acid. *See* Vitamin C
Asparagine, 837
Aspartame, 559
Aspartate, 555, 831
Aspartate transaminase (AST), 601
Aspirin
  case study, 8
  inhibition of prostaglandin synthesis, 746
  role in metabolic acidosis, 312
  as salicylic acid derivative, 8, 532
  solubility in body fluids, 473
  therapeutic dosage, 499
AST (aspartate transaminase), 601
Astatine, 55
Ataxia, 831
Atherosclerosis, 757
Athletes, 875
Atmospheres (atm), 221
Atmospheric pressure, 221, 238–239
Atomic mass, 49, 51
Atomic mass units (amu), **46**, 47
Atomic number (Z), **48**–49, 50, 51, 329
Atomic radius, 54
Atomic theory, **45**–47
Atomic weight, **50**–51
Atoms, **45**
  chemical formula representation, 10
  oxidation numbers, 148
  size, 45–46
  structure, 46–47, 48, 49, 56–58
  in visible light, 66
ATP (adenosine triphosphate), **629**
ATP–ADP conversion
  in covalent modification, 609
  energy transfer in, 609, 632–635
  role of enzymes in, 591, 592
  during running, 712
ATP–cyclic AMP conversion, 849
ATP production. *See also* Citric acid cycle
  blocking and uncoupling, 646–647
  from fatty acid oxidation, 763–764
  glucose metabolism in, 694
  in glycolysis pathway, 696, 697, 698, 703

harmful by-products, 647–649
in hydrothermal vents, 627
in mitochondria, 629
role of energy transport chain in, 643–647
role of liver in, 767
during running, 712
as stage in metabolism, 631
ATP synthase, 644, 645–646
Atropine, 476, 859
Autoclaves, 239
Autoimmune diseases, 708, **882**
Autoradiograms, 817
Auxin, 855
Avian influenza viruses, 799
Avogadro, Amedeo, 164
Avogadro's law, **232**–233
Avogadro's number, 159–162, **161**, 164
Axons, 855
Azidothymidine (AZT), 607, 795

Bacteria, 627, 685
Bacteriophages, 685
Bakelite, 493
Balanced equations, **134**–137. See also Chemical equations
Ball-and-stick models, 109
Banks, William, 879
Barbiturates, 646
Barium, 55, 66
Barium hydroxide, 92
Barometers, 221
Basal metabolic rate, 636–637
Base equivalents, 313–315
Base pairing, 783–785, **784**
Bases, **92**, 291–322. See also Acid–base reactions; Basicity; pH; Strong bases
in aqueous solution, 291–292
base equivalents, 313–315
basic concepts, 91–92
buffer solutions, 308–311
common, 292–293
conjugate, **295**
corrosiveness, 296
definitions, 291–292, 293–296
in DNA, 777, 780
examples, 92
in neutralization reactions, 141–142
nitrogenous, 776, 777
normality of, 314
in RNA, 777, 780
strength, 296–300
strong, 296–300, 307, 322
titration, 317–319
water as, 302
weak, 296–300, **297**, 322, 473
Basicity
of alkoxide ions, 448–450
of amines, 467, 471–472, 522
as denaturation agent, 578
of salt solutions, 321–322
Batteries, 147
BBB (blood–brain barrier), 850, 878–879
B cells, 881
Becquerel (Bq), 346

Becquerel, Henri, 330
Beer, 703, 704
Belladonna, 476, 859
Benedict's reagent, 495–496
Benedict's test, 496
Bent geometry, 114, 115, 116
Benzaldehyde, 488
Benzaldehyde cyanohydrin, 489
Benzalkonium chlorides, 475
Benz[a]pyrene, 420
Benzene, 406–407, 419–421
Benzenes, substituted, 421–423
Benzoquinone, 489
Benzyl methyl imidazolium chloride, 78
Beryllium, 55, 60
Beta-carotene, 406, 613, 614, 648
Beta decay, 341
Beta emission, 334
β-Hexosaminidase A, 737
β-Hydroxyacyl-CoA dehydrogenase, 763
β-Oxidation pathway, 762–763, 764–765
Beta particles, **331,** 342
Beta radiation, 330–331
Beta rays, 330
Beta-sheet structure, **570,** 571
BHA (butylated hydroxyanisole), 449
BHT (butylated hydroxytoluene), 449
Bicarbonate buffer system, 310, 312–313
Bicarbonate ions, 316
"Big bang," 56
Bile, 754
Bile acids, 733, 739, **754,** 767
Binary compounds, **123**–124
Biochemical pathways, 626, 649, 768
Biochemical reactions, 550, 624–626
Biochemistry, 549–550
Bioethics, 821
Biofilm, 701
Biomolecules, 550
Biosynthesis
fatty acid, 759, 768–770
nonessential amino acid, 835–838
protein, 786, 790, 796–799
Biotin, 610
Birth control pills, 853
1,3-Bisphosphoglycerate, 697, 698
Bisphosphonates, 93
"Black rod wax," 385
Black widow spider venom, 859
Bleaching, 145
Blood. See also Blood plasma; Hemoglobin
ammonium ions in, 471
cholesterol levels, 757
citric acid in, 525
components, 876–878
functions, 876–877
pH, 768, 886
sodium urate in, 140
whole, **876**
Blood alcohol concentration, 446
Blood–brain barrier (BBB), 850, 878–879
Blood clots, **883**
Blood clotting, 877, 882–883
Blood-clotting factors, 767, 883

Blood doping, 263
Blood gases, 872, 883–886
Blood glucose, 705, 706, 709
Blood plasma, **871**
composition, 872
electrolyte concentrations in, 274
in extracellular fluid, 871
filtration of, 887
osmolarity, 281, 874, 875
pH, 312, 886
separation from blood cells, 876
in whole blood, 876
Blood pressure, 228, 284
Blood proteins, 871–872
Blood serum, 601, **876**
Blood sugar, 664–665, 679. See also Diabetes mellitus
Blood types, 672
Blood vessels, 195
Blood volume, 338
Blotting, 817
B lymphocytes (B cells), 881
BNCT (boron neutron-capture therapy), 338
Body fat, 35, 623, 636, 647, 760. See also Triacylglycerols
Body fluids, 871–890. See also Blood; Blood plasma
acid–base balance in, 884–886, 888–890
ammonium ions in, 471
blood–brain barrier, 878–879
blood clotting mechanism, 877, 882–883
blood gas transport, 883–886
body water, 871–874
buffer systems, 310, 312–313
carboxylic acids in, 526
charge neutrality, 86
electrolytes, 273–275
fluid balance, 874–876, 890
gas concentrations, 261
ions in, 81, 82, 86
maintenance of, 887–888
organic compounds in, 473
osmolarity, 283, 871, 890
partial pressures in, 261
proteins in, 871–872
red blood cells, 883–884
role in immunity, 879–882
urine, 140, 887–890
water in, 871–874
white blood cells, 879–883
Body mass index (BMI), 35, 760
Body temperature, 195, 877
Boiling point (bp), **215**
alcohols, 433–434, 439
aldehydes, 490
alkanes, 380, 381
alkenes and alkynes, 402
amides, 521
amines, 468
carboxylic acids, 517
common substances, 243
effect of atmospheric pressure on, 238–239
effect of dipole–dipole forces on, 216

effect of hydrogen bonds on, 219, 434
esters, 520
ethers, 433–434
ionic compounds, 77
ketones, 490
molecular compounds, 107
normal, **238**
phenols, 433–434
simple organic compounds, 359
solutions, 277–278
water, 434
Bombardier beetle, 489
Bond angles, **115,** 116, 117
Bond dissociation energies, **180,** 183
Bond length, **100**
Bones, 86, 93
Boron, 12, 60, 102
Boron neutron-capture therapy (BNCT), 338
Botox, 499, 859
Botulinum toxin, 499, 859
Botulism, 859
Bovine spongiform encephalopathy (BSE), 579–580
Boyle's law, 225–227, **226**
Brain
blood–brain barrier, 850, 878–879
capillaries in, 878–879
fuel for, 694, 705, 706, 767
Branched-chain alkanes, **365,** 377–378
Bread, 703, 704
Breathalyzer test, 446
Breathing. See Respiration
Brittle bone disease, 576–577
Bromine
covalent bonds formed by, 101, 103
as halogen, 55
London dispersion forces in, 217
in periodic table, 11
reaction with aluminum, 30
in unsaturation test, 409
Bromotrifluoromethane, 454
Brønsted, Johannes, 293
Brønsted-Lowry acids, **293**–295
Brønsted-Lowry bases, **293**–295, 467
Brown fat, 647
BSE (bovine spongiform encephalopathy), 579–580
Bt corn, 818
Buffering capacity, 320
Buffers, **308**
Buffer solutions, 308–311, 568
Buffer systems, 310, 312–313
Buret, 317
Butadiene, 385
Butane, 216, 372, 380
1,4-Butanediol, 439
Butanol, 439, 440, 441, 703
1-Butene, 105
Butter, 740
Butylated hydroxyanisole (BHA), 449
Butylated hydroxytoluene (BHT), 449
Butyl groups, 376
Butyric acid, 525, 703

Cadaverine, 452
Caffeine, 476, 879

Calcium, 55, 61, 86, 616
Calcium hydroxide, 292
Calomel, 14
Calorie (cal), 31
Calorie, food, 185, 636
Calorimeter, 185
cAMP (cyclic adenosine monophosphate), 848, 849
Camphor, 491
Cancer
 antioxidant benefits, 614
 causes, 420
 diagnosis, 348
 dietary factors, 565, 679–680
 genomic surveys for, 820
 risk factors, 740, 760
 role of free radicals in, 449
 treatment, 338, 342, 790, 879
Cancer cells, 809
Capillaries, 872–873, 878–879
Capillary walls, 872
Caproic acid, 525
Capsid, 794
Captopril, 607
Caraway seeds, 558
Carbamoyl phosphate, 831
Carbocations, 414–415, 441
Carbohydrate metabolism, 693–715
Carbohydrates, 656–686, **657**
 chitin as, 680
 complex, **658**, 679
 connective tissue as, 681
 in diet, 676, 679–680, 728
 digestion of, 693–694
 as fuel for running, 712
 glycoproteins as, 681–682
 handedness of, 659–660
 heparin as, 681
 metabolism of, 679
 as polymers, 118
 polysaccharides as, 681
 on red blood cell surfaces, 672
 role in tooth decay, 701
Carbolic acid, 422, 447
Carbon
 in amino acid metabolism, 828, 834–835
 covalent bonds in, 102, 105, 109
 electron configuration, 60
 as foundation of organic chemistry, 357
 oxidation of, 637
 substitution patterns, 376, 379
Carbonate ions, 316
Carbon dioxide
 covalent bonds in, 104, 122
 as dry ice, 215
 in fermentation, 703, 704
 as greenhouse gas, 224–225
 in hydrothermal vents, 627
 as solvent, 245
 in supercritical state, 245
Carbon dioxide transport, 884–886
Carbonic acid–bicarbonate buffer system, 310, 312–313
Carbonic anhydrase, 885
Carbon monoxide, 113, 382–383
Carbonyl additions, 506–507

Carbonyl compounds, **485**, 486–487. *See also* Aldehydes; Amides; Carboxylic acids; Esters; Ketones
Carbonyl groups, **443**, **485**
 in acetal hydrolysis, 505
 in alcohol oxidation, 443–444
 in aldehydes, 485
 in carbonyl compounds, 485
 in ketones, 485
 polarity, 485–486, 497
 reactivity, 486
 reduction of, 496–498
 substitution reactions, **516**
Carbonyl-group substitution reactions, **516**
Carbonyl hydrates, 507
Carboxylate anions, **526**, 535
Carboxylate groups, 473
Carboxyl family, 486
Carboxyl groups, **517**, 551
Carboxylic acids, **515**. *See also* Amides; Esters
 acidity, 526–527
 in body fluids, 526
 common, 525
 dissociation constants, 526
 functional groups, 362
 naming, 517–519
 odor, 495
 as organic molecule family, 361, 362
 as products of alcohol oxidation, 444
 as products of aldehyde oxidation, 495
 as products of amide hydrolysis, 536
 properties, 517–519, 522–523
 reactions of, 526–527, 528–529
 strength, 527
 structure, 515–516, 527
 water solubility, 527
Carboxylic acid salts, **526**, 527
Carboxyl-terminal amino acids, **561**
Carboxypeptidase A, 588
Carcinogens, 420
Carnitine, 762
Cast iron cookware, 172
Catabolism, **630**
 amino acid, 827, 828–830, 834–835
 glucose, 704, 764
 lipid, 715, 758, 765–766
 in metabolic pathways, 630, 643, 704–705
 products, 643
 protein, 872
Catalase, 588, 648
Catalysis, 587–589, 595–596
Catalysts, **193**
 effect on equilibria, 201, 203, 204
 effect on reaction rates, 193–194
 enzymes as, 587–589
 in hydrogenation of alkenes and alkynes, 407
Catalytic converters, 193–194
Catalytic effect, 596
Cationic detergents, 731

Cations, **74**
 in body fluids, 872
 formation by periodic table groups, 75, 81
 naming, 82–84
 octet rule for, 79–80
 in room temperature ionic liquids, 78
CAT scan, 348
Caustic potash, 92, 732
Caustic soda, 92, 292, 732
Cech, Thomas, 790
Celebrex™, 746
Celera Genomics, 806, 808
Cell-mediated immune response, 880, 881
Cell membranes
 lipids in, 733–734, 739, 741, 742
 properties, 744
 structure, 741–743
 transport across, 743–744
Cell structure, 628–629
Cellulose
 breakdown of, 78
 in cell walls, 685
 digestibility of, 683, 693
 as polysaccharide, 682
 structure, 682, 683
Cell walls, 685–686
Celsius (°C) scale, 15, 32
Centimeters (cm), 18
Centrifugation, 876
Centromeres, **809**
Cerebrosides, 736
Cesium, 54
CFCs (chlorofluorocarbons), 224
Chain elongation, 769, 770
Chain-growth polymers, 416
Chain reactions, 349, **349**, 350, 449
Chair conformation, 384, 683
Change of state, **6**, **214**
 free-energy change in, 214–215
 liquid–gas, 242, 327–328
 solid–liquid, 242–244
Chaperonins, 600
Chargaff's rule, 783
Charge clouds, 114–117
Charging, 796
Charles's law, 228–**229**
Cheese, 703
Chelation therapy, 606
Chemical bonds, 73, 179–180
Chemical change, **4**, 14–15
Chemical compounds, 7
Chemical elements, **6**. *See also* Main group elements; Nonmetals; Periodic table; Transition metals
 atomic numbers, 48–49
 common ions formed by, 80–82
 diatomic, 101
 essential for human life, 13
 inner transition metals, 52, 53
 metalloids, **11**, 53, 55
 naturally occurring, 9–10, 50
 origins, 56
 properties, 56, 62–63
 symbols, 9
Chemical equations, **133**
 balancing, 134–137

 equilibrium, 196–200
 law of conservation of mass, 133–134
 mass relationships in, 167–168
 mole relationships in, 165–166
 net ionic, 150–152
Chemical equilibrium, **196**
 effect of changing conditions on, 200–204
 equations, 196–200
 Henry's Law, 260–262
 in reversible reactions, 196
Chemical formulas, 10
Chemical messengers, 843–866. *See also* Hormones
 antihistamines, 860–861
 eicosanoids, 745
 histamine, 860
 messenger molecules, 843–844
 monoamines, 861–862
 neuropeptides, 863–864
 neurotransmitters, 855–859
 role in drug discovery/design, 864–866
 role in homeostasis, 845
 in sodium balance, 890
 therapeutic drugs, 861–862
Chemical names, 125. *See also* Prefixes; Suffixes; *specific compounds*
Chemical reactions, 7
 classes, 138–139
 conservation of mass, 133–134
 heat changes during, 180–185, 187–189, 214
 limiting reagents, 169–171, 172
 reaction rate, 190–194
 reverse, 196, 626
 vs. nuclear reactions, 330
Chemistry, **4**
 clinical, 845
 combinatorial, 865–866
 inorganic, 357
 organic, **357**, 366
 relationship to other disciplines, 3–4
Chenodeoxycholic acid, 739
Chernobyl nuclear reactor, 350
Chesebrough, Robert, 385
Chime, 866
Chimeric therapeutics, 879
Chiral carbon atoms, **558**
Chiral centers, 558
Chirality, **557**
 of amino acids, 557–559
 in drug development, 663
 of enzymes, 588
 relation to handedness, 659–660
Chitin, 680
Chloride, 616
Chlorine
 covalent bonds formed by, 103
 elemental, as oxidizing agent, 145
 as halogen, 55
 in halogenation of alkenes, 409
 reaction with sodium, 76, 77
Chlorofluorocarbons (CFCs), 224
Chloroform, 453, 454, 864

Chloromethane, 121–122
Chlorophyll, 649
Chloroplasts, 649
Cholesterol
　in cell membranes, 741, 742
　dietary, 728, 740
　regulating drugs, 757
　as sterol, 733
　structure, 738
Cholesterol levels, 757
Cholesterol synthesis, 759, 767
Cholestyramine, 757
Cholic acid, 739, 754
Choline, 734
Cholinergic nerves, 857
Chondroitin sulfates, 681
Chromatin, 775
Chromosomes, **775,** 785, 808–810, 812
Chronic toxicity, 499
Chylomicrons, 754, 756, 758
Chymotrypsinogen, 608
Chymotrypsin, 595
Cimetidine, 860
Cinnamaldehyde, 491
Circulatory system, 263
Cirrhosis, 767
Cis configurations, 400, 401, 727
Cis-trans isomers, 399–402, **401**
Citrates, 640–641, 876, 525
Citric acid, 525
Citric acid cycle, **640**
　dietary triacylglycerols in, 758, 759
　energy generation in, 631, 704–705
　identifying reactants and products in, 642
　outcomes of, 640, 642
　overview, 639–640
　rate of, 642
　redox reactions in, 638, 640–642
　as stage in metabolism, 630–631, 639
　steps of, 640–642
Citronellal, 491
Citronellol, 125
Citrus fruits, 525
Civetone, 491
CJD (Creutzfeldt–Jakob disease), 579–580
Clean Air Act Amendments (1990), 320–321
Cleavage, 696, 698, 763, 831
Clinical chemistry, 845
Clinical toxicology, 478
Clones, **807**
Clostridium, 703, 499
Clothianidin, 859
Cocaine, 862, 864
Codeine, 477, 864, 879
Codons, 791, **793**
Coefficients, **134**
Coenzyme Q-10 (CoQ), 644
Coenzymes, **589,** 611–612, 637–639
Cofactors, **589**
Coffee beans, 819
Cold packs, 256
Colestipol, 757
Collagen, 93, 574–575, 576–577

Colligative properties of solutions, **275**
　boiling point elevation, 277–278
　freezing point depression, 278–279
　osmotic pressure, 279–282
　vapor pressure lowering, 275–277
Collisions, molecular, 190–193
Colloids, **253,** 254
Colon cancer, 760
Color, chemistry of, 406–407
Combinatorial chemistry, 865–866
Combined gas law, **231**
Combustion, 145, 181, 381–383
Common names
　acids, 517–518
　alkenes and alkynes, 398
　aromatic compounds, 422
Compact bone, 93
Competitive inhibition, **605**–606, 609
Complementary proteins, 564, 835
Complete proteins, 564
Complex carbohydrates. *See* Polysaccharides
Concentration, **193**
　converting to mass, 28
　effect of dilution on, 271–272
　effect on enzyme activity, 597
　effect on equilibria, 201–202, 203
　effect on reaction rates, 193
Concentration gradient, **744**
Concentration units, 262–270
　dilution, 270–272
　equivalents, 274
　molarity, 268–270
　osmolarity, 280–281
　percents, 264–269
Condensation, 769, 770
Condensed structures, **109, 367**
　drawing from names, 379–380
　line structure conversion, 369–371
　omission of shape, 372
　writing, 367–368
Conditional amino acids, 836
Cone cells, 406
Conformations, **372**
Conformers, **372**–374
Coniine, 476
Conjugate acid–base pairs, **295,** 296
Conjugate acids, **295**
Conjugate bases, **295,** 297
Conjugated proteins, **573**
Conjugated systems, 406–407
Conjugation, 406
Connective tissue, 681
Conservation of energy, law of, **181,** 185
Conservation of mass, law of, 133–**134**
Constitutional isomers, **366**
Conversion factors, **26**
　equivalents, 274–275
　mass percent, 265
　molarity, 268–270
　volume percent, 265–266
Coordinate covalent bonds, **106**–107
Coordination compounds, 107
Copper, 12, 66, 767
CoQ (coenzyme Q-10), 644
Cori, Carl and Gerti, 713

Cori cycle, 701, 713–714
Corn, genetically modified, 818
Coronary artery disease, 757
"Corpse flower," 460
Corrosion, 144
Corrosiveness, 296, 525
Cortical bone, 93
Cortisol, 739
Cortisone, 739
Cosmic rays, **341**
Coupled reactions, 634–635, 636–637
Covalent bonds, **99**
　coordinate, **106**–107
　electronegativity and, 120–121, 122–123
　formation of, 99–101, 103
　in Lewis structures, 109–112
　multiple, 104–106, 110–112, 358
　octet rule, 99, 103
　in organic compounds, 358, 359
　periodic table patterns, 101–103, 109
　polar, **119**–123, 358
　sulfur-sulfur, 567
Covalent bonds in, 101
Covalent modification, 608–609
Covalent network solids, 241, 242
COX (cyclooxygenase), 746
CPK (creatine phosphokinase), 601
"Cracking," 396
Creatine phosphate, 712
Creatine phosphokinase (CPK), 601
Crenation, 281
Cretinism, 851
Creutzfeldt–Jakob disease (CJD), 579–580
Crick, Francis, 783, 789
Critical mass, **349**
Critical point, 245
Crystalline solids, 240, **241**
Crystallization, 259
CT scan, 348
Cubic centimeter (cm³), 19
Cubic meter (m³), 15, 16, 18–19
Curie (Ci), 345
Curie, Marie Sklodowska, 330
Curie, Pierre, 330
Curved arrow formalism, 382
Cyanide, 646
Cyanidin, 407
Cyanohydrins, 489
Cyclic adenosine monophosphate (cAMP), 848, 849
Cyclic alkanes, 384
Cyclic AMP, 848, 849
Cyclic ethers, 450
Cyclic GMP, 113
Cyclic hemiacetals, 500–501, 666, 667
Cycloalkanes, **383**–384, 386–387
Cycloalkenes, **397**
Cyclobutane, 384
Cyclohexanone, 488
Cyclooxygenase (COX), 746
Cyclopropane, 384
Cysteine, 452–453, 567
Cytidine, 778
Cytochromes, 644

Cytoplasm, 628, **628**
Cytosine, 777, 782, 783–785
Cytosol, 473, **628,** 696, 697
Cytotoxic T cells, 881

Dacron, 538
Dalton, John, 45, 46, 47
Dalton's law, **236**
Damascenone, 125
Danlos, Henri, 338
Darwin, Charles, 855
D-block elements, **62**
Deadly nightshade, 476, 859
Debridement, 588
Decay series, 340–**341**
Degreasing agents, 454
Dehydration, **440**
　of alcohols, 440, 443
　in athletes, 875
　in lipogenesis, 769, 770
Delocalization, 406
Democritus, 45
Denaturation, **578,** 600
Denatured alcohol, 435
Dendrites, 855
Density, **33**
　of alkanes, 381
　of alkenes and alkynes, 402
　of common materials, 33
　measurement, 33–34
　units, 16
Dental amalgam, 14
Dental caries, 701
Dental plaque, 701
Deoxyadenosine, 778
Deoxycytidine, 778
Deoxyguanosine, 778
Deoxyhemoglobin, 884
Deoxyribonucleic acid. *See* DNA
Deoxyribonucleotides, **779**
2-Deoxyribose, 671, 776
Deoxythymidine, 778
Depression, 861–862
Depth of anesthesia, 453
Dermal patches, 284
Designer steroids, 853–854
Detergents, 578, 731, 733
Deuterium, 50
Dextran, 701
Dextrose, 664–665
D-Glucose, 664–665
DHAP (dihydroxyacetone phosphate), 696, 759, 761
Diabetes insipidus, 876
Diabetes mellitus, **707**
　carbohydrate metabolism in, 707–710
　cirrhosis in, 767
　diagnosis and monitoring, 709
　ketoacidosis in, 708, 767–768
　ketosis in, 767–768
　polymorphisms associated with, 813
　role in metabolic acidosis, 312
1,2-Diacylglycerol, 761
Dialcohols, 437
Dialkenes, 417
Dialysis, 283–284

INDEX

Diamond, 241
Diastereomers, **660**
Diastolic pressure, 228
Diatomic elements, 101
Diazepam, 879
Dicarboxylic acids, 517–518
1,2-Dichloroethane, 409
Dichloromethane, 454
Dicumarol, 647
Diet
   carbohydrates in, 676, 679–680, 728, 762
   cholesterol in, 728, 740
   disaccharides in, 676
   essential amino acids in, 835
   fiber in, 679
   iodine in, 454, 850–851
   linoleic acid in, 724
   linolenic acid in, 724
   lipids in, 728, 757–759
   lysine in, 564
   mercury in, 606
   methionine in, 564
   monosaccharides in, 700
   proteins in, 564–565
   Recommended Dietary Allowances (RDAs), 615
   Reference Daily Intakes (RDIs), 615
   salt in, 83
   threonine in, 564
   tryptophan in, 564
Diethyl ether, 450, 451, 864
Diffusion, facilitated, **744**
Diffusion, simple, **744**
Digestion, **693**
   of carbohydrates, 693–694
   of polysaccharides, 683
   of proteins, 825–826
   role of proteases in, 591
   as stage of metabolism, 630, 631
   of triacylglycerols, 753–755
1,2-Dihaloalkane, 409
Dihydrogen phosphate–hydrogen phosphate buffer system, 310
Dihydroxyacetone phosphate (DHAP), 696, 759, 761
Dilution, 270–272
Dilution factor, **271**
Dimethyl ether, 433–434
2,2-Dimethylpropane, 217
Diol epoxide, 420
Diols, 437
Dipeptides, 551, 560–561, 562–563
Diphosphates, 541
Diphosphoric acid, 541
Dipole–dipole forces, **216**, 218, 219
Diprotic acids, 293, 315
Disaccharides, **658**
   in digestive process, 693
   in food, 676
   formation of, 674–675
   hydrolysis of, 675
   naturally occurring forms, 676–678
   structure, 676
Disease, monogenic, 820
Disinfectants, 447
Dissociation, **296**, 302–303
Dissociation constants, 301, 526

Dissolution, 255–257
Disubstituted amides, 522
Disulfide bonds, **567**
Disulfides, 361, 362, **452**–453
DNA (deoxyribonucleic acid), **775**.
   See also Genomes
   α-helix structure of, 217–218
   base pairing in, 783–785
   bases in, 777, 780
   breakdown of, 140
   in cell division, 775–776
   compared to RNA, 776, 780, 789
   double-helix structure, 783–784, 788
   and heredity, 785–786, 811–812, 831
   mutations, 810–813
   noncoding, 809–810
   nucleosides in, 778, 780
   nucleotides in, 779, 780
   as polymer, 118
   polymorphisms, 811, 812–813
   recombinant, **814**, 814–816
   replication, 785, 786–789
   sugars in, 776
   transcription, 790–792
   viral, 794–795
DNA chips, 820–821
DNA fingerprinting, 817
DNA ligase, 788
DNA polymerases, 786–788
Döbereiner, Johann, 52
Dopamine, 861–862, 879
Dose, 499
Double bonds, **104**
Double helix, **783**–784
Double leaflet structure, 741
D-Ribose, 658, 671, 776
Drug addiction, 862, 864
Drugs, 499, **858**, 860, 864–866
Drug therapy, 820–821
Dry cell batteries, 147
Dry ice, 215
D Sugars, **662**, 664
Duodenum, 754

E1 reactions, 441
E2 reactions, 441
E85 (gasohol), 435
Earth's crust, 10, 331
E. coli, 345
EDTA (Ethylenediaminetetraacetic acid), 606
Eicosanoic acids, 745
Eicosanoids, 723, **745**–746
Eight Belles, 853
Electrical conductivity
   alkali metal–halogen compounds, 74
   electrolytes, 272–273
   ionic compounds, 77
   metallic solids, 241
   organic vs. inorganic compounds, 359
Electrolytes, **273**
   in body fluids, 273–275, 871–872, 884

   conductivity, 272–273
   in sports drinks, 276, 875
   strong, **273**
   weak, **273**
Electromagnetic spectrum, 66, 406–407
Electron affinity, **75**
Electron capture, 335–336
Electron carriers, 639, 643, 645
Electron configurations, **59**–64, 79–80
Electron-dot symbols, **65**
   for alkali metals, 79
   in molecular compounds, 103, 106, 110
   overview, 65–67
Electronegativity, **120**–121, 122–123
Electron octet, 79. *See also* Octet rule
"Electron pushing," 382
Electrons, **46**, 49, 56–58
Electron shells, **57**–58
Electron subshells, **57**
Electron transport, 629, 643–645, 704, 705
Electron-transport chain, **643**–647
Electrophilic aromatic substitution reactions, 424
Electrophoresis, 568, 807, 814
Electrostatic forces. *See* Covalent bonds; Intermolecular forces; Ionic bonds
Electrostatic potential maps, 119–120
Elements. *See* Chemical elements
Elimination reactions, **403**
   dehydration of alcohols, 440, 442–443
   mechanism of, 441
   overview, 403–404
   oxidation of alcohols, 440, 443–445
   types of, 441
Elongation, translation, 797, 798
ELSI (Ethical, Legal, and Social Implications) program, 821
Embden–Meyerhoff pathway, 696
Emulsification, 753–754
Emulsifying agents, 734, 735
Enantiomers, **558**, 588, 660
Endergonic reactions, **188**, **626**
   in biochemical processes, 625, 634–635
   coupling with exergonic reactions, 713
   energy diagrams, 191
Endocrine system, **844**–848
Endorphins, 864
Endosymbiosis, 629
Endothelial cells, 878–879
Endothermic reactions, **180**
   in biochemical processes, 625
   changes of state, 214
   and chemical equilibrium, 202–203
   entropy in, 187
   heat of reaction, 180–185
   in solution process, 256
Endotoxins, 816
Endurance athletes, 712, 875
Enediols, 674
Energy, **29**
   chemical bonds, 179–180

   of common fuels, 182
   heat of reaction, 180–185
   units for, 31, 185
Energy, biochemical, 622–649.
   See also ATP production; Citric acid cycle
   in coupled reactions, 634–637
   electron transport chain, 643–647
   from fatty acid oxidation, 764–765
   flow through biosphere, 623–624
   from food, 185
   from glucose metabolism, 704–706, 764
   in metabolism, 629–632, 636
   overview, 623–626
   oxidized/reduced coenzymes in production of, 637–639
   from photosynthesis, 649
Energy, free. *See* Free-energy change (ΔG)
Energy diagrams, 191–192
Energy effect, 596
Energy investment, 698
Energy transfer
Enflurane, 451, 453
Enkephalins, 863–864
Enolase, 697, 699
Enoyl-CoA hydratase, 763
Enteric coatings, 284
Enthalpy (*H*), **181**
Enthalpy change (ΔH), 214–215, 256, 624–625
Entropy (*S*), **187**
Entropy change (ΔS), **187**–189, 214
Environmental toxicology, 478
Enzyme inhibitors, 604–606, 607
Enzymes, **587**
   allosteric, **603**–604
   catalysis by, 587–589, 595–596
   chirality, 588
   in citric acid cycle, 641
   classification, 590–593
   cofactors, 589
   effect of concentration on, 597
   effect of denaturation on, 578
   effect of pH on, 598–599
   effect of temperature on, 598, 599
   extremozymes, 600
   mechanisms, 594–596
   in medical diagnosis, 601
   regulation, 602–609, 698
   role in second messenger release, 847
   specificity, 588, 594
   thermophilic/psychrophilic, 600
   transaminase, 829
Enzyme specificity, **588**, 594
Enzyme–substrate complex, 594–596
Epidemics, 799, 800
Epinephrine
   autoinjection pens, 850
   in body temperature regulation, 195
   function, 848–850
   role in glycogen breakdown, 706
   role in neural control, 844
   synthesis of, 850–851

INDEX I-7

EpiPen™, 850
EPO (erythropoietin), 263
Equilibrium. *See* Chemical equilibrium
Equilibrium constant *(K)*, **197**–200
Equivalent of acid, **314**
Equivalent of base, **314**
Equivalents (Eq), **273**, 274–275, 313–315
Error analysis, 24
Erythrocytes, **876**, 883, 885
Erythropoietin (EPO), 263
Erythrose, 660
Escherichia coli, 814
Essential amino acids, 564, **835**–838
Esterification, **529**
Esters, **515**
 of fatty acids, 722–723, 724–726
 functional groups in, 362
 naming, 520–521, 779
 as organic molecule family, 361, 362
 properties, 520, 522–523
 reactions of, 528, 534–536
 structure, 515–516, 520–521
Estimating answers, 25–29
Estradiol, 739, 852
Estrogens, 739–740
Estrone, 739, 740, 852
Ethane, 109, 117, 380, 381
Ethanol, 435
 in acetal formation, 501
 in hemiacetal formation, 500
 medical uses, 446
 metabolism, 446
 production, 413, 703
 properties, 255, 433–434, 446, 448
 pyruvate conversion to, 702, 703
Ethanolamine, 734
Ethene, 398
Ethers, 362, **433**, 450–451
Ethical, Legal, and Social Implications (ELSI) program, 821
Ethnobotanists, 864
Ethoxy group, 450, 451
Ethyl acetate, 195–196
Ethyl alcohol. *See* Ethanol
Ethylammonium nitrate, 78
Ethyl chloride, 160, 454
Ethylene
 addition mechanism for, 414–415
 as common name, 398
 molecular geometry, 117
 molecular weight, 159
 multiple covalent bonding in, 105
 from petroleum refining, 385
 as plant hormone, 855
 polymerization, 416
 quantity produced, 396
 reaction with hydrogen, 360
 structure, 399
 uses, 385
Ethylenediaminetetraacetic acid (EDTA), 606
Ethylene gas, 855
Ethylene glycol, 435, 437
Ethyl ester, 520

Ethyl group, **375**, 376
Ethyne, 398
Ethynyl estradiol, 853
Eugenol, 448
Eukaryotic cells, 628
Excitement, of atoms, 66
Excretory system, 140
Exercise, 312, 712
Exergonic reactions, **188, 626**
 in biochemical processes, 625, 713, 714
 energy diagrams, 191
Exons, **792**, 806, 810
Exothermic reactions, **180**
 in biochemical processes, 625
 bond formation in, 180, 181–182
 changes of state, 214
 and chemical equilibrium, 202–203
 heat of reaction, 181–185
 law of conservation of energy, 181
 in solution process, 256
Extracellular fluid, **871**
Extremozymes, 600
Extrinsic pathway, 883
Eye, human, 406, 708

Facilitated diffusion, **744**
Factor-label method, **26**, 27–29, 166
FAD (flavin adenine dinucleotide), 639, 702, 762–763
FADH$_2$, 639, 645–646, 762–764
Fahrenheit (°F) scale, 29–30, 32
Fasting, 706–707, 713
Fat, body, 35, 623, 636, 647, 760
Fatal familial insomnia (FFI), 579
Fats, **727**. *See also* Fatty acids; Lipids; Triacylglycerols
 in diet, 728, 758–859
 metabolism of, 706–707, 756
 properties, 727–729
 triacylglycerols from, 729
Fatty acids, **722**
 biosynthesis of, 759, 768–770
 conversion to acetyl-CoA, 758, 759
 conversion to acyl-CoA, 758, 762
 conversion to/from triacylglycerols, 758, 759, 761, 762
 in digestion, 630, 753, 754
 esters of, 722–723, 724–726
 as fuel during running, 712
 in glucose storage, 694
 in lipid structure, 722
 melting point, 725, 727, 729–730
 oxidation of, 762–765, 768
 polyunsaturated, **724**
 saturated, **724**, 727, 743
 storage, 761
 straight-chain, 722
 structure, 722–723, 724–726
 unsaturated, **724**, 727, 743, 757
 water solubility of, 753, 754, 756
Fatty acid synthase, 769
Favorable reactions
 coupled reactions, 636–637
 energy diagrams, 625
 in metabolic pathway, 634–635
 spontaneous reactions as, 624–625

F-block elements, **62**
Feedback control, **602**–603, 609, 642
Feedstock, 385
Female sex hormones, 852
Fermentation, 435, 703–**704**
Fiber, 679–680
Fibrin, **882**
Fibroin, 571
Fibrous proteins, 570–572, **571**
Fight-or-flight hormone. *See* Epinephrine
Filtration (kidney), **887**
Fingerprinting, DNA, 817
Fireworks, 66
Fischer projections, **661**–662, 665–666, 667
Fish, 819, 831
Fish farming, 819
Fission, 349–350
Flammability
 aldehydes, 492, 493
 alkanes, 381
 alkenes and alkynes, 402
 aromatic hydrocarbons, 420
 cycloalkanes, 384
 ethers, 450, 451
 ketones, 492
Flavin adenine dinucleotide (FAD), 639, 702, 762–763
Florigene, 820
Flu, 799–800
Fluid balance, 874–876, 890
Fluid-mosaic model, 741
Fluorapatite, 93
Fluoride, 73
Fluorine, 55, 60, 101, 120
Fluorite, 240
Fluoxetine, 861
Food. *See also* Metabolism
 caloric values, 185, 636
 fermentation, 703
 irradiation, 345
Food and Nutrition Board, 615
Food calorie (Cal), 31, 636
Food labels, 615
Forensic toxicology, 478
Formaldehyde
 molecular geometry, 115–116
 polymers, 493
 properties, 490, 492, 499
 structure, 488
 uses, 492–493
Formalin, 492
Formic acid, 514, 525
Formula units, **87**
Formula weight, **159**
Forward reaction, 196
Francium, 54
Franklin, Benjamin, 164
Free-energy change (Δ*G*), **188**
 in biochemical reactions, 624–625
 in changes of state, 214–215
 in chemical equilibrium, 201–202
 in coupled reactions, 634–635
 effect on reaction rate, 190–192
 in energy diagrams, 191
 in nonspontaneous processes, 186–189

 in phosphate hydrolysis, 633
 in spontaneous processes, 188, 189
Free radicals, 449, **614**, 647, 648
Freezing, 240
Freezing point, 278–279
Fructose
 in food, 670, 671
 metabolism of, 693–694, 700
 structure, 658, 671
Fructose 1,6-bisphosphate, 696, 697, 714
Fructose 6-phosphate, 696, 697, 700, 714
Fruit, ripening of, 345, 396, 592, 855
Fruit sugar. *See* Fructose
Fuels, energy values of, 182
Fukushima nuclear reactor, 350
Fumarase, 592, 593
Fumarate, 641–642, 831
Functional group isomers, **366**
Functional groups, organic, **360**
 in biomolecules, 473
 characteristics, 359–363, 389
 drawing, 363
 in families of organic molecules, 362, 395
 flow scheme, 389
 identifying, 362–363
Fungicides, 454–455
Furosemide, 890
Fusion, nuclear, **349**, 351

Galactose, 670–671, 693–694, 700
Galactosemias, 671, 700
Galileo thermometer, 33
Gallbladder, 754
Gamma emission, 335
Gamma globulins, 882
γ-Glutamyl transferase (GGT), 601
Gamma radiation, 330–331
Gamma rays, 330–331, 335, 342
Gangliosides, 737
Gas constant *(R)*, **234**
Gases, 5. *See also* Change of state; Gas laws
 in body fluids, 261
 ideal, **221**
 kinetic-molecular theory of, 220–221
 particle attraction in, 213
 properties, 220
 solubility of, 259–262
Gas gangrene, 703
Gas laws
 Avogadro's law, 232–233
 Boyle's law, 225–227
 Charles's law, 228–229
 combined gas law, 231
 Dalton's law, 236
 Gay-Lussac's law, 230
 ideal gas law, 233–235
Gasohol (E85), 435
Gastro-esophageal reflux disease (GERD), 299
Gatorade, 875
Gaucher's disease, 737
Gay-Lussac's law, **230**
GDP (guanosine diphosphate), **640**, 848

Geiger counters, 343–344
Gel electrophoresis, 568, 807, 814
Genes, **776**
   expression, 786
   function, 775
   human genome, 788, 805–808, 810
   mapping, 810
   transcription process, 791–792
Gene therapy, 820
Genetically modified organisms (GMOs), 818–820
Genetic code, 791, **793**–794
Genetic control, **609**
Genetic maps, 806, 807
Genetic markers, 806, 807
Genetics, 760, 785–786, 811–812, 831
Genomes, **788**
   human, 788, 805–808, 810
   viral, 794–795
Genomics, 804–821, **818**
   applications, 818–821
   bioethics, 821
   chromosomes, 775, 785, 808–810, 812
   DNA fingerprinting, 817
   genome mapping, 805–808
   mutations, 810–813
   polymerase chain reactions, 814, 815–816, 817
   polymorphisms, 811, 812
   recombinant DNA, 814–816
Genomic screening, 820–821
Genomic surveys, personal, 820
Geometry, molecular, 114–117
Geraniol, 125
GERD (gastro-esophageal reflux disease), 299
Gerhardt, Charles, 8
Gestrinone, 854
GGT (γ-glutamyl transferase), *601*
GHG (greenhouse gases), 224–225
Glacial acetic acid, 525
Global warming, 224–225
Globular proteins, 570–572, **571**, 573
Glomerular filtrate, **887**
Glomerulus, 887
Glucagon, 706
Glucocerebrosidase, 737
Glucocorticoids, 739, 852
Glucogenic amino acids, 835
Gluconate, 709
Gluconeogenesis, **713**
   in fasting and starvation, 706–707
   from non-carbohydrates, 713–715
   pyruvate–glucose conversion by, 701, 706, 714
   in triacylglycerol metabolism, 758, 759
Glucosamine sulfate, 681
Glucose
   in blood, 705, 706, 709, 872
   conversion from glycerol, 706
   conversion to pyruvate, 701
   cyclic form of, 501
   as fuel for brain, 694, 705, 706, 767
   interaction with hexokinase, 594
   metabolic function, 670
   naming of, 657

   storage, 694, 710
   structure, 657, 658, 664–667
Glucose metabolism
   digestion, 693
   energy output, 704–706, 764
   glycogenesis pathway, 694
   glycolysis pathway, 696–699
   overview, 694–695
   pentose phosphate pathway, 694
   regulation, 705–706, 739
Glucose oxidase, 709
Glucose 6-phosphatase, 711, 714
Glucose phosphate
   in glucose metabolism, 694
   in glycogenesis, 710–711, 767
   in glycogenolysis, 711, 767
   in glycolysis pathway, 696, 697, 698
Glucose-tolerance test, 709
Glucosyltransferase, 701
Glutamate, 555, 561, 829, 837
Glutamine, 837
Glutaric acid, 518
Glycemic index, 679
Glyceraldehyde, 659, 660, 662
Glyceraldehyde 3-phosphate, 696, 697, 698, 700, 759
Glycerin. *See* Glycerol
Glycerol
   conversion to glucose, 706, 713
   intestinal absorption of, 754
   as source of pyruvate, 701
   in triacylglyceride metabolism, 759
   uses, 435
   water solubility, 754
Glycerol 3-phosphate, 761
Glyceroneogenesis, 761
Glycerophospholipids, 723, 733, **734**, 743
Glycine, 739, 879
Glycogen
   function, 684–685, 694
   metabolism, 710–713
   as polysaccharide, 682
   storage, 767
   structure, 682, 684, 685
Glycogenesis, 694, **710**–711
Glycogenolysis, 710, **711**, 712
Glycogen phosphorylase, 711, 849
Glycogen synthase, 711
Glycolic acid, 528
Glycolipids, 723, **733**, 736–737, 741, 742
Glycols, **437**
Glycolysis, **696**
   aerobic/anaerobic, 702
   energy generation, 698, 704–705
   energy investment, 698
   in phosphate ester formation, 676
   role of isomerases in, 591–592
   during running, 712
   as source of lactate, 703, 713
   as source of pyruvate, 701
   steps in, 696–699
   sugars in, 700
   in triacylglycerol metabolism, 758, 759
Glycolysis pathway, 696–699, 714
Glycoproteins, **681**–682, 742

Glycosides, **674**–675
Glycosidic bonds, **675**
Goiter, 454, 851
Gold, 12
"Golden rice," 819
Gout, 140, 833
G protein, 847, 848–849
Grain alcohol. *See* Ethanol
Gram–atom conversions, 47
Gram-equivalents (g-Eq), **273**–275
Gram-mole conversions, 163–165
Grams (g), 15, 18
Gray (Gy), 346
Greenhouse effect, 224
Greenhouse gases (GHG), 224–225
Grehlin, 760
Ground state, of atoms, 66
Groups, periodic table, **52**–53, 54–55, 81
Guanine, 777, 783–785
Guanosine, 778
Guanosine diphosphate (GDP), **640**
Guanosine triphosphate (GTP), **640**, 848–849
Guanylyl cyclase, 113
Guar gum, 600
Gums, 670, 679
Gypsum, 240

H_2-receptor blockers, 299
Hair perms/rebonding, 453
Half-life ($t_{1/2}$), **337**–340, 350
Halite, 83
Halogenation, **409**, 425
   of alkanes, 383
   in alkenes, 409
   of aromatic compounds, 425
Halogen-containing compounds, 454–455
Halogens, **55**
   anion formation by, 75, 81
   characteristics, 55
   compounds with alkali metals, 73–74
   covalent bonding in, 109
   electron configurations, 63
   electron-dot symbol, 79
   electronegativity, 120
Halothane, 453, 454
Handedness
   of amino acids, 557–559, 662
   of carbohydrates, 659–660
   concept of, 556–557
   relation to chirality, 659–660
Haptens, 879
Haworth projections, 666
Hazards, chemical, 465–466
HbA (normal adult hemoglobin), 568
HbS (sickle-cell trait hemoglobin), 568
HDLs (high-density lipoproteins), 756, 757
Heart attacks, 601, 746, 757, 760
Heart disease, 740, 757
Heat, **179**, 578
Heating curves, 243
Heat of fusion, **242**, 243, 244
Heat of reaction (ΔH), **181**–185, 187–189, 214

Heat of vaporization, 239, **242**, 243
Heat stroke, 195, 276
Helical secondary structures, 566
Helicases, 786–787
Helium, 55, 60, 65
Helper T cells, 881
Heme, 573–574, 617, 884
Heme groups, 644
Hemiacetal groups, 677, 678
Hemiacetals, **500**
   in acetal formation, 501–502, 674
   cyclic, 666, 667
   formation of, 500–501, 502, 505, 665–666
   identifying, 502
Hemicellulose, 679, 685
Hemiketals, 501
Hemlock, poison, 476
Hemodialysis, 283–284, 831
Hemoglobin (Hb)
   in anemia, 172
   gel electrophoresis of, 568
   iron in, 617
   role in carbon dioxide transport, 883, 884–886
   role in oxygen transport, 263, 883–884
   in sickle-cell anemia, 562, 568
   structure, 573–574
Hemolysis, 281
Hemophilia, 882–883
Hemostasis, **883**
Henderson–Hasselbalch equation, **309**
Henry's law, 260–262, **261**
Heparin, 681, 876
Hepatocytes, 754, 758
1-Heptanol, 439
Herbicides, 454–455
Heredity, 785–786, 811–812, 831
Heroin, 477, 879
Heterocycles, **469**
Heterocyclic amines, 463–464, 471
Heterocyclic nitrogen compounds, 469–470
Heterogeneous mixtures, **6**, **253**, 254
Heterogeneous nuclear RNAs (hnRNAs), *792*
Hexamethylenediamine, 538
Hexokinase, 594, 696, 697
High-density lipoproteins (HDLs), 756, 757
High-density polyethylene, 417
Hippocrates, 8
Histamine, 461, 471, 860, 880–881
Histidine, 836
"Hitting the wall," 712
HIV-1 (human immunodeficiency virus)
   AIDS, 607, 795
   computer modeling of, 865
   drug therapy, 312
   effect of AZT treatment on, 607
   as retrovirus, 795
   virus particles, 16
HMG-CoA lyase, 766
HMG-CoA synthase, 766
Hoffman, Felix, 8
Homeostasis, 761, 767, 845, 874, 886
Homogeneous mixtures, **6**, **253**, 254

Honeybees, death of, 859
Hormones, **844**
  amino acid derivatives, 846, 847, 850–851
  antidiuretic, 845, 875–876, 890
  direct release of, 845
  interaction with receptors, 847
  mechanism, 848–850
  plant, 855
  polypeptides, 846, 847, 850, 851–852
  regulatory, 845–846
  role in endocrine system, 844–848
  sex, 739, 852
  steroid, 739–740, 846, 847, 852–854
  thyroid, 846
Human body. *See also* Blood; Blood plasma; Body fluids; Respiration
  blood pressure, 228, 284
  bones, 86, 93
  carbon monoxide in, 113
  effect of ionizing radiation on, 342, 346
  elemental composition, 10
  essential elements in, 13
  ions required by, 86
  nitric oxide in, 113
  teeth, 86
  temperature regulation, 195
Human genome, 788, 805–808, 810
Human Genome Project, 806–807, 808
Human immunodeficiency virus. *See* HIV-1
Hyaluronate molecules, 681
Hydration, 256, **413**, 763
Hydrocarbons, **360**–362, 382–383
Hydrochloric acid, 92, 119, 292, 410–412
Hydrocortisone, 739
Hydrogen
  covalent bonds in, 101, 102, 109
  electron configuration, 60
  in enzyme-catalyzed redox reactions, 638
  in hydrogenation of alkenes and alkynes, 407–408
  isotopes, 50
  in organic molecules, 358
  oxidation numbers, 148
Hydrogenation, 407–**408**, 730
Hydrogen bonds, **217**
  in aldehydes, 491
  in α-helix structure, 217–218
  in amides, 521–522
  in amines, 467–468
  in carboxylic acids, 517
  in DNA double-helix, 784, 788
  effect on alcohol properties, 439, 447
  effect on boiling point, 219, 434
  effect on protein shape, 565–566
  as intermolecular forces, 217–220
  in ketones, 491
Hydrogen bromide, 410–411, 414
Hydrogen chloride, 160
Hydrogen cyanide, 110–111, 122
Hydrogen ions
  in acids and bases, 91–92
  in electron-transport chain, 644, 645
  in enzyme-catalyzed redox reactions, 638
  in photosynthesis, 649
  in polyatomic ions, 85
Hydrogen peroxide, 145, 647, 709
Hydrogen sulfide, 627
Hydrohalogenation, **410**–412
Hydrolases, 591, 731
Hydrolysis, **505**
  acetal, 504–506, 507
  amide, 534–535, 536–537
  ATP, 632–633
  catalysts for, 591
  in digestion, 693, 825–826
  disaccharide, 675, 700
  enzyme catalysis in, 595–596
  ester, 534–536
  protein, 577
  triacylglycerol, 731–732, 759, 761, 762
  in urea cycle, 831
Hydrometers, 34, 36
Hydronium ion (H$_3$O$^+$), **291**
  in acids and bases, 91, 302–307
  in buffer solutions, 308–309
  molecular shape of, 117
Hydrophilicity, 439, 554
Hydrophobicity
  alcohols, 439
  amino acid side chains, 554, 566–567
  cholesterol, 739
  cholic acid, 754
  hormone–receptor interaction, 847
  triacylglycerols, 727, 729
Hydrothermal vents, 627
Hydroxide anion (OH$^-$), 91–92, 291, 302–305
Hydroxide ions, 316
Hydroxyapatite, 93
Hydroxybenzene, 422
3-Hydroxybutyrate, 765, 766
3-Hydroxybutyrate dehydrogenase, 766
Hydroxyl free radical, 647
Hygroscopic compounds, **258**, 284
Hyperammonemia, 831
Hyperbaric oxygen treatment, 703
Hyperglycemia, **705**
Hyperthermia, 195
Hypertonic solutions, **281**
Hyperventilation, 312, 313
Hypochlorhydria, 299
Hypoglycemia, **705**, 708
Hypothalamus, 844–846
Hypothermia, 195
Hypotonic solutions, **281**

Ibuprofen, 163, 865
Ice, 186–187, 240
Ideal gases, **221**
Ideal gas law, **233**–235
Ideograms, 806, 807
IDLs (intermediate-density lipoproteins), 756
Imaging procedures, 348
IM forces. *See* Intermolecular (IM) forces
Immune response, 877, 879–882, **880**
Immunoglobulin E, 882
Immunoglobulin G antibodies, 882
Immunoglobulins, **880**, 881, 882
Incomplete amino acids, 835
Index number, 396
Indole, 421
Induced-fit model, **594**
Inflammation, 739
Inflammatory response, **879**–881
Influenza, 799–800
Informational strands, 791, 792, 794
Inhibition, enzyme, **602**, 604–607, 609
Initiation, translation, in protein synthesis, 797, 798
Initiator, 416
Inner transition metals, **52**, 53
Inorganic chemistry, 357
Inorganic compounds, 359
Insecticides, 454–455, 606
Insects, 489
Insoluble fiber, 679
Insulin
  appetite regulation by, 760
  in biochemical research, 568
  as chemical messenger, 844
  disulfide bonds in, 567–568
  as polypeptide hormone, 852
  recombinant DNA in manufacture of, 816
  role in blood glucose regulation, 706
Insulin-dependent diabetes, 707–708
Insulin resistance, 708
Insulin shock, 708
Integral proteins, 741, 742
Intermediate-density lipoproteins (IDLs), 756
Intermolecular (IM) forces, **216**. *See also* Covalent bonds; Hydrogen bonds; Ionic bonds
  in biomolecules, 554, 565–567
  during changes of state, 242–243
  dipole–dipole, 216, 219
  effect on properties, 107
  in gases, 220–221
  hydrogen bonds, 217–219
  in liquids, 237–240
  London dispersion, 217, 219
  in solids, 241
Internal environment, 845
Internal radiation therapy, 338
International Union of Pure and Applied Chemistry. *See* IUPAC
International unit (IU), 601
Interstitial fluid, **871**
Intracellular fluid, **871**
Intramolecular forces. *See* Covalent bonds; Ionic bonds
Intrinsic pathway, 883
Introns, 790, **792**, 806
Invert sugar, 678
In vivo nuclear procedures, 338
Iodine
  covalent bonds formed by, 101, 103
  deficiency, 454, 851
  in diet, 454, 850–851
  as halogen, 55
  in periodic table, 12
Ionic bonds, **77**
  effect of electron affinity on, 75
  effect of electron configurations on, 79–80
  effect of ionization energy on, 75
  electronegativity and, 120, 121
  and melting/boiling points, 77
Ionic compounds, 73–93, **77**. *See also* Ionic bonds
  balancing of charges, 76, 86
  common applications, 89
  common elements in, 81
  compared to molecular compounds, 107
  formation of, 73–74
  formulas for, 86–88, 90
  in human bone, 93
  naming, 89–91
  properties, 77–78, 107
  solubility guidelines for, 139–140
  in stalactites and stalagmites, 72
Ionic detergents, 731
Ionic equations, **150**
Ionic formulas, 86–88, 108
Ionic liquids, 78
Ionic solids, **77**, 78, 241, 242, 256
Ionic solubility switches, 473
Ionization energy, **75**, 76, 144, 342
Ionizing radiation, 341–346
Ion-product constant for water (K_w), 302
Ions, **74**. *See also* Ionic compounds
  biologically important, 86
  of common elements, 80–82
  formation of, 73–74, 75–76, 81–82
  naming, 82–84
  octet rule and, 79–80
  oxidation numbers, 148
  polyatomic, **85**, 90
  in solution, 272–273
Iron, 617, 883–884
Iron-deficiency anemia, 172
Irradiation, of food, 345
Irreversible inhibition, 606, 609
Isocitrate, 640–641
Isoelectric point (pI), **555**–556
Isoenzymes, 601
Isoflurane, 451, 453
Isoleucine, 836
Isomerases, 591–592
Isomerization, 697, 698, 699
Isomers, **365**
  cis-trans, 399–402
  drawing, 366
  examples, 365–366
  types, 366
  vs. conformers, 373–374
Isoprene, 398
Isopropyl alcohol, 435
Isopropyl group, **376**
Isotonic solutions, **281**
Isotope dilution, 338

Isotopes, **50**
  atomic mass, 51
  atomic number, 50, 51, 329
  atomic weight, 50–51
  mass number, 50, 329
  stable, 331–332, 347
  unstable, 331–332, 336
  vs. conformations, 373–374
IU (international unit), 601
IUPAC (International Union of Pure and Applied Chemistry) nomenclature
  alcohols, 436
  aldehydes, 488
  alkanes, 374–375
  alkenes and alkynes, 396
  carboxylic acid, 517
  halogen-containing compounds, 454
  ketones, 488

Joint European Torus (JET), 351
Jones, Marion, 853
Joule (J), 31
Juvenile hormone, 451
Juvenile-onset diabetes, 708

Kcalorie (kcal), 185, 636
Kelvin (K), 15, 16, 29–30
Keratin, 217–218
Ketals, 501
Ketoacidosis, 708, 765–**768**
Keto–enol tautomerism, 674
Ketogenesis, **765**
Ketogenesis pathway, 759, 765–766
Ketogenic amino acids, 835
Ketone bodies, **765**
  acetyl-CoA conversion to, 707, 759, 765
  formation in ketogenesis, 766
  as fuel for brain, 767
  in low-carbohydrate diets, 762
  overproduction in diabetes, 767–768
Ketonemia, 767
Ketones, **486**
  aromas of, 485
  carbonyl groups in, 485
  common, 493–494
  functional groups in, 362
  lack of oxidation, 494
  in monosaccharides, 658
  naming, 488
  as organic molecule family, 361, 362
  as product of alcohol oxidation, 444
  properties, 490–492
  reactions of, 496–498, 500–507
  structure, 486–487
  toxicity, 492
Ketonuria, 767
Ketoses, **658**, 674
Ketosis, 494, 762, 767
Kevlar, 539
Kidneys
  as backup to bicarbonate buffer system, 312, 313
  glycerol conversion in, 759
  role in fluid balance, 875, 890
  role in urine formation, 887–888

Kidney stones, 140
Killer T cells, 881
Kilocalorie (kcal), 31
Kilogram (kg), 15, 16, 18
Kinases, 591, 609
Kinetic energy, **179**, 213
Kinetic-molecular theory of gases, **220**–221
Kinetics, 190–194
Krebs cycle. *See* Citric acid cycle
Krypton, 55, 66
Kuru, 579
Kwashiorkor, 564–565
Kwolek, Stephanie, 539
Kyoto Protocol to the United Nations Framework Convention on Climate Change (UNFCCC), 224

Lactate
  conversion of pyruvate into, 701, 702, 712
  conversion to glucose, 713–714
  as product of anaerobic glycolysis, 703, 713
Lactate dehydrogenase (LDH), 601
Lactic acid
  buildup of, 312, 497–498
  naming of, 525
  in skin treatments, 528
  structure, 517
Lactose, 677
Lactose intolerance, 677
Lagging strands, 788
Lamivudine (3TC), 795
Lanthanides, 53
Large calories (Cal), 31
Large subunit, 797
Lasix®, 890
Laudanum, 477
Lauric acid, 724
Law of conservation of energy, **181**, 185
Law of conservation of mass, 133–**134**
LD$_{50}$ (lethal dose), 499
LDH (lactate dehydrogenase), 601
LDLs (low-density lipoproteins), 756, 757
L-Dopa, 879
Leading strands, 788
Lead poisoning, 605–606
Le Châtelier's principle, **200**–204, 261, 507
Lecithin, 734, 735
Length, 15, 16, 18
Leptin, 760
Lethal dose (LD$_{50}$), 499
Leucine, 836
Leu-enkephalin, 863–864
Leukocytes, 876, 880, 881
Leukotrienes, 745, 746
Levulose. *See* Fructose
Lewis bases, **467**
Lewis structures, **108**–112, 117
Ligases, 592–593
Light, polarized, 558, 659–660
Light, visible, 66
Light echoes, 56
Lignin, 679, 685
"Like dissolves like" rule, 255

Limewater, 292
Limiting reagents, **169**–171, 172
Lind, James, 577
Line-angle structures. *See* Line structures
Linear geometry, 114, 115
Line structures, **369**
  converting condensed structures to, 369–370
  converting to condensed structures, 370–371
  for cycloalkanes, 386, 387
  representing shape in, 114–117, 372
  structural formulas, 106, 108
Linoleic acid, 724, 726
Linolenic acid, 724
Lipase, 754, 759
Lipid bilayer, **741**, 742, 744
Lipid metabolism, 752–770
  fatty acid biosynthesis, 768–770
  fatty acid oxidation, 762–765
  ketoacidosis, 765–768
  lipid transport, 755–756
  role of liver in, 767
  TAG mobilization, 762
  triacylglycerol digestion, 753–755
  triacylglycerol metabolism, 758–759
  triacylglycerol synthesis, 761
Lipids, **721**, 721–746. *See also* Fats; Fatty acids; Triacylglycerols
  in cell membranes, 733–734, 739
  classification, 721–723
  density, 756
  in diet, 728
  emulsification of, 753–754
  in glucose synthesis, 701–702, 706
  glycolipids, 723, **733**, 736–737, 741, 742
  identifying components of, 737
  oils, **727**–729
  pathways through villi, 755
  phospholipids, **733**, 734–736, 741, 742
  role in atherosclerosis, 757
  role in energy storage, 721, 767
  sterols, 723, 738–740
  structure, 721–723
Lipid transport, 755–756, 872
Lipogenesis, 759, **768**–770
Lipoproteins, 753–**754**, 755–756, 757
Liposomes, **741**
Liquids, **5**. *See also* Change of state
  intermolecular forces in, 237–240
  ionic, 78
  particle attraction in, 213
  properties, 237–239
Lister, Joseph, 447
Liter (L), 15
Lithium, 54, 60, 66
Lithium–iodine batteries, 147
Litmus, 308
Liver. *See also* Gluconeogenesis
  fructose metabolism in, 700
  functions, 767
  glycerol conversion in, 759
  glycogenolysis in, 711, 713
  lactate–pyruvate conversion in, 701

  in lipid metabolism, 755
  lipogenesis in, 759, 768–770
  pathological conditions, 767
  pyruvate–glucose conversion in, 701
  role in triacylglycerol digestion, 754
  role in urea cycle, 831
Liver function, 312, 446
Lock-and-key model, 594
London dispersion forces, **217**, 219, 380, 402
Lone pairs, **108**, 121
Low-carbohydrate diets, 762
Low-density lipoproteins (LDLs), 756, 757
Low-density polyethylene, 417
Lowry, Thomas, 293
L Sugars, **662**, 664
Lubricants, 385
Lung volume, 226–227. *See also* Respiration
Lyases, 592
Lye (sodium hydroxide), 92, 292, 731
Lymph, 873
Lymphatic system, 873
Lymph capillaries, 873
Lymphocytes, 881
Lysine, 564, 836
Lysozyme, 685

MAC (minimum alveolar concentration), 453
Macrominerals, 616
Mad-cow disease, 579–580
Magnesium, 55, 616
Magnesium hydroxide, 292
Magnetic resonance imaging (MRI), 348
Main group elements, **52**
  covalent bonds formed by, 102
  in ionic compound names, 89
  octet rule for, 79, 82
  in periodic table, 53
Malaria, 562
Malate, 641–642
Male sex hormones, 852
Malleability, 241
Malnutrition, 724
Malonyl-CoA, 768–769
Maltose (malt sugar), 676–677, 703
Mammals, 830
Mannose, 700
Manometers, 221, 222
MAO (monoamine oxidase) inhibitors, 861
Marasmus, 565
Marble, 316
Margarine, 730, 740
Marijuana, 862
Markers, genetic, 806, 807
Markovnikov's rule, 410–412, 415
Mass, **17**. *See also* Mole-mass relationships
  converting from concentration, 28
  converting to volume, 34
  mole–mass conversions, 167, 168
  percent concentration, **264**, 265, 267–268
  units for, 18

Mass–equivalent conversion, 315
Mass–mass conversions, 167, 184
Mass/mass percent concentration (m/m)%, **264**, 265, 267–268
Mass–mole conversions, 167, 168
Mass–mole relationships. *See* Mole–mass relationships
Mass number *(A)*, **49**, 50, 329
Mass ratios, 160
Mass–volume conversion, 34
Mass/volume percent concentration (m/v)%, **264**–265, 266, 267–268
Material Safety Data Sheets (MSDS), 465–466
Matter, **4**, 6–7
Matter, states of, 5–**6**
Mauverine, 407
Measurement, 15–35
   atmospheric pressure, 221
   blood pressure, 228
   body fat, 35
   density, 33–34
   energy, 31
   length, 18
   mass, 17–18
   rounding numbers, 23–24
   scientific notation, 21–23
   significant figures, 19–21
   specific gravity, 34
   specific heat, 31–32
   temperature, 29–32
   volume, 18–19
Medications, 284, 473
Melting point (mp), **215**
   alkanes, 380, 381
   amides, 521
   amino acids, 555
   common substances, 243
   effect dipole–dipole forces on, 216
   fatty acids, 725, 727, 729–730
   ionic compounds, 77
   molecular compounds, 107
   periodicity, 54
   phenols, 447
   simple organic compounds, 359
   triacylglycerols, 727
Memory T cells, 881
Mendeleev, Dmitri, 52
Mercaptans (thiols), 361, 362, 452–453
Mercuric nitrate, 14
Mercury, 14, 606
Mercury poisoning, 14
Messenger molecules, 843–844
Messenger RNAs (mRNAs), **790**–792, 796–799
Metabolic acidosis, 312–313, 886
Metabolic alkalosis, 886
Metabolic pathways, 629–632
Metabolic syndrome, 707–708, 709
Metabolism. *See also* Glucose metabolism; Lipid metabolism
   amino acid, 826–830
   ATP/energy transfer strategies, 632–634
   basal, 636–637
   biochemical energy in, 629–634, 636

carbohydrate, 679, 693–715, 707–710
citric acid cycle in, 630–631, 639
copper, 767
in diabetes mellitus, 707–710
of endurance athletes, 712
ethanol, 446
in fasting and starvation, 706–707, 713
fructose, 693–694, 700
galactose, 693–694, 700
glycogen, 710–713
mannose, 700
metabolic pathway/coupled reaction strategies, 634–635
oxidized/reduced coenzyme strategies, 637–639
protein, 827
role of digestion in, 630, 631
Metal alloys, 254
Metal carbonates, 316
Metal hydroxides, 292, 316
Metallic solids, 241, 242
Metalloids, **11**, 53, 55
Metallurgy, 145
Metals, **11**
   cation formation, 81
   cation naming, 82–84
   electronegativity, 120
   in ionic compound names, 89
   oxidation numbers, 148
   in periodic table, 53, 55
   redox behavior, 144–145
Metarhodopsin II, 406
Met-enkephalin, 863–864
Meters (m), 15, 16
Methane
   covalent bonds in, 101
   as greenhouse gas, 224
   molecular geometry, 116
   physiological effects, 381
   properties, 380
   structure, 399
Methanethiol, 452
Methanol
   acidity, 448
   addition to propionaldehyde, 507
   hydrophobic part of, 439
   production, 385
   properties, 435
   toxicity, 435, 605
   uses, 434
Methanol poisoning, 605
Methionine, 564, 836
Methoxyflurane, 453
Methoxy group, 450
Methyl alcohol. *See* Methanol
Methylamine hydrochloride, 474
Methylammonium chloride, 474
Methylbenzene, 422
3-Methylbutanal, 488
Methyl ester, 520
Methyl ethyl ketone, 488
Methyl group, **375**, 376
Metric units, 15–17
MI (myocardial infarctions), 601, 746, 757, 760
Micelles, **732**, 753, 754

Microcurie (μCi), 345
Microgram (μg), 18
Micronutrients, 589, 616–617
Mifepristone, 853
Milk, fermentation of, 703
Milk of magnesia, 292
Milliequivalents (mEq), 274
Milligram (mg), 16, 18
Millikan, Robert, 48
Milliliter (μL), 19
Millimeter (mm), 18
Millimeter of mercury (mmHg), 221
Millimolar (mM), 268
Millipedes, 489
Mineralocorticoids, 739, 852
Mineral oil, 381
Minerals, 589, 615, 616–617
Minimum alveolar concentration (MAC), 453
Mirror images, 556–557
Miscibility, **258**. *See also* Solubility
Mitochondria, **629**, 643–645, 649, 714
Mitochondrial matrix, **629**, 702, 762
Mixtures, **6**, 254
Mixtures, heterogeneous, **6**, **253**, 254
Mixtures, homogeneous, **6**, **253**, 254
Molar amount *(n)*, 233–235
Molarity (M), **268**–270
Molar mass, **160**
   calculating, 161, 163–165
   in mole-mass conversions, 167, 168
Molecular compounds, 99–125, **101**. *See also* Covalent bonds; Organic compounds
   binary, **123**–124
   compared to ionic compounds, 107
   electron-dot symbols, 103, 106, 110
   formulas, 108, 124
   Lewis structures, 108–110
   multiple covalent bonds in, 104–106
   naming, 123–124, 125
   oxidation numbers, 148–149
   polar covalent bonds in, 119–120
   properties, 107
   shapes, 114–117
Molecular disorder, 187
Molecular formulas, **108**, 124
Molecular geometry, 114–117
Molecular polarity, 121–123, 216, 219–220
Molecular solids, 241, 242
Molecular weight (MW), **159**–160
Molecules, **99**
   diatomic, 101
   nonpolar, 216, 219, 742
   polar, 121–123, 216, 219–220
   shapes of, 114–117
   size of, 118
Mole–gram conversions, 163–165
Mole–mass conversions, 167, 168
Mole–mass relationships, 159–173
   Avogadro's number, 159–162
   gram–mole conversions, 163
   limiting reagents, 169–171, 172
   mass relationships, 167–168
   mole relationships, 165–166
   percent yields, 169–170

Mole–mole conversions, 167
Mole ratios, 166, 167, 184–185
Moles, **160**
Mole/volume concentration, 268–270
Monoamine oxidase, 861
Monoamine oxidase (MAO) inhibitors, 861
Monoamines, 861–863
Monogenic diseases, 820
Monomers, **415**
Monoprotic acids, 293
Monosaccharides, **657**. *See also* Fructose; Glucose
   classifying, 658, 659, 662
   in digestive process, 693
   examples, 669–671
   galactose, 670–671, 693–694, 700
   mannose, 700
   in nucleotides, 776
   properties, 669
   reactions of, 673–676
   structure, 657–658, 664–667
   testing for, 674
Monosubstituted amides, 522
Morphine, 477, 863, 879
Morton, William, 453
Motrin, 865
MRI (magnetic resonance imaging), 348
mRNAs (messenger RNAs), **790**–792, 796–799
MSDS (Material Safety Data Sheets), 465–466
Mullis, Kary, 815
Multiple covalent bonds, 104–106, 110–112, 358
Muriatic acid, 92, 119, 292, 410–412
Muscarine, 478
Muscles
   acetoacetate use by, 767
   fructose metabolism in, 700
   glycogen breakdown in, 706
   glycogenolysis in, 711
   lactate production by, 713
   lactic acid buildup in, 312, 497–498
   neurotransmitter control of, 857
Mushrooms, poisonous, 499
Mutagen, **811**
Mutarotation, 666
Mutations, 810–813, **811**
Mylar, 538
Myocardial infarctions (MI), 601, 746, 757, 760
Myoglobin, 573, 617
MyPlate, 565, 679

N-Acetylglucosamine (NAG), 685
N-Acetylmuraminic acid (NAMA), 685
NAD⁺ (nicotinamide adenine dinucleotide)
   in body fluids, 473
   in ethanol metabolism, 446
   in glycolysis, 697
   in metabolic processes, 497, 638–639, 640, 645
   in pyruvate metabolism, 702

NADH
  in aerobic/anaerobic glycolysis, 702
  in ATP production, 645–646
  as electron carrier, 639, 645
  in fatty acid oxidation, 763–764
  in oxidative phosphorylation, 763–764
  as product of glycolysis, 696
  as reducing agent, 497–498, 638
NADP+ (nicotinamide adenine dinucleotide phosphate), 638, 649
NADPH (nicotinamide adenine dinucleotide phosphate), 638, 649, 694, 769
NAG (N-acetylglucosamine), 685
NAMA (N-acetylmuraminic acid), 685
Naphthalene, 420
Naproxen, 663, 865
Narcotics, 493
National Academy of Sciences-National Research Council, 615
National Human Genome Research Institute, 813, 821
Native proteins, **572**
Natural radioisotopes, 331
N-Butyl group, 376
Necrosis, 703
Negative nitrogen balance, 564
Neon, 55, 60, 66
"Neon" lights, 66
Nephrons, 887
Net ionic equations, **150**–152
Neurons, 855–856
Neuropeptides, 863–864
Neurotoxins, 478, 499
Neurotransmitters, **844**
  histamine as, 860–861
  mechanism of, 857–859
  monoamines as, 861–862
  in neurotoxin mechanisms, 478
  synthesis of, 855–857
Neutralization reactions, 138, **141**–142, 318
Neutrons, **46**, 49
Niacin deficiency, 610
Nicotinamide adenine dinucleotide. See NAD+; NADH
Nicotine, 470, 476, 859, 862, 879
Niemann-Pick disease, 737
Nightshade, 476, 859
Nitration reaction, 425
Nitric acid, 92, 292, 320
Nitric oxide, 113, 320–321
Nitric oxide synthases (NOS), 113
Nitrogen
  amino nitrogen, 828–830
  covalent bonds in, 101, 102, 104, 109
  from dietary protein, 564
  electron configuration, 60
  in organic molecules, 358
  in periodic table, 12
Nitrogen compounds, heterocyclic, 469–470
Nitrogen dioxide, 113, 320–321
Nitrogenous bases, 776, 777
Nitrogen oxide, 828

Nitrous oxide, 453
Noble gases, **55**, 63, 79
Noncoding DNA, 809–810
Noncovalent forces, 554, 572
Nonelectrolytes, **273**
Nonessential amino acids, **835**–838
Non-GMO Project, 819
Non-insulin-dependent diabetes, 707–708
Non-ionic detergents, 731
Nonmetals, **11**
  anion formation, 75, 81
  in coordination compounds, 107
  electronegativity, 120
  oxidation numbers of compounds, 149
  in periodic table, 53, 55
  redox behavior, 144–145
Nonpolar molecules, 216, 219, 742
Nonpolar solvents, 255
Nonspontaneous processes, 186, 187, 188. See also Endergonic reactions
Norepinephrine, 861–862
Norethindrone, 853
Normal boiling point, **238**
Normality (N), **314**
NOS (nitric oxide synthases), 113
N-Propyl group, 376
N-Terminal amino acids, **561**
NTPs (nucleoside triphosphates), 786–788
Nuclear chemistry, 329
Nuclear decay, **332**–336, 337–340, 350
Nuclear equations, 332–333, 334, 336
Nuclear fission, **349**–350
Nuclear fusion, **349**, 351
Nuclear magnetic resonance imaging, 348
Nuclear medicine, 338, 348
Nuclear power, 349–350
Nuclear reactions, **329**
  balanced equations, 332–333, 334, 336, 347
  half-life ($t_{1/2}$), **337**–340, 350
  spontaneous, 329–330
  vs. chemical reactions, 330
Nuclear strong force, 47
Nucleic acid chains, 781–782
Nucleic acids, **776**
  base pairing, 783–785
  complementary sequences, 785
  composition of, 776–778, 780
  function, 775, 782
  naming of, 780
  nucleic acid chains, 781–782
  role in heredity, 785–786
  structure, 781–782
Nucleon, **329**
Nucleosides, 777–778, 780
Nucleoside triphosphates (NTPs), 786–788
Nucleotides, **776**
  in DNA, 779–780, 806
  in RNA, 779–780
  structure, 776, 779–780, 781–782
Nucleus, **46**
Nuclide, **329**

Nutrition Facts labels, 615, 679, 728
Nylons, 537–538

Obesity, 35, 728, 760
Occupational Health and Safety Administration (OSHA), 465–466
Octet rule, **79**–80, 82, 99, 102–103
Odors
  aldehydes, 485, 491, 493
  carboxylic acids, 517
  esters, 520
  ketones, 485, 708, 767
  thiols, 452
  volatile amines, 468
Oils, 727–729
Oil spills, 359
Okazaki fragments, 788
Oleic acid, 360, 724, 726
Oligonucleotides, 815
Oligosaccharides, 672
Open-chain alkanes, 383, 384
Opium, 477
Opsin, 406
Optical activity, 660
Optical isomers, **558**, 663
Orbital-filling diagrams, 61
Orbitals, **57**–58, 59
Organelles, 628
Organic chemistry, **357**, 366
Organic compounds. See also specific compounds
  as conjugated systems, 406–407
  as denaturation agents, 578
  vs. inorganic compounds, 359
  water solubility, 473
Organic molecules. See also Functional groups, organic
  conjugated systems, 406–407
  drawing, 363, 366, 367–371
  families, 359–363
  identifying functional groups, 362–363
  multiple covalent bonding in, 105
  nature of, 357–359
  shapes, 358, 372–374
  structure, 357–358
Organic oxidation, 444
Organic phosphates, 540–542
Organic reactions. See also Addition reactions; Elimination reactions
  alkanes, 381–383
  alkenes, 407–418
  balancing, 383
  oxidation, 444–445
  rearrangement, 404–405
  substitution, 404, 424, 425, 516
Organic reduction, 444
Organophosphorus insecticides, 859
Orientation effect, 596
Origins of replication, 786–787
OSHA (Occupational Health and Safety Administration), 465–466
Osmolarity (osmol), **280, 871**
  of body fluids, 283, 871, 890
  effect of proteins on, 283
  examples, 280–281
Osmosis, **280**, 281, 283

Osmotic pressure (π), 279–282, **280**, 283–284
Osteogenesis imperfecta, 576–577
Osteoporosis, 93
Oxalate ion, 876
Oxalic acid, 518
Oxaloacetate, 640–642
Oxidation, 142. See also Redox reactions
  alcohols, 443–445, 446, 494–495
  aldehydes, 494–496, 673–674
  carbon, 637
  coenzymes, 637–639
  fatty acids, 762–765
  malate, 638
  organic, 444
  pyruvate, 702, 704–705
  thiols, 452–453
Oxidation numbers, **148**–149
Oxidation-reduction reactions, **138**. See also Redox reactions
Oxidative deamination, **829**
Oxidative phosphorylation
  in ATP synthesis, 631, 645–646, 704, 712
  blockers, 646–647
  dietary triacylglycerols in, 758, 759
  energy output, 704, 705, 763–764
  uncouplers, 646–647
Oxidizing agents, 143. See also NAD+
  in bleaching, 145
  characteristics, 144
  in electron-transport chain, 645
  FAD as, 639
  identifying, 146, 148
  monosaccharide reactions with, 673–674
  oxygen as, 495
Oxidoreductases, 590–591
Oxygen
  covalent bonds in, 101, 102, 105, 109
  electron configuration, 60
  harmful by-products of, 647–648
  in organic molecules, 358
  as oxidizing agent, 495
Oxygen masks, 137
Oxygen transport
  in dialysis, 283
  gas concentration in blood, 263
  mechanism, 883–884
  role of iron in, 617
Oxyhemoglobin, 884

Pain relief, 863–864
Palmitic acid, 724, 726, 769
Pancreas, 693, 706, 708
Pancreatic lipases, 754, 759
Pancreatitis, 608
Pandemics, 799
Papain, 588
Para-aramids, 539
Paracelsus, 477, 499
Paraffin wax, 381, 385
Paraformaldehyde, 492–493
Paregoric, 477
Parent compounds
  alcohols, 436
  alkanes, 374–375

alkenes and alkynes, 396–397
amines, 462
halogen-containing compounds, 454
Parkinson's disease, 862, 879
Partial pressure, **234**, 260–262, 263
Parts per billion (ppb), **267**
Parts per million (ppm), **267**–268
Pascals (Pa), 221
Passive transport, **743**–744, 872
Pasteur, Louis, 703
P-Block elements, **62**
PCR (polymerase chain reaction), 814, 815–816, 817
Pectins, 670, 679, 680, 685
Pellagra, 610
PEM (protein-energy malnutrition), 564
Penicillin, 685, 686
Pentane, 217
Pentose phosphate pathway, **694**
PEP (phosphoenolpyruvate), 697, 699
Pepsinogen, 826
Peptide bonds, **551**
Peptidoglycan, 685, 686
Percent concentrations, 264–268
Percent saturation, 283, 884
Percent yield, **169**–170
Periodicity, 54
Periodic table, **11**
  atomic numbers, 52–53
  blocks, 62–63
  covalent bonding trends, 101–103
  electron configuration trends, 62–63
  electronegativity trends, 120
  groups, 52–55, 62–63
  ion formation trends, 80–82
  ionization energy trends, 75, 76
  isotope stability trends, 331, 332
  organization, 11–13
  periods, 52–53
Periods, in periodic table, **52**
Peripheral proteins, 741
Peroxidase, 709
Peroxides, 450
PET (poly(ethylene terephthalate)), 538
PET (positron emission tomography), 348
Petrochemicals, 385
Petroleum jelly, 381, 385
P Function, **304**
pH, **304**. See also Acidity; Acids; Bases; Basicity
  of body fluids, 310, 312
  in buffer solutions, 309–311
  of common substances, 304
  converting to H_3O^+ concentration, 306
  effect on amino acid charge, 555–556
  effect on enzyme activity, 598–599
  effect on protein charge, 568
  measuring, 303–305, 308
  relationship to H^+/OH^- concentrations, 305
  role in tooth decay, 701

Phagocytes, 648, 881
Phase change. See Change of state
Phase diagrams, 245
Phenacetin, 865
Phenelzine, 861
Phenol (hydroxybenzene), 422, 447
Phenolphthalein, 308
Phenols (family of compounds), **433**, 447–450
Phenyl, **422**
Phenylalanine, 837
Phenylephrine hydrochloride, 473
Phenylketonuria (PKU), 838
Phosphatase enzymes, 609
Phosphate diesters, 540–541
Phosphate esters, **540**, 675–676
Phosphate groups, 473, 776
Phosphate ion ($HOPO_3^{2-}$), 697
Phosphate monoesters, 540–541
Phosphates, organic, 540–542
Phosphatidic acid, 761
Phosphatidylcholines, 735
Phosphoenolpyruvate (PEP), 697, 699
Phosphofructokinase, 696, 697
Phosphoglucomutase, 710, 711
Phosphoglycerate kinase, 697, 698
Phosphoglycerate mutase, 697, 699
Phosphoglycerides, 723, 733, 734, 743
Phospholipids, **733**, 734–736, 741, 742
Phosphor, 343
Phosphoric acid, 92, 292, 540–542
Phosphoric acid anhydrides, 540
Phosphoric acid ester, 515
Phosphorus, 102, 103, 616
Phosphorylation, **541**. See also Oxidative phosphorylation
  of ADP, 632–634
  determining type of, 647
  of fructose, 700
  glucose–hexokinase interaction, 594
  in glycolysis pathway, 697, 699
  substrate-level, 699
Phosphoryl groups, **541**, 609
Photosynthesis, 625, 649
Phototropism, 855
Physical change, **4**
Physical maps, 806, 807
Physical quantities, **15**–16
Physiological saline solution, 281
Phytohormones, 855
Pineapple, 520
PKU (phenylketonuria), 838
Plane-polarized light, 659–660
Plant gums, 670, 679
Plant hormones, 855
Plants, cell wall structure in, 685
Plants, drugs derived from, 864
Plaque, arterial, 757
Plaque, dental, 701
Plasma. See Blood plasma
Plasmalogens, 737
Plasma proteins, 767
Plasmids, 814–816
Plastics, 118
Platelets, 883
Pleated sheet secondary structures, 566
Poison hemlock, 476

Poisons, 478. See also Toxic substances
Polar covalent bonds, **119**–123, 358
Polarimeters, 660
Polar molecules, 121–123, 216, 219–220
Polar solvents, 255
Pollutants, 113, 320–321, 492
Polyamides, 537–539
Polyatomic ions, **85**, 90
Polycyclic aromatic compounds, 420
Polyesters, 538
Polyethylene, 118, 396, 416, 417, 538
Polymerase chain reaction (PCR), 814, 815–816, 817
Polymer chains, 118
Polymerization, 416–418, 493
Polymers, **415**–418, 417, 493
Polymorphisms, **811**, 812
Polynucleotides, 776, 783, 784, 788, 789
Polypeptide hormones, 846, 847, 851–852
Polypeptides, 551, 573–574, 826
Polypropylene, 416
Polysaccharides, **658**
  as carbohydrates, 681
  in diet, 679
  in digestive process, 693
  examples, 682–685
  identifying sugars in, 668
  role in tooth decay, 701
Polystyrene, 416
Polyunsaturated fatty acids, **724**
Poly(vinylchloride) plastics (PVC), 409
P Orbitals, 57–58
Positive nitrogen balance, 564
Positron emission, 335
Positron emission tomography (PET), 348
Postsynaptic neurons, 855–856
Post-translational modification, 797
Potash, 731
Potassium, 54, 61, 66, 616
Potassium dichromate, 446
Potassium hydroxide, 92, 732
Potassium iodide, 73, 77, 454
Potatoes, 476–477, 819
Potential energy, **179**
PPI (proton-pump inhibitors), 299
Precipitates, **138**
Precipitation reactions, 138–141
Prefixes
  amino-, 463
  bi-, 85
  in chemical names, 124
  deoxy-, 777–778
  Greek letters, 517
  halo-, 454
  m- (*meta*), 421
  mono-, 124
  o- (*ortho*), 421
  in organic compound names, 374, 375
  p- (*para*), 421
  sec-, 376
  tert-, 376
  in units, 17

Pressors, Angiotensin II as, 607
Pressure *(P)*, 221
  atmospheric, 221
  effect on equilibria, 203
  effect on solubility of gases (Henry's law), 260–262
  relation to temperature of gases (Gay-Lussac's law), 230
  relation to V, T, and n (ideal gas law), 233–235
  relation to volume of gases (Boyle's law), 225–227
  unit conversions, 222–223
Presynaptic neurons, 855–856
Primary (1°) alcohols, 437, 441, 496
Primary (1°) amines, 461, 462, 467–468
Primary (1°) carbon atom, **376**, 379
Primary protein structures, **560**–563, 575
Primers, 815
Prions, 579–580
Products, **7**, 133
  in chemical equations, 133
  in chemical equilibrium, 201–202
  of chemical reactions, 14–15
  major/minor, 441–443
  stability of, 179–180, 195–196
Proelastase, 608
Proenzymes, 608
Progesterone, 852
Progestins, 852
Prokaryotic cells, 628
Propane
  boiling point, 433–434
  chirality, 557–558
  physiological effects, 381
  properties, 380
Propene, 398
Properties, **4**, 5, 62–63
Propylene, 385, 398, 416
Propylene glycol, 435, 437
Propyl group, **376**
Prostaglandins, 745–746
Protease inhibitors, 607
Proteases, 591, 607
Protein analysis, 568
Protein catabolism, 872
Protein-energy malnutrition (PEM), 564
Protein metabolism, 827
Proteins, **551**, 559–580. See also Enzymes
  amino acids in, 552–554
  in body fluids, 871–872
  chemical properties, 577–578
  classes, 552, 576
  coding in DNA, 806, 809, 810
  conjugated, **573**
  conversion to glucose, 706
  denaturation/renaturation of, 578
  in diet, 564–565, 835
  digestion of, 825–826
  effect on osmolarity of body fluids, 283
  fibrous, 570–572
  function, 552
  globular, 570–572, 573

Proteins (continued)
    in glucose synthesis, 701–702, 706
    hydrolysis of, 577
    integral, 741, 742
    misfolding of, 578
    peripheral, 741
    as polymers, 118
    in polymorphisms, 812–813
    primary structures, 560–563, 575
    quaternary structures, 552, **573**, 573–575
    secondary structures, 552, 566, **569**–572, 573, 575, 579
    shape-determining interactions, 565–569
    simple, **573**
    tertiary structures, 552, **572**–573, 575
    translating RNA into, 794
Protein synthesis, 796–799
Protium, 50
Proton-pump inhibitors (PPI), 299
Protons, **46**, 49
Proximity effect, 596
Prozac™, 861
Prusiner, Stanley, 579
Psychrophiles, 600
Puffer fish, 478
Pulse oximetry, 884
Pure substances, **6**
Purine bases, 777
PVC (poly(vinylchloride) plastics), 409
Pyramidal geometry, 114, 116
Pyridine, 421
Pyrimidine bases, 777
Pyrophosphoric acid, 541
Pyruvate
    biochemical transformations of, 701–705, 712
    in glycolysis pathway, 696, 697, 701
Pyruvate dehydrogenase complex, 702
Pyruvate kinase, 697, 699
Pyruvic acid, 497–498

Quantization, 57
Quantum mechanical model, 56
Quaternary ammonium ions, **462**
Quaternary ammonium salts, **475**
Quaternary (4°) carbon atom, **376**, 379
Quaternary protein structures, 552, **573**–575
Quinine, 470, 476, 864
Quinone ring structure, 644

R (organic substituent), 377, 437
R′ (organic substituent), 437
R″ (organic substituent), 437
Racemic mixtures, 663
Rad, 346
Radiation
    detecting, 73, 343–344
    intensity vs. distance, 342
    ionizing, **341**–343, 345, 346
    measuring, 344–346
    types of, 330–331
Radiation therapy, 338
Radicals, 416, 449, **614**, 647, 648
Radioactive decay, 332–336, 337–340, 350

Radioactive decay series, 340–**341**
Radioactive half-life ($t_{1/2}$), **337**, 339–340, 350
Radioactivity, **330**–332, 336–338. *See also* Radiation
Radiocarbon dating, 339
Radioisotopes, **331**, 332, 339, 348
Radiolytic products, 345
Radionuclides, **331**
Radiopharmaceutical agents, 338
Radium, 55
Radium-226, 341
Radon, 55, 341
Radon-222, 341
Rainwater, 320–321
Randomness, 187–189, 214
Ranitidine, 861
Rational drug design, 860
RBE (relative biological effectiveness) factor, 346
RDAs (Recommended Dietary Allowances), 615
RDIs (Reference Daily Intakes), 615
Reabsorption (kidney), **887**–888
Reactants, **7, 133**
    in chemical equations, 133
    in chemical equilibrium, 201–202
    of chemical reactions, 14–15
    concentration of, 193, 194
    representing in organic reactions, 383
    stability of, 179–180, 195–196
Reaction energy diagrams, 191–192
Reaction mechanisms, **415**
Reaction rate, 190–194, **191**
Reagents, 383
Rearrangement reactions, **404**–405
Receptors, **843**, 847
Recombinant DNA, **814**–816
Recommended Dietary Allowances (RDAs), 615
Red blood cells
    blood gas transport by, 883–886
    hemoglobin in, 172, 263, 617
    lactate production by, 703, 713
    osmolarity surrounding, 281
    in sickle-cell anemia, 561–562, 568, 811
    surface of, 672
Redox reactions, **138**. *See also* Oxidation; Reduction
    in batteries, 147
    as class of reactions, 138
    combustion as, 381–383
    in metabolic processes, 637–639, 640–642
    overview, 142–146
    recognizing, 148–149
Reducing agents, **143**, 144, 146, 497–498, 638
Reducing sugars, **674**, 678
Reduction, **143**. *See also* Redox reactions
    of aldehydes and ketones, 496–498
    of carbon, 637
    of coenzymes, 637–639
    in lipogenesis, 769, 770
    organic, 444
Reductive deamination, **837**

Reference Daily Intakes (RDIs), 615
Regular tetrahedrons, **116**
Regulatory hormones, 845–846
Relative atomic mass scale, 46
Relative biological effectiveness (RBE) factor, 346
Relative mass scale, 46
Releasing factors, 797
rem, 346
Renaturation, 578
Reperfusion, 648
Replication, **786**
Replication fork, 788
Residues, **561**
Resonance, **419**
Respiration
    Boyle's law in, 226–227
    in oxygen transport process, 263
    rapid, 708–710
    as redox reaction, 145
Respiratory acidosis, 312–313, 886
Respiratory alkalosis, 886
Respiratory burst, 648
Respiratory chain, 643–647
Restriction endonuclease, 814, 817
Restriction enzymes, 806, 814, 817
Restriction fragment length polymorphism (RFLP), 817
Retinal, 406, 613
Retinoic acid, 613
Retinol, 406, 613
Retroviruses, 795
Reverse reactions, 196, 626
Reverse transcriptase, 794–795
Reverse transcription, 794–795
Reversible inhibition, 195–196, 604–606, 609
Reversible reactions, **196**
RFLP (restriction fragment length polymorphism), 817
R groups, 433, 566–567
Rhodopsin, 406
Ribonuclease, 572–573
Ribonucleic acid. *See* RNA
Ribonucleotides, **779**, 793
Ribose, 658, 671, 776
Ribose 5-phosphate, 694
Ribosomal RNAs (rRNAs), **790**, 791
Ribosomes, **790**, 796–797
Ribozyme, **790**
Rice, genetically modified, 819
Rickets, 610
Ritonavir, 607
RNA (ribonucleic acid), **776**
    bases in, 777, 780
    breakdown of, 140
    compared to DNA, 776, 780, 789
    function, 776
    hnRNA, 792
    mRNA, 790–792, 796–799
    nucleosides in, 778, 780
    nucleotides in, 779, 780
    rRNA, 790, 791
    structure, 789
    sugars in, 776
    synthesis, 790–792
    translating into protein, 794
    tRNA, 790, 791, 796–799
    viral, 794–795

RNA polymerase, 791
Rock salt, 83
Rod cells, 406
Roentgen (R), 345, 346
Room temperature ionic liquids (RTILs), 78
Roses, genetically modified, 820
Rotenone, 646
Rounding off, 23–25, **24**
rRNAs (ribosomal RNAs), **790**, 791
RTILs (room temperature ionic liquids), 78
RU-486, 853
Rubber, synthetic, 385
Rubbing alcohol, 435
Rubidium, 54
Running, biochemistry of, 712
Rutherford, Ernest, 48, 330, 331

SA (salicylic acid), 8, 528, 532–534
Saccharomyces cerevisiae, 703
Salicin, 8
Salicylic acid (SA), 8, 528, 532–534
Saline solution, 281
Saliva, 693
Salmon, genetically engineered, 819
Salmonella, 345
Salt, dietary, 73, 83, 454
Salt bridges, 566
Salts, chemical, **138**, 578
Salt solutions, 321–322
Saponification, **535**, 732
Saquinavir, 795
Sarin, 606
Saturated compounds, 258, 283, **395**–396, 884
Saturated fatty acids, **724**, 727, 743, 757
Saturated solutions, **258**
S-Block elements, **62**
Scanning tunneling microscope (STM), 48
Schizophrenia, 862
Schrödinger, Erwin, 56
Scientific method, **4**, 8, 48
Scientific notation, **21**–23
Scintillation counters, 343
Scopolamine, 864
Scurvy, 575, 576–577, 610
sec-Butyl group, 376
Second (s), 15, 16
Secondary (2°) alcohols, 437, 441, 444, 496
Secondary (2°) amines, 461, 462, 467–468
Secondary (2°) carbon atom, **376**, 379
Secondary protein structures, **569**
    alpha-helix, 569–570, 575
    beta-sheet, 570, 575
    in disease-causing prions, 579
    in fibrous and globular proteins, 570–572
    fibrous proteins, 570–572
    helical, 566
    pleated sheet, 566
    as structural level, 552
Second messengers, 847, 848–849
Secretion (kidney), **887**–888

Selective serotonin re-uptake inhibitors (SSRIs), 861
Selenium, 614
Semiconservative DNA replication, 788
Semimetals, 11, 53, 55
Semipermeability, 279–280
Senescence, 809
S-Enzyme 1, 768–769
Serine, 734
Serotonin, 471, 861–862
Serum cholesterol, 757
Sex hormones, 739, 852
Shell, valence, 63, 64
Shells, electron, 57–58
Shortenings, hydrogenation of, 730
SIADH (syndrome of inappropriate antidiuretic hormone secretion), 876
Sickle-cell anemia, 561–562, 568, 811
Side chains, amino acid, **551**
   acidity, 553, 554
   effect on protein structure, 566–569
   identifying function of, 596
   in nucleotides, 782
   polarity, 554
   in primary protein structure, 560
Side reactions, 169
Sievert (Sv), 346
Significant figures, **20**
   in antilogarithms, 306
   calculations using, 24–25
   in experimental measurement, 19–21
   in scientific notation, 22
Silicon, 12
Silk, 571
Simple diffusion, **744**
Simple proteins, **573**
Simplest formula, 87
Simple sugars. *See* Monosaccharides
Single bonds, **104**
Single-nucleotide polymorphism (SNP), **812**–813
Sister chromatids, 809
SI units, **15**–17, 346
Skin-care treatments, 528
Skunk odor, 452
Slaked lime, 292
Small subunit, 797
Smog, 113, 492
SNP (single-nucleotide polymorphism), **812**–813
Soaps, **731**–733
Sodium
   as alkali metal, 54
   in biological measurement, 66
   in body fluids, 890
   electron configuration, 61
   as macromineral, 616
   reaction with chlorine, 76, 77
Sodium chlorate, 137
Sodium chloride
   dissolution process, 256
   formation of, 76, 77
   properties, 74, 77
   in salt production, 83
   in soap, 732

Sodium fluoride, 73, 93
Sodium hydroxide, 92, 292, 732
Sodium hypochlorite, 145
Sodium iodide, 73
Sodium urate, 140
Solanine, 476–477
Solid hydrates, 257–258
Solids, **5**. *See also* Change of state
   molecular arrangements, 240–241
   particle attraction in, 213
   properties, 241–242, 258–259
   types of, 242
Solubility, **139**, **258**
   acylglycerols, 754
   aldehydes, 490
   alkanes, 381
   alkenes and alkynes, 402
   amides, 522
   amine salts, 474
   amino acids, 555
   aromatic hydrocarbons, 420
   bile acids, 739
   biological problems, 140
   carboxylic acid salts, 527
   citric acid, 525
   cycloalkanes, 384
   effect of denaturation on, 578
   effect of pressure on, 260–262
   effect of temperature on, 258–260
   ethers, 450, 451
   gases, 259–262
   glycerol, 754
   inorganic compounds, 359
   ionic compounds, 78
   ketones, 490
   "like dissolves like" rule, 255
   lipids, 721
   organic compounds, 359, 473
   phenols, 447–448
   proteins, 556, 571
   of solids, 258–259
   strength of particle attractions and, 255
   of water, 255–256
Solubility guidelines, 139–141
Solubility switches, 473
Soluble fiber, 679, 680
Solutes, **254**
Solution process, 255–257. *See also* Solubility
Solutions, 252–284, **253**. *See also* Concentration units; Solubility; Temperature
   boiling point elevation, 277–278
   characteristics, 254
   colligative properties, **275**
   entropy change, 187–189, 214
   examples, 254
   freezing point depression, 278–279
   Henry's law, 260–262
   hypertonic, **281**
   hypotonic, **281**
   ions in, 272–273
   isotonic, **281**
   "like dissolves like" rule, 255
   metal alloys as, 254
   osmotic pressure, 279–282
   saturated, **258**
   supersaturated, **259**

types, 255
   vapor pressure lowering, 275–277
Solvation, **256**
Solvents, **254**
   acetone, 493
   carbon dioxide, 245
   ethers, 450
   halogenated compounds, 454
   ketones, 490
   polarity of, 255
   production by halogenation of alkanes, 383
Somatic cells, 809
s Orbitals, 57–58
Sorbitol, 708
Soybeans, genetically modified, 818
Spanish flu, 800
Spearmint leaves, 558
Specific gravity, **34**
Specific heat, **31**, 32, 239
Specificity, enzyme, **588**, 594
Spectator ions, **150**
Speed, 16
Sphingolipids, **733**, 736, 737
Sphingomyelinase, 737
Sphingomyelins, 723, **733**, 736, 737
Sphingosine, 723, **733**, 736
Sphygmomanometers, 228
Spider webs, 571
Spiral, β-oxidation, 762, 763, 764–765
Spliceosome activity, 790
Spliceosomes, 792
Spongy bone, 93
Spontaneous processes, **186**. *See also* Exergonic reactions
   in biochemical reactions, 624–625
   examples, 186–187
   free energy change in, 188, 189
Sports drinks, 276, 875
SSRIs (selective serotonin re-uptake inhibitors), 861
Stability
   of isotopes, 331–332, 336, 347
   of reactants and products, 179–180, 195–196
   in reversible reactions, 195–196
Stable isotopes, 331–332, 347
Stalactites, 72
Stalagmites, 72
Standard molar volume, **233**
Standard temperature and pressure (STP), **233**
Starches, 682, 683–684, 693
Stars, 351
Starvation, 702, 706–707, 713, 767
States of matter, 5–**6**, 245. *See also* Change of state; Gases; Liquids; Solids
Statins, 757
Stearic acid, 525
Stereochemistry. *See* Chirality
Stereoisomers, **558**, 660, 661
Steroid hormones, 739–740, 846, 847, 852–854
Steroids, 853–854
Sterols, **738**. *See also* Cholesterol
   bile acids, 733, 739, **754**, 767
   characteristics, 723
   overview, 738–740

steroid hormones, 739–740, 846, 847, 852–854
Sticky ends, 814
St. Louis University School of Medicine, 879
STM (scanning tunneling microscope), 48
Stomach acid, 299
Stop codons, 793
STP (standard temperature and pressure), **233**
Straight-chain alcohols, 439
Straight-chain alkanes, **365**, 375–377, 380, 381
Straight-chain fatty acids, 722
Straight-chain polyethylene, 417
Stratosphere, 224
Streptococcus mutans, 701
Streptococcus sanguis, 701
Stress, 201
Stroke, 8, 228, 708, 760
Strong acids, **296**
   calculating pH for, 305, 306–307
   determining strength of, 296–300
   K_a values for, 301
   reaction with strong bases, 322
Strong bases, **297**
   calculating pH for, 307
   determining strength of, 296–300
   reactions with strong acids, 322
   salts of, 322
Strong electrolytes, **273**
Strontium, 55, 66
Structural formulas, 106, **108**. *See also* Line structures
Styrene, 416
Subatomic particles, **46**
Sublimation, 215
Subshells, electron, **57**
Substituents, **374**, 375, 377, 437
Substituted amides, 522, 530
Substitution reactions, **404**, 424, 425, 516
Substrate-level phosphorylation, 699
Substrates, **588**, 594–596, 597
Subunits, 797
Succinate, 641–642
Succinic acid, 518
Succinyl-CoA, 640–641
Sucrose, 670, 676, 677–678, 701
Suffixes
   -al, 488
   -amide, 522
   -amine, 462
   -ammonium, 471
   -ane, 375, 396
   -ase, 592–593
   -ate, 526
   -benzene, 447
   -diene, 396
   -dioic acid, 518
   -ene, 396
   -ide, 454
   -idine, 777
   -ium, 471
   -oic acid, 517
   -one, 488
   in organic compound names, 374, 375

Suffixes (continued)
  -ose, 658
  -osine, 777
  -oyl, 518
  -phenol, 447
  -thiol, 452
  -triene, 396
  -yl, 561
  -yne, 396
Sugar, invert, 678
Sugar molecules, 661–662
Sugars. See also Disaccharides; Monosaccharides; Polysaccharides
  common examples, 670
  in diet, 676, 679
  in DNA and RNA, 776
  families, 661–664
  identifying, 668
  reducing, **674**, 678
  sweetness, 669
Sugar substitutes, 669
Sulfate groups, 731
Sulfides, 361, 362
Sulfonation, 425
Sulfur, 12, 102–103, 567, 616
Sulfur dioxide, 112, 115–116, 320–321
Sulfuric acid, 92, 292
Sulfur trioxide, 320
Sun, 351
Sunlight, 613, 623, 625, 627, 649
Suntory Limited, 820
Supercritical state, 245
Supernovas, 56
Superoxide dismutase, 648
Superoxide ion, 647
Supersaturated solutions, **259**
Supplements, dietary, 681, 826
Surface tension, 239
Surfactants, 731
Sweating, 195
Sweetness, 669, 671
Synapses, **855**
Syndrome of inappropriate antidiuretic hormone secretion (SIADH), 876
Synovial fluid, 681
Synthesis gas, 385
Synthetase, 796
Synthetic polymers, 118
Synthetic rubber, 385
Systolic pressure, 228

Table salt, 73, 83, 454
Tagamet™, 860
TAG mobilization, 762
Taq polymerase, 600, 816
Targets, 843
Taurine, 739
Tay-Sachs disease, 737, 811
TCA (tricarboxylic acid cycle). See Citric acid cycle
T cells, 880, 881
Technetium-99m, 348
Teeth, 86
Telomerase, 809
Telomeres, **809**
Temperature (T), **29**
  body, 195, 877
  in changes of state, 214–215

effect on enzyme activity, 598, 599
effect on equilibria, 202–203
effect on reaction rates, 192, 194
effect on solubility, 258–260
Fahrenheit–Celsius conversion, 32
measurement, 29–32
relation to P, V, and n (ideal gas law), 233–235
relation to pressure of gases (Gay-Lussac's law), 230
relation to volume of gases (Charles's law), 228–229
in spontaneous processes, 188
Temperature-sensitivity, 31
Template strands, 791, 794
Termination, translation, 797–798
tert-Butyl group, 376
Tertiary (3°) alcohols, 437, 444
Tertiary (3°) amines, 461, 462, 467, 468, 530
Tertiary (3°) carbon atom, **376**, 379
Tertiary protein structures, 552, 570–572, 573
Testosterone, 739, 740, 852
Tetrachloromethane, 122
Tetrahedral geometry, 114, 115, 116
Tetrahydrocannabinol (THC), 862
Tetrahydrogestrinone (THG), 853–854
Tetravalent bonding, 357
Tetrodotoxin, 478
TFTR (Tokamak Fusion Test Reactor), 351
THC (tetrahydrocannabinol), 862
Theoretical yield, 169–171
Therapeutic procedures, 338
Thermal "cracking," 396
Thermite reaction, 181
Thermogenin, 647
Thermophiles, 600
THG (tetrahydrogestrinone), 853–854
Thioalcohols. See Thiols
Thiolase, 763, 766
Thiols, 361, 362, **452**–453
Thomson, J. J., 48
Three Mile Island nuclear reactor, 350
Threonine, 564, 837
Threose, 660
Thrombin, 588, 876, 883
Thymine, 777, 782, 783–785
Thyroid gland, 454
Thyroid hormones, 846
Thyroid-stimulating hormone (TSH), 851
Thyrotropin-releasing hormone (TRH), 851
Thyroxine, 454, 850–851
Timed-release medications, 284
Tissue factor, 883
Titration, **317**–319
T lymphocytes (T cells), 880, 881
Tocopherols, 613
Tokamak Fusion Test Reactor (TFTR), 351
Tollens' reagent, 495
Tollens' test, 495, 496
Toluene, 422
Tomography, 348
Tooth decay, 701

Torr, 221
Torricelli, Evangelista, 221
Toxicity, 499
Toxicology, 478
Toxic substances
  acetaldehyde, 493
  acetanilide, 865
  alcohols, 435, 446
  aldehydes
  amines, 468, 476–477
  ammonia, 830
  botulinum, 499, 859
  carbon monoxide, 113, 382–383
  formaldehyde, 492–493, 499, 605
  muscarine, 478
  phenols, 447
Trabecular bone, 93
Trace minerals, 589, 616–617
Tracers, 332
Transaminase enzymes, 829
Transaminases, 591
Transamination, **828**, 829–830
Trans configurations, 400, 401
Transcription, **786**, 790–792, 794–795
Trans-fatty acids, 408, 740
Transferases, 591
Transfer RNAs (tRNAs), **790**, 791, 796–799
Transition group elements, 616–617
Transition metals, **52**
  cation formation in, 81
  cation naming in, 82–84
  in coordination compounds, 107
  inner, **52**, 53
  in periodic table, 53
Translation, **786**, 796–799
Translation elongation, 797, 798
Translation initiation, 797, 798
Translation termination, 797–798
Transmutation, **332**
Transmutation, artificial, **347**
Transuranium elements, 347
Trenbolone, 854
TRH (thyrotropin-releasing hormone), 851
Triacylglycerol lipase, 762
Triacylglycerols, **725**. See also Fats; Fatty acids
  from adipocytes, 759, 760
  catabolism, 715
  as class of lipids, 722, 725–726
  dietary, 728, 758–759
  digestion of, 753–755
  excessive buildup, 767
  from fats and oils, 729
  glycerol from, 759
  hydrogenation of, 730
  hydrolysis of, 731–732, 758, 759, 761, 762
  metabolism of, 758–759
  mobilization of, **761**–762
  oils, 727–729
  storage, 727, 760, 761
  structure, 725, 727
  synthesis of, 759, 761
Triads, 52
Tricarboxylic acid cycle (TCA). See Citric acid cycle
Trichloroacetic acid, 528

Trichloroethylene, 454
Trichloromethane, 454
Tricyclic antidepressants, 861
Triglycerides. See Triacylglycerols
Trigonal planar geometry, 114, 115
Triose phosphate isomerase, 697, 698, 810
Tripeptides, 551
Triphosphates, 541
Triphosphoric acid, 541
Triple bonds, **104**
Triprotic acids, 293
Tritium, 50
tRNAs (transfer RNAs), **790**, 791, 796–799
Tropocollagen, 574–575, 577
Troposphere, 224
Trypsinogen, 608
Tryptophan, 470, 564, 837, 857
TSH (thyroid-stimulating hormone), 851
Tube worms, 627
Tubocurarine, 859
Turnover numbers, **588**–589
Tylenol, 865
Type I diabetes, 707–708
Type II diabetes, 707–708, 760, 813
Type I ionic compounds, 89
Type II ionic compounds, 89
Tyrosine, 448, 838, 851

Ubiquinone, 644
UCLA, 879
UDP (uridine diphosphate), 710
UDP-glucose pyrophosphorylase, 710
Uncompetitive inhibition, **604**–605, 609
Unfavorable reactions, 625, 634–635
UNFCCC (United Nations Framework Convention on Climate Change), 224
Unit conversions, 23, 25–29, 222–223
Units, **15**
  atomic mass, 46
  concentration, 262–270, 274
  conversions, 23, 25–29, 222–223
  density, 16
  energy, 31
  enzyme activity, 601
  food calories, 185
  length, 15, 16, 18
  mass, 18
  metric, 15–17
  prefixes for, 17
  radiation intensity, 344–346
  SI, 15–17
  speed, 16
  temperature, 15
  volume, 15, 16, 18–19
Universal indicator, 308
Unsaturated acids, 517–518, 519
Unsaturated compounds, **396**, 409
Unsaturated fatty acids, **724**, 727, 743, 757
Unstable isotopes, 331–332, 336
Unsubstituted amides, 521, 530
Uracil, 777, 782
Uranium-235, 349
Uranium-238, 341

Urea cycle, 767, **830,** 830–832
Uric acid, 140, 833
Uridine, 778
Uridine diphosphate (UDP), 710
Urine, 140, 887–890
Urinometers, 34
Urushiol, 448
U.S. Department of Agriculture, 565
U.S. Food and Drug Administration
    on atherosclerosis, 757
    fat recommendations, 728
    on food irradiation concerns, 345
    food labeling rules, 615
    health claim investigation by, 680
    on human gene therapy, 820

Vaccines, 795
Valence electrons, **63**–64, 81–82, 110
Valence-shell electron-pair repulsion (VSEPR) models, **114**
Valence shells, **63,** 64
Valine, 561, 836
Vallium™, 879
Vane, John, 8
Vanillin, 484
Vapor, **237**
Vapor pressure, **237,** 238, 275–277
Variable number tandem repeats (VNTRs), 817
Vascular endothelial growth factor, 790
Vaseline™, 381, 385
Vasodilators, 113
Vasopressin, 564, 845, 846, 875–876, 890
Vectors, 820
Vegetable gums, 670, 679
Vegetable oils, 725–728, 730, 731

Venom, 859
Very-low-density lipoproteins (VLDLs), 756
Vesicles, 856
Villi, 693, 754, 755
Vinblastine, 864
Vinegar, 525
Vinyl chloride, 111–112, 122–123
Vinyl group, 417
Vinyl monomers, 417
Vioxx™, 746
Viroids, 790
Viruses
    HIV, 607, 794–795, 865
    influenza, 799–800
    retroviruses, 795
    as vectors in gene therapy, 820
Viscosity, 239
Vision, 406–407, 708
Vitalism, 357
Vitamin A, 406, 612, 613, 648, 819
Vitamin C
    antioxidant properties, 614, 648
    deficiency, 575, 576–577
    in scurvy prevention and treatment, 575, 576–577, 610
    structure, 610, 613
Vitamin D, 610, 612, 613
Vitamin E
    antioxidant properties, 449, 613, 614, 648
    storage, 612
    structure, 449, 613
Vitamin K, 612, 613–614, 882
Vitamins, **610**
    antioxidant, 610, 613, 614, 648
    fat-soluble, 612–614
    food labeling, 615

    role in vision, 406
    water-soluble, 610–612
VLDLs (very-low-density lipoproteins), 756
VNTRs (variable number tandem repeats), 817
Volatility, 242, 420, 451
Volume (V)
    converting from mass, 34
    effect of dilution on, 272
    percent concentrations, 264, 265–266, 267–268
    relation to molar amount of gases (Avogadro's law), 232–233
    relation to P, T, and n (ideal gas law), 233–235
    relation to pressure of gases (Boyle's law), 225–227
    relation to temperature of gases (Charles's law), 228–229
    units for, 15, 16, 18–19
Volumetric flasks, 265
Volume/volume percent concentration (v/v)%, **264,** 265–268
VSEPR (valence-shell electron-pair repulsion) models, **114**

Walden, Paul, 78
Water
    in acid rain, 320–321
    in body fluids, 871–874
    as both acid and base, 302
    covalent bonds in, 101
    dissociation of, 302–303
    in hydration reactions, 413
    hydrogen bonds in, 218
    intake/output of, 874–875
    molecular geometry, 116

    molecular polarity, 121
    properties, 239–240, 243, 434
    solubility, 255–256
Water vapor, 224
Watson, James, 783, 789
Watson–Crick model, 783–784, 786
Wave function, 56
Wavelength, 66
Waxes, 722, **724**–725
Weak acids, **296**
    in buffer solutions, 310
    determining strength of, 296–300
    K_a values for, 301
    salts of, 322
    solubility in body fluids, 473
Weak bases, 296–300, **297,** 322, 473
Weak electrolytes, **273**
Weight, **17**
Wheal-and-flare reactions, 880
White blood cells, 876, 880, 881
Whole blood, **876**
Wilson's disease, 767
Wine, 703, 704
Wöhler, Friedrich, 357
Wood alcohol. *See* Methanol

Xenon, 55
X rays, **341,** 342
X-ray tomography, 348

Yeast, 702

Zaitsev's Rule, 441
Zantac™, 861
Zinc, 12
Zwitterions, **555,** 556
Zymogens, **608,** 609

Functional Groups of Importance in Biochemical Molecules

Functional Group	Structure	Type of Biomolecule
Amino group	$-NH_3^+$, $-NH_2$	Alkaloids and neurotransmitters; amino acids and proteins (Sections 15.1, 15.3, 15.6, 18.3, 18.7, 28.6)
Hydroxyl group	$-OH$	Monosaccharides (carbohydrates) and glycerol: a component of triacylglycerols (lipids) (Sections 17.4, 21.4, 23.2)
Carbonyl group	$\overset{O}{\underset{\|}{-C-}}$	Monosaccharides (carbohydrates); in acetyl group (CH_3CO) used to transfer carbon atoms during catabolism (Sections 16.1, 17.4, 20.4, 20.8, 21.4)
Carboxyl group	$-\overset{O}{\underset{\|}{C}}-OH$, $-\overset{O}{\underset{\|}{C}}-O^-$	Amino acids, proteins, and fatty acids (lipids) (Sections 17.1, 18.3, 18.7, 23.2)
Amide group	$-\overset{O}{\underset{\|}{C}}-N-$	Links amino acids in proteins; formed by reaction of amino group and carboxyl group (Sections 17.1, 17.4, 18.7)
Carboxylic acid ester	$-\overset{O}{\underset{\|}{C}}-O-R$	Triacylglycerols (and other lipids); formed by reaction of carboxyl group and hydroxyl group (Sections 17.1, 17.4, 23.2)
Phosphates: mono-, di-, tri-	$-C-O-P(=O)(O^-)-O^-$ $-C-O-P(=O)(O^-)-O-P(=O)(O^-)-O^-$ $-C-O-P(=O)(O^-)-O-P(=O)(O^-)-O-P(=O)(O^-)-O^-$	ATP and many metabolism intermediates (Sections 17.8, 20.5, and throughout metabolism sections)
Hemiacetal group	$-\overset{\|}{\underset{\|}{C}}\begin{matrix}-OH\\-OR\end{matrix}$	Cyclic forms of monosaccharides; formed by a reaction of carbonyl group with hydroxyl group (Sections 16.7, 21.4)
Acetal group	$-\overset{\|}{\underset{\|}{C}}\begin{matrix}-OR\\-OR\end{matrix}$	Connects monosaccharides in disaccharides and larger carbohydrates; formed by reaction of carbonyl group with hydroxyl group (Sections 16.7, 21.7, 21.9)
Thiols	$-SH$	Found in amino acids cysteine, methionine; structural components of proteins (Sections 14.9, 18.3, 18.8, 18.10)
Sulfides	$-S-$	
Disulfides	$-S-S-$	